THE SOC SOLUTION

Print + Online

SOC⁵ delivers all the key terms and core concepts for the **Introduction to Sociology** course.

SOC Online provides the complete narrative from the printed text with additional interactive media and the unique functionality of **StudyBits**—all available on nearly any device!

What is a StudyBit™? Created through a deep investigation of students' challenges and workflows, the StudyBit™ functionality of **SOC Online** enables students of different generations and learning styles to study more effectively by allowing them to learn their way. Here's how they work:

COLLECT WHAT'S IMPORTANT
Create StudyBits as you highlight text, images or take notes!

WEAK

FAIR

STRONG

UNASSIGNED

RATE AND ORGANIZE STUDYBITS
Rate your understanding and use the color-coding to quickly organize your study time and personalize your flashcards and quizzes.

StudyBit™

TRACK/MONITOR PROGRESS
Use Concept Tracker to decide how you'll spend study time and study YOUR way!

85%

PERSONALIZE QUIZZES
Filter by your StudyBits to personalize quizzes or just take chapter quizzes off-the-shelf.

○ CORRECT
○ INCORRECT
○ INCORRECT
○ INCORRECT

CENGAGE
Learning®

SOC 5
Nijole V. Benokraitis

Vice President, General Manager, 4LTR Press:
   Neil Marquardt

Product Director, 4LTR Press: Steven E. Joos

Product Manager: Clinton Kernen

Content/Media Developer: Sarah Dorger

Product Assistant: Lauren Dame

Marketing Manager: Jennifer Levinduski

Sr. Content Project Manager: Martha Conway

Manufacturing Planner: Ron Montgomery

Production Service: MPS Limited

Sr. Art Director: Bethany Casey

Internal Designer: KeDesign, Mason, OH

Cover Designer: Curio Press, LLC./Lisa Kuhn

Cover Image: © GettyImages, Inc./paci77

Title Page Image: © GettyImages, Inc./paci77

Special Page Images: Computer and tablet
   illustration: © iStockphoto.com/furtaev;
   Smart Phone illustration: © iStockphoto
   .com/dashadima; Feedback image: Rawpixel
   .com/Shutterstock.com

Back Cover Images: Computer and tablet
   illustration: © iStockphoto.com/furtaev;
   Smart Phone illustration: © iStockphoto
   .com/dashadima

Guest Photo Researcher: Vitalius Benokraitis

Intellectual Property Analyst: Alexandra
   Ricciardi

Intellectual Property Project Manager:
   Nick Barrows

Library of Congress Control Number: 2015959315

Student Edition ISBN: 978-1-305-66056-4

Student Edition with Online ISBN: 978-1-305-66062-5

Cengage Learning
20 Channel Center Street
Boston, MA 02210
USA

Cengage Learning is a leading provider of customized learning solutions with employees residing in nearly 40 different countries and sales in more than 125 countries around the world. Find your local representative at **www.cengage.com**

Cengage Learning products are represented in Canada by Nelson Education, Ltd.

To learn more about Cengage Learning Solutions, visit **www.cengage.com**

Purchase any of our products at your local college store or at our preferred online store **www.cengagebrain.com**

Printed in the United States of America
Print Number: 01   Print Year: 2015

BENOKRAITIS
SOC⁵

BRIEF CONTENTS

© GettyImages, Inc./paci77

CONTENTS

Stewart Cohen/Blend Images/Alamy

4 Socialization 60

5 Social Interaction in Everyday Life 82

Jdwfoto/Shutterstock.com

6 Groups, Organizations, and Institutions 100

7 Deviance, Crime, and Social Control 118

8 Social Stratification: United States and Global 138

gpointstudio/Shutterstock.com

AP Images/Paul Sancy

11 The Economy and Politics 200

Ted Thai/Time & Life Pictures/Getty Images

Allison Shelley/Getty Images

1 | Thinking Like a Sociologist

LEARNING OUTCOMES

After studying this chapter, you will be able to...

1-1 Explain what sociology is and how it differs from other social sciences and common sense.

1-2 Explain how and why a sociological imagination helps us understand society.

1-3 Identify and illustrate why it's worthwhile to study sociology.

1-4 Describe and explain the origins of sociology, why sociology developed, and its most influential early theorists.

1-5 Compare, illustrate, and evaluate the four contemporary sociological perspectives.

After you finish this chapter go to **PAGE 19** for **STUDY TOOLS.**

Texting and emailing are associated with the highest risk of car crashes, and headset cell phones aren't much safer than handheld cell phones. Almost 90 percent of U.S. drivers say that talking on a cell phone or texting while driving are a serious threat to safety. However, 70 percent engage in smartphone activities while driving, including surfing the Internet and even video chats (AAA Foundation for Traffic Safety, 2015; AT&T Newsroom, 2015).

Why is there such a disconnection between many Americans' attitudes and behavior? This chapter examines these and other questions. Let's begin by considering what sociology is (and isn't) and how a "sociological imagination" can give us more control over our lives. We'll then look at how sociologists grapple with complex theoretical issues in explaining social life. Before reading further, take the quiz on the next page.

What do you think?

Sociology is basically common sense.

1	2	3	4	5	6	7

strongly agree   strongly disagree

1-1 WHAT IS SOCIOLOGY?

Stated simply, **sociology** is the systematic study of human behavior in society. When sociologists refer to the *systematic* study of behavior, they mean that social behavior is regular and patterned, and that it takes place between individuals, among small groups (such as families), large organizations (such as Apple), and entire

John Lund/Getty Images

societies (such as the United States). But, you might protest, "I'm unique."

1-1a Are You Unique?

Yes and no. Each of us is unique in the sense that you and I are like no one else on earth. Even identical twins, who have the same physical characteristics and genetic matter, often differ in personality and interests. One of my colleagues likes to tell the story about his 3-year-old twin girls who received the same doll. One twin chattered that the doll's name was Lori, that she loved Lori, and would take good care of her. The second twin muttered, "Her name is Stupid," and flung the doll into a corner.

Despite some individual differences, identical twins, you, and I are like other people in many ways. Around the world, we experience grief when a loved one dies, participate in rituals that celebrate marriage or the birth of a child, and want to have healthy and happy lives. Some actions, such as terrorist attacks, are unpredictable. For the most part, however, people conform to expected and acceptable behavior. From the time that we get up until we go to bed, we follow a variety of rules and customs about what we eat, how we drive, how we act in different

TRUEORFALSE?

Everybody Knows That ...

1. The death penalty reduces crime.
2. Women's earnings are now similar to men's, especially in high-income jobs.
3. People age 65 and older make up the largest group of those who are poor.
4. There are more married than unmarried U.S. adults.
5. Divorce rates are higher today than in the past.
6. Latinos are the fastest-growing racial-ethic group in the United States.
7. The best way to get an accurate measure of public opinion is to poll as many people as possible.
8. Illegal drugs are the biggest health hazard.

The answers are at the end of this section.

sociology   the systematic study of human behavior in society.

social situations, and how we dress for work, classes, and leisure activities.

So what? you might shrug. Isn't it "obvious" that we dress differently for classes than for job interviews? Isn't all of this just plain old common sense?

1-1b Isn't Sociology Just Common Sense?

No. Sociology goes well beyond conventional wisdom, what we call common sense, in several ways:

▶ **Common sense is subjective.** If a woman crashes into my car, I might conclude, according to the conventional wisdom, statements that we've heard over the years, that "women are terrible drivers." In fact, most drivers involved in crashes are men—especially teenagers and those age 70 and older (Insurance Institute for Highway Safety, 2013). Thus, *objective* data show that, overall, men are worse drivers than women.

▶ **Common sense ignores facts.** Because common sense is subjective, it ignores facts that challenge cherished beliefs. For example, many Americans are most concerned about street crimes, such as robbery or homicide. FBI and sociological data show, however, that we're much more likely to be assaulted or murdered by someone we know or live with (see Chapters 7 and 12).

▶ **Common sense varies across groups and cultures.** In the United States, it's common sense to date before choosing a marriage partner. In many other societies, in contrast, it's common sense for parents and relatives to select marital mates. Thus, common sense notions about mate selection vary considerably around the world.

▶ **Much of our common sense is based on myths and misconceptions.** A common myth is that living together is a good way to find out whether partners will get along after marriage. Generally, however, couples who live together before marriage have higher divorce rates than those who don't (see Chapter 12).

Sociology, in contrast to conventional wisdom, examines claims and beliefs critically, considers many points of view, and enables us to move beyond established ways of thinking. The *sociological perspective* analyzes how social context influences people's lives. The "sociological imagination" is at the center of the sociological perspective.

Marriage without Love? No Way!

When I ask my students, "Would you marry someone you're not in love with?" most laugh, raise an eyebrow, or stare at me in disbelief. "Of course not!" they exclaim. In fact, the "open" courtship and dating systems common in Western nations, including the United States, are foreign to much of the world. In many African, Asian, Mediterranean, and Middle Eastern countries, marriages are arranged: They forge bonds between families rather than individuals, and preserve family continuity along religious and socioeconomic lines. Love isn't a prerequisite for marriage in societies that value kin groups rather than individual choices (see Chapters 9 and 12).

TRUEORFALSE?

Everybody Knows That ...

All of the answers are false.

1. States without the death penalty have had consistently lower homicide rates than those with death penalties (see Chapter 7).

2. Regardless of education or occupation, women's earnings are lower than men's (see Chapters 9 and 11).

3. Children ages 5 and under make up the largest group of Americans who are poor (see Chapter 8).

4. The number of unmarried U.S. adults outnumbers those who are married (see Chapter 12).

5. Divorce rates are lower today than they were between 1975 and 1990 (see Chapter 12).

6. Latinos are the largest ethnic group in the United States, but Asian Americans are the fastest growing (see Chapter 10).

7. What matters in polling is not the number of people polled, but their representativeness in the population studied (see Chapter 2).

8. In the United States and worldwide, tobacco use is the leading cause of preventable death and disability (see Chapter 14).

WHAT IS A SOCIOLOGICAL IMAGINATION?

According to sociologist C. Wright Mills (1916–1962), our behavior is influenced by social factors such as religion, ethnicity, and politics. Mills (1959) called this ability to see the relationship between individual experiences and larger social influences the **sociological imagination**. The sociological imagination emphasizes the connection between personal troubles (biography) and structural (public and historical) issues.

Consider unemployment. If only a small group of people can't find a job, it's a *personal trouble* that may be due, in part, to an individual's low educational attainment, lack of specific skills that employers want, not searching for work, and so on. If unemployment is widespread, it's a *public issue* because economic problems are also the result of structural factors such as mass layoffs, sending jobs overseas, technological changes, and restrictive hiring policies (see Chapter 11). Thus, people may be unemployed regardless of skills, a college degree, and job searches.

A sociological imagination helps us understand how larger social forces affect individuals and how individuals affect society. It identifies why our personal troubles often reflect larger public issues and policies over which we have little, if any, control. A sociological imagination relies on both micro- and macro-level approaches to understand our social world.

1-2a Microsociology: How People Affect Our Everyday Lives

Microsociology examines the patterns of individuals' social interaction in specific settings. In most of our relationships, we interact with others on a micro, or "small," level (e.g., members of a work group discussing who will perform which tasks). These everyday interactions involve what people think, say, or do on a daily basis.

1-2b Macrosociology: How Social Structure Affects Our Everyday Lives

Macrosociology focuses on large-scale patterns and processes that characterize society as a whole. Macro, or "large," approaches are especially useful in understanding some of the constraints—such as economic forces and public policies.

Microsociology and macrosociology differ conceptually, but are interrelated. Consider the reasons for

AF archive/Alamy

The Wire was a popular television show that aired on HBO between 2002 and 2008. It depicted the lives of poor people and blue-color workers in Baltimore, Maryland. For sociologists, *The Wire* illustrates the connection between macro-level structural constraints and micro-level individual behavior.

divorce. On a micro level, sociologists study factors such as extramarital affairs, substance abuse, arguments about money, and other everyday interactions that fuel marital tension and unhappiness, leading to divorce. On a macro level, sociologists look at how the economy, laws, cultural values, and technology affect divorce rates (see Chapter 12). Examining micro, macro, and micro–macro forces is one of the reasons why sociology is a powerful tool in understanding (and changing) our behavior and society at large (Ritzer, 1992).

1-3 WHY STUDY SOCIOLOGY?

Sociology offers explanations that can greatly improve the quality of your everyday life. These explanations can influence choices that range from your personal decisions to expanding your career opportunities.

sociological imagination the ability to see the relationship between individual experiences and larger social influences.

microsociology examines the patterns of individuals' social interaction in specific settings.

macrosociology examines the large-scale patterns and processes that characterize society as a whole.

1-3a Making Informed Decisions

Sociology can help us make more informed decisions. We often hear that grief counseling is essential after the death of a loved one. In fact, 4 in 10 Americans are better off without it. Grief is normal, and most people work through their losses on their own, whereas counseling sometimes prolongs depression and anxiety (Stroebe et al., 2000).

1-3b Understanding Diversity

The racial and ethnic composition of the United States is changing. By 2025, only 58 percent of the U.S. population is projected to be white, down from 76 percent in 1990 and 86 percent in 1950 (Passel et al., 2011; U.S. Census Bureau, 2012). As you'll see in later chapters, this racial/ethnic shift has already affected interpersonal relationships as well as education, politics, religion, and other spheres of social life.

Recognizing and understand diversity is one of sociology's central themes. Our gender, social class, marital status, ethnicity, sexual orientation, and age—among other factors—shape our beliefs, behavior, and experiences. If, for example, you're a white middle-class male who attends a private college, your experiences are very different from those of a female Vietnamese immigrant who is struggling to pay expenses at a community college.

Increasingly, nations around the world are intertwined through political and economic ties. What happens in other societies often has a direct or indirect impact on contemporary U.S. life. Decisions in oil-producing countries, for example, affect gas prices, spur the development of hybrid cars that are less dependent on oil, and stimulate research on alternative sources of energy.

Suzanne Tucker/Shutterstock.com

1-3c Shaping Social and Public Policies and Practices

Sociology is valuable in applied, clinical, and policy settings because many jobs require understanding society and research to create social change. According to a director of a research institute, sociology increased her professional contributions: "I can look at problems of concern to the National Institutes of Health and say 'here's a different way to solve this problem'" (Nyseth et al., 2011: 48).

Snapshots

"I love our lunches out here, but I always get the feeling that we're being watched."

Jason Love/Cartoonstock.com

1-3d Thinking Critically

We develop a sociological imagination not only when we understand and can apply the concepts, but also when we can think, speak, and write critically. Much of our thinking and decision making is often impulsive and emotional. In contrast, critical thinking involves knowledge and problem solving (Paul and Elder, 2007).

Critical sociological thinking goes even further because we begin to understand how our individual lives, choices, and troubles are shaped by race, gender, social class, and social institutions such as the economy, politics, and education (Eckstein et al., 1995; Grauerholz and Bouma-Holtrop, 2003). *Table 1.1* summarizes some of the basic elements of critical sociological thinking.

1-3e Expanding Your Career Opportunities

A degree in sociology is a springboard for entering many jobs and professions. A national survey of undergraduate sociology majors found that 44 percent were in administrative or management positions, 22 percent were employed in social service and counseling, 18 percent were in sales and marketing, and 12 percent were teachers (Senter et al., 2014).

TABLE 1.1 WHAT IS CRITICAL SOCIOLOGICAL THINKING?

Critical sociological thinking requires a combination of skills. Some of the basic elements include the ability to:

- rely on reason rather than emotion

- ask questions, avoid snap judgments, and examine popular and unpopular beliefs

- recognize one's own and others' assumptions, prejudices, and points of view

- remain open to alternative explanations and theories

- require and examine competing evidence (see Chapter 2)

- understand how public issues affect private troubles

What specific skills do sociology majors learn that are useful in their jobs? Some of the most important are being better able to work with people (71 percent), to organize information (69 percent), to write reports that nonsociologists understand (61 percent), and to interpret research findings (56 percent) (Van Vooren and Spalter-Roth, 2010). In other cases, students major in sociology because it provides a broad liberal arts foundation for professions such as law, education, and social work.

Even if you don't major in sociology, developing your sociological imagination can enrich your job skills. Sociology courses help you learn to think abstractly and critically, formulate problems, ask incisive questions, search for data in the most reliable and up-to-date sources, organize material, and improve your oral presentations (ASA Research Department, 2013; Spalter-Roth et al., 2013).

1-4 SOME ORIGINS OF SOCIOLOGICAL THEORY

During college, most of my classmates and I postponed taking theory courses (regardless of our major) as long as possible. "This stuff is boring, boring, boring," we'd grump, "and has nothing to do with the real world." In fact, theorizing is part of our everyday lives. Every time you try to explain why your family and friends behave as they do, for example, you're theorizing.

As people struggle to understand human behavior, they develop theories. A **theory** is a set of statements that explains why a phenomenon occurs. Theories produce knowledge, guide our research, help us analyze our findings, and, ideally, offer solutions for social problems.

Sociologist James White (2005: 170–171) describes theories as "tools" that don't profess to know "the truth" but "may need replacing" over time as our understanding of society becomes more sophisticated. In effect, theories evolve over time because of cultural and technological changes. You'll see shortly, for example, that sociological theories about behavior changed considerably after the rise of feminist scholarship during the late 1960s.

Sociology and Other Social Sciences: *What's the Difference?*

How would different social scientists study the same phenomenon, such as homelessness? Criminologists might examine whether crime rates are higher among homeless people than in the general population. Economists might measure the financial impact of programs for the homeless. Political scientists

Mitchell Funk/Photographer's Choice/Getty Images

might study whether and how government officials respond to homelessness. Psychologists might be interested in how homelessness affects individuals' emotional and mental health. Social workers are most likely to try to provide needed services (e.g., food, shelter, medical care, and jobs). Sociologists have been most interested in examining homelessness across gender, age, and social class, and explaining how this social problem devastates families and communities.

According to sociologist Herbert Gans (2005), sociologists "study everything." There are currently 43 different subfields in sociology, and the number continues to increase, because sociologists' interests range across many areas.

theory a set of statements that explains why a phenomenon occurs.

that the study of society must be **empirical**. That is, information should be based on observations, experiments, or other data collection rather than on ideology, religion, intuition, or conventional wisdom.

He saw sociology as the scientific study of two aspects of society: social statics and social dynamics. *Social statics* investigates how principles of social order explain a particular society, as well as the interconnections between institutions. *Social dynamics* explores how individuals and societies change over time. Comte's emphasis on social order and change within and across societies is still useful today because many sociologists examine the relationships between education and politics (social statics), as well as how such interconnections change over time (social dynamics).

Creatas Images/Jupiter Images

Theories = Tools

Harriet Martineau

Spencer Arnold/Hulton Archive/Getty Images

1-4b Harriet Martineau

Harriet Martineau (1802–1876), an English author, published several dozen books on a wide range of topics in social science, politics, literature, and history. Her translation and condensation of Auguste Comte's difficult material for popular consumption was largely responsible for the dissemination of Comte's work. "We might say, then, that sociology had parents of both sexes" (Adams and Sydie, 2001: 32). She emphasized the importance of systematic data collection through observation and interviews, and an objective analysis of data to explain events and behavior. She also published the first sociology research methods textbook.

Martineau, a feminist and strong opponent of slavery, denounced many aspects of capitalism as alienating and degrading, and criticized dangerous workplaces that often led to injury and death. Martineau promoted improving women's positions in the workforce through education, nondiscriminatory employment, and training programs. She advocated women's admission into medical schools and emphasized issues such as infant care, the rights of the

Sociological theories didn't emerge overnight. Nineteenth-century thinkers grappled with some of the same questions that sociologists try to answer today: Why do people behave as they do? What holds society together? What pulls it apart? Of the many early sociological theorists, some of the most influential were Auguste Comte, Harriet Martineau, Émile Durkheim, Karl Marx, Max Weber, Jane Addams, and W. E. B. Du Bois.

1-4a Auguste Comte

Auguste Comte (pronounced oh-gust KONT; 1798–1857) coined the term *sociology* and is often described as the "father of sociology." Comte maintained

Hulton Archive/Getty Images

The Father of Sociology—Auguste Comte

empirical information that is based on observations, experiments, or other data collection rather than on ideology, religion, intuition, or conventional wisdom.

aged, the prevention of suicide, and other social problems (Hoecker-Drysdale, 1992).

After a long tour of the United States, Martineau described American women as being socialized to be subservient and dependent rather than equal marriage partners. She also criticized American and European religious institutions for expecting women to be pious and passive rather than educating them in philosophy and politics. Most scholars, including sociologists, ridiculed and dismissed such ideas as too radical.

1-4c Émile Durkheim

Émile Durkheim (1858–1917), a French sociologist and writer, agreed with Comte that societies are characterized by unity and cohesion because their members are bound together by common interests and attitudes. Whereas Comte acknowledged the importance of using scientific methods to study society, Durkheim actually did so by poring over official statistics to test a theory about suicide (Adams and Sydie, 2001).

SOCIAL FACTS

To be scientific, Durkheim maintained, sociology must study **social facts**—aspects of social life, external to the individual, that can be measured. Sociologists can determine *material facts* by examining demographic characteristics such as age, place of residence, and population size. They can gauge *nonmaterial facts*, such as communication processes, by observing everyday behavior and how people relate to each other (see Chapters 3 to 6). For contemporary sociologists, social facts also include collecting and analyzing data on *social currents*, such as collective behavior and social movements (see Chapter 16).

Pictorial Press Ltd/Alamy

Émile Durkheim

DIVISION OF LABOR

One of Durkheim's central questions was how people can be autonomous and individualistic while being integrated in society. **Social solidarity**, or social cohesiveness and harmony, according to Durkheim, is maintained by a **division of labor**—an interdependence of different tasks and occupations, characteristic of industrialized societies, that produces social unity and facilitates change.

As the division of labor becomes more specialized, people become increasingly dependent on others for specific goods and services. Today, for example, many couples who are getting married often contract "experts" such as photographers, florists, deejays, caterers, bartenders, travel agents (for the honeymoon), and even "wedding planners."

SOCIAL INTEGRATION

Durkheim was one of the first sociologists to test a theory using data. In his classic study, *Suicide*, Durkheim (1897/1951) relied on extensive data collection to test his theory that suicide is associated with social integration. He concluded that people who experience meaningful social relationships in families, social groups, and communities are less likely to commit suicide than those who feel alone, helpless, or hopeless. Thus, many seemingly isolated individual acts, including suicide, are often the result of structural arrangements, such as weak social ties.

Are Durkheim's findings on social integration dated? No. We typically hear about high teenage suicide rates, but they're much higher at later ages. As in Durkheim's day, men have higher suicide rates than women across all age groups (see *Figure 1.1*), and white males age 85 and older are the most likely to take their own lives (National Center for Health Statistics, 2015).

Durkheim's connection of social integration to the suicide rate is still relevant today. The high suicide rates of older white men are due to a complex interplay of depression, substance abuse, access to guns, hopelessness because of terminal illnesses, and not being "connected" to family, friends, community groups, and support systems that women tend to develop throughout their lives (American Association of Suicidology, 2009; see also Chapters 9 and 12).

1-4d Karl Marx

Karl Marx (1818–1883), a German social philosopher, is often described as the most influential social scientist

---

social facts aspects of social life, external to the individual, that can be measured.

social solidarity social cohesiveness and harmony.

division of labor an interdependence of different tasks and occupations, characteristic of industrialized societies, that produces social unity and facilitates change.

FIGURE 1.1 U.S. SUICIDE RATES, BY SEX AND AGE

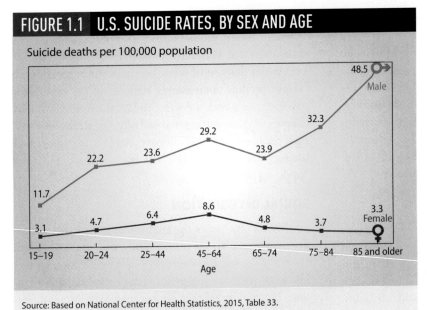

Suicide deaths per 100,000 population

Male: 11.7, 22.2, 23.6, 29.2, 23.9, 32.3, 48.5
Female: 3.1, 4.7, 6.4, 8.6, 4.8, 3.7, 3.3

Age: 15–19, 20–24, 25–44, 45–64, 65–74, 75–84, 85 and older

Source: Based on National Center for Health Statistics, 2015, Table 33.

who ever lived. Marx, like Comte and Durkheim, tried to explain the societal changes that were taking place during the Industrial Revolution.

The Industrial Revolution began in England around 1780 and spread throughout Western Europe and the United States during the nineteenth century. A number of technological inventions—such as the spinning wheel, the steam engine, and large weaving looms—enabled the development of large-scale manufacturing and mining industries over a relatively short period. The extensive mechanization shifted agricultural and home-based work to factories in cities. As masses of people migrated from small farms to factories to find jobs, urbanization and capitalism grew rapidly.

CAPITALISM

Unlike his predecessors and contemporaries, Marx (1867/1967, 1964) maintained that economic issues produce divisiveness rather than social solidarity. For Marx, the most important social change was the development of **capitalism**, an economic system in which the ownership of the means of production—such as land, factories, large sums of money, and machines—is in private hands. As a result, Marx saw industrial society as composed of three social classes:

▶ **capitalists**—the ruling elite who own the means of producing wealth (such as factories)

▶ **petit bourgeoisie**—small business owners and workers who still have their own means of production but might end up in the proletariat because they're driven out by competition or their businesses fail

▶ **proletariat**—the masses of workers who depend on wages to survive, have few resources, and make up the working class

CLASS CONFLICT

Marx believed that society is divided into the haves (capitalists) and the have-nots (proletariat). For Marx, capitalism is a class system in which conflict between the classes is common and society is anything but cohesive. Instead, class antagonisms revolve around struggles between the capitalists, who increase their profits by exploiting workers, and workers, who resist but give in because they depend on capitalists for jobs.

Marx argued that there's a close relationship between inequality, social conflict, and social class. History, he maintained, is a series of class struggles between capitalists and workers. As wealth becomes more concentrated in the hands of a few capitalists, Marx predicted, the ranks of an increasingly dissatisfied proletariat would swell, leading to bloody revolution and eventually a classless society. The Occupy Wall Street movement

Roger Viollet Collection/Getty Images

Karl Marx

Comstock/Stockbyte/Getty Images

capitalism an economic system in which the ownership of the means of production is in private hands.

Occupy Wall Street (OWS) was a protest movement against corporate greed, corruption, and influence on government. It began in mid-September, 2011, in New York City's Wall Street financial district. The OWS slogan, "We are the 99%," referred to U.S. income and wealth inequality between the wealthiest 1 percent and the rest of the population. OWS received global attention and spawned similar movements worldwide but was short-lived (see Chapter 16).

showed that thousands of Americans are very unhappy about the growing inequality between the haves and the have-nots, but there hasn't been a "bloody revolution" in the United States, unlike some countries in the Middle East.

ALIENATION

In industrial capitalist systems, Marx (1844/1964) contended, **alienation**—the feeling of separation from one's group or society—is common across all social classes. Workers feel alienated because they don't own or control either the means of production or the product. Because meaningful labor is what makes us human, Marx maintained, our workplace has alienated us "from the essence of our humanness." Instead of collaboration, a capitalistic society encourages competition, backstabbing, and "looking out for number one."

According to Marx, capitalists are also alienated. They regard goods and services as important simply because they're sources of profit. Capitalists don't care who buys or sells their products, how the workers feel about the products they make, or whether buyers value the products. The major focus, for capitalists, is on increasing profits as much as possible rather than feeling "connected" to the products or services they sell. Every year, for example, companies must recall cars, pharmaceutical items, toys, and food that cause injuries, illness, or death.

1-4e Max Weber

Max Weber (pronounced VAY-ber; 1864–1920) was a German sociologist, economist, legal scholar, historian, and politician. Unlike Marx's emphasis on economics as a major factor in explaining society, Weber focused on social organization, a subjective understanding of behavior, and a value-free sociology.

SOCIAL ORGANIZATION

For Weber, economic factors were important, but ideas, religious values, ideologies, and charismatic leaders were just as crucial in shaping and changing societies. He maintained that a complete understanding of society requires an analysis of the social organization and interrelationships among economic, political, and cultural institutions. In his *Protestant Ethic and the Spirit of Capitalism*, for example, Weber (1920/1958) argued that the self-denial fostered by Calvinism supported the rise of capitalism and shaped many of our current values about working hard (see Chapters 3 and 6).

SUBJECTIVE UNDERSTANDING

Weber posited that an understanding of society requires a "subjective" understanding of behavior. Such understanding, or *verstehen* (pronounced fer-SHTAY-en), requires knowing how people perceive the world in which they live. Weber described two types of *verstehen*. In *direct observational understanding*, the social scientist observes a person's facial expressions, gestures, and listens to his/her words. In *explanatory understanding*, the social scientist tries to grasp the intention and context of the behavior.

If a person bursts into tears (direct observational understanding), the observer knows what the person may be feeling (anger, sorrow, and so on). An explanatory understanding goes a step further by spelling out the

Max Weber

alienation the feeling of separation from one's group or society.

Is It Possible to Be a Value-Free Sociologist?

Max Weber was concerned about popular professors who took political positions that pleased many of their students. He felt that these professors were behaving improperly because science, including sociology, must be "value free." Faculty must set their personal values aside to make a contribution to society. According to a sociologist who agrees with Weber, sociology's weakness is its tendency toward moralism and ideology:

> Many people become sociologists out of an impulse to reform society, fight injustice, and help people. Those sentiments are noble, but unless they are tempered by skepticism, discipline, and scientific detachment, they can be destructive. Especially when you are morally outraged and burning with a desire for action, you need to be cautious (Massey, 2007: B12).

Some argue that being value free is a myth because it's impossible for a scholar's attitudes and opinions to be totally divorced from her or his scholarship (Gouldner, 1962). Many sociologists, after all, do research on topics that they consider significant and about which they have strong views.

Others maintain that one's values should be passionately partisan, should frame research issues,

CAN SOCIOLOGISTS BE VALUE FREE— ESPECIALLY WHEN THEY HAVE STRONG FEELINGS ABOUT MANY SOCIETAL ISSUES? *SHOULD THEY BE?*

and should improve society (Feagin, 2001). Sociologists shouldn't apologize for being subjective in their teaching and research. By staying silent, social scientists "cede the conversation to those with the loudest voices or deepest pockets . . . people with megaphones who spread sensational misinformation" that deprives the public of the best available data (Wang, 2015: A48).

Can sociologists *really* be value free—especially when they have strong feelings about many societal issues? Should they be?

---

reason for the behavior (rejection by a loved one, frustration when your computer crashes, humiliation if a boss yells at you in public).

VALUE-FREE SOCIOLOGY

One of Weber's most lasting and controversial views was the notion that sociologists must be as objective, or "value free," as possible in analyzing society. A researcher who is **value free** is one who separates her or his personal values, opinions, ideology, and beliefs from scientific research.

During Weber's time, the government and other organizations demanded that university faculty teach the "right" ideas. Weber encouraged everyone to be involved as citizens, but he maintained that educators and scholars should be as dispassionate as possible politically and ideologically. The task of the teacher, Weber argued, was to provide students with knowledge and scientific experience, not to "imprint" the teacher's personal political views and value judgments (Gerth and Mills, 1946). "Is It Possible to Be a Value-Free Sociologist?" examines this issue further.

1-4f Jane Addams

Jane Addams (1860–1935) was a social worker who co-founded Hull House, one of the first settlement houses in Chicago that served as a community center for the neighborhood poor. An active reformer throughout her life, Jane Addams was a leader in the women's suffrage movement and, in 1931, was the first American woman to be awarded the Nobel Peace Prize for her advocacy of negotiating, rather than waging war, to settle disputes.

value free separating one's personal values, opinions, ideology, and beliefs from scientific research.

Sociologist Mary Jo Deegan (1986) describes Jane Addams as "the greatest woman sociologist of her day." However, she was ignored by her colleagues at the University of Chicago (the first sociology department established in the United States in 1892) because discrimination against women sociologists was "rampant" (p. 8).

Despite such discrimination, Addams published articles in many popular and scholarly journals, as well as many books on the everyday life of urban neighborhoods, especially the effects of social disorganization and immigration. Much of her work contributed to symbolic interaction, an emerging school of thought that you'll read about shortly. One of Addams' greatest intellectual legacies was her emphasis on applying knowledge to everyday problems. Her pioneering work in criminology included ecological maps of Chicago that were later credited to men (Moyer, 2003).

Jane Addams with a child at Hull House

1-4g W. E. B. Du Bois

W. E. B. Du Bois (pronounced Do-BOICE; 1868–1963) was a prominent black sociologist, writer, editor, social reformer, and orator. The author of almost two dozen books on Africans and black Americans, Du Bois spent most of his life responding to the critics and detractors of black life. He was the first African American to receive a Ph.D. from Harvard University, but once remarked, "I was in Harvard but not of it."

Du Bois helped found the National Association for the Advancement of Colored People (NAACP) and edited its journal, *Crisis*. The problem of the twentieth century, he wrote, is the problem of the color line. Du Bois believed that the race problem was one of ignorance, and advocated a "cure" for prejudice and discrimination. Such cures included promoting black political power and civil rights and providing blacks with a higher education rather than funneling them into techical schools.

These and other writings were unpopular at a time when Booker T. Washington, a well-known black educator, encouraged black people to be patient instead of demanding equal rights. As a result, Du Bois was dismissed as a radical

W. E. B. DuBois

by his contemporaries but was rediscovered by a new generation of black scholars during the 1970s and 1980s. Among his many contributions, Du Bois examined the oppressive effects of race and class, advocated women's rights, and played a key role in reshaping black–white relations in America (Du Bois, 1986; Lewis, 1993).

All of these and other early thinkers agreed that people are transformed by each other's actions, social patterns, and historical changes. They and other scholars shaped contemporary sociological theories.

1-5 CONTEMPORARY SOCIOLOGICAL THEORIES

How one defines "contemporary sociological theory" is somewhat arbitrary. The mid-twentieth century is a good starting point because "the late 1950s and 1960s have, in historical hindsight, been regarded as significant years of momentous changes in the social and cultural life of most Western societies" (Adams and Sydie, 2001: 479). Some of the sociological perspectives had earlier origins, but all matured during this period.

Sociologists typically use more than one theory to explain behavior. The theories view our social world somewhat differently, but all of them analyze why society is organized the way it is and why we behave as we do. Four of the most influential theoretical perspectives are functionalism, conflict theory, feminist theories, and symbolic interaction.

1-5a Functionalism

Functionalism (also known as *structural functionalism*) maintains that society is a complex system of interdependent parts that work together

> **functionalism (structural functionalism)** maintains that society is a complex system of interdependent parts that work together to ensure a society's survival.

to ensure a society's survival. Much of contemporary functionalism grew out of the work of Auguste Comte and Émile Durkheim, both of whom believed that human behavior is a result of social structures that promote order and integration in society.

One of their contemporaries, English philosopher Herbert Spencer (1820–1903), used an organic analogy to explain the evolution of societies. To survive, Spencer (1862/1901) wrote, our vital organs—like the heart, lungs, kidneys, liver, and so on—must function together. Similarly, the parts of a society, like the parts of a body, work together to maintain the whole structure.

SOCIETY IS A SOCIAL SYSTEM

Prominent American sociologists, especially Talcott Parsons (1902–1979) and Robert K. Merton (1910–2003), developed the earlier ideas of structure and function. For these and other functionalists, a society is a system that is composed of major institutions such as government, religion, the economy, education, medicine, and the family.

Each institution or other social group has *structures*, or organized units, that are connected to each other and within which behavior occurs. Education structures such as colleges, for instance, aren't only organized internally in terms of who does what and when, but depend on other structures such as government (to provide funding), business (to produce textbooks and construct buildings), and medicine (to ensure that students, staff, and faculty are healthy).

FUNCTIONS AND DYSFUNCTIONS

Each structure fulfills certain *functions*, or purposes and activities, to meet different needs that contribute to a society's stability and survival (Merton, 1938). The purpose of education, for instance, is to transmit knowledge to the young, to teach them to be good citizens, and to prepare them for jobs (see Chapter 13).

Dysfunctions are social patterns that have a negative impact on a group or society. When one part of society isn't working, it affects other parts by creating conflict, divisiveness, and social problems. Consider religion.

dysfunctions social patterns that have a negative impact on a group or society.

manifest functions purposes and activities that are intended and recognized; they're present and clearly evident.

latent functions purposes and activities that are unintended and unrecognized; they're present but not immediately obvious.

SOCIOLOGISTS TYPICALLY USE MORE THAN ONE THEORY TO EXPLAIN BEHAVIOR AND WHY SOCIETY IS ORGANIZED THE WAY IT IS.

In the United States, the Catholic Church's stance on issues such as not ordaining women to be priests and denouncing abortion and homosexuality has produced a rift between those who embrace or question papal edicts. In other countries, religious intolerance has led to wars and terrorism (see Chapter 13).

MANIFEST AND LATENT FUNCTIONS

There are two kinds of functions. **Manifest functions** are intended and recognized; they're present and clearly evident. **Latent functions** are unintended and unrecognized; they're present but not immediately obvious.

Consider the marriage ceremony. Its primary manifest function is to publicize the formation of a new family unit and to legitimize sexual intercourse and childbirth (even though both might occur outside of marriage). The latent functions of a marriage ceremony include communicating a "hands-off" message to suitors, providing the new couple with household goods and products through bridal showers and wedding gifts, and redefining family boundaries to include in-laws or stepfamily members.

CRITICAL EVALUATION

You'll see in later chapters that functionalism is useful in seeing the "big picture" of interrelated structures and functions. Its influence waned during the 1960s and 1970s, however, because functionalism was so focused on order and stability that it often ignored social change. For example, functionalism couldn't explain the many rapid changes sparked by the civil rights, women's, and gay movements.

A second and related criticism is that functionalism often glosses over the widespread inequality that a handful of powerful people create and maintain. Conflict theorists, especially, have pointed out that what's functional for some privileged groups is dysfunctional for many others.

iStockphoto.com/Blue_Cutler

Among other manifest functions, schools transmit knowledge and prepare children for adult economic roles. Among their latent functions, schools provide matchmaking opportunities. What are some other examples of education's manifest and latent functions?

1-5b Conflict Theory

In contrast to functionalism—which emphasizes order, stability, cohesion, and consensus—**conflict theory** examines how and why groups disagree, struggle over power, and compete for scarce resources (such as property, wealth, and prestige). Conflict theorists see disagreement and the resulting changes in society as natural, inevitable, and even desirable.

SOURCES OF CONFLICT

The conflict perspective has a long history. As you saw earlier, Karl Marx predicted that conflict would result from widespread economic inequality, and W. E. B. Du Bois criticized U.S. society for its ongoing and divisive racial discrimination. Since the 1960s, many sociologists—especially feminist and minority scholars—have emphasized that the key sources of economic inequity in any society include race, ethnicity, gender, age, and sexual orientation.

Conflict theorists agree with functionalists that many societal arrangements are functional. But, conflict theorists ask, who benefits? And who loses? When corporations merge, workers in lower-end jobs are often laid off while the salaries and benefits of corporate executives soar and the value of stocks (usually held by higher social classes) increase. Thus, mergers might be functional for those at the upper end of the socioeconomic ladder, but dysfunctional for those on the lower rungs.

SOCIAL INEQUALITY

Unlike functionalists, conflict theorists see society not as cooperative and harmonious, but as a system of widespread inequality. For conflict theorists, there's a continuous tension between the haves and the have-nots, most of whom are children, women, minorities, people with low incomes, and the poor.

Many conflict theorists focus on how those in power—typically wealthy white Anglo-Saxon Protestant males

"Sometimes the best man for the job isn't."
Author Unknown

(WASPs)—dominate political and economic decision making in U.S. society. This group controls a variety of institutions—such as education, criminal justice, and the media—and passes laws that benefit primarily people like themselves (see Chapters 8 and 11).

CRITICAL EVALUATION

Conflict theory explains how societies create and cope with disagreements. However, some have criticized conflict theorists for overemphasizing competition and coercion at the expense of order and stability. Inequality exists and struggles over scarce resources occur, critics agree, but conflict theorists often ignore cooperation and harmony. Voters, for example, can boot dominant white males out of office and replace them with women and minority group members. Critics also point out that the have-nots can increase their power through negotiation, bargaining, lawsuits, and strikes.

1-5c Feminist Theories

You'll recall that influential male theorists generally overlooked or marginalized early female sociologists' contributions. Until the feminist activism of the 1960s and 1970s, men—who dominated universities and scholarship—were largely "blind to the importance of gender" (Kramer and Beutel, 2015: 17).

Feminist scholars agree with contemporary conflict theorists that much of society is characterized by tension and struggle, but **feminist theories** go a step further because they examine women's social, economic, and political inequality. The theories maintain that women often suffer injustice primarily because of their gender rather than personal inadequacies such as low educational levels or not caring

conflict theory examines how and why groups disagree, struggle over power, and compete for scarce resources.

feminist theories examine women's social, economic, and political inequality.

about success. Feminist scholars assert that people should be treated fairly and equally regardless not only of their sex but also of other characteristics such as their race, ethnicity, national origin, age, religion, class, sexual orientation, or disability. They emphasize that women should be freed from traditionally oppressive expectations, constraints, roles, and behavior (see Reger, 2012).

FOCUSING ON GENDER

Feminist scholars have documented women's historical exclusion from most sociological analyses (see, for example, Smith, 1987, and Adams and Sydie, 2001). Before the 1960s women's movement in the United States, very few sociologists published anything about gender roles, women's sexuality, fathers, or intimate partner violence. According to sociologist Myra Ferree (2005: B10), during the 1970s, "the Harvard social-science library could fit all its books on gender inequalities onto a single half-shelf." Because of feminist scholars, many researchers—both women and men—now routinely include gender as an important research variable on both micro and macro levels.

Globally, except for some predominantly Muslim countries, solid majorities of both women and men support gender equality and agree that women should be able to work outside the home. When jobs are scarce, however, many women and men believe that men should be given preferential treatment (see *Figure 1.2*). Thus, even many equal rights proponents place a higher priority on men's economic rights.

LISTENING TO MANY VOICES

Feminist scholars contend that gender inequality is central to *all* behavior, ranging from everyday interactions to political and economic institutions, but feminist theories encompass many perspectives. For example, *liberal feminism* emphasizes social and legal reform to create equal opportunities for women. *Radical feminism* sees male dominance in social institutions (e.g., as the economy and politics) as the major cause of women's inequality. *Global feminism* focuses on how the

"I myself have never been able to find out precisely what feminism is; I only know that people call me a feminist whenever I express sentiments that *differentiate me from a doormat.*"

Rebecca West, British journalist

intersection of gender with race, social class, and colonization has exploited women in the developing world (see Lengermann and Niebrugge-Brantley, 1992). Most of us are feminists because we endorse equal opportunities for women and men in the economy, politics, education, and other institutions.

CRITICAL EVALUATION

Feminist scholars have challenged employment discrimination, particularly practices that routinely exclude women who aren't part of the "old boy network" (Wenneras and Wold, 1997). One criticism, however, is that many feminists are part of an "old girl network" that hasn't always welcomed different points of view from black, Asian American, American Indian, Muslim, Latina, lesbian, working-class, and disabled women (Lynn and Todoroff, 1995; Jackson, 1998; Sánchez, 2013).

A second criticism is that feminist perspectives tend to downplay social class inequality by focusing on low-income and minority women but not on their male counterparts. Thus, some contend, feminist theories aren't as gender balanced as they claim. Some critics, including feminists, also question whether feminist scholars have lost their bearings by concentrating on personal issues such as greater sexual freedom rather than broader social issues, particularly wage inequality (Rowe-Finkbeiner, 2004; Chesler, 2006; Shteir, 2013).

1-5d Symbolic Interaction

Symbolic interaction theory (sometimes called *interactionism*) is a micro-level perspective that examines people's everyday behavior through the communication of knowledge, ideas, beliefs, and attitudes. Whereas functionalists, conflict theorists, and some feminist theories emphasize structures and large (macro) systems, symbolic interactionists focus on *process* and keep the *person* at the center of their analysis.

There have been many influential symbolic interactionists whom we'll cover in later chapters. In brief, George Herbert Mead's (1863–1931) proposal that the human mind and self arise in the process of social communication became the foundation of the symbolic interaction schools of thought in sociology and social psychology. Herbert Blumer (1900–1987) coined the term

symbolic interaction theory (*interactionism*) examines people's everyday behavior through the communication of knowledge, ideas, beliefs, and attitudes.

FIGURE 1.2 WOMEN SHOULD HAVE THE SAME RIGHTS AS MEN. WELL, MAYBE NOT ALWAYS...

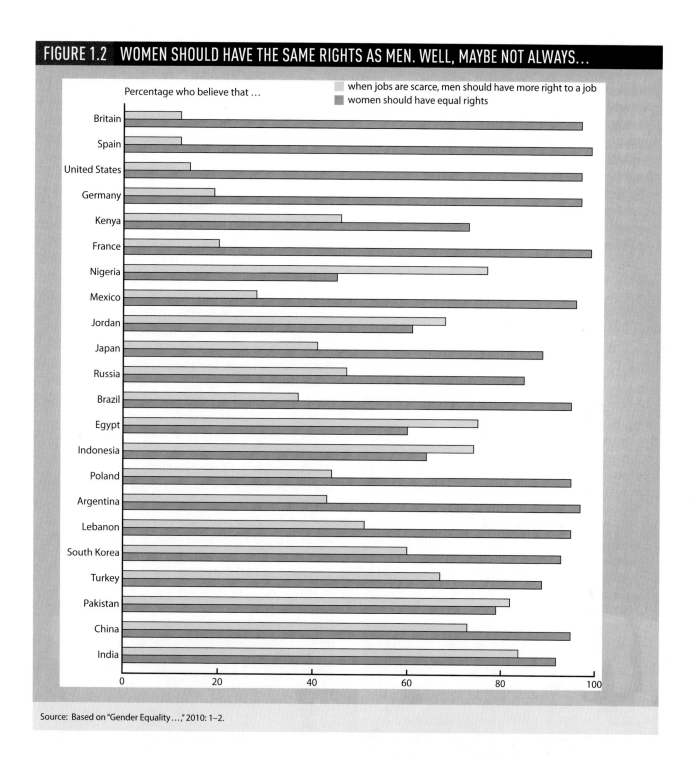

Source: Based on "Gender Equality...," 2010: 1–2.

symbolic interactionism in 1937, developed Mead's ideas, and emphasized that people interpret or "define" each other's actions, especially through symbols, instead of merely reacting to them.

Erving Goffman (1922–1982) enriched these earlier theories by examining human interaction in everyday situations ranging from jobs to funerals. Among his other contributions, Goffman used "dramaturgical analysis" to compare everyday social interaction to a theatrical presentation (see Chapter 5).

CONSTRUCTING MEANING

Our actions are based on **social interaction** in the sense that people take each other into account in their own behavior. Thus, we act differently in different social settings and continuously adjust our behavior, including our body language, as we interact (Goffman, 1959; Blumer, 1969).

social interaction a process in which people take each other into account in their own behavior.

A woman's interactions with her husband differ from those with her children. And she will interact still differently when she is teaching, talking to a colleague in the hall, or addressing an audience at a professional conference.

For symbolic interactionists, society is *socially constructed* through human interpretation (O'Brien and Kollock, 2001). That is, meanings aren't inherent but are created and modified through interaction with others. For example, a daughter who has batting practice with her dad will probably interpret her father's behavior as loving and involved. In contrast, she'll see batting practice with her baseball coach as less personal and more goal-oriented. In this sense, our interpretations of even the same behavior, such as batting practice, vary across situations and depend on the people with whom we interact.

SYMBOLS AND SHARED MEANINGS

Symbolic interaction looks at subjective, interpersonal meanings and at the ways in which we interact with and influence each other by communicating through *symbols*—words, gestures, or pictures that stand for something and that can have different meanings for different individuals.

After the 9/11 terrorist attacks, many Americans displayed the flag on buildings, bridges, homes, and cars to show their solidarity and pride in the United States. In contrast, some groups in the Middle East burned the U.S. flag to show their contempt for U.S. culture and policies. Thus, symbols are powerful forms of communication that show how people feel and interpret a situation.

To interact effectively, our symbols must have *shared meanings*, or agreed-on definitions. One of the most important of these shared meanings is the *definition of the*

situation, or the way we perceive reality and react to it. Relationships often end, for example, because partners view emotional closeness differently ("We broke up because Tom wanted sex. I wanted more communication."). We typically learn our definitions of the situation through interaction with *significant others*—especially parents, friends, relatives, and teachers—who play an important role in our socialization (as you'll see in Chapters 4 and 5).

CRITICAL EVALUATION

Unlike other theorists, symbolic interactionists show how people play an active role in shaping their lives on a micro level. One of the most common criticisms, however, is that symbolic interaction overlooks the widespread impact of macro-level factors such as economic forces, social movements, and public policies on our everyday behavior and relationships. During economic downturns, for example, unemployment and ensuing financial problems create considerable interpersonal conflict among couples and families (see Chapters 11 and 12). Symbolic interaction rarely considers such macro-level changes in explaining everyday behavior.

A related criticism is that interactionists sometimes have an optimistic and unrealistic view of people's everyday choices. Most of us enjoy little flexibility in our daily lives because deeply embedded social arrangements and practices benefit those in power. For example, people are usually powerless when corporations transfer jobs overseas or cut the pension funds of retired employees.

Some also believe that interaction theory is flawed because it ignores the irrational and unconscious aspects of human behavior (LaRossa and Reitzes, 1993). People

For many people, a diamond, especially in an engagement ring, signifies love and commitment. For others, diamonds represent Western exploitation of poor people in Africa who are paid next to nothing for their backbreaking labor in mining these stones.

TABLE 1.2 LEADING CONTEMPORARY PERSPECTIVES IN SOCIOLOGY

THEORETICAL PERSPECTIVE	FUNCTIONALISM	CONFLICT	FEMINIST	SYMBOLIC INTERACTION
Level of Analysis	**Macro**	**Macro**	**Macro and Micro**	**Micro**
Key Points	• Society is composed of interrelated, mutually dependent parts. • Structures and functions maintain a society's or group's stability, cohesion, and continuity. • Dysfunctional activities that threaten a society's or group's survival are controlled or eliminated.	• Life is a continuous struggle between the haves and the have-nots. • People compete for limited resources that are controlled by a small number of powerful groups. • Society is based on inequality in terms of ethnicity, race, social class, and gender.	• Women experience widespread inequality in society because, as a group, they have little power. • Gender, ethnicity, race, age, sexual orientation, and social class—rather than a person's intelligence and ability—explain many of our social interactions and lack of access to resources. • Social change is possible only if we change our institutional structures and our day-to-day interactions.	• People act on the basis of the meaning they attribute to others. Meaning grows out of the social interaction that we have with others. • People continuously reinterpret and reevaluate their knowledge and information in their everyday encounters.
Key Questions	• What holds society together? How does it work? • What is the structure of society? • What functions does society perform? • How do structures and functions contribute to social stability?	• How are resources distributed in a society? • Who benefits when resources are limited? Who loses? • How do those in power protect their privileges? • When does conflict lead to social change?	• Do men and women experience social situations in the same way? • How does our everyday behavior reflect our gender, social class, age, race, ethnicity, sexual orientation, and other factors? • How do macro structures (such as the economy and the political system) shape our opportunities? • How can we change current structures through social activism?	• How does social interaction influence our behavior? • How do social interactions change across situations and between people? • Why does our behavior change because of our beliefs, attitudes, values, and roles? • How is "right" and "wrong" behavior defined, interpreted, reinforced, or discouraged?
Example	• A college education increases one's job opportunities and income.	• Most low-income families can't afford to pay for a college education.	• Gender affects decisions about a major and which college to attend.	• College students succeed or fail based on their degree of academic engagement.

don't always consider the meaning of their actions or behave as reflectively as interactionists assume. Instead, we often act impulsively or say hurtful things without weighing the consequences of our actions or words.

1-5e Other Theoretical Approaches

Table 1.2 summarizes the major sociological perspectives that you've just read about. However, new theoretical perspectives arise because society is always changing. For example, *postmodern theory* attempts to explain social life in contemporary societies that are characterized by postindustrialization, consumerism, and global communications.

Sociology, like other social sciences, has subfields. The subfields—such as socialization, deviance, and social stratification—offer specific theories that reinforce and illustrate functionalist, conflict, feminist, and interactionist approaches. No single theory explains social life completely. Each theory, however, provides different insights that guide sociological research, the topic of the next chapter.

STUDY TOOLS 1

READY TO STUDY? IN THE BOOK, YOU CAN:

☐ Check your understanding of what you've read with the Test Your Learning Questions provided on the chapter review card at the back of the book.

☐ Rip out the chapter review card for a handy summary of the chapter and key terms.

ONLINE AT CENGAGEBRAIN.COM YOU CAN:

☐ Collect StudyBits while you read and study the chapter.

☐ Quiz yourself on key concepts.

☐ Prepare for tests with chapter flash cards as well as flash cards that you create.

☐ Watch videos to explore sociology further.

© Anelina/Shutterstock.com

2 | Examining Our Social World

LEARNING OUTCOMES

After studying this chapter, you will be able to...

2-1 Compare knowledge based on tradition, authority, and research.

2-2 Explain why sociological research is important in our everyday lives.

2-3 Describe the scientific method.

2-4 Describe the basic steps of the research process.

2-5 Compare and illustrate the five most common sociological data collection methods, including their strengths and limitations.

2-6 Explain why ethics are important in scientific research.

After you finish this chapter go to **PAGE 37** for **STUDY TOOLS**.

Spring break is all about beer fests, wet T-shirt contests, frolicking on the beach, and hooking up, right? Maybe not. A national survey found that 70 percent of college students stay home with their parents, and 84 percent of those who throng to vacation spots report consuming alcohol in moderation (The Nielsen Company, 2008). If you suspect that these numbers are too high or too low and wonder how the survey was done, you're thinking like a researcher, the focus of this chapter.

2-1 HOW DO WE KNOW WHAT WE KNOW?

Much of our knowledge is based on *tradition*, a handing down of statements, beliefs, and customs from generation to generation ("The groom's parents should pay for the wedding rehearsal dinner"). Another common source of knowledge is *authority*, a socially accepted source of information that includes "experts," parents, government officials, police, judges, and religious leaders ("My mom says that . . ." or "According to the American Heart Association . . .").

Knowledge based on tradition and authority simplifies our lives because it provides us with basic rules about socially and legally acceptable behavior. The

What do you think?

People can find data to support any opinion they have.

1	2	3	4	5	6	7
strongly agree						strongly disagree

information can be misleading or wrong, however. Suppose a 2-year-old throws a temper tantrum at a family barbecue. One adult comments, "What that kid needs

20   SOC

is a smack on the behind." Someone else immediately disagrees: "All kids go through this stage. Just ignore it."

Who's right? To answer this and other questions, sociologists rely on **research methods**, organized and systematic procedures to gain knowledge about a particular topic. Much research shows, for example, that neither ignoring a problem nor inflicting physical punishment (such as spanking) stops a toddler's bad behavior. Instead, most young children's misbehavior can be curbed by having simple rules, being consistent in disciplining misbehavior, praising good behavior, and setting a good example (see Benokraitis, 2015).

2-2 WHY IS SOCIOLOGICAL RESEARCH IMPORTANT IN OUR EVERYDAY LIVES?

In contrast to knowledge based on tradition and authority, sociological research is important in our everyday lives for several reasons:

1. It challenges overgeneralizations. A common reason for the disconnection between reality and perception is *overgeneralization*, drawing conclusions about behavior or events based on limited observations. U.S. gun homicide rates dropped by 49 percent from 1993 to 2011, but 56 percent of Americans believe that the rate has gone up. This misperception—fueled by mass shootings, the media's focus on crime, and powerful lobby groups such as the National Rifle Association—is partly responsible for the growth of gun ownership in the past 20 years (Cohn et al., 2013). In effect, then, overgeneralization can change behavior.

2. It unmasks myths. According to many newspapers and television shows, suicide rates are highest during the end of the Christmas holidays. In fact, suicide rates are lowest in December and highest in the spring and fall (but the reasons for these peaks are unclear). Another myth is that more women are victims of domestic violence on Super Bowl Sunday than on any other day of the year, presumably because men become intoxicated and abusive. In fact, intimate partner violence rates are high, and consistent, throughout the year (Romer, 2011; "Super Bull Sunday," 2015).

research methods organized and systematic procedures to gain knowledge about a particular topic.

"Facebook Causes 20 Percent of Today's Divorces." What?!

The founder and self-described leader of the United Kingdom's online divorce site (Divorce-Online) sent out a press release titled "Facebook Is Bad for Your Marriage—Research Finds," and claimed that Facebook causes 20 percent of today's divorces. News media around the world ran stories about this press release with headlines such as "Facebook to Blame for Divorce Boom." You'll see in Chapter 12 that there's no divorce boom, so where did the 20 percent number come from?

In 2009, the managing director of Divorce-Online scanned its online divorce petition database for use of the word *Facebook*, and found 989 instances in about 5,000 petitions. Divorce-Online never said that the petitions were only those filed by members of the American Academy of Matrimonial Lawyers, who comprise a very small percentage of all divorce attorneys. Two years later, many Internet sites and blogs were still spreading the fiction that "Facebook Causes Divorce" (see Bialik, 2011). In reality, as you'll see in Chapter 12, there are a number of interrelated macro- and micro-level reasons for divorce; there's no single "cause," much less a website.

3. **It helps explain *why* people behave as they do.** A recent study predicted that older drivers, particularly those age 70 and older, would be more likely than younger drivers to have fatal crashes, but found the opposite (Braitman et al., 2011). The researchers couldn't explain why their prediction turned out to be false. Sociologists posit that older Americans are less likely to have fatal crashes because many avoid driving at night or during bad weather, and they're much less likely than younger drivers to use cell phones or text while driving (Halsey, 2010).

4. **It influences social policies.** In the past, many psychologists attributed school bullying to students with emotional and behavioral problems. Recently, however, sociologists have found that bullying is due largely to jockeying for social status rather than to mental or emotional problems (Faris and Felmlee, 2011). That is, the most aggressive bullies are trying to prevent their rivals from becoming members of a higher-ranking group. Because of these findings, some educators have enlisted the support of students at the top of the high school social ladder to prevent or decrease bullying (Parker-Pope, 2011).

5. **It sharpens critical thinking skills.** Many Americans, particularly women, rely on talk shows for information on a number of topics. During 2009 alone, Oprah Winfrey featured and applauded guests who maintained, among other things, that children contract autism from the measles, mumps, and rubella (MMR) vaccinations they receive as babies; that fortune cards can help people diagnose their illnesses; and that people can wish away cancer (Kosova and Wingert, 2009). *All of these claims are false.*

Such misinformation can be dangerous. Because of the "MMR vaccinations can cause autism" scare, about 30 percent of U.S. parents are hesitant to vaccinate their children (Kennedy et al., 2011). Partly because of such fears, by mid-2015, the United States was experiencing the largest increase in measles cases since 1996. Up to three of every 1,000 measles cases are fatal (Clemmons et al., 2015; see also Gibson, 2012, and "Clueless," 2014, for other recent examples of "celebrity bogus science").

2-3 THE SCIENTIFIC METHOD

Sociologists rely on the **scientific method**, a body of objective and systematic techniques to investigate phenomena, acquire knowledge, and test hypotheses and theories. The techniques include careful data collection, exact measurement, accurate recording and analysis of the findings, thoughtful interpretation of results, and, when appropriate, generalization of the findings to a larger group. Before collecting any data, however, social scientists must grapple with a number of research-related issues. Let's begin with concepts, variables, and hypotheses.

2-3a Concepts, Variables, and Hypotheses

A basic element of the scientific method is a **concept**—an abstract idea, mental image, or general notion that represents some aspect of the world. Some examples of concepts are "blood pressure," "religion," and "marriage."

scientific method a body of objective and systematic techniques for investigating phenomena, acquiring knowledge, and testing hypotheses and theories.

concept an abstract idea, mental image, or general notion that represents some aspect of the world.

Because concepts are abstract and may vary among individuals and cultures, scientists rely on variables to measure (*operationalize*) concepts. A **variable** is a characteristic that can change in value or magnitude under different conditions. Variables can be attitudes, behaviors, or traits (e.g., ethnicity, age, and social class).

An **independent variable** is a characteristic that has an effect on the **dependent variable**, the outcome. A **control variable** is a characteristic that is constant and unchanged during the research process.

Scientists can simply ask a research question ("Why are people poor?"), but they usually begin with a **hypothesis**—a statement of the expected relationship between two or more variables—such as "Unemployment increases poverty." In this example, "unemployment" is the independent variable and "poverty" is the dependent variable.

Researchers might also use control variables, such as education, to explain the relationship between unemployment and poverty. For example, people with at least a college degree generally have lower poverty rates than those with lower educational levels because the former are less likely to experience long periods of unemployment.

2-3b Deductive and Inductive Reasoning

Deduction and induction are two different but equally valuable approaches in examining the relationship between variables. Generally, **deductive reasoning** begins with a theory, prediction, or general principle that is then tested through data collection. An alternative mode of inquiry, **inductive reasoning**, begins with specific observations, followed by data collection, a conclusion about patterns or regularities, and the formulation of hypotheses that can lead to theory construction (see *Figure 2.1*).

FIGURE 2.1   DEDUCTIVE AND INDUCTIVE APPROACHES

DEDUCTIVE* — GENERAL — START

Theory → Hypothesis → Data collection → Confirmation or rejection of hypothesis/theory — SPECIFIC

INDUCTIVE* — SPECIFIC

Observation → Data collection → Conclusion about patterns or regularities → Hypothesis → Theory — GENERAL

* END

*Note that the components are very similar in both approaches.

theory
+ hypothesis
+ variables
+ data
+ explanation of results
―――――――――――
= foundation of scientific method

variable a characteristic that can change in value or magnitude under different conditions.

independent variable a characteristic that has an effect on the dependent variable.

dependent variable the outcome that may be affected by the independent variable.

control variable a characteristic that is constant and unchanged during the research process.

hypothesis a statement of the expected relationship between two or more variables.

deductive reasoning an inquiry process that begins with a theory, prediction, or general principle that is then tested through data collection.

inductive reasoning an inquiry process that begins with a specific observation, followed by data collection, a conclusion about patterns or regularities, and the formulation of hypotheses that can lead to theory construction.

Taking a deductive approach, you might decide to test a theory of academic success using the following hypothesis: "Students who study in groups perform better on exams than those who study alone." You would collect the data, ultimately confirming or rejecting your hypothesis (or theory).

Alternatively, you might notice that your classmates who participate in study groups seem to get higher grades on exams than those who study alone. Using an inductive approach, you would collect the data systematically and formulate hypotheses (or suggest a theory) that could then be tested deductively. Most social science research involves both inductive and deductive reasoning.

2-3c Reliability and Validity

Sociologists are always concerned about reliability and validity. **Reliability** is the *consistency* with which the same measure produces similar results time after time. If, for example, you ask "How old are you?" on two subsequent days and a respondent gives two different answers, such as 25 and 30, there's either something wrong with the question or the respondent is lying. Respondents might lie, but scientists must make sure that their measures are as reliable as possible.

Validity is the degree to which a measure is accurate and *really* measures what it claims to measure. Consider student course evaluations. The measures of a "good" professor often include items such as "The instructor is interesting" and "The instructor is fair."

Because we don't know what students mean by "interesting" and "fair," how accurate are such measures in differentiating between "good" and "bad" professors? A study at two large public universities found that a third of the students admitted being dishonest in end-of-semester course evaluations. Some fibbed to make their instructors look good, but most lied to "punish"

Are *The Voice* voters an example of a probability or nonprobability sample of the show's fans?

Tyler Golden/NBC/NBCU Photo Bank via Getty Images

professors they didn't like, especially when they received lower grades than they thought they deserved (Clayson and Haley, 2011). Such research findings raise questions about the accuracy and usefulness of student course evaluations in measuring an instructor's actual performance.

2-3d Sampling

Early in the research process, sociologists decide what sampling procedures to use. Ideally, researchers would like to study all the units of the population in which they're interested—say, all adolescents who use drugs. A **population** is any well-defined group of people (or things) about which researchers want to know something. Obtaining information about and from populations is problematic, however. The population may be so large that it would be too expensive and time consuming to conduct the research. In other cases—such as all adolescents who use drugs—it's impossible even to identify the population.

As a result, researchers typically select a **sample**, a group of people (or things) that's representative of the population they wish to study. In obtaining a sample, researchers decide whether to use probability or nonprobability sampling. A **probability sample** is one in which each person (or thing, such as an email address) has an equal chance of being selected because the selection process is *random*. The most desirable characteristic of a probability sample is that the results can be generalized to the larger population because all the people (or things) have had an equal chance of being selected.

In a **nonprobability sample**, there's little or no attempt to get a representative cross section of the population. Instead, researchers use sampling criteria such as convenience or the availability of respondents

reliability the consistency with which the same measure produces similar results time after time.

validity the degree to which a measure is accurate and really measures what it claims to measure.

population any well-defined group of people (or things) about whom researchers want to know something.

sample a group of people (or things) that's representative of the population researchers wish to study.

probability sample each person (or thing) has an equal chance of being selected because the selection is random.

nonprobability sample there is little or no attempt to get a representative cross section of the population.

or information. Nonprobability samples are especially useful when sociologists are exploring a new topic or want to get people's insights on a particular topic before launching a larger study (Babbie, 2013).

Television news programs, newsmagazines, and entertainment shows often provide a toll-free number, a texting number, or a Web address and encourage viewers to vote on an issue (such as whether marijuana should be legal in all states). How representative are these voters of the general population? And how many enthusiasts skew the results by voting more than once?

But, you might think, if as many as 100,000 people respond, doesn't such a large number indicate what most people think? No. Because the respondents are self-selected and don't comprise a random sample, they're not representative of a population.

2-3e The Time Dimension

Researchers have two principal ways of dealing with the time issue: cross-sectional studies and longitudinal studies. The data can be *longitudinal* (collected at two or more points in time from the same or different samples of respondents) or *cross-sectional* (collected at one point in time). *Figure 2.2* shows a change over time in Americans' attitudes toward same-sex marriage; this is an example of a longitudinal study. If the researchers had collected data at only one point in time (2015, 2000, or 1996), this would have been a cross-sectional study. Cross-sectional studies provide valuable information, but longitudinal studies are especially useful in examining trends in behavior or attitudes; a researcher can compare similar populations across different years or follow a particular group of people over time.

2-3f Qualitative and Quantitative Approaches

In **qualitative research**, sociologists examine nonnumerical material that they then interpret. In a study of grandfathers who were raising their grandchildren, for example, the researcher tape-recorded in-depth interviews and then analyzed the responses to questions about financial issues and daily parenting tasks (Bullock, 2005).

In **quantitative research**, sociologists focus on a numerical analysis of people's responses or specific characteristics, studying a wide range of attitudes, behaviors, and traits (e.g., homeowners versus renters).

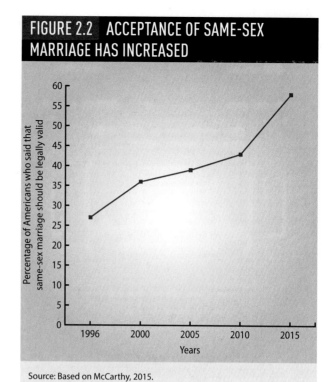

FIGURE 2.2 ACCEPTANCE OF SAME-SEX MARRIAGE HAS INCREASED

Percentage of Americans who said that same-sex marriage should be legally valid

Years

Source: Based on McCarthy, 2015.

Stockbyte/Getty Images

C Squared Studios/ Photodisc/Getty Images

In one national probability study, for example, the researchers surveyed almost 7,000 respondents to understand the influence of grandparents who live with their children and grandchildren (Dunifon and Kowaleski-Jones, 2007).

Which approach should a researcher use? It depends on her or his purpose. Consider college attrition. Quantitative data provide information on characteristics such as national college graduation rates. Qualitative data, in contrast, yield in-depth descriptions of why some college students drop out whereas others graduate. In many studies, sociologists use both approaches.

2-3g Correlation Is Not Causation

Ideally, researchers would like to determine **causation**, a relationship in which one variable is the direct consequence of another. For example, the more groceries you

qualitative research examines nonnumerical material and interprets it.

quantitative research focuses on a numerical analysis of people's responses or specific characteristics.

causation a relationship in which one variable is the direct consequence of another.

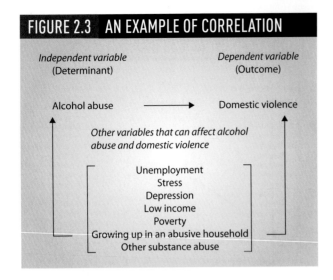

FIGURE 2.3   AN EXAMPLE OF CORRELATION

Independent variable
(Determinant)

Dependent variable
(Outcome)

Alcohol abuse ⟶ Domestic violence

Other variables that can affect alcohol abuse and domestic violence

Unemployment
Stress
Depression
Low income
Poverty
Growing up in an abusive household
Other substance abuse

buy, the more money you spend (buying groceries causes you to spend money). Most human behavior is much more complex. Because causation is difficult to prove and nothing in life (except death) is certain, sociologists and other scientists calculate a **correlation**, the relationship between two or more variables. For example, much research shows that there's an association between alcohol abuse and domestic violence. Moreover, the more frequent the alcohol abuse, the greater the likelihood of domestic violence (see Chapter 12).

Alcohol abuse and domestic violence often occur together, but this doesn't mean that one *causes* the other. Domestic violence also occurs when people don't drink, and not all people who abuse alcohol become violent or aggressive towards others. Instead, there may be other factors that affect alcohol abuse, domestic violence, or both variables (see *Figure 2.3*). As a result, sociologists rarely use the term *cause* when interpreting their results because they can't *prove* that there's a cause-and-effect relationship. Instead, a researcher might conclude that alcohol abuse is "associated (or correlated) with," "contributes to," or "increases the likelihood of" rather than "causes" domestic violence.

(2-4) THE RESEARCH PROCESS: THE BASICS

Hypotheses construction, deductive and inductive reasoning, establishing reliability and validity, and sampling are some of the preliminary and often most challenging steps in the research process. *Figure 2.4* outlines the scientific

method, using a deductive approach that begins with an idea and ends with writing up (and sometimes publishing) the results. Later in this chapter, we'll examine some studies that use an inductive approach.

1. **Choose a topic to study.** The topic can be general or very specific. Some sociologists begin with a new question or idea; others extend or refine previous research findings. A topic can generate new information, replicate a previous study, or propose an intervention (e.g., a new substance abuse program).

2. **Summarize the related research.** In what is often called a *literature review*, a sociologist summarizes the pertinent research, shows how her or his topic is related to previous and current research, and indicates how the study will extend the body of knowledge. If the research is applied, a sociologist also explains how the proposed service or program will improve people's lives.

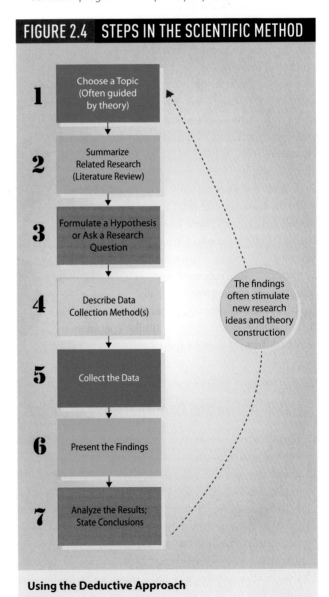

FIGURE 2.4   STEPS IN THE SCIENTIFIC METHOD

1 Choose a Topic (Often guided by theory)

2 Summarize Related Research (Literature Review)

3 Formulate a Hypothesis or Ask a Research Question

4 Describe Data Collection Method(s)

5 Collect the Data

6 Present the Findings

7 Analyze the Results; State Conclusions

The findings often stimulate new research ideas and theory construction

Using the Deductive Approach

correlation the relationship between two or more variables.

3. Formulate a hypothesis or ask a research question. A sociologist next states a hypothesis or asks a research question. In either case, she or he has to be sure that the measures of the variables are as reliable and valid as possible.

4. Describe the data collection method(s). A sociologist describes which method or combination of methods (sometimes called *methodology, procedure,* or *research design*) is best for testing a hypothesis or answering a research question. This step also describes sampling, the sample size, and the respondents' characteristics.

5. Collect the data. The actual data collection might rely on fieldwork, surveys, experiments, or existing sources of information such as Census Bureau data.

6. Present the findings. After coding (tabulating the results) and running statistical tests, a sociologist presents the findings as clearly as possible.

7. Analyze and explain the results. After analyzing the data, a sociologist explains why the findings are important. This can be done in many ways. She or he might show how the results provide new information, enrich our understanding of behavior or attitudes that researchers have examined previously, or refine existing theories or research approaches.

In drawing conclusions about the study, sociologists typically discuss its implications. For instance, does a study of juvenile arrests suggest that new policies should be implemented, that existing ones should be changed, or that current police practices may be affecting the arrest rates? That is, the researcher answers the question "So what?" by showing the importance and usefulness of the study.

2-5 SOME MAJOR DATA COLLECTION METHODS

Sociologists typically use one or more of the following major data collection methods: surveys, field research, content analysis, experiments, and secondary analysis of existing data. Because each method has strengths and weaknesses, researchers must decide which will provide the most accurate information, given time and budget constraints.

"How long have you been dead? Do you have any complaints about your treatment here? Do you have any suggestions?"

Farris, Joseph/CSL, CartoonStock Ltd

2-5a Surveys

Many sociologists use **surveys** that include questionnaires, face-to-face or telephone interviews, or a combination of these techniques. Two important elements in survey research are sampling and constructing a series of questions for *respondents*, the people who answer the questions.

SELECTING A SAMPLE

Random sample surveys are preferred because the results can be generalized to a larger population. Researchers can obtain representative samples through *random digit dialing*, which involves selecting area codes and exchanges (the next three numbers) followed by four random digits. In the procedure called *computer-assisted telephone interviewing* (CATI), the interviewer uses a computer to select random telephone numbers, reads the questions to the respondent from a computer screen, and then enters the answers in precoded spaces, saving time and expense by not having to reenter the data after the interview.

Two types of electronic surveys are becoming increasingly popular. The first type is a survey sent via email, either as text in the body of the message or as an attachment. The second type is the more familiar survey that is posted on a website. Electronic surveys are cost effective and can represent a broad spectrum of the general population. In addition, Web surveys provide

survey a data collection method that includes questionnaires, face-to-face or telephone interviews, or a combination.

respondents with visual material, including videos, to look at and respond to (Keeter, 2009, 2010).

QUESTIONNAIRES AND INTERVIEWS

A survey must have a specific plan for asking questions and recording answers. The most common way to do this is to give respondents a **questionnaire**, a series of written questions that ask for information. The questions can be *closed-ended* (the researcher provides a list of answers that a respondent chooses), *open-ended* (the researcher asks respondents to answer questions in their own words) (see *Table 2.1*), or a combination. Questionnaires can be mailed, used during an interview, or self-administered (e.g., student course evaluations, Web surveys).

The **interview**, in which a researcher directly asks respondents a series of questions, is another way to collect survey data. Interviews can be conducted face-to-face or by telephone. *Structured interviews* use closed-ended questions, whereas *unstructured interviews* use open-ended questions that allow respondents to answer as they wish.

STRENGTHS

Surveys are usually inexpensive, simple to administer, and have a fast turnaround. Because the results are anonymous, respondents are generally willing to answer questions on sensitive topics such as income, sexual behavior, and drug usage (Hamby and Finkelhor, 2001).

Researchers often simplify surveys to increase response rates. During the 2010 census, for example, the Census Bureau used only a short form and slogans such as "10 Questions, 10 Minutes" to encourage people to mail back the forms. And, for the first time, the Census Bureau used six different languages for questionnaires, and hired staff who spoke over 150 different languages. Doing so generated higher response rates, reduced the need for expensive follow-up interviews, and saved millions of dollars (Groves and Vitrano, 2011; Mather et al., 2011; Wozniak and Groves, 2011).

TABLE 2.1	ONE QUESTION—TWO FORMATS

Open-Ended Version

How would you describe your current financial situation?

Closed-Ended Version

How would you describe your current financial situation?

[ ] Excellent	[ ] Fair	[ ] Terrible
[ ] Very good	[ ] Poor	[ ] Not sure
[ ] Good		

Suppose that you're the researcher on an income study. Do you think that you'd use the open- or closed-ended version? Why?

Now, suppose that you're the respondent. Do you think that you'd be more likely to answer the open- or closed-ended version? Why?

Telephone interviews are popular because they're a relatively inexpensive way to collect data. Face-to-face interviews have high response rates (often up to 85 percent) because they involve personal contact. People are more likely to discuss sensitive issues in an interview than via a mailed questionnaire, a phone survey, or an electronic survey. If respondents don't understand a question, are reluctant to answer, or give incomplete answers, interviewers can clarify, keep respondents from digressing, or *probe* by asking respondents to elaborate on an answer (Babbie, 2013).

With the innovation of "robo-polls," the entire interview is conducted by a programmed recording that interprets the respondent's spoken answers, records them, and determines how to continue the interview. This method is cost effective because it cuts out the cost of hiring people, but respondents may be more reluctant to answer sensitive questions (Babbie, 2013).

LIMITATIONS

A major limitation of surveys that use mailed questionnaires is a low response rate, often only about 10 percent (Gray et al., 2007). If the questions are unclear, complicated, or seen as offensive, a respondent may simply throw the questionnaire away or give answers on subjects they know nothing about (Babbie, 2013).

A survey's wording can also substantially affect people's answers. According to a recent Gallup poll, for instance, 51 percent of Americans agreed that doctors should be allowed to "assist the patient to commit suicide," but a whopping 70 percent said that doctors should be allowed to "end the patient's life by some painless means." Thus, not using the word *suicide* was much

questionnaire a series of written questions that ask for information.

interview a researcher directly asks respondents a series of questions.

more acceptable even though the patient's outcome was the same (Saad, 2013).

Another concern is a *social desirability bias*—the tendency of respondents to give the answer that they think they "should" give or that will cast them in a favorable light. For example, the proportion of Americans who say they voted in a given election is always much higher than the actual number of votes cast (Taylor and Lopez, 2013).

People also underreport (or lie about) socially undesirable behaviors and traits (such as using illicit drugs or mental health problems), but exaggerate attendance at religious services, donating to charity, and washing their hands after using a restroom. In a recent survey, 28 percent of the respondents admitted that they had lied to their doctors about smoking, drinking alcohol, using illicit drugs, not exercising, having unsafe sex with multiple partners, and other risky behavior (Reddy, 2013; Bharadwaj et al., 2015; Keeter et al., 2015).

Spurred by the rising cost of telephone surveys and plummeting participation rates, *The New York Times* and CBS News recently announced that they'll use online-only surveys for their public-opinion polling. The announcement drew sharply divided views. Opponents maintain that because the surveys are based on nonprobability samples, the respondents are self-selected, and people who don't use the Internet tend to be older, poorer, and less educated than Internet users. Thus, the results of online surveys aren't accurate and nationally representative. Proponents argue, on the other hand, that the online surveys will increase the pool of people who are increasingly difficult to reach. Moreover, the potential bias from excluding non-Internet users is getting smaller—84 percent of U.S. adults were online in mid-2015 and the numbers are steadily increasing (DeSilver, 2014; Voosen, 2014; Perrin and Duggan, 2015). Future research will determine whether online surveys are as accurate as traditional telephone surveys.

Unlike questionnaires and telephone surveys, face-to-face interviews can be very expensive. They can also be dangerous. During the 2010 census, a number of Americans, frustrated with the economy and government, took their anger out on census takers. Hundreds of census workers were shot at with pellet guns; hit by baseball bats; spat at; and confronted with shotguns, packs of snarling pit bulls, pickaxes, crossbows, and hammers (Morello, 2010).

Because the survey is the research approach you'll encounter most often in sociology and many other disciplines, it's important to be an informed consumer. You can't simply assume that a survey, including a public opinion poll, is accurate or representative of

TABLE 2.2 CAN I TRUST THESE NUMBERS?

To determine a survey's credibility, ask:

- Who sponsored the survey? A government agency, a political organization, a business, or a group that's lobbying for change?

- What's the purpose of the survey? To provide objective information, to promote an idea or a political candidate, or to get attention through sensationalism?

- How was the sample drawn? Randomly? Or was it a "self-selected opinion poll" (a SLOP, according to Tanur, 1994)?

- How were the questions worded? Were they clear and objective, or loaded and biased? If the survey questions are not provided, why not?

- How did the researchers report their findings? Were they objective, or did they make value judgments?

a larger population. Asking a few basic questions about a survey, such as those in *Table 2.2*, will help you assess its credibility.

2-5b Field Research

In **field research**, sociologists collect data by systematically observing people in their natural surroundings. In *participant observation*, researchers interact with the people they're studying; they may or may not reveal their identities as researchers. If you recorded interaction patterns between students and professors during your classes, you would be engaging in participant observation. In *nonparticipant observation*, researchers study phenomena without being part of the situation (observing young children in classrooms through one-way mirrors). Researchers sometimes combine participant and nonparticipant observation.

Some field research studies are short term (e.g., observing, over a few weeks or months, whether and how parents discipline their unruly children in grocery stores). Others, called *ethnographies*, require a considerable amount of time in the field. Sudhir Venkatesh (2008), while a graduate student at the University of Chicago, spent more than 6 years studying the culture and members of the Black Kings, a crack-selling gang in Chicago's inner city.

Observational studies are usually highly structured and carefully designed projects in which data are recorded and then converted to quantitative summaries. These studies may examine complex communication

field research data collected by observing people in their natural surroundings.

Field researchers study a variety of topics, including how customers, cocktail waitresses, bouncers, and paid consultants engage in deception, hustling, and bribes to help people hook up at a nightclub (see Grazian, 2008, for a description of this study).

patterns, measure the frequency of acts (e.g., the number of head nods or angry statements), or note the duration of a particular behavior (e.g., the length of eye contact) (Stillars, 1991). Thus, observational studies are much more complex and sophisticated than they appear to the general public.

STRENGTHS

Unlike other data collection methods, field research provides rich detail in describing and understanding attitudes and behavior in the "real world." The researcher can explore new topics from the respondents' viewpoint using observation or in-depth interviews. Because observation usually doesn't disrupt the natural surroundings, the researcher doesn't directly influence the subjects.

Counting the homeless is an ongoing research problem. Communities must provide accurate numbers to qualify for federal funds, but census takers have difficulty getting such counts because safety rules prohibit them from entering private property (such as warehouses) or dark alleys. In one approach, the YWCA in Trenton, New Jersey, gave free haircuts at a fair for the homeless to attract the city's homeless people to a place where they could be counted (Jonsson, 2007). Is such field research unethical because it violates people's privacy? Or innovative in helping a community get more federal funding for the homeless?

Field research is also more flexible than some other methods. The researcher can modify the research design, for example, by deciding to interview (rather than just observe) key people after the research has started.

Validity is usually a strength of field research because researchers can determine how participants define and interpret concepts such as "success," "inequality," or "marital infidelity." Compared with other data collection methods, field research is also more flexible if a situation changes. Studying gang behavior, for example, may require shifting the focus of the analysis if members are killed or imprisoned (see Venkatesh, 2008). Finally, field research is a powerful data collection method because it describes a phenomenon using the subjects' own words.

LIMITATIONS

Observation can be expensive if a researcher needs elaborate recording equipment, must travel far or often, or has to live in a different society or community for an extended period. Researchers who study other cultures must often learn a new language, a time-consuming task.

A second limitation is the problem of *reactivity*: If people know they're being observed, they might behave differently. For example, a study of 392 restaurants from five different chains (such as the Olive Garden) found that employee theft decreased by 22 percent after management installed surveillance cameras (Pierce et al., 2013). In the case of participant observation, anything that the researcher does or doesn't do can affect what's being observed because the researcher's presence changes a group's dynamics, behavior, and outcomes.

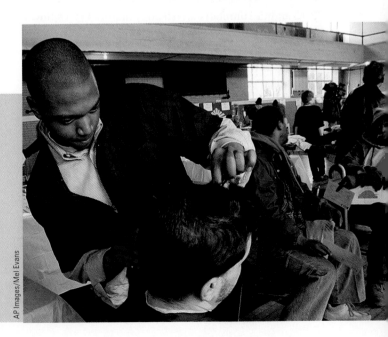

A field researcher may encounter other data collection challenges. Homeless and battered women's shelters, for example, are usually—and understandably—wary of researchers intruding on their residents' privacy. Even if the researcher has access to such groups, it's often difficult to be objective while collecting and interpreting the data because poverty and domestic violence can evoke strong emotional reactions (e.g., anxiety, anger against perpetrators, sympathy for victims). Finally, the findings can't be generalized because the data come from nonprobability samples.

2-5c Content Analysis

Content analysis systematically examines some form of communication. Researchers can apply this unobtrusive data collection method to almost any form of written or oral communication: speeches, TV programs, online blogs, advertisements, office emails, songs, diaries, advice columns, poems, or Facebook chatter, to name just a few.

The researcher develops categories for coding the material, sorts and analyzes the data in terms of frequency, intensity, or other characteristics, and draws conclusions about the results. Look at the two baby cards in *Figure 2.5*. What messages do they send about gender roles? These are only two cards, but we could do a content analysis of all birth congratulations at a particular store or from a number of stores or online sites to determine whether such cards reinforce stereotypical gender-role expectations.

Sociologists have used content analysis to examine a number of topics. A few examples are images of women and men in video games and music videos, changes in child-rearing advice in popular parenting magazines, gender and ethnic differences in yearbook photographs, and gender biases in job advertisements (Downs and Smith, 2010; Zhang et al., 2010; Wallis, 2011; Lindner, 2012; Wondergem and Friedlmeier, 2012).

STRENGTHS

Content analysis is usually inexpensive and often less time consuming than other data collection methods, particularly field research. If, for example, you wanted to examine the content of television commercials during football games, you wouldn't need fancy equipment, a travel budget, or a research staff.

A second advantage is that researchers can recode errors fairly easily. This isn't the case with surveys. If you mail a questionnaire with poorly constructed items, it's too late to change anything.

Third, content analysis is unobtrusive. Because researchers aren't dealing with human subjects, they don't need permission to do the research, and don't have to worry about influencing the respondents' attitudes or behavior.

Finally, researchers can use content analysis to compare societies at single or different points in time. In one study, the researchers analyzed gender portrayals in the 101 top-grossing G-rated films in the United States and Canada between 1990 and 2005 (Smith et al., 2010; see also Joshi et al., 2011). It would be very difficult, using most other data collection methods, to analyze such material over a 15-year period.

LIMITATIONS

Content analysis can be very labor intensive, especially if a project is ambitious. In one study, the researchers examined the amount and intensity of violence in children's animated movies that were released between 1938 and 1999 (Yokota and Thompson, 2000). It took several years to code one or more of the major characters' words, expressions, and actions.

A second disadvantage is that the coding may be subjective. Having several researchers on a project can increase coding objectivity, but only one researcher often codes the content.

FIGURE 2.5 HOW DO VIEWS OF FEMALE AND MALE BABIES DIFFER?

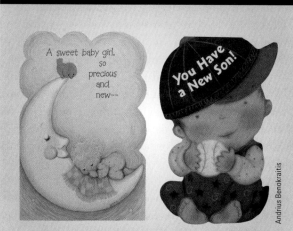

Andrius Benokraitis

As these cards illustrate, girls and boys are viewed differently from the time they're born. In these and other baby cards, girls, but not boys, are typically described as "sweet" or "precious." Also, the images usually show boys as active but girls as passive.

content analysis a data collection method that systematically examines some form of communication.

An Experiment That Went Awry

In a well-known 1971 experiment, psychologist Philip Zimbardo created a mock prison, using 24 undergraduate volunteers as prisoners and guards. He stopped the experiment after 6 days because some of the participants experienced intense negative reactions. For more information, go to Stanford Prison Experiment, prisonexp.org. We'll examine this study, including the ethical issues it raised, in more detail in Chapter 6.

AP Images/The Oklahoman/Chris Landsberger

A third limitation is that content analysis often reflects social class bias. Because most books, articles, speeches, films, and so forth are produced by people in upper socioeconomic levels, content analysis rarely captures the behavior or attitudes of working-class people and the poor. Even when documents created by lower-class individuals or groups are available, it's difficult to determine whether the coding reflects a researcher's social class prejudices.

Finally, content analyses can't always tell us *why* people behave as they do. We'd have to turn to studies that use different data collection methods—such as surveys and field research—to understand why, for example, people buy stereotypically feminine or masculine baby cards or produce video games that portray men, but not women, as competitive and aggressive (see Lindner, 2012).

2-5d Experiments

An **experiment** is a controlled artificial situation that allows researchers to manipulate variables and measure the effects. The classic experimental design includes two equal-sized groups that are similar on characteristics such as gender age, ethnicity, race, and education.

Suppose our hypothesis is "Watching a film on race discrimination reduces prejudice." In the **experimental group**, the participants are exposed to the independent variable (they watch the film on race discrimination). In the **control group**, participants don't watch the film because they're not exposed to the independent variable. Before the experiment, we'll measure the dependent variable (prejudice) in both groups using a *pretest* (a prejudice scale). After the experimental group watches the film, we'll measure both groups again using a *posttest* (the same prejudice scale). If we find a difference in the scores of the dependent variable (prejudice), we might conclude that the independent variable affects the dependent variable (e.g., watching the film on race discrimination reduced or increased prejudice). *Figure 2.6* illustrates this basic experimental design.

STRENGTHS

Experiments come closer than other data collection methods in suggesting a possible cause-and-effect relationship. They're also usually less expensive and time consuming than other data collection techniques (especially large surveys and multiyear field research), and there's often no need to purchase special equipment. A second and related strength is that because researchers recruit students or other volunteers (as in medical research), participants are usually readily available and don't expect much, if any, monetary compensation.

A third advantage is that experiments can be *replicated* (repeated) many times with different participants. Such replication strengthens the researchers' confidence in the validity and reliability of the measures and the study's results. For example, doctors stopped prescribing hormone pills for menopausal women because better-designed experiments that replicated earlier studies found that the pills increased the risk for breast cancer and strokes (Ioannidis, 2005).

experiment a controlled artificial situation that allows researchers to manipulate variables and measure the effects.

experimental group the participants who are exposed to the independent variable.

control group the participants who aren't exposed to the independent variable.

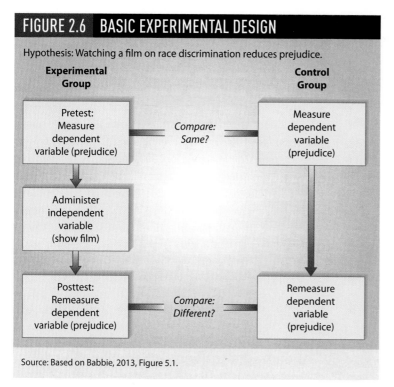

FIGURE 2.6 BASIC EXPERIMENTAL DESIGN

Hypothesis: Watching a film on race discrimination reduces prejudice.

Experimental Group

- Pretest: Measure dependent variable (prejudice)
- Administer independent variable (show film)
- Posttest: Remeasure dependent variable (prejudice)

Control Group

- Measure dependent variable (prejudice)
- Remeasure dependent variable (prejudice)

Compare: Same?

Compare: Different?

Source: Based on Babbie, 2013, Figure 5.1.

LIMITATIONS

One disadvantage of laboratory experiments is their reliance on student volunteers or paid respondents. Students often feel obligated to participate as part of their grade, or they may fear antagonizing an instructor who's conducting a study. Participants might also give the answers that they think the researcher expects (the social desirability bias described earlier). In the case of paid subjects, those who are the busiest, don't need the extra cash, move, or get sick may not participate fully or drop out of the study.

Another limitation is that experiments, especially those conducted in laboratories, are artificial. People *know* that they're being observed and may behave very differently than they would in a natural setting (Gray et al., 2007; see also Chapter 6 on the Hawthorne effect).

Third, the results of experimental studies can't be generalized to a larger population because they come from small, nonrepresentative, or self-selected samples. For example, college students who participate in experiments aren't necessarily representative of other college students, much less of people who aren't in college.

Fourth, experiments aren't suitable for studying large groups of people, a major focus in sociology. As you saw in Chapter 1, many sociologists examine macro-level issues (such as employment trends) that can't be observed in a laboratory or other experimental setting.

Finally, even if a researcher finds an association between variables (e.g., watching a film reduces prejudice), it doesn't mean that the former "causes" the latter. The control group's responses in the posttest may have been influenced by other factors, including communicating with the members of the experimental group (Cook and Campbell, 1979) or living in a racially mixed neighborhood, which tends to diminish racial and ethnic prejudice (see Chapter 10). That is, experiments have few control variables that might be explaining the relationship between the independent and dependent variables.

Because of such limitations, consumers are often confused and frustrated by contradictory findings and advice. Recently, different experiments have concluded that caffeine can relieve *or* lead to chronic headaches, inhibit *or* improve memory, increase *or* reduce depression, and that some high-fat foods are healthier than low-fat foods (Bartlett, 2011; Mozzafarian and Ludwig, 2015).

2-5e Secondary Analysis of Existing Data

Sociologists rely heavily on **secondary analysis**, an unobtrusive data collection method that examines information collected by someone else. The data may be historical materials (e.g., court proceedings), personal documents (e.g., letters and diaries), public records (e.g., state archives on births, marriages, and deaths), or official statistics (e.g., U.S. Bureau of Labor reports). Researchers with slightly different interests could analyze or reanalyze the data or examine only specific variables. For example, the federal government compiles information on a large number of variables—including population, age, and education—that social scientists, marketing analysts, health practitioners, and other researchers analyze for their own purposes.

STRENGTHS

Secondary analysis is usually accessible, convenient, and inexpensive. A staggering amount of data on topics such as income and employment—for the United States and worldwide—are readily available at libraries, colleges, and online.

secondary analysis examination of data that have been collected by someone else.

Second, there's a huge savings in time. Because many of the data sets are stored electronically, researchers can spend most of their time analyzing, rather than collecting, the data.

A third advantage of secondary analysis is the data's high quality. The federal government and nationally known survey organizations have budgets and well-trained staff equipped to address any data collection problems. Because the samples are representative of national populations, researchers can be more confident about generalizing the findings.

Fourth, data sets often contain hundreds of variables. Thus, researchers can access national, international, and historical information that they themselves can't collect (Boslough, 2007). Finally, because many of the data sets are longitudinal, researchers can look at trends and changes over time.

LIMITATIONS

Secondary analysis has several drawbacks. First, it may be difficult to gain access to historical materials because the documents are deteriorating, housed in only a few libraries in the country, or part of private collections. It may also be difficult to determine the accuracy and authenticity of historical data.

Second, existing sources may not have the exact data a researcher needs. In 2000, the Census Bureau changed how it counted race and ethnicity to include the growing numbers of recent immigrants and mixed-race Americans. Doing so increased the accuracy of gauging racial and ethnic diversity, but created 63 possible racial-ethnic combinations. As a result, today's numbers are different from those before 2000, making comparisons on race and ethnicity across time problematic (Saulny, 2011).

A third and related limitation is that even the same source may modify its measures of variables because of legal and social changes. Regarding marital and family status, for example, the Census Bureau added "separated" only in 1950, "unmarried partner" in 1990, and started categorizing same-sex married couples as families only in 2013 (Bates and DeMaio, 2013; Morello, 2014). Such changes are necessary. They can be frustrating, however, if, for instance, you wanted to compare the number of separated Americans between 1940 and 2015 or to determine the number of same-sex married couples since 2004, when Massachusetts became the first state to legalize same-sex marriage.

Finally, because the researcher didn't collect the data, she or he has no control over the specific variables used or how they were measured, and may not know how seriously problems such as low response rates, validity, and reliability affect the data. Most federal and other established data sets provide such information, but researchers can't simply assume that all data sets are problem-free.

Table 2.3 summarizes the data collection methods described in this chapter. Often, researchers use a combination of methods because "the social world is a multifaceted and multi-layered reality that reveals itself only in part with any single method" (Jacobs, 2005: 4). You should also be familiar with evaluation research, an increasingly important research tool.

TABLE 2.3 SOME DATA COLLECTION METHODS IN SOCIOLOGICAL RESEARCH

Method	Example	Advantages	Disadvantages
Surveys	Sending questionnaires and/or interviewing students on why they succeeded in college or dropped out	Questionnaires are fairly inexpensive and simple to administer; interviews have high response rates; findings are often generalizable	Mailed questionnaires may have low response rates; respondents tend to be self-selected; interviews are usually expensive
Secondary analysis	Using data from the National Center for Education Statistics (or similar organizations) to examine why students drop out of college	Usually accessible, convenient, and inexpensive; often longitudinal and historical	Information may be incomplete; some documents may be inaccessible; some data can't be collected over time
Field research	Observing classroom participation and other activities of first-year college students with high and low grade-point averages (GPAs)	Flexible; offers deeper understanding of social behavior; usually inexpensive	Difficult to quantify and to maintain observer/subject boundaries; the observer may be biased or judgmental; findings are not generalizable
Content analysis	Comparing the transcripts of college graduates and dropouts on variables such as gender, race/ethnicity, and social class	Usually inexpensive; can recode errors easily; unobtrusive; permits comparisons over time	Can be labor intensive; coding is often subjective (and may be distorted); may reflect social class bias
Experiments	Providing tutors to some students with low GPAs to find out if such resources increase college graduation rates	Usually inexpensive; plentiful supply of subjects; can be replicated	Subjects aren't representative of a larger population; the laboratory setting is artificial; findings can't be generalized

2-5f Evaluation Research: Pulling It All Together

Evaluation research studies whether a social intervention has produced the intended result. *Social interventions* are typically programs and strategies that seek to prevent or change negative outcomes (e.g., teenage pregnancy and substance abuse).

Evaluation research "refers to a research purpose rather than a specific method" (Babbie, 2013: 192). Thus, sociologists use a variety of data collection methods, such as secondary analysis and content analysis to examine an organization's records, surveys to gauge employee and client satisfaction, and interviews with the staff and program recipients for in-depth information on an organization's processes.

Evaluation research is *applied* because it's intended to have some real-world effect (Weiss, 1998). Examples include *needs assessment studies* (e.g., to improve or initiate a new service such as an after-school program), *cost-benefit studies* (e.g., to determine whether funding teen abstinence programs reduces pregnancies), and *monitoring studies* (e.g., to examine the effect of new policing policies on crime rates). The findings are generally used to improve the efficiency and effectiveness of a particular policy or group.

STRENGTHS

Evaluation research is important for several reasons. First, because local and state governments have been cutting their budgets, social service agencies often rely on evaluation research to streamline their programs, and to achieve the best results at the lowest possible cost (Kettner et al., 1999).

Second, evaluation research is versatile because it includes qualitative and quantitative approaches. It can address almost any topic such as the effectiveness of driver education programs, foreign aid policies, job training programs, and premarital counseling, to name just a few. The research costs can also be low if, for example, researchers can use secondary analysis rather than collect new data.

Third, evaluation research addresses real-life problems that confront many families and communities. Because state and local resources are shrinking and there are hundreds of intervention strategies, evaluation research can be invaluable to program directors or agency heads in deciding which programs to keep, improve, or cut (Peterson et al., 1994).

LIMITATIONS

Evaluations can usually address only one or a few of the many factors that affect behavior. For instance, adolescent substance abuse is not only due to individual high-risk decisions, but also to multiple interrelated

DARE programs have been popular with schools, police departments, parents, and politicians across the country even though evaluation research has found that the curriculum doesn't reduce students' drug use.

variables such as peers, school policies, and cultural values (Bandy, 2012).

Another limitation is that the social context affects evaluation research because of politics, vested interests, and conflicts of interest. Groups that fund the evaluation may pressure researchers to present only positive findings. Also, the results may not be well received if they contradict deeply held beliefs, challenge politicians' pet projects, or conflict with official points of view (Olson, 2010; Babbie, 2013).

Consider the popular DARE (Drug Abuse Resistance Education) program, introduced in 1983, to teach elementary school children about the dangers of drug use. Social scientists evaluated DARE in 1994, 1998, and 1999, and found no significant difference in the incidence of drug use among students who had and hadn't completed the DARE curriculum. When DARE funding was threatened, its promoters dismissed the research as "voodoo science" (Miller, 2001).

2-6 ETHICS AND SOCIAL RESEARCH

Researchers today operate under much stricter guidelines than they did in the past. In conducting their research, what ethical and political dilemmas do sociologists (and other social scientists) encounter?

evaluation research studies whether a social intervention has produced the intended result.

2-6a What Is Ethical Research?

Because so much research relies on human participants, the federal government, university institutional review boards (IRBs), and many professional organizations have formulated codes of ethics to protect participants. Regardless of the discipline or the research methods used, all ethical standards have at least three golden rules:

▶ First, *do no harm* by causing participants physical, psychological, or emotional pain.

▶ Second, the researcher must get the participant's *informed consent* to be in a study. This includes the participant's knowing what the study is about and how the results will be used.

▶ Third, researchers must always protect a participant's *confidentiality* even if the participant has broken a law that she or he discloses to the researcher. *Table 2.4* lists some of the basic ethical principles in doing sociological research.

2-6b Scientific Misconduct

Some disciplines are more susceptible to ethical violations than others. Medical researchers, especially, have been accused of considerable scientific dishonesty. Some of the alleged violations have included the following: changing

TABLE 2.4 SOME BASIC PRINCIPLES OF ETHICAL SOCIOLOGICAL RESEARCH

In its most recent update, the American Sociological Association (1999) has reinforced its previous ethical codes and guidelines for researchers. Researchers must:

- Obtain all subjects' consent to participate in the research and their permission to quote from their responses and comments.

- Not exploit subjects or research assistants involved in the research for personal gain.

- Never harm, humiliate, abuse, or coerce the participants in their studies, either physically or psychologically.

- Honor all guarantees to participants of privacy, anonymity, and confidentiality.

- Use the highest methodological standards and be as accurate as possible.

- Describe the limitations and shortcomings of the research in their published reports.

- Identify the sponsors who funded the research.

- Acknowledge the contributions of research assistants (usually underpaid and overworked graduate students) who participate in the research project.

research results to please the corporation (usually tobacco or pharmaceutical companies) that sponsored the research; being paid by companies to deliver speeches to health practitioners that endorse specific drugs, even if the medications don't reduce health problems; allowing drug manufacturers to ghostwrite their articles that are published in prestigious medical journals (and even draft textbooks); and falsifying data. A research team examined more than 2,000 published articles that were withdrawn from some of the most prestigious medical journals, including the *Journal of the American Medical Association* (*JAMA*). Almost 68 percent were withdrawn because of misconduct and fraud rather than errors (Project on Government Oversight, 2010; Shamoo and Bricker, 2011; Fang et al., 2012; McNeill, 2012).

In the social sciences, some data collection methods are more prone to ethical violations than others. Surveys, secondary analysis, and content analysis are less vulnerable than field research and experiments because the researchers typically don't interact directly with participants or affect them. In contrast, experiments and field research can raise ethical questions because of deception or influencing the participants' feelings, attitudes, or behavior.

In a recent experiment, Facebook and several academic researchers randomly selected 689,000 subscribers. For two of the groups, Facebook filtered the users' posts, providing either a positive or negative flow of comments, videos, pictures, and Web links by other people in their social network. The study—finding that users who saw fewer positive posts were less likely to post something positive, and vice versa—concluded that similar emotional language on social networks could be contagious (Kramer et al., 2014).

Some researchers questioned the conclusion, but many criticized the experiment for breaching ethical guidelines. Facebook users weren't informed about the experiment, hadn't given their consent to participate, and didn't know that they were being manipulated. Non-researchers described the emotional manipulation as "scandalous," "spooky," and "disturbing." Even so, the university IRB had approved the study because the researchers themselves didn't collect the data (Booth, 2014; Voosen, 2014).

All scholars face emerging ethical questions when they do research using social networks and other online environments. IRBs lack experience with Web research and sometimes unintentionally approve social science research that might violate ethics guidelines.

2-6c Do People Believe Scientific Findings?

Not always. For example, 98 percent of scientists say that humans and other living things have evolved over time, compared with 65 percent of U.S. adults. And although

Fisher-Price and other toy companies claim that, based on their research, their iPhone and iPad apps can teach infants as young as 9 months old to read and count (Singer, 2013). In fact, studies have found that such products do more harm than good because infants who watch the apps learn fewer words than those whose parents talk or read to them (see Chapter 12).

there has been no scientific evidence of paranormal events, 60 percent of Americans believe that people have seen or been in the presence of ghosts (Speigel, 2013; Funk and Rainie, 2015).

Why do many people reject scientific findings? One reason may be that when the media publicize the relatively rare occurrences of medical misconduct, readers generalize the few cases to all research and become more skeptical of all scientific studies.

Second, research findings often challenge personal attitudes and beliefs that people cherish. For example, a number of conservative Christians believe that global warming is a myth because only God, not humans, can affect the climate (Kahan et al., 2011; Sheppard, 2011; Smith and Leiserowitz, 2013).

Third, many Americans, especially minorities, are suspicious of the scientific community. Between 1906 and 1932, for example, medical researchers purposely infected unknowing subjects with fatal venereal diseases to test the effectiveness of penicillin. The subjects were almost always black men, prison inmates, mental patients, or soldiers (McNeil, 2010; Sanburn, 2010).

Finally, a large majority of scientists, including social scientists, spend little or no time engaging in public policy debates or explaining their research to the public and journalists (Rainie et al., 2015). As a result, "lobbyists and activists can promote their ideological agendas . . . over those of good science and public policy" (Lynas, 2015).

STUDY TOOLS 2

READY TO STUDY? IN THE BOOK, YOU CAN:

☐ Check your understanding of what you've read with the Test Your Learning Questions provided on the chapter review card at the back of the book.

☐ Rip out the chapter review card for a handy summary of the chapter and key terms.

ONLINE AT CENGAGEBRAIN.COM YOU CAN:

☐ Collect StudyBits while you read and study the chapter.

☐ Quiz yourself on key concepts.

☐ Prepare for tests with chapter flash cards as well as flash cards that you create.

☐ Watch videos to explore sociology further.

3 | Culture

LEARNING OUTCOMES

After studying this chapter, you will be able to...

3-1   Describe and illustrate a culture's characteristics.

3-2   Explain the significance of symbols, language, values, and norms.

3-3   Discuss and illustrate cultural similarities.

3-4   Discuss and illustrate cultural variations.

3-5   Differentiate between high culture and popular culture.

3-6   Explain how and why technology affects cultural change.

3-7   Compare and evaluate the theoretical explanations of culture.

After you finish
this chapter go to
PAGE 59 for
STUDY TOOLS.

Once when I returned a set of exams, a student who was unhappy with his grade blurted out an obscenity. A voice from the back of the classroom snapped, "You ain't got no culture, man!" Is it true that people who use vulgar language "ain't got no culture"?

3-1 CULTURE AND SOCIETY

As popularly used, *culture* often means an appreciation of the finer things in life, such as Shakespeare's sonnets, the opera, and using civil language. In contrast, sociologists use the term in a much broader sense: **Culture** refers to the learned and shared behaviors, beliefs, attitudes, values, and material objects that characterize a particular group or society. Thus, culture shapes a people's total way of life.

Most human behavior isn't random or haphazard. Among other things, culture influences what you eat; how you were raised and will raise your own children; if, when, and whom you will marry; how you make and spend money; and what you read. Even people who

What do you think?

I shape my culture; my culture doesn't shape me.

1	2	3	4	5	6	7

strongly agree   strongly disagree

pride themselves on their individualism conform to most cultural rules. The next time you're in class, for example, count how many students are *not* wearing jeans, T-shirts, sweatshirts, or sneakers—clothes that are the prevalent uniform of adolescents, college students, and many U.S. adults.

---

culture the learned and shared behaviors, beliefs, attitudes, values, and material objects that characterize a particular group or society.

A **society** is a group of people who share a culture and a defined territory. Society and culture go hand in hand; neither can exist without the other. Because of this interdependence, social scientists sometimes use the terms *culture* and *society* interchangeably.

3-1a Some Characteristics of Culture

All human societies, despite their diversity, share some cultural characteristics and functions (Murdock, 1940; see also Aguilar and Stokes, 1995). We don't see culture directly, but it shapes our attitudes and behaviors.

1. **Culture is learned.** Culture isn't innate but learned, and affects how we think, feel, and behave. If a child is born in one region of the world but raised in another, she or he will learn the customs, attitudes, and beliefs of the host culture.

2. **Culture is transmitted from one generation to the next.** We learn many customs, habits, and attitudes informally through parents, relatives, friends, and the media. We also learn culture formally in such settings as schools, workplaces, and community organizations. Whether our learning is formal or informal, culture is cumulative because each generation transmits cultural information to the next one.

3. **Culture is shared.** Culture brings members of a society together. We have a sense of belonging because we share

similar beliefs, values, and attitudes about what's right and wrong. Imagine the chaos if we did what we wanted (as physically assaulting an annoying neighbor) or if we couldn't make numerous daily assumptions about other people's behavior (as stopping at red traffic lights).

4. **Culture is adaptive and always changing.** Culture changes over time. New generations discard technological aspects of culture that are no longer practical, such as replacing typewriters with personal computers, and landlines with cell phones. Attitudes can also change over time. Compared with several generations ago, for example, many Americans now say that nonmarital sex is acceptable (see *Figure 3.1*).

Culture reflects who we are, but remember that it's people who create culture. As a result, culture changes as people adapt to their surroundings. Since the attacks on September 11, 2001, most people worldwide, including Americans, don't complain at airports when we have to pass through a metal detector, when our baggage is X-rayed or searched, when items are confiscated, and when airports use full-body scanners. Thus, as the dangers in society have grown, people have passed laws or

society a group of people who share a culture and defined territory.

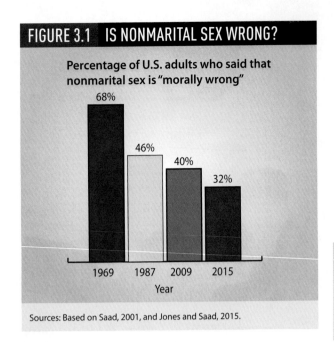

FIGURE 3.1 IS NONMARITAL SEX WRONG?

Percentage of U.S. adults who said that nonmarital sex is "morally wrong"

- 1969: 68%
- 1987: 46%
- 2009: 40%
- 2015: 32%

Year

Sources: Based on Saad, 2001, and Jones and Saad, 2015.

Communication symbols vary across societies and change over time. The earliest known map of the world (700–500 B.C.E) used cuneiform script; modern iPads are loaded with icons that provide a variety of information, including maps.

implemented rules that have increased their security but decreased their privacy.

3-1b Material and Nonmaterial Culture

Cultures that people construct are both material and nonmaterial (Ogburn, 1922). **Material culture** consists of the physical objects that people make, use, and share. These objects include such things as buildings, furniture, music, weapons, jewelry, hairstyles, and the Internet. **Nonmaterial culture** consists of the ideas that people create to interpret and understand the world. Beliefs about the supernatural, customs, and rules of behavior are examples of nonmaterial culture.

Because material and nonmaterial culture are created by people, the physical objects and ideas can vary considerably across societies (e.g., the farming tools people create, how they form governments, or raise their children). Frequently, as you'll see shortly, there's a "cultural lag" because material culture (invention of the automobile and the cell phone) changes faster than nonmaterial culture (laws about pollution and texting while driving).

material culture the physical objects that people make, use, and share.

nonmaterial culture the ideas that people create to interpret and understand the world.

symbol anything that stands for something else and has a particular meaning for people who share a culture.

3-2 THE BUILDING BLOCKS OF CULTURE

Recently, 74-year-old James Davis granted his dying wife's wish to be buried in their front yard. The Alabama Supreme Court agreed with a lower court judge that family members can't bury people anywhere they like. Davis refused to budge because the town doesn't have an ordinance governing burials, and his family members and neighbors supported his decision. After four years of court hearings and appeals, a city crew exhumed the body (Reeves, 2013).

Who's right—Davis, his family and neighbors, or the judges? To answer such questions, we must understand the building blocks of culture, particularly symbols, language, values, norms, and rituals.

3-2a Symbols

A **symbol** is anything that stands for something else and has a particular meaning for people who share a culture. In most societies, for example, a handshake communicates friendship or courtesy, a wedding ring signals that a person isn't a potential dating partner, and a siren denotes an emergency. People influence each other through the use of symbols. A smile and a frown communicate different information and elicit different responses. Through symbols, we engage in *symbolic interaction* (see Chapter 1).

SYMBOLS TAKE MANY FORMS

Written words are the most common symbols, but we also communicate by tattooing our bodies or not, celebrating or disregarding wedding anniversaries, and purchasing or avoiding goods and services to increase our social status. Gestures (e.g., raised fists, hugs, and stares) convey important messages about people's feelings and attitudes.

SYMBOLS DISTINGUISH ONE CULTURE FROM ANOTHER

In Islamic societies, many women wear a head scarf (*hijab* or *hejab*) that covers their hair and neck, a veil that hides part of the face from just below the eyes (*nijab*), or a garment that covers the entire body (*chador* or *burqa*). Whatever form it takes, the veil is just a piece of cloth, but it means different things to people across and within cultures.

For non-Muslims, especially Westerners, veils are often symbols of male dominance over women and religious zealotry; they conjure up images of terrorism and connote hiding something. For many Muslims, on the other hand, veiling symbolizes religious commitment, women's modesty, and an Islamic identity. Other Muslims—including those who live in the Middle East or have emigrated to Western countries—view most veiling, especially *nijabs* and *burqas*, not as a religious requirement but a custom that discourages face-to-face interaction (Murphy, 2009).

SYMBOLS CAN UNIFY OR DIVIDE A SOCIETY

Symbols unify people culturally. About 75 percent of Americans say that they display the nation's flag in their home, workplace, car, or clothing ("75%—A Nation of Flag Wavers," 2011). Every Fourth of July, many Americans celebrate independence from Britain with parades, firecrackers, barbecues, and speeches by local and national politicians. All of this symbolic behavior signifies freedom and democracy. For this reason, some immigrants purposely choose July 4th to be naturalized.

Symbols can also be divisive. Recently, South Carolina's governor, with the legislature's approval, decided to stop flying the Confederate flag on statehouse grounds because it represents slavery and white domination of African

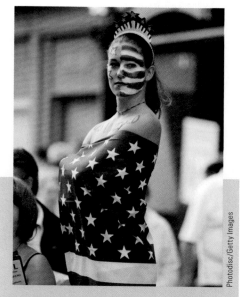

The U.S. Flag Code describes the flag as a "living thing" and forbids its display on clothing, bedding, drapery, or in advertising (Luckey, 2008). Is this woman disrespecting a cherished symbol? Or expressing her patriotism?

Americans. Protesters denounced the flag's removal because they see it as a proud emblem of their Southern heritage.

SYMBOLS CAN CHANGE OVER TIME

Symbols communicate information that varies across societies and may change over time. In 1986, the International Red Cross changed its name to the International Red Cross and the Red Crescent Movement to encompass a number of Arab branches, and adopted a crescent emblem in addition to the well-known cross. Israel wanted to use a red Star of David, rejecting the cross as a Christian symbol and the crescent as Islamic. Red Cross officials offered a red diamond as the new shape, but some countries rejected the diamond because it represents bloody conflicts in many African countries that mine diamonds (Whitelaw, 2000). The red cross and crescent continue to be the organization's emblems because the issue hasn't been resolved.

The meaning of even simple symbols varies across groups and can change over time. The # is commonly called the pound and number sign (#2 pencil). For musicians, # denotes a note that's sharp. For proofreaders, # means "insert a space." We're used to seeing # on telephone pads, but since the ascent of social media, # is a hashtag used to mark keywords or topics in a tweet (#culture).

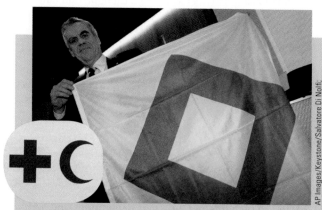

Left: The two emblems of the International Movement of the Red Cross and the Red Crescent. Right: The proposed red diamond emblem.

3-2b Language

Perhaps the most powerful of all human symbols is **language**, a system of shared symbols that enables people to communicate with one another. Language is a human invention that communities of people have endowed with meaning. In every society, children begin to grasp the essential structure of their language at a very early age and without any instruction. Babbling rapidly leads to uttering words and combinations of words. The average child knows approximately 900 words by the age of 2 and 8,000 words by age 6 (Hetherington et al., 2006). Each word is a symbol that represents an idea, feeling, or physical object.

Anthropological linguist Edward Sapir (1929) and later his student, Benjamin Whorf (1956), theorized that thoughts and behavior are determined by language. Their theory, known as the *Sapir-Whorf hypothesis*, posited that language not only is a tool for interpersonal communication but also provides people with a framework for interpreting social reality and the world around them. Thus, as languages vary, so do interpretations of social reality.

Subsequent research found little evidence to support the Sapir-Whorf hypothesis. Some critics maintained that language affects but does not "determine" culture. Others argued that language influences some thinking processes, but not all (e.g., Berlin and Kay, 1969; Heider and Olivier, 1972; Schlesinger, 1991). Still, the Sapir-Whorf hypothesis has generated considerable research across disciplines such as philosophy, sociology, and psychology in exploring the relationship between language, thought, and culture.

WHY LANGUAGE IS IMPORTANT

Language makes us human: It conveys our ideas, transmits information, and influences people's attitudes and behavior. Language also directs our thinking, controls our actions, and shapes our expression of emotions. The meaning of words change over time, but some variations are more controversial than others. Both birds and humans now tweet. Spam is a canned meat product, but also an unsolicited electronic message. Hallmark generated considerable controversy, however, when it changed gay to fun on a line of Christmas decorations and accessories, including a sweater (see photo).

language a system of shared symbols that enables people to communicate with one another.

AP Images/Hallmark Inc.

Every year, Hallmark produces "Keepsake" Christmas ornaments and apparel, many of which are collectibles. In 2013, one of the new pieces paid tribute to "Deck the Halls," a traditional Christmas carol. Hallmark changed the lyrics from "Don we now our gay apparel" to "Don we now our fun apparel" on this sweater because the meaning of gay has changed since the song's publication in the 1800s (Wong, 2013). Because of some negative reactions, Hallmark apologized for the change (Murphy, 2013). Why do you think that one word generated so much controversy?

Language gives us a sense of belonging to a group, but can also spark anger and conflict. Recognizing the connection between language and behavior, in 1990 Brazil passed a law criminalizing speech that incites racial, religious, or ethnic tensions. The penalties—including jail, fines, and community service—have been imposed on a variety of offenders: People who produced an online video that disparaged deities at an Afro-Brazilian spiritual center, an Argentine soccer player who shouted racist slurs during a game, and a law student who sent racist comments on Twitter (Barnes, 2012).

LANGUAGE AND GENDER

Many professors routinely use phrases such as "Okay, guys, in class today…" and no one objects. One of my colleagues illustrates the linkage between language and gender, and how it affects our thinking, by saying "Okay, gals, in class today…." "I always get a reaction of gaping mouths, laughs, and bewildered looks," he says. The students react differently to "guys" and "gals" because we've internalized male terms (e.g., *guys, policeman,* and *maintenance man*) as normal and acceptable.

Language has a profound influence on how we think about and act toward people. Those who adhere to traditional language contend that nouns such as *fireman, chairman, mailman,* and *mankind* and pronouns such as *he* refer to both women and men, and that women who object to such usage are too sensitive. Suppose, however, that all of your professors used only *she, her,* and *women* when referring to all people. Would the men in the class feel excluded? One solution is to use sentences with plural, gender-neutral pronouns (e.g., *they, them*) and gender-neutral nouns (e.g., *firefighter, chair, police officer,* and *mail carrier*).

Wanted: Someone Who Can Translate Federal Gibberish into Plain English

In the United States, it's almost impossible to find a high-level government document that people can understand. To increase the readability of federal publications, in 2010 President Obama signed the Plain Writing Act, which requires federal documents to be rewritten in plain English. Here's a "before" and "after" example from the Department of Health and Human Services Handbook that assures health care for low-income HIV/AIDS patients:

Before: Title I of the CARE Act creates a program of formula and supplemental competitive grants to help metropolitan areas with 2,000 or more reported AIDS cases meet emergency care needs of low-income HIV patients. Title II of the Ryan White Act provides formula grants to States and territories for operation of HIV service consortia in the localities most affected by the epidemic, provision of home and community-based care, continuation of insurance coverage for persons with HIV infection, and treatments that prolong life and prevent serious deterioration of health. Up to 10 percent of the funds for this program can be used to support Special Projects of National Significance.

After: Low income people living with HIV/AIDS gain, literally, years, through the advanced drug treatments and ongoing care supported by HRSA's Ryan White Comprehensive AIDS Resources Emergency (CARE) Act.

Language is supposed to help people communicate. However, the Plain Writing Act isn't enforced, and many federal agencies, including the Internal Revenue Service, are exempted. Does a government have more control if people don't understand its rules and regulations?

Sources: Based on Health Resources and Services Administration (2010), and "Government Resolves…," (2011).

LANGUAGE, RACE, AND ETHNICITY

Words—written and spoken—create and reinforce both positive and negative images about race and ethnicity. Someone might receive a *black mark*, and a *white lie* isn't "really" a lie. We *blackball* someone, *blacken* someone's reputation, denounce *blackguards* (villains), and prosecute people in the *black market*. In contrast, a *white knight* rescues people in distress, the *white hope* brings glory to a group, and the good guys wear *white hats*.

Racist or ethnic slurs, labels, and stereotypes demean and stigmatize people. Derogatory ethnic words abound: *honky, hebe, kike, spic, chink, jap, polack, wetback*, and many others. Self-ascribed racial epithets are as harmful as those imposed by outsiders. When Italians refer to themselves as *dagos* or African Americans call each other *nigger*, they tacitly accept stereotypes about themselves and legitimize the general usage of such derogatory ethnic labels (Attinasi, 1994).

Through language, children learn about their cultural heritage and develop a sense of group identity. Spanish is the fastest-growing and most spoken non-English language in the United States today. Some demographers predict, however, that about 34 percent of Latinos will speak only English at home by 2020, up from 25 percent in 2010 (Lopez and Gonzalez-Barrera, 2013). In many Latino and other immigrant families, the native language begins to disappear after just one generation. Assimilation has positive effects, but can also lead to communication problems between Americanized children and their non-English-speaking neighbors and kin in the homeland (Tobar, 2009; see also Chapter 10).

LANGUAGE AND SOCIAL CHANGE

Language is dynamic and changes over time. The most recent edition of the New American Bible, an annual best seller, has changed the word "booty," which now has a sexual connotation, to "spoils of war." "Cereal," which many think of as breakfast food, is now "grain" to reference loads of wheat (*Huffington Post*, 2011).

U.S. English is composed of hundreds of thousands of words borrowed from many countries and groups that were

TABLE 3.1 U.S. ENGLISH IS A MIXED SALAD

Here are a few examples of English words borrowed from other languages, showing our mixture of heritages. What words would you add to the list?

☐ Africa: apartheid, Kwanzaa, safari	☐ Spain: anchovy, bizarre
☐ Alaska and Siberia: husky, igloo, kayak	☐ Thailand: Siamese
☐ Bangladesh: bungalow, dinghy	☐ Turkey: baklava, caviar, kebob
☐ Hungary: coach, goulash, paprika	☐ France: bacon, police, ballet
☐ India: bandanna, cheetah, shampoo	☐ Japan: geisha, judo, sushi
☐ Iran and Afghanistan: bazaar, caravan, tiger	☐ Norway: iceberg, rig, walrus
☐ Israel: kosher, rabbi, Sabbath	☐ Mexico: avocado, chocolate, coyote
☐ Italy: fresco, spaghetti, piano	☐ Germany: strudel, vitamin, sauerkraut, kindergarten

in the Americas before the colonists arrived (Carney, 1997). *Table 3.1* provides a few examples of English words borrowed from other languages.

In response to cultural and technological changes, our vocabulary now includes *sexting*, *ebook*, *tweet*, *staycation*, *twerk*, *unfriend*, and *selfie*, among other new words. And to some people's dismay and others' delight, some writers are now substituting *partner* for the traditional *spouse*, *wife*, or *husband*.

3-2c Values

Values are the standards by which people define what is good or bad, moral or immoral, proper or improper, desirable or undesirable, beautiful or ugly. These widely shared standards provide *general guidelines* for everyday behavior rather than concrete rules for specific situations. Honesty, for example, is an important but abstract U.S. value. Thus, faculty typically define in course syllabi what they mean by cheating, such as using material from the Internet or other sources without proper acknowledgment, signing an attendance sheet for another student, or getting help on an assignment.

MAJOR U.S. VALUES

Sociologist Robin Williams (1970: 452–500) has identified a number of core U.S. values. All are central to the American way of life because they're widespread, have endured over time, and indicate what we believe is important in life.

values the standards by which people define what is good or bad, moral or immoral, proper or improper, desirable or undesirable, beautiful or ugly.

1. **Achievement and success.** U.S. culture stresses personal achievement, especially occupational success. Many Americans believe that if they work hard, apply themselves, and save their money, they'll be successful in the future.

2. **Activity and Work.** Americans want to "make things happen." They respect people who are focused and disciplined, and assume that hard work will be rewarded. Journalists and others often praise those who work past their retirement age.

3. **Humanitarianism.** U.S. society emphasizes concern for others, helpfulness, kindness, and offering comfort and support. During natural disasters—such as earthquakes, floods, fires, and famines—at home and abroad, many Americans are enormously generous (see Chapter 5).

4. **Efficiency and practicality.** Americans emphasize technological innovation, up-to-dateness, practicality, and getting things done. Many American colleges now tout their programs or courses as being practical instead of emphasizing knowledge and intellectual growth.

5. **Progress.** Americans focus on the future rather than the present or the past. The next time you walk down the aisle of a grocery store, note how many products are "new," "improved," and "better than ever."

6. **Material possessions.** Americans "have more stuff now than any society in history" (even if they can't afford the products). When the stuff accumulates, we buy stuff to stuff it into. Consequently, an entire industry, self-storage, has emerged to house our extra belongings (Sanburn, 2015: 46).

7. **Freedom and equality.** Countless U.S. documents affirm freedom of speech, freedom of the press, and freedom of worship. U.S. laws also tell Americans that they have been "created equal" and have the same legal rights regardless of race, ethnicity, gender, religion, disability, age, or social class.

8. **Conformity.** Most people don't want to be labeled as "strange," "peculiar," or "different." They conform because they want to be accepted, liked, hired, and promoted. As a result, we "toe the line," "go along to get along," and try to "fit in."

9. **Democracy.** Democracy promises the average citizen equal political rights. It emphasizes equality, freedom, and faith in the people rather than giving power to a monarch, dictator, or emperor.

10. Individualism. U.S. culture values each person's development. Thus, we often encourage children to be independent, creative, self-directed, self-motivated, and spontaneous.

VALUES SOMETIMES CONFLICT

Williams acknowledged that these core American values are sometimes contradictory and don't always mesh with reality. For example,

▶ We value equality but are comfortable with enormous gaps in wealth and power, and continue to discriminate against people because of their ethnicity, race, sexual orientation, gender, or age (see Chapters 10 and 11).

▶ We prize individualism, but reward people who conform to a group's expectations (see Chapter 6).

▶ We encourage responsibility, but often blame popular culture rather than parenting for children's bad behavior (see Chapter 12).

▶ We rail against waste, rising food prices, and the growing number of landfills, but throw away up to 40 percent of food every year (Gunders, 2012).

VALUES VARY ACROSS CULTURES AND CHANGE OVER TIME

Cultural values can change because of technological advances, immigration, and contact with outsiders. For example, the Japanese parliament passed a law making love of country a compulsory part of school curricula. The lawmakers and their numerous supporters hope that teaching patriotism will counteract the American-style emphasis on individualism and self-expression that they believe has undermined Japanese values of cooperation, self-discipline, responsibility, and respect for others (Wallace, 2006).

In the United States, surveys of first-year college students show a shift in values. Between 1968 and 2013, for example, "developing a meaningful philosophy of life" plummeted in importance whereas "being very well off financially" surged (Eagan et al., 2014). Consistent with our humanitarian values, however, in 2013 a sizeable majority (63 percent) of the students said that "helping others who are in difficulty" is essential or very important, but women were more likely to feel this way (see *Figure 3.2*).

The share of U.S. adults ages 25 and older who have never been married is at an historic high—19 percent in 2013 compared with only 9 percent in 1960. You'll see in Chapter 12 that marriage rates vary by social class, among

Every year, Americans throw out more food than plastic, paper, metal, and glass. The French government fines large supermarkets up to $82,000 for throwing away food that's considered edible (Ferdman, 2015).

iStockphoto.com/violinconcertono3/ David Henderson

other factors, but values also affect marital status. For example, 50 percent of Americans, another historic high, say society is just as well off if people have priorities other than marriage and children (Wang and Parker, 2014; U.S. Census Bureau News, 2014).

3-2d Norms

Values are general standards, whereas **norms** are a society's specific rules of right and wrong behavior. Norms tell us what we *should*, *ought*, and *must* do, as well as what we *should not*, *ought not*, and *must not* do: don't

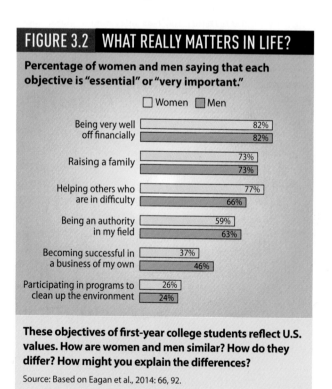

FIGURE 3.2   WHAT REALLY MATTERS IN LIFE?

Percentage of women and men saying that each objective is "essential" or "very important."

☐ Women ■ Men

Being very well off financially	82% (Women), 82% (Men)
Raising a family	73% (Women), 73% (Men)
Helping others who are in difficulty	77% (Women), 66% (Men)
Being an authority in my field	59% (Women), 63% (Men)
Becoming successful in a business of my own	37% (Women), 46% (Men)
Participating in programs to clean up the environment	26% (Women), 24% (Men)

These objectives of first-year college students reflect U.S. values. How are women and men similar? How do they differ? How might you explain the differences?

Source: Based on Eagan et al., 2014: 66, 92.

norms   specific rules of right and wrong behavior.

talk in church, stand in line, and so on. We may not like many of the rules, but they make our everyday lives more orderly and predictable. Here are some general characteristics of norms:

▸ Most are *unwritten*, passed down orally from generation to generation (use the good dishes and tablecloth on special occasions).

▸ They're *instrumental* because they serve a specific purpose (get rid of garbage because it attracts roaches and rats).

▸ Some are *explicit* (save your money "for a rainy day"), whereas others are *implicit* (be respectful at a wake or funeral).

▸ They *change* over time (it's now more acceptable to have a child out of wedlock but much less acceptable to smoke or drink alcohol while pregnant).

▸ Most are *conditional* because they apply in specific situations (slipping out of your smelly shoes may be fine at home, but not on an airplane).

▸ Because they're situational, norms can be *rigid* ("You *must* turn in your term paper at the beginning of class on May 5") or *flexible* ("You can have another week to finish the paper").

Norms regulate our behavior, but can differ because of characteristics such as age, social class, race, and gender. For example, a female professor's students sometimes comment on her clothes in their course evaluations and want her to "dress like a professor." In contrast, her husband, also a professor, "has yet to hear a single student comment about his wardrobe." So she changes her businesslike outfits every day while her husband usually wears khaki pants (or jeans) and a polo shirt day in and day out (Johnston, 2005). Thus, "her" and "his" norms differ.

Sociologists differentiate three types of norms—folkways, mores, and laws—that vary because some rules are more important than others. As a result, a society punishes some wrongdoers more severely than others.

FOLKWAYS

Folkways are norms that involve everyday customs, practices, and interaction. Etiquette rules are good examples of folkways: Cover your mouth when you cough or sneeze, say "please" and "thank you," knock before entering someone's office, and don't pick your ear wax or toenails in public.

We automatically practice numerous folkways because we've internalized them since birth. We often don't realize that we conform to folkways until someone violates one, such as picking one's nose or popping pimples in public, or talking loudly on a cell phone.

Folkways vary from one country to another. Punctuality is important in many Western countries but not in much of Latin America and the Mediterranean. Japan is notable for its gift giving among businesspeople, but this may be viewed as bribery in Western countries. Austrians, as well as Germans and the Swiss, consider chewing gum in public vulgar. There are also many differences in table manners: Europeans keep their hands above the table at all times, not in their laps, and Koreans frown on sniffling or blowing one's nose at the table (Axtell et al., 1997).

Folkways can change in response to macro-level changes. The growth of technology has altered many campus folkways. In the past, a syllabus was typically a few pages long, outlining course requirements and deadlines for exams and papers. Now, many syllabi—some as

folkways norms that involve everyday customs, practices, and interaction.

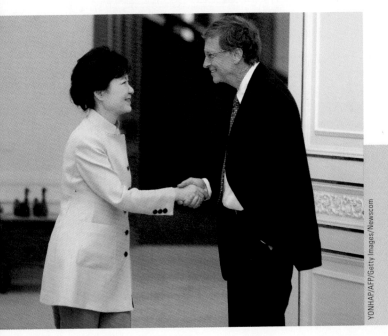

South Korea's national media criticized Microsoft founder Bill Gates for his "rude" handshake when meeting President Park Geun-Hye in 2013. As one person tweeted, "How can he put his hand in his pocket when meeting a leader of a state?!" A one-hand shake is disrespectful in South Korea and other parts of Asia, and is usually reserved for someone younger or a good friend (Agence France-Presse, 2013; Evans, 2013). Which types of norms did Gates violate— folkways, mores, or laws?

YONHAP/AFP/Getty Images/Newscom

long as 20 pages—look more like legal documents. They describe, often in great detail, a variety of rules on Internet plagiarism, unacceptable online sources for papers and projects, and laptop and cell phone use during class, even forbidding videos of professors that may appear on YouTube (Wasley, 2008).

MORES

Mores (pronounced "MORE-ayz") are norms that people consider very important because they maintain moral and ethical behavior. According to U.S. cultural mores, people must be sexually faithful to their spouse or partner, loyal to their country, and must not kill another person (except during war or in self-defense).

Folkways emphasize *ought to* behavior whereas mores define *must* behavior. The Ten Commandments are a good example of both civil and religious mores. Other mores include ethical guidelines (don't cheat, don't lie), expectations about behavior (don't use illicit drugs), and rules about sexual partners (don't have sexual intercourse with children or family members).

Mores, like folkways, can change. Until about the late 1960s, 80 percent of U.S. nonmarital babies were given up for adoption. This rate has dropped to about 1 percent because most unwed mothers, who are no longer stigmatized for having nonmarital babies, keep their infants (Benokraitis, 2015).

One of the most powerful types of mores are **taboos**, strong prohibitions of any act that's forbidden because it's considered extremely offensive. Taboos reflect social customs, religious or moral beliefs, or laws. Cannibalism, incest, and infanticide (killing infants) are examples of tabooed behavior.

Taboos define and reinforce mores, but there are many violations. For instance, 92 percent of Americans say that having an extramarital affair is morally wrong. Over a lifetime, however, 21 percent of husbands and 15 percent of wives admit to having had extramarital sex ("Adultery in New England . . .," 2014; Dugan, 2015).

LAWS

Laws are formally defined norms about what is legal or illegal. Unlike folkways and most mores, laws are deliberate, formal, "precisely specified in written texts," and "enforced by a specialized bureaucracy," usually police and courts, which have the power to punish violators (Hechter and Opp, 2001: xi).

Laws change over time. In 2014 alone, states passed more than 40,000 bills, ranging from drones to school sports. So far, 23 states have legalized medical marijuana; Colorado, Oregon, Alaska, Washington, and the District of Columbia have legalized marijuana for recreational use. Also, every state permits gambling (e.g., lotteries, casinos, racetrack betting). Some people believe that legalizing marijuana and gambling violates mores, but taxing both of these activities has greatly augmented state coffers.

Laws also vary across societies. You'll see shortly and throughout this textbook that countries' laws differ considerably regarding, for example, alcohol, adultery, elections, health care, religious worship, freedom of speech, and adoption.

SANCTIONS

Most people conform to norms because of **sanctions**, rewards for good or appropriate behavior and/or penalties for bad or inappropriate behavior. Children learn norms through both *positive* sanctions (praise, hugs, smiles,

mores norms that people consider very important because they maintain moral and ethical behavior.

taboos strong prohibitions of any act that's forbidden because it's considered extremely offensive.

laws formally defined norms about what is legal or illegal.

sanctions rewards for good or appropriate behavior and/or penalties for bad or inappropriate behavior.

China's new "Elderly Rights Law" allows aging parents to sue children who don't visit them "often." The judge ordered the daughter of this 77-year-old woman to visit her mother every two months and on at least two of the country's major holidays (Makinen, 2013).

AFP/Getty Images

new toys) and *negative* sanctions (frowning, scolding, spanking, withdrawing love) (see Chapter 4).

Negative sanctions vary in the degree of punishment. When we violate folkways, the sanctions are relatively mild (e.g., gossip, raised eyebrows, ridicule) because folkways aren't critical for a society's survival. If you don't bathe or brush your teeth, you may not be invited to parties because others will see you as crude, but not sinful or evil.

The sanctions for violating mores and some laws can be severe: loss of employment, expulsion from college, whipping, torture, banishment, imprisonment, and even execution. The punishment is usually harsh because the unacceptable behavior threatens a society's moral foundations. For example, Iran's public floggings of women who associate with men who aren't relatives punish offenders for their "immorality" and send a warning to others to follow the rules (see Chapter 9).

Sanctions aren't always consistent, despite universally held norms. Someone who can hire a good attorney often receives no penalty or a light one for a serious crime. In some cultures, young girls and women may be lashed for engaging in premarital sex, but men aren't punished at all for the same behavior, even though it violates Islamic law (Peter, 2012). Also, laws are sometimes enforced selectively or not at all. In the United States and elsewhere, offenders who violate rape, sexual harassment, and domestic violence laws are rarely punished (Jewkes et al., 2013; see also Chapters 7 and 12).

3-2e Rituals

Rituals are formal and repeated behaviors that unite people. They include giving gifts during Christmas, oaths of allegiance, wedding and funeral ceremonies, worship rites, dressing up for Halloween, graduations, veterans' parades, and even shaking hands. What all rituals have in common is the transmission and reinforcement of norms that unite us and strengthen relationships (Rossano, 2012).

Rituals are also outward symbols of values. During the playing of our national anthem, for example, everyone stands and faces the flag (or music). Many people place their hands over their hearts, and men remove their caps and hats. Such ritualistic behavior demonstrates respect and reinforces our shared values about patriotism and democracy.

rituals formal and repeated behaviors that unite people.

cultural universals customs and practices that are common to all societies.

Hindus, who make up more than 80 percent of India's 1.2 billion population, consider cows sacred. In some Indian states, violators who possess or sell beef can be punished by up to five years in prison (Bagri and Najar, 2015).

NOAH SEELAM/Getty Images

3-3 SOME CULTURAL SIMILARITIES

You've seen that there's considerable diversity across societies in symbols, language, values, and norms. There are also some striking similarities across and within cultures because of cultural universals, ideal versus real culture, ethnocentrism, and cultural relativism.

3-3a Cultural Universals

Cultural universals are customs and practices that are common to all societies. Anthropologist George Murdock and his associates studied hundreds of societies and compiled a list of 88 activities that they found among all cultures (see *Table 3.2* for some examples).

There are many cultural universals, but specific behaviors vary across cultures, from one group to another in the same society, and over time. For example, all societies have food taboos, but they differ across societies. About 75 percent of the world's people eat insects. In Thailand, locusts, crickets, silkworms, grasshoppers, ants, and other insects are a big part of the diet (Stolley and Taphaneeyapan, 2002). Vietnamese restaurants offer cat on the menu, and many poor people in rural China eat dog meat because they can't afford poultry, pork, and beef. Increasingly, however, urban, middle-class Chinese households "have learned to love dogs at the end of a leash rather than on a skewer" ("Dog-lovers . . . ," 2015: 46; see also Wan, 2011).

TABLE 3.2 SOME CULTURAL UNIVERSALS

Athletics	Games	Kin terminology	Property rights
Cooking	Gift giving	Language	Religious rituals
Courtship	Greetings	Magic	Sexual restrictions
Dancing	Hairstyles	Marriage	Status differentiation
Division of labor	Housing	Medicine	
Food taboos	Inheritance rules	Music	

Source: Based on Murdock, 1945.

3-3b Ideal versus Real Culture

The **ideal culture** comprises the beliefs, values, and norms that people say they hold or follow. In every society, however, ideal culture differs from **real culture**, or people's actual everyday behavior.

Consider parent-child relationships. Americans say that they love and cherish their children, but every year hundreds of thousands of children experience abuse and neglect on a daily basis. Indeed, 82 percent of the offenders are parents (U.S. Department of Health and Human Services, 2015; see also Chapter 12).

In China, the 2,500-year-old Confucian ideal of *filial piety* (respecting and taking care of one's parents) still runs deep. However, nearly half of the country's older people now live apart from their children, most of whom live and work in urban areas—"a phenomenon unheard of a generation ago"—leaving elders to fend for themselves (Mencher, 2013). Thus, ideal culture and actual behavior are often inconsistent.

3-3c Ethnocentrism and Cultural Relativism

Ethnocentrism is the belief that one's culture, society, or group is inherently superior to others. Thus, "*our* customs, *our* laws, *our* food, *our* traditions, *our* music, *our* religion, *our* beliefs and values, and so forth, are somehow better than those of other societies" (Smedley, 2007: 32). Because people internalize their culture and tend to see their way of life as the best and the most natural, they often disparage those with differing attitudes and behavior as inferior, wrong, or backward. How many of you, for example, are repelled at the thought of eating insects, cats, or dogs?

Some of my black students argue that it's impossible for African Americans to be ethnocentric because they suffer much prejudice and discrimination. *Any* group can be ethnocentric, however (Rose, 1997). An immigrant from Nigeria who believes that native-born African Americans are lazy and criminal is just as ethnocentric as a native-born African American who assumes that Nigerians are arrogant and "uppity."

Ethnocentrism can be functional. Appreciating one's own country and heritage promotes loyalty and cultural unity. When children learn their country's national anthem and customs, they have a sense of belonging. Ethnocentrism also reinforces conformity and maintains stability. Members of a society become committed to their particular values and customs, and transmit them to the next generation. As a result, life is (generally) orderly and predictable.

Ethnocentrism is also dysfunctional because viewing others as inferior generates hatred, discrimination, and conflict. Many current wars and battles—as those in some African nations and between Israelis and Palestinians—are due to religious, ethnic, or political intolerance (see Chapter 13). Thus, ethnocentrism discourages intergroup understanding and cooperation.

Kevin Foy/Alamy

About 75 percent of the world's people consume insects, which are high in protein, vitamins, and fiber, and usually low in fat. More than two out of every three American adults are overweight or obese, the highest percentage on the planet (see Chapter 14). Would we be healthier if we ate insects instead of gulping down hamburgers and french fries?

ideal culture the beliefs, values, and norms that people say they hold or follow.

real culture people's actual everyday behavior.

ethnocentrism the belief that one's own culture, society, or group is inherently superior to others.

The opposite of ethnocentrism is **cultural relativism**, the belief that no culture is better than another and should be judged by its own standards. Most Japanese mothers stay home with their children, whereas many American mothers work outside the home. Is one practice better than another? No. Because Japanese fathers are expected to be the breadwinners, it's common for many Japanese women to be homemakers. In the United States, in contrast, some mothers choose to stay home, but economic recessions and the loss of many high-paying jobs have catapulted many married women into the labor force to help support their families (see Chapters 10, 11, and 12). Thus, Japanese and American parenting may differ, but one culture isn't better or worse than the other.

3-4 SOME CULTURAL VARIATIONS

There's considerable cultural variation not only *across* societies but also *within* the same society. *Subcultures* and *countercultures* account for some of the complexity within a society.

3-4a Subcultures

A **subculture** is a group within society that has distinctive norms, values, beliefs, lifestyle, or language. A subculture is part of the larger, dominant culture but has its own particular ways of thinking, feeling, and behaving. U.S. society contains thousands of subcultures based on a variety of characteristics, interests, or activities:

▶ **Ethnicity** (Irish, Mexican, Vietnamese)

▶ **Religion** (evangelical Christians, atheists, Mormons)

▶ **Politics** (Maine Republicans, Southern Democrats, independents)

▶ **Sex and gender** (heterosexual, lesbian, transgender)

▶ **Age** (older widows, middle schoolers, college students)

▶ **Occupation** (surgeons, prostitutes, truck drivers)

▶ **Music and art** (jazz aficionados, country music buffs, art lovers)

cultural relativism the belief that no culture is better than another and should be judged by its own standards.

subculture a group within society that has distinctive norms, values, beliefs, lifestyle, or language.

counterculture a group within society that openly opposes and/or rejects some of the dominant culture's norms, values, or laws.

Dozens of rules and customs govern what an American first lady should and shouldn't wear while visiting other countries. In keeping with diplomatic protocol for Western women, Michelle Obama wore a headscarf during a recent trip to Indonesia (Davis, 2015).

▶ **Social class** (billionaires, working poor, middle class)

▶ **Recreation** (mountain bikers, poker players, dancers)

As you read this list, you'll realize that most of us are members of numerous subcultures. We usually participate in subcultures without having much commitment to any of them (e.g., basketball fans, guitar players, collectors of ceramic frogs). On the other hand, members of racial or ethnic subcultures may intentionally live in the same neighborhoods, associate with each other, have close personal relationships, and marry others who are similar to themselves.

To fit in, members of most subcultures adapt to the larger society, but they may also maintain some of their traditional customs. Among Ghanaian immigrants, for instance, funerals are lavish celebrations rather than sad gatherings. The funerals are "all-night affairs with open bars and window-rattling music." As in Ghana, the guests need not know the deceased or even the family. Many people attend simply to meet other Ghanaians and to "cut loose on the dance floor" (Dolnick, 2011: A1; see also Bax, 2013).

In many instances, subcultures arise because of technological or other societal changes. After the Internet emerged, subcultures appeared that identified themselves as hackers, techies, and computer geeks.

3-4b Countercultures

A **counterculture** is a group within society that openly opposes and/or rejects some of the dominant culture's norms, values, or laws. Countercultures usually emerge when people believe they can't achieve their goals within

FIGURE 3.3 THE GEOGRAPHY OF U.S. HATE GROUPS, 2014

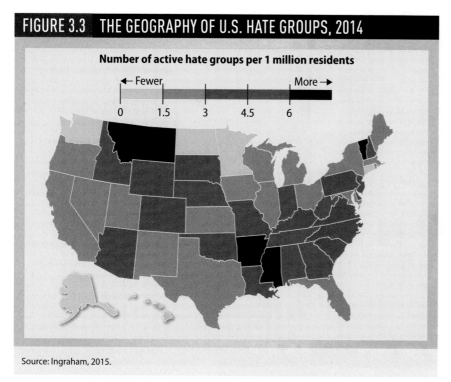

Number of active hate groups per 1 million residents

← Fewer   More →

0   1.5   3   4.5   6

Source: Ingraham, 2015.

the existing society. As a result, such groups develop values and practices that run counter to those of the dominant society. Some countercultures are small and informal, but others, such as religious militias, have millions of members and are highly organized (see Chapter 11).

Most countercultures are law-abiding. Some, however, are violent and extremist, such as the 784 active hate groups across the United States that intimidate ethnic groups and gays (Southern Poverty Law Center, 2015). Hate groups are concentrated most in the Deep South and in the Montana/Idaho region (see *Figure 3.3*). The reasons for the clustering include factors such as racist and anti-gay attitudes, religious intolerance, low educational and income levels, and right-wing political views (Florida, 2011; Frohlich et al., 2015; Ingraham, 2015).

Some skinheads have killed gays; antigovernment militia adherents bombed a federal building in Oklahoma, killing dozens of adults and children; and antiabortion advocates have murdered physicians and bombed abortion clinics. Radical anti-government groups or white supremacists are responsible for most terrorist acts in the United States. The FBI is particularly concerned by this violence because it's carried out by individuals who aren't affiliated with any countercultural movement and are, therefore, hard to detect (Kurzman and Schanzer, 2015).

3-4c Multiculturalism

Multiculturalism (sometimes called *cultural pluralism*) refers to the coexistence of several cultures in the same geographic area, without one culture dominating another. Many applaud multiculturalism because it encourages intercultural dialogue (e.g., U.S. schools offering programs in African American, Latino, Arabic, and Asian studies). Supporters hope that emphasizing multiculturalism—especially in schools and the workplace—will decrease ethnocentrism, racism, sexism, and other forms of discrimination.

Despite its benefits, not everyone is enthusiastic about multiculturalism. Not learning the language or following the customs of the country where one lives and works can spur tension and conflict. European political leaders in Germany, Great Britain, and France have described multiculturalism, particularly integrating Muslims, as a "failed policy" and a "nightmare" that's had "disastrous results." Instead of adopting the dominant culture's values and beliefs, according to critics, some ethnic and religious subcultures "build tiny nations" within the host nation, thereby jeopardizing a country's national identity (Marquand, 2011; Kymlicka, 2012; Tharoor, 2015).

3-4d Culture Shock

People who travel to other countries often experience **culture shock**—confusion, disorientation, or anxiety that accompanies exposure to an unfamiliar way of life. Conventional cues about how to behave are missing or have a different meaning. Culture shock affects some people more than others, but the most stressful changes involve differences of food, clothes, punctuality, ideas about what offends people, language, the general pace of life, and a lack of privacy (Spradley and Phillips, 1972; Pedersen, 1995).

To some degree, everyone is *culture bound* because we've internalized cultural norms and values. For example, an American journalist who works in Mexico City says that perpetual lateness is common: A child's birthday party may start 2 or 3 hours late, a wedding may begin an hour

multiculturalism (sometimes called *cultural pluralism*) the coexistence of several cultures in the same geographic area, without one culture dominating another.

culture shock confusion, disorientation, or anxiety that accompanies exposure to an unfamiliar way of life.

Does Pop Culture Make Kids Fat?

The prevalence of obesity among American children ages 2 to 19 years has more than tripled, increasing from 5 percent in the 1960s to 17 percent in 2012 (National Center for Health Statistics, 2014). There are many reasons for the increase, but physicians and researchers lay much of the blame on popular culture, especially the advertising industry. The advertising, particularly on television, uses cartoon characters such as SpongeBob SquarePants and Scooby-Doo to sell sugary cereals, cookies, candy, and other high-calorie

iStockphoto.com/Juanmonino

snacks. In contrast, many European countries forbid advertising on children's television programs. After considerable pressure from the federal government, the nation's largest food companies agreed to curb advertising, but children are watching more fast food ads on TV than ever before. Only six companies are responsible for more than 70 percent of all TV fast-food ads viewed by children age 18 and younger ("Food Companies Propose...," 2011; Harris et al., 2013).

---

after the announced time, and interviews with top Mexico City officials may be up to 2 hours late or the official never arrives. These norms are changing because of the growth of Mexico's global economy, but the more relaxed attitude toward time in Mexico City (and elsewhere) can be a culture shock to Americans who have grown up in a "clock-obsessed" culture (Ellingwood, 2009).

3-5 HIGH CULTURE AND POPULAR CULTURE

High culture refers to the cultural expression of a society's highest social classes. Examples include opera, ballet, theatre, and classical music. In contrast, **popular culture** refers to beliefs, practices, activities, and products that are widespread among a population. Examples include television, music, social media, advertising, sports, hobbies, fads, fashions, movies, the food we eat, the gossip we share, and the jokes we tell (Levine, 1998; Gans, 1999).

high culture the cultural expression of a society's highest social classes.

popular culture the beliefs, practices, activities, and products that are widespread among a population.

cultural capital resources such as knowledge, verbal and social skills, education, and other assets that give a group advantages.

Social class affects our participation in high culture and popular culture. Members of the upper class can pursue the fine arts and similar activities because they have **cultural capital**—resources such as knowledge, verbal and social skills, education, and other assets that give a group advantages (Bourdieu, 1986). Resources alone, however, don't determine our cultural tastes and interests. For example, in Baltimore (as in many other cities), tickets for the symphony, opera, or ballet are usually much less expensive than those for a Ravens football game or rock concert.

Cultural capital sets up boundaries between social classes, but our cultural expressions and products often contain elements of both popular culture and high culture. For instance, African American jazz players have inspired European classical music composers and themselves perform classical music (Salamone, 2005; Lawn, 2013).

The Queen of England has conferred knighthood—one of the highest honors given to an individual in the United Kingdom—on pop singers such as Elton John and Paul McCartney. Since 1998, the John F. Kennedy Center for the Performing Arts in Washington, D.C., has presented an annual Mark Twain Prize for American Humor to popular culture comedians such as Richard Pryor, Will Ferrell, and Tina Fey. Millions of Americans who think of themselves as "cultured" watch "trashy TV" like *Hoarders, The Bachelor/The Bachelorette, Two and a Half Men,* and *Keeping Up with the Kardashians.* Doing so allows them to feel superior to the shows' conventional "low-brow" viewers ("Thank God I'm not like that!") (McCoy and

American fast-food restaurants are popular in many Asian countries. Because of greater competition from local fast-food chains, U.S. companies now offer a large number of items geared specifically to Asian tastes. At one of China's Pizza Huts, the menu includes shrimp pizza, fried squid, and green tea ice cream. In Latin America, McDonald's sells "McMollettes," English muffins with refried beans, cheese, and salsa (York, 2012). In India, McDonald's offers an entirely meat-free menu.

iStockphoto.com/Holger Mette

Scarborough, 2015). Thus, the divide between popular culture and high culture isn't as rigid as we might think.

3-5a The Influence of Mass Media

People don't always believe everything they read or see on television or online. Nonetheless, many are highly influenced by a popular culture that is largely controlled and manipulated by newspapers and magazines, television, movies, music, and ads (see Chapter 4). These **mass media**, or forms of communication designed to reach large numbers of people, have enormous power in shaping public attitudes and behavior.

Many of my students who claim that they "don't pay any attention to ads" come to class wearing branded apparel: Budweiser caps, Old Navy t-shirts, and Nike footwear. Advertising affects our self-image and self-esteem. As you'll see in later chapters, many women and men spend considerable time and money to meet ideal standards of femininity and masculinity—generated by ads—that are unrealistic and impossible to attain. (We'll examine fashion and fads, two important components of popular culture, in Chapter 16.)

Because companies are driven by the pursuit of profit, considerable mass media content is basically marketing. Many television morning shows and MTV have become "a kind of sophisticated infomercial" (O'Donnell, 2007: 30). In effect, a third of the content on morning shows is essentially selling something (a book, music, a movie, or another television program) that the corporation owns. One of the most lucrative alliances is between Hollywood and toy manufacturers (see Chapter 4). And U.S. mass media exert a powerful influence abroad.

3-5b The Globalization of Popular Culture

Because the United States is the biggest producer of popular culture, many countries have become inundated with America's movies, music, television shows, newspapers, satellite broadcasts, fast food, clothing, and other consumer goods. Overseas corporations contribute to the spread of U.S. popular culture because they profit by distributing the products of popular musicians, authors, and filmmakers (The Levin Institute, 2013).

Some observers have described the spread of American popular culture as **cultural imperialism**, a process by which the cultural values and products of one society influence or dominate those of another. A great deal of cultural imperialism is voluntary. According to a British journalist, much of England's population has embraced American pop music, films, television programs, architecture, and is now experiencing an "obesity epidemic" because of the "bread-potatoes-and-lard-based fast food on which America nourishes itself" (Heffer, 2010).

Other countries complain that U.S. cultural imperialism displaces authentic local culture and results in cultural loss. For instance, Iran's government has

mass media forms of communication designed to reach large numbers of people.

cultural imperialism the cultural values and products of one society influence or dominate those of another.

denounced Batman, Spider-Man, and Harry Potter toys as a form of "cultural invasion" that challenges the country's conservative and religious values. The curvaceous and often scantily clad Barbie dolls with peroxide-blond hair have been especially singled out as "destructive culturally and a social danger." Nonetheless, many of the girls who watch foreign television and (illegal) satellite TV want the dolls (Peterson, 2008: 4).

3-6 CULTURAL CHANGE AND TECHNOLOGY

Why do cultures persist? How and why do they change? And what happens when technology changes faster than cultural norms, values, and laws?

3-6a Cultural Persistence: Why Cultures Are Stable

In many ways, culture is a conservative force. As you saw earlier, values, norms, and language are transmitted from generation to generation. Such **cultural integration**, or the consistency of various aspects of society, promotes order and stability. Even when new behaviors and beliefs emerge, they commonly adapt to existing ones. Recent immigrants may speak their native language at home and celebrate their own holy days, but they're expected to gradually absorb the host country's values, to obey its civil and criminal laws, and to adopt its national language. Life would be chaotic and unpredictable without such cultural integration.

3-6b Cultural Dynamics: Why Cultures Change

Cultural stability is important, but all societies change over time. Some of the major reasons for cultural change include diffusion, invention and innovation, discovery, and external pressures.

DIFFUSION

A culture may change because of *diffusion*, the process through which components of culture spread from one society to another. Such borrowing may have occurred so long ago that the members of a society consider their culture to be entirely their own creation. However, anthropologist Ralph Linton (1964) has estimated that 90 percent of the elements of any culture are a result of diffusion (see *Figure 3.4*).

Diffusion can be direct and interpersonal, occurring through trade, tourism, immigration, intermarriage, or the invasion of one country by another. Diffusion can also be indirect and largely impersonal, as in the Internet transmissions that zip around the world.

INVENTION AND INNOVATION

Cultures change because people are continually finding new ways of doing things. *Invention*, the process of creating new things, brought about products such as toothpaste (invented in 3000 B.C.E.), eyeglasses (262 C.E.), flushable toilets (the sixteenth century), fax machines (1843—that's right, invented in 1843!), credit cards (1920s), computer mouses (1964), Post-it™ notes (1980), and DVDs (1995).

Innovation—turning inventions into mass-market products—also sparks cultural changes. An innovator is someone who markets an invention, even if it's someone else's idea. For example, Henry Ford invented nothing new but "assembled into a car the discoveries of other men behind whom were centuries of work," an innovation that changed people's lives (Evans et al., 2006: 465).

Inventors and innovators are rarely isolated geniuses who work alone. Instead, they depend on collaborative relationships and social networks for ideas and help. Thomas Edison's laboratory, which had a staff of 35 inventors, produced the light bulb. The Human Genome Project included 110 authors from 20 universities, research centers, institutes, and hospitals. Even Michelangelo used twelve assistants to paint the renowned Sistine Chapel in Rome (Dahlin, 2011).

DISCOVERY

Like invention, *discovery* involves exploration and investigation, and results in new products, insights, ideas, or behavior. The discovery of penicillin prolonged lives, which, in turn, means that more grandparents (as well as great-grandparents) and grandchildren get to know each other. However, longer life spans also mean that many adult children and grandchildren now care for older family members over many years (see Chapter 12).

Discovery usually requires dedicated work and years of commitment, but some discoveries occur by chance, called the *serendipity effect*. George de Mestral, a Swiss electrical engineer, was hiking through the woods and was annoyed by burrs that clung to his clothing. Why were they so difficult to remove? A closer examination showed that the burrs had hooklike arms that locked into the open weave of his clothes. The discovery led de Mestral to invent a hook-and-loop

cultural integration the consistency of various aspects of society that promotes order and stability.

FIGURE 3.4 THE 100 PERCENT AMERICAN

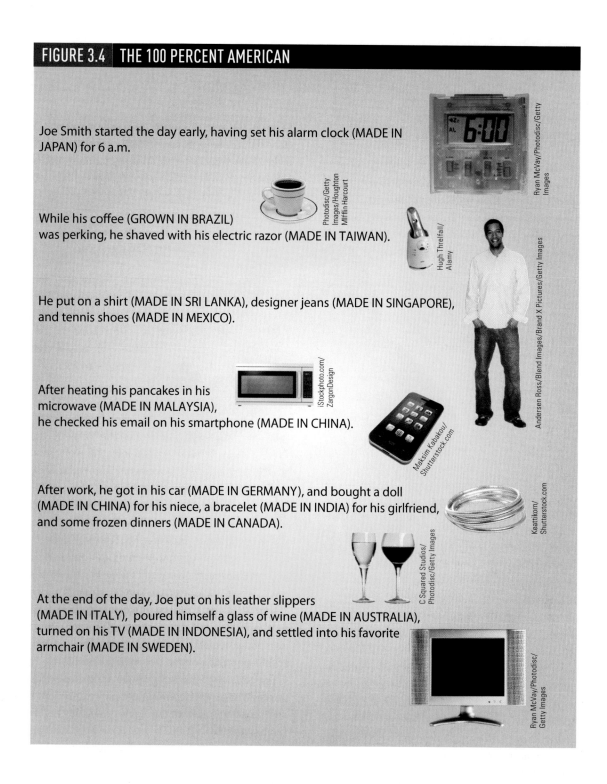

Joe Smith started the day early, having set his alarm clock (MADE IN JAPAN) for 6 a.m.

While his coffee (GROWN IN BRAZIL) was perking, he shaved with his electric razor (MADE IN TAIWAN).

He put on a shirt (MADE IN SRI LANKA), designer jeans (MADE IN SINGAPORE), and tennis shoes (MADE IN MEXICO).

After heating his pancakes in his microwave (MADE IN MALAYSIA), he checked his email on his smartphone (MADE IN CHINA).

After work, he got in his car (MADE IN GERMANY), and bought a doll (MADE IN CHINA) for his niece, a bracelet (MADE IN INDIA) for his girlfriend, and some frozen dinners (MADE IN CANADA).

At the end of the day, Joe put on his leather slippers (MADE IN ITALY), poured himself a glass of wine (MADE IN AUSTRALIA), turned on his TV (MADE IN INDONESIA), and settled into his favorite armchair (MADE IN SWEDEN).

fastener. His invention, Velcro—derived from the French words *velour* (velvet) and *crochet* (hooks)—can now be found on everything from wallets to spacecraft.

EXTERNAL PRESSURES

External pressure for cultural change can take various forms. In its most direct form—war, conquest, or colonization—the dominant group uses force or the threat of force to change an existing culture. When the Soviet Union invaded and occupied many small countries (e.g., Lithuania, Latvia, Estonia, Ukraine, and Armenia) after World War II, it forbade citizens to speak their native languages, banned traditions and customs, and turned churches into warehouses.

Pressures for change can also be indirect. Some countries (e.g., Thailand, Vietnam, China, and Russia) have reduced their prostitution and international sex trafficking because of widespread criticism by the

United Nations and some European countries. The United Nations has no power to intervene in a country's internal affairs but can embarrass nations by publicizing human rights violations (Farley, 2001).

3-6c Technology and Cultural Lag

Some parts of culture change more rapidly than others. **Cultural lag** refers to the gap that occurs when material culture changes faster than nonmaterial culture.

There are numerous examples of cultural lag in modern society because our nonmaterial culture (e.g., norms, values, and laws) hasn't kept up with material culture (e.g., technological advances). Texting while driving has resulted in thousands of fatal car crashes (see Chapters 1 and 2), but many states have neither banned this practice nor enforced the laws they've passed.

Because technology is moving faster than laws to regulate and ensure Internet-based communication privacy, there are many breaches. The National Security Agency (NSA), one of the largest U.S. intelligence organizations, has secretly collected phone records of millions of Americans, and spied on leading U.S. technology companies (e.g., Facebook, Google, Apple, Yahoo) as well as European political heads of state who are allies (Timberg and Nakashima, 2013).

Cultural lags can create confusion, ambiguity about what's right and wrong, conflict, and a feeling of helplessness. They also expose contradictory values and behavior. U.S. technology executives have railed against the NSA's mass compilation of data, but the companies themselves collect information on users that they sell to advertisers (Miller, 2013; Sengupta, 2013). Some consumers complain that retailers track their behavior, but have no problem with cookies, profiles, apps, and other online tools that let e-commerce sites know who they are, how they shop, and what they purchase (Clifford and Hardy, 2013). And as you'll see in several later chapters, even when mores, values, and laws catch up with technological innovations and inventions, they can create unexpected problems such as pollution, drug shortages, and high medical costs.

---

cultural lag the gap that occurs when material culture changes faster than nonmaterial culture.

3-7 SOCIOLOGICAL PERSPECTIVES ON CULTURE

What is the role of culture in modern society? And how does culture help us understand ourselves and the world around us? Functionalist, conflict, feminist, and symbolic interaction scholars offer important insights in answering these and other questions. *Table 3.3* summarizes the perspectives.

3-7a Functionalism

Functionalists focus on society as a system of interrelated parts (see Chapter 1). Similarly, in their analysis of culture, functionalists emphasize the social bonds that attach people to society.

KEY POINTS

For functionalists, culture is the cement that binds society. As you saw earlier, norms and values shape our lives, provide guideposts for our everyday behavior, and promote cultural integration and societal stability. Especially in developed countries such as the United States that have high immigration rates, cultural norms and values help newcomers adjust to the host society.

All societies have similar strategies for meeting human needs. Cultural universals, such as religious rituals, may play out differently in different countries, but all known societies have religious rituals. Critics often blame popular culture for societal problems, including violence and crime, but popular culture also serves important functions in uniting people with similar interests and producing new technologies (e.g., DVDs, YouTube, Instagram) (Kidd, 2007).

Functionalists also note that culture can be dysfunctional. For example, some countercultures, particularly right-wing extremist groups, can create chaos by bombing federal buildings and killing or injuring hundreds of people.

CRITICAL EVALUATION

Functionalism shows that shared norms and values create cultural solidarity and stability. In emphasizing culture's role in meeting people's daily needs, however, functionalism often overlooks diversity and social change. A number of influential functionalists have proposed, for instance, that immigration should be restricted because

THEORETICAL PERSPECTIVE	FUNCTIONALIST	CONFLICT	FEMINIST	SYMBOLIC INTERACTIONIST

TABLE 3.3 SOCIOLOGICAL PERSPECTIVES OF CULTURE

THEORETICAL PERSPECTIVE	FUNCTIONALIST	CONFLICT	FEMINIST	SYMBOLIC INTERACTIONIST
Level of Analysis	**Macro**	**Macro**	**Macro and Micro**	**Micro**
Key Points	• Similar beliefs bind people together and create stability. • Sharing core values unifies a society and promotes cultural solidarity.	• Culture benefits some groups at the expense of others. • As powerful economic monopolies increase worldwide, the rich get richer and the rest of us get poorer.	• Women and men often experience culture differently. • Cultural values and norms can increase inequality because of sex, race/ ethnicity, and social class.	• Cultural symbols forge identities (that change over time). • Culture (like norms and values) helps people merge into a society despite their differences .
Examples	• Speaking the same language (English in the United States) binds people together because they can communicate with one another, express their feelings, and influence one another's attitudes and behaviors.	• Much of the English language reinforces negative images about gender ("slut"), race ("honky"), ethnicity ("jap"), and age ("old geezer") that create inequality and foster ethnocentrism.	• Using male language (e.g., "congressman," "fireman," and "chairman") conveys the idea that men are superior to and dominant over women, even when women have the same jobs.	• People can change the language they create as they interact with others. Many Americans now use "police officer" instead of "policeman," and "single person" instead of "bachelor" or "old maid."

it dilutes shared U.S. values. Such proposals overlook the many contributions that newcomers make to society (see Chapter 10).

3-7b Conflict Theory

Unlike functionalists, conflict theorists argue that culture can generate considerable inequality instead of unify society. Because the rich and powerful determine economic, political, educational, and legal policies for their own benefit and control the mass media, the average American has little power to change the culture—whether

it's low wages, unpopular wars, or corporate corruption (see Chapters 8 and 11).

KEY POINTS

Conflict theorists maintain that many cultural values and norms benefit some members of society more than others. We are taught to work hard, for example, but who profits from the average worker's efforts? U.S. taxpayers have paid billions of dollars to bail out financial industries that collapsed because of greedy top executives. The executives then used the money to give themselves and their staff higher salaries and bonuses than ever before.

India's high-income families usually have lavish weddings that follow centuries-long traditions (left). For functionalists, the customs and rituals reinforce cultural identity and bind people. For conflict theorists, India's inequality also creates a large population of poor people (right) who are excluded from the traditions and ceremonies that reinforce social bonds.

FotoFlirt/Alamy

Anne-Marie Palmer/Alamy

Very rarely, moreover, are wealthy people incarcerated for crime (see Chapters 7 and 8). Thus, those at the top of the socioeconomic ladder can violate laws and mores, including honesty, that other Americans are expected to observe.

Conflict theorists also point out that technology serves primarily the rich. Desperate low-income women use online sites to sell their hair, breast milk, and ovarian eggs. Selling kidneys is illegal in the United States, but the organ can fetch at least $15,000 on the Internet black market (Stillwell, 2013).

CRITICAL EVALUATION

According to some critics, conflict theorists place too much emphasis on societal discord and downplay a culture's benefits. For instance, cultural values integrate members of a society and decrease divisiveness.

Conflict theorists maintain that mass media conglomerates control popular culture and promote goods and services that most people don't need and/or can't afford (e.g., big screen TVs, the most recent smartphones, and designer clothes). Critics counter that people aren't mindless puppets but can make choices, including not living on credit. Thus, according to some conflict theory critics, conglomerates shouldn't be blamed for people's irrational behavior.

3-7c Feminist Theories

Feminist scholars agree with conflict theorists that culture creates considerable inequality, but they focus on gender differences. Feminists are also more likely than other theorists to examine multicultural variations across groups.

KEY POINTS

Culture affects gender roles. When media portrayals of women are absent or stereotypical, we get a distorted view of reality (Neuhaus, 2010). Feminist scholars also emphasize that subcultures—for example, female students or single mothers—may experience culture differently than their male counterparts do because women typically have fewer resources, particularly income and power (see Chapters 9 and 11).

Unequal access to resources often results in women having fewer choices than men, and living under laws and customs that subordinate women. In the United States, a nonmarital birth typically impoverishes women

but not men. In Japan, where rape is often kept hush-hush, police and judges rarely take the offense seriously (Makino, 2009). In most Islamic societies, men dictate women's attire and behavior (Heath, 2008; Zahedi, 2008). In many Latin American countries, violence against women is not only "firmly entrenched" but has recently spiked (Llana and Brodzinsky, 2012).

In India, gang rapes are a "routine and a largely invisible crime." Also, young women who move to a city are policed by a network of males. If the women violate village moral codes, including owning a cell phone, they disappear or are quickly married to village men of their parents' choosing (Barry, 2013; Barry and Choksi, 2013; Srivastava and Mehrotra, 2013).

CRITICAL EVALUATION

Feminist analyses expand our understanding of cultural components that other theoretical perspectives ignore or gloss over. Like conflict theorists, however, feminist theorists often emphasize discord rather than how culture integrates women and men into society. Another weakness is that feminist scholars tend to downplay social class inequality by focusing on the cultural experiences of low-income and minority women but not of their male counterparts.

3-7d Symbolic Interaction

Unlike functionalists and conflict theorists, symbolic interactionists examine culture through micro lenses. They're most interested in understanding how individuals—in contrast to groups, organizations, or societies—create, maintain, and modify culture.

KEY POINTS

Interactionists study how culture influences our everyday lives. Language, you'll recall, shapes our views and behavior. People within and across societies create a variety of symbols that may change over time. Technology has also changed many people's communication patterns. Some complain, for example, that texting dominates our lives. A much viewed YouTube clip shows the bride pulling out her cell phone from under her veil and texting with one hand as she walks down the aisle with her father (Pflaumer, 2011). On the other hand, students become more engaged when they're required to use technology, such as tweeting weekly observations that illustrate their reading assignments (Lang, 2013).

Which sociological perspectives help explain how children's birthday parties reproduce culture?

Symbolic interactionists also note that our values and norms, like other components of culture, aren't superimposed by some unknown external force. Instead, as people construct their perception of reality, they create, change, and reinterpret values and norms through interaction with others. Peer pressure, for example, can either encourage or discourage inappropriate language or behavior.

CRITICAL EVALUATION

Micro approaches are useful in understanding what culture means to people and how these meanings differ across societies. However, symbolic interactionists don't offer a systematic framework that explains *how* people create and shape culture or develop shared meanings of reality. Why, for instance, are some of us less polite than others even though we share the same cultural values and norms? Another weakness is that interactionists don't address the linkages between culture and subcultures. Language bonds people, but interactionists say little about how organized groups (such as the English-only movement) try to maintain control over language to promote their beliefs.

STUDY TOOLS 3

READY TO STUDY? IN THE BOOK, YOU CAN:

☐ Check your understanding of what you've read with the Test Your Learning Questions provided on the chapter review card at the back of the book.

☐ Rip out the chapter review card for a handy summary of the chapter and key terms.

ONLINE AT CENGAGEBRAIN.COM YOU CAN:

☐ Collect StudyBits while you read and study the chapter.

☐ Quiz yourself on key concepts.

☐ Prepare for tests with SOC5 Flash Cards as well as those you create.

☐ Watch videos to explore sociology further.

4 | Socialization

LEARNING OUTCOMES

After studying this chapter, you will be able to…

4-1 Define and illustrate socialization and explain its importance.

4-2 Describe the nature versus nurture debate.

4-3 Compare social learning and symbolic interaction theories of socialization.

4-4 Describe and illustrate five socialization agents.

4-5 Explain how socialization changes throughout life.

4-6 Explain when and how resocialization occurs.

After you finish
this chapter go to
PAGE 80 for
STUDY TOOLS.

Two of MTV's most popular reality shows are *16 and Pregnant* and *Teen Mom*. The producers say that the goal of the programs is to decrease teen pregnancy by showing the harmful effects of not using protection or birth control, and the resulting struggles in raising a baby. Critics argue that the shows glamorize teen pregnancy, make the teen moms instant celebrities, and ignore the long-term negative impact on the mothers' and children's healthy social development (Thompson, 2010). Both sides are talking about **socialization**, the lifelong process through which people learn culture and become functioning members of society.

4-1 SOCIALIZATION: ITS PURPOSE AND IMPORTANCE

Socialization is critical in all societies. To understand why, let's begin by looking at the purpose of socialization.

4-1a What Is the Purpose of Socialization?

Socialization—from childhood to old age—can be relatively smooth or bumpy, depending on factors such as age, gender, race/ethnicity, and social class. Generally,

What do you think?

I'm always the same person, no matter where I am or who's around.

1	2	3	4	5	6	7

strongly agree   strongly disagree

Comstock Images/Getty Images

however, socialization has four key functions that range from providing us with a social identity to transmitting culture to the next generation.

SOCIALIZATION ESTABLISHES OUR SOCIAL IDENTITY

Have you ever thought about how you became the person you are today? Sociology professors sometimes ask their students to give 20 answers to the question "Who am I?" (Kuhn and McPartland, 1954). How would you respond? You would probably include a variety of descriptions such as college student, single or married, female or male, and son or daughter. Your answers would show a sense of your *self* (a concept we'll examine shortly). You are who you are largely because of socialization.

SOCIALIZATION TEACHES US ROLE TAKING

Why do you act differently in class than with your friends? Because we play different roles in different settings. A *role* is the behavior expected of a person in a particular social position (see Chapter 5). The way we interact with a parent is typically very different from the way we talk to an employer, a child, or a professor. We all learn appropriate roles through socialization.

SOCIALIZATION CONTROLS OUR BEHAVIOR

In learning appropriate roles, we absorb values and a variety of rules about how we should (and shouldn't) interact in everyday situations. If we follow the rules, we're usually rewarded. If we break the rules, we may be punished. Socialization maintains the social order and controls our behavior by teaching us to conform to social expectations.

We conform because we've internalized societal norms and values. **Internalization** is the process of learning cultural behaviors and expectations so deeply that we accept them without question. Obeying laws, paying bills on time, and respecting teachers are examples of internalized behaviors.

SOCIALIZATION TRANSMITS CULTURE TO THE NEXT GENERATION

Each generation passes its culture on to the next generation. The culture that is transmitted, as you saw in Chapter 3, includes language, beliefs, values, norms, and symbols.

socialization the lifelong process through which people learn culture and become functioning members of society.

internalization the process of learning cultural behaviors and expectations so deeply that we accept them without question.

The Harlow Studies and Emotional Attachment

In the early 1960s, psychologists Margaret and Harry Harlow (1962) conducted several studies on infant monkeys. In one group, a "mother" made of terry cloth provided no food, while a "mother" made of wire did so through an attached baby bottle containing milk. In another group, the cloth mother provided food but the wire mother didn't. Regardless of which mother provided milk, when both groups of monkeys were frightened, they clung to the cloth mother. The Harlows concluded that physical contact and comfort were more important to the infant monkeys than nourishment.

Since then, some sociologists have cited the Harlow studies to argue that emotional attachment may be even more critical than food for human infants. Do you see any problems with sociologists' generalizing the results of animal studies to humans?

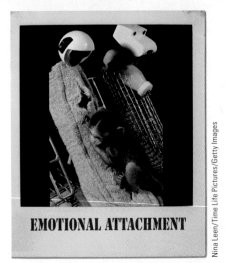

EMOTIONAL ATTACHMENT

Nina Leen/Time Life Pictures/Getty Images

4-1b Why Is Socialization Important?

Social isolation can be devastating. An example of its negative effects is Genie. When Genie was 20 months old, her father decided she was "retarded," put her in a wire mesh cage, and locked her away in a back room with the curtains drawn and the door shut. Genie's father frequently beat her with a wooden stick and no one in the house was allowed to speak to her. When she was discovered in 1970 at age 13, a psychiatrist described Genie as "unsocialized, primitive, and hardly human." Except for high-pitched whimpers, she never spoke. Genie had little bowel control, experienced rages, and tried to hurt herself. After living in a rehabilitation ward and a foster home, she learned to eat normally, was toilet-trained, and gradually developed a vocabulary, but her language use never progressed beyond that of a 3- or 4-year-old (Curtiss, 1977).

The research on children who are isolated (such as Genie) or institutionalized (such as orphans) demonstrates that socialization is critical to our development. Behaviors such as talking, eating with utensils, and controlling our bowel movements don't come naturally. Instead, we learn to do all of these things beginning in infancy. When children are deprived of social interaction, they don't develop the characteristics that most of us see as normal and human.

4-2 NATURE AND NURTURE

Biologists focus on the role of heredity (or genetics) in human development. In contrast, most social scientists, including sociologists, underscore the importance of learning, socialization, and culture. This difference of opinion is often called the *nature–nurture debate* (see *Table 4.1*).

4-2a How Important Is Nature?

Those who argue that nature (biology) shapes behavior point to two kinds of evidence—developmental and health differences between males and females, and cases of unsuccessful sex reassignment.

DEVELOPMENTAL AND HEALTH DIFFERENCES

Boys mature more slowly than girls, get sick more often, and are less likely to have mastered the self-control and fine-motor skills necessary for a successful start in school. Boys are also at greater risk than girls for most of the major learning and developmental disorders—as much as four times more likely to suffer from autism, attention deficit disorder, and dyslexia. Girls, however, are at least twice as likely as boys to suffer from depression, anxiety, and eating disorders (Eliot, 2012).

Among adults, the senses of smell and taste are more acute in women than in men, and hearing is better and lasts longer in women than in men. Women, however, have a higher risk of developing diabetes. Some conditions (such as migraine headaches and breast cancer) are more common in women, whereas others (such as hemophilia and skin cancer) are more common in men (McDonald, 1999; Kreeger, 2002, 2002).

TABLE 4.1 THE NATURE–NURTURE DEBATE

Nature	Nurture
Human development is . . .	Human development is . . .
Innate	Learned
Biological, physiological	Psychological, social, cultural
Due largely to heredity	Due largely to environment
Fairly fixed	Fairly changeable

UNSUCCESSFUL SEX REASSIGNMENT

Scientists who believe that nature (not nurture) molds behavior also point to unsuccessful sex reassignment attempts. In the 1960s, John Money, a highly respected psychologist, published numerous articles and books in which he maintained that gender identity is determined as much by culture and nurture as by hormones (see, for example, Money and Ehrhardt, 1972). His views were based on the case of David Reimer, a child who underwent sex reassignment after his penis was mutilated in a botched circumcision. His parents, upon Money's recommendation, raised David as a girl (Brenda). Money maintained that the reassignment was successful. A biologist and a psychiatrist who followed up on David/Brenda's case in the 1990's, however, concluded that the sex reassignment had failed (Diamond and Sigmundson, 1997).

4-2b How Important Is Nurture?

Most sociologists maintain that nurture is more significant than nature because socialization and culture shape human behavior. They point to successful sex reassignment cases (see Vitello, 2006) and two types of data to support their argument: cross-cultural variations in male violence, and the environment's effect on biology.

CROSS-CULTURAL VARIATIONS IN MALE VIOLENCE

We often hear that males are "naturally" more aggressive than females because their glands produce more testosterone, the dominant male hormone. If men were innately aggressive due to biology, they would be equally violent across all societies. This isn't the case. The proportion of women who have ever suffered physical violence by a male partner varies considerably: 16 percent in East Asia, 21 percent in North America, 41 percent in western Latin America, and 66 percent in central sub-Saharan Africa (World Health Organization, 2013).

All mass murderers (those who have killed a large number of people during one incident) have been men, but most such murders have occurred in the United States (Christakis, 2012; Kluger, 2014). Such variations reflect cultural laws and practices and other environmental factors (nurture) rather than biology or genetics (nature) (Chesney-Lind and Pasko, 2004).

HOW ENVIRONMENT AFFECTS BIOLOGY

Much research shows that environment (nurture) influences children's genetic makeup (nature). For example, birth defects associated with prenatal alcohol exposure can occur in the first 3 to 8 weeks of pregnancy, before a woman even knows she's pregnant. A woman's single drinking "binge"—lasting four hours or more—can permanently damage an unborn child's brain. A man who drinks heavily may have genetically damaged sperm that also leads to birth defects. Thus, alcohol abuse (an environmental factor) by either a woman or a man can have a devastating, irreversible, and lifelong negative biological impact on a child (Denny et al., 2009; Warren, 2012).

In a growing field known as "fetal origins," scientists are finding that the nine months of pregnancy constitute the most consequential period of our lives, permanently influencing the wiring of the brain, the functioning of organs such as the heart and liver, and behavior. That is, the kind and quantity of nutrition you received in the womb; the pollutants, drugs, and infections you were exposed to during gestation; your parents' health and eating habits; and your mother's stress level during pregnancy—all these factors have lasting effects in infancy, childhood, and adulthood (Begley, 2010; Paul, 2010).

Adult behavior can influence a child's biological makeup in other ways. Physical, psychological, or sexual abuse can impair a child's developing brain, and these changes can trigger disorders such as depression in adulthood (see Chapter 12). Thus, childhood mistreatment can blunt biological development and lead to behavioral and emotional problems throughout life. Even when identical twins look alike because they have the same genetic makeup, socialization can affect their interests, personalities, lifestyles, and life outcomes.

4-2c What Can We Conclude About the Nature–Nurture Debate?

A growing number of sociologists maintain that the nature *or* nurture debate is outdated in understanding

Nature or nurture?

Denys Kurbatov/Shutterstock.com; Graphixmania/Shutterstock.com; Lichtmeister/Shutterstock.com

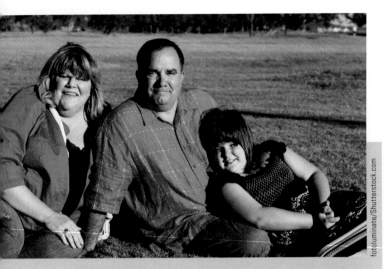

Is children's obesity due to nature, nurture, or both?

socialization processes and outcomes (Ledger, 2009). They believe that genes shape our lives and could help explain why there's so much variation across families and other groups.

Humans have hundreds of traits—such as height, metabolism, aggression, leadership traits, and cognitive ability—that are partly inherited (Freese, 2008). A team of researchers analyzed 50 years of twin studies and more than 300 characteristics. They found that, on average, 49 percent of the individual differences were genetic and 51 percent were environmental, but some traits (such as cleft lip) were more likely to be inherited than others (such as anti-social behavior) (Polderman et al., 2015).

Our social environment can enhance or dampen biological characteristics. For example, children who are genetically predisposed to obesity don't always become overweight if parents discourage overeating

The Case of Brenda/David

In 1963, twin boys were being circumcised when the penis of one of the infants, David Reimer, was accidentally burned off. Encouraged by John Money, a highly respected psychologist, the parents agreed to raise David as "Brenda." The child's testicles were removed, and surgery to construct a vagina was planned. Money reported that the twins were growing into happy, well-adjusted children, setting a precedent for sex reassignment as the standard treatment for 15,000 newborns with similarly injured genitals (Colapinto, 1997, 2001).

In the mid-1990s, a biologist and a psychiatrist followed up on Brenda's progress and concluded that the sex reassignment had failed. Almost from the beginning, Brenda refused to be treated like a girl. When her mother dressed her in frilly clothes as a toddler, Brenda tried to rip them off. She preferred to play with boys and stereotypical boys' toys such as machine guns (Diamond and Sigmundson, 1997).

At age 14, Brenda rebelled and stopped living as a girl: She refused to wear dresses, urinated standing up, refused vaginal surgery, and decided she would either commit suicide or live as a male. When his father finally told David the true story of his birth and sex change, David recalls that "all of a sudden everything clicked. For the first time things made sense and I understood who and what I was" (Diamond and Sigmundson, 1997: 300).

David had a mastectomy (breast removal surgery) at age 14 and underwent several operations to reconstruct a penis. At age 25, he married an older woman and adopted her three children. At age 38, he committed suicide. Most suicides have multiple reasons, but some speculated that David committed suicide because of the "physical and mental torments he suffered in childhood that haunted him the rest of his life" (Colapinto, 2004: 96). David's experience suggests to some scientists that nature outweighs nurture in shaping a person's gender identity.

David Reimer was raised as a girl, Brenda, until he was 14.

and encourage physical recreational activities (Martin, 2008). Thus, research that combines both biological and social factors helps us understand human behavior (D'Onofrio and Lahey, 2010; Shanahan et al., 2010; Bryant, 2014).

 4-3 ## SOCIOLOGICAL EXPLANATIONS OF SOCIALIZATION

Functionalism provides a foundation for understanding the purposes of socialization described at the beginning of this chapter, but doesn't tell us *how* socialization works on a micro level. There are well-known psychological and psychosocial theories of human development (e.g., Sigmund Freud, Erik Erikson, Jean Piaget, and Lawrence Kohlberg). In sociology, two influential approaches that explain socialization on a micro level are *social learning theories* and *symbolic interaction theories* (*Table 4.2* summarizes these perspectives).

4-3a Social Learning Theories

The central notion of **social learning theories** is that people learn new attitudes, beliefs, and behaviors through social interaction, especially during childhood. The learning is direct and indirect, and a result of observation, reinforcement, and imitation (Bandura and Walters, 1963; Mischel, 1966; Lynn, 1969).

DIRECT AND INDIRECT LEARNING

Reinforcement refers to direct or indirect rewards or punishments for particular behaviors. A little girl who puts on her mother's makeup may be told that she's cute, but her brother will be scolded ("boys don't wear makeup"). Children also learn through indirect reinforcement. If a little boy's male friends are punished for crying, he'll learn that "boys don't cry."

Children also learn through *observation* and *imitation*. Even when children aren't directly rewarded or punished for "behaving like boys" or "behaving like girls," they learn about gender by watching who does what in their families. A father who is rarely at home because he's always working sends the message that men are supposed to earn money. A mother who is always complaining about being overweight or old sends the message that women are supposed to be thin and young. Because parents are emotionally important to their children, they're

typically a child's most powerful **role models**, people we admire and whose behavior we imitate. Other role models, as you'll see shortly, include siblings, grandparents, teachers, friends, and even celebrities.

LEARNING AND PERFORMING

Social learning theorists also distinguish between *learning* and *performing* behavior. Children and adults learn norms and roles through observation, but they don't always imitate the behavior. For example, children may see their friends cheat in school but don't do so themselves. Adults, similarly, may see their coworkers steal office supplies but buy their own. We behave as we do, then, because our society teaches us what's appropriate and inappropriate.

CRITICAL EVALUATION

Social learning theories help us understand why we behave as we do, but much of the emphasis is on early socialization rather than on what occurs throughout life. Social learning theories don't explain why reinforcement and modeling affect some children more than others, especially those in the same family. There may be personality differences, but if learning is as effective as social learning theorists maintain, siblings' attitudes and behavior should be more similar than different. This is often not the case, even with identical twins.

Most social learning theories ignore factors such as birth order, which brings different advantages and disadvantages. Except for affluent families, larger families have

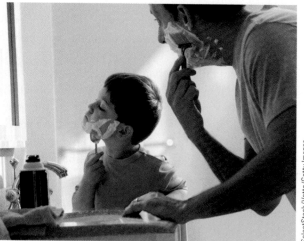

SelectStock/Vetta/Getty Images

social learning theories maintain that people learn new attitudes, beliefs, and behaviors through social interaction, especially during childhood.

role models people we admire and whose behavior we imitate.

TABLE 4.2 KEY ELEMENTS OF SOCIALIZATION THEORIES

Social Learning Theories	Symbolic Interaction Theories
• Social interaction is important in learning appropriate and inappropriate behavior.	• The self emerges through social interaction with significant others.
• Socialization relies on direct and indirect reinforcement.	• Socialization includes role taking and controlling the impression we give to others.
Example: Children learn how to behave when they are scolded or praised for specific behaviors.	*Example:* Children who are praised are more likely to develop a strong self-image than those who are always criticized.

fewer resources for the second, third, and later children. The first child may be disciplined and monitored more, but also has higher grades than her or his younger siblings, and is more likely to attend college. In contrast, later-born children often have fewer rules, less parental supervision, lower grades, and are less likely to receive financial support to attend college (Conley, 2004; Hotz and Pantano, 2013).

4-3b Symbolic Interaction Theories

Symbolic interaction theories have had a major impact in explaining social development. Sociologists Charles Horton Cooley (1864–1929), George Herbert Mead (1863–1931), and Erving Goffman (1922–1982) were especially influential in showing how social interaction shapes socialization.

CHARLES HORTON COOLEY: EMERGENCE OF THE SELF AND THE LOOKING-GLASS SELF

Newborn infants lack a sense of **self**, an awareness of their social identity. Gradually, they begin to differentiate themselves from their environment and to develop a sense of self.

After carefully observing the development of his young daughter, Charles Horton Cooley (1909/1983) concluded that children acquire a sense of who they are through their interactions with others, especially by imagining how others view them. The sense of self, then, isn't innate but emerges out of social relationships. Cooley called this social self the *reflected self,* or the **looking-glass self**, a self-image based on how we think others see us. He proposed that the looking-glass self develops in an ongoing process of three phases:

▶ **Phase 1: Perception.** We imagine how we appear to other people and how they *perceive* us ("She thinks I'm attractive" or "I bet he thinks I'm fat").

▶ **Phase 2: Interpretation of the perception.** We imagine how others *judge* us ("She's impressed with me" or "He's disgusted with the way I look").

▶ **Phase 3: Response.** We experience *self-feelings* based on how we think other people judge us. If we think they see us in a favorable light, we may feel proud, happy, or self-confident ("I'm terrific"). If we think others see us in a negative light, we may feel angry, embarrassed, or insecure ("I'm pathetic").

Our interpretation of others' perceptions (phase 2) may be totally wrong. The looking-glass self, remember, refers to how we *think* others see us rather than how they *actually* see us. However, our perceptions of other people's views—whether we're right or wrong—affect our self-image, which then affects our behavior.

Cooley focused on how children acquire a sense of who they are through their interactions with others, but he noted that the process of forming a looking-glass self doesn't end in childhood. Instead, our self-concept may change over time because we reimagine ourselves as we

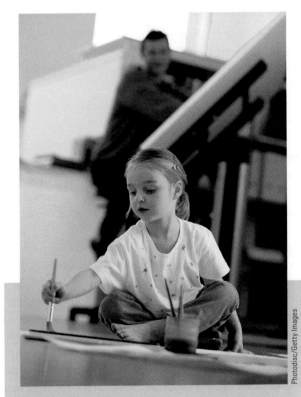

Parents who encourage their children to express themselves help them develop a positive self-image.

self an awareness of one's social identity.

looking-glass self a self-image based on how we think others see us.

think others see us—attractive or ugly, interesting or boring, intelligent or stupid, graceful or awkward, and so on.

Sometimes we're aware of the process shaping our looking-glass self. Consider your own self-concept: If others keep telling you what a great student you are, you'll likely see yourself as intelligent. In many settings, however—such as public places, large classrooms, and sports events—the looking-glass self is irrelevant because there's relatively little interpersonal interaction and the other people are strangers who aren't very significant in our lives.

GEORGE HERBERT MEAD: DEVELOPMENT OF THE SELF AND ROLE TAKING

Cooley described how an individual's sense of self emerges, but not how it develops. George Herbert Mead, one of Cooley's colleagues, took up this task. For Mead (1934), the most critical social interaction occurs in the family, the foundation of socialization.

Our self develops, according to Mead, when we learn to differentiate the *me* from the *I*—two parts of the self. The *I* is creative, imaginative, impulsive, spontaneous, self-centered, and sometimes unpredictable. The *me* that has been successfully socialized is aware of the attitudes of others, has self-control, and has internalized social roles. Instead of impulsively and selfishly grabbing another child's toys, for example, as the *I* would do, the *me* asks for permission to use someone's toys and shares them with others.

For Mead, the *me* forms as children engage in **role taking**, learning to take the perspective of others. Children gradually acquire this ability early in the socialization process through three sequential stages:

1. **Preparatory stage (roughly birth to 2 years).** An infant doesn't distinguish between the self and others. The *I* is dominant, while the *me* is forming in the background. In this stage, children learn through imitation. They may mimic daddy's shaving or mommy's angry tone of voice without really understanding the parent's behavior. In this exploratory stage, children engage in behavior that they rarely associate with words or symbols, but they begin to understand cause and effect (e.g., crying leads to being picked up). Gradually, as the child begins to recognize others' reactions and forms a looking-glass self, he or she develops a self.

2. **Play stage (roughly 2 to 6 years).** The child begins to use language and to understand that words (such as *dog* and *cat*) have a shared cultural meaning. Through play,

children begin to learn role taking in two ways. First, they emulate significant others. A **significant other** is someone whose opinions we value and who influences our thinking, especially about ourselves. A significant other can be anyone: usually a parent, grandparent, or sibling in early childhood; a teacher, girlfriend/ boyfriend, spouse, best friend, or employer in later childhood and adulthood. The child learns that he or she has a self that is distinct from those of others, that others behave in many different ways, and that others expect her or him to behave in specific ways. In other words, the child learns social norms (see Chapter 3).

In the play stage, the child moves beyond imitation and acts out imagined roles ("I'll be the mommy and you be the daddy"). The play stage involves relatively simple role taking because the child plays one role at a time but doesn't yet understand the relationships between roles. This stage is crucial, according to Mead, because the child *is learning to take the role of the other*. For the first time, the child tries to imagine how others behave or feel. The *me* grows stronger because the child is concerned about the judgments of significant others.

Also in the play stage, children experience **anticipatory socialization**, learning how to perform a role that they'll occupy. By playing "mommy" or "daddy," children prepare themselves for eventually becoming parents. Anticipatory socialization continues in later years when, for instance, expectant parents attend childbirth classes, job seekers practice their skills in mock interviews, and high school students visit campuses and attend orientation meetings to prepare for college life.

3. **Game stage (roughly 6 years and older).** This stage involves acquiring the ability to understand connections between roles. The child must "not only take the role of the other … but must assume the various roles of all participants in the game, and govern his action accordingly" (Mead, 1964: 285).

Mead used baseball to illustrate this stage. In baseball (or other organized games and activities), the child plays one role at a time (batter) but understands and anticipates the actions of other players (pitcher, shortstop,

Vahan Shirvanian/Cartoonstock.com

role taking learning to take the perspective of others.

significant other someone whose opinions we value and who influences our thinking, especially about ourselves.

anticipatory socialization learning how to perform a role that a person will occupy.

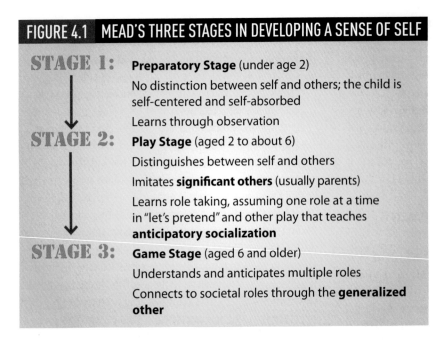

FIGURE 4.1   MEAD'S THREE STAGES IN DEVELOPING A SENSE OF SELF

STAGE 1:   **Preparatory Stage** (under age 2)

No distinction between self and others; the child is self-centered and self-absorbed

Learns through observation

STAGE 2:   **Play Stage** (aged 2 to about 6)

Distinguishes between self and others

Imitates **significant others** (usually parents)

Learns role taking, assuming one role at a time in "let's pretend" and other play that teaches **anticipatory socialization**

STAGE 3:   **Game Stage** (aged 6 and older)

Understands and anticipates multiple roles

Connects to societal roles through the **generalized other**

runners on bases) on both teams. To successfully participate in a game of baseball, the individual must be able to anticipate the roles of others. The same is true of participating in society. Thus, as children grow older and interact with a wider range of people, they learn to anticipate, respond to, and fulfill a variety of social roles.

At the game stage, children connect to societal roles through the **generalized other**, the norms, values, and expectations of society as a whole that affect a person's behavior (see *Figure 4.1*).

This concept suggests that when an individual acts, she or he takes into account people in general. For example, a child is honest not merely to please a particular parent, but because parents, teachers, and others endorse honesty as an important social value. For sociologists, the development of the generalized other is a central feature of socialization because the *me* becomes an integral part of the self (see *Figure 4.1*).

Who's engaging in impression management? How? Why?

Rob Marmion/Shutterstock.com

After the generalized other has developed, the *me* never fully controls the *I*, even in adulthood. We sometimes break rules or act impulsively, even though we know better. When we're frustrated, sick, tired, or angry, we may lash out against other people even though they haven't done anything wrong.

ERVING GOFFMAN: STAGING THE SELF IN EVERYDAY LIFE

Cooley and Mead described how the self and role taking emerge and develop during early socialization. Erving Goffman (1959, 1969) extended these analyses by showing that we interact differently in different settings throughout adulthood. Goffman proposed that social life mirrors the theatre because we're like actors: We engage in "role performances," want to influence an "audience," and can have considerable control over the image that we project while we're "on stage." (We'll explore these concepts in greater detail in Chapter 5.)

In a process that Goffman called **impression management**, we provide information and cues to others to present ourselves in a favorable light while downplaying or concealing our less appealing characteristics. Being successful in this presentation of the self requires managing three types of expressive resources. First, we try to control the *setting*—the physical space, or "scene," where the interaction takes place. In the classroom, "a professor may use items such as chalk, lecture notes, computers, videos, and desks to facilitate a class and show that he or she is an excellent teacher" (Sandstrom et al., 2006: 105).

A second expressive resource is controlling *appearance,* such as clothing and titles that convey information about our social status. When physicians or professors use the title "Doctor," for example, they're telling the "audience" that they expect respect.

The third expressive resource is *manner*—the mood or style of behavior we display that sends important messages to the audience. Faculty members regularly manage their manner when interacting with students. When you email or text a professor a question about your course paper, the response is usually encouraging

generalized other the norms, values, and expectations of society as a whole that affect a person's behavior.

impression management the process of providing information and cues to others to present oneself in a favorable light while downplaying or concealing one's less appealing characteristics.

("Have you also thought about . . .") even though the professor may actually be annoyed ("You haven't thought about this very much; stop wasting my time").

Sometimes, no matter how much we try to manage setting, appearance, and manner, we slip up in our role performances. This leads to discomfort and embarrassment for everyone involved, but the audience is typically tactful:

> *Imagine that you are giving a presentation in one of your classes and, just as you make a key point, a large droplet of saliva lands near someone sitting in the front row. Although several of your classmates cannot help but notice it, none of them are likely to shout out, "Hey, you just about spit on someone in the first row!" Instead, they will probably act as if nothing unusual or embarrassing took place* (Sandstrom et al., 2006: 109).

As we move from one situation to another, according to Goffman, we maintain self-control, for example, by avoiding emotional outbursts and altering our facial expressions and verbal tones. Thus, all of us engage in impression management practically every day.

CRITICAL EVALUATION

Symbolic interactionists have provided major insights about socialization, particularly of young children, and have shown that fitting into our social world is a complex process that continues through adulthood. Like other theories, however, symbolic interaction has its limitations.

First, it's not clear why some children have a positive looking-glass self and are successful later in life even when the cues are consistently negative. Over the years, some of my best students have included those who grew up in abusive households or attended low-quality schools.

Second, essential concepts such as *self, me,* and *I* are vague. Because the concepts are imprecise, it's difficult, if not impossible, to measure them.

Third, stage theories, such as Mead's, maintain that children automatically pass through play and game stages as they get older. Some critics argue that the stages aren't as rigid as symbolic interactionists claim, may overlap or vary depending, for example, on whether parents and other primary caregivers are actively involved in the child's upbringing and provide enriching interaction. Even then, some children never develop the role of the generalized other (see Ritzer, 1992).

Fourth, some have questioned the value of the concept of the generalized other in understanding early socialization processes. Because children interact in many social contexts (home, preschool, play groups), they don't simply assume the role of one generalized other, but many. That is, as children get older, they may have several **reference groups**—groups of people who shape an individual's self-image, behavior, values, and attitudes in different contexts (Merton and Rossi, 1950; Shibutani, 1986).

Fifth, interactionists credit people with more free will than they have. Individuals don't always have the power or ability to affect others' reactions or their own lives. For example, impression management is less common among lower than higher socioeconomic groups, whose members have internalized such skills to be successful in jobs. Also, how people react to us may be due to characteristics (e.g., gender, age, race/ethnicity) that we can't change (Powers, 2004).

Finally, a major criticism of Cooley, Mead, Goffman, and other interactionists is that they tend to downplay or ignore macro-level factors that affect people's development. As the next two sections show, institutions such as the family, education, economy, and popular culture have a major impact on socialization. Let's begin by looking at socialization agents who play important roles in our lives.

4-4 PRIMARY SOCIALIZATION AGENTS

By fretting, crying, and whining, infants teach adults when to feed, change, and pick them up. Babies are actively engaged in the socialization process, but family, friends, peer groups, teachers, and the media are some of the primary **socialization agents**—the individuals, groups, or institutions that teach us how to participate effectively in society.

4-4a Family

Parents, siblings, grandparents, and other family members play a critical role in our socialization. Parents, however, are the first and most influential socialization agents.

Gelpi JM/Shutterstock.com

---

reference groups groups of people who shape an individual's self-image, behavior, values, and attitudes in different contexts.

socialization agents the individuals, groups, or institutions that teach us how to participate effectively in society.

TABLE 4.3 PARENTING STYLES

Parenting Style	Characteristics	Example
Authoritarian	Very demanding, controlling, punitive	"You can't borrow the car because I said so."
Authoritative	Demanding, controlling, warm, supportive	"You can borrow the car, but be home by curfew."
Permissive	Not demanding, warm, indulgent, set few rules	"Borrow the car whenever you want."
Uninvolved	Neither supportive nor controlling	"I don't care what you do; I'm busy."

HOW PARENTS SOCIALIZE CHILDREN

The purpose of socialization is to enable children to regulate their behavior and to make responsible decisions. In teaching their children social rules and roles, parents rely on several learning techniques, such as reinforcement, that you've just read about. Parents also manage many aspects of the environment that influence a child's social development: They often choose the neighborhood (which then determines what school a child attends), decorate the child's room in a masculine or feminine style, provide her or him with particular toys and books, and arrange enriching social events and other activities (e.g., sports, art, music).

PARENTING STYLES

Four common parenting styles affect socialization (see *Table 4.3*). In *authoritarian parenting*, parents tend to be harsh, unresponsive, and rigid, using their power to control a child's behavior. *Authoritative parenting* is warm, responsive, and involved, but parents set limits and expect appropriately mature behavior from their children. *Permissive parenting* is lax: Parents set few rules but are usually warm and responsive. *Uninvolved parenting* is indifferent and neglectful: Parents focus on their own needs, spend little time interacting with the children, and know little about their interests or whereabouts (Baumrind, 1968, 1989; Maccoby and Martin, 1983; Aunola and Nurmi, 2005).

Healthy child development is most likely in authoritative homes because parents are consistent in combining warmth, monitoring, and discipline. Authoritative parenting tends to produce children who are self-reliant, achievement-oriented, and successful in school. Adolescents in households with authoritarian, permissive, or uninvolved parents tend to have poorer psychosocial development, lower school grades, lower self-reliance, higher levels of delinquent behavior, and

are more likely to be swayed by harmful peer pressure (e.g., to use drugs and alcohol) (Hillaker et al., 2008; Meteyer and Perry-Jenkins, 2009).

Such findings vary by social class and race/ethnicity, however. For many Latino and Asian immigrants, authoritarian parenting produces positive outcomes, such as academic success. This parenting style is also effective in safeguarding children who are growing up in low-income neighborhoods with high crime rates and drug peddling (Pong et al., 2005).

SIBLINGS

Siblings, like parents, are important socialization agents. Older siblings may transmit beliefs that having sex at an early age, smoking, drinking, and marijuana use "aren't a

AP Images/Yonhap/Shin Young-kuyn

During the last decade, numerous summer camps have sprung up in South Korea where children ages 7 to 19 live in military-style barracks, undergo drills and exercise from dawn to dusk, eat very simple meals, and have no access to computers, television, or cell phones. Most are sent by parents "who realize that their pampered offspring need more discipline to become better students and grow into conscientious adults" (Glionna, 2009: A1). Some of the kids' offenses have included getting low grades, accidentally breaking a window, and, most commonly, talking back to their parents.

big deal" (Altonji et al., 2010). On the other hand, older siblings are positive role models when they encourage their younger brothers and sisters to do well in school, to stay away from friends who get in trouble, and to get along with people. Siblings can also have a positive impact by helping their younger brothers and sisters with their homework and protecting them from neighborhood and school bullies. During adulthood, siblings often act as confidantes and offer financial and emotional assistance during stressful times (Whiteman et al., 2011; McHale et al., 2012).

GRANDPARENTS

In many ways, grandparents are the glue that keeps a family close. Some grandparents and grandchildren live far apart and rarely see each other, but the majority see their grandchildren often, frequently do things with them, and offer them emotional and instrumental support (e.g., listening and giving them money). Many recent Asian immigrants live in extended families and are more likely to do so than any other group, including Latinos. In such co-residence, grandparents are often "historians" who transmit values and cultural traditions to their grandchildren even if there are language barriers (Benokraitis, 2015; see also Chapter 12).

Grandparents are sometimes *surrogates* because they provide regular care or raise the grandchildren. Whether custody is legal or informal, about 10 percent of all U.S. children live with a grandparent, up from 5 percent in 1990. About 2.7 million grandparents are raising their children; 33 percent of these households have no parent present (Ellis and Simmons, 2014; "National Grandparents Day...," 2014).

The most common pattern is *living-with* grandparents who have the grandchild in their own home or, less often, live in the home of a grandchild's parents. Living-with grandparents take on child-rearing responsibilities either because their children haven't yet moved out of the house or because teenage or adult parents can't afford to live on their own. These living arrangements have increased the number of **multigenerational households**, homes in which three or more generations live together. In effect, then, grandparents are important socialization agents.

4-4b Play, Peer Groups, and Friends

As you saw earlier, Mead believed that play and games are important in children's development. They help us become more skilled in using language and symbols, learn role playing, and internalize roles that we don't necessarily enact (the generalized other).

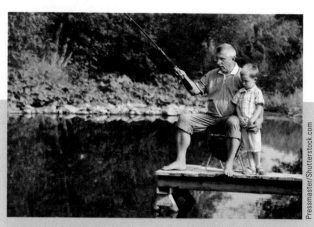

Grandparents often provide stability in family relationships and a continuity of family rituals and values. When a teacher asked her class of 8-year-olds what they liked best about a grandparent, the essays included comments such as "They don't say 'Hurry up'" and "When they take us for a walk, they slow down for things like pretty leaves and caterpillars."

A **peer group** consists of people who are similar in age, social status, and interests. All of us are members of peer groups, but such groups are especially influential until about our mid-20s. After that, coworkers, spouses, children, and close friends are usually more important than peer groups in our everyday lives.

PLAY

Play serves several important functions. First, it promotes cognitive development. Whether it's doing simple five-piece puzzles or tackling complex video games such as *Sim City* and *Roller Coaster Tycoon*, play encourages children to think, formulate strategies, and budget and manage resources. From an early age, however, play is generally gender typed. Among preschoolers, when playing alone, girls prefer feminine activities (such as playing with dolls) whereas boys are more likely to play with balls and transportation toys. However, when girls and boys are in mixed-sex play groups, they're as likely to play with "masculine" as with "feminine" toys. Thus, gender-typed play breaks down and promotes a greater range of activities (Goble et al., 2012).

American children make up only 3 percent of the world's children, but U.S. families buy more than 40 percent of the toys purchased globally (Sanburn, 2015).

multigenerational households homes in which three or more generations live together.

peer group people who are similar in age, social status, and interests.

Girls' sections of catalogs and toy stores are swamped with cosmetics, dolls and accessories, arts and crafts kits, and housekeeping and cookware. In contrast, boys' sections feature sports equipment, building sets, workbenches, construction equipment, and toy guns. Some family scholars propose that children's toys should be more gender neutral to discourage gender stereotyping and to develop a "wider repertoire of skills" (Orenstein, 2011; Kapp, 2013).

The second function of play—especially when it's structured—is to keep children out of trouble and to enhance their social development. Children who spend more of their free time in structured and supervised activities—such as hobbies and sports—rather than just hanging out with their friends, perform better academically, are emotionally better adjusted, and have fewer problems at school and at home. In effect, sports and hobbies provide children with constructive ways to channel their energy and intelligence (McHale, 2001).

Third, play can strengthen peer relationships. Beginning in elementary school, few things are more important to most children than being accepted by their peers. Even if children aren't popular, belonging to a friendship group enhances their psychological well-being and ability to cope with stress (Rubin et al., 1998; Scarlett et al., 2005).

PEERS AND FRIENDS

Peer influence usually increases as young children get older. Especially during the early teen years, friends often reinforce desirable behavior or skills that enhance a child's self-image ("Wow, you're really good in math!"). Thus, to apply Cooley's concept, peers can help each other develop a positive looking-glass self.

Peers also serve as good role models. Children acquire a wide array of information and knowledge by observing their peers. Even during the first days of school, children learn to imitate their peers at standing in line, raising their hands in class, and being quiet while the teacher is speaking.

Among teens and young adults who are lesbian or gay, heterosexual friends can be especially supportive in accepting one's homosexuality and disclosing it to family. Peers can also have a significant impact in teen dating violence by encouraging their friends to leave a relationship, to report the abuse, and to seek protection orders (Shilo and Savaya, 2011; Oudekerk et al., 2014).

Not all peer or friend influence is positive, however. Among seventh- to twelfth-graders, having a best friend who engages in sexual intercourse, is truant, joins a gang, and uses tobacco and marijuana increases the probability of imitating such behavior (Card and Giuliano, 2011; Haas and Schaefer, 2014). Among adolescents and college students, poor physical fitness, obesity, and eating "junk food" increases the likelihood of adopting or maintaining an unhealthy diet and developing health problems (Carrell et al., 2010; De La Haye et al., 2013).

4-4c Teachers and Schools

Oprah Winfrey often praised her fourth-grade teacher, who recognized her abilities, encouraged her to read, and inspired her to excel academically. Like family and peer groups, teachers and schools affect our socialization.

In Sweden, Top-Toy Group, a licensee of the Toys "R" Us brand, published a "gender-blind" catalog for the 2012 Christmas season. "On some pages, girls brandish toy guns and boys wield blow-dryers and cuddle dolls" (Molin, 2012: D12). The United States is going in the opposite direction (see Auster and Mansbach, 2012).

Jekaterina Nikitina/Getty Images

GODONG/BSIP SA/Alamy

THE SCHOOL'S ROLE IN SOCIALIZATION

By the time children are 4 or 5, school fills an increasingly large portion of their lives. The school's primary purpose is to instruct children and enhance their cognitive development. Schools don't simply transmit knowledge; they also teach children to think about the world in different ways. Because of the emphasis on multiculturalism, for example, children often learn about other societies and customs. Even outside of classes, schools affect children's daily activities through homework assignments and participation in clubs and other extracurricular activities.

Because many parents are employed, schools have had to devote more time and resources to topics—such as sex education, abusive adolescent relationships, and drug abuse prevention—that were once the sole responsibility of families (Miller et al., 2015). In many ways, then, schools play an increasingly important role in socialization.

TEACHERS' IMPACT ON CHILDREN'S DEVELOPMENT

Teachers are among the most important socialization agents. From kindergarten through high school, teachers play numerous roles in the classroom—instructor, role model, evaluator, moral guide, and disciplinarian, to name just a few. Once children enter school, their relationships with their teachers are important for academic success. Kindergarten and elementary school teachers' reports of behavioral problems (such as unexcused absences) and poor grades often follow students into middle school (Hamre and Pianta, 2001; see also Chapter 13).

Siede Preis/Getty Images

4-4d Popular Culture and the Media

Because of tablets, smartphones, YouTube, and social networking sites, young people are constantly connected to electronic media (see Chapter 5). How does such technology affect socialization?

ELECTRONIC MEDIA

The American Academy of Pediatrics (AAP) advises parents to avoid television entirely for children younger than 2. The Academy also counsels parents to limit the viewing time of elementary school children to no more than 2 hours a day to encourage more interactive activities such as talking, playing, singing, and reading together "that will promote proper brain development." Still, 66 percent of children under age 2 spend, on average, an hour a day using screen media, and 16 percent of 1-year-olds have a TV in their bedrooms. Among parents of children age 8 and younger, 55 percent use media as a babysitter while they do chores around the house. Indeed, 71 percent of all children under age 18 have a TV in their bedroom (Rideout, 2013; Strasburger and Hogan, 2013).

The average young American now spends more time with electronic media than in school. The typical 8- to 10-year-old spends nearly 8 hours a day using some type of electronic media (which is more time than many adults spend in a full-time job), and older children and teenagers spend more than 11 hours a day. Too much time on the Internet has been linked with violence, cyberbullying, trouble in school, obesity, lack of sleep, and other problems (Strasburger and Hogan, 2013). Generally, youths who spend more time with media have lower grades and

moodboard - Mike Watson Images/Getty Images

Class field trips enhance children's socialization. Among other benefits, the field trips expand learning beyond the classroom, enrich students' understanding of subject matter, and encourage them to think about occupational choices and careers that they haven't considered.

TABLE 4.4 HOW DO ELECTRONIC MEDIA AFFECT CHILDREN?

Among all 8- to 18-year-olds, percentage who said that they are . . .

	Heavy Users (More Than 16 Hours/Day)	Moderate Users (3–16 Hours/Day)	Light Users (Less Than 3 Hours/Day)
Get good grades (A's and B's)	51	65	66
Get fair/poor grades (C's or lower)	47	31	23
Have been happy at school this year	72	81	82
Are often bored	60	53	48
Get into trouble a lot	33	21	16
Are often sad or unhappy	32	23	22

Source: Based on Rideout et al., 2010: 4.

lower levels of personal contentment (see *Table 4.4*). These findings are similar for girls and boys across all ages, race/ethnicity, parents' social class, and single- and two-parent households (Rideout et al., 2010).

The prevalence of gun violence in PG-13 movies has more than tripled since the rating was introduced in the mid-1980s, and now surpasses the violence in R-rated movies. (PG-13-rated movies caution parents that the material isn't appropriate for children under age 13. R-rated movies require children age 17 and younger to be accompanied by a parent or adult guardian.) Even if youth don't use guns, "because of the increasing popularity of PG-13-rated films, youth are exposed to considerable gun violence . . . that may increase aggressive behavior" (Bushman et al., 2013: 1017).

There's also a growing concern that violent video games make violence seem normal. Playing violent video games (e.g., *Grand Theft Auto, Call of Duty,* and *Halo*) can increase a person's aggressive thoughts, feelings, and behavior both in laboratory settings and in real life. Violent video games also encourage male-to-female aggression because much of the violence is directed at women (Carnagey and Anderson, 2005; Brody, 2015).

Still, it's not clear why violent video games affect people differently. Many young males who enjoy playing such games aren't any more aggressive, vicious, or destructive than those who aren't violent video game enthusiasts (Ferguson and Olson, 2013). Instead, factors such as angry and aggressive personalities, indifference to other people's feelings, and the quantity of violent gaming increase the likelihood of violent behavior (see Dennis, 2013; Markey et al., 2014; Przybylski and Mishkin, 2015).

ADVERTISING

In 2006, dozens of companies promised to limit child-directed TV ads that used cartoons and other characters to promote unhealthy food. Some companies have run shorter commercials, but the content hasn't changed very much (Kunkel et al., 2015; see also Chapter 3).

Advertisers are targeting children at increasingly younger ages. A new form of advertising called

What? Adolescents Can Buy Ultraviolent Video Games, but Not a Magazine with an Image of a Nude Woman?!

In 2011, the Supreme Court ruled (7–2 in *Brown v. Entertainment Merchants Association*) that video games, even ultraviolent ones, that are sold to minors are protected by the First Amendment's guarantee of free speech. The majority said that none of the scientific studies *prove* that violent games *cause* minors to act aggressively, and that parents, not the government, should decide what is appropriate for their children (Schiesel, 2011; Walls, 2011).

Hot Property/Shutterstock.com

Not surprisingly, the video game industry was jubilant over the decision. Many parents and lawmakers, on the other hand, agreed with one of the dissenting justices that it makes no sense to forbid selling a 13-year-old boy a magazine with an image of a nude woman, but not an interactive video game in which the same boy "actively . . . binds and gags the woman, then tortures and kills her" (Barnes, 2011: A1).

Many men's health and fitness magazines routinely feature models who have undergone several months of extreme regimens, including starvation and dehydration, to tighten their skin and make their muscles "pop." The magazines also use camera and lighting tricks and Photoshop to project an idealized image of hypermasculinity that, in reality, is impossible to attain (Christina, 2011; see also Ricciardelli et al., 2010).

advergaming combines free online games with advertising. Advergaming is growing rapidly. Sites such as Wonka.com, operated by Nestlé, and Barbie.com, operated by Mattel, attract millions of young children and provide marketers with an inexpensive way to "draw attention to their brand in a playful way, and for an extended period of time" (Moore, 2006: 5).

Young people see 45 percent more beer ads and 27 percent more ads for hard liquor in teen magazines than adults do in their magazines (Jernigan, 2010). Girls ages 11 to 14 are subjected to about 500 advertisements a day on the Internet, billboards, and magazines in which the majority of models are "nipped, tucked, and airbrushed to perfection" (Bennett, 2009: 43).

What effect do such ads have on girls' and women's self-image? About 43 percent of 6- to 9-year-old girls use lipstick or lip gloss, and 38 percent use hairstyling products. In addition, 8- to 12-year-old girls spend more than $40 million a month on beauty products, 80 percent of 10-year-old girls have been on a diet, and 80 percent of girls ages 13 to 18 list shopping as their favorite activity (Hanes, 2011; Seltzer, 2012). Many girls and young women believe that they have to be gorgeous, thin, and almost perfect to be loved by their parents and boyfriends (Schwyzer, 2011).

Many women, especially white women, who are unhappy with their bodies, turn to cosmetic surgery. In 2014 women had 90 percent of all cosmetic procedures; the number (almost 9.7 million) has increased 429 percent from 1997. Men had more than 1 million procedures, an increase of 273 percent from 1997. Almost 78 percent of all patients were white. The most common cosmetic surgery was breast enlargement or breast lift for women and liposuction for men (American Society for Aesthetic Plastic Surgery, 2015).

4-5 SOCIALIZATION THROUGHOUT LIFE

Biological aging comes naturally, but social aging is a different matter. As we progress through the life course—from infancy to death—we are expected to learn culturally approved norms, values, and roles.

4-5a Infancy and Toddlerhood

Infancy (between birth and 12 months) and toddlerhood (ages 1 to 3) encompass only a small fraction of the average person's life span, but are periods of both extreme helplessness and enormous physical and cognitive growth. Some scientists describe healthy infants' brains as "small computers" because of their enormous capacity for learning (Gopnik et al., 2001: 142).

The quality of relationships with adults and other caregivers has a profound effect on infants' and toddlers' development. Engaging infants in talk increases their vocabulary. By age 2, toddlers who know more words have better language skills that, in turn, help them control their behavior at age 3 and later (Vallotton and Ayoub, 2011).

More than half (53 percent) of children ages 1 and 2 are read to seven or more times a week by a family member ("Half of Young Children . . . ," 2011). The biggest change is children's media use. In 2013, 38 percent of all children under age 2 were using mobile devices (e.g., smartphone, tablet), up from only 10 percent in 2011. You saw earlier that the AAP has advised parents to limit their infants' and toddlers' screen time. Only 28 percent of parents say that media decreases face-to-face family time, but they often underestimate the amount of time that children under age 2 spend with electronic media (Rideout, 2013).

That's Not All . . .

There are many socialization agents other than family, play and peer groups, teachers and schools, and the media. If you think about the last few years, which groups or organizations have influenced who you are? What about the military, a company you've worked for, college and athletic clubs, religious organizations, support groups, athletic memberships, community groups, or others? All affect our socialization throughout life.

Childhood—viewed as a distinct stage of development—is a fairly recent phenomenon (Ariès, 1962). Until 1938, children made up a large segment of the U.S. labor force, especially in factories. They worked 6 days a week, 12 or more hours a day, received very low wages, and were often injured because of dangerous equipment.

4-5b Childhood Through Adolescence

Childhood and adolescence (roughly ages 4 to 12) are marked by considerable physical, emotional, social, and cognitive growth. Parents and teachers, particularly, reinforce norms and values through rewards and punishments (see Chapter 3), but are U.S. children and adolescents being coddled? For example, children usually get trophies and prizes simply for participating in extracurricular activities, and are constantly assured that they're winners, but "collapse at the first experience of difficulty" (Merryman, 2013: A29; see also Twenge, 2007).

Self-help books and self-proclaimed child experts routinely instruct parents to praise their kids to bolster their self-esteem. Such advice is misguided. A recent study of parents with children ranging from 7 to 12 years old found that constantly praising kids as "special" or "exceptional" and applauding every little achievement results in narcissism. These "self-centered brats" often demand special treatment, can't tolerate setbacks, are more violent than other people, and are at high risk for drug addiction and depression (Brummelman et al., 2015). A recent review of 41 studies of discipline strategies concluded that, in the long run, consistent and well-deserved (but not constant) praise strengthens the parent–child relationship, teaches children to overcome setbacks, and improves their problem-solving skills (Owen et al., 2012; Petersen, 2012).

Whereas some parents are accused of smothering their kids with never-ending attention and devotion, others are criticized for "free-range parenting," giving young children greater freedom and responsibility as they get older. Recently, a suburban Maryland couple allowed their 10- and 6-year-olds to play outside and walk home from a park, unsupervised. As a result, the police picked up the siblings, drove them to a crisis center, and child protective services investigated the parents. The couple's child-raising approach generated national debates on what constitutes neglect, when the state should intervene in parenting decisions, and whether constant adult surveillance or giving children greater age-appropriate independence and responsibility is in their best interest (Meitiv, 2015; Paulson, 2015; St. George, 2015).

Most U.S. children enjoy happy and healthy lives, but many don't. Almost 3 million children younger than 18 (almost 4 percent of all children in this age group) have an incarcerated parent. Another 400,000 are in the foster care system. Almost 13 percent of all children live in households where a parent or other adult uses, manufactures, or distributes illicit drugs. And, every year, at least 686,000 children experience neglect or physical and sexual abuse (U.S. Department of Justice, 2011; Children's Bureau, 2013; Reilly, 2013; "Child Maltreatment," 2014). Thus, for millions of U.S. children, the socialization process is shaky at best.

4-5c The Teenage Years

Ages 13 to 19 is a period of tremendous change. Most societies have distinct *rites of passage*, public rituals that mark the transition from one social position to another. Dating and getting a driver's license are the most common rites of passage in the United States that ease adolescents into adulthood (see Chapters 10 and 13 for religious and ethnic rites of passage).

Teenagers establish their own identity and become more independent as they mature and break away from parental supervision, a healthy development process. The most difficult part of raising adolescents, according to some parents, is dealing with their changing moods and behavior.

For many years, people attributed such dramatic changes to "raging hormones." Scientists now believe that there's a link between a teen's baffling behavior and the

Best Attendance

Nationally, 57 percent of 6- to 17-year olds participate in at least one after-school extracurricular activity. The benefits include learning time management, exploring diverse interests, and building social skills, but some family practitioners believe that today's children's are too overbooked and overscheduled (Laughlin, 2014; Lawrence, 2015). Do you agree?

fact that his or her brain may be changing far more than was thought previously. During adolescence, the brain matures at different rates. Areas involved in basic functions such as processing information and controlling physical movement mature first. The parts of the brain responsible for controlling impulses, avoiding risky behavior, and planning ahead—the hallmarks of adult behavior—are among the last to mature (National Institute of Mental Health, 2011).

Parental involvement is usually beneficial in a child's development, but "helicopter parents," who hover over their kids, are hyperinvolved, intrusive, and overcontrolling. Helicopter parenting can occur at any stage, but usually refers to parents of preteens, teenagers, and young adults. Anecdotal examples include parents verbally attacking teachers about their children's low grades, demanding that their child be moved to another

class before the school year has even begun, completing difficult homework assignments, showing up in the guidance counselor's office with college applications that they have filled out for their children, and haggling with college professors over the student's grade on an exam or paper (Krache, 2008; Weintraub, 2010; Rochman, 2013).

Helicopter parenting diminishes teens' and young adults' ability to develop decision-making and problem-solving skills. However well-intentioned, helicopter parents increase their own stress and decrease their children's feelings of autonomy and competence. In turn, not believing in one's ability to successfully accomplish tasks and achieve goals on one's own leads to teenagers' and young adults' feeling anxious, depressed, and more dissatisfied with life (Shellenbarger, 2013; Schiffrin et al., 2014).

4-5d Adulthood

Socialization continues throughout adulthood, the period roughly between ages 21 and 65. Most adults adopt a series of new roles that may include work, marriage, parenthood, divorce, remarriage, buying a house, and experiencing the death of a child, parent, or grandparent. We'll examine these transitions in later chapters. Two of the most important roles in adulthood are work and parenthood, but let's begin with young adults who are reluctant to leave their parents' nest.

THE CROWDED EMPTY NEST

During the 1960s and 1970s, sociologists almost always included the "empty nest" in describing the family life course. This is the stage in which parents, typically in their 50s, find themselves alone after their children have married, gone to college, or found jobs and moved out.

Today, young adults are living at home longer than they did in the past. The terms *boomerang children* and *boomerang generation* refer to young adults who move back in with their parents. Many people in their 30s, 40s, and older—often with a spouse, girlfriend or boyfriend, and children in tow—are moving back into their parents' home.

Most young adults leave the parental nest by age 23, but the proportion of those ages 25 to 34 who are living with parents increased from 9 percent in 1960 to 12 percent in 2013 (Fields and Casper, 2001; U.S. Census Bureau, Current Population Survey . . . , 2013). Among young adults ages 18 to 34, men are more likely than women to live with their parents (see *Figure 4.2*).

Hosting an adult child who lives at home can cost between $8,000 and $18,000 a year, depending on how much parents are paying for extras such as travel and entertainment. About 26 percent of parents have taken on debt to pay for their adult children's expenses (Grind, 2013).

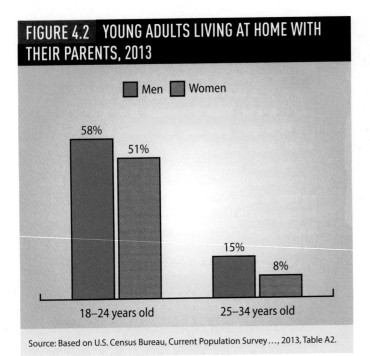

FIGURE 4.2 YOUNG ADULTS LIVING AT HOME WITH THEIR PARENTS, 2013

Men Women

58%
51%
15%
8%

18–24 years old   25–34 years old

Source: Based on U.S. Census Bureau, Current Population Survey..., 2013, Table A2.

Macro-level factors have been the major reason for the delay in many Americans' transition to adulthood, especially living independently. Financial insecurity, student loan debt, low wages, divorce, credit card debt, and going from job to job until they find work they enjoy have made it harder for young middle-class adults to maintain the lifestyles that their parents created. And the transition to independence gets tougher the lower a person's occupation and education. Nonetheless, and despite an improved job market since the 2007–2009 recession, the number of 18- to 34-year-olds living independently has declined—from 71 percent in 2007 to 67 percent in 2015 (Parker, 2012; Fry, 2013, 2015).

On a micro level, almost 33 percent of today's parents, compared with 19 percent in 1993, say that children shouldn't be expected to be on their own financially until age 25 or later (Taylor et al., 2012). Among those ages 18 to 24, 87 percent live at home because parents "make it easy for me to stay" (Payne and Copp, 2013). They enjoy the comforts of the pampering parental nest. According to one of my male students in his late 20s, for example, "My mom enjoys cooking, cleaning my room, and just having me around. I don't pitch in for any of the expenses, but we get along great because she doesn't hassle me about my comings and goings" (Benokraitis, 2015: 354).

Many young adults, as you'll see in Chapter 12, are delaying marriage and don't feel a need to establish their own homes. Perhaps most important, the stigma traditionally linked to young adults' living at home has faded as the practice grew. Among those ages 25 to 34 who live at home, 61 percent say they have friends or family

members who have moved back in with their parents (Parker, 2012). The implication (and perhaps justification) of moving back home or not leaving in the first place is that "a lot of people are doing it."

WORK ROLES

The average American who was born between 1957 and 1964 has held 11 jobs from ages 18 to 46 alone; 26 percent have held 15 jobs or more ("Number of Jobs Held...," 2012). Such job changing means that our occupational socialization is an ongoing process.

Learning work roles can be difficult because each job has different demands and expectations. Even when there's formal job training, we must also learn the subtle rules that are implicit in many job settings. A supervisor who says, proudly and loudly, "I have an open door policy. Come in and chat whenever you want," may *really* mean "I have an open door policy if you're not going to complain about something." Even if we remain in the same job for many years, we must often acquire new skills, especially technological ones, or risk being laid off.

Being laid off can be stressful at any age. In the case of midlife men (generally defined as those ages 45 to 60), such changes are especially traumatic: They must learn new job-hunting skills and often adjust to as much as a 70 percent decrease in their earnings (see Chapter 11). Instead of enjoying the security of a job in their midlife years, these men must undergo occupational socialization that assaults their self-image as good employees and family providers.

PARENTING ROLES

Like workplace roles, parenting does *not* come naturally. Most first-time parents muddle through by trial and error. Family sociologists often point out that we get more training for driving a car or a fishing license than for parenting.

Most couples don't realize that children are expensive. Middle-income couples, with an average income of $82,800 a year, spend about 16 percent of their earnings on a child during the first 2 years (see *Figure 4.3*). These couples will spend almost $246,000 for each child from birth through high school. Child-rearing costs are much higher for single parents and especially for low-income families if a child is disabled, chronically ill, or needs specialized care that welfare benefits don't cover. Expenses are also much higher for parents whose children attend private schools (Lino, 2014).

You'll see in Chapter 12 that arguments over finances and child-rearing strategies are two of the major reasons for divorce. All in all, adjusting to marital and

FIGURE 4.3 HOW MUCH DOES IT COST TO RAISE A CHILD?

In 2013, middle-income married couples—those with before-tax income between $61,530 and $106,540—spent $12,940 per year for a child under age 2. This amount doesn't include the costs of prenatal care or delivery.

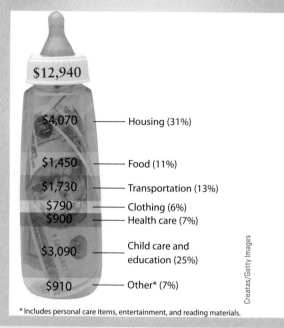

$12,940

$4,070 — Housing (31%)

$1,450 — Food (11%)

$1,730 — Transportation (13%)

$790 — Clothing (6%)
$900 — Health care (7%)

$3,090 — Child care and education (25%)

$910 — Other* (7%)

Creatas/Getty Images

*Includes personal care items, entertainment, and reading materials.

Source: Based on Lino, 2014, Table 1.

parental roles during adulthood requires considerable patience, effort, and work.

About 23 percent of Americans are "sandwiched" between their children and their parents. That is, they're caring for at least one parent age 65 or older, raising a child under age 18, and/or providing economic assistance to a grown child. Pulled in many directions, these midlife adults must learn to deal with the stress of juggling a full-time job and caregiving responsibilities (Parker and Horowitz, 2015; see also Chapter 12).

4-5e Later Life

Socialization continues in later life. People age 65 and older often remarry and must learn to play new roles such as stepparent or step-grandparent. Many others undergo unwanted rites of passage such as widowhood, divorce or re-divorce, becoming a custodial parent, and dealing with the death of an adult child or grandchild.

The adult child–aging parent relationship has mixed outcomes. Adult children who receive parental financial and emotional support are happier, not surprisingly, than their counterparts who don't receive such assistance.

However, older parents often experience stress when their adult children are doing poorly and have to rely on their parents for financial and emotional support (Fingerman et al., 2012).

Because we live longer, we may spend 20 percent of our adult life in retirement. When retired people are unhappy, it's usually because of health or income problems rather than loss of the worker role (see Chapter 14). If retirement benefits don't keep up with inflation or if a retiree isn't covered by a pension plan, people can plunge into poverty soon after retirement. Poverty is devastating at any age, but sliding down the economic ladder during one's 60s and 70s is especially traumatic because there's little chance of recouping one's income (see Chapters 12 and 14).

Older people who are in poor health and experience continuous pain sometimes welcome death. Among those who are healthy, some reenter the labor force or volunteer in various organizations, many are involved with their grandchildren, and others forge new relationships after widowhood (see Chapter 12). Thus, even in the later years, many people continue to learn new roles.

4-6 RESOCIALIZATION

Throughout life, many people undergo **resocialization**, the process of unlearning old ways of doing things and adopting new attitudes, values, norms, and behavior. Resocialization ranges from mild to intense, and can be voluntary or involuntary.

4-6a Voluntary Resocialization

Much resocialization is voluntary, as when an American wife moves to the Middle East with her Iranian husband and follows local customs about veiling, women's submissive roles, and staying out of public life. Other examples of voluntary resocialization include entering a religious order, joining a religious cult, seeking treatment in a drug rehabilitation facility, or serving in the military.

The process of voluntary resocialization is usually mild. First-year college students, new employees, and members of support groups, for instance, must learn new rules and live up to an individual's, a group's, or an organization's expectations. Such changes, however, are usually short-term and may even be enjoyable.

resocialization the process of unlearning old ways of doing things and adopting new attitudes, values, norms, and behavior.

On the other hand, some voluntary resocialization can be long, difficult, and intense. New parents' lives change dramatically as they learn parenting roles. In divorce, people may experience a sense of relief and liberation, but must also cope with emotional distress, financial changes, and child custody disputes. People who experience a permanent physical disability must cope with physical limitations, psychological issues, and stigma. Soldiers who return from Iraq or Afghanistan, even if they haven't suffered physical injuries, describe coming home and living a normal life as daunting. They must reconnect with their children, adjust to spouses who have become more independent, and deal with posttraumatic stress disorders (PTSD) that include depression and nightmares (Jones, 2013; Taormina-Weiss, 2013; van Agtmael, 2013).

4-6b Involuntary Resocialization

Resocialization is also involuntary, as when children are sent to a foster home or a detention camp, or when people are imprisoned or committed to a psychiatric hospital. Such settings are **total institutions** where people are isolated from the rest of society, stripped of their former identities, and required to conform to new rules and behavior (Goffman, 1961).

Resocialization in total institutions is an intense two-step process. First, the staff uses *degradation ceremonies*, humiliating rituals that publicly stigmatize people (Garfinkel, 1956). The inmates are strip-searched, deloused, and fingerprinted; their heads are shaved; they're issued a uniform and are given a serial number; they have no privacy, limited access to family and friends, and little communication with the outside world; and they're told what to do and when. In the military, soldiers get short haircuts, wear matching uniforms, and follow rigid schedules. The purpose of these practices and restrictions is to destroy any sense of individualism, autonomy, and past identity, and to produce a more compliant person (Goffman, 1961; see also Chapter 7).

In the second resocialization step, the staff tries to build a new identity that conforms to the institution's expectations. Inmates who conform are rewarded with simple privileges (e.g., making phone calls, more visits with family) and an earlier release. Soldiers often bond with one another, become more self-disciplined, may use their new skills in civilian jobs, and are proud of having served in the military ("Once a Marine, always a Marine").

As this chapter shows, socialization is a powerful force in shaping who we are. Socialization doesn't produce robots, however, because people are creative, adapt to new environments, and change as they interact with others.

total institutions settings where people are isolated from the rest of society, stripped of their former identities, and required to conform to new rules and behavior.

STUDY TOOLS 4

READY TO STUDY? IN THE BOOK, YOU CAN:

☐ Check your understanding of what you've read with the Test Your Learning Questions provided on the chapter review card at the back of the book.

☐ Rip out the chapter review card for a handy summary of the chapter and key terms.

ONLINE AT CENGAGEBRAIN.COM YOU CAN:

☐ Collect StudyBits while you read and study the chapter.

☐ Quiz yourself on key concepts.

☐ Prepare for tests with chapter flash cards as well as flash cards that you create.

☐ Watch videos to explore sociology further.

Most resocialization occurs in institutions ranging from monastic orders to the military.

5 Social Interaction in Everyday Life

LEARNING OUTCOMES

After studying this chapter, you will be able to…

5-1 Explain the importance of social interaction and its relationship to social structure.

5-2 Define and illustrate status set, ascribed and achieved statuses, master status, and status inconsistency.

5-3 Explain how and why social roles differ, and how people cope with role conflict and role strain.

5-4 Compare and illustrate symbolic interaction, social exchange, and feminist explanations of social interaction.

5-5 Describe and illustrate nonverbal communication, its importance, and cross-cultural variations.

5-6 Summarize the benefits and costs of online interaction.

After you finish this chapter go to **PAGE 99** for **STUDY TOOLS**.

In mid-2011, a Boeing 767 took off from an airport in Washington, D.C., bound for Ghana, Africa. Shortly after the flight was airborne, a passenger reclined his seat, which was close to the lap of the man sitting behind him. There was a smack to the head, angry words, and a fist fight broke out. A flight attendant and another passenger jumped in between. The pilot returned to the airport to determine the scope of the problem (Halsey, 2011: A1; see also Grossman, 2011).

This incident of air rage is an example of **social interaction**, the process by which we act toward and react to people around us. The scuffle began with a typical passenger annoyance, escalated into an aggressive nonverbal reaction (a smack to the head), burst into an angry argument and fight, and quickly involved others—a third passenger, the airline attendant, and the pilot. This incident illustrates four critical components of social interaction (Maines, 2001; Schwalbe, 2001):

▶ It is central to all human social activity, and includes both nonverbal behavior (a physical smack) and words (an argument).

▶ People interact differently depending on what's at stake for them ("I'm not going to put up with this guy all the way to Ghana").

▶ Social interaction affects people's behavior ("No one can hit me on the head and get away with it").

What do you think?

I spend more time on social media than I do talking face to face with my family and friends.

1	2	3	4	5	6	7

strongly agree   strongly disagree

David Sacks/Stone/Getty Images

▶ Elements of social structure, such as rules about proper behavior, affect all social interaction but can produce different personal outcomes (e.g., most airline passengers tolerate cramped seats quietly whereas others explode).

5-1 SOCIAL STRUCTURE

Social structure is an organized pattern of behavior that governs people's relationships (Smelser, 1988). Because social structure shapes our actions, life is typically orderly and predictable rather than chaotic or random. We're often not aware of the impact of social structure until we violate cultural rules, formal or informal, that dictate our daily behavior—whether in class, at work, or in an airplane.

Every society has a social structure that encompasses statuses, roles, groups, organizations, and institutions (Smelser, 1988). We'll examine groups, organizations, and institutions in later chapters. Let's take a closer look here at statuses and roles, two building blocks of social structure.

5-2 STATUS

For most people, the word *status* signifies prestige: An executive, for example, has more status than a receptionist, and a physician has a higher status than a nurse. For sociologists, **status** is a social position that a person occupies in a society (Linton, 1936). Thus, executive, secretary, physician, and nurse are all social statuses. Other examples of statuses are musician, voter, sister, parent, police officer, and friend.

Sociologists don't assume that one position is more important than another. A mother, for example, isn't more important than a father, and an adult isn't more important than a child. Instead, all statuses are significant because they determine social identity, or who we are.

social interaction the process by which we act toward and react to people around us.

social structure an organized pattern of behavior that governs people's relationships.

status a social position that a person occupies in a society.

FIGURE 5.1   IS THIS YOUR STATUS SET?

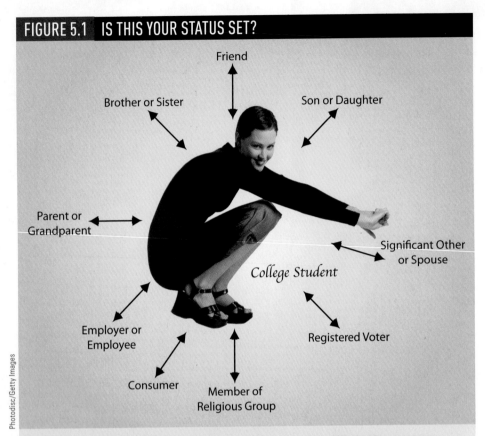

Photodisc/Getty Images

Being a college student is only one of your current statuses. What other statuses comprise your status set?

5-2a Status Set

Every person has many statuses (see *Figure 5.1*). Together, they form her or his **status set**, a collection of social statuses that a person occupies at a given time (Merton, 1968). Dionne, one of my students, is female, African American, 42 years old, divorced, mother of two, daughter, aunt, Baptist, voter, bank supervisor, volunteer at a soup kitchen, and country music fan. All of these socially defined positions (and others) make up Dionne's status set.

Status sets change throughout the life course. Because she'll graduate next year and is considering remarrying and starting an after-school program, Dionne will add at least three more statuses to her status set

status set   a collection of social statuses that a person occupies at a given time.

ascribed status   a social position that a person is born into.

achieved status   a social position that a person attains through personal effort or assumes voluntarily.

and will also lose the statuses of divorced and college student. As Dionne ages, she'll continue to gain new statuses (grandmother, retiree) and lose others (supervisor at a bank, or wife, if she is widowed).

Statuses are *relational*, or complementary, because they're connected to other statuses: A *husband* has a *wife*, a *real-estate agent* has *customers*, and a *teacher* has *students*. No matter how many statuses you occupy, every status is linked to those of other people. These connections between statuses influence our behavior and relationships.

5-2b Ascribed and Achieved Status

Status sets include both ascribed and achieved statuses. An **ascribed status** is a social position that a person is born into. We can't control, change, or choose our ascribed statuses, which include sex (male or female), age, race, ethnicity, and family relationships. Your ascribed statuses, for example, might include *male*, *Latino*, and *brother*. Some argue that sex isn't really an ascribed status because people can have sex-change operations. However, a sex-change operation doesn't change the fact that someone was born a male or a female (except in a minority of cases, as you'll see in Chapter 9).

An **achieved status**, in contrast, is a social position that a person attains through personal effort or assumes voluntarily. Your achieved statuses might include college graduate, mother, and employee. Unlike our ascribed statuses, our achieved statuses can be controlled and changed. We have no choice about being a son or daughter (an ascribed status), but we have an option to become a parent (an achieved status).

Students sometimes think that religion and social class are ascribed rather than achieved statuses. It's true that someone may be born into a family that practices a certain religion or one that is poor, middle class, or wealthy. Because we can change these statuses through our own actions, however, neither religion nor social class is an ascribed status. A Catholic might convert to Judaism (or vice versa), and thousands of Americans born into poor or working-class families have become millionaires (see Chapter 8).

5-2c Master Status

A **master status** overrides other statuses and forms an important part of a person's social identity (Hughes, 1945; Becker, 1963). A master status is usually immediately apparent, makes the biggest impression, affects others' perceptions, and, consequently, often shapes a person's entire life.

Some master statuses—such as sex, age, and race—are ascribed. Whatever you do, for example, people will see you as male or female and react accordingly. Other master statuses are achieved. For many people, their occupation is their master status: Occupation tends to be the first thing we ask about when we meet someone; conveys much information about income, education, skills, and achievements; and lingers after we've left our jobs (e.g., someone is introduced as a former CEO or ex-military officer). Other achieved master statuses can be based on religion, community leadership, family wealth, or deviant behavior.

A physical disability can become a master status. One of the military's major tasks is helping severely injured soldiers to become as independent as possible.

5-2d Status Inconsistency

Because we hold many statuses, some clash. **Status inconsistency** refers to the conflict that arises from occupying social positions that are ranked differently. Examples include a teacher who works as a bartender or a skilled welder who stocks shelves at Walmart because neither can find a better job in a stagnant economy. We'll cover status inconsistency in more detail in Chapter 7. For now, you should be aware that you occupy many statuses, and some of them may clash now and in the future.

5-3 ROLE

Each status is associated with one or more roles. A **role** is the behavior expected of a person who has a particular status. We *occupy* a status but *play* a role. In this sense, a role is the dynamic aspect of a status (Linton, 1936).

College student is a status, but the role of a college student requires many *formal* behaviors such as going to class, reading, thinking, completing weekly assignments, writing papers, and taking exams. *Informal* behaviors may include joining a student club, befriending classmates, attending football games, and even abusing alcohol on weekends.

Like statuses, roles are relational. Playing the role of professor requires teaching, advising students, answering their email, and grading exams and assignments. Most professors are also expected to serve on committees, do research, publish articles and/or books, and perform services such as giving talks to community groups. Thus, the status of college student or professor involves numerous role requirements that govern who does what, where, when, and how.

Roles can be rigid or flexible. A person who occupies the status of secretary typically plays a role that's defined by rules about when to come to work, how to answer the phone, and when to submit the necessary work and in what format. A boss, on the other hand,

master status a status that overrides other statuses and forms an important part of a person's social identity.

status inconsistency the conflict that arises from occupying social positions that are ranked differently.

role the behavior expected of a person who has a particular status.

usually enjoys considerable flexibility: She or he has more freedom to come in late or leave early, to determine which projects should be completed first, and to decide when to hire or fire employees.

Because roles are based on mutual obligations, they ensure that social relations are fairly orderly. We know what we're supposed to do and what others expect of us. If professors fail to meet their role obligations by missing many classes or coming to classes unprepared, students may respond by studying very little, cutting classes, and submitting negative course evaluations. If students miss classes, don't turn in the required work, or cheat, professors can fail them or assign low grades.

5-3a Role Performance

Roles define how we're *expected* to behave in a particular status, but people vary considerably in fulfilling the responsibilities associated with their roles. Many college students succeed, whereas others fail; some professors inspire their students, whereas others put them to sleep. These differences reflect **role performance**, the *actual* behavior of a person who occupies a status. For example, a professor may demand more of graduate students than of undergraduates. An instructor may also interact differently with a 19-year-old student than a 40-year-old student who's anxious about returning to college.

5-3b Role Set

We occupy many statuses and play many roles associated with each status. A **role set** is the array of roles attached to a particular status. The status of a college student, for example, may include at least six different roles (see *Figure 5.2*). Each of these roles is linked to people with different statuses (e.g., professor, registrar, librarian). Because a different set of norms governs each

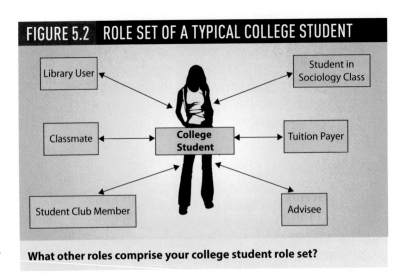

FIGURE 5.2 ROLE SET OF A TYPICAL COLLEGE STUDENT

Library User — College Student — Student in Sociology Class

Classmate — College Student — Tuition Payer

Student Club Member — College Student — Advisee

What other roles comprise your college student role set?

of these relationships, the student interacts differently with a classmate than with a reference librarian or a professor.

Figure 5.2 shows only one status—that of college student. If you think about a college student's other statuses (employee, son or daughter, parent, girlfriend or boyfriend, husband or wife), you can see that meeting the expectations of numerous role sets can create considerable role conflict and role strain.

5-3c Role Conflict and Role Strain

Playing many roles often leads to **role conflict**, difficulties in playing two or more contradictory roles. College students who have a job, especially if it's full-time, often experience role conflict because both employers and professors expect them to excel, but it's very difficult to meet these expectations. The role conflict increases if the student has young children or cares for an aging parent or grandparent.

Whereas role conflict arises because of the expectations of contradictory roles, **role strain** is due to the conflicting demands within the same role. Students experience role strain when several exams are scheduled on the same day and they're also involved in time-consuming extracurricular activities. Faculty members experience role strain when the requirements of being a professor—teaching, research, and community service—sap their energy and time. Military chaplains report role strain in preaching about peace while blessing those about to go to war.

Almost all of us experience role strain because many inconsistencies are built into our roles. *Table 5.1* gives examples of some of the reasons for role conflict and role strain.

role performance the actual behavior of a person who occupies a status.

role set array of roles attached to a particular status.

role conflict difficulties in playing two or more contradictory roles.

role strain difficulties due to conflicting demands within the same role.

TABLE 5.1 WHY DO WE EXPERIENCE ROLE CONFLICT AND ROLE STRAIN?

Reason	Example
Because many people are overextended, some roles are bound to conflict with others.	Students may study less than they want because employers demand that they work overtime or on weekends.
People have little or no training for many roles.	Parents are expected to turn out "perfect" kids even though they receive training for driving a car but none for parenting.
Some role expectations are unclear or contradictory.	Some employers pride themselves on having family-friendly policies but expect employees to work 12 hours a day, travel on weekends, and use vacation days to care for a sick child.
Highly demanding jobs often create difficulties at home.	Some jobs (being in the military, policing, and firefighting) require people to be away from their families for extended periods of time or during crises.

Like many other people, college students can reduce their role conflict and role strain by setting priorities and compartmentalizing their roles.

5-3d Coping with Role Conflict and Role Strain

Role conflict and role strain can produce tension, hostility, aggression, and stress-related physical problems that include insomnia, headaches, ulcers, eating disorders, anxiety attacks, chronic fatigue, nausea, weight loss or gain, and drug and alcohol abuse (Weber et al., 1997). Some role conflict and role strain may last only a few weeks, but others are long-lived (e.g., working in a stressful or low-paying job).

To deal with role conflict and role strain, some people *deny that there's a problem.* Employed mothers, especially those who are divorced, often become supermoms—providing home-cooked meals, attending their children's sports activities, and volunteering for a school's fund-raising campaign even though they're exhausted, must do laundry at midnight, and neglect their own interests and well-being. Supermoms may succeed over a number of years in managing role conflict and role strain. Eventually, however, they may become angry and resentful, or experience health or emotional problems (Douglas and Michaels, 2004).

There are five more effective ways to minimize role conflict and role strain. First, we can reduce role conflict through *compromise* or *negotiation.* To decrease the conflict between work and family roles, many couples draw up schedules that require fathers to do more of the housework and child rearing (see Chapter 12).

Second, we can *set priorities.* If extracurricular activities interfere with studying, which is more important? Succeeding in college always requires making sacrifices, such as attending fewer parties or not seeing friends as often as we'd like.

Third, we can *compartmentalize* our roles. Many college students take courses in the morning, work part-time during the afternoon or evening, and devote part of the weekend to leisure activities. It's not always easy but usually possible to separate our various roles.

Fourth, we can decide *not to take on more roles.* One of the most effective ways to avoid role conflict is to just say "no" to requests to do volunteer work, pressure from family or friends to take on unwanted tasks (e.g., babysitting), or pleas to become involved in college or community activities.

Finally, we can *exit* a role or status. Withdrawing from community activities and club offices, for example, can decrease role conflict and role strain considerably. There's always pressure to remain in a role, but none of us is indispensable, and there are many people who are eager to replace us. Although necessary, some role exits are painful and long-lived. Divorce, for instance, isn't a quick event but a process that goes on many years (sometimes decades), during which two people (and their children) must redefine their expectations and adjust to a different household structure (see Chapter 12).

5-4 EXPLAINING SOCIAL INTERACTION

Statuses and roles shape our everyday relationships, but social interaction provides the basis of these relationships: It affects who you are, how you behave, and what you say. Social interaction seems natural and simple, but why do we interact as we do? And why do people sometimes interpret the same words differently? Three micro-level perspectives—symbolic interaction, social exchange theory, and feminist theories—provide answers to these and other questions. These perspectives offer different insights, but each explains how people interact in their daily lives. (*Table 5.2* summarizes these approaches.)

5-4a Symbolic Interaction

For interactionists, the most significant characteristic of all human communication is that people take each other and the context into account (Blumer, 1969). When your professors ask, "How are you?" they expect a "Fine, thanks" and probably barely look at you. A doctor, on the other hand, usually looks you in the eye when asking "How are you?" and takes notes as soon as you start to reply. Thus, "How are you?" has different meanings in different social contexts and elicits different responses ("I'm doing okay" vs. "I've been having a lot of headaches lately"). In this sense, people construct reality.

Doublespeak

1.	automotive internist	A.	a lie
2.	reducing costs	B.	bombing
3.	misstatement	C.	undertaker
4.	auto dismantler and recycler	D.	cutting salaries and/ or firing people
5.	air support	E.	fake
6.	previously owned	F.	car mechanic
7.	genuine imitation	G.	junk dealer
8.	bereavement care expert	H.	stock market crash
9.	equity retreat	I.	used
10.	revenue enhancement	J.	tax increase

Answers: 1-F; 2-D; 3-A; 4-G; 5-B; 6-I; 7-E; 8-C; 9-H; 10-J

SOCIAL CONSTRUCTION OF REALITY

"Human reality is socially constructed reality" (Berger and Luckmann, 1966: 172). That is, we produce, interpret, and share the reality of everyday life with others. This social construction of reality typically evolves through direct, face-to-face interaction, but the interaction can also be indirect, as in watching television or participating in social media.

Businesspeople, advertisers, politicians, educators, and even social scientists use words deliberatively to shape or change our perceptions of reality. For example, doublespeak is "language that pretends to communicate but really doesn't. [It] makes the bad seem good, the

TABLE 5.2 SOCIOLOGICAL EXPLANATIONS OF SOCIAL INTERACTION

Perspective	Key Points
Symbolic Interactionist	• People create and define their reality through social interaction. • Our definitions of reality, which vary according to context, can lead to self-fulfilling prophecies.
Social Exchange	• Social interaction is based on a balancing of benefits and costs. • Relationships involve trading a variety of resources, such as money, youth, and good looks.
Feminist	• Females and males act similarly in many interactions but may differ in communication styles and speech patterns. • Men are more likely to use speech that's assertive (to achieve dominance and goals), whereas women are more likely to use language that connects with others.

negative appear positive, the unpleasant appear attractive or at least tolerable" (Lutz, 1989: 1).

There are several kinds of doublespeak: *Euphemisms* are words or phrases that avoid a harsh, unpleasant, or distasteful reality; *gobbledygook* (or *bureaucratese*) overwhelms the listener with big words and long sentences; and *inflated language* makes simple everyday things seem complex (Lutz, 1989; see also Beard and Cerf, 2015). Take the quiz on the previous page to see how much doublespeak you understand or use.

SOCIAL INTERACTION AND SELF-FULFILLING PROPHECIES

Our perceptions of reality shape our behavior. In an oft-cited statement, known as the *Thomas Theorem*, sociologists W. I. Thomas and Dorothy Thomas (1928: 572) observed, "If men define situations as real, they are real in their consequences." That is, our behavior is a result of how we interpret a situation ("My mother-in-law hates me, so I go shopping when she drops in").

Carrying this idea further, sociologist Robert Merton (1948/1966) proposed that our definitions of reality can result in a **self-fulfilling prophecy**: If we define something as real and act on it, it can, in fact, become real. For example, physical education teachers who publicly humiliated students often turned them off physical fitness for good. One man said, "To this day I feel totally inadequate in team-related activities and have a natural reflex to AVOID THEM AT ALL COSTS" (Strean, 2009: 217; capitalization in original). Thus, gym teachers' negative comments made students feel inadequate in sports, regardless of their ability, and changed their behavior during adulthood.

Our perceptions of reality shape our behavior, but how do people define that reality? Interactionists use two research tools—*ethnomethodology* and *dramaturgical analysis*—to help answer this question.

ETHNOMETHODOLOGY

A term coined by sociologist Harold Garfinkel (1967), **ethnomethodology** is the study of how people construct and learn to share definitions of reality that make everyday interactions possible. That is, we base our interactions on common assumptions about what makes sense in specific situations (Schutz, 1967; Hilbert, 1992).

Cory Thoman/Shutterstock.com

People make sense of their everyday lives in two ways. First, by observing conversations, people discover the general rules that we all use to interact. Second, people can understand interaction rules by breaking them. Over a number of years, Garfinkel instructed his students to purposely violate everyday interaction rules and then to analyze the results. In these exercises, some of his students went to a grocery store and insisted on paying more than was asked for a product. Others were instructed, in the course of an ordinary conversation "to bring their faces up to the subject's until their noses were almost touching" (Garfinkel, 1967: 72).

When students violated such everyday interaction rules, they were sanctioned. Grocery clerks became hostile when the students insisted on paying more than the marked price for a product, and people backed off when their noses were almost touching: "Reports were filled with accounts of astonishment, bewilderment, shock, anxiety, embarrassment, and anger" (Garfinkel, 1967: 47).

Violating interaction rules, even unspoken ones, can trigger anger, hostility, and frustration. College students become upset if professors ignore teaching norms by using sarcasm and putdowns, coming to class unprepared or consistently late, constantly reading from the book, or discouraging students' comments and questions (Berkos et al., 2001).

DRAMATURGICAL ANALYSIS

Dramaturgical analysis is a research approach that examines social interaction as if occurring on a stage where people play different roles and act out scenes for the "audiences" with whom they interact. According to sociologist Erving Goffman (1959, 1967), life is similar to a play in which each of us is an actor, and our social interaction is much like theatre because we're always on stage and always performing. In our everyday "performances", we present different versions of ourselves to people in different settings (audiences).

---

self-fulfilling prophecy a situation in which if we define something as real and act on it, it can, in fact, become real.

ethnomethodology the study of how people construct and learn to share definitions of reality that make everyday interactions possible.

dramaturgical analysis examines social interaction as if occurring on a stage where people play different roles and act out scenes for the audiences with whom they interact.

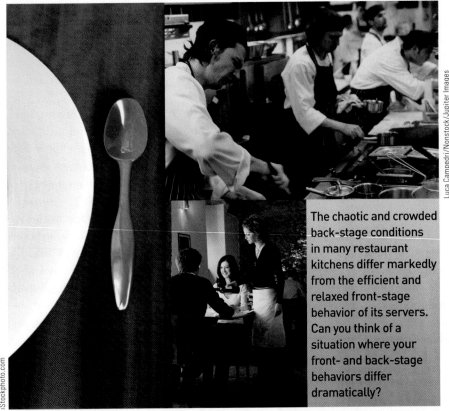

The chaotic and crowded back-stage conditions in many restaurant kitchens differ markedly from the efficient and relaxed front-stage behavior of its servers. Can you think of a situation where your front- and back-stage behaviors differ dramatically?

iStockphoto.com

Luca Campedri/Nonstock/Jupiter Images

Thinkstock Images/Stockbyte/Getty Images

Because most of us try to present a positive image of ourselves, much social interaction involves *impression management*, a process of suppressing unfavorable traits and stressing favorable ones (see Chapter 4). To control information about ourselves, we often rely on props to convey or reinforce a particular image. For example, physicians, lawyers, and college professors may line their office walls with framed diplomas, medical certificates, or community awards to give the impression that they're competent, respected, and successful.

For Goffman, a performance involves front- and back-stage behaviors. The *front stage* is an area where an actual performance takes place. In front-stage areas, such as living rooms or restaurants, the setting is clean and the servers or hosts are typically polite and deferential to guests. The *back stage,* an area concealed from the audience, is where people can relax. Bedrooms and restaurant kitchens are examples of back stages. After guests have left, the host and hostess may kick off their shoes and gossip about their company. In restaurants, cooks and servers may criticize the guests, use vulgar language, and yell at each other. Thus, the civility and decorum of the front stage may change to rudeness in the back stage.

Another example of front- and back-stage behavior involves faculty and students. Professors want to create the impression that they're well prepared, knowledgeable, and hardworking. In back-stage areas such as their offices, however, faculty may complain to their colleagues that they're tired of teaching a particular course or grading terrible exams and papers, or they may confess that they don't always prepare for classes as well as they should. Students often plead with instructors for higher grades: "But I studied very hard" or "I'm an A student in my other courses." In back-stage areas such as dormitories, student lounges, or libraries, however, students may admit to their friends that they barely studied or that they have a low grade-point average.

5-4b Social Exchange Theory

Social exchange theory proposes that individuals seek through their interactions to maximize their rewards and minimize their costs. An interaction that elicits a reward, such as approval or a smile, is more likely to be repeated than an interaction that evokes censure, such as disapproval or criticism (Thibaut and Kelley, 1959; Homans, 1974; Blau, 1986). Because interactions are most satisfying when there's a balance between giving and taking, individuals must decide whether and which resources to exchange.

People bring various tangible and intangible resources to a relationship (money, status, intelligence, good looks, youth, power, affection). Any of a person's resources can be traded for more, better, or different resources. Marriages between older men and younger women often reflect an exchange of the man's power, money, and/or fame for the woman's youth, physical attractiveness, and ability to bear children.

Many of our cost–reward decisions are conscious and deliberate, but others are passive or based on long-term negative interactions. For example, much of the domestic violence research shows that women stay in

social exchange theory proposes that individuals seek through their interactions to maximize their rewards and minimize their costs.

Tyler Olson/Shutterstock.com

abusive relationships because their self-esteem has eroded after years of criticism and ridicule from both their parents and their spouses or partners ("You'll be lucky if anyone marries you," "You're dumb," "You're ugly," and so on). In effect, the abused women (and sometimes men) believe that they have nothing to exchange in a relationship, that they don't have the right to expect benefits (especially when the abuser blames the victim for provoking the anger), or that enduring an abusive relationship is better than being alone (see Chapter 12).

5-4c Feminist Theories

Cultural norms and gender role expectations shape our interaction patterns. Generally, women are socialized to be more comfortable talking about their feelings, whereas men are socialized to be dominant and take charge, especially in the workplace. Women often ask questions that probe about feelings ("Were you glad it happened?"). Women are also more likely than men to do conversational "maintenance work," such as asking questions that encourage conversation ("What happened at the meeting?") (Lakoff, 1990; Farley et al., 2010).

Women's speech tends to be supportive ("You must've felt terrible") and tentative ("I kind of feel that . . ."). And, in many social situations, women—particularly those ages 40 to 60—often apologize for the way they look (weight, hair, wrinkles). Men don't do so, "even when they look terrible" (Wolven, 2014; Wood, 2015).

Compared with female speech, men's speech often reflects *conversational dominance*, speaking more frequently, and for longer periods. Men also show dominance by interrupting or ignoring others, reinterpreting the speaker's meaning, changing the topic, or giving unsolicited and unwanted advice (Tannen, 1990; Toth, 2011; Gurkoff and Ranieri, 2012).

Such interaction differences aren't innate because of one's sex, however. Instead, much interaction is contextual—that is, it varies depending on the situation. For example, fathers who do much of the parenting have communication styles that are similar to those of mothers. In the workplace, women and men who occupy high-level decision-making positions have similar interaction styles with superiors and subordinates (Cameron, 2007).

You've seen, so far, that much of our interaction is verbal and face-to-face. Nonverbal communication is also important in shaping our everyday relationships.

5-5 NONVERBAL COMMUNICATION

Nonverbal communication refers to messages sent without using words. This silent language, a "language of behavior" that conveys our real feelings, can be more potent than our words (Hall, 1959: 15). For example, sobbing sends a much stronger message than saying "I feel very sad." Some of the most common nonverbal messages are silence, visual cues, touch, and personal space.

5-5a Silence

Silence expresses a variety of emotions: Agreement, apathy, awe, confusion, disagreement, embarrassment, regret, respect, sadness, thoughtfulness, and fear, to name just a few. In various contexts, and at particular points in a conversation, silence means different things to different people. Sometimes silence saves us from embarrassing ourselves. Think, for example, about the times you fired off an angry email or text message, regretted doing so an hour later, and then dreaded getting a fuming response.

nonverbal communication messages that are sent without using words.

"This concludes my lecture on non-verbal communication. Any comments or questions?"

Chris Wildt/CartoonStock.com

Many of us have experienced the pain of getting "the silent treatment" from friends, family members, coworkers, or lovers. Not talking to people who are important to us builds up anger and hostility. Initially, the "offender" may work hard to make the close-mouthed person discuss a problem. Eventually, the target of the silent treatment gets fed up, gives up, or ends a relationship.

5-5b Visual Cues

Visual cues, another form of nonverbal communication, include gestures, facial expressions, and eye contact. Let's consider a few examples of such body language in our daily interactions.

GESTURES

Most of us think we know what certain gestures mean—folding your arms across your chest indicates a closed, defensive attitude; leaning forward often shows interest; shrugging your shoulders signals indifference; narrowing your eyes and setting your jaw shows defiance; and smiling and nodding shows agreement.

Finger-pointing is usually a gesture that directs attention outward, placing blame or responsibility on someone else. The common "talk to the hand" gesture sends a stronger message: "Go away!" or "I'm not listening to you." Few gestures, however, clearly convey a meaning by themselves. Instead, they must be interpreted in context. Because of habit or hearing problems, for instance, a coworker may always lean forward when listening, regardless of whether he or she is interested.

The same gesture may have different meanings in different countries:

▶ Tapping one's elbow several times with the palm of one's hand indicates that someone is sneaky (in Holland), stupid (in Germany and Austria), or mean or stingy (in South America).

▶ Twisting an index finger into one cheek means "She's beautiful" in Libya, and that something tastes good in Italy. The same gesture in southern Spain means that a man is effeminate, and "You're crazy!" in Germany.

▶ In many Middle Eastern countries, people view the shoes and the soles of one's feet as unclean. Thus, stretching out one's legs with the feet pointing at someone or crossing one's legs so that a sole faces another person is considered rude (Morris, 1994; Lynch and Hanson, 1999; Jandt, 2001; Donadio, 2013).

In a remarkable tarmac ritual, and regardless of weather, the ground crews at Japanese airports use many gestures as a jet pushes back from the gate. First, the crew members line up, snap to attention, and then, "in perfect unison," they bow. This gesture is directed toward the passengers ("Thank you for visiting Japan"), the plane's

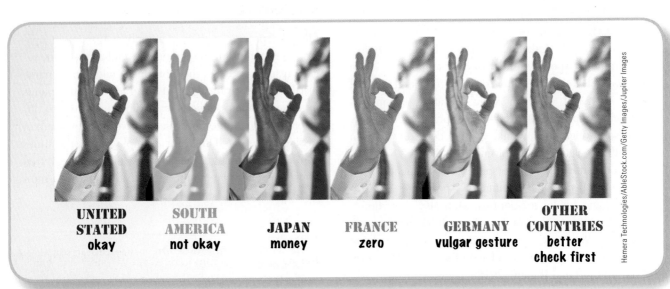

UNITED STATED	SOUTH AMERICA	JAPAN	FRANCE	GERMANY	OTHER COUNTRIES
okay	not okay	money	zero	vulgar gesture	better check first

Hemera Technologies/AbleStock.com/Getty Images/Jupiter Images

Departing jets get a wave from the ground crew in Tokyo.

flight crew ("We respect your expertise and dedication"), and, ultimately, as recognition of the ground crew's commitment to service ("We have fulfilled our duties to the best of our abilities"). Then they straighten up, smile broadly, and wave goodbye to show the passengers that they had been welcome to Japan (Gottlieb, 2011: 44).

FACIAL EXPRESSIONS

Facial expressions are visual cues of feelings, but they can be deceptive. First, our facial expressions don't always show our true emotions. Parents tell their children "Don't you roll your eyes at me!" or "Look happy when Aunt Minnie hugs you." Thus, children learn that displaying their real feelings—especially when they're negative—is unacceptable.

Second, faces can lie about feelings. Parents often know when children are lying because the children avoid eye contact, cry, or blink and swallow frequently. Some doctors say that they can tell if their patients are lying because of signs such as avoiding eye contact, pausing, or voice inflections, but many adults can control their facial expressions. They can deceive, successfully and over many years, because they've rehearsed the lies in their heads, are smooth talkers with a reputation for being trustworthy, or

Is she being friendly? Or flirting?

have gotten away with lying in the past. In nonverbal behavior, people can correctly distinguish truth from lies only about 54 percent of the time (Reddy, 2013; Shea, 2014).

Third, facial expressions can be misleading because of cultural variations. American businesspeople have grumbled that Germans are cool and aloof, whereas many German businesspeople have complained that Americans are excessively friendly and hide their true feelings with grins and smiles. The Japanese, who believe that it's rude to display negative feelings in public, smile more than Americans do to disguise embarrassment, anger, and other negative emotions (Jandt, 2001).

The only consistent research finding, across cultures, is that women smile more than men. This difference may be due to cultural norms that socialize women to hide negative feelings that might make people uncomfortable, and to show deference to men (Szarota, 2010). Regardless of the reason, men are more likely than women to mistake being friendly as being flirtatious and sexually seductive (La France et al., 2009; Rutter and Schwartz, 2012).

EYE CONTACT

Eye contact serves several social purposes (Eisenberg and Smith, 1971; Ekman and Friesen, 1984; Siegman and Feldstein, 1987). First, we get much information about other people by looking at their eyes. Eyes open wide show surprise, fear, or a flicker of interest. When we're angry, we stare in an unflinching manner. When we're sad or ashamed, our eyes may be cast down.

Second, appropriate eye contact depends on the social context. Especially during job interviews, eye contact conveys attentiveness, confidence, and respect. Managers expect eye contact from subordinates, but subordinates often complain that a manager's prolonged eye contact is domineering or intimidating. Looking at a coworker when speaking shows trust and interest, but an overly intense gaze can be disturbing (Chen et al., 2013). When two people like each other, they establish eye contact more often and for longer durations than when there's tension in the relationship.

Finally, cultural norms affect eye contact. In many Asian cultures, including Japan and some Caribbean cultures, meeting another's eyes can be rude. Asians are more likely than Westerners to regard a person who makes frequent eye contact as angry, unapproachable, and unpleasant (Akechi et al., 2013). In effect, then, cross-cultural communication breaks down when we don't understand cultural rules about acceptable body language.

5-5c Touch

Touching, another important form of nonverbal communication, sends powerful messages about our feelings

Pictured are former President Bush and Saudi Crown Prince Abdullah at Bush's Texas ranch in 2002. Many U.S. journalists raised questions about two men holding hands. In response, Arab Americans were quick to point out that Arab society sees the outward display of affection between male friends as an expression of respect and trust. Thus, government officials and military officers often hold hands as they walk together or converse with one another.

and attitudes. Parents worldwide communicate with their infants through touch—stroking, holding, patting, rubbing, and cuddling.

Touching can be positive (hugging, embracing, kissing, and holding hands) or negative (hitting, shoving, pushing, spanking). Other forms of touching are controlling. One of my students, for example, left a boyfriend because "He said he loved me and trusted me but he gripped my arm tightly every time I talked to another guy and pulled me away." Between intimate partners, a decline in the amount of touching may signal that feelings are cooling off.

GENDERED TOUCHING

Whether touching is viewed as positive or negative depends on the situation and one's gender. In higher education, even when the faculty member is well liked, male and female students perceive touching differently. When female professors touch male students on the arm while talking to them, the students view the gesture as friendly. When a male professor touches a female student on the arm, she may get nervous because she's afraid that the touching may escalate (Lannutti et al., 2001; Fogg, 2005). Generally, women are more likely than men to initiate hugs and touches that express support, affection,

and comfort. In contrast, men more often use touching to assert power or show sexual interest (Wood, 2015).

CROSS-CULTURAL VARIATIONS IN TOUCHING

As with other forms of nonverbal communication, the interpretation of touching varies from culture to culture. In some Middle Eastern countries, people don't offer anything to another with the left hand because it's used to clean oneself after using the toilet. And among many Chinese and other Asian groups, hugging, backslapping, and handshaking aren't as typical as they are in the United States because such touching is seen as too intimate (Lynch and Hanson, 1999).

According to one scholar, Americans—even strangers—seem to be hugging more. Nonetheless, she maintains, the United States is a "medium touch" culture: "More physically demonstrative than Japan, where a bow is the all-purpose hello and goodbye, but less demonstrative than Latin or Eastern European cultures, where hugs are robust and can include a kiss on both cheeks" (Drexler, 2013: C3).

5-5d Personal Space

In the example of airline seats at the beginning of this chapter, you saw that the distance that people establish between themselves is an important aspect of nonverbal communication. Americans are more annoyed by a passenger who invades their seat space (47 percent) than skimpy airline leg room (39 percent), long security lines (37 percent), or weather delays (20 percent) (Aguila, 2014). Personal space plays a significant role in our everyday nonverbal interactions, reflects power and status, and varies across societies.

WHEN IS OUR PERSONAL SPACE VIOLATED?

Our living space is public or private. In the public sphere, which is usually formal, we have clearly delineated spaces: "This is your locker," "That's her office," or "You're parking in my spot." We usually decorate our public spaces with businesslike artifacts such as awards or framed photos of an institution's accomplishments.

After a snowstorm, people in densely populated cities who shovel parking spots "mark" their personal spaces with lawn chairs, strollers, and even a table set for two, complete with a bottle of wine. Some of the most aggressive (and locally accepted) retaliation occurs in South Boston, where residents punish violators by slashing their tires or smashing their car windows (Goodnough, 2010).

Private spaces send a different message. They convey informality and a relaxed feeling. Private spaces—homes, apartments, and dorm rooms—are often mini-museums that reflect people's interests, hobbies, hygiene, and personalities.

IMAGEMORE Co., Ltd./Alamy

members often share office space, even though many of their discussions with students are confidential. Generally, the higher the socioeconomic status, the greater the consumption of space, including large cars, reserved parking spaces, private dining areas, first-class airline seats, and luxurious skyboxes at sports stadiums.

CROSS-CULTURAL VARIATIONS AND SPACE

Cultural norms and values determine how we use space. Americans not only stand in line but also have strict queuing rules: "Hey, the end of the line is back there" and "That guy is trying to cut into the line." In contrast, "along with Italians and Spaniards, the French are among the least queue-conscious in Europe" (Jandt, 2001: 109). In these countries and others, people routinely push into the front of a group waiting for taxis, food, and tickets, and nobody objects.

Americans maintain personal space in an elevator by moving to the corners or to the back. In contrast, an Arab male may stand right next to or touching another man even when no one else is in the elevator. Because most Arab men don't share American concepts of personal space in public places and in private conversations, they consider it offensive if the other man steps or leans away. In many Middle Eastern countries, however, there are stringent rules about

People, particularly men, sometimes invade our personal space by standing too close to us in a line, leaning against us, using all of the shared armrest in tight airline seats, or plopping their feet on the chair next to us in a classroom, airport, or movie theatre. Not all space intrusions are physical encroachments. Loud cell phone conversations are a good example of an auditory intrusion into our personal space.

In intimate relationships (as between family members), our personal space is often 2 feet or less because we're at ease with close physical proximity. In contrast, in public situations (someone speaking to a large audience or faculty in large classrooms), the personal space is often 12 feet or more because the speaker, who has a higher social status than the audience, has a formal relationship with the listeners and avoids close physical contact (Hall, 1966).

SPACE AND POWER

Space usage signifies who has privilege, status, and power. Wealthy people can afford enormous apartments in the city or houses in the suburbs, whereas the poor are crowded into the most undesirable sections of a city or town or in trailer parks. Executives, including college presidents, usually have huge offices (even entire suites), whereas faculty

public | social | private

—12+— —4— >1 1/2 –2–
feet feet feet

George Doyle & Ciaran Griffin/Stockbyte/Getty Images

women and men separating themselves spatially during religious ceremonies and about women avoiding any physical contact with men in public places (Falah and Nagel, 2005; Office of the Deputy Chief . . . , 2006; see also Chapter 3).

5-6 ONLINE INTERACTION

Many of us now interact in *cyberspace*, an online world of computer networks. In cyberspace, **social media** are websites that enable users to create, share, and/or exchange information and ideas. Social media include social networking sites (e.g., Facebook and Twitter), gaming sites (e.g., *Second Life*), video sites (e.g., YouTube), and blogs (websites maintained by people who provide commentary and other material). Who's online and why? And what are the benefits and costs of online interaction?

5-6a Who's Online and Why?

The percentage of Americans age 18 and older who are connected to the Internet increased from practically zero in 1994 to 85 percent in 2015. Fully 91 percent of U.S. adults have a cell phone. More than half (64 percent) of all cell phone users have smartphones, giving them online access all the time (Anderson and Perrin, 2015; Smith, 2015).

DEMOGRAPHIC VARIATIONS

About 85 percent of women and men use the Internet. Only 58 percent of Americans aged 65 and older are online compared with 93 percent of those aged 30 to 49, and 96 percent of those aged 18 to 29. Of the 15 percent who don't use the Internet, nearly half are 65 and older (Perrin and Duggan, 2015).

Besides age, race, ethnicity, and social class also affect a person's likelihood of being online. As *Figure 5.3* shows, Asian Americans are the most connected group in the United States. This group's greater "connectivity" is due primarily to high education and income levels. Many Asian American parents are professionals who can afford computers and online service, and they encourage their children to use technology for education, a major avenue of upward mobility. Moreover, Asian Americans tend to live in urban and suburban areas where high-speed Internet connections are readily available (File and Ryan, 2014; see also Chapters 10 and 13).

social media websites that enable users to create, share, and/or exchange information and ideas.

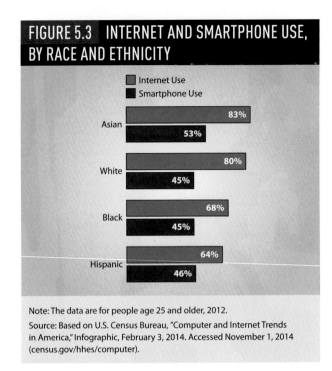

FIGURE 5.3 INTERNET AND SMARTPHONE USE, BY RACE AND ETHNICITY

Note: The data are for people age 25 and older, 2012.

Source: Based on U.S. Census Bureau, "Computer and Internet Trends in America," Infographic, February 3, 2014. Accessed November 1, 2014 (census.gov/hhes/computer).

The biggest digital divide is between social classes. The higher a family's income, the greater the likelihood that its members are Internet users (see *Figure 5.4*). Affluent families can buy home computers, pay for Internet service, and send their children to schools that provide computer-related instruction. Thus, the children from the poorest families are the most likely to lack technological skills.

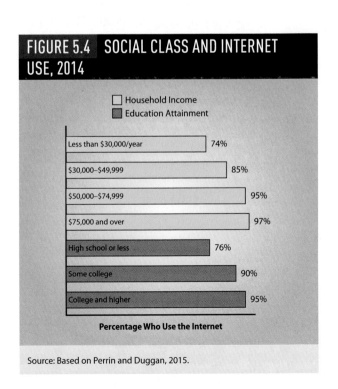

FIGURE 5.4 SOCIAL CLASS AND INTERNET USE, 2014

Source: Based on Perrin and Duggan, 2015.

WHAT WE DO ONLINE

Among our many online activities, the most popular are **social networking sites**, websites that connect people who have similar personal or professional interests. About 74 percent of adults use social networking sites, up from only 8 percent in 2005 ("Social Networking Fact Sheet," 2015). They spend more time, on average, perusing social networks than any other online activity (see *Table 5.3*).

Among adults ages 18 and older, 70 percent use Facebook at least once a day, ahead of Instagram (49 percent) and Twitter (36 percent), and 52 percent—up from 42 percent in 2013—use two or more social media sites. Besides social networking, more than half of online users access healthcare information, seek parenting advice, do banking, learn about community events or activities, use navigation devices while driving, and follow breaking news events (Duggan, Ellison, et al., 2015; Duggan, Lenhart, et al., 2015; Smith, 2015). Thus, people go online for a variety of reasons.

5-6b What Are the Benefits and Costs of Online Interaction?

You'll see in Chapter 16 that technological changes bring both benefits and costs. The same is true of online interaction, ranging from family ties to privacy issues.

FAMILY TIES

Social media has had a mixed effect on families. By age 3, many children have learned technological skills such as creating a password, navigating some websites, and using tablets. Video games like *Minecraft* encourage school age children to build, explore, collaborate, and improve hand-eye coordination and problem solving (Dockterman, 2013; Lewin, 2013). On the other hand, screen time erodes the quality and quantity of interaction between parents and young children: "Some parents are perpetually tuned into their own devices, responding to every ping of their cellphones and tablets" (Brody, 2015: D7).

Cell phones have increased the frequency of interaction between parents, especially if both are employed, to coordinate schedules and to chat with their children. Also, parents report spending more time with their children by playing home video games. On the other hand, some children complain that they rarely receive their parents' full attention because a parent is often immersed in email, texting, or being online even when pushing a swing, driving, or eating meals.

A team of researchers observed 55 parents and their young children at fast-food restaurants. Forty parents

TABLE 5.3 HOW AMERICANS SPEND THEIR ONLINE TIME	
Average minutes per day spent on . . .	
Social networks	37
Email	29
Online video, search	23 minutes each
Online games	19
Blogs, online radio	8 minutes each
Other	39

Note: These data are based on a national survey of adults ages 18–64. "Other" includes activities ranging from reading online newspapers and magazines to downloading pornography.

Source: Based on Statista, 2014.

immediately took out mobile devices and used them throughout most of the meal (Radesky et al., 2014). By paying more attention to their electronic gadgets than their children, parents are sending their children a powerful nonverbal message: "Whatever I'm doing on my phone is far more important than you are."

Among adults who are married or in committed relationships, 74 percent say that the Internet, smartphones, and social media have strengthened their communication and emotional closeness. On the other hand, 20 percent believe that the technology has had a negative effect on their relationships (e.g., a cell phone distracts a spouse or partner when the couple is together, they argue about the amount of time one of them spends online) (Lenhart and Duggan, 2014).

ONLINE RELATIONSHIPS

An increasing number of people use social networking to discuss important matters, get social support, and keep up or revive dormant relationships. Some ethnic groups, especially Asian Americans, who are rarely featured on television, have become successful on YouTube. They've found "millions of eager fans" who follow their comedy sketches and discuss topics, particularly sex and race, that are taboo among older generations (Considine, 2011: ST6; see also Pipkin, 2013).

Millions of Americans turn to the Web to find romance. About 11 percent of Internet users (representing 9 percent of all American adults) have used online dating sites (e.g., Match.com, plentyoffish.com). Seventy percent of online daters believe that online dating is a good way to meet prospective marriage mates, but

social networking site a website that connects people who have similar personal or professional interests.

only 5 percent of Americans who are currently married or in a long-term relationship met their partner online. Some 54 percent of online daters, particularly women, say that people have lied about their age, height, weight, income, occupation, and marital status. And, in 2014 alone, the Federal Bureau of Investigation (2015) received nearly 6,000 complaints about romance scams from people who reported being swindled out of a total of over $82 million (Smith and Anderson, 2015; see also Benokraitis, 2015).

CONNECTIONS AND ANXIETIES

Online interaction develops networking skills, helps people find jobs, and encourages political engagement. On the other hand, there's a "constant compulsion to connect" (Suddath, 2013: 80). For example:

▸ The average worker checks his or her email "an extraordinary" 74 times a day (Roberts, 2014).

▸ About 67 percent of cell phone owners check their phone, even when it's not ringing or vibrating (Smith, 2012).

iStockphoto.com/pearleye

▸ Among smartphone owners, 46 percent say they couldn't live without their devices, 41 percent check their smartphones several times an hour, and 11 percent do so every few minutes (Anderson, 2015; Newport, 2015).

Fully 57 percent of U.S. teenagers have met a new friend online while browsing social networks like Instagram or playing games like *Call of Duty* (Lenhart et al., 2015). The average Facebook-using teen has 300 "friends," 33 percent of whom she or he has never met in person. According to some critics, social networking sites, especially Facebook, are superficial, give people a false sense of connection to others, inflate egos, and diminish time to develop relationships with the few friends that we have (DiSalvo, 2010; Madden et al., 2013).

The more often young adults use Facebook, according to several studies, the unhappier they become, and regardless of gender, level of loneliness, or self-esteem. The root cause of the unhappiness is envy, even though Facebook users suspect that their "friends" Photoshop images and exaggerate their achievements, job success, vacations, and love life. Jealousy, anger, and loneliness are especially likely if Facebook users receive fewer positive comments, "likes," and general feedback than their friends do (Krasnova et al., 2013; Kross et al., 2013). Even adults admit that social media makes them feel bad about themselves when their friends describe awards, promotions, and "living a more fulfilling, fabulous life than I am" (Melton, 2013).

ONLINE HARASSMENT AND HUMILIATION

Nationally, about 24 percent of adolescents have experienced *cyberbullying*—deliberately using digital media to communicate false, embarrassing, or hostile information about someone. A greater percentage of females (79 percent) than males (60 percent) report being cyberbullied during the school year. Cyberbullying is less common than traditional bullying, but it has more profound negative outcomes that include depression, anxiety, severe isolation, and, most tragically, suicide (Patchin, 2013; Robers et al., 2013).

Harassment—from name calling to threatening behavior—is becoming a common part of online life. Fully 73 percent of adult Internet users have seen someone harassed online, and 40 percent have experienced it personally. Half of the latter didn't know the person who had most recently attacked them. Men are more likely to experience name-calling and embarrassment, whereas women are targets of more intense and serious forms of online assaults like sexual harassment and cyberstalking (Duggan, 2014).

Cyberstalking, which is more common among adults than adolescents, is threatening behavior or unwanted advances using email, texting, social networking sites, and other electronic devices. About 80 percent of the victims are women, and 87 percent of the offenders are men. Cyberstalking may evolve into offline stalking, including abusive or harassing phone calls, vandalism, threatening or obscene mail, trespassing, and physical assault (National White Collar Crime Center, 2013).

Tweeters can be engaging, funny, and supportive. Many of these faceless strangers can also be racist, sexist, abusive, and hateful. Because tweets are short and easy to read, public accusations and vile, ugly things said about anyone can quickly get a huge audience's attention.

Sylvie Bouchard/Shutterstock.com

Many Americans, especially teens and young adults, willingly give out personal information on social media sites—photos, phone number, address, age, education and work background, political and religious views, and so on (Madden et al., 2013; see also Fleming, 2015). It takes a private investigator just a few clicks to get a composite picture of someone from public records, email messages, websites, blogs, and networking sites.

Also, smartphones know everything—where people go, what they search for, what they buy, what they do for fun, and when they go to bed. Tech companies such as Google and Facebook can track people on their cell phones and reach them with "individualized, hypertargeted ads." Thus, "if you research a Hawaiian vacation on your work desktop, you could see a Hawaii ad that night on your personal cell phone" (Miller and Sengupta, 2013: A1). Besides collecting data on its own users, Facebook has partnerships with several companies that collect behavioral data—from store transactions and customer email lists to divorce and Web browsing records—that it then sells to advertisers (Sengupta, 2013). Neither state nor federal laws prohibit the collection or sharing of such data by third parties.

PRIVACY ISSUES

A major cost of online interaction is jeopardizing our privacy because email and text messages are neither anonymous nor confidential. Snapchat—an app that allows people to take a picture, send it to someone, and the photo is automatically deleted after 10 seconds—doesn't guarantee that the deletion always occurs (McLane, 2013). Many companies monitor and preserve their employees' email messages and can use them to discipline or fire people. It's also becoming increasingly common for college admissions officers and employers to search networking sites and Twitter before deciding whether to accept a student or to make a job offer (Belkin and Porter, 2012; Singer, 2013).

People often text and say things in emails they'd never say in person, especially when they're flirting, angry, frustrated, or tired. Emails, text messages, and Snapchat photos don't disappear after being deleted but may last indefinitely in cyberspace. Divorce lawyers have successfully retrieved deleted emails, Facebook posts, tweets, and Instagram photos to convince judges that a divorcing spouse shouldn't get financial support or custody of the children (Hillin, 2014).

STUDY TOOLS 5

READY TO STUDY? IN THE BOOK, YOU CAN:

☐ Check your understanding of what you've read with the Test Your Learning Questions provided on the chapter review card at the back of the book.

☐ Rip out the chapter review card for a handy summary of the chapter and key terms.

ONLINE AT CENGAGEBRAIN.COM YOU CAN:

☐ Collect StudyBits while you read and study the chapter.

☐ Quiz yourself on key concepts.

☐ Prepare for tests with chapter flash cards as well as flash cards that you create.

☐ Watch videos to explore sociology further.

© Anelina/Shutterstock.com

Groups, Organizations, and Institutions

LEARNING OUTCOMES

After studying this chapter, you will be able to...

6-1 Compare and illustrate the different types of social groups, explain why people conform to group pressure, and discuss the impact of social networks.

6-2 Describe and illustrate three types of formal organizations, summarize the strengths and shortcomings of bureaucracies, and explain how informal groups affect organizations.

6-3 Compare the theoretical explanations of groups and organizations, including their contributions and limitations.

6-4 Explain why social institutions are important and how they're interconnected.

After you finish this chapter go to **PAGE 117** for **STUDY TOOLS.**

Every year, about a dozen of our neighbors, all fervent Ravens football fans, get together to buy season tickets and to plan elaborate tailgating parties. They wear Ravens caps and purple jerseys, and most of the drivers attach pennants to their car antennas. These football enthusiasts are a social group.

6-1 SOCIAL GROUPS

A **social group** is two or more people who share some attribute and interact with one another. They have a sense of belonging or "we-ness." Friends, families, work groups, religious congregations, clubs, athletic teams, and World War II veterans are all examples of social groups. Each of us is a member of many social groups, but we identify more closely with some groups than others.

6-1a Types of Social Groups

Some groups are small and personal (families); others are large and impersonal (financial organizations). Some are

highly organized and stable (political parties); others are fluid and temporary (high school classmates). The most basic types of social groups are primary and secondary groups.

social group two or more people who share some attribute and interact with one another.

What do you think?

It's a good idea for companies to use video surveillance, including in restrooms.

| 1 | 2 | 3 | 4 | 5 | 6 | 7 |

strongly agree   strongly disagree

Jdwfoto/Shutterstock.com

PRIMARY AND SECONDARY GROUPS

A **primary group** is a a relatively small group of people who engage in intimate face-to-face interaction over an extended period (Cooley, 1909/1983). Primary groups, such as families and close friends, are our emotional glue. They have a powerful influence on our social identity because we interact with them on a regular basis over many years, usually throughout our lives. Because primary group members genuinely care about each other, they contribute to one another's personal development, security, and well-being. Our family and close friends, particularly, stick with us through good and bad, and we feel comfortable being ourselves with them.

Lew Robertson/Spirit/Corbis

In contrast to primary groups, a **secondary group** is a large, usually formal, impersonal, and temporary collection of people who pursue a specific goal or activity. Your sociology class is a good example of a secondary group. You might have a few friends in class, but students typically interact infrequently and impersonally. When the semester (or quarter) is over and you've accomplished your goal of passing the course, you may not see each other again (especially if you're attending a large college or university). Other examples of secondary groups include sports teams, labor unions, and a company's employees.

Unlike primary groups, secondary groups are usually highly structured: There are many rules and regulations, people know (or care) little about each other personally, relationships are formal, and members are expected to accomplish specific tasks. Whereas primary groups meet our *expressive* (emotional) needs, secondary groups fulfill *instrumental* (task-oriented) needs. Once a task or activity is completed—whether it's earning a grade, turning in a committee report, or building a bridge—secondary groups split up and become members of other secondary groups.

primary group a small group of people who engage in intimate face-to-face interaction over an extended period.

secondary group a large, usually formal, impersonal, and temporary collection of people who pursue a specific goal or activity.

TABLE 6.1 CHARACTERISTICS OF PRIMARY AND SECONDARY GROUPS

	Characteristics of a Primary Group	Characteristics of a Secondary Group
Interaction	• Face-to-face • Usually small	• Face-to-face or indirect • Usually large
Communication	• Emotional, personal, and satisfying	• Emotionally neutral and impersonal
Relationships	• Intimate, warm, and informal • Usually long-term • Valued for their own sake (expressive)	• Typically remote, cool, and formal • Usually short-term • Goal-oriented (instrumental)
Individual Conformity	• Relatively free to stray from norms and rules	• Expected to adhere to rules and regulations
Membership	• Members aren't easily replaced	• Members are easily replaced
Examples	• Family, close friends, girlfriends and boyfriends, self-help groups, street gangs	• College classes, political parties, professional associations, religious organizations

Table 6.1 summarizes the characteristics of primary and secondary groups. These characteristics are **ideal types**—general traits that describe a social phenomenon rather than every case. Ideal types provide composite pictures of how structures, events, and behaviors differ rather than specific descriptions of reality. Because primary and secondary groups are ideal types, their characteristics can vary. Thus, primary group members may sometimes devote themselves to meeting instrumental needs (running a family-owned business), and secondary group members (military units and athletic teams) can develop lasting ties.

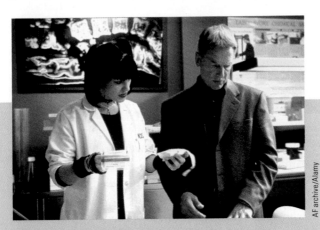

Is the cast of characters on the television series *NCIS* an example of a primary group? A secondary group? Both? Neither?

IN-GROUPS AND OUT-GROUPS

Rudyard Kipling's poem "We and They" captures the essence of in-groups and out-groups. Members of an **in-group** share a sense of identity and "we-ness" that typically excludes and devalues outsiders. **Out-group** members are people who are viewed and treated negatively because they're seen as having values, beliefs, and other characteristics different from those of an in-group. For example, "we" vegetarians are healthier than "you" meat eaters, "we" computer nerds are smarter than "you" fraternity and sorority "types," and so on.

Almost everyone sees others as members of in-groups and out-groups. From ancient times to the present, people in various parts of the world have made "we" and "they" distinctions based on race, ethnicity, gender, sexual orientation, religion, age, social class, and other social and biological characteristics (Coser, 1956; Tajfel,

ideal types general traits that describe a social phenomenon rather than every case.

in-group people who share a sense of identity and "we-ness" that typically excludes and devalues outsiders.

out-group people who are viewed and treated negatively because they're seen as having values, beliefs, and other characteristics different from those of an in-group.

> All good people agree,
> And all good people say,
> All nice people, like Us, are We
> And everyone else is They.
> —from "We and They," by Rudyard Kipling

1982; Hinkle and Schopler, 1986). Such distinctions can promote in-group solidarity and cohesion. They can also create conflict and provoke inhumane actions, including wars, massacres of out-group members, and civil wars (see Chapter 3).

One person's in-group is another person's out-group. The general public and criminal justice system see gangs as dangerous out-groups. Many young people who join gangs, however, experience in-group benefits. Attractions include making money, although illegally; close relationships with family members and friends who are already involved in gang life; protection against violent family or community members and rival gangs; a sense of self-worth, support, and belonging; and the status of being an "outlaw" that the entertainment industry often glamorizes as sexy and exciting (Taylor and Smith, 2013).

REFERENCE GROUPS

An in-group or an out-group can become a **reference group**, people who shape our behavior, values, and attitudes (Merton and Rossi, 1950). Reference groups influence who we are, what we do, and who we'd like to be in the future. Unlike primary groups, however, reference groups rarely provide personal support or face-to-face interaction over time.

Reference groups might be people with whom we already associate (a college club or a recreational athletic team). They can also be groups that we admire and want to be part of (teachers or doctors). Each person has many reference groups. Your sociology professor, for example, may be a member of several professional associations, a golf enthusiast, a parent, and a homeowner. Identification with each of these groups influences her or his everyday attitudes and actions.

Like in-groups and primary groups, reference groups can have a strong impact on our self-identity, self-esteem, and sense of belonging because they shape our current and future attitudes and behavior. We typically add or drop reference groups throughout the life course. If you aspire to move up the occupational ladder, your reference group may change from entry-level employees to managers, vice presidents, and CEOs.

6-1b Group Size and Structure

Two important characteristics of a social group are its size and leadership. Both affect a group's interaction and dynamics (behavior over time).

GROUP SIZE: DYADS, TRIADS, AND BEYOND

German sociologist Georg Simmel (1858–1918) pioneered the study of **dyads**, groups with two members, and **triads**, groups with three members. A dyad (e.g., parent–child, lovers, husband–wife, close friends) is the most cohesive of all groups because its members tend to have a personal relationship and to interact more intensely than do people in larger groups. Dyads can be unstable, however, because both persons must cooperate. If either member doesn't fulfill her or his responsibilities, the dyad collapses (Simmel, 1902; Mills, 1958).

Adding only one member to a dyad has important consequences because it changes a group's dynamics. A triad is more stable than a dyad because the group can continue if a person drops out, and a member may patch up an argument that erupts between the other two. On the other hand, a third person who enters a dyad may intentionally create conflict to break up a relationship or to attain a dominant position. In any triad, two of the members can gang up against or leave out the third (Simmel, 1902; Mills, 1958). Triads also tend to interact less often and intensely than do dyads (e.g., think about a third roommate who moves in, another person who joins you and a coworker at lunch, or someone who tags along on a date).

As more members are added—a fourth, fifth, and so on—interaction changes even more and alliance possibilities increase. At about seven or eight members, the group may break down into dyads and triads because it's difficult to have a single conversation about the task at hand (Becker and Useem, 1942). As a result, larger groups often require some kind of leadership to keep the members focused.

GROUP LEADERSHIP

Several classic studies (Lewin et al., 1939; Bales, 1950) identified three basic types of group leaders. An **authoritarian leader** gives orders, assigns tasks, and makes all major decisions. There's a clear division between leaders and followers, and leaders focus on

reference group people who shape our behavior, values, and attitudes.

dyad a group with two members.

triad a group with three members.

authoritarian leader gives orders, assigns tasks, and makes all major decisions.

completing *instrumental* tasks (e.g., marketing a new smartphone). Authoritarian leadership is most effective when there's little time for group decision making and the leader is the most knowledgeable member of the group. On the other hand, authoritarian leaders tend to stifle creativity and may be viewed as controlling and bossy.

By contrast, a **democratic leader** encourages group discussion and includes everyone in the decision-making process. Democratic leaders have the final say, but tend to be concerned about meeting a group's *expressive* needs (e.g., maintaining harmony, high morale, cohesiveness). Democratic leaders engage group members and encourage their creativity, but have also been blamed for not taking charge in times of crisis, and lower productivity than groups with authoritarian leaders.

A **laissez-faire** (pronounced les-ey-fair) **leader** offers little or no guidance to group members and allows them to make their own decisions. Laissez-faire leaders are similar to the permissive parents you met in Chapter 4: There are few rules, few demands, and people do pretty much what they want. Laissez-faire leadership is effective in situations where group members are highly qualified in an area of expertise. However, this leadership style often leads to poorly defined roles, low productivity, a lack of motivation, little cooperation between group members, and not meeting a group's goals.

It's difficult for a leader to meet both a group's instrumental and expressive needs because involving everyone in decision making may result in not accomplishing a group's objectives. Also, leadership styles are more effective in some settings than others. Imagine, for example, if your college professors were laissez-faire ("Take the exams when you think you're ready") or if your employers were always democratic ("Let's vote on who'll do what today").

6-1c Group Conformity

Most Americans see themselves as rugged individualists who have minds of their own (see the discussion of values in Chapter 3). A number of studies have shown, however, that many of us are profoundly influenced by group pressure. Four of the best known of these studies are by Solomon Asch, Stanley Milgram, Philip Zimbardo, and Irving Janis.

democratic leader encourages group discussion and includes everyone in the decision-making process.

laissez-faire leader offers little or no guidance to group members and allows them to make their own decisions.

U.S. government shutdowns occur when political leaders don't compromise on legislation that funds federal agencies and operations.

ASCH'S RESEARCH

In a now classic study of group influence, social psychologist Solomon Asch (1952) told subjects that they were taking part in an experiment on visual judgment. After seating six to eight male undergraduates around a table, Asch showed them the line drawn on card 1 and asked them to match the line to one of three lines on card 2 (see *Figure 6.1*). The correct answer, clearly, is line C.

All but one of the subjects—who usually sat in the last chair—were Asch's confederates, or accomplices. In the first test, all the confederates selected the correct matching line. In the other tests, each of them, one by one, deliberately chose an incorrect line. Thus, the nonconfederates faced a situation in which seven other group members

FIGURE 6.1 CARDS IN ASCH'S EXPERIMENT

Card 1   Card 2

A   B   C

had unanimously agreed on a wrong answer. Averaged over all of the trials, 37 percent of the nonconfederate subjects ended up agreeing with the group's incorrect answers. When they were asked to judge the length of the lines alone, away from the influence of the group, they made errors only 1 percent of the time.

Alexandra Milgram

Asch's research demonstrated the power of groups over individuals. Even when we know that something is wrong, we may go along with the group to avoid ridicule or exclusion. Remember that these experiments were done in a laboratory and with people who didn't know each other. A group's influence on a person's attitudes and behavior can be even stronger when it's a real-life situation.

MILGRAM'S RESEARCH

In a well-known laboratory experiment on obedience, psychologist Stanley Milgram (1963, 1965) asked 40 volunteers to administer electric shocks to other study participants. In each experimental trial, one participant was a "teacher" and the other a "learner," one of Milgram's accomplices. The teachers were businessmen, professionals, and blue-collar workers.

The learner was strapped into a realistic-looking chair that supposedly regulated electric currents. The teacher read aloud pairs of words that the learner had to memorize. Whenever the learner gave a wrong answer, the teacher was told to apply an electric shock from a low of 15 volts to a high of 450 volts. The learner didn't actually receive a shock, but was told to fake pain and fear. When the learners shrieked in agony, the majority of the teachers, although distressed, obeyed the study supervisor and administered the shocks when told to do so.

Milgram's study was controversial. Ordering electric shocks raised numerous ethical questions about the participants' suffering extreme emotional stress (see Chapter 2, and Perry, 2013, for a recent critique). However, the results showed that an astonishingly large proportion of ordinary people obeyed an authority figure's instructions to inflict pain on others. It's easy to sit back and say "I'd never do something like that," but we do. For example, many workers obey employers when told to ignore evidence that their product is unsafe (Tavris, 2013).

ZIMBARDO'S RESEARCH

The Stanford Prison Experiment conducted by social psychologist Philip Zimbardo also underscores the influence of groups on behavior (Haney et al., 1973; Zimbardo, 1975). Zimbardo recruited volunteers through a local California newspaper for an experiment on prison life. He then selected 24 young men, most of them college students.

On a Sunday morning, nine of the men were "arrested" at their homes as neighbors watched. The men were booked and transported to a mock prison that Zimbardo and his colleagues had constructed in the basement of the psychology building at Stanford University. The "prisoners" were searched, issued an identification number, and outfitted in a dresslike shirt and heavy ankle chains. Those assigned to be "guards" were given uniforms, billy clubs, whistles, and reflective sunglasses. The guards were told that their job was to maintain control of the prisoners but not to use violence.

All the young men quickly assumed the roles of either obedient and docile prisoners or autocratic and controlling guards. The guards became increasingly more cruel and demanding. The prisoners complied with dehumanizing demands (such as eating filthy sausages) to gain the guards' approval and bowed to their authority.

Zimbardo's study was scheduled to run for 2 weeks but was stopped after 6 days because the guards became increasingly aggressive. Among other things, they forced the prisoners to clean out toilet bowls with their bare hands, locked them in a closet, and made them stand at attention for hours. Instead of simply walking out or rebelling, the prisoners became withdrawn and depressed. Zimbardo ended the experiment because of the prisoners' stressed-out reactions.

The experiment raised numerous ethical questions about the harmful treatment of participants. It demonstrated, however, the powerful effect of group conformity: People exercise authority, even to the point of hurting others, or submit to authority if there's group pressure to conform (Zimbardo et al., 2000).

JANIS'S RESEARCH

Sometimes intelligent people, including those in highly responsible positions, make disastrous and irrational decisions. Why? Social psychologist Irving Janis (1972, 1982) cautioned presidents and other heads of state to be wary of **groupthink**—a situation in which in-group members make faulty decisions because of group pressures, rather than critically testing, analyzing, and evaluating ideas and evidence. To demonstrate group

groupthink a situation in which in-group members make faulty decisions because of group pressures, rather than critically testing, analyzing, and evaluating ideas and evidence.

Many social scientists have used Milgram's findings to explain the Abu Ghraib prison in Iraq, where U.S. soldiers humiliated prisoners because "I was just following orders."

SOMETIMES INTELLIGENT PEOPLE, EVEN THOSE IN HIGHLY RESPONSIBLE POSITIONS, MAKE DISASTROUS AND IRRATIONAL DECISIONS.

loyalty, Janis argues, individuals don't raise controversial issues, don't question weak arguments, or probe "soft-headed thinking." As a result, influential leaders often make decisions, based on their advisors' consensus, that turn out to be political and economic disasters.

Examples of groupthink "fiascoes" that Janis studied included U.S. failure to anticipate the attack on Pearl Harbor, the Bay of Pigs invasion, the escalation of the Vietnam War, and the ill-fated hostage rescue in Iran. More recently, a United States Senate study (2004) concluded that U.S. leaders' decision to invade Iraq in 2003 was based on a groupthink dynamic that relied on unproven and inaccurate assumptions, inadequate or misleading sources, and a dismissal of conflicting information showing that Iraq had no weapons of mass destruction.

Janis and other researchers have focused on high-level decision making, but groupthink is common in all kinds of groups—student clubs, PTAs, search committees, juries, and community activists, for example (see Hansen, 2010). If group members are aware of groupthink, they can avoid some of the pitfalls by hammering out disagreements and seeking advice from informed and objective people outside the group.

6-1d Social Networks

A **social network** is a web of social ties that links individuals or groups to one another. It may involve as few as three people or millions.

social network a web of social ties that links individuals or groups to one another.

Network links between people or groups can be strong or weak. Some of our social networks, such as our primary and secondary groups, may be tightly knit, involve interactions on a daily basis, and have clear boundaries about who belongs and who doesn't. In other cases, our social networks connect us to large numbers of people whom we don't know personally, with whom we interact only rarely or indirectly, and the boundaries change as people come and go. Examples of distant networks include members of the American Sociological Association and LinkedIn subscribers.

The Internet, in particular, has generated numerous social networks that unite people who have similar interests (see *Table 6.2*). Almost 75 percent of all adult Americans are active in some kind of voluntary group or organization. Internet users (80 percent) are more likely to participate in groups than non-Internet users (56 percent) (Rainie et al., 2011).

Do social networks affect divorce rates? A research team that analyzed three decades of divorce data found that people were 75 percent more likely to get divorced if a friend was divorced, and 33 percent more likely to end a marriage if a friend of a friend was divorced (McDermott et al., 2013; Chapter 12 discusses some of the micro- and macro-level reasons for divorce).

TABLE 6.2 HOW THE INTERNET AFFECTS SOCIAL NETWORKS

Percentage of U.S. adults who said the Internet has a "major impact" on their groups' ability to ...

	Internet Users	Non-Internet Users
Communicate with members	75	44
Draw attention to an issue	68	43
Connect with other groups	67	40
Organize activities	65	41
Impact society at large	64	45
Recruit new members	55	38

Source: Based on Rainie et al., 2011: 31.

6-2 FORMAL ORGANIZATIONS

A **formal organization** is a complex and structured secondary group designed to achieve specific goals in an efficient manner. We depend on a variety of formal organizations to provide goods and services in a stable and predictable way. Examples include water departments, food producers and grocery chains that stock our favorite items, and garment industries and retailers that produce and sell the clothes we wear.

6-2a Characteristics of Formal Organizations

Formal organizations share some common characteristics:

▸ Social statuses and roles are organized around shared expectations and goals.

▸ Norms governing social relationships among members specify rights, duties, and sanctions.

▸ A formal hierarchy includes leaders or individuals who are in charge.

Modern, complex societies rely on formal organizations, including businesses and schools, to accomplish complex tasks. Organizations differ, however, in their goals, membership, and degree of hierarchy.

6-2b Types of Formal Organizations

Sociologist Amitai Etzioni (1975) identified three types of formal organizations—utilitarian, normative, and coercive—based on their purpose and people's reasons for participating.

UTILITARIAN ORGANIZATIONS

Most people belong to a *utilitarian organization*, one that provides an income or other specific current or future material reward. Examples include government agencies, factories, corporations, and schools. Colleges are workplaces, but also offer students the opportunity to earn degrees.

NORMATIVE ORGANIZATIONS

People join *normative organizations* (also called *voluntary organizations* or *voluntary associations*) because of shared interests and to pursue goals that they consider personally worthwhile or rewarding. Examples include political parties, civic organizations (e.g., Habitat for Humanity), religious organizations, choirs, cultural groups, occupational groups (e.g., American Accounting Association), and groups that have a recreational focus (e.g., Beer Collectors of America).

Helping others is a strong American value (see Chapter 3). Thus, every year, millions of Americans—many of whom have full-time jobs—are volunteers in a variety of local, regional, and national organizations, saving government agencies billions of dollars by providing needed services (see *Table 6.3*). In addition, 83 percent of Americans donate money to charitable organizations (Gallup Editors, 2013).

COERCIVE ORGANIZATIONS

Membership in *coercive organizations* is largely involuntary. People are pushed or forced to join these organizations because of punishment (prisons) or treatment (psychiatric hospitals, drug rehabilitation centers). Most coercive organizations are total institutions characterized by strict rules, the members' isolation, and resocialization (see Chapter 4).

TABLE 6.3 VOLUNTEERING IN THE UNITED STATES, 2014

Percentage of adults who volunteer	25
Total number of volunteers age 16 and older	63 million
Median annual hours per volunteer	50
Total dollar value of volunteer time	$173 billion

Sources: Based on Corporation for National & Community Service, 2015, and BLS News Release, 2015.

formal organization a complex and structured secondary group designed to achieve specific goals in an efficient manner.

Utilitarian, normative, and coercive organizations differ in their general characteristics and membership, but a single formal organization can fall into all three categories. For example, the military is a type of coercive organization because of its rigid rules, resocialization, and members' isolation from the rest of the population. It's a utilitarian organization that provides millions of jobs for both soldiers and civilians. The military is also a normative organization: Some people join during wartime because of civic responsibility; many volunteer their time and resources to support soldiers and their families during deployment; and others help veterans find jobs, housing, and health care for war-related injuries.

6-2c Bureaucracies

A **bureaucracy** is a formal organization designed to accomplish goals and tasks in an efficient and rational way. Bureaucracies aren't a modern invention but existed thousands of years ago in ancient Egypt, China, and Africa. Some bureaucracies function more smoothly than others, and some formal organizations are more bureaucratic than others. Whether they're relatively small (a 100-bed hospital) or huge (the Social Security Administration), bureaucracies have some common characteristics.

IDEAL CHARACTERISTICS OF BUREAUCRACIES

Max Weber (1925/1947) identified six key characteristics of the ideal type of bureaucracy. In Weber's model, the following characteristics describe what an efficient and productive bureaucracy *should* be like:

▸ **High degree of division of labor and specialization.** People perform very specific tasks.

▸ **Hierarchy of authority.** Workers are arranged in a hierarchy in which each person is supervised by someone in a higher position. The resulting pyramids—often presented in organizational charts—show who has authority over whom and who is responsible to whom. Thus, there's a chain of command, stretching from top to bottom, that coordinates decision making.

▸ **Explicit written rules and regulations.** Detailed written rules and regulations cover almost every possible kind of situation and problem that might arise, including hiring, firing, salary scales, rules for sick pay and absences, and everyday operations.

▸ **Impersonality.** There's no place for personal likes, dislikes, or tantrums. Workers are expected to behave professionally. An impersonal workplace in which all employees are treated equally minimizes conflict and favoritism and increases efficiency.

▸ **Qualifications-based employment.** People are hired based on objective criteria such as skills, education, experience, and standardized test scores. If workers perform well and have the necessary credentials and technical competence, they'll move up the career ladder.

▸ **Separation of work and ownership.** Neither managers nor employees own the offices they work in, the desks they sit at, the technology they use, or the products that they assemble, invent, or design.

A "rational matter-of-factness," Weber maintained, made bureaucracies more productive by "eliminating from official business love, hatred, and all purely personal, irrational, and emotional elements" (Weber, 1946: 216). Weber viewed bureaucracies as superior to other forms of organization because they're more efficient and predictable. He worried, however, that bureaucracies could become "iron cages" because people become trapped in them, "their basic humanity denied."

BUREAUCRATIC DYSFUNCTIONS

Weber described the ideal characteristics of a productive and efficient bureaucracy, but what's the reality? As you

bureaucracy a formal organization designed to accomplish goals and tasks in an efficient and rational way.

PzAxe/Shutterstock.com

read through the following list of problems, think about the ones you've experienced while working in a bureaucracy or dealing with one.

▸ *Weak reward systems* reduce the motivation to do a good job and are thus a major source of inefficiency and lack of innovation (Barton, 1980). Besides low wages or salaries, weak reward systems include few or no health benefits, little recognition, unsafe equipment and work environments, and few incentives to be creative.

▸ *Rigid rules* discourage creativity and can lead to **goal displacement**, a preoccupation with rules and regulations rather than achieving the organization's objectives (Merton, 1968). Instead of questioning whether all the red tape is really necessary, bureaucrats are usually more concerned that people follow established procedures because "we've always done it this way."

▸ Rigid rules and goal displacement often lead to **alienation**, a feeling of isolation, meaninglessness, and powerlessness. Alienation—at all levels—may result in high turnover, tardiness, absenteeism, stealing, sabotage, stress, health problems, and in some cases, whistle-blowing (reporting organizational misconduct to legal authorities).

▸ *Communication problems* are common in bureaucracies. Because communication typically flows down rather than up the hierarchy, employees (and many managers below the highest echelons) rarely know what's going on. Supervisors and their subordinates may be reluctant to discuss problems or offer suggestions because they fear being criticized, demoted, or fired (Blau and Meyer, 1987; Fletcher and Taplin, 2002). Communication problems also waste time and resources. An employee prepared a monthly report for nearly three years because no one ever mentioned that the company no longer needed it (Harnish, 2014).

▸ The **iron law of oligarchy** is the tendency of a bureaucracy to become increasingly dominated by a small group of people (Michels, 1911/1949). A handful of people can control and rule a bureaucracy because the top officials and leaders monopolize information and resources. As a result, those at the top maintain their power and privilege.

Brand X Pictures/Getty Images

$65 VISIT

Health Care on Aisle 7!

To avoid medical and health insurance bureaucracies, more people are turning to retail clinics (also called "convenient care clinics"). The first one opened in 2000; there were about 1,400 in 2012; and they're expected to increase to at least 3,000 by 2016 (Robert Wood Johnson Foundation, 2015).

The clinics—usually staffed by nurse practitioners—offer low-cost treatment for more than 25 common conditions (e.g., strep throat and ear infections), and provide health screening tests, immunizations, and physicals. The clinics are located in pharmacies, grocery stores, and "big box" stores such as Target and Walmart. There's no need to make an appointment and minimal (if any) waiting, the patients are in and out in about 15 minutes, and most of the clinics are open evenings and weekends (Spetz et al., 2013; *Consumer Reports*, 2014).

goal displacement a preoccupation with rules and regulations rather than achieving the organization's objectives.

alienation a feeling of isolation, meaninglessness, and powerlessness.

iron law of oligarchy the tendency of a bureaucracy to become increasingly dominated by a small group of people.

▶ The cumulative effect of these and other bureaucratic dysfunctions can result in *dehumanization* because weak reward systems, rigid rules, and other problems limit organizational creativity and freedom. As a result, work becomes more automated and impersonal (see "The McDonaldization of Society").

Bureaucracies are plagued by other problems, such as favoritism and dishonest employee evaluations (Kilmann, 2011). They function, however, largely because of the internal development of informal groups and recent changes in performance reviews.

6-2d Getting Around Bureaucratic Dysfunctions

Weber's model overlooked the many ways that workers get around rigid bureaucratic regulations and impersonality. Since around 2000, self-managing work teams have become more common, and many large corporations are rethinking age-old performance reviews that alienate employees. Let's begin with a brief look at some of the early studies' missteps and contributions regarding the importance of informal group networks.

HISTORIC STUDIES OF INFORMAL GROUP NETWORKS

For many years, and well into the 1930s, experts concentrated on organizational efficiency based on the principles of *scientific management* developed by Frederick Winslow Taylor, a mechanical engineer who wanted to improve industrial productivity. Taylor (1911/1967) considered workers—especially those in factories and on assembly lines—as mere adjuncts to machines, assumed that people don't like to work, maintained that employees need close supervision, and believed that management had to enforce cooperation because most workers aren't capable of handling even the simplest tasks.

A number of studies conducted by industrial psychologists and sociologists between 1927 and 1932 shook up Taylor's views. The research teams studied employees at the Western Electric Company's Hawthorne plant in Chicago. The *Hawthorne studies*, as they're often called, found that informal groups were critical to the organization's functioning (Roethlisberger and Dickson, 1939/1942; Mayo, 1945; Landsberger, 1958).

Informal social groups can promote an organization's goals if they collaborate, are cohesive, and motivate each other. They can also resist an organization's goals and formal rules. In one of the Hawthorne studies, Roethlisberger and Dickson (1939/1942) spent 6 months observing a group of 14 men who wired telephone switchboards in what the company called the "bank-wiring room."

The bank-wiring room work group consisted of nine wiremen, three solderers, and two inspectors. Management offered financial incentives for higher productivity, but the work group pressured people to limit output. The wiremen developed the following norms that controlled the group's behavior:

1. **You shouldn't turn out too much work.** If you do, you're a "rate-buster" and a "speed king."

2. **You shouldn't turn out too little work.** If you do, you're a "chiseler."

3. **You shouldn't tell a supervisor anything that will be detrimental to a coworker.** If you do, you're a "squealer."

4. **You shouldn't "act officious."** If you're an inspector, for example, you shouldn't act like one.

If individuals violated any of these norms, the other workers used "binging" (striking a person on the shoulder) to punish them.

Why did the men control productivity rather than take advantage of the management's promise of higher wages? They feared that if they produced at a high level, they'd be required to do so regularly. They also worried that high productivity would lead to layoffs because supervisors would conclude that fewer workers could achieve the same output. In addition, the bank-wiring room men experienced high morale and job satisfaction because they felt they had some control over their work. In contradiction to Taylor's scientific management perspective, then, the Hawthorne studies concluded that there's an important relationship between formal and informal organization.

MODERN WORK TEAMS

Since the Hawthorne studies, many companies have recognized that informal groups affect workers' productivity. As a result, numerous organizations have implemented alternative management strategies.

Today, *self-managing work teams* are the dominant model in most large organizations (Cloke and Goldsmith, 2002; Neider and Schriesheim, 2005). Contrary to Taylor's notion that workers do and managers think, self-managing work teams, sometimes referred to as *postbureaucratic organizations*, involve groups of 10 to 15 people who take on the duties of their former supervisors. Instead of being told what to do, self-managing workers gather and interpret information, act on the information, and take collective responsibility for their actions—whether it's designing a new refrigerator or handling food service for a university. Because they're not held back by unresponsive managers, effective self-managing teams are focused on their goals and are committed to the organization and its success (Colvin, 2012).

The McDonaldization of Society

According to sociologist George Ritzer (1996: 1), the organizational principles that underlie McDonald's, the well-known fast-food chain, are beginning to dominate "more and more sectors of American society as well as of the rest of the world." McDonaldization has four components: efficiency, calculability, predictability, and control.

1. *Efficiency* means that consumers have a quick way of getting meals. In a society where people rush from one place to another and where both parents are likely to work or single parents are pressed for time, McDonald's (and similar franchises) offers an efficient way to satisfy hunger and avoid much "fuss and mess" (see also Jargon, 2013). Like their customers, McDonald's workers function efficiently: The menu is limited, the registers are automated, and employees perform their tasks rapidly and easily.

2. *Calculability* emphasizes the quantitative aspects of the products sold (portion size and cost) and the time it takes to get the products. Customers often feel that they're getting a lot of food for what appears to be a nominal sum of money and that a trip to a fast-food restaurant will take less time than eating at home. In reality, soft drinks are sold at a 600 percent markup because most of the space in the cup is taken up by ice. It costs at least twice as much and takes twice as long to get to the restaurant, stand in line, and pick up the food than to prepare a similar meal at home.

3. *Predictability* means that products and services will be the same over time and in all locales. "There's great comfort in knowing that McDonald's offers no surprises" (p. 10). The Quarter Pounder is the same regardless where customers order. Outside of the United States, "homesick American tourists in far-off countries can take comfort in the knowledge that they'll likely run into those familiar golden arches and the restaurant they have become so accustomed to" (p. 81). Workers, like customers, behave in predictable ways: "There are, for example, six steps to window service: greet the customer, take the order, assemble the order, present the order, receive payment, thank the customer and ask for repeat business" (p. 81).

4. *Control* means that technology shapes behavior. McDonald's controls customers subtly by offering limited menus and uncomfortable seats that encourage diners to eat quickly and leave. "Consumers know that they're supposed to line up, move to the counter, order their food, pay, carry the food to an available table, eat, gather up their debris, deposit it in the trash receptacle, and return to their cars. People are moved along in this system not by a conveyor belt, but by the unwritten, but universally known, norms for eating in a fast-food restaurant" (p. 105).

McDonald's controls employees more openly and directly. Workers are trained to do a limited number of things in precisely the same way, managers and inspectors make sure that subordinates toe the line, and McDonald's has steadily replaced human beings with technologies that include soft-drink dispensers that shut off when the cup is full and a machine that rings and lifts the French fries out of the oil when they're crisp. Such technology increases the corporation's control over workers because employees don't have to use their own judgment or need many skills to prepare and serve the food.

Efficiency, calculability, predictability, and control reflect a rational system (remember Weber?) that increases a bureaucracy's efficiency. On the other hand, Ritzer contends, McDonaldization reflects the "irrationality of rationality" because the results can be harmful. The huge farms that now produce "uniform potatoes to create those predictable French fries" of the same size rely on the extensive use of chemicals that then pollute water supplies. And the enormous amount of nonbiodegradable trash that McDonald's produces wastes our money because we—and not McDonald's—pay for landfills.

McDonald's is such a powerful model that many businesses have cloned its four dimensions of efficiency, calculability, predictability, and control. Examples include "McDentists" and "McDoctors," drive-in clinics designed to deal quickly with minor dental and medical problems; "McChild" care centers like KinderCare; and "McPaper" newspapers, such as *USA Today* (Ritzer, 2008).

iStockphoto.com/EllenMoran

TABLE 6.4 SOCIOLOGICAL PERSPECTIVES ON GROUPS AND ORGANIZATIONS

Theoretical Perspective	Level of Analysis	Main Points	Key Questions
Functionalist	Macro	Organizations are made up of interrelated parts and rules and regulations that produce cooperation in meeting a common goal.	• Why are some organizations more effective than others? • How do dysfunctions prevent organizations from being rational and effective?
Conflict	Macro	Organizations promote inequality that benefits elites, not workers.	• Who controls an organization's resources and decision making? • How do those with power protect their interests and privileges?
Feminist	Macro and micro	Organizations tend not to recognize or reward talented women and regularly exclude them from decision-making processes.	• Why do many women hit a glass ceiling? • How do gender stereotypes affect women in groups and organizations?
Symbolic Interactionist	Micro	People aren't puppets but can determine what goes on in a group or organization.	• Why do people ignore or change an organization's rules? • How do members of social groups influence workplace behavior?

PERFORMANCE REVIEWS

Since the early 1980s, many companies have been using "stack" or "forced" ranking to evaluate employees. These annual performance reviews grade workers against one another and rank them on a scale of one to five. The fours and fives are designated as underperformers and fired, demoted, or not promoted. Stack ranking creates a sense of failure and helplessness. The evaluations are done infrequently and become "management hammers" rather than reviews that provide workers frequent feedback on improving their performance (Suddath, 2013).

Some companies are adopting "calibration" to evaluate employees." This is a process in which managers across different divisions or departments ensure that their review processes are similar to each other regarding criteria and scoring. Calibration is based on facts and measurable tasks, not feelings, greatly reducing the likelihood of a leader's biases and inconsistent scoring (Culbert and Rout, 2010; Ovide and Feintzeig, 2013).

6-3 SOCIOLOGICAL PERSPECTIVES ON SOCIAL GROUPS AND ORGANIZATIONS

How do organizations operate? And how do social groups affect massive bureaucracies? In answering such questions, functionalist, conflict, feminist, and interactionist theories offer different insights. Taken together, these perspectives provide a multifaceted understanding of groups and organizations (*Table 6.4* summarizes these perspectives).

6-3a Functionalism: Social Groups and Organizations Benefit Society

Functionalist perspectives emphasize that social groups and organizations are composed of interrelated, mutually dependent parts. As Weber pointed out, when a bureaucracy operates rationally and efficiently, workers and bosses cooperate to turn out a final product, whether it's an automobile or a can of soup.

Effective leadership is a primary factor in job satisfaction. Even in our sprawling government bureaucracies, and across all organizational levels, the happiest workers are those whose bosses reward motivation and commitment, encourage integrity, manage people fairly, and promote the employees' professional development and creativity (Partnership for Public Service, 2011).

A recent study of a large company with technology-based service jobs (e.g., retail clerks, airline gate agents, call-center workers) found that replacing a supervisor from the bottom 10 percent of the pool with one from the top 10 percent increased output almost as much as adding a tenth worker to a nine-worker team. The researchers concluded that productivity increases when the most promising workers are paired with top bosses who provide frequent feedback and teach better work methods and skills (Lazear et al., 2013). Thus, good bosses don't threaten or punish employees, but provide concrete training to get the job done.

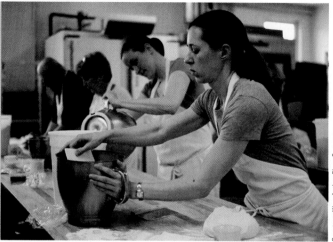

Since 1975, the number of employee-owned companies in the United States has grown from 1,600 to more than 11,500, and now represents about 12 percent of the private sector workforce. These firms are profitable because the workers are highly motivated: They have a stake in the business, control the decision making, contribute original ideas and cost-saving solutions, and depend on each other. Pictured here are people learning to bake bread at the King Arthur Flour Company's Baking Education Center in Norwich, Vermont, an employee-owned company that has "a ferocious focus on treating employees well." (Koba, 2013; Marikar, 2015).

Courtesy King Arthur Flour Company

Some companies have boosted productivity by encouraging workers' interaction. Examples include providing group rather than individual breaks, and carving out meeting spaces that accommodate three or four people instead of 12 or more (Silverman, 2013).

Groups, organizations, and bureaucracies can also be dysfunctional. Workers may be alienated because of weak reward systems, favoritism, and incompetent supervisors. Only 30 percent of U.S. workers and 35 percent of their managers are engaged in their jobs. Among employees who feel unchallenged, bored, or dissatisfied, 75 to 80 percent spend more than 2 hours a day surfing the Net or on Twitter, Facebook, and other social networking sites. Wasting time—which results in low productivity, unhappy customers, and lower profits—costs organizations up to $398 billion annually (Adkins, 2015; OfficeTime, 2015).

Since 2008, business spending on employees has grown only 2 percent compared with 26 percent on equipment and software (Rampell, 2011). Many people contend that doing so increases unemployment rates. From a functionalist perspective, however, investing more in equipment than workers is a rational bureaucratic response to the rising costs of health care benefits and economic competition from other countries (see Chapter 11).

CRITICAL EVALUATION

Functionalism is useful in understanding how groups and organizations fulfill essential functions, including motivating people to get work done and achieving a common goal. For critics, especially conflict theorists, functionalists exaggerate cooperation and tend to gloss over dysfunctions such as worker dissatisfaction and alienation. At some of the largest and most successful companies like Amazon, current and former employees have complained of 80-hour work weeks, interrupted vacations, sending negative feedback about coworkers to bosses, and little tolerance for those struggling with life-threatening illnesses or family crises (Kantor and Streitfeld, 2015).

Another issue is whether informal social networks improve worker morale and control as much as functionalists claim. After all, a tedious and monotonous job is tedious and monotonous regardless of the degree of informal coworker interaction.

6-3b Conflict Theory: Some Benefit More Than Others

Conflict theorists contend that organizations promote inequality that benefits those at the top of the hierarchy, not workers. In many companies, those at higher levels are more comfortable hiring and promoting people like themselves (male, white, and middle class or higher). Phrases such as "fitting the mold" and "having common interests" are often code words for belonging to an in-group.

Inequality in income, status, and other rewards means that owners and managers can easily exploit workers. Those at the top dictate to those at the middle and the bottom. Because organizations serve elites, conflict theorists argue, they routinely ignore workers' needs and interests.

What about performance reviews in rewarding the best employees? According to a management professor, "I've examined scores of empirical studies since the early 1980s and have not found convincing evidence that performance reviews are fair, accurate or consistent across managers, or that they improve organization effectiveness." Instead, subjective evaluations "are intimidating tools that make employees too scared to speak their minds" (Culbert, 2011: A25).

There's also considerable incompetence, waste, and corruption in many government agencies and programs. For example:

▶ In 2014 alone, the federal government paid more than $125 billion to ineligible Medicare, Medicaid, and low-income recipients ("Government Efficiency and Effectiveness," 2015).

▶ Between 2012 and 2014, the Department of Housing and Urban Development (HUD) spent $37 million per month on tens of thousands of public-housing tenants who weren't eligible for assistance (Hicks, 2015).

▶ Since at least 2012, the Veterans Administration has ignored the agency's competitive bidding laws, wasting at least $6 billion a year on high prices for medical goods and services (Rein and Wax-Thibodeaux, 2015).

▶ In 2012, the FBI's "spotty" database allowed 3,000 unauthorized people with criminal and mental health histories to purchase firearms (Schuck, 2013).

▶ Between 2008 and 2012, the Agriculture Department paid $22 million to farmers who had been dead for at least two years.

▶ In 2012, the Defense Department paid $13.4 million for helicopter parts that should have cost $4.4 million, but no one knows why; at least 40 percent of all U.S. weapons given to the Afghan and Iraqi armies "can't be traced" (Capaccio and Salant, 2014; Vicens, 2015).

Federal workers' job satisfaction has been dropping since 2003. Over the years, the major complaints have included poor leadership, incompetent employees, few opportunities for advancement, and meager rewards for good performance (Partnership for Public Service, 2013; see also U.S. Office of Personnel Management, 2014).

According to many federal employees, besides wasting billions of taxpayer dollars, the agencies have numerous (and unnecessary) levels of executives and managers rather than people working in the field. One employee complained, for instance, "The layers of management are insane. . . . It takes 13 steps and five layers to get a signature from our office director" (Rein, 2011: B4).

Many private organizations, especially big businesses, are also deficient. Consider the airline industry. Because of mergers, four major carriers (American, Delta, Southwest, and United) now control 80 percent of all domestic travel. Because there's little competition, and despite lower fuel and other costs, airlines have limited flights and seats, increased prices, and enjoy larger profit margins than ever before. Most carriers have upgraded amenities for first- and business-class travelers, but have received the worst customer ratings since 2008. The most common complaints include getting bumped from flights; lost, stolen, or delayed baggage; and long lines at ticket counters (Bowen and Headley, 2015; Stewart, 2015).

TABLE 6.5 MANY AMERICANS HAVE LITTLE CONFIDENCE IN ORGANIZATIONS

Percentage who said, in 2015, that they have "a great deal" or "quite a lot" of confidence in ...

Organization	Percentage
The church or organized religion	42
The medical system	37
The presidency	33
The U.S. Supreme Court	32
The public schools	31
Banks	28
Newspapers and organized labor	24 (each)
The criminal justice system	23
Big business and television news	21 (each)
Congress	8

Note: A majority of Americans had high confidence only in the military (72 percent), small business (67 percent), and the police (52 percent).

Source: Based on Jones, 2015.

Because of these and similar problems, many Americans are disillusioned with a number of both public and private organizations (see *Table 6.5*).

CRITICAL EVALUATION

Conflict theory has underscored the inequality within groups and organizations that saps workers' motivation and limits their economic success, but has several weaknesses. First, conflict theory assumes that greater equality leads to a more successful and productive organization. However, even self-managing work teams can fail because of ineffective team leaders or management interference (Barker, 1993; Lencioni, 2002).

Second, conflict theorists rarely credit many organizations' supporting controversial social issues that promote greater equality. Regardless of motives (e.g., improving a tarnished public image, boosting profits, becoming more socially responsible), a number of the nation's most powerful CEOs and businesses—including Walmart, NASCAR, Apple, AT&T, and Southwest Airlines—have supported controversial gay marriage laws (see Chapters 7, 9, and 11).

Third, do conflict theorists focus too much on organizational deficiencies? There are, after all, many organizations that are efficient, profitable, and that don't exploit their workers (Kaplan, 2010).

6-3c Feminist Theories: Men Benefit More Than Women

Feminist scholars emphasize that across all social classes, women (and especially minority women) consistently fare worse than men, especially in leadership roles. Women have enjoyed increased success in organizations since the 1980s, but rarely at the highest levels. For example, women hold more than half of all management and professional positions, but only 4.6 percent of *Fortune* 500 and *Fortune* 1000 CEO positions (Catalyst, 2014).

Large numbers of women still hit a **glass ceiling**—workplace attitudes or organizational biases that prevent them from advancing to leadership positions. Examples of obstacles include men's negative attitudes about women in the workplace, women's placement in staff positions that aren't on the career track to the top, a lack of mentoring, biased evaluations by male supervisors, and little or no access to highly visible committees or task forces (Barreto et al., 2009; Sandberg, 2013; see also Chapter 11).

In contrast, many men who enter female-dominated occupations (e.g., nursing and teaching) receive higher wages and faster promotions, a phenomenon known as a **glass escalator**. In 2014, for example, men made up only 11 percent of all full-time registered nurses. However, the average female nurse earned $55,952, about 10 percent less than the $61,880 that the average male nurse earned. Men on glass escalators also move up to supervisory positions more quickly than do females (Goudreau, 2012; U.S. Bureau of Labor Statistics, 2015).

CRITICAL EVALUATION

Feminist theories have expanded conflict explanations of groups and organizations by showing that many talented women are treated like outsiders. One weakness of feminist theories, however, is that much of the emphasis is still on white and black women in professional and managerial positions, even though Latinas and Asian American women comprise a large segment of these workers and those in the general labor force.

A second limitation is that even when feminist scholars say that both sexes suffer from organizational stereotypes, they offer little data on how such stereotypes affect men. Also, female supervisors rarely affect an organization's glass ceiling because they neither support nor promote women workers (see Huffman, 2013).

6-3d Symbolic Interaction: People Define and Shape Their Situations

Whereas functionalist, conflict, and (some) feminist theorists examine groups and organizations on a macro level,

The Army's Ranger School is a grueling and prestigious program that develops elite fighters and leaders. Pictured here is one of the two female soldiers who, in 2015, became the first women to graduate. So far, however, women aren't allowed to serve in the infantry or other ground combat, including the Ranger regiment.

symbolic interactionists focus on micro-level behavior. They emphasize that an individual's perception and definition of a situation shape group dynamics and, consequently, organizations.

Group leaders or members can create or reinforce conformity (as shown in the Asch and Milgram studies). Informal groups can also determine what goes on in an organization by refusing to obey the rules and implementing their own (as in the Hawthorne studies). Thus, according to symbolic interactionists, individuals make choices, change rules, and mold their own identities instead of being manipulated (Kivisto and Pittman, 2001).

Symbolic interactionists also note that people's outcomes depend on how coworkers and bosses interpret the *same* behavior. For example, whether women are supervisors or clerical workers, if they lose their temper, they're overwhelmingly seen as too emotional, incompetent, out of control, weak, and worth less pay. Their angry male counterparts, in contrast, are often viewed as authoritative and in control (Brescoll and Uhlmann, 2008).

glass ceiling workplace attitudes or organizational biases that prevent women from advancing to leadership positions.

glass escalator men who enter female-dominated occupations and receive higher wages and faster promotions than women.

CRITICAL EVALUATION

Symbolic interaction explains how members of groups and organizations interpret the world around them and, as a result, affect what goes on. Social networking sites, especially Facebook, have empowered employees to discuss everyday problems despite management's control (Kirkpatrick, 2010).

Despite these contributions, because symbolic interaction is a micro-level theory, it ignores macro-level factors that exploit workers and consumers. About 51 percent of Americans seek out "all natural" products, but much of the food contains genetically modified organisms, chemically processed vitamins, and artificial ingredients. The average consumer has little knowledge of and control over such false advertising, a macro-level factor (Esterl, 2013). In addition, and in contrast to interactionists' claims, most people can't shape or change their situations. Instead, formal organizations (the U.S. government and large companies) often invade the privacy of citizens, workers, or consumers by collecting information and monitoring people's behavior.

Fully 93 percent of Americans say that controlling who gets access to their private information is important to them, but half believe that they have little or no control over their data. In reality, few of us have control over our personal data because we trade privacy for the convenience of shopping and other online activities (Madden and Rainie, 2015; Scheer, 2015).

We've looked at social groups and formal organizations, both of which are part of larger structures that sociologists call *social institutions*. The final section introduces you to this important concept and lays the foundation for analyses of social institutions in later chapters.

6-4 SOCIAL INSTITUTIONS

A **social institution** (or simply, *institution*) is an organized and established social system that meets one or more of a society's basic needs. According to functionalists, there are five major social institutions, worldwide, that ensure a society's survival:

▶ The *family* replaces a society's members through procreation, socializes children, and legitimizes sexual activity between adults.

▶ The *economy* determines how a society produces, distributes, and consumes goods and services.

▶ The *government* (which includes police, military, and courts) maintains law and order, creates and

---

social institution an organized and established social system that meets one or more of a society's basic needs.

Five major social institutions

- family
- economy
- government
- education
- religion

---

enforces laws, regulates elections, and protects people's civil liberties.

▶ *Education* helps to socialize children, transmits values and knowledge, and teaches work-related skills.

▶ *Religion* involves shared beliefs, values, and practices related to the supernatural.

Besides these core institutions (which we'll examine in Chapters 7 and 11–14), people develop new institutions in response to cultural changes. In the past, for example, the family and religion addressed most health problems. Only in relatively recent times have *medicine* and *health care systems* emerged as distinct institutions that diagnose, treat, and try to prevent illness and disease. Because of technological and economic changes, the *mass media* is another relatively recent institution. It disseminates information to a large audience, creates desires for goods and services, influences our attitudes toward social issues, and reinforces a society's values and beliefs.

6-4a Why Social Institutions Are Important

People create social institutions to address basic needs or problems. Institutions have an organized purpose, weave together norms and values, and, consequently, guide behavior. They may differ across countries, but each institution carries out certain tasks in a particular society and contributes to its overall functioning and stability. No two

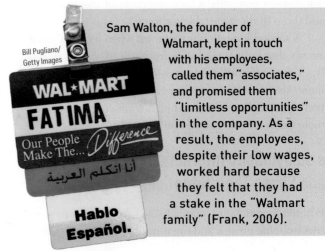

Bill Pugliano/Getty Images

Sam Walton, the founder of Walmart, kept in touch with his employees, called them "associates," and promised them "limitless opportunities" in the company. As a result, the employees, despite their low wages, worked hard because they felt that they had a stake in the "Walmart family" (Frank, 2006).

On average, a 12-ounce can of sugared soda contains at least 9 teaspoons of sugar. Over the course of a year, drinking one soda a day can make you 16 pounds fatter (Loumarr, 2015). Which social institutions affect your beverage choices?

families are exactly alike, but the family institution reflects broadly shared cultural agreements about what a family should be and do. In the same vein, no two grocery stores are exactly alike, but the economy establishes and maintains a variety of rules that make food shopping predictable.

Institutions aren't just "somewhere out there." We create, live inside, change, resist, and reproduce them, and participate simultaneously in multiple institutions. Understanding institutions can tell us a lot about how a society functions and how we're connected to one another because institutions are highly interdependent.

6-4b How Social Institutions Are Interconnected

Social institutions are linked to one another. Consider, for example, the linkages among seven institutions in explaining U.S. obesity, a major indicator of health. *Medicine* tells us that obesity is the second top cause of preventable disease, disability, and death. (The top cause is smoking.) High obesity rates have resulted in negative health outcomes (e.g., heart disease, strokes, diabetes, several types of cancer) and steadily rising *health care* costs (see Chapter 14).

The *mass media* (such as television, magazines, and online sites) routinely broadcast such results and publish medical advice.

In *education*, particularly K–12, many schools now offer nutritious meals that are lower in salt and fat and higher in whole grains. However, school vending machines still sell high-fat snacks and sugary drinks. Children may not choose healthier food because, by the time they enter school, their *families* have already shaped their eating habits and preferences (see Chapter 4).

The *government* has had mixed effects on weight issues. First Lady Michelle Obama has convened medical professionals, business and government leaders, grassroots activists, and nutrition advocates to combat childhood obesity. Also, some towns have refused permits to build more fast food restaurants. On the other hand, Congress hasn't passed laws to rid school vending machines of sugary snacks and drinks. A major reason is that almost all politicians rely on the food industry to help finance their elections and to hire them as lobbyists after they leave public office (see Chapter 11).

The *economy* has other far-reaching effects on obesity through its linkages to other institutions. For example, food corporations like PepsiCo, Kellogg, and Kraft regularly attend annual meetings where they make presentations to dietitians and other nutrition professionals, give seminars and parties, and provide free samples of their products. The mass media relies on ads from food companies even though the businesses create and deliver unhealthy products to customers. And, because of advertising and packaging claims, many parents think that fruit and sports drinks and flavored waters—all of which have high amounts of sugar—are healthy choices for children (Munsell et al., 2015; Strom, 2015).

You've seen that social groups, formal organizations, and social institutions shape our behavior. Nonetheless, and despite numerous constraints, because people create all these structures, they can also change them.

STUDY TOOLS 6

READY TO STUDY? IN THE BOOK, YOU CAN:

☐ Check your understanding of what you've read with the Test Your Learning Questions provided on the chapter review card at the back of the book.

☐ Rip out the chapter review card for a handy summary of the chapter and key terms.

ONLINE AT CENGAGEBRAIN.COM YOU CAN:

☐ Collect StudyBits while you read and study the chapter.

☐ Quiz yourself on key concepts.

☐ Prepare for tests with chapter flash cards as well as flash cards that you create.

☐ Watch videos to explore sociology further.

7 | Deviance, Crime, and Social Control

LEARNING OUTCOMES

After studying this chapter, you will be able to...

7-1 Differentiate between deviance and crime, and describe the key characteristics of deviance.

7-2 After evaluating the two major crime measures, identify and illustrate the different types of crime.

7-3 Describe, illustrate, and evaluate functionalist perspectives on deviance.

7-4 Describe, illustrate, and evaluate conflict perspectives on deviance.

7-5 Describe, illustrate, and evaluate feminist perspectives on deviance.

7-6 Describe, illustrate, and evaluate symbolic interaction perspectives on deviance.

7-7 Identify and evaluate the criminal justice system's social control methods.

After you finish this chapter go to **PAGE 137** for **STUDY TOOLS**.

Recently, a group of hackers broke into AshleyMadison.com, a website that arranges extramarital affairs, and released more than 35 million of the subscribers' names, addresses, and personal information. The online profiles included thousands of American and European military and government personnel and high-level executives. Besides cheating on their spouses, these subscribers had also violated rules that forbade using work computers for personal reasons. The hackers wanted to expose deceitful spouses, and many people believed that the site users got what they deserved, but Ashley Madison's CEO denounced the hackers as "criminals" (Mason and Sharp, 2015; Weise and Vanden Brook, 2015).

Hacking is usually illegal, but why is an infidelity site, marital dishonesty, or misusing workplace computers considered deviant? And what, exactly, is deviance? Before reading further, take the quiz on the next page to see how much you know about U.S. deviance and crime.

7-1 WHAT IS DEVIANCE?

Have you ever driven above the speed limit? Cheated on an exam? Engaged in underage drinking? All are examples of **deviance**—violations of social norms. Unlike the

What do you think?

There would be less crime if the punishments were more severe.

| 1 | 2 | 3 | 4 | 5 | 6 | 7 |

strongly agree   strongly disagree

Blair Seitz/Alamy

general public, sociologists use the concept *deviance* nonjudgmentally: They're interested in understanding and explaining why all of us violate norms from time to time, and why people judge some acts (hacking) more negatively than others (subscribing to an infidelity site).

Deviance becomes **crime**, a violation of society's formal laws, when it breaks rules that have been written into law and enforced by a political authority. For example, trafficking children under age 18 for prostitution, pornography, or sex is deviant. It's not a crime, however, until a country enacts laws that punish offenders.

7-1a Some Key Characteristics of Deviance

Deviance is universal because it exists in every society. Still, deviance can vary quite a bit over time, from situation to situation, from group to group, and from culture to culture (Sumner, 1906; Schur, 1968):

▶ **Deviance can be a trait, a belief, or a behavior.** People usually do something to be considered deviant. We can also be treated as outsiders simply

---

---

deviance a violation of social norms.

crime a violation of society's formal laws.

because of our appearance, skin color, religious beliefs, or sexual orientation (Becker, 1963; see also Chapters 9, 10, and 13).

▸ **Deviance is accompanied by social stigmas.** A **stigma** is a negative label that devalues a person and changes her or his self-concept and social identity. Stigmatized individuals may react in many ways: They may alter their appearance (as through cosmetic surgery), associate with others like themselves who accept them (as in gangs), hide information about some aspect of their deviance (an ex-convict who doesn't reveal that status), or divert attention from a stigma by excelling in some area (as music or sports) (Goffman, 1963).

▸ **Deviance varies across and within societies.** What's accepted or tolerated in one society may be deviant in another. The United States comprises only 5 percent of the world's population, but accounts for 31 percent of global public *mass shootings* (incidents that result in killing four or more people outside homes and that aren't gang-related, hostage situations, or robberies). Between 1966 and 2012, among 171 countries, and those with the most such shootings, there were 90 in the United States compared with 18 in the Philippines and 15 in Russia. A high firearm ownership rate is a major reason for the high prevalence of U.S. mass shootings (Lankford, 2015).

Within a country, public killings, including mass shootings, vary by location. Between 2000 and 2013, the United States had 160 shooting incidents:

More than 800 women in Ghana live in exile in witch camps. Most are poor widows or older women who are blamed for outbreaks of disease in their villages and relatives' or neighbors' illness or death (Küntzle and Blondé, 2013). Murders of women and children accused of sorcery have risen since 2009. Saudi Arabia, Nepal, India, Papua New Guinea, Tanzania, and Uganda still kill people suspected of witchcraft (Hoffman, 2014).

Nyani Quarmyne/Aurora Photos (RM)/Corbis

stigma a negative label that devalues a person and changes her or his self-concept and social identity.

46 percent occurred in or near businesses, and 24 percent took place in schools (including colleges), compared with 4 percent in religious buildings (Blair and Schweit, 2014).

▸ **Deviance varies across situations.** What's acceptable in one context may be stigmatized in another. Among U.S. adults, between 74 and 77 percent say it's generally okay to use a cellphone on public transportation and while walking down the street or waiting in line. In contrast, only 4 to 5 percent approve of using cellphones at meetings, movie theaters, and religious services (Rainie and Zickuhr, 2015).

▸ **Deviance is formal or informal.** *Formal deviance* is behavior that violates laws. A major example is crime, a topic we'll examine shortly. In contrast, *informal deviance* is behavior that disregards accepted social norms, such as picking one's nose or teeth or scratching one's private parts in public, belching loudly, and not dressing appropriately (e.g., wearing jeans to a wedding reception or job interview).

▸ **Perceptions of deviance can change over time.** Many behaviors that were acceptable in the past are now seen as deviant. Only during the 1980s and 1990s did U.S. laws define date rape, marital rape, stalking, and child abuse as crimes. Smoking—widely accepted in the past—has been prohibited in many public places. In 2011, for the first time, a majority of Americans (59 percent) supported a smoking ban in *all* public places, a considerable increase from 31 percent in 2003 (Dugan, 2013). Also, increasing numbers of employers don't hire people who fail tests for nicotine usage.

On the other hand, most Americans now shrug off behaviors that were stigmatized in the past. Cohabitation, seen as sinful and immoral 40 years ago, is now widespread and considered normal. And Americans' support for legalizing marijuana for recreational use increased from only 12 percent in 1969 to 53 percent in 2015 (Doherty and Tyson, 2015).

7-1b Who Decides What's Deviant?

Because deviance is culturally relative and standards change over time, who decides what's right or wrong? An important group is those who have authority or power. During our early years, parents and teachers specify acceptable and

Are any of these people deviant? Why or why not?

Alfred Wekelo/Shutterstock.com; MATTHIAS BEIN/Corbis Wire/Corbis; © Dmitry Naumov/Shutterstock.com

unacceptable behavior (see Chapter 4). As we get older, laws also define what's deviant (e.g., being forbidden to drive until age 16 or to purchase or consume alcohol until age 21).

Many employers—especially those in businesses and federal agencies—don't hire people with visible "body art" ("Tattoos in the Workplace . . .," 2014). Lawmakers pass numerous laws that people must obey, but often exempt themselves. Examples include having immunity from employment discrimination and sexual harassment lawsuits, and passing laws that allow guns in schools, restaurants, public buildings, and churches, but not in state legislative chambers (Skoning, 2013; Deprez and Selway, 2014).

Public attitudes and behavior also affect definitions of deviance. Many people, including health practitioners, now use the word "overweight" instead of "fat" or "obese" to decrease the stigma or being fat. On the other hand, there has been considerable *normalization of deviance*, the gradual process through which unacceptable practices or standards become acceptable. Examples include the growing acceptance of same-sex marriage, legalized gambling, and recreational marijuana (see Chapter 3). Normalizing some types of deviance generates considerable income, particularly at the state level.

7-2   TYPES OF DEVIANCE AND CRIME

Noncriminal deviance—such as suicide, alcoholism, lying, mental illness, and adult pornography—is common. Sociologists are especially interested in *criminal deviance*, behavior that violates laws, because most of the transgressions are visible, publicized by the media, and threaten people's lives or property. Measuring criminal deviance may seem straightforward, but isn't as simple as it appears.

7-2a Measuring Crime

Two of the most important sources of crime statistics are the FBI's *Uniform Crime Report* (UCR) and the U.S. Department of Justice's *National Crime Victimization Survey* (NCVS). Each has its strengths and weaknesses.

The UCR data are crimes reported to the police and arrests made each year. The statistics are useful in examining trends over time, but the UCR doesn't include offenses such as corporate crime, kidnapping, and Internet crimes; simple assaults (those not involving weapons or serious injury); nor the 56 percent of all violent crimes (e.g., rape, robbery, and aggravated assault) and 66 percent of property crimes (e.g., burglary and motor vehicle theft) that aren't reported to the police (Truman et al., 2013). Because of such limitations, many sociologists also use victimization surveys to measure the extent of crime.

A **victimization survey** interviews people about being crime victims. Because the response rates are at least 90 percent, the NCVS offers a more accurate picture of many offenses than does the UCR. The NCVS also includes both reported and unreported crime, and isn't affected by police discretion in deciding whether to arrest an offender.

Still, many victims don't report a crime to either the police or interviewers for a variety of reasons: They fear reprisal for getting the offender in trouble, believe that the police can't or won't do anything to help, or the crime, particularly rape, is too painful to discuss. Some people don't want to admit having been victims—especially when the perpetrator is a family member or friend. Others may be too embarrassed to tell an interviewer that they were victimized while drunk or engaged in drug sales (Hagan, 2008; Truman and Langton, 2015).

No one knows the extent of U.S. crime because even the best sources (the UCR and NCVS) provide only estimates, and data for higher-income lawbreakers are more elusive than for lower-income offenders. Let's begin with the latter, often called *street crimes*, offenses committed by ordinary people, usually in public places.

victimization survey interviews people about being crime victims.

One of the biggest costs for retail stores is *wardrobing*, returning expensive clothes after wearing them. Such "borrowing" has also become prevalent in fine jewelry, high-priced cameras for weddings, men's clothes (such as business suits for special occasions), and even household tools. Wardrobing has increased 40 percent since 2009, and cost retailers almost $9 billion in 2013 (National Retail Federation, 2013; Timberlake, 2013). Who do you think pays for wardrobing fraud?

ESLINE/Shutterstock.com

Becky Stares/Shutterstock.com

7-2b Street Crimes

A majority of Americans (55 percent) say crime is an "extremely" or "very serious" problem in the United States (Dugan, 2013). They're usually referring to *violent crimes* (e.g., murder, rape, robbery, aggravated assault) that use force or the threat of force against others. *Property crimes* (e.g., burglary, motor-vehicle theft, arson) involve the destruction or theft of property, but offenders don't use force or the threat of force.

INCIDENCE OF VIOLENT AND PROPERTY CRIMES

Of the almost 10 million crimes reported to the police in 2013, 88 percent were property crimes. There has been a much higher rate of property to violent crimes since at least 1993 (see *Figure 7.1*). The media are most likely to cover violent crimes because they inspire the greatest fear, but Americans are much more likely to be victimized by theft or burglary than to be murdered, raped, robbed, or assaulted with a deadly weapon. In fact, the average person is 433 times more likely to experience a theft than to be murdered (Glassner, 2010; Federal Bureau of Investigation, 2014).

Despite occasional crime spikes in some large U.S. cities, between 2004 and 2013 violent crimes decreased by 15 percent and property crimes by 16 percent (Federal Bureau of Investigation, 2014). The odds of being murdered or robbed are now less than half of what they were in the early 1990s, when violent crimes peaked (Oppel, 2011).

OFFENDERS AND VICTIMS

Most offenders are never caught, but street crimes aren't random. Of those arrested in 2013, for example:

▶ 10 percent were age 17 and younger, 44 percent were ages 18 to 29, and 13 percent were ages 30 to 34. In contrast, only 5 percent were 55 or older, and many of the crimes involved drug abuse violations and drunk driving rather than violent and property crimes.

▶ 74 percent were males. They accounted for 80 percent of people arrested for violent crime and 62 percent of those arrested for property crime.

▶ 69 percent were white, 28 percent were black, and the remaining 3 percent were American Indian or Asian American. (According to the agencies that provided ethnicity data, 17 percent of those arrested were Latino.) Proportionately, African Americans make up about 13 percent of the general population but account for 28 percent of street crimes (Federal Bureau of Investigation, 2014).

Men are more likely than women to be crime victims, but the type of victimization differs. Males experience higher rates of homicide, robbery, and aggravated assault, whereas females are more likely to be victims of rape, sexual assault, and intimate partner violence.

Violent victimization rates are lower for Latinos (16 per 1,000 persons age 12 and older) and whites (20) than for all other groups (23). The rates are also much higher for people ages 12 to 17 and 18 to 24 (30 and 27, respectively) than those age 65 and older (3) (Truman and Langton, 2015). Crimes against older people are especially traumatic, however, because the victims are usually physically unable to fight back, and it takes them longer to recover from a crime both physically and emotionally (Madigan, 2009; Morgan and Mason, 2014).

Crime rates are higher in low-income areas, but does this mean that poor people are more deviant? Because police devote more resources to poor neighborhoods, those at the lower end of the socioeconomic ladder are more likely to be caught, arrested, prosecuted,

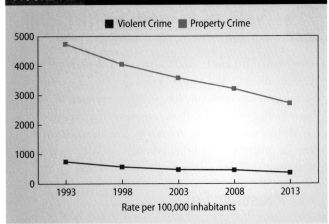

FIGURE 7.1 U.S. CRIME RATES HAVE DECREASED

■ Violent Crime ■ Property Crime

Rate per 100,000 inhabitants

Source: Based on Federal Bureau of Investigation, 2014, Table 1.

Men are much more likely than women to be victims of violent crime.

and incarcerated. *Intellectual property theft*—which includes offenses such as software piracy, bootlegging musical recordings and movies, selling company trade secrets, and copyright violations—is a primarily middle-class crime, but the offenders are difficult to catch.

7-2c Hate Crimes

A **hate crime** is a criminal act motivated by hostility toward a particular race, religion, ethnicity, sexual orientation, gender, or disability. The number of hate crimes decreased from nearly 7,500 in 2002 to 5,928 in 2013, but about 65 percent aren't reported to the police. In 2013, 65 percent of hate crimes were against people and 35 percent against property. Among the former, 74 percent were directed at individuals; the remainder targeted businesses or financial institutions, the government, and religious organizations. Almost half of the incidents were racially motivated (see *Table 7.1*). A majority (52 percent) of the offenders were white, 24 percent were black, and 15 percent were of unknown races (Federal Bureau of Investigation, 2014).

7-2d White-Collar Crime

White-collar crime refers to illegal activities committed by high-status people in the course of their occupations (Sutherland, 1949; Cressey, 1953). The offenses don't involve violence because white-collar criminals use their

status and powerful positions to enrich themselves or others. Typical white-collar crimes include embezzlement (stealing an employer's or client's money); identity theft; bribery; insider trading (illegal stock manipulation); and mass marketing, mortgage, and investment fraud (Federal Bureau of Investigation, 2014).

Comstock/Stockbyte/Getty Images

White-collar crimes cause significant public harm. The financial cost of white-collar crime is unknown, but estimated between $300 and $660 billion a year (Huff et al., 2010). In 2012, Americans who experienced identity theft lost about $25 billion—almost twice as much as the $14 billion for all property crimes (Harrell and Langton, 2013).

Citing the high fraud level in Utah, the state's legislators recently passed a bill to build the nation's first white-collar offender registry. The website contains offenders' names, aliases, a photograph, physical description, and crimes for which they were convicted (Harvey, 2015).

White-collar crimes are lucrative because most offenders aren't caught. According to one estimate, there's only one FBI investigator per almost 4,000 financial services sales agents, personal financial advisers, and financial analysts. Thus, the odds of being arrested are very small (Rohrlich, 2010; see fbi.gov/wanted/wcc for annual updates of the FBI's most wanted white-collar criminals).

Even when white-collar criminals are arrested and convicted, few are incarcerated. Of the almost 3,800 suspected embezzlers, for example, only 525 (14 percent) have been arrested and sentenced; of those arrested and sentenced, only half (7 percent) served any prison time (Motivans, 2013). Because so few of the most affluent white-collar offenders are punished, they enjoy lavish lifestyles that include yachts, private jets, and $2 million birthday parties (Lublin, 2013; Spitznagel, 2013).

7-2e Corporate Crimes

Corporate crimes (also known as *organizational crimes*) are illegal acts committed by executives to benefit

TABLE 7.1 HATE CRIMES: TYPES AND TARGETS, 2013		
Source of Hostility	Percentage of All Cases	Largest Target Group
Race	49	Blacks (66%)
Sexual Orientation	20	Gay Men (61%)
Religion	17	Jews (60%)
Ethnicity/National Origin	12	Latinos (53%)
Other*	2	
Total	100	

*Includes mental and physical disabilities, gender, and gender identity.

Source: Based on Federal Bureau of Investigation, 2014, Table 1 of hate crime statistics.

hate crime a criminal act motivated by hostility toward a particular race, religion, ethnicity, sexual orientation, gender, or disability.

white-collar crime illegal activities committed by high-status people in the course of their occupations.

corporate crimes (also called *organizational crimes*) illegal acts committed by executives to benefit themselves and their companies.

themselves and their companies. Corporate crimes include a vast array of illicit activities such as conspiracies to stifle free market competition, price-fixing, tax evasion, and false advertising. The target of the crime can be the general public, the environment, or even a company's own workers (see Chapters 11 and 15). Often, corporate offenders commit multiple crimes (fraud, insider trading, and perjury).

Corporate crimes are fairly common. For example, General Motors (G.M.) knew about a malfunctioning ignition switch as far back as 2001. It started recalling cars only in 2014, however, after the ignition switch was linked to at least 16 deaths, a $35 million federal fine, and dozens of wrongful death and injury lawsuits. As this book goes to press, G.M. has approved payments for 100 deaths and 184 injuries, and is reviewing claims for 37 deaths and 589 injuries (Hirsch, 2014; Vlasic, 2015).

Corporate crimes are profitable. Fraudulent billings to health care programs are estimated between 3 and 10 percent of the nation's total health care expenditures of $2.4 trillion a year (Federal Bureau of Investigation, 2012; see also "Health-Care Fraud...," 2014). The collapse of Enron, a major U.S. energy company, put more than 5,000 people out of work, wiped out more than $2 billion in employee pensions, and investors lost $60 billion in stock. Jeffrey Skilling, Enron's CEO, was convicted on 19 counts of conspiracy, securities fraud, insider trading, and other illegal business deals. His prison sentence was reduced from 24 to 14 years, and could be cut to 12 years for good behavior (McCoy, 2013; Fowler, 2013).

7-2f Cybercrime

Cybercrime (also called *computer crime*) refers to illegal activities that are conducted online. These high-tech crimes include defrauding consumers with bogus financial investments, embezzling, sabotaging computer systems, hacking, and stealing confidential business and personal information. During just the first six months of 2015, U.S. companies and federal agencies experienced data breaches that involved almost 118 million people. Some of the major cyberattacks have targeted the government (e.g., Internal Revenue Service), the healthcare industry (e.g., CareFirst BlueCross BlueShield), and businesses (e.g., Target, Sony, Home Depot, Staples). The most technologically sophisticated hackers are in China, Russia, and Eastern Europe, but cybercriminals also include people living in Brazil, Nigeria, Vietnam, and the United States (Rayman, 2014;

In 2009, Bernard Madoff—a highly respected stockbroker and investment advisor—pleaded guilty to defrauding thousands of investors of almost $18 billion between 1991 and 2008. Madoff was sentenced to 150 years in a federal prison. So far, it's cost almost $1 billion to recover 59 percent of the lost money (Bloomberg News, 2014).

Granville, 2015; Urrico, 2015). The theft of personal information from large companies and government agencies has become so routine, according to one observer, "that most such breaches are quickly forgotten" and are "merely an occasional inconvenience" ("The Ashley Madison Hack," 2015: 14).

In 2014 alone, the Internet Crime Complaint Center received almost 300,000 complaints that resulted in nearly $801 million in financial losses. The most common offenses were selling cars online for a full or partial payment and not delivering the vehicle; impersonating the FBI and other government agencies to get information about addresses, Social Security numbers, and other personal and financial data; and romance scams that manipulated victims who sent money or expensive gifts to their new "lovers" (Federal Bureau of Investigation, 2015).

Cybercrime and cyber espionage cost the United States about $100 billion a year. Preventing or dealing with cybercrime, particularly the theft of intellectual property, slows research and development (which results in less hiring), increases a business's or agency's operating costs, and results in considerable revenue loss (Center for Strategic and International Studies, 2014).

7-2g Organized Crime

Organized crime refers to activities of individuals and groups that supply illegal goods and services for profit. Organized crime includes drug distribution, loan-sharking (lending money at illegal rates), illegal gambling, pornography, theft rings, hijacking cargo, and laundering money obtained from illegal activities (e.g., prostitution) by

cybercrime (also called *computer crime*) illegal activities that are conducted online.

organized crime activities of individuals and groups that supply illegal goods and services for profit.

depositing the money in legitimate enterprises (e.g., banks) (Federal Bureau of Investigation, 2014).

One of the most profitable activities, *organized retail crime*, involves stealing billions of dollars in merchandise that is then sold online, a practice known as "e-fencing." Organized retail crime results in an estimated $15 to $30 billion in annual financial loss (Larence, 2011).

7-2h Victimless Crimes

Offenses that are the least likely to be reported—such as illicit drug use, prostitution, drunkenness, and illegal gambling—are called *public order crimes*, or *victimless crimes*. **Victimless crimes** are acts that violate laws but those involved don't consider themselves victims. For example, prostitutes argue that they're simply providing services to people who want them, and substance abusers claim that they aren't hurting anyone but themselves.

Some contend that this term is misleading because victimless crimes often lead to property and violent crimes, as when addicts commit burglary, rob people at

Drug abuse involves consensual or solitary acts that don't directly violate another individual's rights. Why, then, do many people argue that drug abuse is not a victimless crime?

VladimirFLoyd/Getty Images

gunpoint, and engage in identity theft to get money for drugs. In a national survey, 70 percent of the respondents said that a family member's drug abuse had a negative effect on the emotional or mental health of at least one other family member, and 39 percent experienced financial problems because they went into debt to pay for an uninsured addict's treatment (Saad, 2006). In 2014, 36 percent of Americans—up from 15 percent in 1947—said that alcohol was a major cause of family problems. Drug use is also often a reason for divorce and parenting problems (White and Witkus, 2013; Jones, 2014).

The next sections examine four important sociological perspectives that help us understand why people are deviant. *Table 7.2* summarizes these theories.

7-3 FUNCTIONALIST PERSPECTIVES ON DEVIANCE

For functionalists, deviance is a normal part of society. Functionalists don't endorse undesirable behavior, but view deviance as both functional and dysfunctional.

7-3a Dysfunctions and Functions

Dysfunctions are the negative consequences of behavior, but what's dysfunctional for one group or individual may be functional for another. For example, gangs are functional because they provide members with a sense of belonging, identity, and protection. Gangs are also dysfunctional because they commit violent and property crimes (Egley and Howell, 2013).

victimless crimes acts that violate laws but those involved don't consider themselves victims.

TABLE 7.2 SOCIOLOGICAL EXPLANATIONS OF DEVIANCE

Theoretical Perspective	Level of Analysis	Key Points
Functionalist	Macro	• Deviance is both functional and dysfunctional. • Anomie increases the likelihood of deviance. • People are deviant when they experience blocked opportunities to achieve the culturally approved goal of economic success.
Conflict	Macro	• There's a strong association between capitalism, social inequality, power, and deviance. • The most powerful groups define what's deviant. • Laws rarely punish the illegal activities of the powerful.
Feminist	Macro and micro	• There's a large gender gap in deviant behavior. • Women's deviance reflects their general oppression due to social, economic, and political inequality. • Many women are offenders or victims because of patriarchal beliefs and practices and living in a rape culture.
Symbolic Interactionist	Micro	• Deviance is socially constructed. • People learn deviant behavior from significant others such as parents and friends. • If people are labeled or stigmatized as deviant, they're likely to develop negative self-concepts and engage in criminal behavior.

Deviance is *dysfunctional* because it:

▶ **Creates tension and insecurity.** Any violation of norms—a babysitter who cancels at the last minute or the theft of your smartphone or tablet—makes life unpredictable and increases anxiety.

▶ **Erodes trust in personal and formal relationships.** Crimes such as date rape and stalking make many women suspicious of men, and victims of identity theft have had problems obtaining banking services or credit cards because financial organizations don't trust them (see Harrell and Langton, 2013).

▶ **Decreases confidence in institutions.** In the 2008 stock market crash, taxpayers had to pay for the financial industry's corporate fraud and mismanagement. Since then, millions of people, even those who didn't lose money, worry that their retirement funds may disappear in the future (see Chapter 12).

▶ **Is costly.** Besides personal costs to victims (e.g., fear, emotional trauma, physical injury), deviance is expensive. All of us pay higher prices for consumer goods and services (e.g., auto and property insurance), as well as higher taxes for prosecuting criminals and for building and maintaining prisons. And as you saw earlier, white-collar, corporate, cyber, and organized crime cost the U.S. economy hundreds of billions of dollars every year.

Deviance can also be *functional* because it provides a number of societal benefits (Durkheim, 1893/1964; Erikson, 1966; Sagarin, 1975). For example, deviance

▶ **Affirms cultural norms and values.** Negative reactions, such as expelling a college student who's caught cheating or incarcerating a robber, reinforce a society's norms and values about being honest and law-abiding.

▶ **Provides temporary safety valves.** Some deviance is accepted under certain conditions, enabling people to "blow off steam." Typically, a community and police will tolerate noise, underage drinking, loud music, and obnoxious behavior during college students' spring break because it's a short-lived nuisance.

▶ **Increases social solidarity.** Some deviant behavior unites a group, community, or society, especially when there's a common enemy. After the U.S. Navy Seals killed Osama bin Laden in 2011, most

Americans—regardless of age, political views, social class, or race and ethnicity—applauded his death.

▶ **Bolsters the economy.** Deviance can benefit a community financially. Some towns welcome new prisons because they generate jobs and stimulate the economy (Semuels, 2010). Unlike a manufacturing plant that might close down or move to another country, prisons are rarely dismantled.

▶ **Triggers social change.** An estimated 450,000 people were killed or injured in 2009 because of "distracted driving" collisions. Consequently, by early 2014, 41 states and the District of Columbia banned text messaging for all drivers (Hanes, 2010; Insurance Institute for Highway Safety, 2010; Governors Highway Safety Association, 2014).

7-3b Anomie and Social Strain Theory

Functionalists offer many explanations of deviance, but two of the most influential are anomie and strain theories. Both analyze why so many people engage in deviant behavior even though they share many of the same goals and values as those who conform to social norms.

DURKHEIM'S CONCEPT OF ANOMIE

Émile Durkheim (1893/1964, 1897/1951) introduced the term **anomie** to describe the condition in which people are unsure of how to behave because of absent, conflicting, or confusing social norms. During periods of rapid social change, such as industrialization in Durkheim's time, societal rules may break down. As many young people moved to the city to look for jobs in the nineteenth century, norms about proper behavior that existed in the countryside crumbled. Even today, as you'll see in Chapter 15, many urban newcomers experience anomie and miss the neighborliness that was common at home.

MERTON'S CONCEPT OF SOCIAL STRAIN

Robert Merton (1938) elaborated on the concept of anomie to explain how social structure helps to create deviance. According to Merton, Americans are socialized to believe that anyone can realize the American dream of accumulating wealth and being successful economically. To achieve the *cultural goal* of economic success, society emphasizes legitimate and *institutionalized means* such as education, hard work, saving, starting at the bottom and working one's way up, and making sacrifices instead of seeking instant gratification.

anomie the condition in which people are unsure how to behave because of absent, conflicting, or confusing social norms.

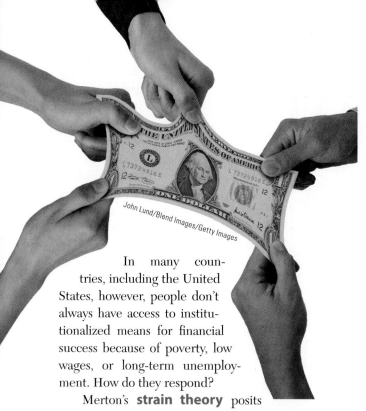

John Lund/Blend Images/Getty Images

In many countries, including the United States, however, people don't always have access to institutionalized means for financial success because of poverty, low wages, or long-term unemployment. How do they respond?

Merton's **strain theory** posits that people may engage in deviant behavior when they experience a conflict between goals and the means available to obtain the goals. Not all people turn to deviance to resolve social strain (see *Table 7.3*). Most of us *conform* by working harder and longer to become successful. The fact that you're reading this text shows that, using Merton's language, your mode of adaptation to strain is conformity—one of achieving success through an institutionalized means such as a college education.

Merton's other four modes of adaptation signify deviance. *Innovation* occurs when people endorse the cultural goal of economic success but turn to illegitimate means, especially crime, to achieve the goal. For many people living in inner-city ghettos, education, hard work, and deferred gratification are often unachievable. Thus, deviance is functional for those who turn to crime to make money.

In *ritualism*, people don't expect to be rich but get the necessary education and experience to obtain or keep a job. In many bureaucratic positions, employees don't make decisions and can't move up the ladder, but their jobs are relatively secure. Ritualists aren't criminals, but they're deviant because they've given up on becoming financially successful. Instead, they do what they're told and "go along to get along."

In *retreatism*, people reject both the goals and the means for success. Merton used examples such as vagrants, alcoholics, and drug addicts, some of whom give up because they believe it's impossible to succeed.

In *rebellion*, people feel so alienated that they want to change the social structure entirely by substituting new goals and means for the current ones. Contemporary examples are terrorists and U.S. militia groups that oppose the federal government.

7-3c Critical Evaluation

A major contribution of functionalism is showing how social structure, not just individual attitudes and behavior, produces or reinforces deviance and crime. For instance, computer-related crimes have increased for a number of structural reasons such as Americans' greater use of the Internet and offenders' overcoming security technology.

Functionalist theories also have weaknesses: (1) The concepts of *anomie* and *strain theory* are limited because they overlook the fact that not everyone in the United States embraces financial success as a major goal in life; (2) the theories don't explain why women's crime rates are much lower than men's, especially since women generally have fewer legitimate opportunities for financial success than men; (3) crime rates have declined since 2000 despite chronic poverty and unemployment; and (4) functionalism

strain theory posits that people may engage in deviant behavior when they experience a conflict between goals and the means available to obtain the goals.

TABLE 7.3 MERTON'S STRAIN THEORY OF DEVIANCE

Mode of Adaptation	Accept Cultural Goals?	Accept Institutionalized Means to Achieve Goals?	Examples
Conformity	Yes	Yes	College graduate
Innovation	Yes	No	Thief
Ritualism	No	Yes	Low-level bureaucrat
Retreatism	No	No	Alcoholic
Rebellion	No—seeks to replace goals	No—seeks to replace means for achieving goals	Antigovernment militia member

Source: Based on Merton, 1938.

doesn't explain why people commit some crimes (setting fires just for kicks or murdering an intimate partner) that have nothing to do with being successful (Anderson and Dyson, 2002; Williams and McShane, 2004).

The most consistent criticism is that functionalism typically focuses on lower-class deviance and crime. Conflict theorists have filled this gap by examining middle- and upper-class crime.

A ritualist? Retreatist? Rebel?

7-4 CONFLICT PERSPECTIVES ON DEVIANCE

Functionalists ask, "Why do some people commit crimes and others don't?" Most conflict theorists focus on capitalism, social inequality, power differences, who makes the laws, and ask, "Why are some acts defined as criminal while others aren't?" (Akers, 1997).

7-4a Capitalism and Social Inequality

For conflict theorists, there's a strong association between capitalism, social inequality, and deviance. That is, people and groups with economic and political power view deviance as any behavior that threatens their own interests (Vold, 1958; Turk, 1969, 1976; Kraska, 2004). As a result, newspapers and TV news stations (which are owned by the affluent) routinely focus on "street crimes" rather than "suite crimes."

The United States, like many other countries, is a capitalist economy based on the private ownership of property (see Chapter 11). Because capitalism is based on personal gain rather than collective well-being, it emphasizes the maximization of profits, an accumulation of wealth, and economic self-interest.

From a conflict perspective, capitalism is a major cause of crime for several interrelated reasons (Chambliss, 1969; Quinney, 1980; Snider, 1993). First, greed and self-interest perpetuate deviance. For example, a study of 36 countries found that 42 percent of board directors and senior managers—pressured to deliver growth and fearing personal pay cuts—"cooked the books" by understating costs, overstating revenue, encouraging customers to buy unnecessary stock, and bribing powerful people to win or retain business (Ernst & Young, 2013).

Second, capitalism is a competitive system that encourages corporate crime. When only a few firms

dominate an industry, they can easily eliminate competitors, conspire to fix prices, falsify financial statements, and commit fraud (see Eavis, 2015).

Third, because capitalism creates considerable social inequality, those at the top can control and exploit lower-level workers. In a recent study of the financial services/banking industry, almost half of the workers believed that illegal or unethical activity was "part and parcel of succeeding in this highly competitive field." Among the most senior employees, 28 percent of those earning $500,000 or more per year said that their company's confidentiality policies and procedures bar them from reporting wrongdoing (Tenbrunsel and Thomas, 2015). People who are the most vulnerable and powerless—unemployed and low-income earners—may turn to street crimes for survival.

7-4b Laws, Power, and Social Control

From a conflict perspective, criminal laws serve the interests of the capitalist ruling class. The most powerful groups in society control the law, which defines what's

Investigation Discovery, which debuted in 2008, is one of the country's fastest growing cable television networks, and airs in 157 countries and territories (Steel, 2015). The shows re-enact true crime incidents, primarily homicides, but also kidnappings, stalkings, sexual assaults, and intimate partner violence (e.g., *Scorned: Love Kills, Deadly Affairs, Fatal Vows, Wives with Knives*). Why do you think this network is so popular?

deviant and who will be punished (Chambliss and Seidman,1982; Reiman and Leighton, 2010).

Why do so many high-status people commit crimes? Because they can, according to conflict theorists. First, most white-collar and corporate crimes are *not criminalized* (Turk, 1969; Lilly et al., 1995). Generally, the wealthy and powerful make and enforce laws that protect their property. The laws against higher-status criminals are relatively lenient and seldom enforced, whereas laws against lower-status offenders, particularly those who commit property crimes, are harsher and enforced more often (Thio et al., 2012).

Second, the powerful have more *opportunities for deviance*. As you saw earlier, white-collar and corporate criminals have access to resources and the necessary occupational skills to defraud customers and employees. Compared with street criminals who rob banks, for example, bankers can steal huge amounts of money without leaving their desks.

Third, it's usually easy for powerful people to *hide their deviance*. Even though corporate fraud is a "widespread problem," it's very difficult to uncover and prosecute offenders in multi-layered corporate structures: The evidence may have been destroyed, and suspects deny knowledge of the criminal activities or maintain that the "mistakes" were unintentional (Federal Bureau of Investigation, 2012).

Fourth, there are usually *few penalties* for high-status crimes because lawmakers and defendants share a common cultural background. Of the senior Wall Street bank executives who played leading roles in the 2008 financial crisis, only one has been convicted of fraud. Even when international banks admit that they promote crimes like tax evasion, there are few penalties. Fraud by the rich wipes out 40 percent of the world's wealth, but no one goes to jail (Taibbi, 2014; "Financial Crime," 2015).

When U.S. prosecutors bring civil charges, not a single individual has to pay even a dime of her or his own money in damages, and the fines often represent less than half of 1 percent of annual revenue. In 2009, top officials at Goldman Sachs agreed to pay a $550 million fine to settle accusations of securities fraud that made almost $52 billion in profits. The fine represented four days of the bank's revenues (Summers, 2010; see also Prins, 2011). If, using a comparable percentage, you made $100,000 a year illegally and had to pay a $1,000 fine, would there be a financial incentive for honesty?

7-4c Critical Evaluation

Conflict theories explain the linkages between capitalism, laws, power, and social class that benefit those at the top. Critics, however, point to several weaknesses. First, some contend that conflict theory exaggerates the importance of capitalism in explaining white-collar and corporate crime. Other capitalist societies—Japan, Finland, Norway, Sweden, and Switzerland—have much lower business crime rates because of greater social solidarity and enforcement of laws that ban corruption and fraud (Leonardsen, 2004; Ernst & Young, 2013). In effect, some critics say, capitalism isn't a major reason for crime.

A second criticism is that conflict theory deemphasizes the crimes committed by low-income people. Street crimes are costly to society because of legal expenses, victim injuries, wage losses, incarceration, and crime prevention programs. A related criticism is that conflict theories tend to ignore the fact that some affluent people, including corporate executives, don't get away with their crimes (Federal Bureau of Investigation, 2012).

Third, some critics maintain that conflict theory ignores the ways that crime is functional for society. As you

Rich Defendants, Lenient Sentences In 2009, 36-year-old Ryan LeVin (pictured here), a super-rich heir to a family costume jewelry fortune, killed two men in Florida when he lost control of his Porsche while hurtling down a street at more than 100 miles an hour, then fled the scene. In 2011, he pleaded guilty to two counts of vehicular homicide and faced 30 years in prison. Because LeVin's attorney convinced the widows to accept a financial settlement (the amount is unknown), LeVin was sentenced to only 2 years of house arrest, which he served at one of his parents' two luxury seaside condos. LeVin never apologized to the victims' families. His major concern was getting back his Porsche (Shepherd, 2011; Mayo, 2012). In another recent case, a great grandson of the wealthy du Pont family was convicted of raping his three-year-old daughter. The felony requires a 10-year mandatory sentence, but a Superior Court judge gave him probation because the offender "will not fare well in prison" (Alman, 2014).

Sun Sentinel/Getty Images

saw earlier, deviance provides jobs and affirms law-abiding cultural norms and values. Finally, many contend, the most influential conflict theories focus almost entirely on men as both victims and offenders (Moyer, 2001; Belknap, 2007). Feminist theories have been filling this gap.

7-5 FEMINIST PERSPECTIVES ON DEVIANCE

Until recently, most sociological research on deviance focused almost entirely on males. Nearly all sociologists were men who saw women as not worthy of much analytical attention or assumed that explanations of male behavior were equally applicable to females (Flavin, 2001; Belknap, 2007). Feminist scholars' analyses of female offenders and victims have greatly enriched our understanding of deviance.

7-5a Women as Offenders

Crime statistics show a large gender gap. Crime (at least reported crime) is still generally a male activity. Men constitute about half of the U.S. population, but they make up 72 percent of those who are arrested. The gender gap is lowest for embezzlement, but prostitution is the only crime for which more women (67 percent) are arrested than men (see *Figure 7.2*).

Some believe that women's lower arrest rates reflect their socialization to take fewer risks and "to an ethic of care" not to hurt others (England, 2005). Others say that women have fewer opportunities to commit crimes. For example, only 9 percent of people involved in high-level corporate crimes are women: They're generally excluded from conspiracies because of their subordinate position in companies, profit less because they play minor roles, and are rarely ringleaders who initiate or orchestrate a fraud. When it comes to lower-level corporate crime, however—which is often

committed by individual bookkeepers, bank tellers, and accountants rather than groups—45 percent of the offenders are women (Association of Certified Fraud Examiners, 2012; Steffensmeier et al., 2013).

Men are much more likely than women to commit crimes, especially violent crimes, but girls' and women's arrest rates have increased for some offenses. Between 2004 and 2013, men's arrests dropped across all offenses, whereas female arrests rose 17 percent for both gambling and robbery, 20 percent for larceny-theft, and 12 percent for drunken driving (Federal Bureau of Investigation, 2014).

Why? Some posit that women are more likely to be employed in jobs that present opportunities for crimes such as fraud, property theft, and embezzlement. Because women on average are in low-income jobs, some resort to crime—especially shoplifting, petty theft, and prostitution—to survive financially or to support a family. Women's offenses are highest in cities, where their economic oppression and poverty are greatest (Heimer et al., 2006; Flower, 2010).

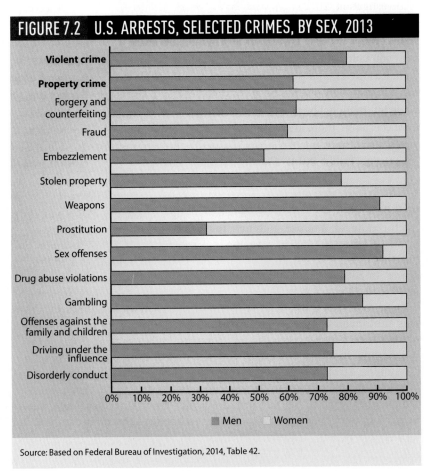

Siede Preis/Getty Images

FIGURE 7.2 U.S. ARRESTS, SELECTED CRIMES, BY SEX, 2013

Violent crime
Property crime
Forgery and counterfeiting
Fraud
Embezzlement
Stolen property
Weapons
Prostitution
Sex offenses
Drug abuse violations
Gambling
Offenses against the family and children
Driving under the influence
Disorderly conduct

0% 10% 20% 30% 40% 50% 60% 70% 80% 90% 100%

■ Men □ Women

Source: Based on Federal Bureau of Investigation, 2014, Table 42.

In 2013, adolescents under age 18 represented 10 percent of all arrests. Like adults, boys accounted for more arrests (71 percent) than girls, but girls' arrests have increased dramatically—from 17 percent in 1980 to 29 percent in 2013 (Federal Bureau of Investigation, 2014; Sickmund and Puzzanchera, 2014).

The higher arrest rates are due to a combination of factors: Policy changes, including more aggressive policing; parents' and school officials' greater likelihood of calling the police to deal with girls' delinquent behavior; and a substantial increase in juvenile females committing crimes that range from drug abuse to assaulting family members or intimate partners (Chesney-Lind and Jones, 2010; Zahn et al., 2010; Sickmund and Puzzanchera, 2014).

7-5b Women as Victims

Between 2002 and 2004, Ariel Castro, a Cleveland bus driver, kidnapped three women. The victims—ages 14, 16, and 20 at the time—were kept in Castro's house, chained in rooms and sometimes in the basement, and repeatedly sexually abused. One of the women gave birth to a child fathered by Castro. The women were rescued in 2013 when one of them broke through a door and called for help.

The Castro case is a recent example of the physical and sexual victimization of women and children. Many of the serious crimes (murder and robbery) are committed by men against men. Women and girls, however, are almost always the victims of sexual assault, rape, intimate partner violence, stalking, sexual exploitation, sexual harassment, female infanticide, and other crimes that degrade and terrorize women. Even in neighborhoods where the household income is $75,000 or higher, 36 percent of women, compared with only 13 percent of men, fear walking alone at night near their homes (Saad, 2010).

Feminist scholars offer several explanations for women's victimization. One is **patriarchy**, a hierarchical system in which cultural, political, and economic structures are controlled by men. In patriarchal societies, including the United States, because women have less access to power, they're at a disadvantage in creating and implementing laws that protect females (Price and Sokoloff, 2004; DeKeseredy, 2011). Such inequity diminishes women's control over their lives and increases their invisibility as victims.

A second reason for female victimization is the effect of culture on gender roles. Many girls and women have been socialized to be victims of male violence because of societal images of women as weaker, less intelligent, and less valued than men: "Girls are rewarded for passivity and feminine behavior, whereas boys are rewarded for

Holding money exchanged for sex, the child of a prostituted woman stands in the doorway of a brothel in Phnom Penh, Cambodia. Children, especially girls, who are raised in brothels are highly vulnerable to sexual exploitation. Many of the customers are Western men.

aggressiveness and masculine behavior" (Belknap, 2007: 243). In effect, then, both sexes internalize the belief that female victimization is normal. Because patriarchal societies rarely punish men who victimize women, or give them light sentences, many female victims feel trapped. For example, a Maryland judge sentenced a 20-year-old man to 3 years in prison for killing his girlfriend's puppy and 3 years in prison for beating the girlfriend over a 2-year period (Siegel, 2009).

A third reason for women's victimization is living in a **rape culture**, an environment in which sexual violence is prevalent, pervasive, and perpetuated by the media and popular culture (Parrott and Parrott, 2015). Here are a few examples of the pervasiveness of rape and sexual assaults in U.S. culture:

▶ During their lifetimes, 3 percent of men and 17 percent of women have been victims of an attempted or completed rape (Rape Crisis Center, 2014);

patriarchy a hierarchical system in which cultural, political, and economic structures are controlled by men.

rape culture an environment in which sexual violence is prevalent, pervasive, and perpetuated by the media and popular culture.

- 27 percent of college women have experienced an attempted or completed rape (Lauerman, 2013);

- 26,000 members of the military were sexually assaulted in 2012, 42 percent more than in 2010 (Department of Defense, 2013).

The actual number of violations is much higher. In 2014, for example, only 34 percent of victims reported rapes and sexual assaults to the police (Truman and Langton, 2015).

7-5c Critical Evaluation

Feminist sociologists have been at the forefront of studying female deviance and increasing awareness of crimes such as date rape, stalking, intimate partner violence, and the international sex trafficking of women and girls. Some critics contend, however, that because concepts such as patriarchy are difficult to measure, feminist research has yet to show, specifically, how patriarchy victimizes women. Others criticize feminist analyses for emphasizing male but not female violence, focusing on men's but not women's crimes, and not addressing the simultaneous effects of gender, social class, and race/ethnicity (Friedrichs, 2009; Dao, 2013).

7-6 SYMBOLIC INTERACTION PERSPECTIVES ON DEVIANCE

For symbolic interactionists, deviance is socially constructed because it's in the eye of the beholder. For instance, only 47 percent of French people believe that extramarital affairs are immoral, compared with 84 percent of Americans and 94 percent of Palestinians and Turks ("Extramarital Affairs," 2014). Interactionists offer many explanations of deviance, but two of the best known are differential association theory and labeling theory.

differential association theory asserts that people learn deviance through interaction, especially with significant others.

labeling theory posits that society's reaction to behavior is a major factor in defining oneself or others as deviant.

7-6a Differential Association Theory

Differential association theory asserts that people learn deviance through interaction, especially with significant others such as family members and friends. Dale "Rooster" Bogle's family in Oregon is an example of this theory. Even though Rooster, the father, served time in prison for theft, he taught his sons and daughters to survive by stealing: "By the time the boys were 10 years old they were breaking into liquor stores for their dad or stealing tractor-trailer trucks, hundreds of them. The girls turned to petty crimes to support their drug addictions" (Butterfield, 2002: 1).

People become deviant, according to sociologist Edwin Sutherland, if they have more contact with significant others who violate laws than with those who are law-abiding. In effect, people learn techniques for committing criminal behavior, along with the values, motives, rationalizations, and attitudes that reinforce such behavior (Sutherland and Cressey, 1970). Thus, in Rooster's clan, all of the children wound up in prison because they grew up in an environment that taught deviance rather than conformity.

Sutherland emphasized that differential association doesn't occur overnight. Instead, people are most likely to engage in crime if they're exposed to deviant values (1) early in life, (2) frequently, (3) over a long period of time, and (4) by significant others and reference groups (e.g., parents, siblings, close friends, and business associates).

Considerable research supports differential association theory. Almost 47 percent of state prisoners have a parent or other close relative who has also been incarcerated. Even before age 13, children who associate with peers who smoke, use drugs, or commit crimes are more likely to do so themselves because they learn these behaviors from their friends (Adamson, 2010; Roman et al., 2012; Huizinga et al., 2013). Thus, according to differential association theory, we're products of our socialization.

7-6b Labeling Theories

Have you ever been accused of something wrong that you didn't do? What about getting credit for something that you and others know you didn't deserve? In either case, did people start treating you differently? The reactions of others are the crux of **labeling theory**, which holds that society's reaction to behavior is a major factor in defining oneself or others as deviant. Some of the earliest sociological studies found that teenagers who misbehaved were tagged as delinquents. Such labeling changed the child's self-concept and resulted in more deviance and criminal behavior (Tannenbaum, 1938). During the 1950s and 1960s, two influential sociologists—Howard Becker and Edwin Lemert—extended labeling theory.

BECKER: DEVIANCE IS IN THE EYES OF THE BEHOLDER

According to Howard Becker (1963: 9), "The deviant is one to whom that label has successfully been applied; *deviant behavior is behavior that people so label*" (emphasis in original). That is, it's not an act that determines deviance, but whether and how others react.

Some people are never caught or prosecuted for crimes they commit, and thus aren't labeled deviant. In other cases, people may be falsely accused (as cheating on taxes) and stigmatized. In effect, then, deviance is in the eye of the beholder because societal reaction, rather than an act, labels people as law-abiding or deviant. Moreover, labeling can lead to secondary deviance.

LEMERT: PRIMARY AND SECONDARY DEVIANCE

Edwin Lemert (1951, 1967) developed labeling theory by differentiating between primary and secondary deviance. **Primary deviance** is the initial act of breaking a rule. Primary deviance can range from relatively minor offenses, such as not attending a family member's funeral, to serious offenses, such as rape and murder.

Even if people aren't guilty of primary deviance, labeling can lead to **secondary deviance**, rule-breaking behavior that people adopt in response to others' reactions. A teenager who's caught trying marijuana may be labeled a "druggie." If the individual is rejected by others, he or she may accept the label, associate with drug users, and become involved in a drug-using subculture. According to Lemert, a single deviant act will rarely result in secondary deviance. The more times a person is labeled, however, the higher the probability that she or he will accept the label and engage in deviant behavior.

There's considerable evidence that labeling affects people's lives. For example, those who have been jobless for a year or longer report stigmas in job searches: Employers assume that they're lazy, don't really want to work, or there's something wrong with them. Some companies have explicitly barred the long-term unemployed from certain job openings, telling them outright in job ads that they need not apply (see Chapter 11).

THE MEDICALIZATION OF DEVIANCE

In 1952, the American Psychiatric Association published the first edition of the *Diagnostic and Statistical Manual of Mental Disorders (DSM)*, which describes mental health disorders. The number of disorders increased from 106 in 1952 to nearly 400 in 2013, and increased in size from "a spiral-bound pamphlet" to "a tome of nearly 1,000 pages" (Tavris, 2013: C5). Some of the new disorders include "binge eating" and "bereavement" (American Psychiatric Association, 2013).

The *DSM*'s ever-changing diagnoses and labels are an example of the **medicalization of deviance**, diagnosing and treating a violation of social norms as a medical disorder. The *DSM* has far-reaching effects, according to one of the past editors: "Anything you put in that book … has huge implications not only for psychiatry but for pharmaceutical marketing, research, for the legal system, for who's considered to be normal or not, for who's considered disabled. *And it has huge implications for stigma because the more disorders you put in, the more people get labels*" (Carey, 2010: 1, emphasis added; Chapter 14 discusses the medicalization of deviance costs).

iStockphoto.com/Darren Mower

Do you think that this arrest is the result of primary or secondary deviance? Why?

7-6c Critical Evaluation

Interaction theories have helped to explain deviance, but critics point to several weaknesses. Differential association theory doesn't explain impulsive crimes triggered by rage or fear, those committed by people who have grown up in law-abiding families, or why only a minority of children who grow up in poor, urban, disadvantaged communities join gangs and/or commit crimes. This theory also ignores the possibility that deviant values and behaviors can be unlearned, as when young children model significant others whom they respect (Williams and McShane, 2004; Graif and Sampson, 2009; Guerra et al., 2013).

Critics have faulted labeling theory on other points: The theory exaggerates the importance of judgments in

> **primary deviance** the initial act of breaking a rule.
>
> **secondary deviance** rule-breaking behavior that people adopt in response to others' reactions.
>
> **medicalization of deviance** diagnosing and treating a violation of social norms as a medical disorder.

Trailer Trash? About 20 million low-income Americans live in mobile homes because they can't afford other housing. Most realize that they're stigmatized as "trailer trash," but maintain their dignity despite the labeling. They point out that they're homeowners, they take care of their property, many are employed, and they differentiate themselves from people who commit crimes (Kusenbach, 2009; Geoghegan, 2013).

changing a person's self-concept, doesn't explain why crime rates are higher in the South than in other parts of the United States or at particular times of the year (as before holidays), and doesn't tell us *why* people commit crimes. Some critics also contend that social reactions are the *result* rather than the "cause" of deviant behavior (Schurman-Kauflin, 2000; Benson, 2002). Conflict theorists, in particular, criticize symbolic interactionists for ignoring structural factors—such as poverty and low-paid jobs—that create or reinforce deviance and crime (Currie, 1985).

7-7 CONTROLLING DEVIANCE AND CRIME

Most of us conform to laws and cultural expectations. We usually conform, however, not because we're innately good, but because of **social control**, techniques and strategies that regulate people's behavior in society. The purpose of social control is to eliminate, or at least reduce, deviance.

social control the techniques and strategies that regulate people's behavior in society.

control theory proposes that deviant behavior decreases when people have strong social bonds with others.

criminal justice system government agencies that are charged with enforcing laws, judging offenders, and changing criminal behavior.

7-7a Control Theory and Types of Social Control

Control theory proposes that deviant behavior decreases when people have strong social bonds with others. The bonds that strengthen our self-control are based on (1) *attachment* to other people (e.g., family, friends, and coworkers), (2) *commitment* to conformity (e.g., participating in community activities), (3) *involvement* in legitimate activities (e.g., getting an education, holding a job), and (4) *belief* in social norms and values (e.g., being moral, obeying laws) (Hirschi, 1969).

Most conformity is due to the internalization of norms during the powerful socialization process. Because most of us care about the opinions of family members, friends, and teachers, we try to live up to their expectations (see Chapter 4). Thus, we conform because of *informal social controls* that we learn and internalize during childhood.

Formal social control, which also regulates social behavior, exists outside of the individual. For example, a college dean might threaten a student with suspension or expulsion, a business might install security cameras, or a police officer might arrest someone.

7-7b The Criminal Justice System and Social Control

Social institutions such as the family, education, and religion try to maintain *social* control over behavior; the criminal justice system has the *legal* power to control crime and punish offenders. The **criminal justice system** is made up of government agencies—including the police, courts, and prisons—that enforce laws, judge offenders, and try to decrease criminal behavior. The criminal justice system relies on three major strategies to control crime: prevention and intervention, punishment, and rehabilitation. Punishment, as you'll see, is the least effective approach.

PREVENTION AND INTERVENTION

Crime prevention is much less expensive than punishment. States spend an average of $241 per day to incarcerate an adolescent. In contrast, it costs about $75 a day for community services that provide counseling, drug treatment, and other support that reduces the risk of re-offending (Khadaroo, 2015).

Among adults, many crimes involve substance abuse, but the average annual cost per outpatient for alcohol or drug abuse treatment is $4,700, compared with $24,000 for incarceration (National Institute on Drug Abuse, 2012). Nearly a quarter-million inmates in state and federal prisons are age 50 and older, and many have serious health problems. An older offender's annual medical costs can exceed $100,000 (Johnson and Beiser, 2013).

Some crimes have plummeted because of technology. Car theft, for example, has decreased 40 percent since 2003 because most new automobiles are almost impossible to hot-wire (Klein and White, 2011). Technology, however, hasn't stopped other crimes such as burglary because many offenders have become more sophisticated in avoiding home security systems.

Nationally, states with the most firearms laws have the lowest gun-related homicide and suicide rates (Fleegler et al., 2013; Anestis and Anestis, 2015), but gun control laws are controversial. Large majorities of Americans support background checks (85 percent) and laws that prevent mentally ill people from buying guns (79 percent). Since 2013, however, Americans have been almost evenly divided on whether it's more important to control or protect gun ownership (Doherty, 2015). A recent study of top-selling movies from 1950 to 2012 found that gun violence in PG-13 rated films now surpass the violence in R-rated films (Bushman et al., 2013). Thus, violence sells.

Social Service Agencies and Community Outreach Programs.

Many organizations—composed of law enforcement professionals, social workers, and nonprofit groups—try to prevent crime. Because of high prison costs, the U.S. Department of Justice, among other federal agencies, funds numerous initiatives that try to prevent and decrease crime. The National Institute of Justice's CrimeSolutions.gov uses research to rate the effectiveness of criminal justice programs and practices.

Andrey Burmakin/Shutterstock.com

Some programs are more successful than others. Compared with police-only interventions, when police and local social service workers respond together to an incident of intimate partner violence, victims experience significantly less violence immediately after the episode and months later. Why? The collaborative partnership between law enforcement and local social service agencies provides victims with more protective strategies and resources (e.g., getting medical attention and protection orders, going someplace where a partner can't find them) (Messing et al., 2014).

The purpose of programs like "Scared Straight," which involve organized visits to prisons and face-to-face interaction with adult inmates, is to deter juvenile delinquency. Evaluations of scared-straight programs have concluded, however, that they may increase rather than deter adolescent deviance. The reasons are unclear, but may be due to the program participants' imitating the inmates' "brutal and vulgar" language and crimes. Still, the tours continue to be popular (Robinson and Slowikowski, 2011; Petrosino et al., 2013).

Police. The primary role of the police is to enforce society's laws, but can the police prevent crimes? Police can head off some crimes by cruising high-risk areas in patrol cars or by having more officers on foot patrols who return to the same street several times during a shift ("Policing Philadelphia . . . ," 2013). Concentrating on *hot spots*, areas of high criminal activity, reduces crime only temporarily, however, because criminals simply move to other parts of the city (U.S. Department of Justice, 1996; Braga, 2003).

Compared with 57 percent of whites and 52 percent of Latinos, only 30 percent of blacks have "a great deal" of confidence in the police (Jones, 2015). Police actions in some cities—including Ferguson, Missouri; North Charleston, South Carolina; and Baltimore, Maryland—have recently come under scrutiny after black men were killed while being apprehended by white police officers. There has been a longstanding distrust between police and many poor and minority neighborhoods, but the deaths of the black men spawned protests in some cities, and violence and riots in Baltimore (which we'll examine in Chapter 16).

Minorities, particularly African Americans, accuse police of discrimination, *racial profiling* (using race to stop and detain someone suspected of having committed a crime), and brutality. Police officers argue that they resort to violence partly because "We've never had a population who are so well-armed," that racial profiling is justified because a high proportion of blacks and Latinos commit crimes, and that they're scapegoated for social problems ("Police Culture," 2015: 28). Police can't prevent most crimes because they have no control over macro-level factors—poverty, unemployment, limited educational opportunities, and neighborhood deterioration—that are associated with high crime rates.

PUNISHMENT

Those who endorse a **crime control model** believe that crime rates increase when offenders don't fear apprehension or punishment. This perspective emphasizes protecting society and supports a tough approach toward criminals in sentencing, imprisonment, and capital punishment.

Sentencing. After someone has been found guilty of a crime or has pleaded guilty, a judge (and sometimes a jury) imposes a *sentence,* or penalty. A sentence can be a fine, *probation* (supervision instead of serving time in jail), incarceration, or the death penalty. Those who receive *parole* are released from prison before the end of

crime control model proposes that crime rates increase when offenders don't fear apprehension or punishment.

FIGURE 7.3 THE NUMBER OF U.S. PRISONERS HAS NEARLY TRIPLED SINCE 1987

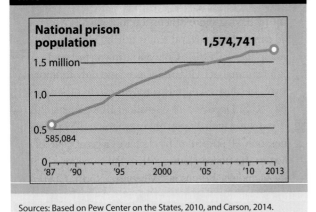

National prison population

1,574,741

Sources: Based on Pew Center on the States, 2010, and Carson, 2014.

IT'S TEMPTING TO ASSUME THAT THE MORE PEOPLE BEHIND BARS, THE LOWER THE CRIME RATE.

their full sentence on condition that they check in regularly with an officer and obey the law.

About 57 percent of those sentenced get probation; the others serve a term in a local jail or a state or federal prison (Glaze and Kaeble, 2014). Many people question the fairness of sentencing because those convicted of similar crimes can receive widely different sentences depending, among other factors, on race/ethnicity, social class, variations in state sentencing laws, and how juries and judges determine the seriousness of a crime. Nationwide, of those serving life-without-parole sentences for nonviolent offenses (e.g., possession, sale, or distribution of drugs; shoplifting; attempting to cash a stolen check), 65 percent are black. And, across all racial-ethnic groups, the vast majority are low-income and don't have a high school diploma (Turner, 2013).

As state and local budgets shrink, even states with reputations for being tough on crime are embracing more lenient policies. Some states have begun to shorten the average number of years on probation and parole, which decreases supervision costs. Other states are reducing the number of people sent to prison: It costs an average of $79 a day to keep an inmate in prison but only about $3.50 a day to monitor the same person on probation (Richburg, 2009; see also Maruschak and Bonczar, 2013).

Incarceration. The number of U.S. prisoners has surged since 1987 (see *Figure 7.3*). Since 2008, and for the first time in history, 1 in every 100 American adults is in prison. The United States has less than 5 percent of the world's population, but almost 25 percent of the planet's prisoners, the majority of whom are

incarcerated for nonviolent drug offenses. Besides the sheer number of inmates, the United States is also the global leader in per capita inmates (716 per 100,000 population), well ahead of nations such as Russia (475), Cuba (510), and the United Kingdom (148) (Walmsley, 2013).

Of all U.S. state and federal prisoners, 92 percent are men, but imprisonment rates vary by race/ethnicity. As *Figure 7.4* shows, incarceration rates are highest for African Americans, both women and men. If current trends continue, 1 of every 3 black males born in 2013 can expect to go to prison in his lifetime, as can 1 of every 6 Latino males—compared with 1 of every 17 white males (The Sentencing Project, 2013).

The U.S. prison population decreased by more than 2 percent between 2010 and 2013 (Carson, 2014). Why the drop? Arrest rates have declined, as you saw earlier, but prison counts have dropped primarily because of economic reasons. Driven by overcrowding and budget crises, many states have pursued a variety of strategies such as giving probation or lighter sentences for nonviolent crimes, releasing low-level offenders before they serve their sentences, and accelerating the release of inmates who have completed educational, vocational, and substance abuse treatment

FIGURE 7.4 U.S. PRISONERS BY RACE/ETHNICITY AND SEX, 2013
Imprisonment Rate Per 100,000 U.S. Population

	Imprisonment Rate
MEN	
White	466
Black	2,805
Latino	1,134
Other*	963
WOMEN	
White	51
Black	113
Latina	66
Other*	90

*Includes American Indians, Alaska Natives, Asians, Native Hawaiians, Pacific Islanders, and persons of two or more races.

Source: Based on Carson, 2014, Table 8.

programs. Releasing prisoners early has been controversial, but state officials maintain that their budgets can't afford the high incarceration costs (Goode, 2013; Kearney et al., 2014).

It's tempting to assume that the more people behind bars, the lower the crime rate. It costs the American taxpayer about $34,000 per inmate per year. State spending on corrections surged from $15 billion in 1982 to $49 billion in 2010 (Kyckelhahn, 2013; "The Right Choices," 2015). However, *recidivism* (being arrested for committing another offense after being released) has barely changed since 1999: Of those released from state prisons, 68 percent are arrested for a new crime within 3 years, and 77 percent within 5 years (Durose et al., 2014).

Among other services, Homeboy Industries, based in Los Angeles, provides previously incarcerated gang members with job training and placement. Pictured here, Dr. Mark Benor uses a laser machine to remove a former gang member's tattoo at a free clinic funded by Homeboy Industries.

ROBYN BECK/Getty Images

Capital Punishment. About 60 percent of Americans support *capital punishment* (the death penalty), down from 80 percent in the mid-1990s (Swift, 2015). By mid-2015, more than half of the world's nations, including many developing countries, had abolished the death penalty or suspended executions. The United States executes more people than any other country except China, Iran, Iraq, and Saudi Arabia. Thirty-one states have the death penalty, but the number of executions fell from 98 in 1999 to 35 in 2014, Since 1976, 56 percent of all executions have been carried out in the South. Just three states—Texas, Missouri, and Florida—accounted for 82 percent of all executions in 2014 (Snell, 2014; Amnesty International, 2015; "Capital Punishment," 2015; Death Penalty Information Center, 2015).

There are several reasons for fewer executions. First, and as you saw earlier, homicide rates have declined sharply since 1980. Second, public support for the death penalty has waned. Third, the nation's police chiefs believe that the death penalty doesn't deter crime. Fourth, there's considerable research which shows racial disparities in executions (e.g., black males, especially if the victims are white, are substantially more likely to be executed than white males who murder blacks). Fifth, most juries can now impose life sentences without the possibility of parole. Finally, some argue that the death penalty is a waste of money because inmates can spend 15 to 20 years appealing the sentence. In California, for example, it costs taxpayers $137 million per year for inmates who appeal a death sentence compared with $12 million per year for those serving life sentences ("Capital Punishment…," 2014; Death Penalty Information Center, 2015; Ghandnoosh, 2015).

REHABILITATION

Rehabilitation, a third approach to controlling deviance, maintains that appropriate treatment can change offenders into productive, law-abiding citizens. Nationally, about 64,000 inmates produce more than $2 billion worth of products a year. They make everything from redwood canoes to specialty motorcycles, fishing poles, and saddles. They also produce apple juice, raise fish, milk cows and goats, grow flowers, teach dogs obedience, and manage vineyards at prison-run facilities (Alsever, 2014).

Are rehabilitation programs successful? Only if they provide employment after release. Other effective rehabilitation efforts include learning a trade while in prison, earning a high school diploma or college degree while in prison or after release, and receiving services that address several needs (e.g., housing, employment, and medical services) rather than just one (e.g., drug abuse counseling) (Cumberworth, 2010; Flower, 2010).

STUDY TOOLS 7

READY TO STUDY? IN THE BOOK, YOU CAN:

☐ Check your understanding of what you've read with the Test Your Learning Questions provided on the chapter review card at the back of the book.

☐ Rip out the chapter review card for a handy summary of the chapter and key terms.

ONLINE AT CENGAGEBRAIN.COM YOU CAN:

☐ Collect StudyBits while you read and study the chapter.

☐ Quiz yourself on key concepts.

☐ Prepare for tests with chapter flash cards as well as flash cards that you create.

☐ Watch videos to explore sociology further.

© Anelina/Shutterstock.com

8 | Social Stratification: United States and Global

LEARNING OUTCOMES

After studying this chapter, you will be able to...

8-1 Explain and illustrate social stratification systems and bases.

8-2 Describe the U.S. class structure and explain how and why social classes differ.

8-3 Describe poverty and explain why people are poor.

8-4 Compare the different types of social mobility, describe recent trends, and explain what factors affect mobility.

8-5 Describe global stratification, its variations and consequences, and the theoretical models that explain why inequality is universal.

8-6 Compare and evaluate the theoretical explanations of social stratification.

After you finish
this chapter go to
PAGE 157 for
STUDY TOOLS.

In 2012, American multi-billionaire Larry Ellison—founder of Oracle, the world's biggest database company—bought the Hawaiian island of Lanai for $500 million. The same year, 17 percent of Hawaii's residents were living in poverty (*Hawai'i Free Press*, 2012; Kroll and Dolan, 2013). Many people in the United States and around the world are struggling to survive while others enjoy astonishing wealth. This chapter examines social stratification and why there are haves and have-nots.

8-1 SOCIAL STRATIFICATION SYSTEMS AND BASES

Social stratification is a society's ranking of people based on their access to valued resources such as wealth, power, and prestige. All societies are stratified, but some more than others. Sociologists distinguish among stratification systems based on the extent to which they're open or closed. Before reading further, take the True/False quiz to see how much you know about U.S. economic inequality.

social stratification a society's ranking of people based on their access to valued resources such as wealth, power, and prestige.

8-1a Closed and Open Stratification Systems

In a *closed stratification system*, movement from one social position to another is limited by ascribed

What do you think?

Americans who are poor just aren't working hard enough.

| 1 | 2 | 3 | 4 | 5 | 6 | 7 |

strongly agree strongly disagree

GMB Akash/Panos

statuses. An *open stratification system* allows movement up or down because people's achievements affect mobility. Closed stratification systems are considerably more fixed than open ones, but neither is completely open or closed.

Two closed stratification systems exist today: slavery and castes. In a **slavery system**, people own others as property and have almost total control over their lives. The slaves have been kidnapped, inherited, or given as gifts to pay a debt. They're bought and sold as commodities, sometimes multiple times. The United Nations banned all forms of slavery worldwide in 1948, but almost 36 million people are living as slaves across the globe. About 61 percent are in just five countries—India, China, Pakistan, Uzbekistan, and Russia—but modern slavery is also prevalent in parts of Africa, the Middle East, and Asia (Walk Free Foundation, 2014).

In a **caste system**, a second type of closed stratification system, people's positions are ascribed at birth and largely fixed. Their places in the hierarchy are primarily determined by inherited characteristics such as race, skin color, gender, family background, or nationality (see Chapter 5). People must marry within their caste; there are severe restrictions in their choice of occupation,

slavery system people own others as property and have almost total control over their lives.

caste system people's positions are ascribed at birth and largely fixed.

CHAPTER 8: Social Stratification: United States and Global **139** residence, and social relationships; and people rarely move from one caste to another.

A good example is India, where a caste system has existed for more than 3,000 years. At the top were the Brahmins. On the bottom rung were the Dalits (formerly called "untouchables"), who were very poor and performed the most menial and unpleasant jobs, including collecting human waste and cleaning streets. Fearing being "polluted," higher castes wouldn't interact with the Dalits (Mendelsohn and Vicziany, 1998).

India outlawed the caste system in 1949, but social distinctions are deeply entrenched, particularly in rural areas, and about 90 to 95 percent marry within their own castes ("Marriage in India," 2015). In 1950, India set up a quota system that reserves nearly half of government jobs and public college slots for members of the most disadvantaged castes. Some claim that the quota system is unfair because it benefits less qualified candidates. Others maintain that such complaints come primarily from groups who fear losing some of their deep-rooted privileges (Barstow and Raj, 2015; McCarthy, 2015).

In a **class system**—a relatively open stratification structure—people's positions are based on both birth and achievement. Because achieved characteristics (e.g., education, work skills, occupation) can change, people in open stratification systems can move from one social class to another. A **social class** is a group of people who have a similar standing or rank in a society based on wealth, education, power, prestige, and other valued resources. A large majority of Americans (60 percent) believe that hard work leads to success (Kohut, 2015), but is this really the case?

8-1b The Bases of Stratification

Conventional wisdom says that money can't buy happiness, but a study of 155 countries, including the United States, found that the richer people are, the happier they are (Stevenson and Wolfers, 2013). Income is an important stratification factor, but it's not the only one. Instead, sociologists use a multidimensional approach that includes wealth, prestige, and power.

class system people's positions are based on both birth and achievement.

social class people who have a similar standing or rank in a society based on wealth, education, power, prestige, and other valued resources.

wealth economic assets that a person or family owns.

income the money a person receives, usually through wages or salaries, but can also include other earnings.

WEALTH

Wealth refers to the economic assets that a person or family owns. It includes money, property (e.g., real estate), stocks and bonds, retirement and savings accounts, personal possessions (e.g., cars and jewelry), and income. **Income** is the money a person receives, usually through wages or salaries, but it can also include rents, interest on savings accounts, stock dividends, royalties, or business proceeds. Income and wealth differ in several important ways:

▸ Wealth is *cumulative*. It increases over time, especially through investment, whereas income is usually spent on everyday expenses.

▸ Because wealth is accumulated over time, much of it can be *passed on to the next generation*. Inheritances and other monetary gifts offer opportunities to start a business, buy a first (or vacation) home, pursue college without debt, provide a cushion when a job is lost or a health emergency arises, and increase assets (such as savings and stocks).

▸ Passing on wealth to the next generation *preserves privilege*. Wealthy people tend to have the greatest amount of *economic capital* (income and other monetary assets such as property), *cultural capital* (advanced degrees and assets such as style of speech, table manners, and physical appearance), and *social capital* (networks comprised of influential people). Differential access to all three types of capital reinforces and reproduces the existing class structure and inequality (Bourdieu, 1984).

U.S. wealth and income inequality is staggering. As *Figure 8.1* shows, the top 1 percent of Americans own 37 percent of all wealth and 20 percent of all income—the highest levels recorded since the government began collecting such data, and income inequality is the highest it's been since 1928 (Lowrey, 2013; Saez, 2013). We can get a better sense

India's Dalits still perform unpleasant tasks such as burning corpses and removing rubbish and human waste.

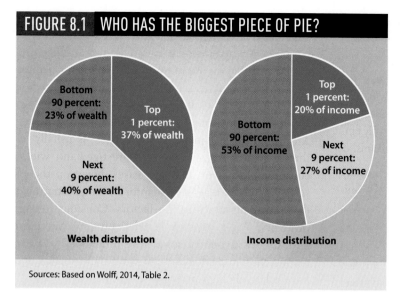

FIGURE 8.1   WHO HAS THE BIGGEST PIECE OF PIE?

Bottom 90 percent: 23% of wealth

Top 1 percent: 37% of wealth

Next 9 percent: 40% of wealth

Wealth distribution

Top 1 percent: 20% of income

Bottom 90 percent: 53% of income

Next 9 percent: 27% of income

Income distribution

Sources: Based on Wolff, 2014, Table 2.

These super-rich Americans average $3.1 billion per person and control 37 percent of the nation's wealth (Wolff, 2012; Wealth-X and UBS, 2014).

PRESTIGE

A second basis of social stratification is **prestige**—respect, recognition, or regard attached to social positions. Prestige is based on many criteria, including wealth, family background, power, and accomplishments. Regarding accomplishments, for example, every college convocation acknowledges students who graduate *cum laude*, *magna cum laude*, and *summa cum laude*.

We typically evaluate people according to the kind of work they do. Higher prestige occupations include physicians, lawyers, pharmacists, college professors, architects, dentists, and teachers. Legislators, police officers, actors, librarians, realtors, and firefighters are examples of jobs at a medium prestige level. Occupations at the lower prestige level consist of janitors, carpenters, bartenders, garbage collectors, truck drivers, and food servers (Smith et al., 2011).

The most prestigious occupations have several characteristics:

▸ They *require more formal education* (college or post-graduate degrees) and/or extensive training. Physicians, for example, must fulfill internship and residency requirements after receiving a medical degree.

▸ They're primarily *nonmanual and require more abstract thought*. All jobs have some physical

of income disparities by looking at the average household income. As *Figure 8.2* shows, there's a significant income gap between even the top 5 and 20 percent of households.

Both wealth and income gaps have *increased*. In 2013, the wealthiest 10 percent of households had 77 of the country's wealth, up from 67 percent in 1989. The wealth of the bottom 90 percent fell by more than 3 percent; in 2013, this group owned only 23 percent of all U.S. wealth, about as much as in 1940 (Levine, 2012; Saez and Zucman, 2014).

In 2003, the net wealth of the typical household, adjusted for inflation, was $87,992. Ten years later, it was down to $56,335. Thus, between 2003 and 2013, ordinary Americans became 36 percent poorer; those in the top 5 percent got 14 percent richer (Pfeffer et al., 2014).

Five years after the Great Recession (late 2007 to mid-2009) officially ended, few families enjoyed an economic recovery, but there are stark racial and ethnic disparities. Although racial wealth gaps have widened since the 1980s, Asians' wealth has increased significantly since 1989. By 2013, the median wealth of Asian families ($91,000) was closer to that of whites ($134,000) than Latinos ($14,000) and blacks ($11,000). Researchers predict that Asian wealth levels will soon surpass even those of whites: Asian families already have the highest household income, high educational levels, diversified assets, little or no credit card debt, and save money every year (Emmons and Noeth, 2015; see also Chapter 10).

Regardless of how the economy is doing, the rich keep getting richer. Between 1983 and 2010, the number of U.S. households that had $10 million or more increased by 426 percent. In 2014, the United States had 571 billionaires, up from 413 in 2011, who comprised nearly 25 percent of the world's billionaires.

prestige respect, recognition, or regard attached to social positions.

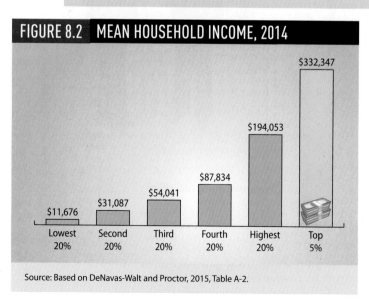

FIGURE 8.2   MEAN HOUSEHOLD INCOME, 2014

$11,676	$31,087	$54,041	$87,834	$194,053	$332,347
Lowest 20%	Second 20%	Third 20%	Fourth 20%	Highest 20%	Top 5%

Source: Based on DeNavas-Walt and Proctor, 2015, Table A-2.

activity, but an architect must use more imagination in designing a building than a carpenter, who usually performs very specific tasks.

▶ They *are paid more*, even though there are some exceptions. A realtor or a truck driver may earn more than a registered nurse, but registered nurses are likely to earn more over a lifetime because they have steady employment, health benefits, and retirement programs.

▶ They're *seen as socially more important*. An elementary school teacher may earn less than the school's janitor, but the teacher's job is more prestigious because teachers contribute more to a society's well-being.

▶ They *involve greater self-expression, autonomy, and freedom from supervision*. A dentist has considerably more freedom in performing her or his job than a dental hygienist. In effect, then, higher prestige occupations provide more privileges.

POWER

A third basis of social stratification is **power**—the ability to influence or control the behavior of others despite

"New money or old money?"

Joseph Farris/Cartoonstock.com

power the ability to influence or control the behavior of others despite opposition.

socioeconomic status (SES) an overall ranking of a person's position in society based on income, education, and occupation.

opposition. Wealthy people generally have more power than others in dominating top government and economic positions. Even if they don't hold a political office or a corporate position, the most powerful people make the nation's important decisions.

People who experience *status consistency* are about equal in terms of wealth, prestige, and power. A Supreme Court justice, for instance, is usually affluent, enjoys a great deal of prestige, and wields considerable power. In many cases, however, there's *status inconsistency* (see Chapter 5) if a person ranks differently on stratification factors. Consider funeral directors. Their prestige is relatively low, but most have higher incomes than college professors, who are among the most educated people in U.S. society and have relatively high prestige.

We've looked at stratification systems and their bases, but how, specifically, do people in different social classes act? And how does social class affect our behavior?

8-2 SOCIAL CLASS IN AMERICA

To measure social class, sociologists typically use **socioeconomic status (SES)**—an overall ranking of a person's position in society based on income, education, and occupation. Some researchers ask people to identify the social classes in their communities (*reputational approach*), some ask people to place themselves in one of a number of classes (*subjective approach*), but most use SES indicators (*objective approach*).

In measuring social class, national surveys tend to use the subjective approach and closed-ended questions (see Chapter 2). Because the choices differ, so do respondents' self-identification. In a Gallup poll that offered five categories, 1 percent of Americans identified themselves as upper class, 13 percent as upper-middle class, 38 percent as middle class, 33 percent as working class, and 15 percent as lower class (Newport, 2015). In another national survey, however, where three of the five categories included the words "middle class," fully 87 percent said that they were upper-middle, middle, or lower-middle class (Doherty and Tyson, 2015).

Because there are different ways of measuring social class, sociologists don't always agree on the number of social classes. There's consensus, however, that there are four general classes: upper, middle, working, and lower. Except for the working class, many sociologists divide these classes further into more specific strata. There aren't rigid dividing lines, but *Figure 8.3* provides an overview of the U.S. social class structure. Besides income, education, and occupation, social classes also

FIGURE 8.3 THE AMERICAN CLASS STRUCTURE

Upper Class
3%

Middle
Class
44%

Working Class
38%

Lower Class
15%

Class	Typical Occupations	Typical Annual Income
Upper	**Upper-Upper (1%)**	
	Heirs of "old money"	Billions
	Lower-Upper (2%)	
	Executives, high-level managers, owners of corporations	Millions
Middle	**Upper-Middle (14%)**	
	Upper-level managers, professionals, medium-sized business owners	$150,000+
	Lower-Middle (30%)	
	Lower-level managers, nurses, teachers, social workers, skilled/unionized craftspeople, nonretail sales	$60,000–$80,000
Working	**Working Class (38%)**	
	Semiskilled workers, clerical workers, retail sales, truck drivers	$40,000–$50,000
Lower	**Working Poor (3%)**	
	Low-paid manual and retail workers, service workers (maids, health care aides)	$25,000
	Underclass (12%)	
	Unemployed, part-time low-wage earners, public assistance recipients	Below $12,000

Sources: Based on BLS Reports, 2013; U.S. Census Bureau, Current Population Survey…, 2013; Wysong et al., 2014.

differ in values, power, prestige, social networks, and *lifestyles* (tastes, preferences, and ways of living).

8-2a The Upper Class

You saw earlier that very rich Americans control a vastly disproportionate amount of the total wealth and income. This group comprises the *upper-upper class* and the *lower-upper class*.

THE UPPER-UPPER CLASS

Upper-upper-class people are the *old rich* who have been wealthy for generations. Because they value their privacy, upper-upper-class members rarely appear on the lists of wealthiest individuals published by *Forbes* or other sources.

An inherited fortune brings power. Upper-upper-class white males, in particular, shape the economic and political climate through a variety of mechanisms: Dominating the upper levels of business and finance, holding top political positions in the federal government, underwriting thousands of think tanks and research institutes that formulate national policies, and shaping public opinion through the mass media (Zweigenhaft and Domhoff, 2006; Hacker and Pierson, 2010).

THE LOWER-UPPER CLASS

The lower-upper class—which is much more diverse than the upper-upper class—is the *nouveau riche*, those with "new money." Some, like the Kennedys, amassed fortunes several generations ago. Others, like politician and businessman Jeb Bush, are augmenting their vast inheritances through a combination of speaking fees, corporate board memberships, investments, and consulting contracts that pay millions each year (O'Keefe et al., 2015). Many of the new rich—Oprah Winfrey, President Bill Clinton, and Mark Zuckerberg—worked for their income rather than inherited it. Besides business entrepreneurs, the lower-upper class also includes high-level managers of international corporations, those who earn at least a million dollars a year, and some highly paid athletes and actors, but the lifestyles vary considerably (Wood, 2012). Some lower-upper class members live modestly, but many flaunt their new wealth. Sociologist Thorstein Veblen (1899/1953) coined the term *conspicuous consumption*, a lavish spending on goods and services to display one's social status and enhance one's prestige.

Most of the *nouveau riche* are unbelievably wealthy. If Bill Gates cashed in all his wealth and spent $1 million every single day, it would take him 218 years to spend all

Conspicuous consumption isn't limited to the super-rich. Many upper-middle-class people, in particular, use status symbols to show that they've made it by buying "almost rich" cars (like a Mercedes Benz that starts at $32,000), upscale kitchen appliances, designer handbags, expensive jewelry, and by taking lavish vacations.

Hemera Technologies/Getty Images

his money (Oxfam, 2014). Because they lack the "right" ancestry and have usually made their money by working for it, however, lower-uppers aren't accepted into "old money" circles that have strong feelings of in-group solidarity. Still, they engage in lifestyles and rituals that parallel those of the upper-upper class, such as having personal chefs, taking exotic vacations (often in private planes), and joining country clubs (Sherwood, 2010).

8-2b The Middle Class

In the two national surveys mentioned earlier, between 38 and 47 percent of Americans identified themselves as middle class. There's no agreement on exactly where the middle class begins and ends, but there are some major differences between the *upper-middle class* and the *lower-middle class*.

THE UPPER-MIDDLE CLASS

Upper-middle-class members, although rich, live on earned income rather than accumulated or inherited wealth. The occupations of this group usually require a Ph.D. or advanced professional degree. People in this class include corporate executives and managers (but not those at the top), high government officials, owners of large businesses, physicians, and successful lawyers and stockbrokers. A higher level of education is associated with greater occupational prestige and autonomy, as well as job quality and security, but people in these occupations are three times more likely than those in the general population to work 50 or more hours per week (Dewan and Gebeloff, 2012).

THE LOWER-MIDDLE CLASS

The lower-middle class, more diverse than the upper-middle class, is composed of people in nonmanual occupations that require some training beyond high school, and professional occupations that require a college degree. Nonmanual jobs include office staff, low-level managers,

working poor people who work at least 27 weeks a year but whose wages fall below the official poverty level.

owners of small businesses, medical and dental technicians, secretaries, police officers, and sales workers (e.g., insurance and real estate agents). Examples of professional occupations are nursing, social work, and teaching. Most families in the lower-middle class rely on two incomes to maintain a comfortable standard of living.

Unlike the upper-middle class, those in the lower-middle class have less autonomy and freedom from supervision, and there's little chance for advancement. Except for some retirement funds, most have only modest savings to cover emergencies. Many buy used or inexpensive late-model cars, eat out at middle-income restaurants, and take occasional vacations, but they rarely have the income to buy luxury products without going deeply into debt.

8-2c The Working Class

About 33 percent of Americans identify themselves as working class (Newport, 2015). This group consists of skilled and semiskilled laborers (e.g., construction and assembly-line workers, truck drivers, auto mechanics, carpenters, electricians). The semiskilled jobs typically require little training, are mechanized, and closely supervised. Most of the occupations are blue-collar, but some—clerks and retail sales workers—are white collar. The jobs don't require a college education, but offer little or no opportunity for advancement.

Working-class homeowners may experience foreclosure because of delinquent payments. Many use credit cards but then can barely pay the monthly minimum, don't have enough money to cover a $500 repair bill, and all of their savings may amount to only about four months of income. Debts become overwhelming when borrowers suffer setbacks like divorce, illness, or job loss (Bell, 2015; PEW Charitable Trusts, 2015).

8-2d The Lower Class

Between 10 and 15 percent of Americans identify themselves as lower class (Doherty and Tyson, 2015; Newport, 2015). People in this group are at the bottom of the economic ladder because they have little education, few occupational skills, work in minimum wage jobs, or are often unemployed. Most of the lower class is poor, but sociologists often distinguish between the *working poor* and the *underclass*.

THE WORKING POOR

More than 23 percent of America's poor are the **working poor**—people who work at least 27 weeks a year but whose wages fall below the official poverty level. Women are more likely than men to be among the working poor. Blacks and Latinos are more than twice as likely as

Among America's working poor are hotel housekeepers. On average, those employed even at expensive hotels earn less than $19,000 a year working full time. The working poor do dirty and often demeaning work for next to nothing, barely scrape by, and live in constant fear of being fired (Wing and Schwartz, 2014).

Asians and whites to be in this group. Full-time workers are less likely than part-time workers to be working poor but, in 2013, more than 4 percent of those usually employed full time were among the working poor, primarily because of low earnings (BLS Reports, 2015).

THE UNDERCLASS

The **underclass**, which occupies the bottom rung of the economic ladder, consists of people who are persistently poor, residentially segregated, and relatively isolated from the rest of the population. Most are chronically unemployed or drift in and out of jobs. They may work erratically or part time, but their lack of skills, low education levels, and mental or physical disabilities make it difficult for them to find full-time jobs. Members of the underclass may have some earnings, but are often dependent on government programs, including Social Security, public assistance, and veterans' benefits. Some also have income from criminal activities (Gilbert, 2011; see also Chapter 7).

8-2e How Social Class Affects Us

Social class, more than any other single variable, affects practically all aspects of our lives. Max Weber referred to the consequences of social stratification as **life chances**—the extent to which people have positive experiences and can secure the good things in life (food, housing, education, good health) because they have economic resources. Regarding health, for example:

▶ Poor children are less healthy than those at higher SES levels and are more likely to be unhealthy in adulthood. Unhealthy adults earn less, spend less time in the labor force, and must often retire earlier (see Chapter 14).

▶ Living in a chronically disadvantaged neighborhood and experiencing violence increases anxiety and stress, accelerating the onset of diseases (Bird et al., 2010; Harrell et al., 2014).

▶ Compared with the upper class, the lower class is four times more likely to report being in poor health, three times more likely to be unhappy, and twice as likely to be frequently stressed and depressed (Morin and Motel, 2012; Graham, 2015).

8-3 POVERTY

There are two ways to define poverty: absolute and relative. **Absolute poverty** is not having enough money to afford the basic necessities of life, such as food, clothing, and shelter ("what I need"). **Relative poverty** is not having enough money to maintain an average standard of living ("what I want").

8-3a The Poverty Line

The **poverty line** (also called the *poverty threshold*) is the minimal income level that the federal government considers necessary for basic subsistence. To determine the poverty line, the Department of Agriculture (DOA) estimates the annual cost of food that meets minimum nutritional guidelines and then multiplies this figure by three to cover the minimum cost of clothing, housing, health care, and other necessities. The poverty line, which in 2014 was $24,008 for a family of four (two adults

underclass people who are persistently poor and seldom employed, residentially segregated, and relatively isolated from the rest of the population.

life chances the extent to which people have positive experiences and can secure the good things in life because they have economic resources.

absolute poverty not having enough money to afford the basic necessities of life.

relative poverty not having enough money to maintain an average standard of living.

poverty line the minimal income level that the federal government considers necessary for basic subsistence (also called the *poverty threshold*).

and two children), is adjusted every year to include cost-of-living increases. If a family makes more than the poverty line, it's usually not eligible for public assistance (De-Navas-Walt and Proctor, 2015).

Is the poverty line realistic? Some contend that the poverty line is too high. They argue, for example, that many people aren't as poor as they seem because the poverty threshold doesn't include the value of noncash benefits such as food stamps, medical services (primarily Medicare and Medicaid), public housing subsidies, and unreported income (Eberstadt, 2009; Haskins, 2015).

Others claim that the poverty line is too low because it doesn't include child care and job transportation costs or the cost-of-living expenses, particularly housing, that vary considerably across states, regions, and urban-rural areas. According to some economists, a family of four needs almost $60,000 a year to cover basic necessities, a considerably higher amount than the official poverty line of just over $24,000. Some also argue that poverty estimates exclude millions of Americans who live above the poverty line but rely on food banks, soup kitchens, and clothing thrift stores to survive (Fremstad, 2012; DeGraw, 2014).

8-3b Who Are the Poor?

In 2014, the official U.S. poverty rate was almost 15 percent (46.7 million Americans), up from 11 percent in 2000. Between the ages of 25 and 60, 62 percent of Americans will experience at least one year of poverty (DeNavas-Walt and Proctor, 2015; Rank and Hirschl,

One in three poor Americans live in the suburbs. Compared with central cities, many suburbs offer safer streets, better schools, and a greater number of low-paying jobs (e.g., retail sales, landscaping, restaurants). Rents are rising, however, and many of the suburban poor worry about the costs of maintaining a car to get to work ("Broke in the 'Burbs," 2013; Kneebone and Berube, 2014).

2015). Both historically and currently, the poor share some of the following characteristics.

AGE

Children under age 18 make up only 23 percent of the U.S. population, but 33 percent (almost 16 million) of the poor, up from a low of 16 percent in 2001. Americans aged 65 and older make up about 14 percent of the total population, but 10 percent (4.6 million) of the poor. The poverty rate of older Americans is at an all-time low, and lower than that of any

Six Ways the Poor Pay More

Food: Not having a car to get to a supermarket chain or discount store means getting groceries at the corner store where staples are more expensive, produce is limited, and some products are stale.

Transportation: About 45 percent of American households don't have access to public transportation. Old cars break down, can be expensive to fix, use a lot of gas, and can cost people a job.

Doing Business: Most neighborhood stores charge more for products because real estate and insurance are more expensive. Poor people also pay more for financial services. Banks are scarce and "check-cashing stores" charge up to 10 percent of a check's value to cash it.

Credit: Because poor people don't have credit, they must rely on "cash advance" stores if they can prove that they get a regular paycheck. If poor people qualify for such a loan, they pay an annual percentage rate of about 825 percent.

Everyday Hassles: The poor are hassled almost every day by bill collectors because of overdue payments, deal with Laundromat trips, carry groceries long distances, and contend with violence in high-crime areas.

Waiting: Poor people spend much of their time waiting—in food pantry lines, at blood banks to sell their blood, at bus stops to get to work, and for apartments in safer neighborhoods.

Sources: Brown, 2009; Bass and Campbell, 2013; Johnson, 2014 "Tackling Poverty," 2015.

other age group, because government programs, particularly Medicare and Medicaid, have generally kept up with the rate of inflation. In contrast, many programs for poor children have been reduced or eliminated since 1980 (Children's Defense Fund, 2014; DeNavas-Walt and Proctor, 2015).

GENDER AND FAMILY STRUCTURE

Women's poverty rates are higher than men's (16 and 13 percent, respectively), but family structure is an important factor. In 2014, 6 percent of married-couple families, 16 percent of male-headed families, and 31 percent of female-headed families lived in poverty (DeNavas-Walt and Proctor, 2015).

The term **feminization of poverty** refers to the disproportionate number of the poor who are women. Because of increases in divorce, nonmarital childbearing, and low-paying jobs, single-mother families are four to five times more likely to be poor than married-couple families that have two wage earners, and more likely to be extremely poor over many years (Schaefer and Edin, 2014; Shriver, 2014; Irving and Loveless, 2015).

RACE AND ETHNICITY

In absolute numbers, there are more poor whites (19.7 million) than poor Latinos (13.1 million), blacks (10.8 million), or Asians (2.1 million) (DeNavas-Walt and Proctor, 2015). Proportionately, however, and as *Figure 8.4* shows, whites and Asian Americans are less likely to be poor than other racial-ethnic groups.

EDUCATION

Increasingly, education separates the poor from the nonpoor. In 2013, just 5 percent of college graduates were poor, compared with 14 percent of high school graduates and 28 percent of those without a high school diploma (Mather and Jarosz, 2014). The more risk factors, the more likely that someone will experience poverty. For example, being a young, black, unmarried female with young children and little education increases the odds of both poverty and extreme poverty (living below the bottom 10 percent of the income distribution) (Rank and Hirschl, 2015; The Annie E. Casey Foundation, 2015).

8-3c Why Are People Poor?

Social scientists have a number of theories about poverty (see Cellini et al., 2008, for a nontechnical summary). Regarding two of the most common explanations, one blames the poor themselves, and the other emphasizes societal factors.

BLAMING THE POOR: INDIVIDUAL FAILINGS

About 60 percent of Americans believe that people can be successful if they're more industrious. And, in a recent

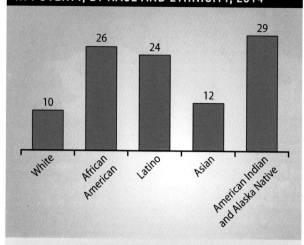

FIGURE 8.4 PERCENTAGE OF AMERICANS LIVING IN POVERTY, BY RACE AND ETHNICITY, 2014

Sources: Based on DeNavas-Walt and Proctor, 2015, Table 3, and American FactFinder, 2013, TableS0201 (accessed September 10, 2015, factfinder2.census.gov.in).

national survey, 45 percent said that "poor people today have it easy because they can get government benefits without doing anything in return" (Keeter and Doherty, 2015).

Such attitudes reflect a still influential *culture of poverty* perspective, which contends that people are poor because they're personally "deficient" or "inadequate." If they worked harder, took initiative, were ambitious, and planned for the future instead of seeking immediate gratification, according to this perspective, poor people could succeed. Instead of being responsible and getting jobs, the poor rely on generous government handouts that erode an incentive to work (Lewis, 1966; Banfield, 1974; Murray, 1984, 2012). Thus, the most effective remedy to poverty is to change people's attitudes and behavior.

BLAMING SOCIETY: STRUCTURAL CHARACTERISTICS

In contrast to blaming the poor, most sociologists contend that macro-level structural factors create and sustain poverty. Technology, automation, industry relocation, and the growth of a low-wage service sector produce conditions of poverty that are difficult for individuals to overcome. As the number of inner-city manufacturing jobs declined, neighborhood businesses collapsed, middle classes—both white and black—moved to the suburbs, and the poor became more isolated and

feminization of poverty the disproportionate number of the poor who are women.

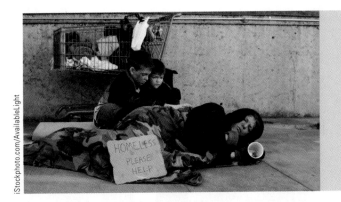

In 2014, almost 579,000 Americans were homeless on a given night, down by 11 percent since 2007. Most (69 percent) stayed in residential programs, 37 percent were people in families, and 23 percent were children under the age of 18. Unemployment, poverty, and a lack of affordable housing are the main causes of homelessness (Henry et al., 2014; National Alliance to End Homelessness, 2015).

welfare-dependent (Oliver and Shapiro, 1995; Wilson, 1996; Conley, 1999). Thus, the most effective remedy to poverty is changing structural forces like unemployment and very low wages.

Which perspective is more accurate, blaming the poor or blaming society? Some people are poor because they're lazy and would rather get a handout than a job. Most people are poor, however, because of economic factors (low wages, job loss, lack of affordable housing, inability to afford health insurance), which, in turn, can result in acute health and employment problems.

Millions of poor Americans don't claim benefits to which they're entitled. They don't seek temporary financial assistance for food, shelter, or utilities, for instance, because they don't know about the programs, don't know where to apply, or have transportation and child-care difficulties in meeting some of the job-training requirements. Moreover, and despite personal failings and deficiencies, millions of middle class Americans get government assistance through programs for the disabled, veterans, college students, older people, and the unemployed, and wealthy farmers receive generous federal subsidies (Levin et al., 2014; Rampell, 2014; Eckhoff, 2015; Rodrigue and Reeves, 2015).

8-4 SOCIAL MOBILITY

Social mobility is movement from one social class to another. Is the American dream of moving up the economic ladder a reality? Only 54 percent of Americans—down from 81 percent in 1998—believe that there's

plenty of opportunity to get ahead, 59 percent say that it's almost impossible to achieve the American dream, and 26 percent expect to move down in social class over the next few years (Dugan and Newport, 2013; Morello and Clement, 2014). Are they right?

8-4a Types of Social Mobility

Intragenerational mobility is movement up or down a social class over one's lifetime. If Ashley begins as a nurse's assistant, becomes a registered nurse, and then a physician's assistant (PA), she experiences intragenerational mobility. **Intergenerational mobility** is movement up or down a social class over two or more generations. If Ashley's parents were blue-collar workers, her upward movement to the middle class is an example of intergenerational mobility. Intragenerational and intergenerational mobility can be downward or upward.

8-4b Recent Trends in Social Mobility

From the 1940s to the 1970s, U.S. upward mobility increased. Since then, many Americans have been less likely to rise to a higher social class. A child raised in a household in the poorest fifth of the income distribution has a 40 percent chance of being stuck at the bottom as an adult, and just a 9 percent chance of making it to the top 20 percent by adulthood. Those raised at the top of the 20 percent have a 37 percent of staying there, and only an 8 percent chance of tumbling to the bottom of the ladder (Reeves and Grannis, 2013; Chetty et al., 2014).

Americans have become considerably less likely to identify themselves as middle class (see *Figure 8.5*). Some have moved up the economic ladder, but many have been downwardly mobile. In 2014, for example, 40 percent of Americans said they were lower-middle or lower class, up from 25 percent in 2008 (Morin and Motel, 2012; Kochhar and Morin, 2014). Because of such changes, some researchers and analysts describe the middle class as "eroding," "sinking," "shrinking," and

social mobility movement from one social class to another.

intragenerational mobility movement up or down a social class over one's lifetime.

intergenerational mobility movement up or down a social class over two or more generations.

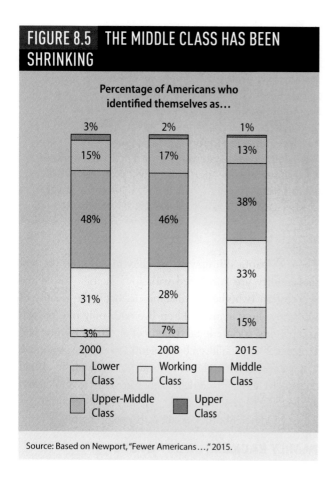

FIGURE 8.5   THE MIDDLE CLASS HAS BEEN SHRINKING

Percentage of Americans who identified themselves as...

	2000	2008	2015
	3%	2%	1%
	15%	17%	13%
	48%	46%	38%
	31%	28%	33%
	3%	7%	15%

- ☐ Lower Class
- ☐ Working Class
- ◼ Middle Class
- ◼ Upper-Middle Class
- ◼ Upper Class

Source: Based on Newport, "Fewer Americans...," 2015.

an "exclusive club" (see, for example, Galston, 2013; and Porter, 2013).

When social mobility occurs, most moves are short because social classes are pretty rigid (Winship, 2011; Wysong et al., 2014). A child from a working-class family, for example, is more likely to move up to the lower-middle class than to jump to the lower-upper class. The same is true of downward mobility: Someone from the middle class is more likely to slide into the working class than to drop to the underclass. Moreover, upward intergenerational mobility has become more limited because the rungs of the economic ladder have grown further apart (Chetty et al., 2014). As you saw earlier, income inequality has increased, making it more difficult for even upper-middle class people to move up.

A study of 22 high-income countries found that the United States ranked only 15th for up-ward intergenerational mobility. There's less upward mobility in the United States than in many other industrialized countries as well as some developing countries (e.g.,

Ecuador and Peru) (Corak, 2013). In effect, countries with the greatest income inequality, such as the United States, are also the most likely to have limited upward mobility (Greeley, 2013).

8-4c What Affects Social Mobility?

Upward mobility doesn't always reflect people's talents, intelligence, or hard work. Instead, much social mobility depends on structural, demographic, and family background factors, all of which are interrelated.

STRUCTURAL FACTORS

Macro-level variables, over which people have little or no control, affect social mobility in several ways. First, *changes in the economy* spur upward or downward mobility. During an economic boom, the number of jobs increases, and many people have an opportunity to move up. During recessions, such as the Great Recession, long-term unemployment leads to downward mobility (see Chapters 11 and 12).

Second, *government policies and programs* affect social mobility. Unlike the United States, European countries that have promoted equality (e.g., Canada, Denmark, Finland, Norway) have the highest upward mobility rates. They have funded universal health care, which reduces the chance of people falling into poverty because of medical emergencies or poor health; have provided affordable housing; and have invested in technical schools where young people learn a high-paying trade (Foroohar, 2011; Deparle, 2012).

Third, *immigration* fuels upward mobility. Because many recent immigrants take low-paying jobs, groups that are already in a country move into higher-paying occupations (Haskins, 2008).

DEMOGRAPHIC FACTORS

Demographic factors also affect social mobility. Four of the most important are education, gender, race, and ethnicity, but geography also has an effect.

AP Images/Kirsty Wigglesworth

J. K. Rowling worked as a teacher. After a divorce, she lived on public assistance, writing *Harry Potter and the Philosopher's Stone* during her daughter's naps. The book was published in 1997; by 2007, Rowling, a billionaire and one of the world's richest women, was considerably wealthier than the Queen of England.

MCT/Tribune News Service/Getty Images

Children raised in poor families can be upwardly mobile. Sonia Sotomayor, whose parents were born in Puerto Rico, is the first Latina to serve on the U.S. Supreme Court. Her family first lived in a South Bronx tenement and later a working-class housing project. Sotomayor's father, an alcoholic, had a third-grade education, didn't speak English, and was a tool and die worker. He died when Sotomayor was nine years old. Her mother—a telephone operator and later a practical nurse—stressed the value of education. Sotomayor excelled in school, received a full scholarship to Princeton University, and put in long hours to catch up with her peers' academic knowledge.

Education. Especially when the economy is slumping, people with college and graduate degrees fare better than those with a high school education or less. The latter often face long and frequent bouts of unemployment, must get by with temporary work, and may move down the socioeconomic ladder because employment opportunities are greatest in high-skill and low-skill rather than middle-skill jobs (Steuerle et al., 2013; Kochhar and Morin, 2014).

Gender. Women's massive entry into the labor force since the 1980s has increased family income and many single women's upward mobility. In 1960, only 42 percent of women aged 25 to 54 were in the labor force compared with almost 75 percent in 2014 (U.S. Bureau of Labor Statistics, 2014). You'll see in later chapters that men's labor participation rates have been dropping, but men's mobility is less affected than women's by a divorce, nonmarital children, or widowhood.

Race and ethnicity. African Americans and Latinos, especially those from low-income families, usually experience little upward mobility. Black and Latino middle classes have grown since the 1970s, but both groups still lag significantly behind whites and Asian Americans in median family income (see Chapters 10 and 11). During the Great Recession, even a college degree didn't protect middle-class black and Latino families from downward mobility. Because the households had little inherited wealth, they had to borrow heavily for down payments on a home and their children's college tuition (Emmons and Noeth, 2015).

Geography. Where one lives can stimulate or dampen upward mobility. For example, the probability of a child born into the poorest fifth of the population in San Jose, California, making it to the top fifth is 13 percent compared with only 4 percent for a child born in Charlotte, North Carolina.

These disparities in upward mobility in different parts of the country are correlated with four factors: residential segregation (whether by income or race), the quality of schooling, how many children live with only one parent, and parents' social capital (Chetty et al., 2014; Berube and Holmes, 2015). Children below age 13 whose parents move to lower-poverty neighborhoods are more likely to attend college, to have substantially higher incomes as adults, and are less likely to become single parents (Chetty et al., 2015). Thus, good neighborhoods can nurture success.

FAMILY BACKGROUND FACTORS

You saw earlier that high-income households have more economic, cultural, and social capital than do low-income households. A study tracked 790 Baltimore children from the year they entered first grade until they turned 29 years old. The researchers found that the children's social class in adulthood was pretty much the same as the family into which they were born. For example, only 4 percent of the children from low-income families had a college degree by age 30 compared with 45 percent of those from higher-income backgrounds (Alexander et al., 2014).

Only 7 of the 44 U.S. presidents came from the lower-middle class or below. Abraham Lincoln, although born in a log cabin, had a father who was one of the wealthiest people in the community. And about a third of the students with low grades at Ivy League universities wouldn't be there if their parents weren't celebrities, well-known politicians, or others who donate millions to the school (see Chapters 11 and 13).

According to one policy analyst, the most important divide in America today is social class, not race, and the growing gulf between how high-income and low-income parents raise children has widened. Unlike children growing up in low SES households, those in high-income families benefit from high-quality schools, helpful mentoring connections, enriching extracurricular activities, and so on (Putnam, 2015; see also Chapter 4). High- and low-income parents spend about 10 percent of their annual household

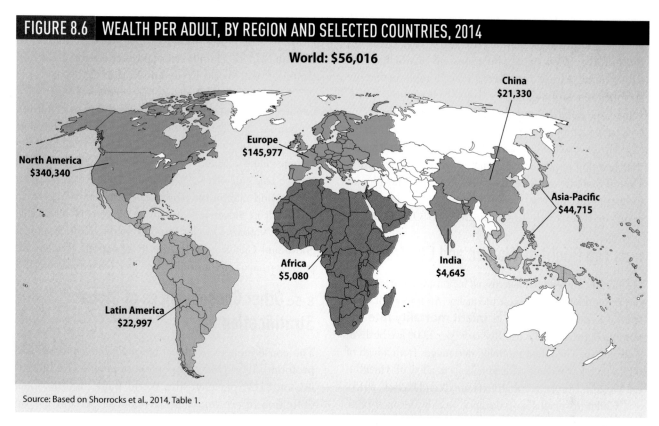

FIGURE 8.6 WEALTH PER ADULT, BY REGION AND SELECTED COUNTRIES, 2014

World: $56,016

China $21,330

Europe $145,977

North America $340,340

Asia-Pacific $44,715

Africa $5,080

India $4,645

Latin America $22,997

Source: Based on Shorrocks et al., 2014, Table 1.

income (almost $12,900 and $5,800, respectively) to help their adult children pay for bills, rent, tuition, and other expenses. However, the higher the parents' SES, the greater the financial assistance and children's opportunities to succeed (Wightman et al., 2012; Mitnik and Grusky, 2015).

Our socialization affects what French sociologist Pierre Bourdieu (1984) called *habitus*—the habits of speech and lifestyle that determine where a person feels comfortable and knowledgeable. Upper-class parents cultivate flexibility, autonomy, and creativity because they expect their children to step into positions that require such characteristics. Poor and working-class parents stress obedience, honesty, and appearance—traits that many employers expect. High-income parents have social connections to jobs and admission to particular schools or colleges. Because of such resources, children of top earners are likely to grow up to be top earners themselves (Khan, 2012; Corak, 2013).

8-5 GLOBAL STRATIFICATION

Global stratification refers to worldwide inequality patterns that result from differences in wealth, power, and prestige. All societies are stratified, but more than 75 percent of the world's population lives in countries where economic inequality has widened, increasing the possibility of social class tension and conflict (United Nations Development Program, 2013; World Economic Forum, 2013).

8-5a Wealth and Income Inequality

In 2014, global wealth reached a new all-time high of $263 trillion, but how is it distributed?

▶ The richest 1 percent of the world's adults own almost half of the world's wealth.

▶ The top 10 percent of the world population owns 86 percent of global wealth.

▶ The 85 richest people in the world have as much wealth as the 3.5 billion poorest (Shorrocks et al., 2014; Hardoon, 2015).

There are enormous wealth disparities across regions (see *Figure 8.6*), and the inequalities have swelled. The incomes of the top 1 percent have increased 60 percent since 1990, and the income growth of the ultra-rich 0.01 percent has been even greater. Even in more economically egalitarian countries such as Norway and Sweden, the share of income going to the richest 1 percent increased by more than 50 percent between 1980 and 2012 (Oxfam, 2013; Fuentes-Nieva and Galasso, 2014).

Worldwide, income inequality is also pervasive. Across 131 countries, the richest 3 percent hold

global stratification worldwide inequality patterns that result from differences in wealth, power, and prestige.

20 percent of all income. The median per capita incomes in the wealthiest populations are more than 50 times those in the 10 poorest populations, all of which are in sub-Saharan Africa, but there's considerable economic stratification within countries. In oil-rich Nigeria, for instance, the richest 6 percent hold 40 percent of the country's total income (Phelps and Crabtree, 2013, 2014).

Moreover, living in a wealthy country doesn't mean that people enjoy income equality. Of 34 countries, the United States ranks 31st in income equality—far behind industrialized nations and just above some developing countries that include Turkey and Mexico (OECD, 2014).

8-5b The Plight of Women and Children

Historically, currently, and across *all* nations, women and children experience the greatest inequality. An important measure of a country's health is its **infant mortality rate**, the number of babies under age 1 who die per 1,000 live births in a given year. The infant mortality rate ranges from a high of 92 in some low-income African nations to a low of 3 in affluent Western European countries (Kaneda and Bietsch, 2015).

Children's deaths have declined worldwide since 1990, but every year nearly 7 million poor children die before their fifth birthday; most of the deaths continue to be concentrated in sub-Saharan Africa and Southern Asia (e.g., Afghanistan, Bangladesh, Iran, Pakistan). Across and within countries, 87 percent of the under-five deaths are due to poverty (United Nations Development Program, 2013; You et al., 2013).

There's a strong association between high infant and child mortality rates and women's low education levels. Investing in women's education reduces unwanted and

infant mortality rate the number of babies under age 1 who die per 1,000 live births in a given year.

unplanned births; lowers infant, child, and maternal mortality; increases women's labor force participation; and can help lift a family out of poverty (World Bank and Collins, 2013; World Economic Forum, 2013).

Education doesn't guarantee women's greater economic equality, however. In both rich and poor countries, women are less likely than men to be employed because of cultural attitudes about women's roles and child rearing responsibilities. When women work outside the home, the *gender pay gap*—the overall income difference between women and men in the workplace—ranges from a low of 3 percent in Slovenia to a high of 46 percent in Zambia (United Nations Development Program, 2013; Fuentes-Nieva and Galasso, 2014). And in all nations, women are paid less than men for similar work (see Chapters 9 and 12).

8-5c Other Consequences of Global Stratification

The consequences of global inequality are massive and profound. Worldwide, 22 percent of people (1.2 billion) live on $1.25 per day—the World Bank's international definition of *extreme poverty*—and 34 percent live on no more than $2.00 per day. The extreme poverty rate fell by 25 percent between 1981 and 2010—from 2.1 to 1.9 million. During the same period, however, the extreme poverty rate in low-income countries (where the annual per capita income is $1,035 or less) increased by 33 percent (from 249 to 352 million people). Across regions, 54 percent of the residents of sub-Saharan Africa and 38 percent of those in South Asia—compared with 1 percent in Canada, Europe, and the United States—live in extreme poverty (Olinto et al., 2013; Phelps and Crabtree, 2013; see also Edin and Shaefer, 2015).

Because of severe destitution, almost one in eight people worldwide suffer from chronic hunger and undernutrition. In some impoverished regions, women and children must walk seven or more miles each day to get clean water. In Australia, Canada, and Germany—three of the most affluent countries—about 10 percent struggle to pay for food, compared with 50 to 70 percent of those in poor African countries (Office of the United Nations…, 2010; FAO, IFAD, and WFP, 2013).

People in poor countries report high levels of unhappiness, physical and mental

We hear about the booming economies of India and China, but not everyone has benefited. A tent city in India houses many of the workers who earn about $1.30 a day building new office towers for the affluent (left). In China, a man in Shanghai begs as wealthier residents pass by (right).

iStockphoto.com/jacus

REUTERS/Aly Song/Landov

El Greco/Shutterstock.com

health problems, and have little chance to escape poverty. Economic inequality has other costs: The world's richest 1 percent use their wealth to get political favors, including bypassing environmental laws; have a global network of tax havens worth almost $19 trillion within their own countries and offshore; and harm democracy by reducing opportunities for upward mobility (Helliwell et al., 2013; Kirkland, 2014; Oxfam, 2014).

8-5d Why Is Inequality Universal?

Many theories try to explain why inequality is universal, but three of the most influential have been modernization theory, dependency theory, and world-system theory.

Modernization theory claims that low-income countries are poor because their leaders don't have the attitudes and values that lead to experimentation and using modern technology. Instead, policy makers adhere to traditional customs that isolate and prevent them from competing in a global economy. In effect, modernization theory blames poor nations for their poverty and other problems. After the key foundations of modernity and capitalism are in place, this perspective maintains, low-income countries will prosper.

Dependency theory contends that the main reason why low-income countries are poor is because they're pawns that high-income countries exploit and dominate. Rich nations wield an enormous amount of power by exporting jobs overseas, manipulating foreign aid, draining

less powerful countries of their resources, penetrating other countries with multinational corporations, and coercing national governments to comply with their interests (e.g., by not passing environmental laws). In effect, according to dependency theorists, high-income countries benefit because the poor provide cheap labor and aren't powerful enough to protest even though they work in hazardous conditions and earn less than $1 a day.

More recently, *world-system theory*, similar to dependency theory, argues that "the economic realities of the world system help rich countries stay rich while poor countries stay poor" (Bradshaw and Wallace, 1996: 44). That is, countries such as the United States that dominate the world economy control the economies of low-income countries because their workers depend on external markets for jobs. High-income countries can extract raw materials (e.g., diamonds and oil) with little cost. They can also set prices, regardless of market values, for agricultural products that low-income countries export. Doing so forces many small farmers to abandon their fields because they can't pay for labor, fertilizer, and other costs (Carl, 2002; Alvaredo et al., 2013).

8-6 SOCIOLOGICAL EXPLANATIONS: WHY THERE ARE HAVES AND HAVE-NOTS

Why are societies stratified? *Table 8.1* summarizes the key points of the four theoretical perspectives. Let's begin by looking at a long-standing debate between functionalists and conflict theorists on why there are haves and have-nots.

TABLE 8.1 SOCIOLOGICAL EXPLANATIONS OF SOCIAL STRATIFICATION

Perspective	Level of Analysis	Key Points
Functionalist	Macro	• Fills social positions that are necessary for a society's survival • Motivates people to succeed and ensures that the most qualified people will fill the most important positions
Conflict	Macro	• Encourages workers' exploitation and promotes the interests of the rich and powerful • Ignores a wealth of talent among the poor
Feminist	Macro and micro	• Constructs numerous barriers in patriarchal societies that limit women's achieving wealth, status, and prestige • Requires most women, not men, to juggle domestic and employment responsibilities that impede upward mobility
Symbolic Interactionist	Micro	• Shapes stratification through socialization, everyday interaction, and group membership • Reflects social class identification through symbols, especially products that signify social status

8-6a Functionalist Perspectives: Stratification Benefits Society

Functionalists see stratification as both necessary and inevitable: Social class provides each individual a place in the social world, and motivates people to contribute to society. Without a system of unequal rewards, functionalists argue, many important jobs wouldn't be performed.

THE DAVIS–MOORE THESIS

Sociologists Kingsley Davis and Wilbert Moore (1945) developed one of the most influential functionalist perspectives on social stratification that persists today. The **Davis–Moore thesis**, as it's commonly called, asserts that social stratification benefits society. The key arguments of the Davis–Moore thesis can be summarized as follows:

1. **Every society must fill a wide variety of positions and ensure that people accomplish important tasks.** Societies need teachers, doctors, farmers, trash collectors, plumbers, police officers, and so on.

2. **Some positions are more crucial than others for a society's survival.** Doctors, for example, provide more critical services to ensure a society's continuation than do lawyers, engineers, or bankers.

3. **The most qualified people must fill the most important positions.** Some jobs require more skill, training, or intelligence than others because they're more demanding, and it's more difficult to replace the workers. Pilots, for example, must have more years of training and aren't replaced as easily as flight attendants.

4. **Society must offer greater rewards to motivate the most qualified people to fill the most important positions.** People won't undergo many years of education or training unless they're rewarded by money, power, status, and/or prestige. If doctors and nurses earned the same salaries, there wouldn't be much incentive for people to spend so many years earning a medical degree.

According to the Davis–Moore thesis and other functionalist perspectives, then, stratification and inequality are necessary to motivate people to work hard and to succeed. For functionalists, social stratification is based on **meritocracy**—people's accomplishments.

> **Davis–Moore thesis** the functionalist view that social stratification benefits a society.
>
> **meritocracy** a belief that social stratification is based on people's accomplishments.

Singer Lady Gaga had an estimated net worth of $220 million in 2015. Do her earnings reflect her contribution to society, especially compared with teachers, physicians, trash collectors, computer scientists, and others?

ChinellatoPhoto/Shutterstock.com

That is, people are rewarded for what they do and how well, rather than their ascribed status.

CRITICAL EVALUATION

Sociologist Melvin Tumin (1953) challenged the Davis–Moore thesis. First, he argued, societies don't always reward the positions that are the most important for the members' survival. If the multi-millionaire professional athletes, actors, and pop musicians went on strike, many of us would probably barely notice. If, on the other hand, those with much lower income—whether they're doctors, garbage collectors, teachers, truck drivers, or mail carriers—refused to work, society would grind to a halt. Thus, according to Tumin, there's little association between earnings and the jobs that keep a society going.

Second, Tumin claimed, Davis and Moore overlook the many ways that stratification limits upward mobility. Where wealth is differentially distributed, access to education, especially higher education, depends on family wealth. As a result, large segments of the population are likely to be deprived of the chance to discover what their talents are, and society loses.

Third, Tumin criticized Davis and Moore for ignoring the critical role of inheritance. In upper classes, sons and daughters don't have to work because their inherited wealth guarantees a lifetime income and perpetuates privileges over generations.

Functionalist theories also don't explain why (1) upward social mobility is more limited in the United States than in other industrialized (and even some developing) countries; (2) so many college graduates can find only low-paying jobs; and (3) racial/ethnic income and wealth gaps persist across all social classes (Legatum Prosperity Index, 2013; McKernan et al., 2013).

8-6b Conflict Perspectives: Stratification Harms Society

Like Tumin, conflict theorists maintain that social stratification is dysfunctional because it hurts individuals and societies. Karl Marx's (1934) analysis of social class and inequality has had a profound influence on modern sociology, especially conflict theory.

CAPITALISM BENEFITS THE RICH

Marx was aware that a diversity of classes can exist at any one time, but he predicted that capitalist societies would ultimately be reduced to two social classes: the capitalist class, or bourgeoisie; and the working class, or proletariat. The **bourgeoisie**, those who own and control capital and the means of production (e.g., factories, land, banks, and other sources of income), can amass wealth and power. The **proletariat**, workers who sell their labor for wages, earn barely enough to survive.

For conflict theorists, the economic struggles of the U.S. middle and working classes since the late 1970s weren't primarily the result of globalization and technological changes but, instead, a long series of government policies that overwhelmingly favored the rich. This policy of **corporate welfare** consists of an array of subsidies, tax breaks, and assistance that the government has created for businesses. For example:

▶ Taxpayers are paying $7 trillion for bailing out mismanaged financial institutions. The companies' executives still receive multimillion-dollar annual salaries and benefits (Ivry et al., 2015; Kiel and Nguyen, 2015; see also Chapter 11).

▶ According to one estimate, each household pays $10,000 per year to subsidize corporations. Companies benefit enormously from the taxpayer-supported highways they use, law enforcement and judicial systems that protect their intellectual and physical property, public education that provides a workforce, and military personnel who safeguard corporate assets abroad. Other costs include paying for corporations' unpaid taxes, cleaning up oil producers' pollution and environmental disasters, and funding research that results in patent-protected technology that reaps massive corporate profits (Buchheit, 2015).

▶ Because of many tax havens and legal loopholes, many giant corporations (e.g., Apple, Amazon, Google, Walmart) pay very low taxes—less than 2 percent compared with the average U.S. employee's 33 to 41 percent. In 2014, 11 of the largest companies (like Goodyear Tire & Rubber) that reported billions in profits paid no taxes at all. In the same year, the Internal Revenue Service awarded $19 million in contracts to 17 corporations that owed back taxes, despite federal bans on such awards (Clemente and Auerbach, 2015; Krantz, 2015; McCoy, 2015; Tsukayama, 2015).

Considerable data support conflict theorists' contention that severe and growing inequality is dysfunctional for society. Among other social problems, ongoing economic inequality intensifies poverty, undermines people's trust in political and economic institutions, and, consequently, erodes national solidarity. Whether their concern is driven by self-interest or a sense of justice, even powerful financiers and conservative economists have begun to worry about economic inequality. They fear that the increasing income and wealth gaps and low wages have made the United States less productive, have narrowed the consumer spending base, have discouraged many people from entering or staying in the labor force, and that the huge gulf between the top 1 to 5 percent and the rest of society may lead to "oppressive taxes" on the superrich, widespread social unrest, and even a revolution (Dumaine, 2015; Hightower, 2015).

CRITICAL EVALUATION

Conflict theories have their limitations. First, and despite Marx's prediction, even though corporate wealth has greatly increased during the past 100 years and many people have become poorer, there have been some protests but no revolutions in capitalist countries. Americans aren't demanding greater economic equality, and many go into debt to purchase luxury items such as TVs, jewelry, and pet supplies, and to replace expensive smartphones and other high-tech hardware every few years. Thus, according to critics, conflict theorists are exaggerating the existence and effects of economic inequality.

Second, do conflict theorists underrate people's ability to be upwardly mobile? Indeed, 60 percent of Americans are satisfied with their opportunities to succeed (Newport, 2015).

Third, some critics point out that conflict theorists ignore the fact that government programs have cut poverty by 40 percent since the 1960s. About 46 percent of American households pay no federal income taxes (and these tax breaks have nearly doubled since 1975) because

bourgeoisie those who own and control capital and the means of production.

proletariat workers who sell their labor for wages.

corporate welfare subsidies, tax breaks, and assistance that the government has created for businesses.

they receive public assistance, get deductions for raising children under age 18, or don't report income. Moreover, the United States spends far more, per capita, than do many other wealthy countries on pensions, health care, family support, unemployment, housing assistance, and similar benefits (Rosenberg, 2014; Kirkegaard, 2015).

8-6c Feminist Perspectives: Women Are Almost Always at the Bottom

For feminist scholars, functionalist and conflict theories are limited because they typically focus on men in describing and analyzing stratification and social class. As a result, women are largely invisible.

STRATIFICATION BENEFITS PRIMARILY MEN, NOT WOMEN

An estimated 70 percent of the world's poor are women. Often, in both rich and poor countries, gender inequality results in discrimination in the economy, politics, and access to medical services. Especially in low-income countries, almost 4 million girls and women are "missing" each year because of high death rates. About 20 percent are aborted because of a preference for sons, and more than a third die from complications of childbirth and pregnancy. Besides excessive deaths, many women are blocked from entering higher-paying occupations (e.g., finance and business), earn less than men in all jobs, have less access to land and credit, and have little decision-making power in households and politics (World Bank, 2012; Independent Television Service, 2014).

More than 100 countries ban women from certain jobs, accessing finance, owning businesses, or conducting legal affairs. And a "global pandemic" of violence against females affects one in three worldwide, including young girls who become sex slaves (UN Women, 2014).

Feminist theorists contend that, in a patriarchal system, men shape the stratification system: They control a disproportionate share of wealth, prestige, and power, and the feminization of poverty results in women's downward mobility. In addition, women often have to overcome economic inequities as well as juggle domestic and workplace responsibilities (see Chapters 11 and 12). The gender gaps in wealth, income, and household burdens make it harder for women to access resources and opportunities to build a strong future for themselves and their families.

CRITICAL EVALUATION

Some critics point out that feminist theorists often focus only on poor women in showing how patriarchy affects stratification and social class (see Kendall, 2002). Another criticism is that many feminist scholars don't explain why so many women succeed despite patriarchal barriers. Third, feminist theories don't account for some striking cross-cultural variations. In a study of 165 countries, for example, Canada, Australia, the Netherlands, and the United States ranked in the top ten regarding women's legal treatment, workplace participation, and

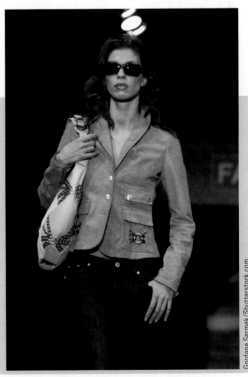

In 2013, more than 1,100 workers were killed and 2,000 were injured when an unsafe garment factory collapsed in Bangladesh. The country exports $20 billion in clothes annually to European and U.S. retailers, including Walmart. The $38-a-month minimum wage (for a six-day workweek) barely covers living expenses. More than 80 percent of the 4 million workers in Bangladesh's garment industry are women, mostly young and with little schooling (Al-Mahmood, 2013; Hammadi, 2013; Srivastava, 2013).

access to education and health care, but well behind developing countries such as Cuba, Rwanda, and Burundi in women's representation in political offices (Ellison, 2011). Thus, feminist theories don't explain why women have greater political power in some patriarchal and developing countries than in many industrialized nations.

8-6d Symbolic Interaction Perspectives: People Create and Shape Stratification

Symbolic interactionists focus on how people create, change, and reproduce social classes. They address micro-level issues such as how people learn their social positions in everyday life and how such learning affects their attitudes, behavior, and lifestyles.

SOCIAL CONTEXTS SHAPE SOCIAL CLASS

People across social classes interact with and socialize their children differently. By age 2, children from higher-income homes know 30 percent more words than those from low-income homes. Early vocabularies increase reading comprehension and proficiency as early as kindergarten (Fernald et al., 2013).

Children as young as 4 years old are aware of and enact their social class. Among preschoolers, upper-middle-class children speak, interrupt, ask for help, and argue more often than do working-class children. Such behavior receives more adult attention and gives upper-middle-class children more opportunities to develop their language skills (Streib, 2011).

Social contexts also affect mobility. The social classes into which children are born affect their aspirations, the skills they value and can access, and the networks and resources on which they can draw. Doctors' children, for example, are more likely to become physicians themselves because they're exposed to medical discussions at home, are encouraged to pursue medicine as an occupation, and are embedded in social networks that provide medical school contacts (Jonsson et al., 2009).

Few people are demanding greater economic equality because social stratification is internalized. About 65 percent of Americans say that the U.S. economic system favors the wealthy, that the income gap between the rich and poor has widened since the early 2000s, and that not everyone has an equal chance to succeed. However, 43 percent believe that the rich are more hardworking and intelligent than the average person, only 47 percent think that the rich–poor gap is a big problem, and 63 percent say that the country benefits from having a rich class of people (Newport, 2012; Drake, 2013; Stokes, 2013; DeSilver, 2014).

Such attitudes may be reflecting many Americans' hopes that they, too, will be wealthy someday. On the other hand, a recent analysis of 40 developed countries found "widespread ignorance and misperceptions" of inequality. Most ordinary people, including Americans, know very little about income inequality in their own nations, whether inequality is increasing or decreasing, where they are personally on the income spectrum, and what they think they know is often wrong (Gimpelson and Treisman, 2015). Regardless of the reasons, many Americans don't see the connection between structural economic inequality and their own financial situation.

CRITICAL EVALUATION

Symbolic interaction theories are important in understanding the everyday processes that underlie social stratification, but there are several weaknesses. First, interactionism, unlike the other perspectives, doesn't help us understand why stratification exists. Second, the theories don't explain why—despite the same family background, resources, and socialization—some siblings are considerably more upwardly mobile than their brothers and sisters. Third, conflict theorists, especially, fault symbolic interactionists for ignoring structural factors—such as the economy, government policies, and educational institutions—that create and reinforce inequality.

STUDY TOOLS 8

READY TO STUDY? IN THE BOOK, YOU CAN:

☐ Check your understanding of what you've read with the Test Your Learning Questions provided on the chapter review card at the back of the book.

☐ Rip out the chapter review card for a handy summary of the chapter and key terms.

ONLINE AT CENGAGEBRAIN.COM YOU CAN:

☐ Collect StudyBits while you read and study the chapter.

☐ Quiz yourself on key concepts.

☐ Prepare for tests with chapter flash cards as well as flash cards that you create.

☐ Watch videos to explore sociology further.

9 Gender and Sexuality

After you finish this chapter go to **PAGE 177** for **STUDY TOOLS.**

Women pay more than men, and sometimes twice as much, for many things, including cars, mortgages, health care, high-end jeans from the same designer, and similar grooming products like moisturizers, deodorants, and even razors (Browne, 2015; Hill, 2015; Newman, 2015). Are the higher costs due to discrimination? This chapter examines how gender and sexuality affect our lives. First, however, take the short quiz on the next page to see how much you know about these topics.

9-1 SEX, GENDER, AND CULTURE

Some differences between women and men are biological; others are social creations. Both biology and social factors shape a person's identity, but sex and gender aren't synonymous.

9-1a How Sex and Gender Differ

Many people use *sex* and *gender* interchangeably, but they have different meanings. **Sex** refers to the biological characteristics with which we are born—chromosomes, anatomy, hormones, and other physical and physiological

What do you think?

Today, women have more employment advantages than men.

1	2	3	4	5	6	7

strongly agree   strongly disagree

gary718/Shutterstock.com

feminine or masculine depends on gender, a more complex concept than sex.

Gender refers to learned attitudes and behaviors that characterize women and men. Gender is based on social and cultural expectations rather than on physical traits. Thus, most people are *born* either male or female, but we *learn* to be women or men because we internalize behavior patterns expected of each sex. In many societies, for example, women are expected to look young, thin, and attractive, and men are expected to amass as much wealth as possible.

9-1b Sex: Our Biological Component

Physical characteristics such as breasts and beards indicate whether someone is a male or female, but sex isn't always clear-cut. Our cultural expectations dictate that

---

TRUE OR FALSE?

How Much Do You Know About Gender and Sexuality?

1. About 10 percent of the U.S. population is gay.

2. If Americans could have only one child, a majority would prefer a boy.

3. Among spouses and partners, men are more likely than women to lie about financial transactions.

4. Americans would rather work for a male than a female boss.

5. U.S. abortion rates have declined since 1990.

6. Almost a third of Americans who are married or in a committed relationship met online.

The answers are on p. 161.

---

sex the biological characteristics with which we are born.

gender learned attitudes and behaviors that characterize women and men.

---

attributes. These attributes influence our behavior (e.g., shaving beards, wearing bras), but *don't determine* how we think or feel. Whether we see ourselves and others as we are female or male, but a number of people are "living on the boundaries of both sexes" (Lorber and Moore, 2007: 141). For example, **intersexuals** (this term has replaced *hermaphrodites*) are people whose medical classification at birth isn't clearly either male or female. About 1 in 2,000 to 4,000 children born each year are classified as intersex because they're born with both male and female external genitals or an incomplete development of internal reproductive organs. Some parents seek surgery; others wait until a child is old enough to decide what to do (Bendavid, 2013).

SEXUAL IDENTITY AND SEXUAL ORIENTATION

Our **sexual identity** is an awareness of ourselves as male or female and how we express our sexual values, attitudes, and feelings. Our sexual identity incorporates a **sexual orientation**—a preference for sexual partners of the same sex, of the opposite sex, of both sexes, or neither sex:

▸ **Homosexuals** (from the Greek root *homo*, meaning "same") are sexually attracted to people of the same sex. Male homosexuals prefer to be called *gay*, female homosexuals are called *lesbians*, and both gay men and lesbians are often referred to, collectively, as *gays*. *Coming out* is a person's public announcement of a gay or lesbian sexual orientation.

▸ **Heterosexuals**, often called *straight*, are attracted to people of the opposite sex.

▸ **Bisexuals**, sometimes called *bis*, are attracted to both sexes.

▸ **Asexuals** lack any interest in or desire for sex.

---

intersexuals people whose medical classification at birth isn't clearly either male or female.

sexual identity an awareness of ourselves as male or female and how we express our sexual values, attitudes, and feelings.

sexual orientation a preference for sexual partners of the same sex, of the opposite sex, of both sexes, or neither sex.

homosexuals those who are sexually attracted to people of the same sex.

heterosexuals those who are sexually attracted to people of the opposite sex.

bisexuals those who are sexually attracted to both sexes.

asexuals those who lack any interest in or desire for sex.

FIGURE 9.1 SEXUAL ORIENTATION CONTINUUM

0 Exclusively heterosexual	4 Usually homosexual
1 Predominantly heterosexual	5 Predominantly homosexual
2 Usually heterosexual	6 Exclusively homosexual
3 Bisexual	7 Asexual

Sources: Kinsey et al., 1948, p. 638; and Kinsey Institute, 2011.

Sexual orientation, like biological sex, isn't as clear-cut as many people believe. Alfred Kinsey (1948) and his associates' classic study found that most people weren't exclusively heterosexual or homosexual. Instead, they fell somewhere along a continuum in terms of sexual desire, attractions, feelings, fantasies, and experiences. Researchers have recently added *asexual* to Kinsey's classification (see *Figure 9.1*).

Most people's sexual identity corresponds with their biological sex, sexual attraction, and sexual behavior, but not always. Among Americans ages 15 to 44, for example, 86 percent *identify themselves* as straight, but 23 percent have had *same-sex experiences* (Moore, 2015).

PREVALENCE OF HOMOSEXUALITY

How many Americans are gay men, lesbians, and bisexuals? No one knows for sure because, among other reasons, researchers define and measure sexual orientation differently. For example, are people gay if they have ever engaged in same-sex behavior? What about heterosexuals who are attracted to people of the same sex? Moreover, respondents may not be willing to disclose this information, and people who engage in same-sex behavior often identify themselves as straight (Gates, 2011).

Researchers measure the prevalence of homosexuality by simply asking people to identify their sexual orientation. Americans estimate that 23 percent of the population is gay (Newport, 2015). In fact, only about 3 percent of Americans identify themselves as lesbian, gay, and/or bisexual (Ward et al., 2014), and women are more likely than men to report being bisexual (see *Figure 9.2*).

WHAT DETERMINES OUR SEXUAL ORIENTATION?

A Hong Kong billionaire offered $65 million to any man who succeeded in marrying his daughter after

FIGURE 9.2   SEXUAL ORIENTATION AMONG U.S. ADULTS AGED 18 AND OVER

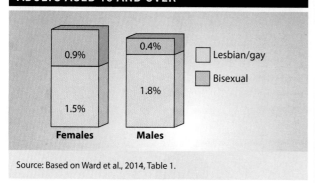

0.9%
0.4%
1.5%
1.8%

☐ Lesbian/gay
☐ Bisexual

Females   Males

Source: Based on Ward et al., 2014, Table 1.

she eloped with her female partner. The offer attracted 20,000 suitors, but none were successful (Nichols, 2014). Like this father, 30 percent of Americans (but down from 56 percent in 1978) believe that being gay or lesbian is a "personal choice" or due to a person's upbringing (Jones, 2015).

Culture shapes people's sexual attitudes and behavior, but no one knows why we're straight, gay, bisexual, or asexual. Sexual orientation must have biological roots, according to some researchers, because homosexuality exists in all societies and, across cultures, the gay population is roughly the same—about 5 percent (Barash, 2012).

There's also growing scientific consensus that biological factors, particularly the early influence of sex hormones after conception and around childbirth, have a strong effect on sexual orientation. For example, pre-gay boys and pre-lesbian girls are gender-nonconforming at an early age compared with their heterosexual counterparts in play and choice of clothes (see LeVay, 2011, for a comprehensive summary of these and other studies). Other researchers speculate that a combination of genetic and cultural factors influence our sexual orientation (Slater, 2013).

9-1c Gender: Our Cultural Component

Gender doesn't occur naturally, but is socially constructed. This means that gender aspects may differ across time, cultures, and even groups within a society. Let's begin with gender identity.

GENDER IDENTITY

People develop a **gender identity**, a perception of themselves as either masculine or feminine, early in life. Many Mexican baby girls but not boys have pierced ears, for example, and hairstyles and clothing for

TRUEORFALSE?

How Much Do You Know About Gender And Sexuality?

Answers to the quiz:

1. False; see p. 160.

2. False; 40 percent of men compared with 28 percent of women say that if they could have only one child, they'd prefer a boy (Newport, 2011).

3. True; 8 percent of men, compared with 5 percent of women, have a hidden bank account, and men (26 percent) are more likely than women (14 percent) to have spent $500 or more without telling their spouses/partners (Mecia, 2015).

4. True; 46 percent have no preference, but of the rest, 33 percent prefer a male boss to a female boss (20 percent) (Riffkin, 2014).

5. True; see p. 169.

6. False; only 5 percent met online. The vast majority of relationships begin offline (Smith, 2014).

American toddlers differ by sex. Gender identity, which typically corresponds to a person's biological sex, is part of our self-concept and usually remains relatively fixed throughout life.

Transgender is an umbrella term for people whose gender identity and behavior differ from the sex to which they were assigned at birth. They comprise about 0.3 percent (700,000) of U.S. adults (Gates, 2011).

Because transgender is independent of sexual orientation, people may identify as heterosexual, gay, bisexual, or asexual. Facebook users can now choose their gender identity from more than 50 possibilities, but here are some of the most common transgender categories (American Psychological Association, 2014):

▶ *Transsexuals* are people whose gender identity differs from their assigned sex. Some, but not all,

gender identity a perception of oneself as either masculine or feminine.

transgender an umbrella term for people whose gender identity and behavior differ from the sex to which they were assigned at birth.

undergo hormone treatment or surgery to change their physical sex to resemble their gender identity.

▶ *Cross-dressers* wear clothing that is traditionally or stereotypically worn by another gender in their culture. People who cross-dress are usually comfortable with their assigned sex and don't wish to change it.

▶ *Genderqueer* are people who identity their gender as falling somewhere on a continuum between female and male, or a combination of gender identities and sexual orientations.

Lesbians, gay men, bisexuals, and transgender people (LGBT) are becoming increasingly more visible, but, as you'll see shortly, their acceptance is mixed.

Gender expression is the way a person communicates gender identity to others through behavior, clothing, hairstyles, voice, or body characteristics. Cross-dressing, girls' frilly dresses, and men's business suits are all examples of gender expression. Even if a person's gender identity is constant, gender expression can vary from situation to situation and change over time. For example, between 2010 and 2014, the number of men's eyelift, facelift, and breast reduction cosmetic surgeries increased by 33 to 44 percent; half of American men now routinely use moisturizers, facial creams, or self-tanning lotions and sprays; and some of the National Basketball Association's "toughest players" have promoted products for Dove, La Mer, and other skin-care companies (Boyle, 2013; Holmes, 2013; American Society for Aesthetic Plastic Surgery, 2015).

GENDER ROLES, GENDER STEREOTYPES, AND SEXISM

Gender roles are the characteristics, attitudes, feelings, and behaviors that society expects of females and males. As you saw in Chapter 4, one of the purposes of socialization, which begins at birth, is to teach people appropriate gender roles. As a result, we learn to become

In 1976, Bruce Jenner won an Olympic gold medal in the men's decathlon. In 2015, at age 65, he became openly transgender, changing his name to Caitlyn Jenner.

Tibrina Hobson/Getty Images

male or female through interactions with family members, teachers, friends, and the larger society.

Americans are more likely now than in the past to pursue jobs and other activities based on their ability and interests rather than their sex. For the most part, however, our society still has fairly rigid gender roles and widespread **gender stereotypes**—expectations about how people will look, act, think, and feel based on their sex. We tend to associate stereotypically female characteristics with weakness and stereotypically male characteristics with strength. Consider, for example, how often we describe the same behavior differently for women and men:

▶ He's firm; she's stubborn.

▶ He's careful about details; she's picky.

▶ He's honest; she's opinionated.

▶ He's raising good points; she's "bitching."

▶ He's a man of the world; she's "been around."

Gender stereotypes fuel **sexism**, an attitude or behavior that discriminates against one sex, usually females, based on the assumed superiority of the other sex. In the late 1990s, and after receiving numerous rejections, a publisher finally accepted J. K. Rowling's manuscript of her Harry Potter book. Rowling followed the publisher's advice to sell the book under her initials, not her first name, Joanne. Even today, particularly for new science fiction and mystery authors, publishers instruct women to use male pseudonyms because "men prefer books written by men" (Cohen, 2012: D9).

---

gender expression the way a person communicates gender identity to others through behavior, clothing, hairstyles, voice, or body characteristics.

gender roles the characteristics, attitudes, feelings, and behaviors that society expects of females and males.

gender stereotypes expectations about how people will look, act, think, and feel based on their sex.

sexism an attitude or behavior that discriminates against one sex, usually females, based on the assumed superiority of the other sex.

At an early age, sex-appropriate activities prepare girls and boys for future adult roles. As a result, there are few male ballet dancers and female auto mechanics.

Belinda Images/Superstock

Men also experience sexism. Here's what one of my students wrote during an online discussion of gender roles:

Some parents live their dreams through their sons by forcing them to be in sports. I disagree with this but want my [9-year-old] to be "all boy." He's the worst player on the basketball team at school and wanted to take dance lessons, including ballet. I assured him that this was not going to happen. I'm going to enroll him in soccer and see if he does better.

Is this mother suppressing her son's natural dancing talent? We'll never know because she, like many parents, expects her son to fulfill sexist gender roles that meet with society's approval.

9-1d Societal Reactions to LGBTs

People's attitudes toward LGBTs are mixed. There's been greater acceptance in some countries, but considerable repression in others.

GREATER ACCEPTANCE

Australian passports and birth certificates now designate male, female, and transgender. In India, the 2011 national census for the first time offered three options: male, female, or a "third sex" that includes LGBTs (Cohn, 2011). In Thailand, which has the world's biggest transsexual population, an airline has recently recruited "third sex" flight attendants (Mutzabaugh, 2011).

In the United States, 92 percent of LGBT adults say society is more accepting of them than 10 years ago (Taylor et al., 2013). Many municipal jurisdictions, corporations, and small companies now extend more health care and other benefits to gay employees and their partners than to unmarried heterosexuals who live together, and the Boy Scouts of America has lifted its ban on gay adult leaders. The U.S. Supreme Court and a growing number of states have legalized same-sex marriages, and large numbers of Americans

support policies that give lesbians and gays equal rights in the workplace and elsewhere (Von Drehle, 2014).

There are many other indicators of greater LGBT acceptance. In 2012, the Army promoted the first openly gay female officer to brigadier general. In 2013, the Pentagon added benefits for same-sex partners, including services on U.S. military bases. Federal workers and Medicare recipients are now eligible for sex-change operations (Mach and Cornell, 2013; "Transgender Rights," 2015). And, since the mid-1990s, a number of LGBT characters have appeared in leading and supporting roles on prime-time TV in popular programs such as *The Simpsons*, *The Good Wife*, *Modern Family*, *Two and a Half Men*, and *American Dad* (GLAAD, 2013).

GREATER INTOLERANCE

According to one scholar, what makes gay people different from others is that "we are discriminated against, mistreated, [and] regarded as sick or perverted" (Halperin, 2012: B17). **Heterosexism**, a belief that heterosexuality is the only legitimate sexual orientation, pervades societal practices, laws, and institutions. Heterosexism can trigger **homophobia**, a fear and hatred of lesbians and gays.

Homophobia often takes the form of *gay bashing*: threats, assaults, or acts of violence directed at LGBTs. Of the nearly 7,300 hate crime incidents reported to the police in 2013, 20 percent of the victims were LGBT, but much gay bashing isn't reported (Federal Bureau of Investigation, 2014; Wilson, 2014). Among transgender

Jonathan Daniel/Getty Images

In mid-2013, the Washington Wizards' Jason Collins appeared on the cover of *Sports Illustrated*. "I'm a 34-year-old NBA center, I'm black, and I'm gay," he announced (Collins and Lidz, 2013: 34). Why do you think his coming out received national coverage?

heterosexism belief that heterosexuality is the only legitimate sexual orientation.

homophobia a fear and hatred of lesbians and gays.

people, 41 percent have attempted suicide sometime in their lives—nearly nine times the national average—due primarily to rejection by family and friends; harassment, discrimination, and violence at school, work, and by police; and health care providers' refusing treatment (Haas et al., 2014).

9-2 CONTEMPORARY GENDER STRATIFICATION

A recent study concluded that "it will take until 2085 for women to reach parity with men in leadership roles in government/politics, business, entrepreneurship, and nonprofit organizations" (Klos, 2013). Why? There's still widespread **gender stratification**—people's unequal access to wealth, power, status, prestige, opportunity, and other valued resources because of their sex. *Gendered institutions* are social structures that enable gender stratification. Let's begin with the family, remembering that institutions intersect and affect each other (see Chapter 6).

9-2a Gender and Family Life

Among parents with children under age 18, fathers spend more hours each week in paid work than do mothers, do less child care and housework, and have more leisure time (see *Figure 9.3*). Many household chores are still gendered. On average, men are four times more likely to do home repairs and maintenance (e.g., repairing cars, lawn care); women are three times more likely to do the never-ending jobs like cooking, cleaning, and laundry (Wang, 2013; "American Time Use Survey…," 2015).

Women do less housework than they used to, but devote twice as much time to child care as men, and there are gender differences in parenting tasks. Fathers tend to do more of the enjoyable tasks (e.g., reading to children, playing with them, taking them to games), and are much more likely to "join in" with child care than to "take over" from mothers. In contrast, mothers do most of the daily, nonstop tasks like picking

For many women, there's nothing sexier than a man who does housework.

gender stratification people's unequal access to wealth, power, status, prestige, and other valued resources because of their sex.

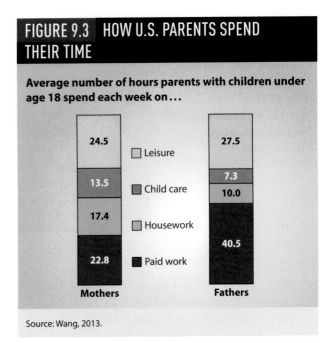

FIGURE 9.3 HOW U.S. PARENTS SPEND THEIR TIME

Average number of hours parents with children under age 18 spend each week on…

Mothers:
- Leisure: 24.5
- Child care: 13.5
- Housework: 17.4
- Paid work: 22.8

Fathers:
- Leisure: 27.5
- Child care: 7.3
- Housework: 10.0
- Paid work: 40.5

Source: Wang, 2013.

children up from school or day care and feeding, bathing, and putting them to bed (Parker and Wang, 2013; Craig, 2015).

Every year, the media feature and applaud stay-at-home dads, but their numbers are negligible. In 2014, 211,000 fathers (3 percent of all fathers) cared for children younger than age 15 while their wives worked outside the home. Being a stay-at-home dad is usually a temporary role that's due to unemployment or health problems. The greatest increase, however, has been among white, college-educated, middle-upper class men who choose to be stay-at-home dads because their partners or wives have high incomes ("Father's Day…," 2015; Kane, 2015).

9-2b Gender and Education

U.S. girls and women have made substantial educational progress in the last few decades. Despite the headway, there are gender differences at all educational levels. In public K–12 schools, as rank and pay increase, the number of women decreases. Among all full-time teachers, 89 percent at the elementary level are women; the number falls to 56 percent in high school. Among principals, the number of women drops from 64 percent at elementary schools to 30 percent in high schools (Bitterman et al., 2013; Goldring et al., 2013; U.S. Bureau of Labor Statistics, 2015).

TABLE 9.1 AS RANK INCREASES, THE NUMBER OF FEMALE FACULTY DECREASES

Rank	Percentage of Female Faculty Members
Instructor	56
Assistant Professor	49
Associate Professor	42
Professor	29

Note: Of the almost 762,000 full-time faculty in 2011, 44 percent were women.

Source: Based on Snyder and Dillow, 2013, Table 291.

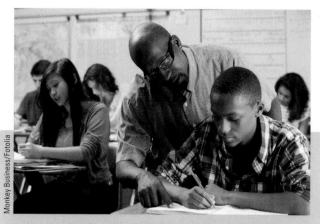

Of the nearly 2.8 million U.S. teachers at elementary and middle schools, only 20 percent are men. A major reason is low salaries compared with other occupations, but gender stereotypes are also a factor (Bonner, 2007; U.S. Bureau of Labor Statistics, 2015).

Because women across all racial and ethnic groups are more likely than men to finish college, some observers have described this phenomenon as "the feminization of higher education." Women have sailed past men in earning associate's, bachelor's, and master's degrees, but earn only 37 percent of professional and doctoral degrees (see Chapter 13). Such data contradict the description of higher education as feminized.

Even when women earn doctoral degrees in male-dominated STEM (science, technology, engineering, and math) fields, they're less likely than men to be hired (Milan, 2012). Once hired, women faculty are less likely to be promoted. Since 2000, 45 percent of all Ph.D. degree recipients have been women (Snyder and Dillow, 2013), but as the academic rank increases, the number of female faculty decreases (see *Table 9.1*).

9-2c Gender and the Workplace

There has been progress toward greater workplace equality, but we still have a long way to go. In the United States (as around the world), many jobs are segregated by sex, there are ongoing gender pay gaps, and numerous women experience sexual harassment.

OCCUPATIONAL SEX SEGREGATION

Occupational sex segregation (sometimes called *occupational gender segregation*) is the process of channeling women and men into different types of jobs. As a result, a number of U.S. occupations are filled almost entirely by either women or men. Between 92 and 98 percent of all child care workers, secretaries, dental hygienists, and preschool and kindergarten teachers are women. Between 96 and 99 percent of all pilots, mechanics, plumbers, and firefighters are men. Women have made progress in a number of the higher-paying occupations, but 74 percent of chief executives, 80 percent of software developers, and 90 percent of engineers are

men (U.S. Bureau of Labor Statistics, 2015). The issue isn't women and men working in different spaces or locations, but that male-dominated occupations usually pay higher wages. And, as in education, women are much less likely than men to move up the occupational ladder (see Chapters 8 and 11).

THE GENDER PAY GAP

On average, full-time, year-round working women earn 77 to 78 cents for every dollar paid to their male counterparts (Hegewisch and Hartmann, 2015). To state this differently, *the average woman must work almost nine extra weeks every year to make the same wages as a man.*

This overall income difference between women and men is the **gender pay gap** (also called the *wage gap, pay gap,* and *gender wage gap*). The average woman who works full-time year-round for 47 years loses a significant amount of money because of the gender pay gap: $700,000 for high school graduates, $1.2 million for college graduates, and more than $2 million for women with a professional degree (e.g., medicine, law). Lower wages and salaries reduce women's savings, purchasing power, and quality of life, and bring them lower Social Security income after retirement (Soguel, 2009; Glynn and Powers, 2012).

occupational sex segregation (sometimes called *occupational gender segregation*) the process of channeling women and men into different types of jobs.

gender pay gap the overall income difference between women and men in the workplace (also called the *wage gap, pay gap,* and *gender wage gap*).

FIGURE 9.4 GENDER PAY GAP, BY EDUCATION

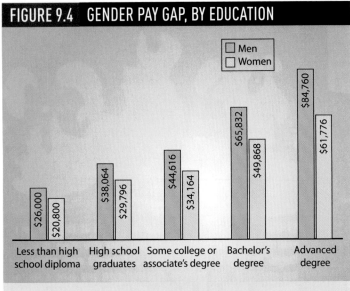

☐ Men
☐ Women

$26,000 / $20,800 — Less than high school diploma
$38,064 / $29,796 — High school graduates
$44,616 / $34,164 — Some college or associate's degree
$65,832 / $49,868 — Bachelor's degree
$84,760 / $61,776 — Advanced degree

Note: These are the median annual earnings of year-round full-time workers age 25 and older in late 2013.

Source: Based on BLS News Release, "Usual Weekly Earnings...," 2014, Table 9.

Not only do women earn less than men at all educational levels, the higher the education level, the bigger the gender pay gap (see *Figure 9.4*). Female college graduates earn 76 percent of what their male counterparts earn, but women's earnings drop to 73 percent of men's for those with advanced degrees. (Note that, as a group, women with advanced degrees earn less than men with a college degree.)

Why is there a gender pay gap? Women tend to choose fields with lower earnings (such as health care and education), whereas men are more likely to major in higher paying fields (such as engineering and computer science). Women on average also work fewer hours than men, usually to care for children or other family members. However, there's a 7 percent pay gap after accounting for a number of variables—including college major, occupation, hours worked, GPA, age, geographical region, and marital status—and the pay gap increases to 12 percent 10 years after college graduation (AAUW, 2013; Frueh, 2014; Goldin, 2014). Only 4 to 5 percent of men, compared with 15 to 17 percent of women, believe that they've been passed over for a promotion or denied a raise because of their gender (Saad, 2015).

SEXUAL HARASSMENT

Sexual harassment is any unwanted sexual advance, request for sexual favors, or other conduct of a sexual

sexual harassment any unwanted sexual advance, request for sexual favors, or other conduct of a sexual nature that makes a person uncomfortable and interferes with her or his work.

nature that makes a person uncomfortable and interferes with her or his work. It includes *verbal behavior* (e.g., pressure for dates, demands for sexual favors in return for hiring and promotion, the threat of rape), *nonverbal behavior* (e.g., indecent gestures, display of sexually explicit posters, photos, or drawings), and *physical contact* (e.g., pinching, touching, or rape). Between 1997 and 2014, the Equal Employment Opportunity Commission (EEOC) received almost 228,000 formal sexual harassment complaints, 83 percent of them from females (Equal Employment Opportunity Commission, 2011, 2015). The statistics would be much higher, but many companies now require new employees to agree to arbitrate complaints, including sexual harassment, as a condition of being hired (Green, 2011).

Nearly two-thirds of Americans say that sexual harassment is a serious problem in this country. About 25 percent of women have been sexually harassed at work. Only 40 percent of them reported the incident, however, because they believed that doing so wouldn't do any good or feared being fired or demoted (Clement, 2011).

9-2d Gender and Politics

Unlike a number of other countries (including Great Britain, Germany, India, Israel, Pakistan, Argentina, Chile, and Philippines), the United States has never had a woman serving as president or even vice president. In the U.S. Congress, 81 percent of the members are men. In other important elective offices (governor, mayor, state legislator), only a handful of the decision makers are women (see *Table 9.2*). These numbers haven't changed much since the early 1990s.

Women's voting rates in the United States have been higher than men's since 1984 (see Chapter 11). Why, in contrast, are there so few women in political office? There's a combination of reasons: (1) Women run for office at a far lower rate than men with similar credentials because women don't consider themselves qualified; (2) they have to do more than their male counterparts to prove themselves; (3) U.S. presidents appoint more men than women to important positions; and (4) there's a lingering sexism, among both men and women, that female politicians are both less feminine and compassionate than the average woman, and lack the leadership traits associated with male politicians (e.g., confident, assertive) (Luling, 2013; Schneider and Bos, 2014; Parker and Horowitz, 2015; see also Chapter 11).

TABLE 9.2 U.S. WOMEN IN ELECTIVE OFFICES, 2015

Political Office	Total Number of Office Holders	Percentage Who Are Women
Senate	100	20%
House of Representatives	435	19
Governor	50	12
State Legislator	7,383	24
Attorney General	50	18
Secretary of State	50	28
State Treasurer	50	14
State Comptroller	50	4
Mayor (100 largest cities)	100	17

Source: Based on material at the Center for American Women and Politics, 2015.

9-3 SEXUALITY

In the movie *Annie Hall*, a therapist asks two lovers how often they have sex. The man rolls his eyes, and complains, "Hardly ever, maybe three times a week!" The woman exclaims, "Constantly, three times a week!" *Sexuality* is considerably more complex than just having sex, however, because it's a product of our sexual identity, sexual orientation and sexual scripts, and includes desire, expression, and behavior.

9-3a Contemporary Sexual Attitudes and Practices

Sex doesn't "just happen." It typically progresses through a series of stages such as approaching, flirting, touching, or asking directly for sex. Sexual attitudes and behavior can vary from situation to situation and change over time, including why we have sex.

WHY WE HAVE SEX

People have sex to reproduce and to experience physical pleasure, but there are other reasons. The majority of teenagers ages 15 to 19 have their first sexual intercourse with someone that they describe as "going steady," but the percentage is higher for females (70 percent) than males (56 percent). Nationwide, 5 percent of male and 12 percent of female high school students have been physically forced to have unwanted sexual intercourse. Teenagers are also more likely to engage in sex at any early age if they use alcohol or other drugs or experience domestic violence (Martinez et al., 2011; Eaton et al., 2012). A

study of nearly 2,000 college students identified 237 reasons for having sex that ranged from the physical (stress reduction) to the spiritual (to get closer to God) and from the altruistic (to make the other person feel good) to the spiteful (to retaliate against a partner who had cheated) (Meston and Buss, 2007).

SEXUALITY THROUGHOUT THE LIFE COURSE

Contrary to some stereotypes, adolescents aren't sexually promiscuous and older people aren't asexual. Among teenagers aged 15 to 19, the percentage who ever had sexual intercourse declined from 51 percent in 1988 to 43 percent in 2010. By 2008, however, almost half of teens in this age group had had oral but not vaginal sex (Chandra et al., 2011). Adolescents who have oral sex prior to vaginal intercourse may be doing so because they view oral sex as "not really sex." Rather, they see it as a way to delay vaginal intercourse, to maintain one's virginity (especially among those who are religious), and to avoid the risk of pregnancy and STDs (Regnerus and Uecker, 2011; Copen et al., 2012).

By age 44, 98 percent of Americans have had vaginal intercourse; by the same age, 6 percent of men and 12 percent of women have had oral, anal, or vaginal contact with a same-sex partner. Fewer than 4 percent of Americans identify themselves as LGBT, but 15 percent of women and 5 percent of men have been sexually attracted to the same sex. Among those aged 15 to 21 who identify themselves as straight, 11 percent of females and 4 percent of males have had same-sex sexual contact (Chandra et al., 2011; McCabe et al., 2011). Thus, as noted earlier, sexual identity, attraction, and behavior overlap.

Christopher Weyant The New Yorker Collection/The Cartoon Bank

"Don't you have any sexual fantasies that don't involve me cleaning?"

A majority of adults ages 45 and older agree that a satisfying sexual relationship is important, but it's not their top priority. Marital sexual frequency may decrease because concerns about earning a living, making a home, and raising a family become more pressing than love-making. Others may be going through a divorce, dealing with unemployment, helping to raise grandchildren, or caring for aging parents—all of which sap people's interest in sex (Jacoby, 2005; ConsumerReports.org, 2009).

As people age, they experience lower levels of sexual desire and some sexual activities, but a third of men and women age 70 and older report having sex at least twice a month. Poor health and relationship problems, rather than just advancing age, are more closely linked to declining sexual activity (Lee et al., 2015). Many couples in their seventies and eighties emphasize emotional intimacy and companionship, and are satisfied with kissing, cuddling, and caressing (Heiman et al., 2011; Lodge and Umberson, 2012).

9-3b Sexual Scripts and Double Standards

We like to think that our sexual behavior is spontaneous, but all of us have internalized sexual scripts. A **sexual script** specifies the formal and informal norms for acceptable or unacceptable sexual behavior. Social scripts can change over time and across groups, but are highly gendered in two ways—women's increasing hypersexualization and a persistent sexual double standard.

THE "SEXY BABES" TREND

Sexualized social messages are reaching ever younger audiences, teaching or reinforcing the idea that girls and women should be valued for how they look rather than their personalities and abilities. For example, there are "bikini onesies" for infant girls, sexy lingerie for girls 3 months and older, and padded bras for 7- and 8-year olds (that's right, for 7- and 8-year-olds!).

Many girls are obsessed about their looks, and from an early age, for a variety of reasons, including their mothers' role modeling. By age 9, girls start imitating the clothes, makeup, and behavior of mothers who dress and act in highly sexualized ways (Starr and Ferguson, 2012).

Media images also play a large role in girls' hypersexualization. Girls and boys see cheerleaders (with increasingly sexualized routines) on TV far more than they see female basketball players or other athletes. A study of the 100 top-grossing films in 2008 found that 13- to 20-year-old females were far more likely than their male counterparts to be depicted in revealing clothes (40 percent vs. 7 percent), partially naked (30 percent vs. 10 percent), and physically attractive (29 percent vs. 11 percent). In television shows, women are now represented in more diverse roles—as doctors, lawyers, and criminal investigators—but they're often sexy ("hot"). Also, top female athletes regularly pose naked or semi-naked for men's magazines (Hanes, 2011; Smith and Choueiti, 2011).

Who benefits from girls' and women's hypersexualization? Marketers who convince girls (and their parents) that being popular and "sexy" requires the right clothes, makeup, hair style, and accessories, create a young generation of shoppers and consumers who will increase business profits more than ever before (Lamb and Brown, 2007; Levin and Kilbourne, 2009).

THE SEXUAL DOUBLE STANDARD

Some believe that the **sexual double standard**—a code that permits greater sexual freedom for men than women—has faded. Others argue that it persists. Among U.S. adolescents, the higher the number of sexual partners, the greater the boy's popularity. In contrast, girls who have more than eight partners are far less popular than their less-experienced female peers. By age 44, many more men (21 percent) than women (8 percent) report having had at least 15 sex partners. And, over a lifetime, men are more likely than women to have sex outside of marriage (19 percent and 14 percent, respectively) (Kreager and Staff, 2009; Chandra et al., 2011; Drexler, 2012).

Another example of the sexual double standard is *hooking up*—which can mean anything from kissing to sexual

Galia Slayen built a life-sized Barbie to show what she would look like if she were a real woman. (Slayen used a toy for the head because she wasn't able to create a proportional head) ("Life Size Barbie...," 2011).

sexual script specifies the formal and informal norms for acceptable or unacceptable sexual behavior.

sexual double standard a code that permits greater sexual freedom for men than women.

intercourse. The prevalence of hooking up has increased only slightly since the late 1980s, but is now more common than dating at many high schools and colleges (Kalish and Kimmel, 2011; Monto and Carey, 2014).

Hooking up has its advantages. Many men prefer hookups because they're inexpensive compared with dating. For women, hook-ups offer sex without becoming involved in time-consuming relationships that compete with schoolwork, dealing with boyfriends who become demanding or controlling, and experiencing breakups (Bogle, 2008; Rosin, 2012).

Hooking up also has disadvantages, especially for women, because it reinforces a sexual double standard. For example, men are more likely than women to perform sexual acts that a partner doesn't like; more than twice as many men as women experience an orgasm because the men typically show no affection and aren't concerned about what pleases women sexually; and women who hook up generally get a reputation as "sluts" (England and Thomas, 2009; Armstrong et al., 2010, 2012). About half of women, compared with only 25 percent of men, have regretted having casual sex (Galperin et al., 2013).

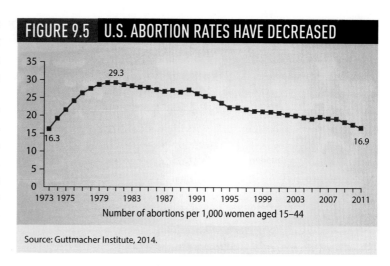

FIGURE 9.5 U.S. ABORTION RATES HAVE DECREASED

Number of abortions per 1,000 women aged 15–44

Source: Guttmacher Institute, 2014.

SOME CURRENT SOCIAL ISSUES ABOUT SEXUALITY

9-4

Most Americans see sex as a private act, but others believe that that there should be government control of some sexual expressions and decisions. People disagree about social policies on sex-related topics such as teenagers' birth control, prostitution, reproductive technologies that include genetic engineering (see Chapter 16), and teen pregnancy (see Chapter 12). Two of the most controversial and politically contested issues continue to be abortion and same-sex marriage.

9-4a Abortion

Abortion is the expulsion of an embryo or fetus from the uterus. It can occur naturally—in *spontaneous abortion* (miscarriage)—or be induced medically. Abortion was outlawed in the nineteenth century, but has been legal since the U.S. Supreme Court's *Roe v. Wade* ruling in 1973.

TRENDS

Every year, 40 percent of unintended pregnancies end in abortion. Over a lifetime, 33 percent of women have

an abortion by age 45 (Guttmacher Institute, 2014). The *abortion rate*, or the number of abortions per 1,000 women ages 15 to 44, increased during the 1970s, then decreased, and has dropped to its lowest point since 1973 (see *Figure 9.5*).

Why have abortion rates decreased? At least half of the states have restricted access to abortion, but most of the decline has been due to an overall drop in pregnancy rates, delaying childbearing, more effective usage of contraceptives, and greater access to emergency contraception that prevents pregnancy (Jones and Jerman, 2014).

Abortion is most common among women who are young (in their twenties), white, and never married (see *Figure 9.6*). About 20 percent of women who get abortions have at least a college degree, but most are poor: 69 percent have incomes below or near the poverty level, and 74 percent are financially unable to support a baby. Low-income women are also much more likely than higher-income women to have experienced intimate partner violence that included being impregnated against their will. Abortion cuts across all religious groups, but 28 percent of women are Catholic and 37 percent are Protestant (including born-again/evangelical Christians) (Jones et al., 2013; Guttmacher Institute, 2014; Reeves and Venator, 2015).

WHY IS ABORTION CONTROVERSIAL?

Since its legalization, abortion has been one of the most persistently contentious issues in U.S. politics and culture. More Americans describe themselves as "pro-choice" (50 percent) than "pro-life" (44 percent), and 6 percent aren't sure. Both anti- and pro-abortion groups agree on some issues, such as

abortion the expulsion of an embryo or fetus from the uterus.

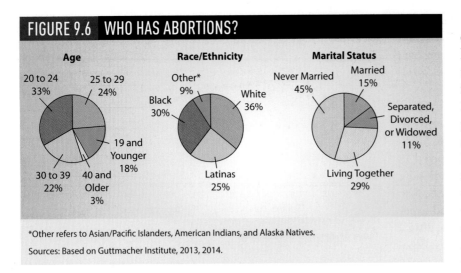

FIGURE 9.6 WHO HAS ABORTIONS?

Age
- 20 to 24 33%
- 25 to 29 24%
- 19 and Younger 18%
- 30 to 39 22%
- 40 and Older 3%

Race/Ethnicity
- Other* 9%
- White 36%
- Black 30%
- Latinas 25%

Marital Status
- Never Married 45%
- Married 15%
- Separated, Divorced, or Widowed 11%
- Living Together 29%

*Other refers to Asian/Pacific Islanders, American Indians, and Alaska Natives.

Sources: Based on Guttmacher Institute, 2013, 2014.

A strong majority (63 percent) of Americans want to keep abortion legal, but it's virtually impossible for many women to get legal abortions in 89 percent of all U.S. counties. Between 2001 and mid-2015, states passed 471 laws restricting access to abortions. Some of the restrictions have included cutting public funding for abortions for low-income women, passing licensing requirements that make it impossible for abortion clinics to remain open, and limiting access to "abortion pills" (medicines that legally end a pregnancy in the first nine weeks). Antiabortion advocates have also set up thousands of "crisis pregnancy centers," usually next to abortion clinics, that pressure women to continue a pregnancy, telling them that an abortion causes breast cancer, leads to suicide, and so on (Bahadur, 2014; Nash et al., 2014; Guttmacher Institute, 2015).

requiring a patient's informed consent, but 19 percent want abortion to be illegal under all circumstances (Saad, 2015).

Antiabortion groups believe that the embryo or fetus isn't just a mass of cells but a human being from the time of conception and, therefore, has a right to life. In contrast, abortion rights advocates point out that, at the moment of conception, the organism lacks a brain and other specifically and uniquely human attributes, such as consciousness and reasoning, and that a pregnant woman—not legislators—should decide whether or not to bear children.

Antiabortion groups maintain that abortion is immoral and endangers a woman's physical, mental, and emotional health. Whether abortion is immoral is a religious and philosophical question. On a physical level, a legal abortion in the first trimester (up to 12 weeks) is safer than driving a car, playing football, motorcycling, getting a penicillin shot, or continuing a pregnancy. There's also no evidence that having an abortion increases the risk of breast cancer or causes infertility (Sheppard, 2013; Pazol et al., 2014; Holloway, 2015).

What about mental and emotional health? Antiabortion activists argue that abortion leads to postabortion stress disorders, depression, and even suicide. National studies conducted since the 1980s have consistently found that abortion poses no hazard to an adolescent or adult woman's mental health, doesn't increase emotional problems such as depression or low self-esteem, and doesn't lead to drug or alcohol abuse or suicide. An unwanted pregnancy, not abortion, increases the risk of mental health problems (Major et al., 2008; Academy of Medical Royal Colleges, 2011; Steinberg and Finer, 2011).

9-4b Same-Sex Marriage

Same-sex marriage (also called *gay marriage*) is a legally recognized marriage between two people of the same biological sex and/or gender identity. Although still controversial, same-sex marriage is becoming more acceptable in the United States and some other countries.

TRENDS

In 2015, the U.S. Supreme Court issued a landmark ruling (*Obergefell v. Hodges*) granting same-sex couples a constitutional right to marry. The 5-4 decision gave married gay couples nationwide the same legal rights and benefits as married heterosexual couples. With the Supreme Court's decision, the United States joined 20 other countries (so far) that allow same-sex marriage.

Prior to the ruling, gay marriage was illegal in 13 states. Some of these states' lawmakers have urged their constituents to accept the new law. Others are pressing their residents to "stand and fight by seeking a constitutional amendment banning gay marriage" (de Vogue and Diamond, 2015).

Opponents sometimes invoke religion to defy gay marriage laws. Shortly after the Supreme Court ruling, for example, Kim Davis, a county clerk in Kentucky, refused to issue marriage licenses to same-sex couples because of "God's moral law," her conscience, and "religious freedom." A judge jailed Davis briefly for contempt, but she vowed to keep fighting (Muskal, 2015). We'll examine same-sex marriages and families in Chapter 12, but why is gay marriage such a contentious issue?

same-sex marriage (also called *gay marriage*) a legally recognized marriage between two people of the same biological sex and/or gender identity.

TABLE 9.3 WHY DO AMERICANS FAVOR OR OPPOSE SAME-SEX MARRIAGES?

What do *you* think? What other reasons can you add for each side of the debate?

Same-sex marriage should be legal because . . .	Same-sex marriage should be illegal because . . .
• Gay marriages strengthen families and long-term unions that already exist. Children are better off with parents who are legally married.	• Children need a mom and a dad, not two dads or two moms.
• There are no scientific studies showing that children raised by gay and lesbian parents are worse off than those raised by heterosexual parents.	• There are no scientific studies showing that children raised by gay and lesbian parents are better off than those raised by heterosexual parents.
• Every person should be able to marry someone that she or he loves.	• People can love each other without getting married.
• Gay marriages are good for the economy because they boost businesses such as restaurants, bakeries, hotels, airlines, and florists.	• What's good for the economy isn't necessarily good for society, especially its moral values and religious beliefs.

Sources: Bennett and Ellison, 2010; Olson, 2010; Sullivan, 2011; Sprigg, 2011; Whitehead, 2011; Bogage, 2015; Kaufman, 2015.

WHY IS SAME-SEX MARRIAGE CONTROVERSIAL?

A large majority (60 percent) of Americans support same-sex marriage (up from 37 percent in 2006). Most of the opposition is coming from people who are Republican, white, male, regularly attend religious services, live in the South, are 55 and older, and hold conservative views on family issues (Doherty and Kiley, 2015; Masci and Motel, 2015; Newport, 2015).

Those who favor same-sex marriage argue that people should have the same rights regardless of sexual orientation. Those who oppose same-sex marriage contend that such unions are immoral, weaken traditional notions of marriage, and are contrary to religious beliefs. *Table 9.3* summarizes some of the major pro and con arguments in this ongoing debate.

CREATISTA/Shutterstock.com

9-5 GENDER AND SEXUALITY ACROSS CULTURES

There's considerable variation worldwide regarding gender inequality and sexual oppression. Such variations show that our behavior is learned, not innate.

9-5a Gender Inequality

A recent United Nations (2015) report concluded that women continue to face discrimination in access to work, economic assets, and participation in private and public decision making. They're also more likely than men to live in poverty, to be illiterate, and to experience violence.

In many countries, women's progress toward equality has been mixed. For example, Saudi Arabia, one of the wealthiest countries in the world, and which has some of the most educated women in the world (including STEM college and advanced degrees), ranks near the bottom in women's economic and political participation (Hausmann et al., 2014).

Of 142 countries (which represent more than 90 percent of the world's population), Iceland, Norway, Finland, Sweden, and New Zealand typically have the highest equality scores in women's quality of life and status in the economy, politics, and education. The United States ranks only 20th, and behind some poor countries such as Burundi, Lesotho, Nicaragua, and Rwanda. The greatest gender inequality is in the Middle East, North Africa, and sub-Saharan Africa; the five bottom countries in gender equality are Chad, Mali, Pakistan, Syria, and Yemen. In *all countries and regions*, the greatest gender gaps are in economic opportunity and participation and political leadership (Hausmann et al., 2014).

ECONOMIC PARTICIPATION

Women remain at a disadvantage in the labor market. Globally, about 75 percent of working-age men participate in the labor force, compared with 50 percent of working-age women, and women earn 24 percent less than men. In 85 percent of countries, women with advanced degrees have higher rates of unemployment than men with similar levels of education (United Nations, 2015).

Countries that have closed education gaps and have high levels of women's economic participation—the Scandinavian countries, United States, Canada, New Zealand, and Australia—demonstrate strong economic growth

(Worley, 2014). However, gender gaps still persist in senior positions, wages, and leadership. For example, Germany is Europe's No. 1 economy, but also has the largest pay gap between men and women in the European Union. Of 191 executives on the management boards of Germany's 30 biggest companies, only 12 are women, a 20 percent decrease from a year before (de Pommereau, 2013; Webb, 2013).

POLITICAL LEADERSHIP

Worldwide, only 22 percent of national legislators are women. Rwanda has 64 percent, followed by nine countries in which women hold 40 to 46 percent of the high-level political positions. In contrast, women make up only 19 percent of the U.S. Congress (Center for American Women and Politics, 2015; UN Women, 2015).

Of the 197 world leaders who are presidents or prime ministers, only 13 percent are women. Worldwide, women occupy only 22 percent of the positions in decision-making bodies that are comparable with our Congress. The United States ranks 80th in women's political leadership, well below many African, European, and Asian countries, and even below most of the Arab countries that many Westerners view as repressing women (Inter-Parliamentary Union, 2015).

9-5b Sexual Inequality

Globally, women have fewer rights and opportunities than men. There's been more acceptance of homosexuality in some countries, but heterosexism prevails.

VIOLENCE AGAINST FEMALES

Violence against women is a persistent global problem. Worldwide, 35 percent of women have endured physical and/or sexual violence by an intimate partner or another male (World Health Organization, 2013; UN Women, 2014). The rates are much higher in many countries. For example,

▸ In Afghanistan and some African countries, about 80 percent of girls—some as young as 8 years old—are forced into marriages; 87 percent of Afghan women have experienced physical, psychological, or sexual abuse (Peter, 2012; "Child Brides...," 2014).

▸ In Pakistan, 90 percent of women undergo domestic violence in their lifetimes. As many as 5,000 females are victims of "honor killings" every year. An *honor killing* is the murder of a family member, almost always a female, who is considered to have shamed the family by being a rape victim or has been of suspected of engaging in premarital

or extramarital sex (Sahgal and Townsend, 2014; "Human Rights Violations," 2015).

▸ In the Democratic Republic of Congo, approximately 1,100 women are raped every day by soldiers, strangers, and intimate partners ("Human Rights Violations," 2015).

▸ In India, rape and gang rape are epidemic, but less than a quarter of reported crimes end in conviction. As many as 100,000 women a year are killed over dowry disputes (the money or goods that a wife brings to her husband at marriage) (Harris, 2013; "Ending the Shame...," 2013).

Female genital mutilation/cutting (FGM/C) is a partial or total removal of the female external genitalia. Most of the 140 million girls and women who have undergone FGM/C live in 29 African countries, Indonesia, and some Middle Eastern countries. Most of the girls are between 3 and 12 years old. The operator is typically an elderly village woman who uses a knife or other sharp object and doesn't administer an anesthetic. Countries justify FGM/C on the grounds that it controls a girl's sexual desires and preserves her virginity, a prerequisite for marriage (UNICEF, 2013; Feldman-Jacobs and Clifton, 2014).

Although Nigeria, Egypt, and other countries have outlawed FGM/C, the practice remains widespread and widely accepted. In the United States, both the procedure and sending a girl abroad to undergo the process are illegal. Because of a surge of African immigrants in the last two decades, however, about half a million girls are at risk (Mather and Feldman-Jacobs, 2015).

Jean-Marc Bouju/Impact/HIP/The Image Works

Little girls like this one scream and writhe in pain during FGM/C (see text). Complications include hemorrhaging to death, a rupture that causes continual dribbling of urine or feces for the rest of the woman's life, severe pain during sexual intercourse, and death during childbirth if the baby can't emerge through the mutilated organs.

VIOLENCE AGAINST MALES

You saw earlier that Americans are more accepting of homosexuality, and that more nations are legalizing gay marriage. In contrast, many countries in Asia, Africa, and the Middle East don't tolerate LGBTs. A vast majority of Africans (e.g., 98 percent in Ghana, 93 percent in Uganda, 88 percent in Kenya) say homosexuality is unacceptable. Gay sex is illegal in 78 nations, including 34 of Africa's 54 countries. Gay men, particularly, may be legally tortured, stoned, imprisoned, or killed ("Deadly Intolerance," 2014; "Global Views on Morality," 2014; Pflanz, 2014; Baker, 2015).

Russia's parliament recently banned LGBT relationships and forbade distributing material on gay rights. Russians are more accepting of extramarital affairs, gambling, and drinking alcohol (a major cause of men's death before age 55) than homosexuality (Council for Global Equity, 2014; Poushter, 2014).

In 2014, 20,000 U.S. service members, representing nearly 5 percent of all active-duty women and 1 percent of active-duty men, experienced at least one incident of unwanted sexual contact. Nearly 35 percent of the assaults reported by men included rape and penetration with an object. Only 25 percent reported the incident; the victims feared retaliation from a higher-ranking perpetrator and ostracism by peers (Department of Defense, 2015).

9-6 SOCIOLOGICAL EXPLANATIONS OF GENDER AND SEXUALITY

Gender and sexuality affect all people's lives, but why is there so much variation over time and across cultural groups? The four sociological perspectives answer this and other questions somewhat differently (*Table 9.4* summarizes these theories).

9-6a Functionalism

Functionalists view women and men as having distinct roles that ensure a family's and society's survival. These roles help society operate smoothly, and have an impact on the types of work that people do.

DIVISION OF GENDER ROLES AND HUMAN CAPITAL

Some of the most influential functionalist theories, developed during the 1950s, proposed that gender roles differ because women and men have distinct roles and responsibilities. A man (typically a husband and father) plays an *instrumental role* of economic provider; he's competitive and works hard. A woman (typically a wife and mother) plays an *expressive role*; she provides the emotional nurturance that sustains the family unit and supports the father/husband (Parsons and Bales, 1955; Betcher and Pollack, 1993).

Instrumental and expressive roles are complementary, and each person knows what's expected: If the house is clean, she's a "good wife"; if the bills are paid, he's a "good husband." The duties are specialized, but both roles are equally important in meeting a family's needs and ensuring a society's survival.

Such traditional gender roles help explain occupational sex segregation because people differ in the amount of human capital that they bring to the labor market. *Human capital* is the array of competencies—including education, job training, skills, and experience—that have economic value and increase productivity.

From a functionalist perspective, what individuals earn is the result of the choices they make and, consequently, the human capital that they accumulate to meet labor market demands. Women diminish their human capital because they choose lower paying occupations (social work rather than computer science), as well as

TABLE 9.4 SOCIOLOGICAL EXPLANATIONS OF GENDER AND SEXUALITY

Theoretical Perspective	Level of Analysis	Key Points
Functionalist	Macro	• Gender roles are complementary, equally important for a society's survival, and affect human capital. • Agreed-on sexual norms contribute to a society's order and stability.
Conflict	Macro	• Gender roles give men power to control women's lives. • Most societies regulate women's, but not men's, sexual behavior.
Feminist	Macro and micro	• Women's inequality reflects their historical and current domination by men, especially in the workplace. • Many men use violence—including sexual harassment, rape, and global sex trafficking—to control women's sexuality.
Symbolic Interactionist	Micro	• Gender is a social construction that emerges and is reinforced through everyday interactions. • The social construction of sexuality varies across cultures because of societal norms and values.

postpone or leave the workforce for childbearing and child care. When they return to work, women have lower earnings than men because, even in higher paying occupations, their human capital has deteriorated or become obsolete (Kemp, 1994).

WHY IS SEXUALITY IMPORTANT?

For functionalists, sexuality is critical for reproduction, but people should limit sex to marriage and forming families. Functionalists view sex outside of marriage as dysfunctional because most unmarried fathers don't support their children. The offspring often experience poverty and a variety of emotional, behavioral, and academic problems (Avellar and Smock, 2005).

You might be tempted to dismiss the functionalist view of limiting sex to marriage as outdated. Worldwide, however, sex outside of marriage is prohibited and *arranged marriages*—in which parents or relatives choose their children's future mates—are the norm. Most children agree to arranged marriages because of social custom and out of respect for their parents' wishes. The matches solidify relationships with other families and ensure that the woman's sexual behavior will be confined to her husband, avoiding any doubt about the offspring's parentage (see Benokraitis, 2015).

Some functionalists encourage marrying in one's mid-to-late twenties instead of delaying marriage. The benefits include enjoying more frequent sex, having an easier time getting pregnant, and being able to have more than one child than people who marry in their thirties or later (Wilcox, 2015).

CRITICAL EVALUATION

Critics fault functionalist perspectives on four counts. First, even during the 1950s, white middle-class male sociologists ignored almost a third of the labor force that was composed of working-class, immigrant, and minority women who played *both* instrumental and expressive roles. Second, functionalists tend to overlook the fact that many people don't have a choice of playing only instrumental or expressive roles because most families rely on two incomes for economic survival. Third, the human capital model assumes that women have lower earnings than men because they "choose" lower paying occupations. As you saw earlier, however, there's a gender pay gap across *all* occupations, even those that require advanced degrees. Fourth, whether they occur within or outside of marriage, unintended pregnancies cost taxpayers at least $21 billion each year for pregnancy-related care, abortions, and miscarriages (Sonfield and Kost, 2015).

Functionalists tend to reject sexual relationships outside of marriage. Compared with married couples, for example, those who cohabit have poorer quality relationships and lower happiness levels (Wilcox and Marquardt, 2011; Sassler et al., 2012). As you'll see in Chapter 12, however, marriage doesn't guarantee long or happy relationships.

9-6b Conflict Theory

Conflict theorists see gender inequality as built into the social structure, both within and outside the home. In both developing and industrialized countries, men control most of a society's resources and dominate women. Like functionalists, conflict theorists see sexuality as a key component of a society's organization, but they view sexuality as reflecting and perpetuating sexism and discrimination.

CAPITALISM AND GENDER INEQUALITY

For conflict theorists, capitalism, not complementary roles, explains gender roles and men's social and economic advantages. Conflict theorists argue that women's inequality is largely due to economic exploitation—both as underpaid workers in the labor force and doing most of the unpaid domestic work, especially housework and caring for children and aging family members. In effect, gender roles are profitable for business. Companies can require their male employees to work long hours or make numerous overnight business trips and not worry about workers demanding payment for child care services that can cost up to $63,000 a year (Carey and Trap, 2014).

Women comprise 51 percent of the U.S. population, but Congress "looks much like the face of corporate America—overwhelmingly wealthy, white, and male" (Weathers, 2015). Women's underrepresentation in political institutions can help explain why occupational sex segregation, gender pay gaps, and sexual harassment persist.

IS GENDER INEQUALITY LINKED TO SEXUAL INEQUALITY?

From a conflict perspective, gender inequality gives men economic, political, and/or interpersonal power to control or dominate women's sexual lives. Most of the victims of domestic violence and rape, in the United States and around the world, are women and girls. In workplace sexual harassment cases, the offender is typically a male supervisor. For example, 66 percent of American female restaurant workers experience sexual harassment by a manager, owner, or supervisor at least once a month. In prostitution and sex trafficking, almost all of the victims worldwide are poor women and girls (U.S. Department of State, 2013; Restaurant Opportunities Center United et al., 2014).

A group of men recently surrounded a woman at a busy bus station in Nairobi, Kenya. They violently tore off her clothes because she was wearing a miniskirt, accused her of being "indecently dressed" and a jezebel, and left her naked on the street. Bystanders watched. A week later, about 1,000 women (some dressed in miniskirts) demonstrated against the ongoing and rising violence against girls and women, and vowed to wear whatever clothes they want. Some men joined the protestors; others threw rocks or shouted lewd comments.

AP Images/Ben Curtis

In some societies, particularly in the Middle East and some African countries, men dictate how women should dress and whether they can travel, work, receive health care, attend school, or start a business; dismiss women's charges of sexual assaults (including gang rapes); and blame girls for child rape because they're "seducing" older men (Neelakantan, 2006). Many women have internalized such sexism. For example, 38 percent of Ugandan women believe that wife beating is acceptable if the wife leaves the house without informing her husband (Kaneda and Bietsch, 2015).

CRITICAL EVALUATION

Conflict theory shows how social structures reinforce men's domination of women, but critics point out several limitations. First, women aren't as powerless as some conflict theorists claim. Like men, women often barter with their employers to get what they want ("I'll work late all this week if I can have Friday off"). Second, conflict theory emphasizes the differences between women and men rather than their common goals, similar attitudes, and comparable power strategies to increase their economic and political influence. Third, conflict theory often ignores women's exploitation of other women. In the United States, women comprise 19 percent of those involved in the sex trafficking industry. The thriving underground sex economy relies heavily on nannies, secretaries, and escort services and brothels owned or operated by females (Banks and Kyckelhahn, 2011; Dank et al., 2014).

9-6c Feminist Theories

Feminist scholars agree with conflict theorists that gender stratification benefits men and capitalism, but emphasize that women's subordination also includes their daily vulnerability to male violence (Katz, 2006). Feminist scholars, more than any other group of theorists, are especially concerned about men's controlling women's sexual lives.

LIVING IN A GENDERED WORLD

During the second Republican presidential debate in 2015, the moderator asked the 11 contenders what prominent woman they'd choose for the redesigned $10 bill. The responses ranged from caregiving moms and wives to ineligible non-Americans; only four candidates were able to name an outstanding American woman. To much applause, the only female candidate, Carly Fiorina, opposed a redesign because "women are the majority of this nation" (see Beckwith, 2015).

Men comprise 48 percent of the U.S. population age 18 and older, but they account for 98 percent of directors of the 700 top-grossing films since 2007, 97 percent of the televised news media's and ESPN's sports broadcasts, 95 percent of *Fortune* 500 CEOs, 94 percent of venture-capital firms' partners, 81 percent of the 500 largest companies' board members, 83 percent of the largest private law firm partners, 80 percent of Congress, 75 percent of college/university presidents, 73 percent of federal and state judges, and only five women have been promoted to the rank of four-star general (Catalyst, 2015; Cooky et al., 2015; June, 2015; Peters and Lagon, 2015; Smith et al., 2015; Triedman, 2015; "Valley of the Dudes," 2015; Weathers, 2015; see also *Table 9.2* on p. 167). In many of these sectors, women's representation has decreased or remained about the same for at least a decade.

Sexism is prevalent in almost all workplaces and online environments. The digital media has amplified feminist voices, but also silenced them. Once a writer is singled out by an anti-feminist men's group, she's deluged with hateful and threatening messages, both public and private. Males with the fewest skills and lowest status are the most likely to be threatened by and hostile toward an influx of talented women. Regardless of the reasons, some women no longer participate online because of the harassment (Goldberg, 2015; Kasumovic and Kuznekoff, 2015).

Feminist theorists emphasize that gender, race, and social class intersect to form a hierarchical stratification system that shapes women's and men's attitudes, experiences, and behavior. Privileged women have less status than privileged men, but upper-class white men *and* women subordinate lower-class women *and* minority men (Andersen and Collins, 2010). Understanding the interconnections between gender, race/ethnicity, and social class provides a more comprehensive picture of living in a gendered world than does any single variable.

SEXUALITY, SOCIAL CONTROL, AND COMMERCIALIZING SEX

Men assert their power and control—across cultures and over time—through rape, intimate partner violence, sexual harassment, exploiting women through prostitution and pornography, forcing women to marry against their will, honor killings, FGM/C, and punishing women, but not men, for seeking a divorce or committing adultery (Magnier, 2009; World Health Organization, 2013).

Sex is big business. From 2001 to 2007, Internet pornography in the United States alone increased from a $1 billion to a $3-billion-a-year industry (Covenant Eyes, Inc., 2013). Men's average testosterone levels decline naturally with aging, but are also due to obesity and alcohol. Pharmaceutical companies "have seized on the decline in testosterone levels as pathological and applicable to every man." As a result, sales of testosterone-boosting drugs are estimated to soar from $2 billion in 2012 to $5 billion by 2017. Prescriptions are surging even though medical studies have shown that the drugs increase the risk of heart attacks (La Puma, 2014: A21; O'Connor, 2014).

The Food and Drug Administration recently approved Addyi, a drug to increase premenopausal women's sexual desire. Some women's groups lobbied for the drug, but the "little pink pill" has been controversial: Only about 10 percent of the women who used it saw an improvement (probably because depression, alcohol, drugs, stress, and dissatisfaction with one's partner can decrease a sex drive); it has side effects like dizziness and fainting; only about 7 percent of premenopausal women have low sexual desire;

and the drug might be used to control women ("Give her some pills, Doc; there's something wrong with her"). Despite the physical hazards, observers are predicting "blockbuster sales" for Addyi (Palermo, 2015; Pollack, 2015).

Pornography, "sex drugs," and the earlier discussion of the "sexy babes" trend are just a few examples of the increasing *commercialization of sex*, making sexuality a commodity that can be sold for financial gain. Commercializing sex demeans both women and men, but the "products" are usually women.

CRITICAL EVALUATION

Feminist perspectives provide many insights on gender inequality, but there are several limitations. First, the explanations are sometimes too narrow; for example, men aren't universally violent or sexually exploitative. Second, feminist analyses are inclusive, but this strength can also be a weakness. Regarding the gender pay gap, for example, should race, ethnicity, or social class be given priority in implementing change? Third, are feminist scholars overstating women's underrepresentation in some sectors? Since 2010, for instance, both cable and network shows have featured women, not men, in strong leading roles (e.g., *Girls, How to Get Away with Murder, Scandal, Madam Secretary, Nurse Jackie*) (O'Keefe, 2014; Blay, 2015). And, like conflict theorists, feminist scholars are sometimes accused of glossing over women's exploitation of others. In Iraq, for example, sex traffickers are often women who target the youngest girls because virgins bring the highest prices (Naili, 2011).

9-6d Symbolic Interaction

Whereas functionalist, conflict, and some feminist theories are macro level, symbolic interactionists focus on the everyday processes that produce and reinforce gender roles. We "do" gender, sometimes consciously and sometimes unconsciously, by adjusting our behavior and our perceptions depending on the sex of the person with whom we're interacting (West and Zimmerman, 2009). Our expression of our sexuality, similarly, isn't inborn but a product of socialization, and what families and other societal groups deem as appropriate and inappropriate behavior (Hubbard, 1990).

GENDER IS A SOCIAL CONSTRUCTION

For interactionists, gender is a social creation, and gender inequality is shaped by learning gender roles through daily social interaction. For example, when teachers tell girls and boys that both are equally capable in math and science, "the difference in performance essentially disappears" (Hill et al., 2010: 2). When, on the other hand, teachers discourage girls from pursuing math and science, girls fare worse than boys on exams, enroll in fewer advanced math and science

Many sexually healthy men and women in their 20s, 30s, and 40s now use drugs like Viagra and Addyi. Are the drugs an example of commercializing sex?

Sean Nel/Shutterstock.com

Activists supporting the "kiss of love" campaign in New Delhi, India.

courses in high school and college, and choose lower-paying careers (Lavy and Sand, 2015). Thus, teachers' gender bias can affect occupational choices and earnings at adulthood.

Believing in gender differences can actually *produce* differences. Because men are self-confident about their worth, they're four times more likely than women to ask for a raise. When women do ask for a raise, they ask for 30 percent less than do men. Men in higher-level positions who aren't promoted often threaten to quit and are offered retention bonuses to stay. Women typically decide to work harder and try again next year (Lipman, 2015).

SEXUALITY IS ALSO A SOCIAL CONSTRUCTION

For interactionists, sexuality is also socially constructed. As with gender roles, we *learn* to be sexual and to express our sexuality differently over time and across cultures because the people around us affect our sexual behavior. For example, students who attend conservative religious high schools are less likely than their public school counterparts to report same-sex attraction or an LGBT self-identity in young adulthood (Wilkinson and Pearson, 2013).

In many Middle Eastern countries, men have premarital sex but don't marry women who aren't virgins (Fleishman and Hassan, 2009). In the United States, as you saw earlier, men who have casual sex are "studs," whereas women are "sluts." Thus, sexual double standards are socially constructed.

Because sexuality is socially constructed, attitudes and behavior can change. Many Americans, for instance, have become more comfortable with bisexuals, lesbians, and gay men.

In China, a university that planned to publicly shame students who engaged in "uncivilized behavior" (e.g., hugging or kissing in public) withdrew the policy after widespread protests. In much of India, people may be beaten for kissing on the street or even holding hands. To protest such moral policing by right-wing groups, two college students started a "Kiss of Love" campaign through a Facebook page. Despite arrests, hundreds of people joined the demonstration. Similar protests quickly spread to other cities (Ming, 2011; Bhardwaj, 2014).

CRITICAL EVALUATION

Despite their contributions, interactionists have been criticized for ignoring the social structures that create and maintain gender inequality. A large majority of childless 18- to 32-year-olds hope to share earning and household/caregiving responsibilities equally with their future partners. An equal division of labor is unlikely, however, because current U.S. workplace and government policies don't support women's and men's balancing work and family life. By the end of 2015, all of the military services are required to lift a ban on women serving in combat roles, an important criterion in promotions. The Marines, however, are trying to get a waiver ("The Marines Need…," 2015; Pedulla and Thébaud, 2015; see also Chapters 11 and 12).

Because behavior is socially constructed, sexual attitudes and practices vary across cultures. However, interactionism doesn't explain why siblings, even identical twins—who are socialized similarly—may have different sexual orientations. A related weakness is that interactionists don't explain why, historically and currently, women around the world are considerably more likely than men to be subjected to sexual control and exploitation. Such analyses require macro-level analyses that examine family, religious, political, and economic institutions.

STUDY TOOLS 9

READY TO STUDY? IN THE BOOK, YOU CAN:

☐ Check your understanding of what you've read with the Test Your Learning Questions provided on the chapter review card at the back of the book.

☐ Rip out the chapter review card for a handy summary of the chapter and key terms.

ONLINE AT CENGAGEBRAIN.COM YOU CAN:

☐ Collect StudyBits while you read and study the chapter.

☐ Quiz yourself on key concepts.

☐ Prepare for tests with chapter flash cards as well as flash cards that you create.

☐ Watch videos to explore sociology further.

10 | Race and Ethnicity

LEARNING OUTCOMES

After studying this chapter, you will be able to...

10-1 Describe how and explain why U.S. racial and ethnic diversity has changed.

10-2 Define and give examples of race, ethnicity, and racial-ethnic group.

10-3 Show how immigration patterns have changed, and describe Americans' reactions to legal and undocumented immigrants.

10-4 Distinguish between dominant and minority groups, and describe the most common patterns of dominant-minority group relations.

10-5 Describe and illustrate the most common sources of racial-ethnic friction.

10-6 Compare the five major minority groups in terms of origins, social class, and achievements.

10-7 Compare and evaluate the theoretical explanations of race and ethnicity.

10-8 Describe how and explain why interracial and interethnic relationships are changing.

After you finish this chapter go to **PAGE 198** for **STUDY TOOLS.**

In 2014, Coca-Cola aired an ad during the Super Bowl that portrayed U.S. ethnic diversity by featuring "America the Beautiful" sung in several languages. After the ad, Twitter lit up with criticism. Some of the commenters said that it's disrespectful to sing "America the Beautiful" in any language other than English, and many disparaged immigrants who don't learn English. Others pointed out that the song isn't our national anthem, perhaps it should've been sung in one of the American Indian tongues that were here before Europeans arrived, and that we don't have an official national language (Sahgal, 2014).

Such discourse illustrates our diversity and divisions. This chapter examines the impact of race and ethnicity on our lives, why racial-ethnic inequality is still widespread, and the growth of interracial and interethnic relationships. First, however, take the short quiz on the next page to see how much you know about these topics.

What do you think?

The United States is a melting pot.

1	2	3	4	5	6	7

strongly agree   strongly disagree

Henryk T. Kaiser/Getty Images

10-1 U.S. RACIAL AND ETHNIC DIVERSITY

The United States is the most multicultural country in the world, a magnet that draws people from hundreds of nations and is home to millions of Americans who are bilingual or multilingual. Of the nearly 322 million U.S. population, 14 percent are foreign born (up from only 5 percent in 1965), and this group is expected to reach a historic 19 percent (78.2 million people) in 2065. As a result, America's multicultural umbrella includes at least 150 ethnic or racial groups and 381 languages (Passel and Cohn, 2008; "Modern Immigration Wave . . . ," 2015).

By 2025, only 58 percent of the U.S. population is projected to be white—down from 86 percent in 1950 (see *Figure 10.1*). By 2044, whites may make up less than half of the total population because Latino and Asian populations are expected to double in size, and persons who identify themselves as being two or more races will more than triple (Frey, 2014).

TRUE OR FALSE?

How Much Do You Know About U.S. Racial and Ethnic Groups?

1. Race is determined biologically.

2. The United States has the most foreign-born residents of any country in the world.

3. People who are prejudiced also discriminate.

4. Among Asian Americans, the most economically successful are Asian Indians.

5. About 85 percent of Americans report being only one race.

6. Minority groups are small in number.

The answer to #4 is true; the others are false. You'll see why as you read this chapter.

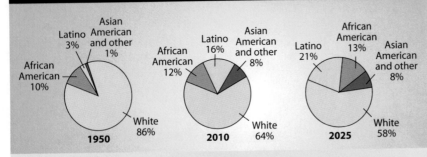

Note: "Asian American and other" includes American/Indian/Alaskan Native, Native Hawaiians and Pacific Islanders, some other race, and those who identify themselves with with two or more races.

Sources: U.S. Census Bureau, 2008; Passel et al., 2011.

10-2 THE SOCIAL SIGNIFICANCE OF RACE AND ETHNICITY

All of us identify with some groups and not others in terms of sex, age, social class, and other factors. Two of the most common and important sources of self-identification, as well as labeling by others, are race and ethnicity.

10-2a Race

A **racial group** refers to people who share visible physical characteristics, such as skin color and facial features, that members of a society consider socially important. Contrary to the popular belief that race is determined biologically, it's a *social construction*, a societal invention that labels people based on physical appearance, social class, or other characteristics (see Daniel, 2014).

Possibly only six of the human body's estimated 35,000 genes determine the color of a person's skin. Because all human beings carry 99.9 percent of the same genetic material (DNA), the "racial" genes that make us look different are miniscule compared with the genes that make us similar (Graves, 2001; Pittz, 2005).

If our DNA is practically identical, why are we so obsessed with race? People react to the physical characteristics of others, and those reactions have consequences. Skin color, hair texture, and eye shape, for example, are easily

observed and mark groups for unequal treatment. As long as we act on the basis of these characteristics, our life experiences will differ in access to jobs and other resources, how we treat people, and how they treat us.

10-2b Ethnicity

An **ethnic group** (from the Greek word *ethnos*, meaning "nation") refers to people who identify with a common national origin or cultural heritage. Cultural heritage includes language, geographic roots, food, customs, traditions, and religion. Ethnic groups in the United States include Puerto Ricans, Chinese, Serbs, Arabs, Swedes, Hungarians, Jews, and many others. Like race, ethnicity can be a basis for unequal treatment, as you'll see shortly.

10-2c Racial-Ethnic Group

People who have distinctive physical and cultural characteristics are a **racial-ethnic group**. Some people use the terms *racial* and *ethnic* interchangeably, but remember that *race* refers to physical characteristics with which we are born, whereas *ethnicity* describes cultural characteristics that we learn. The term *racial-ethnic* includes both physical and cultural traits.

Describing racial-ethnic groups has become more complex because the U.S. government allows people to

There's much variation in skin color across and within groups. People of African descent have at least 35 different shades of skin tone (Taylor, 2003). So, can you determine someone's race simply by looking at her or him?

racial group people who share visible physical characteristics that members of a society consider socially important.

ethnic group people who identify with a common national origin or cultural heritage.

racial-ethnic group people who have distinctive physical and cultural characteristics.

identify themselves in terms of both race and ethnicity. In the 2000 and 2010 censuses, for example, Latinos could check off "black" or "white" for race and "Cuban" for ethnic origin. Such choices generate dozens of racial-ethnic categories. People prefer some "labels" to others, and racial and ethnic self-identification varies.

10-2d What We Call Ourselves

In 1976, the U.S. government began to use the word *Hispanic* or *Latino* to categorize Americans who trace their roots to Spanish-speaking countries. About 70 percent say that it doesn't matter if they're referred to as *Latino* or *Hispanic*. If given a choice, however, 51 percent prefer to identify themselves by the family's country of origin (e.g., Mexican, Cuban, Salvadoran) (Pew Hispanic Center, 2012; Jones, 2013).

Since 1900, the Census Bureau has used *Negro*, *black*, and *African American*. Currently, 65 percent of this group say it doesn't matter to them whether they're called *black* or *African American*, and this trend hasn't changed much since 1991 (Jones, 2013). Many people, including African American scholars, use *black* and *African American* interchangeably.

We see similar variations in the usage of *Native American* and *American Indian*. These groups prefer their tribal identities (such as Cherokee, Apache, and Lumbi) to being lumped together under a single term. Among white Americans, even some third and fourth generations identify themselves as Irish, German, Ukrainian, and so on.

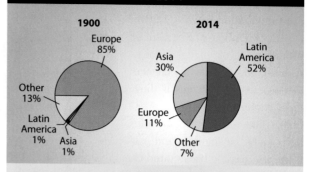

JStaley401/Getty Images

(10-3) OUR CHANGING IMMIGRATION MOSAIC

The United States, a nation of immigrants, has historically and currently both welcomed immigration and feared its consequences. What fuels public debate is the rising number of foreign-born people in the United States.

10-3a The Foreign-Born Population

The United States has one of the highest foreign-born populations in the world, but there's been a significant shift in the immigrants' country of origin. In 1900, almost 85 percent of immigrants came from Europe compared with only 11 percent in 2014 (see *Figure 10.2*). Today,

FIGURE 10.2 ORIGINS OF U.S. FOREIGN-BORN POPULATION: 1900 AND 2014

1900
- Europe 85%
- Other 13%
- Latin America 1%
- Asia 1%

2014
- Asia 30%
- Latin America 52%
- Europe 11%
- Other 7%

Note: Latin America includes the Caribbean, Central America (including Mexico), and South America.

Sources: Based on U.S. Department of Commerce, 1993, and U.S. Census Bureau, 2014 American Community Survey, Table S0201. Accessed October 1, 2015 (factfinder.census.gov).

the foreign born come primarily from Asia (mainly China and the Philippines) and Latin America (mainly Mexico) ("Modern Immigration Wave . . . ," 2015).

The United States admits more than 1 million immigrants every year—more than any other nation. A major change has been the rise of *undocumented* (also called *unauthorized*) *immigrants*—from 180,000 in the early 1980s to 11.3 million in 2014. Undocumented immigrants make up 27 percent of all foreign-born U.S. residents, and 4 percent of the nation's population. About 58 percent are from Mexico, 20 percent are from Central and Latin America, 13 percent are from Asia, and 9 percent are from other countries, including Canada, Europe, and Africa (Passel and Cohn, 2015; Zong and Batalova, 2015). Most immigrants—legal and undocumented—come to the United States for the same reasons as in the past—religious and political freedom, economic and educational opportunities, and escape from wars and natural disasters (see Chapter 15).

10-3b U.S. Reactions to Immigration

American attitudes toward immigrants have grown more positive since the mid-1990s, but are still mixed: 45 percent say immigrants strengthen the country because of their hard work and talents; 37 percent believe that immigrants weaken the United States because "they take our jobs, housing, and health care"; and 6 percent say that immigrants don't have much of an effect one way or the other. Regarding the contentious issue of immigration levels, 65 percent believe the current levels should remain the same or be increased, but 34 percent advocate reducing the stream of immigrants (Dugan, 2015; "Modern Immigration Wave...," 2015).

Between 65 to 72 percent of Americans favor giving undocumented immigrants a chance to become U.S. citizens if they meet certain requirements over time. The conditions include having jobs, paying fines and back taxes, and passing background checks (Steinhauser, 2014; "Broad Public Support ...," 2015; Jones, 2015).

Tim Boyle/Bloomberg/Getty Images

Some cities, towns, and counties—including those in states with hostile laws—have passed ordinances and resolutions that welcome undocumented immigrants. The residents believe that newcomers, both legal and illegal, are vital for economic growth and breathing new life into small rural towns and cities that have been "staggering toward the grave" (Mertens, 2015; Preston, 2015).

Allowing undocumented immigrants to work legally would increase state and local taxes by $2 billion a year because about half work "off the books" and don't pay income taxes. Citizenship would augment the health-care workforce, reducing waiting time for health services, and boost the ratio of retirees to workers. And, whether legal or undocumented, unskilled immigrants are more willing than their native-born counterparts to take grueling but low-paying jobs in service occupations (e.g., child

dominant group any physically or culturally distinctive group that has the most economic and political power, the greatest privileges, and the highest social status.

care, maids, cooks), agriculture, and the seafood industry (Institute on Taxation and Economic Policy, 2013; Kenny, 2015; Mirabella, 2015; Susman, 2015).

On a number of measures, immigrants and their children—both legal and undocumented—outperform native-born Americans. Children of immigrants are more likely than their U.S.-born peers to live in a two-parent household; to have a parent with a secure job; to earn a college degree; to attend religious services regularly; and to be "prodigious job creators." They're also less likely than their native-born counterparts to be obese, disabled, or to suffer from chronic illnesses; to commit crimes; and to have nonmarital children ("Second-Generation Americans," 2013; Alsever, 2014; Waters and Pineau, 2015).

10-4 DOMINANT AND MINORITY GROUPS

Race and ethnicity often make people feel like outsiders, or "the other." Otherness is being different or having characteristics that set you apart from the dominant group (Thorpe-Moscon and Pollack, 2014).

10-4a What Is a Dominant Group?

A **dominant group** is any physically or culturally distinctive group that has the most economic and political power, the greatest privileges, and the highest social

iStockphoto.com/Glenn Frank

Alexey Stiop/Shutterstock.com

The U.S. government issues about 66,000 temporary visas each year for seasonal workers because of labor shortages, especially in agriculture. After millions of undocumented immigrants fled to other states from Arizona, Alabama, and Georgia because of harsh laws, many employers complained that couldn't find enough laborers—in construction and landscaping—to clean up and rebuild communities in the aftermath of devastating tornados (Shepherd, 2011).

status. As a result, it can treat other groups as subordinate. In most societies, for instance, men are a dominant group because they have more status, resources, and power than women (see Chapters 8 and 9).

Dominant groups aren't necessarily the largest in number. From the seventeenth century until 1994, about 10 percent of the population in South Africa was white and had almost complete control of the black population. Because of *apartheid*, a formal system of racial segregation, the black residents couldn't vote, lost their property, and had minimal access to education and politics. Apartheid ended in 1994, but most black South Africans are still a minority because whites "hold the best jobs, live in the most expensive homes, and control the bulk of the country's capital" (Murphy, 2004: A4).

10-4b What Is a Minority?

Sociologists describe Latinos, African Americans, Asian Americans, Middle Eastern Americans, and American Indians as minorities. A **minority** is any group that may be treated differently and unequally because of their physical, cultural, or other characteristics. The characteristics include gender, age, sexual orientation, religion, ethnicity, or skin color. Minorities may be larger in number than a dominant group, but they have less power, privilege, and social status. For example, middle-class African Americans are much more likely than their white counterparts to be called by debt collectors even though both groups have similar debt levels and repayment rates (Ruetschlin and Asante-Muhammad, 2013). *Table 10.1* offers other examples of everyday privileges associated with skin color.

10-4c Some Patterns of Dominant-Minority Group Relations

To understand some of the complexity of dominant-minority group relations, think of a continuum. At one end of the continuum is genocide; at the other end is pluralism (see *Figure 10.3*).

TABLE 10.1 AM I PRIVILEGED?

Most white people don't feel privileged because they aren't wealthy. Nonetheless, they enjoy everyday benefits, and take them for granted, simply because they're members of the dominant group. Would you add other advantages of being white?

1. I can go shopping and feel fairly sure that I won't be followed or harassed by store detectives.

2. If a traffic cop pulls me over, I can be sure that I haven't been singled out because of my race.

3. I can be late to a meeting without having the lateness reflect on my race.

4. I can turn on the television and see people who look like me represented in positive ways and in a wide range of roles.

5. I live in a safe neighborhood with good schools.

6. I never think twice about calling the police when trouble occurs.

7. I can wear a hoodie and not worry about others thinking that I'm up to no good.

8. I don't get dirty looks if I listen to loud music at a gas station.

Sources: McIntosh, 1995; Independent Television Service, 2003; Williams, 2014.

GENOCIDE

Genocide is the systematic effort to kill all members of a particular ethnic, religious, political, racial, or national group. By 1710, the colonists in America had killed thousands of Indians, poisoned others, and promoted scalp bounties. In 1851, the governor of California officially called for the extermination of all Indians in the state (de las Casas, 1992; Churchill, 1997). Worldwide, and

minority a group of people who may be treated differently and unequally because of their physical, cultural, or other characteristics.

genocide the systematic effort to kill all members of a particular ethnic, religious, political, racial, or national group.

FIGURE 10.3 CONTINUUM OF SOME DOMINANT-MINORITY GROUP RELATIONS

INTOLERANCE
INEQUALITY

ACCEPTANCE
EQUALITY

Genocide
Systematic efforts to destroy minorities (e.g., American Indians)

Segregation
Physical and social separation of dominant and minority groups (e.g., housing segregation)

Acculturation
A minority group adopts the language, values, and other characteristics of the dominant group (e.g., learning and speaking English)

Pluralism
There is no dominant group because all groups share power and other resources fairly equally (e.g., possibly Switzerland)

between 1915 to 1995 alone, well over 74 million people have been victims of genocide in Africa, Cambodia, China, Eastern Europe, and Turkey (United Human Rights Council, 2004).

SEGREGATION

Segregation is the physical and social separation of dominant and minority groups. In 1954, the Supreme Court ruling in *Brown v. Board of Education* declared *de jure*, or legal, segregation unconstitutional. It was followed by the passage of a variety of federal laws that prohibited racial segregation in public schools, as well as discrimination in employment, voting, and housing.

De facto, or informal, segregation has replaced *de jure* segregation in the United States and many other countries. Some *de facto* segregation may be voluntary, as when minorities prefer to live among their own racial or ethnic group. In most cases, however, *de facto* segregation is due to discrimination. In the case of equally qualified homeseekers, for instance, realtors show minorities fewer homes and apartments than they do whites. Because minorities have fewer choices, the time and cost of their housing searches rise (Turner et al., 2013; see also Rugh and Massey, 2010).

ACCULTURATION AND ASSIMILATION

Many minority group members blend into U.S. society through **acculturation**, the process of adopting the language, values, beliefs, and other characteristics of the host culture (e.g., learning English, celebrating Thanksgiving). Acculturation doesn't include intermarriage, but the newcomers merge into the host culture in most other ways. **Assimilation** involves conforming to the dominant group's culture, adopting its language and values, and intermarrying with that group. Journalists often use *assimilation* and *acculturation* interchangeably, but the former absorbs, rather than just changes, a group culturally.

---

segregation the physical and social separation of dominant and minority groups.

acculturation the process of adopting the language, values, beliefs, and other characteristics of the host culture.

assimilation conforming to the dominant group's culture, adopting its language and values, and intermarrying with that group.

pluralism minority groups maintain many aspects of their original culture while living peacefully with the host culture.

racism a set of beliefs that one's own racial group is inherently superior to other groups.

PLURALISM

In **pluralism**, sometimes called *multiculturalism*, minority groups maintain many aspects of their original culture—including using their own language and marrying within their own racial or ethnic group—while living peacefully with the host culture. Pluralism is especially evident in urban areas with large racial and ethnic communities (e.g., "Little Italy," "Greek Town," "Little Korea," "Spanish Harlem"). U.S. minorities have numerous ethnic newspapers and radio stations, and the same constitutional rights (such as freedom of speech) as the dominant group. Nonetheless, people of various skin colors and cultures don't always experience the same social standing, and there can be considerable racial-ethnic friction.

10-5 SOME SOURCES OF RACIAL-ETHNIC FRICTION

During a recent trip to Switzerland, Oprah Winfrey asked a clerk at an expensive shop to show her a $37,000 purse. The clerk refused: "That one will cost too much, you won't be able to afford that." Winfrey didn't know for sure whether or not the snub was motivated by racism, but said that the incident was what "people with black or brown skin experience every day" (Goyette, 2013). Ahmed Mohamed, a 14-year-old Sudanese-American student in Texas, took a homemade digital clock to school to show his engineering teacher. The principal, concerned that the clock was a bomb, notified the police, who handcuffed Mohamed and transported him to a juvenile detention facility. Would people have reacted differently if Winfrey and Mohamed had been white? And were the reactions racist?

10-5a Racism

Racism is a set of beliefs that one's own racial group is inherently superior to other groups. Using this definition, anyone can be racist if she or he believes that another group is inferior. It's a way of thinking about racial and ethnic differences that justifies and preserves the social, economic, and political interests of dominant groups.

Many whites believe that racism has been pretty much "solved." After all, we've elected an African American president twice, have a Latina Supreme Court justice, and minorities now own many businesses. In fact, racism is widespread. A few recent examples—and all involving young white men—include the murder of nine blacks in a historic black church in Charleston, South Carolina, and,

A number of colleges and professional teams have replaced their logos, nicknames, and mascots because many American Indians have denounced them as racist. Some exceptions are the Cleveland Indians in baseball (left) and the Washington Redskins in football (right). Why do many people see the Indian images and team names as demeaning? Why do you think that the Minnesota Vikings and University of Notre Dame's Fighting Irish are acceptable?

at several universities, racist slurs and videos about lynching blacks (King, 2015; Schmidt, 2015). Racism fuels both prejudice and discrimination.

10-5b Prejudice

Prejudice is an *attitude* that prejudges people, usually in a negative way, who are different from "us" in race, ethnicity, or religion. If an employer assumes, for example, that white workers will be more productive than blacks or Latinos, she or he is prejudiced. Prejudice isn't one-sided because *anyone* can be prejudiced ("White people can't be trusted" or "Black women can't afford expensive purses"). Prejudice is most evident in stereotypes and scapegoating.

A **stereotype** is an oversimplified or exaggerated generalization about a group of people (see, for example, www.stuffwhite-peoplelike.com). Stereotypes can be positive ("African Americans are great athletes") or negative ("African Americans are violent"). Whether positive or negative, stereotypes distort reality. Some blacks are great athletes and some are violent, just like people in other groups. Whatever the intention, stereotypes lump people together and reinforce the belief that many traits are biological and fixed. Once established, stereotypes are difficult to change because people often dismiss any evidence to the contrary as an exception ("For an Asian, Jeremy Lin is a terrific basketball player").

Stereotypes can lead to a displacement of anger and aggression on **scapegoats**, individuals or groups whom people blame for their own problems or shortcomings ("They didn't hire me because the company wants blacks" or "I didn't get into that college because Asians are at the top of the list"). Minorities are easy targets because they typically differ in physical appearance and are usually too powerless to strike back.

At one time or another, almost all newcomers to the United States have been scapegoats. Especially in times of economic hardship, the most recent immigrants often become scapegoats ("Latinos are replacing Americans in construction jobs"). Prejudice, stereotypes, and scapegoating are attitudes, but they often lead to discrimination.

10-5c Discrimination

Discrimination is *behavior* that treats people unequally or unfairly because of their group membership. It encompasses all sorts of actions, ranging from social slights (e.g., not inviting minority coworkers to lunch) to rejection of job applications and racially motivated hate crimes. Discrimination can be subtle (e.g., not sitting next to someone) or blatant (e.g., racial slurs), and it occurs at individual and institutional levels.

Individual discrimination is harmful action on a one-to-one basis by a dominant group member against someone in a minority group. Nationally, 74 percent of blacks, compared with only 31 percent of whites, believe that they have been discriminated against because of their race (Romano and Samuels, 2012).

---

prejudice an attitude that prejudges people, usually in a negative way.

stereotype an oversimplified or exaggerated generalization about a group of people.

scapegoats individuals or groups whom people blame for their own problems or shortcomings.

discrimination behavior that treats people unequally or unfairly because of their group membership.

individual discrimination harmful action on a one-to-one basis by a dominant group member against someone in a minority group.

In **institutional discrimination** (also called *institutionalized discrimination*, *systemic discrimination*, and *structural discrimination*), minority group members experience unequal treatment and fewer opportunities because of a society's everyday laws, policies, practices, and customs. Institutional discrimination is widespread, but being challenged. In a recent 5-4 decision, the Supreme Court ruled that housing discrimination is illegal even if it's unintentional. Individuals and groups can now sue banks and insurance companies, for example—even without proof of intentional discrimination—if their policies or practices have a discriminatory *effect* on minorities (Hurley, 2015).

Discrimination, both individual and institutional, also occurs *within* racial-ethnic groups. For example, a study of New York City's Korean-owned nail salons found an "ethnic caste system." Among Asian manicurists, Koreans earned twice as much as their non-Korean counterparts. The latter frequently worked longer hours, were often forced into less desirable salons that had fewer customers and paltry tips, were told not to chat during their entire 12-hour shifts, and were assigned male customers that manicurists often dread because of the men's "thick toenails and hair-covered knuckles" (Nir, 2015).

10-5d Relationship Between Prejudice and Discrimination

Sociologist Robert Merton (1949) created a model showing how the relationship between prejudice and discrimination can vary. His model includes four types of people and their possible response patterns (see *Table 10.2*).

TABLE 10.2 RELATIONSHIP BETWEEN PREJUDICE AND DISCRIMINATION

		Does the Person Discriminate?	
		Yes	No
Is the Person Prejudiced?	Yes	Prejudiced discriminator (e.g., a prejudiced person who attacks minority group members verbally or physically)	Prejudiced nondiscriminator (e.g., a prejudiced person who goes along with equal employment opportunity policies)
	No	Unprejudiced discriminator (e.g., an unprejudiced person who joins a club that excludes minorities)	Unprejudiced nondiscriminator (e.g., an unprejudiced employer who hires minorities)

Source: Based on Merton, 1949.

institutional discrimination unequal treatment and fewer opportunities because of a society's everyday laws, policies, practices, and customs.

"You look like this sketch of someone who's thinking about committing a crime."

David Sipress/The New Yorker Collection/Cartoon Bank.Com

Unprejudiced nondiscriminators aren't prejudiced and don't discriminate. They believe in the American creed of freedom and equality for all and cherish egalitarian values. They may not do much, however, individually or collectively, to change discrimination. In contrast, but equally consistent in attitude and action, are *prejudiced discriminators* who are both prejudiced and discriminate. They're willing to defy laws, such as not renting to minorities, to express their beliefs.

Unprejudiced discriminators aren't prejudiced, but they discriminate because it's expedient or in their own self-interest to do so. If, for example, an insurance company charges higher automobile insurance premiums to people with low occupational and educational levels (often minorities), agents will implement these policies even though they themselves aren't prejudiced.

Prejudiced nondiscriminators are prejudiced, but don't discriminate. Despite their negative attitudes, they hire minorities and are civil in everyday interactions because they believe they must conform to antidiscrimination laws or situational norms. If, for example, most of their neighbors or coworkers don't discriminate, prejudiced nondiscriminators will go along with them.

10-6 MAJOR RACIAL AND ETHNIC GROUPS IN THE UNITED STATES

Four states (California, Hawaii, New Mexico, and Texas) and the District of Columbia are now "majority-minority," meaning that minorities make up more than half of the population ("Millennials Outnumber...," 2015). Of the major racial-ethnic groups, some encounter more barriers than others, but all have numerous strengths that enhance U.S. society. *Figure 10.4* shows the racial-ethnic makeup of the U.S. population. Let's begin with white ethnic groups whose ancestors are from Europe.

FIGURE 10.4 MAJOR U.S. RACIAL AND ETHNIC GROUPS

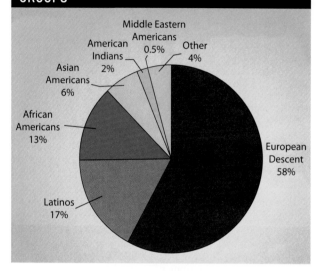

10-6a European Americans: A Declining Majority

During the seventeenth century, English immigrants settled the first colonies in Massachusetts and Virginia. Other white Anglo-Saxon Protestants (WASPs), who included people from Wales and Scotland, quickly followed. Most of these groups spoke English. Some of the immigrants were affluent, but many were poor or had criminal backgrounds.

DIVERSITY

About 58 percent of the U.S. population has a European heritage. The largest groups have ancestors from Germany, Ireland, England, Italy, Poland, France, and the Scandinavian countries (see *Figure 10.5*).

CHARACTERISTICS AND CHANGES

WASPs generally looked down on later waves of immigrants from southern and eastern Europe. They viewed the newcomers as inferior, dirty, lazy, and uncivilized because they differed in language, religion, and customs. New England, which was 90 percent Protestant, was particularly hostile to Irish Catholics, characterizing them as irresponsible and shiftless (Feagin and Feagin, 2008).

All the later waves of European immigrants faced varying degrees of hardship in adjusting to the new land because the first English settlers had a great deal of power in shaping economic and educational institutions. In response to prejudice and discrimination, many of the immigrants founded churches, schools, and recreational activities that maintained their language and traditions (Myers, 2007).

Despite stereotypes, prejudice, and discrimination, European immigrants began to prosper within a few generations. They surmounted numerous obstacles and became influential in all sectors of U.S. life. Overall, they now fare much better financially than most other groups in the United States. This doesn't mean that all are rich. In fact, in absolute numbers, poor whites outnumber those of other racial-ethnic groups (see Chapter 8).

10-6b Latinos: The Largest Minority

About 1 in 3 Americans is a member of a racial or ethnic minority, but Latinos are the largest group, constituting 17 percent of the nation's population. The size and growth of the Latino population this century is due mainly to births in the United States, not recent immigration. On average, Latinas have three births each, one more than white, black, and Asian or Pacific Islander women (Krogstad and Lopez, 2014; "Hispanic Heritage Month . . . ," 2015).

DIVERSITY

Worldwide, only Mexico has a larger Latino population (120 million) than the United States (55 million). More persons of Puerto Rican origin now live in the 50 states and the District of Columbia than in Puerto Rico (Cohn et al., 2014; "Hispanic Heritage Month . . . ," 2015).

FIGURE 10.5 EUROPEAN AMERICANS BY ORIGIN, 2014

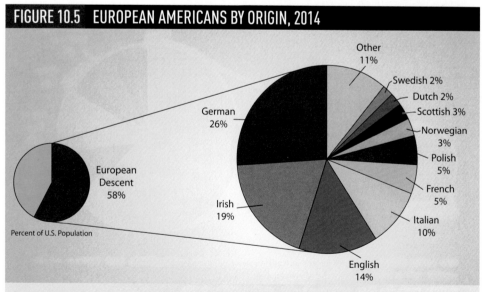

Source: Based on U.S. Census Bureau, Selected Social Characteristics in the United States, 2014 American Community Survey 1-Year Estimates, Table DP02. Accessed October 5, 2015 (factfinder.census.gov).

Dominican-born Alfredo Rodriguez is one of numerous successful Latino businessmen who are rebuilding neglected inner-city neighborhoods. In 1985, Rodriguez bought his first grocery store in Queens, New York, with the $25 a week his mother had been setting aside for him for a decade. In 2002, he purchased a supermarket in Newark, New Jersey, to meet the needs of local Latino shoppers. Five years later, his Xtra Supermarket had annual sales of $9 million (Rayasam, 2007).

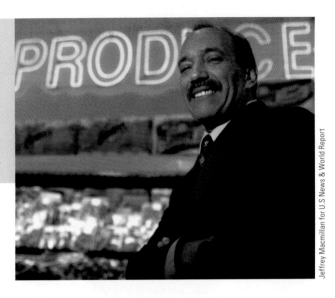

Jeffrey Macmillan for U.S News & World Report

Some Latinos trace their roots to the Spanish and Mexican settlers who established homes and founded cities in the Southwest before the arrival of the first English settlers on the East Coast. Others are recent immigrants or children of the immigrants who arrived in large numbers at the beginning of the twentieth century. Of the Latinos living in the United States, most are from Mexico (see *Figure 10.6*), but Spanish-speaking people from different countries vary widely in their customs, cuisines, and cultural practices.

CHARACTERISTICS AND CHANGES

About 21 percent of Americans age 5 and older speak a language other than English at home. In this group, 73 percent speak Spanish. The number of U.S. residents who speak Spanish at home increased by 120 percent since 1990, whereas the number who spoke Italian, French, German and other Indo-European languages declined during this period (Ryan, 2013; "Hispanic Heritage Month...," 2015).

The median household income of Latinos is lower than the national median of $53,657, but higher than that of American Indians and African Americans (see *Figure 10.7*). About 44 percent of Latino families earn $50,000 a year or more, up considerably from only 7 percent in 1972. However, 25 percent of Mexican Americans, Dominicans, and Puerto Ricans and 16 percent of Cuban Americans live below the poverty line (Macartney et al., 2013; DeNavas-Walt and Proctor, 2015).

Latinos have among the lowest education levels (see *Figure 10.8*), but there's considerable variation across subgroups. Almost half of Venezuelans have a bachelor's degree or higher compared with only 8 to 9 percent of those from Guatemala, Mexico, and Salvador (Ogunwole et al., 2012).

As with other groups, the socioeconomic status of Latinos reflects a number of interrelated factors, particularly education, occupation, English language proficiency, and recency of immigration. For example, 21 percent of second-generation Latinos have at least a college degree compared with 11 percent of foreign-born Latinos, and higher median household incomes—$48,400 and $34,600, respectively ("Second-Generation Americans," 2013).

Almost 78 percent of second-generation Latinos, compared with 58 percent of all U.S. adults,

FIGURE 10.6 LATINOS BY ORIGIN, 2014

Dominican 3%
Cuban 4%
South American 6%
*Other 4%
Puerto Rican 10%
Central American 9%
Mexican 64%

Latinos 17%

Percent of U.S. Population

Note: Central American includes countries such as El Salvador, Honduras, and Guatemala; South American includes countries such as Argentina, Bolivia, and Venezuela.

*Includes people from Spain and those who didn't specify a country of origin.

Source: Based on U.S. Census Bureau. 2015. "Hispanic Roots." Accessed October 1, 2015 (www.census.gov).

FIGURE 10.7 U.S. MEDIAN HOUSEHOLD INCOME, BY RACE AND ETHNICITY, 2014

$74,297

$60,256

$53,657

$42,491

$36,252 $35,398

All Households

Asian

White

Latino

American Indian/ Alaska Native

African American

Sources: Based on "American Indian and Alaska Native...," 2014, and DeNavas-Walt and Proctor, 2015, Table 1.

believe that most people can get ahead if they're willing to work hard ("Second-Generation Americans," 2013). Such attitudes help explain why many Latinos are successful. They own 12 percent of all businesses, up from 8 percent in 2007. Since 2005, many Latino immigrants have earned better hourly wages than their predecessors because they tended to be older, better educated, and more likely to be employed in construction than in agriculture (Kochhar, 2007; "Hispanic Heritage Month ...," 2015; "2012 Survey of Business Owners," 2015).

10-6c African Americans: A Changing Minority

The 45 million African Americans are the second largest minority group in the United States, making up 13 percent of the population ("Black (African-American)...," 2015). This percentage includes those of more than one race.

DIVERSITY

Most African Americans share a common characteristic: They're members of the only group ever brought to the United States involuntarily and legally enslaved. The term *African American* encompasses tremendous diversity, including native-born Americans with black, white, American Indian, and/or Latino ancestors, as well as recent immigrants from Africa and elsewhere. Of the nearly 4 million foreign-born blacks, 50 percent are from the Caribbean, 36 percent from Africa, and 9 percent from a Central or South American country. The largest African-born populations are from Nigeria and Ethiopia (Anderson, 2015).

CHARACTERISTICS AND CHANGES

African Americans make up 13 percent of the U.S. population but 26 percent of people living in poverty (DeNavas-Walt and Proctor, 2015). The median family income of African Americans is the lowest of all racial-ethnic groups (see *Figure 10.7*). Between 2010 and 2014, the number of black households with annual incomes of $50,000 or more dropped from 36 to 28 percent (DeNavas-Walt and Proctor, 2015).

Why? About 20 percent of blacks work for the government (e.g., as teachers, mail carriers, bus drivers), and are more likely to do so than whites or Latinos. During and since the Great Recession, many of these positions declined, increasing unemployment and erasing job stability and promotion opportunities. White households recovered from the downturn more quickly than other groups because they had more wealth to fall back on. Black households were also hit harder because of home foreclosures caused, in part, by greedy and discriminatory lending practices (Cohen, 2015; Burd-Sharps and Rasch, 2015).

You saw in Chapter 8 that there's a high correlation between occupation, education, and income. Some cities have implemented mentoring programs that are increasing black high school graduation and college enrollment rates, but blacks have much lower education levels than

Few people know that Madame C. J. Walker (1867–1919), who manufactured hair care products for black women, was one of the first American female millionaires. Or that Dr. Charles R. Drew (1904–1950) was a renowned surgeon, teacher, and researcher. He founded two of the world's largest blood banks, saving untold lives during and since World War II.

FIGURE 10.8 PERCENTAGE WITH A BACHELOR'S DEGREE OR HIGHER, BY RACE AND ETHNICITY, 2014

52% Asian
36% White
22% Black
18% American Indian/ Alaska Native
15% Latino

Sources: Based on U.S. Census Bureau, "Educational Attainment in the United States: 2014 Detailed Tables," Table 1, accessed October 2, 2015 (census.gov), and "American Indian and Alaska Native...," 2015.

Asians or whites (see *Figure 10.8*). Nonetheless, some economists estimate that at least 30 percent of the black–white wage gap is due to differential treatment in hiring, firing, and pay rather than differences in formal schooling or specific job skills (Fryer et al., 2011).

Despite discrimination, many African Americans are successful. They own almost 9 percent of U.S. businesses, up from 6 percent in 2007. The largest shares are in health care/social assistance and transportation/warehousing businesses (20 percent and 15 percent, respectively) (DeSilver, 2015; "2012 Survey of Business Owners," 2015).

10-6d Asian Americans: A Model Minority?

The nearly 20 million Asian Americans comprise almost 6 percent of the U.S. population. In 2012, Asians became the fastest-growing racial or ethnic group in the country; over 60 percent of the growth came from international migration. Hawaii is the only state with an Asian majority of 64 percent ("Asians Fastest-Growing...," 2013; "Asian/Pacific American...," 2015).

DIVERSITY

Asian Americans encompass a broad swath of cultural groups. They come from at least 26 countries in East and Southeast Asia (e.g., China, Korea, Vietnam, Cambodia, the Philippines) and South Asia (especially India, Pakistan, and Sri Lanka), and speak at least 19 languages in the United States (Ryan, 2013; "Asian/Pacific American...," 2015).

These diverse origins mean that there are huge differences in languages and dialects (and even alphabets), religions, cuisines, and customs. Chinese are the largest Asian American group, followed by Filipinos and Asian Indians (see *Figure 10.9*). At least 83 percent of Asian Americans trace their roots to only six countries—China, India, Japan, Korea, the Philippines, and Vietnam (Taylor et al., 2012).

CHARACTERISTICS AND CHANGES

Compared with whites and other racial-ethnic groups, many Asian Americans are doing well. They have the highest education levels (see *Figure 10.8*), and almost 21 percent, compared with 11 percent of the general population, have a graduate or professional degree. Because of their high education levels, almost half of Asian Americans are in highly skilled and high-paying occupations such as information technology, science, engineering, and medicine ("Asian/Pacific American...," 2015). Consequently, they have the highest median household income (see *Figure 10.7*).

Because of their educational and economic success, Asian Americans are often hailed as a "model minority."

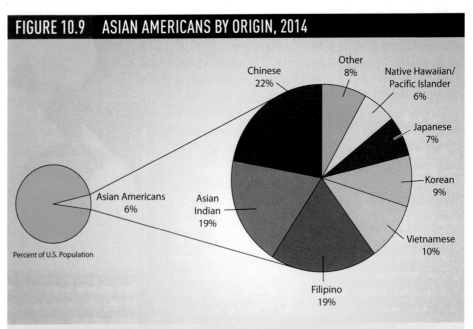

FIGURE 10.9 ASIAN AMERICANS BY ORIGIN, 2014

Percent of U.S. Population

Asian Americans 6%

Chinese 22%
Other 8%
Native Hawaiian/ Pacific Islander 6%
Japanese 7%
Korean 9%
Vietnamese 10%
Filipino 19%
Asian Indian 19%

Note: "Other" includes people from at least 15 countries, including Laos, Cambodia, and Sri Lanka.

Source: Based on U.S. Census Bureau, 2014, American Community Survey, Tables B02018 and B02019. Accessed October 2, 2015 (factfinder.census. gov).

Asian Indian students have won the Scripps National Spelling Bee eight years in a row, and all but four of the last 16 years. In 2015, 13-year-old Vanya Shivashankar and 14-year-old Gokul Venkatachalam were co-champions. The winning words were *scherenschnitte* and *nuntak*. (Am I the only one who had to look them up?) Asian Indians' extraordinary performance has been attributed to factors like hard work, sacrifice, pursuing a championship over many years, and growing up in households where one or both parents are well-educated professionals who set high academic goals (Heim, 2015).

Such labels are misleading, however, because there's considerable variation across subgroups. For example, more than 70 percent of those from India and Taiwan have a bachelor's degree or higher compared with only 12 to 14 percent of Hmong, Laotians, and Cambodians. More than 63 percent of Hmong, Laotians, and Cambodians and 50 percent of Native Hawaiians and Vietnamese haven't attended college, compared with only 23 percent of Filipinos and Asian Indians (Teranishi, 2011; Ogunwole et al., 2012).

Fully 93 percent of Asian Americans describe members of their country of origin group as "very hardworking"; just 57 percent say the same about Americans as a whole (Taylor et al., 2012). You'll see in Chapter 13 that many Asian American college students succeed, and despite discrimination, because they work harder than most of their peers. Stereotypes of Asian-Americans as hard-working and high-achieving can enhance performance: Teachers and guidance counselors place the students in advanced classes, the honors track, and give them extra help with homework and college applications. On the other hand, the model minority stereotype has negative effects: It pigeonholes Asian students into specific careers and doesn't them offer the chance "to see themselves as something other than doctors, engineers, or accountants"; assumes that all minorities can overcome institutionalized discrimination; and holds all

Asian Americans to unrealistically high expectations (Kay et al., 2013; Leung, 2013; Lee and Zhou, 2015).

10-6e American Indians: A Growing Nation

American Indians used to be called the "vanishing Americans," but they've "staged a surprising comeback" due to higher birth rates, a longer life expectancy, and better health services (Snipp, 1996: 4). The 5.2 million American Indians and Alaska Natives (AIANs) make up almost 2 percent of the U.S. population, and are expected to increase to almost 3 percent by 2060 ("American Indian and Alaska Native...," 2014).

DIVERSITY

Like Asian Americans, American Indians and Alaska Natives are a heterogeneous group. Of the 566 federally recognized tribes, 8 have more than 100,000 members. The Cherokee, with almost 820,000 members, is the largest, followed by the Navajo and Choctaw (Norris et al., 2012). Tribes speak 150 native languages, although many are quickly vanishing. For example, only about 25 people nationwide speak Comanche. Tribes also vary widely in their religious beliefs and cultural practices. Thus, a Comanche-Kiowa educator cautions, "Lumping all Indians together is a mistake. Tribes ... are sovereign nations and are as different from another tribe as Italians are from Swedes" (Pewewardy, 1998: 71; Ashburn, 2007; Mangan, 2013).

CHARACTERISTICS AND CHANGES

AIANs are a unique minority group because they're not immigrants and have been in what is now the United States longer than any other group. They have experienced centuries of subjugation, exploitation, and political exclusion (e.g., not having the right to vote until 1924). Some tribes are still trying to reclaim billions of dollars that the federal government has squandered or mismanaged (Wilkinson, 2006; Volz, 2012; Kindy, 2013).

Many AIANs are better off today than they were a decade ago, but long-term institutional discrimination has been difficult to shake. For example, 29 percent live below the poverty line compared with 15 percent of the general population. The median household income of AIANs is slightly higher than that of African Americans, but lower than that of other racial-ethnic groups (see *Figure 10.7*).

Their educational levels have increased, but only 18 percent of AIANs have a bachelor's degree or higher compared with 29 percent of the general population. One of four civilian-employed AIANs works in a management or professional occupation. Even when AIANs are similar to whites in age, gender, education level, marital status, state of residence, and other factors, their odds of being

Mohegan Sun in southern Connecticut is one of the largest casinos in the United States. It has spent some of its profits on college scholarships, a $15 million senior center, and health insurance for tribal members. In contrast, some of the poorest tribes, such as the Navajo and Hopi, who have rejected gaming for religious reasons, have many members who live in poverty without kitchen facilities (like stoves or refrigerators) or indoor plumbing.

employed are 31 percent lower than those of whites. This suggests, according to some researchers, that AIANs are experiencing discrimination in the labor market (Austin, 2013; "American Indian and Alaska Native...," 2014).

Despite numerous obstacles, AIANs have made considerable economic progress by insisting on self-determination and the rights of tribes to run their own affairs. For example, 244 tribes now generate about 43 percent of all U.S. casino gaming revenue. The Justice Department allows tribes to grow or sell marijuana on their lands, even in states that ban the practice. Some tribes, "mindful of the painful legacy of alcohol abuse in their communities," are opposed to selling or using marijuana on their territory. Others see marijuana sales as a huge source of future income, similar to cigarette sales and casino gambling. The 400 members of the Santee Sioux tribe in South Dakota run a successful casino, a 120-room hotel, and a 240-head buffalo ranch. Because of greater competition and the lingering effects of the Great Recession, however, the tribe is opening the nation's first marijuana resort on its reservation. The project could generate up to $2 million a month in profit (Phelps, 2014; Cano, 2015; "2013 Indian Gaming Industry Report," 2015).

10-6f Middle Eastern Americans: An Emerging Minority

The Middle East is "one of the most diverse and complex combinations of geographic, historical, religious, linguistic, and even racial places on Earth" (Sharifzadeh, 1997: 442). It encompasses about 30 countries, including Armenia, Turkey, Israel, Iran, Afghanistan, Pakistan, and 22 Arab nations (e.g., Algeria, Iraq, Kuwait, Saudi Arabia, and the United Arab Emirates).

DIVERSITY

Of the 57 million people in the United States who speak a language other than English at home, more than 3 percent speak Middle Eastern languages such as Armenian, Arabic, Hebrew, Persian, or Urdu (U.S. Census Bureau,

Steve Jobs—the late co-founder, chairman, and CEO of Apple Inc.—was born to unmarried university students. His father was a Syrian-born Muslim, and his mother was a U.S.-born Catholic of Swiss descent. The father put the baby up for adoption because his girlfriend's family wouldn't allow her to marry an Arab ("Steve Jobs...," 2011).

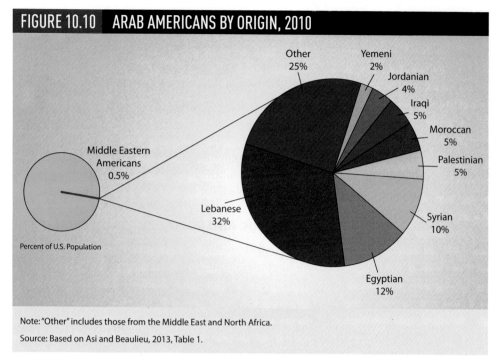

FIGURE 10.10 ARAB AMERICANS BY ORIGIN, 2010

Middle Eastern Americans 0.5%
Percent of U.S. Population

Other 25%
Yemeni 2%
Jordanian 4%
Iraqi 5%
Moroccan 5%
Palestinian 5%
Lebanese 32%
Syrian 10%
Egyptian 12%

Note: "Other" includes those from the Middle East and North Africa.

Source: Based on Asi and Beaulieu, 2013, Table 1.

2012). There are more than 1.8 million Americans of Arab descent, accounting for about 0.5 percent of the total population (Asi and Beaulieu, 2013; Krogstad, 2015). Those who identify themselves as Arab Americans come from many different countries (see *Figure 10.10*). As in the case of Asian American families, Middle Eastern families make up a heterogeneous population that is a "multicultural, multiracial, and multiethnic mosaic" (Abudabbeh, 1996: 333).

CHARACTERISTICS AND CHANGES

Middle Eastern Americans tend to be better educated and wealthier than other Americans. About 46 percent have a college degree or higher compared with 29 percent of the general population; 20 percent have a post-graduate degree, which is nearly twice the U.S. average of 10 percent. In 2010, the median income of Arab American households was almost $57,000 compared with $52,000 for all U.S. households. As in other groups, however, there are wide variations. Lebanese have higher median family incomes ($67,300) than Iraqis ($32,000) (Asi and Beaulieu, 2013; Motel and Patten, 2013).

Not all Middle Eastern Americans are successful, of course. Lebanese and Syrians have the lowest poverty rates, 11 percent, compared with more than 26 percent of Iraqis, and 23 percent of all foreign-born Middle Easterners (Arab American Institute Foundation, 2012; Motel and Patten, 2013).

10-7 SOCIOLOGICAL EXPLANATIONS OF RACIAL-ETHNIC INEQUALITY

As with other topics, the four major sociological theories help us understand racial-ethnic relations. (*Table 10.3* summarizes the key points of each perspective.)

TABLE 10.3 SOCIOLOGICAL EXPLANATIONS OF RACIAL-ETHNIC INEQUALITY

Theoretical Perspective	Level of Analysis	Key Points
Functionalist	Macro	Immigration provides needed workers; acculturation and assimilation increase social solidarity; racial-ethnic inequality can be dysfunctional, but benefits dominant groups.
Conflict	Macro	There's ongoing strife between dominant and minority groups; powerful groups maintain their advantages primarily through economic exploitation; race is a more important factor than social class in perpetuating racial-ethnic inequality.
Feminist	Macro and micro	Minority women suffer from the combined effects of racism and sexism; gendered racism occurs within and across racial-ethnic groups.
Symbolic Interactionist	Micro	Because race and ethnicity are socially constructed, social interaction can increase or reduce racial and ethnic hostility; antagonistic attitudes toward minorities, which are learned, can be lessened through cooperative interracial and interethnic contacts.

10-7a Functionalism

Those who criticize immigrants for not becoming Americanized quickly enough reflect a functionalist view of racial-ethnic relations. That is, if a society is to work harmoniously, newcomers must adopt the dominant group's values, goals, and particularly language. Doing so increases a society's cultural solidarity.

STABILITY AND COHESION

Immigration is functional for the host nation if it gains needed workers. Highly educated and skilled immigrants fill important positions in medicine, science, and business. Many employers also rely on immigrants to work in fields, orchards, and vineyards at low wages; others are actively recruiting immigrants for decent-paying but "dirty" jobs at meat and fish factories that native-born Americans avoid (Newkirk and Douban, 2012; Davey, 2014).

Functionalists view racial-ethnic inequality as dysfunctional, but attribute much of the inequity to individual failings and a lack of acculturation. In a meritocratic society, people should be rewarded for their accomplishments rather than ethnicity or skin color (see Chapter 8). There are incentives for not acculturating, however. For example, hundreds of Puerto Ricans get federal disability benefits, and regardless of work experience or level of education, because they're considered less employable if they can't speak English. About 76 percent of immigrant households with children receive at least one kind of welfare benefit (e.g., Medicaid, food stamps, housing assistance) compared with 52 percent of their native-born counterparts (Camarota, 2015; Office of the Inspector General, 2015).

CRITICAL EVALUATION

Functionalist explanations are limited in several ways. First, acculturation and assimilation increase social solidarity, but can also have negative outcomes. For example, second-generation immigrants are more likely than their foreign-born peers to join gangs, commit crimes, and experience obesity and other health problems (Bersani, 2014; see also Chapter 14).

Second, by focusing on order and stability, functionalists ignore racial-ethnic inequalities that often spawn tension and discord. Although, as you saw earlier, immigrants have revitalized many communities, their presence has been controversial and divisive. Third, many whites—particularly middle-to-upper class men—assume that their privileges are universal

(Killerman, 2014). Functionalists acknowledge that inequality is dysfunctional, but this isn't their major focus. Thus, they seem to accept discrimination as inevitable (Chasin, 2004).

10-7b Conflict Theory

Conflict theorists see ongoing strife between dominant and minority groups. Dominant groups try to protect their power and privilege, whereas subordinate groups struggle to gain a larger share of societal resources.

ECONOMIC AND SOCIAL CLASS INEQUALITY

For conflict theorists, capitalism creates and sustains racial-ethnic inequality. According to a classic explanation, there's a "split labor market." Jobs in the *primary labor market*, held primarily by white workers, provide better wages, health and pension benefits, and some measure of job security. In contrast, workers in the *secondary labor market* (e.g., fast-food employees) are largely minorities and easily replaced. Their wages are low, there are few fringe benefits, and working conditions are generally poor (Doeringer and Piore, 1971; Bonacich, 1972).

Such economic stratification pits minorities against each other and low-income whites. Because these groups compete with each other instead of uniting against exploitation, capitalists don't have to worry about increasing wages or providing safer work environments.

Social class doesn't always protect minorities from economic inequality. Because of residential segregation, for instance, middle-income black and Latino households are much more likely than white ones to live in poor neighborhoods, exposing children to weaker schools, more crime, and bad influences. At leading technology companies like Facebook and Google, blacks and

Latinos make up only 2 to 3 percent of college-educated officials and managers (Andrews, 2015; Guynn, 2015; Kang, 2015; Reardon et al., 2015).

CRITICAL EVALUATION

First, large majorities of both blacks and Latinos say that immigration and income, not race or ethnicity, are the primary sources of social conflict in the United States (Morin, 2009). Such data suggest a substantial consensus among some minorities that economic inequality isn't due entirely to racism.

Second, discrimination isn't always as conscious and deliberate as some conflict theorists claim. To increase diversity, Google has for some time recruited at historically black colleges, trains employees about unconscious bias, and expects them to take action when they see prejudice or discrimination (McGregor, 2015). Third, economic inequality reinforces racial-ethnic inequality, but racism existed hundreds of years before the development of capitalism.

10-7c Feminist Theories

Walk through almost any hotel, large discount store, nursing home, or fast-food restaurant in the United States. You'll notice two things: Most of the low-paid employees are women, and predominantly minority women. For feminist scholars, such segregation of minority women is due to gendered racism.

GENDERED RACISM

Gendered racism refers to the overlapping and cumulative effects of inequality due to racism *and* sexism. Many white women encounter discrimination on a daily basis (see Chapter 9). Minority women, however, are also members of a racial-ethnic group, bringing them a double dose of inequality. If social class is included, some minority women experience *triple oppression.* Many affluent women, in particular, have no qualms about exploiting recent immigrants, especially Latinas, who perform demanding housework at very low wages (Hondagneu-Sotelo, 2001).

Gendered racism also occurs *within* racial-ethnic groups. According to a black male sociologist, scholars rarely discuss black male privilege, which is characterized by having advantages over black women. Examples include being promoted more often and getting higher pay than their black female counterparts who are equally skilled and educated (National Public Radio, 2010).

CRITICAL EVALUATION

Feminist perspectives can be faulted on three counts. First, because all of us have internalized institutional

In 2015, Viola Davis became the first African-American woman to win an Emmy for best actress in a drama series (*How to Get Away with Murder*). In her acceptance speech, Davis said, "The only thing that separates women of color from anyone else is opportunity."

discrimination, minority group members are also guilty of reinforcing gendered racism in schools, workplaces, and elsewhere. African American girls as young as 15 have to cope with black men's sexual harassment, stereotypes that black girls and women are sexually promiscuous, and many black men's preference for women with lighter complexions, facial features, and hair that are closer to European standards of beauty (Friedman, 2011; Thomas et al., 2011). Second, feminist scholars seldom explore women's contribution to gendered racism (e.g., affluent black women and Latinas who exploit domestic and farm workers).

A third weakness is that minority scholars themselves don't always address gendered racism. In a recent study of Hollywood diversity, for example, the researchers examined nearly 350 of the top-grossing films and more than 1,000 television programs (Hunt and Ramón, 2015). The report compared dominant/minority groups, males/females, and minority subgroups (e.g., Latino, Asian,

gendered racism the overlapping and cumulative effects of inequality due to racism *and* sexism.

Looting

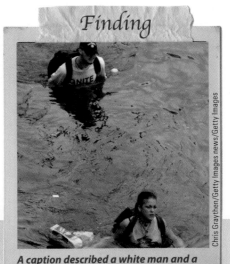
Finding

After Hurricane Katrina in 2005, the caption on a photo of a young black man wading through water described him as "looting."

A caption described a white man and a light-skinned woman as "finding" goods at a flooded local grocery store.

For symbolic interactionists, images shape our perceptions of racial and ethnic groups.

mixed), but omitted gender, race, and ethnic interconnections (e.g., share of acting roles by race *and* gender).

10-7d Symbolic Interaction

According to symbolic interactionists, we learn attitudes, norms, and values throughout the life course. Because race and ethnicity are constructed socially, labeling, selective perception, and social contact can have powerful effects on everyday intergroup relations.

LABELING, SELECTIVE PERCEPTION, AND THE CONTACT HYPOTHESIS

We learn attitudes toward dominant and minority groups through labeling and selective perception, both of which can increase prejudice and discrimination. For example, a comprehensive study of major U.S. news magazines (e.g., *Time, U.S. News & World Report*) concluded that labeling immigrants as a "problem" and a "menace" ignored "the broader array of Latino roles and contributions to American communities" (Gavrilos, 2006: 4; see also "Media Coverage of Hispanics," 2009).

Selective perception also affects racial attitudes and behavior. For example, when white people are given proof of being privileged, they may insist that *other* people, not

contact hypothesis posits that the more people get to know members of a minority group personally, the less likely they are to be prejudiced against that group.

they themselves, benefit personally from privilege (Phillips and Lowery, 2015). People who see what they want and ignore any evidence that challenges their perceptions are unlikely to support policies to reduce racial inequality.

Media critics have lambasted some televisions shows (e.g., *Love & Hip Hop, Real Housewives of Atlanta, Married to Medicine*) for reinforcing negative stereotypes of African Americans (Riley, 2014). Others have similarly criticized popular television shows such as *The Big Bang Theory* and *2 Broke Girls* for stereotyping South Asian men as nerdy scientists, asexual, speaking broken English, and being socially awkward (Deggans, 2012; Dholia, 2014).

Negative images create and reinforce racial and ethnic stereotypes, but people can decrease labeling and selective perception. The **contact hypothesis** posits that the more people get to know members of a minority group personally, the less likely they are to be prejudiced against that group. Such contacts are most effective when dominant and minority group members have approximately the same ability and status (e.g., both are coworkers or bosses), when they share common goals (e.g., work on a project), when they cooperate rather than compete, and if an authority figure supports intergroup interaction (e.g., an employer requires white supervisors to mentor minority workers) (Allport, 1954; Kalev et al., 2006; Carrell et al., 2015).

CRITICAL EVALUATION

Interactionism is limited for two reasons. First, it's not clear why labeling, selective perception, and racial bias are more common among some people than others, especially when they're similar on a number of variables such as social class, gender, age, religion, race, and ethnicity (Dovidio, 2009).

Another weakness is that symbolic interaction tells us little about the social structures that create and maintain racial-ethnic inequality. For instance, people who aren't prejudiced can foster discrimination by simply going along with the inequitable policies that have been institutionalized in education, the workplace, and other settings (i.e., the unprejudiced discriminators described earlier).

10-8 INTERRACIAL AND INTERETHNIC RELATIONSHIPS

President Obama, the son of a black father from Kenya and a white mother from Kansas, identified himself as "black" on the 2010 census questionnaire, although he could have checked off both black and white or "other." The president is just one of the growing number of Americans who are biracial or multiracial.

10-8a Growing Multiracial Diversity

The 2000 U.S. Census for the first time allowed people to mark more than one race, which generated about 126 categories and combinations. In 2010, 97 percent of Americans reported being only one race, but almost 3 percent (9 million people) self-identified as being two or more races, up from 2.4 percent in 2000. Every state saw its multiple-race population jump by at least 8 percent, but Native Hawaiians/Pacific Islanders were the most likely to identify themselves as belonging to two or more races (see *Figure 10.11*).

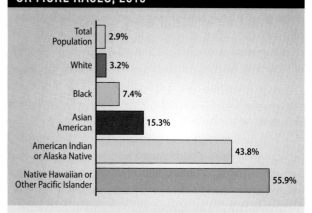

FIGURE 10.11 PERCENTAGE OF AMERICANS WHO IDENTIFIED THEMSELVES AS BEING TWO OR MORE RACES, 2010

Total Population	2.9%
White	3.2%
Black	7.4%
Asian American	15.3%
American Indian or Alaska Native	43.8%
Native Hawaiian or Other Pacific Islander	55.9%

Note: Of the almost 309 million U.S. population in 2010, only 0.3 percent said they were three or more races.

Source: Based on Jones and Bullock, 2012, Figure 10.

A recent survey found that 7 percent of U.S. adults could be considered multiracial based on how people describe themselves as well as their parents' and grandparents' racial backgrounds. The percentage may be even higher because 61 percent of adults with a mixed racial background don't consider themselves "multiracial" (Parker et al., 2015).

10-8b Interracial Dating and Marriage

Laws against **miscegenation**, marriage or sexual relations between a man and a woman of different races, existed in America as early as 1661. It wasn't until 1967, in the U.S. Supreme Court's *Loving v. Virginia* decision, that antimiscegenation laws were overturned nationally, but not everyone is happy about interracial relationships. Recently, for example, a Baptist congregation in Kentucky voted to ban interracial couples from becoming members or participating in worship activities (Waldron, 2011).

In 2013, 87 percent of Americans approved of black–white marriages, up from only 4 percent in 1958 (Newport, 2013). A large majority (63 percent) of Americans say that they'd be fine with a family member's marriage to someone of a different racial or ethnic group (Wang, 2012).

White supremacist Craig Cobb made headlines worldwide when he tried, unsuccessfully, to establish a white-only town in North Dakota. Shortly after that, Cobb's DNA test showed that he was 14 percent black. How awkward, a journalist observed ("DNA Reveals...," 2013: 8).

miscegenation marriage or sexual relations between a man and a woman of different races.

Attitudes about interracial relationships are changing, but what about behavior? Racial-ethnic intermarriages have increased slowly—from only 0.7 percent of all marriages in 1970 to nearly 10 percent in 2010. In 2010, also, 18 percent of heterosexual unmarried couples were of different races and 21 percent of same-sex couples were interracial. A record 15 percent (1 in 7) of new U.S. marriages were between people of different races or ethnicities. Asian Americans have the highest intermarriage rates and whites and blacks the lowest (Lofquist et al., 2012; Wang, 2012).

The increase in intermarriage is due to many interrelated factors—both micro and macro level—that include everyday contact and changing attitudes. For example, we tend to date and marry people we see on a regular basis. The higher the educational level, the greater the potential for intermarriage, because educated minority group members often attend integrated colleges, and their workplaces and neighborhoods are more racially mixed than in the past (Chen and Takeuchi, 2011; Wang, 2012).

People often marry outside of their racial-ethnic groups because of a shortage of potential spouses within their own group. Because the Arab American population is so small, 80 percent of U.S.-born Arabs have non-Arab spouses. Acculturation also affects intermarriage rates. Intermarriage is more common among second-generation immigrants (15 percent) than those in the first generation (8 percent). By the third generation, 31 percent of Latinos and Asian Americans have a spouse of a different race or ethnicity from themselves (Kulczycki and Lobo, 2002; "Second-Generation Americans," 2013).

STUDY TOOLS 10

READY TO STUDY? IN THE BOOK, YOU CAN:

☐ Check your understanding of what you've read with the Test Your Learning Questions provided on the chapter review card at the back of the book.

☐ Rip out the chapter review card for a handy summary of the chapter and key terms.

ONLINE AT CENGAGEBRAIN.COM YOU CAN:

☐ Collect StudyBits while you read and study the chapter.

☐ Quiz yourself on key concepts.

☐ Prepare for tests with chapter flash cards as well as flash cards that you create.

☐ Watch videos to explore sociology further.

© Anelina/Shutterstock.com

SOC
ONLINE
STUDY YOUR WAY
WITH STUDYBITS!

WEAK

FAIR

STRONG

UNASSIGNED

Rate and Organize StudyBits

Collect What's Important

Create Flashcards From Your StudyBits

CORRECT

INCORRECT

INCORRECT

INCORRECT

Personalize Your Quizzes

Track/Monitor Your Progress

85%

Access SOC **ONLINE** at www.cengagebrain.com

11 The Economy and Politics

After you finish
this chapter go to
PAGE 223 for
STUDY TOOLS.

LEARNING OUTCOMES

After studying this chapter, you will be able to…

11-1 Compare and illustrate the different types of global economic systems.

11-2 Describe corporations and explain how corporate and political power are interwoven.

11-3 Explain how and why macro-level variables have changed the U.S. economy and jobs.

11-4 Compare and illustrate the different types of global political systems.

11-5 Explain the relationship between power, authority, and politics.

11-6 Describe the U.S. political system, and explain who votes, who doesn't, and why.

11-7 Compare and evaluate the theoretical explanations of the economy and politics.

About 42 percent of Americans say that they're worse off financially than in the past, but a whopping 92 percent believe that Congress is too dysfunctional to address such issues (Dugan, 2014; McCarthy, 2015). Such national data show that many Americans see a close linkage between their personal lives, the economy, and politics.

The **economy** determines how a society produces, distributes, and consumes goods and services. In **politics** individuals and groups acquire and exercise power and authority and make decisions. Societies worldwide differ in the kinds of economic and political systems they develop because of many factors such as globalization and technology. Let's begin by looking at global economic systems.

economy determines how a society produces, distributes, and consumes goods and services.

politics individuals and groups acquire and exercise power and authority and make decisions.

What do you think?

The country would be better off if there were more women in political office.

1 2 3 4 5 6 7

strongly agree   strongly disagree

11-1 GLOBAL ECONOMIC SYSTEMS

The two major economic systems around the world are capitalism and socialism. A handful of countries endorse communism, but in actual practice, economies are usually some mixture of both capitalism and socialism.

11-1a Capitalism

Capitalism is an economic system based on private ownership of property and competition in producing and selling goods and services. Ideally, capitalism has four essential characteristics (Smith, 1776/1937; Heilbroner and Thurow, 1998):

▶ **Private ownership of the means of production.** Property (e.g., real estate, banks, utilities) belongs to individuals or organizations rather than the state or the community.

▶ **Market competition.** Owners of production compete in deciding what goods and services to produce and setting prices that offer consumers the greatest value in price and quality.

▶ **Profit.** Selling something for more than it costs to produce generates profits and an accumulation of wealth for individuals and companies.

▶ **Investment.** By investing profits, capitalists can increase their own wealth. Workers, too, can save and invest their money.

In reality, capitalism doesn't function ideally because of abuses, greed, and worker exploitation. It usually results in monopolies and oligopolies rather than a free market that encourages competition. Federal laws prohibit the establishment of a **monopoly**, the domination of a particular market or industry by one person or company. With little or no competition, a company would be able to dictate prices and lower the quality of goods and services.

capitalism an economic system based on private ownership of property and competition in producing and selling goods and services.

monopoly domination of a particular market or industry by one person or company.

An **oligopoly**, which is legal, is the domination of a market by a few large producers or suppliers. In the mid-1980s, 50 companies controlled 90 percent of all U.S. media. Today, it's six companies—including CBS, Time Warner, and the Walt Disney Corporation. There were hundreds of airlines in the mid-1980s, but four giant airlines (United, Delta, American, and Southwest) now control 85 percent of the U.S. market. CEOs maintain that merging companies increases efficiency and safety. Critics say, however, that oligopolies raise prices, stifle competition, and that consumers have fewer choices in buying goods and services (Fox, 2013; Yglesias, 2013; Koenig and Mayerowitz, 2015). Moreover, entrepreneurs have little chance of breaking into an industry dominated by an oligopoly.

11-1b Socialism

Socialism is an economic system based on the public ownership of the production of goods and services. Ideally, socialism is the opposite of capitalism and has the following characteristics:

▶ **Collective ownership of property.** The community, rather than the individual, owns property. The state owns utilities, factories, land, and equipment, but distributes them equally among all members of a society.

▶ **Cooperation.** Working together and providing social services to all people are more important than competition.

▶ **No profit motive.** Private profits that are fueled by greed and exploitation of workers are forbidden.

▶ **Collective goals.** The state is responsible for all economic planning and programs. People are discouraged from accumulating individual profits and investments, and are expected to work for the greater good.

There have been many socialist governments during the last 150 years, but none has reflected pure socialism. Competition and individual profits are officially forbidden, but government officials, top athletes, and high-ranking party members enjoy more freedom, larger

Walt Disney Company is an example of an oligopoly. Among other holdings, it owns publishing companies, web portals, 19 major television and cable stations, at least 60 radio stations, numerous magazines, and 14 theme parks and resorts around the world. All of these outlets promote and sell Disney products, increasing the corporation's profits.

apartments, higher incomes, and greater access to education and other resources than the rank and file.

11-1c Communism

Communism is a political and economic system in which property is communally owned and all people are considered equal. Whereas socialism allows some free market economy, and people receive resources according to how hard they work, communism demands that all production be owned by the public, and that people receive resources according to their need.

Karl Marx (1867/1967) believed that socialist states would evolve into communist societies. This hasn't happened and most formerly communist countries have collapsed. Today, only China, Cuba, Laos, North Korea, and Vietnam practice some form of communism, but have also incorporated capitalism. In contrast to its principles, communism has failed, according to historians, because of widespread corruption and mismanagement, oppression, imprisonment or execution of people who questioned coercive policies, economic inefficiency that resulted in shortages of food and other goods, and few rewards for working hard (Brown, 2010; Johnson, 2010; Pollick, 2014).

11-1d Mixed Economies

Welfare capitalism (also called *state capitalism*) is an economic system that combines private ownership of property, market competition, and a government's

oligopoly domination of a market by a few large producers or suppliers.

socialism an economic system based on the public ownership of the production of goods and services.

communism a political and economic system in which property is communally owned and all people are considered equal.

welfare capitalism (also called *state capitalism*) an economic system that combines private ownership of property, market competition, and a government's regulation of many programs and services.

China is an example of a country with a mixed economy: It espouses communism, practices socialism, and has endorsed many aspects of capitalism. China's economic boom hasn't benefited all of its citizens. Many urban centers are thriving—offering those in upper and middle classes high-rise apartments, stores, and restaurants. In contrast, millions of Chinese, including many older people, like the one pictured here, survive by scouring trash bins for plastic bottles to recycle.

regulation of many programs and services. Many countries, including the United States, have some form of welfare capitalism. Most industry is private, but the government owns or operates some of the largest industries and services (e.g., education, transportation, postal services) that benefit the entire population (Jacoby, 1997; Esping-Andersen, 1990; Hacker, 2002).

Crony capitalism is an economy based on close relationships between business people and the government. In this system, the government gives wealthy people and large corporations preferential treatment through special tax breaks (called *tax credits* and *deductions*), direct payment or loans (called *subsidies*, not welfare), and provides grants that rarely require competition in an open market. From 2000 to 2012, for example, all but one of the *Fortune* 100 companies received almost $1.3 trillion from the federal government in tax credits, contracts, grants, loans, and various subsidies (Andrzejewski, 2014; see also the discussion of corporate welfare in Chapter 8).

Large companies drive both capitalism and mixed economies. Corporations, in particular, wield enormous influence both at home and abroad.

11-2 CORPORATIONS AND THE ECONOMY

When Bob Thompson sold his construction company for $442 million, he distributed almost a third of the profits to his 550 workers, as well as to some retirees and widows. About 80 became instant millionaires ("Boss Sells His Company...," 1999). Unlike Thompson, most corporate heads have luxurious lifestyles, eliminate employees' pensions, retire as billionaires, and pass on their massive wealth to their heirs (see Chapter 8). Corporate and political power are interwoven, but let's begin by looking at some of the characteristics of corporations.

11-2a Corporations

A **corporation** is an organization that has legal rights, privileges, and liabilities apart from those of its members. Until the 1890s, there were only a few U.S. corporations—in textiles, railroads, and the oil and steel industries. Today, there are almost 6 million, most created for profit, but a mere 0.5 percent bring in 90 percent of all corporate income (U.S. Census Bureau, 2012).

Whereas many industrialized nations have closed corporate tax loopholes, the United States has expanded them. As a result, many corporations exercise more power than do governments, and have amassed enormous wealth. A number of U.S. corporations use legal tax loopholes to avoid paying tens of billions of dollars every year on overseas income; 29 companies alone have more cash than the U.S. Treasury Department; and the profits of the *Fortune* 500 corporations have soared (Durden, 2011; Jilani, 2011; Niquette and Rubin, 2014).

11-2b Conglomerates

A **conglomerate** is a corporation that owns a collection of companies in different industries. Conglomerates emerged during the 1960s and grow by acquiring companies through mergers. Mergers might increase the value of shareholders' stock, but typically make chief executives "truly, titanically, stupefyingly rich" (Morgenson, 2004: C1). An example of a conglomerate is Kraft Foods, which owns companies that produce snacks, beverages, pet foods, a variety of groceries and convenience foods, and has ties with other corporations such as Starbucks (see www.kraft.com/brands).

Conglomerates diversify business risk by participating in a number of different markets, and can ward off

corporation an organization that has legal rights, privileges, and liabilities apart from those of its members.

conglomerate a corporation that owns a collection of companies in different industries.

emerging competition. Conglomerates waned after the 1980s, but are making a comeback. Google, for example, has set up companies to create driverless cars, extend human lifespans, and develop online education ("From Alpha to Omega," 2015).

11-2c Interlocking Directorates

In an **interlocking directorate**, the same people serve on the boards of directors of several companies or corporations. An estimated 15 to 20 percent of corporate directors sit on two or more corporate boards. Some interlocking directorates are especially powerful because they include past U.S. presidents and past members of Congress who provide connections in Washington, D.C. Interlocking directors develop similar economic perspectives; those who sit on two or more corporate boards are especially likely to receive appointments to influential government advisory committees (Domhoff, 2013). Thus, a very small group of people has the power to shape the national agenda.

11-2d Transnational Corporations and Conglomerates

Interlocking directorates have become more influential than ever because of the proliferation of transnational corporations. A **transnational corporation** (also called a *multinational corporation* or an *international corporation*) is a large company that's based in one country but operates across international boundaries. By moving production plants abroad, large U.S. corporations can avoid trade tariffs, bypass environmental regulations, and pay low wages. The political leaders of many countries welcome transnational corporations to stimulate their economies, create jobs, and enrich their personal bank accounts (Caston, 1998).

The most powerful are **transnational conglomerates** (also called *multinational conglomerates*), corporations that own a collection of different companies in various industries in a number of countries. General Electric is the world's biggest transnational conglomerate, with 70 percent of its businesses and more than half of its employees based abroad. It owns hundreds of companies in the United States, and has subsidiaries in at least 27 countries and on every continent ("Biggest Transnational Companies," 2012). Of the 25 most profitable conglomerates worldwide, 17 are based in the United States. In effect, then, a very small group of corporations holds considerable global economic power (McGregor and Hamm, 2008).

11-3 WORK IN U.S. SOCIETY TODAY

Poll after poll shows that financial hardship is a top issue for many Americans. Many worry about the economy in general, losing their job, and paying off debt (Dugan, 2014; Wilke and Newport, 2014). Economic problems vary by social class, but much of the hardship has to do with changes in **work**, a physical or mental activity that

Many U.S. factories today need only a few skilled workers to manage the high-tech machines that have replaced people.

---

interlocking directorate the same people serve on the boards of directors of several companies or corporations.

transnational corporation (also called a *multinational corporation* or an *international corporation*) a large company that's based in one country but operates across international boundaries.

transnational conglomerate (also called a *multinational conglomerate*) a corporation that owns a collection of different companies in various industries in a number of countries.

work a physical or mental activity that produces goods or services.

produces goods or services. Even those who work hard have lost economic ground because of macro-level variables such as deindustrialization, globalization, offshoring, and weakened labor unions.

11-3a Deindustrialization and Globalization

Many Americans have been casualties of **deindustrialization**, a process of social and economic change resulting from the reduction of industrial activity, especially manufacturing. Since 2000, 32 percent of U.S. manufacturing jobs have disappeared (Hargrove, 2011). In mid-2012, 304,000 manufacturing plants were operating in the United States, 27,000 fewer than at the end of 2007 (Philips, 2012). A major reason for deindustrialization is that, beginning in the early 1960s, employers easily replaced workers with the lowest skill levels, usually those on assembly lines, with automation. Machines and robots have also replaced many blue collar jobs that provided a comfortable standard of living in the past.

The number of manufacturing jobs plummeted from a peak of almost 20 million in 1979 to less than 12 million by the end of the 2007–2009 recession (Barker, 2011). Since then, manufacturing has been rebounding. It was the fourth largest U.S. employer in 2011, behind only health care, retail trade, and food services, and the average wage is almost $30.00 an hour ("Manufacturing Day…," 2013).

Some of the work has changed. Manufacturers complain about a shortage of skilled candidates for many jobs, particularly operating or programming computer-controlled cutting tools, repairing sophisticated machinery, and designing software that runs a factory's operations (Hagerty, 2013; Phillips, 2014).

The manufacturing industry is also facing a major shortage of skilled workers like welders, ironworkers, and brick masons. An estimated two million such jobs will go unfilled over the next decade because employers can't find enough people with the required skills, and despite above-average wages. Entry-level welders, for example, can move quickly from $17 to $30 an hour (Cohen, 2015; Payne and Somerville, 2015).

Deindustrialization was hastened by **globalization**, the growth and spread of investment, trade, production, communication, and new technology around the world. One example of globalization is a car that's assembled in the United States with practically all of its parts manufactured and produced in Germany, Japan, South Korea, or developing countries.

Proponents argue that everyone benefits from globalization because it creates millions of jobs and brings affordable goods and services (e.g., cell phones) to millions of people around the world. Critics contend that while globalization has improved the living standards of millions of workers in developing countries such as China and India, it has reduced the number of workers in Western economies. They also maintain that, by and large, globalization benefits only the world's most powerful transnational corporations: It increases their profits by expanding their worldwide base of consumers, exploits poor people in developing countries, creates greater income inequality between workers and the global elite, and gives corporations unprecedented political power (Schaeffer, 2003; Porter, 2014).

11-3b Offshoring and Labor Unions

Offshoring refers to sending work or jobs to another country to cut a company's costs at home. Sometimes called *international outsourcing* or *offshore outsourcing*, the transfer of manufacturing jobs overseas has been going on since at least the 1970s.

Between 2001 and 2011, U.S. companies moved more than 2.7 million jobs to China, 77 percent of them in manufacturing and primarily blue-collar (Scott, 2012). During the same period, U.S. firms also offshored 28 percent of high-level, well-paid information technology (IT) jobs, including those in accounting, computer science, and engineering (National Science Board, 2012). Most of the offshored jobs go to India and China, but many have also moved to Canada, Hungary, Mexico, Poland, Russia, Egypt, Venezuela, Vietnam, and South Africa.

Companies can get accounting services in India that cost much less than in the United States (see *Figure 11.1*). In Mexico, the average General Motors worker earns wages and benefits that cost less than $4 an hour compared with $55 an hour in the United States. Because of such large wage differences,

iStockphoto.com/P_Wei

---

deindustrialization a process of social and economic change resulting from the reduction of industrial activity, especially manufacturing.

globalization the growth and spread of investment, trade, production, communication, and new technology around the world.

offshoring sending work or jobs to another country to cut a company's costs at home.

FIGURE 11.1 AVERAGE ANNUAL EARNINGS OF AN ACCOUNTANT IN THE UNITED STATES AND INDIA

$70,000

$23,000

American consumers can purchase goods and services at low prices, and the majority doesn't care where products are made (Black, 2010; Guarino, 2013; Newport et al., 2014).

Deindustrialization, globalization, and offshoring have weakened *labor unions*, organized groups that seek to improve wages, benefits, and working conditions. Union membership has dropped sharply—from 35 percent of the workforce in the mid-1950s to 11 percent in 2014 ("Union Members—2014," 2015). In the 1950s, 75 percent of Americans approved of labor unions compared with 58 percent in 2015 (Saad, 2015). Five states prohibit unionization; in 2012, 24 states significantly restricted collective bargaining rights; and anti-union groups have helped defeat efforts to unionize autoworkers in Tennessee and other states (Trottman and Maher, 2013; Whitesides, 2014).

Do we still need unions? Opponents argue that unions have too much influence, that their members are overpaid, and that unions drain state resources because of high pensions and salaries of public sector employees (e.g., teachers, nurses, sanitation workers, police). Critics also contend that unions have limited employers' flexibility in hiring and firing decisions, and that ever-increasing labor and health care costs have forced employers to move their operations overseas to remain competitive (McKinnon, 2011; Schlesinger, 2011).

Proponents argue that states have experienced budget deficits because of the housing crisis and a recession that was due to Wall Street greed and not overpaid union members. Others point out that union members have made numerous concessions like decreasing their wages and benefits. Most important, historically, unions have benefited almost all workers by insisting on paid holidays and vacations, greater workplace safety, overtime pay, and by challenging corporations that are becoming more dominant and powerful in both the economy and politics (Stepan-Norris, 2015).

11-3c How Americans' Work Has Changed

Only 33 percent of U.S. workers are "completely satisfied" with their pay (McCarthy, 2015). The U.S. economy started adding more jobs in 2012, but millions of Americans still have only low-paying and part-time jobs. If these strategies fail, they find themselves among the unemployed.

LOW-WAGE JOBS

A recent study of 702 U.S. occupations found that automation and globalization have wiped out millions of middle-income jobs (e.g., sales, accounting, office and administrative support). By 2025, 230 million white collar jobs and 47 percent of all U.S. employees are expected to be replaced by computers (Frey and Osborne, 2013; Manyika et al., 2013).

Except for registered nurses, whose median annual salary is about $57,000, seven of the occupations expected to add the most jobs through 2020 are low paying, up to a maximum of about $24,000 a year. These occupations include home health aides and personal care aides who assist older and disabled people. Other fast-growing but low-paying jobs—from about $22,000 to $31,000 a year—include retail salespersons, general office clerks, and nurses' aides (Sommers and Franklin, 2012; "Labor Day 2015...," 2015).

The federal minimum wage, which rose from $6.55 to $7.25 an hour in 2009, increased the wages of less than 4 percent of the workforce. The District of Columbia and 29 states now exceed the federal minimum wage. If it had been raised since 1968 at the same rate as the average worker's growth in productivity, the federal minimum wage would be nearly $18.50 an hour. In 2015, Walmart, the nation's largest low-wage employer, increased the hourly wage to $9 an hour. Some other large retailers (e.g., Target, TJ Maxx, Marshalls) followed suit. If Walmart paid its U.S. workers $12 per hour and passed every penny of the costs to consumers, the average customer would pay just 46 cents more per shopping trip, or about $12 a year. Costco, a discount retailer, has huge profits yet pays employees an average of $21 per hour, not including overtime (Jacobs et al., 2011; "Wages Maximus," 2014; Cooper, 2015; Tabuchi, 2015). Thus, many Americans are being paid less for doing more.

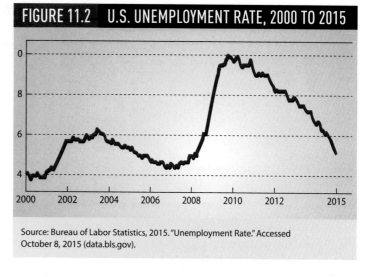

FIGURE 11.2 U.S. UNEMPLOYMENT RATE, 2000 TO 2015

Source: Bureau of Labor Statistics, 2015. "Unemployment Rate." Accessed October 8, 2015 (data.bls.gov).

PART-TIME WORK

Of the almost 27 million part-timers (those who work less than 35 hours a week), 31 percent are involuntary because people can't find suitable full-time employment or employers have reduced their hours (BLS News Release, 2015). Part-time work—traditionally parceled out to lower-level hourly employees like cashiers and administrative assistants—has spread to white-collar and professional sectors and now includes marketing directors, engineers, and financial officers. Employers save on costs—including vacations and health and unemployment insurance—by hiring more (or all) part-time workers. They admit, however, that the downside includes high turnover and employees who are less committed to turning out a better product or providing a better service (Davidson, 2011).

UNEMPLOYMENT AND DISCOURAGED WORKERS

The U.S. unemployment rate surged between 2008 and 2010, but has dropped since then (see *Figure 11.2*). Despite the drop, only 63 percent of Americans age 16 and older have a job or are looking for one—the lowest share of the population participating in the labor force since 1978. Before and since the recession, unemployment rates have been highest for teenagers, single female heads of households, blacks, those with less than a college degree, and people employed in mining, construction, and leisure and hospitality industries (BLS News Release, 2015).

About 36 percent of the jobless are *long-term unemployed*, people who have searched for but not found a job for 27 weeks or longer (Bureau of Labor Statistics, 2014). The longer that people are unemployed, the harder it is to find a job: Only 11 percent find steady, full-time employment a year later. Many of the long-term unemployed become **discouraged workers**,

people who stop searching for employment because they believe that job hunting is futile. About 35 percent of Americans are discouraged workers who aren't included in the Census Bureau's unemployment rate (Krueger et al., 2014).

UNDEREMPLOYED WORKERS

Besides not counting discouraged workers, unemployment rates exclude the **underemployed**—people who have part-time jobs but want full-time work or whose jobs are below their experience, skill, and education levels. In 2014, an estimated 17 percent of Americans were underemployed compared with only 8 percent in 2007 (Mishel et al., 2012; Gallup, 2014).

Among the underemployed, almost 33 percent of those who graduated from college between 2006 and 2010 said that their postcollege job was below their education and skill level. Underemployed Americans are better off financially than the unemployed, but both groups experience similar levels of sadness, stress, anger, worry, and depression (Godofsky et al., 2011; Mishel et al., 2012).

JOB SATISFACTION AND STRESS

A whopping 69 percent of Americans report being "not engaged" or "actively disengaged" in their jobs, meaning that they're not emotionally involved in or enthusiastic

discouraged workers people who stop searching for employment because they believe that job hunting is futile.

underemployed people who have part-time jobs but want full-time work or whose jobs are below their experience, skill, and education levels.

In good times and in bad, African American men have the highest unemployment rates. And compared with their white counterparts, black workers are unemployed longer after their unemployment insurance benefits end. Pictured here are laid-off workers examining listings at a job fair.

TABLE 11.1 VACATION DAYS, ON AVERAGE, THAT WORKERS RECEIVE AND USE

	Receives	Uses
France	30	30
Denmark	30	29
Spain	30	26
Italy	28	20
Japan	18	7
United States	14	10
Malaysia	12	10
Thailand	11	8
South Korea	10	7

Source: Based on "Vacation Habits…," 2013.

about their work (Gallup, 2013). Worker engagement varies by job type: Service workers are less engaged than managers and executives (28 percent and 40 percent, respectively), but 60 percent of the latter aren't engaged or are actively disengaged. Across generations, people born after 1980 are the least engaged, primarily because they believe that their jobs don't allow them to use their talents and strengths (Adkins, 2015).

In most European countries, the typical work week is 35 hours—comparable with the U.S. definition of part-time work. Americans work more hours (1,804 per year) and more weeks per year (48 weeks) than people in any other industrialized country except Japan, and they have the highest productivity rate in the world. In 2014, full-time employees worked an average of 47 hours a week, and nearly 40 percent worked at least 50 hours a week (Begala, 2011; Saad, 2014).

The United States is the only industrialized nation in the world that doesn't legally guarantee workers a paid vacation; 25 percent of Americans have no paid vacation *and* no paid holidays. In contrast, the typical worker, especially in Western Europe, has at least 6 weeks of paid vacation, regardless of job seniority or the number of years worked (Ray et al., 2013; Greenfield, 2015). U.S. workers have the least vacation days and don't use all of them (see *Table 11.1*). Workers in France receive and take 30 days,

but 90 percent complain that they're "vacation deprived" ("Vacation Habits…," 2013).

In 2014, 41 percent of Americans didn't take any vacation days. The rest shrank their vacations to a few days or long weekends because they couldn't afford a vacation, had too much to do on the job, wanted to impress their boss with their dedication, or feared being replaced while away. Employees who forfeit paid time off don't get more raises or bonuses and report higher stress than those who take all of their vacation time, but employers save almost $53 billion a year when people work for free (U.S. Travel Association, 2014; Dickey, 2015; Mudallal, 2015).

11-3d Women and Minorities in the Workplace

One of the most dramatic changes in the United States during the twentieth century was the increase of women in the labor force (see *Table 11.2*). Many factors have contributed to the surge in women's employment, especially since the 1970s, including the growth in the number of college-educated women (who, consequently, had more job opportunities), an increase in the number of working single mothers, and the higher costs of homeownership that require two incomes.

TABLE 11.2 WOMEN AND MEN IN THE LABOR FORCE, 1890–2014

	Percentage of Men and Women in the Labor Force		Women as a Percentage of All Workers
Year	Men	Women	
1890	84	18	17
1900	86	20	18
1920	85	23	20
1940	83	28	25
1960	84	38	33
1980	78	52	42
1990	76	58	45
2014	69	57	47

Source: Based on Labor Force Statistics from the Current Population Survey, Table 3, "Employment Status," 2014. Accessed October 8, 2015 (bls.gov/cps/tables.htm).

Largely because of their higher educational attainment, 29 percent of women in two-income marriages bring home the bigger paycheck, up from only 4 percent in 1970 ("Wives Who Earn More . . . ," 2014). As a group, however, women have lower earnings than men in both the highest and lowest paying occupations, and the wage gaps are greater in high-income jobs (see *Figure 11.3*).

There are earnings disparities across racial-ethnic groups, but the differences are especially striking by sex. As you examine *Figure 11.4*, note two general characteristics. First, earnings increase—across all racial-ethnic groups and for both sexes—as people go up the occupational ladder. But across all occupations, men have higher earnings than women of the same racial-ethnic group. At the bottom of the occupational ladder are African American women and Latinas, with the latter faring worse than any of the other groups. Thus, both gender *and* race-ethnicity affect earnings. (See Hegewisch et al., 2015, for a detailed analysis of the gender wage gap, by race and ethnicity, since 1955.)

FIGURE 11.3 WOMEN EARN LESS THAN MEN WHETHER THEY'RE CEOs OR COOKS

These are five of the highest and lowest paid occupations of full-time, year-round U.S. workers in 2014. How might you explain why the earnings differ by sex, especially in the highest paid jobs?

Men Make More Than Women in Some of the Highest Paying Jobs

Chief executives
$2,246
$1,572
Pharmacists
$2,176
$1,902
Physicians and surgeons
$2,002
$1,246
Lawyers
$1,915
$1,590
Computer/information systems managers
$1,763
$1,529

Median Weekly Earnings

. . . and in Some of the Lowest Paying Jobs

Bartenders
$634
$459
Personal care aides
$465
$425
Cashiers
$412
$387
Waiters and waitresses
$501
$415
Cooks
$416
$405

Median Weekly Earnings

Men   Women

Note: Some of the differences between men's and women's median weekly earnings may seem small, but multiply each number by 52 weeks. Thus, in annual earnings, male physicians and surgeons average more than $104,000 compared with only $65,000 for females.

Source: Based on Labor Force Statistics from the Current Population Survey, 2015, Table 39, "Median Weekly Earning of Full-Time Wage and Salary Workers by Detailed Occupation and Sex." Accessed October 8, 2015 (bls.gov/cps).

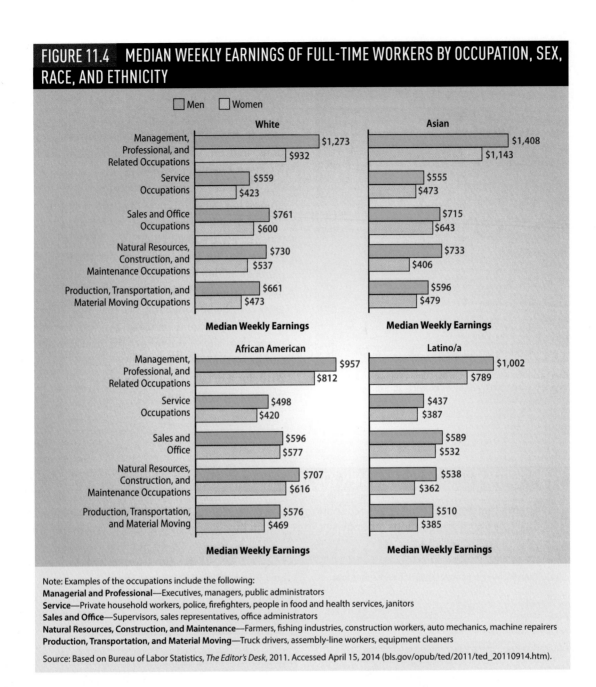

FIGURE 11.4   MEDIAN WEEKLY EARNINGS OF FULL-TIME WORKERS BY OCCUPATION, SEX, RACE, AND ETHNICITY

☐ Men   ☐ Women

White

Occupation	Men	Women
Management, Professional, and Related Occupations	$1,273	$932
Service Occupations	$559	$423
Sales and Office Occupations	$761	$600
Natural Resources, Construction, and Maintenance Occupations	$730	$537
Production, Transportation, and Material Moving Occupations	$661	$473

Median Weekly Earnings

Asian

Occupation	Men	Women
Management, Professional, and Related Occupations	$1,408	$1,143
Service Occupations	$555	$473
Sales and Office Occupations	$715	$643
Natural Resources, Construction, and Maintenance Occupations	$733	$406
Production, Transportation, and Material Moving Occupations	$596	$479

Median Weekly Earnings

African American

Occupation	Men	Women
Management, Professional, and Related Occupations	$957	$812
Service Occupations	$498	$420
Sales and Office	$596	$577
Natural Resources, Construction, and Maintenance Occupations	$707	$616
Production, Transportation, and Material Moving	$576	$469

Median Weekly Earnings

Latino/a

Occupation	Men	Women
Management, Professional, and Related Occupations	$1,002	$789
Service Occupations	$437	$387
Sales and Office	$589	$532
Natural Resources, Construction, and Maintenance Occupations	$538	$362
Production, Transportation, and Material Moving	$510	$385

Median Weekly Earnings

Note: Examples of the occupations include the following:
Managerial and Professional—Executives, managers, public administrators
Service—Private household workers, police, firefighters, people in food and health services, janitors
Sales and Office—Supervisors, sales representatives, office administrators
Natural Resources, Construction, and Maintenance—Farmers, fishing industries, construction workers, auto mechanics, machine repairers
Production, Transportation, and Material Moving—Truck drivers, assembly-line workers, equipment cleaners

Source: Based on Bureau of Labor Statistics, *The Editor's Desk*, 2011. Accessed April 15, 2014 (bls.gov/opub/ted/2011/ted_20110914.htm).

11-4 SOCIOLOGICAL EXPLANATIONS OF WORK AND THE ECONOMY

Sociological theories offer differing perspectives on the economy and its impact on society. *Table 11.3* summarizes each perspective's key points.

11-4a Functionalism

Functionalists typically emphasize the benefits of work and the economy. They also see capitalism as bringing prosperity to society as a whole.

KEY CHARACTERISTICS

For functionalists, work is necessary for a society's survival and defining its members' roles. Work is also important because it connects people to each other: As jobs become more specialized, a small group of workers is responsible for getting the work done, and coworkers can get to know one another (Parsons, 1954, 1960; Merton, 1968). Such networks increase workplace solidarity and enhance a sense of belonging to a group where people listen to each other's ideas, interact, and "share a vision for the work [they] do together" (Gardner, 2008: 16).

Besides providing income, work has social meaning. For many Americans, jobs offer a sense of accomplishment and feeling valued; stability, order, and a daily rhythm; and

TABLE 11.3 SOCIOLOGICAL EXPLANATIONS OF WORK AND THE ECONOMY

Theoretical Perspective	Level of Analysis	Key Points
Functionalist	Macro	Capitalism benefits society; work provides an income, structures people's lives, and gives them a sense of accomplishment.
Conflict	Macro	Capitalism enables the rich to exploit other groups; most jobs are low-paying, monotonous, and alienating; productivity isn't always rewarded.
Feminist	Macro and micro	Gender roles structure women's and men's work experiences differently and inequitably.
Symbolic Interactionist	Micro	How people define and experience work in their everyday lives affects their workplace behavior and relationships with coworkers and employers.

a network of interesting social and professional contacts (Katzenbach, 2003; Brooks, 2007). Functionalists also maintain that wage inequities spur people to persevere and to set higher goals. Low-paying jobs, for example, can motivate people to work harder or obtain further education to move up the economic ladder (see Chapter 8).

CRITICAL EVALUATION

Critics point out that work is a significant source of stress for 69 percent of Americans and often leads to myriad health problems (American Psychological Association, 2014). Many people have unstimulating and low-paid jobs that require repetitive and routine tasks, instead of providing a connection to a product or a group (see the discussion of McDonaldization in Chapter 6). Even in higher paying jobs, only a third of physicians and nurses report being engaged in their work (Gallup, 2013).

Functionalists also gloss over many U.S. corporations' disinterest in workers' well-being. In the private sector, 30 percent of full-timers and 74 percent of part-timers don't have paid sick days (Williams and Gault, 2014). Even when business improves, many employers convert full-time jobs to part-time or temporary positions, and have been cutting health benefits (Schultz, 2011).

AP Images/Connecticut Post, Ned Gerard

11-4b Conflict Theory

Whereas functionalists emphasize the economy's benefits, conflict theorists argue that capitalism creates social problems. They contend, for example, that globalization has led to job insecurity, and that a handful of transnational conglomerates have enormous power.

KEY CHARACTERISTICS

For conflict theorists, low wages alienate employees rather than motivate them to work harder, as functionalists claim. Many people are stuck in low-paying, monotonous jobs and don't always get promotions regardless of their skills, attitudes, perseverance, or productivity (Clements, 2012).

Conflict theorists also argue that capitalism enables the rich and powerful to exploit other groups, resulting in huge wealth disparities. In most industrialized countries, CEOs earn 10 to 25 times more than the average worker. In the United States, CEOs—including those responsible for the Great Recession's financial crisis—earn 303 times more than the typical worker. In some of the largest companies, CEOs earn 1,000 times more than their employees. Between 1978 and 2014, CEO pay increased by 937 percent compared with only 10 percent for workers (Krantz, 2015; Mishel and Davis, 2015; Weaver, 2015).

Why are there such astounding differences between CEO and worker incomes? Because of interlocking directorates, CEOs give each other "stunning" compensation packages, whereas the typical male worker earned less in 2014 than in 1973. Fueled by worker productivity but stagnant wages, CEO salaries and bonuses have

For functionalists, the economy provides order and stability. Many unemployed men—including recent Asian and Latino immigrants—are joining the military because it offers steady work, college education benefits, and an opportunity for career advancement.

skyrocketed. And, unlike other industrialized countries, the U.S. government doesn't cap corporate pay (Bivens and Mishel, 2015; Kim et al., 2015; Wessel, 2015).

CRITICAL EVALUATION

Critics fault conflict theorists for emphasizing economic constraints rather than choices. For example, almost a third of households making more than $75,000 a year live paycheck-to-paycheck because they enjoy "lifestyle purchases" like dining out and entertainment ("SunTrust: Many Higher Income...," 2015).

Also, according to functionalists, conflict theorists underestimate people's ability to get ahead. Many young people are thronging schools that teach high-tech skills in welding and other trades, fast-food workers are unionizing

Jordin Althaus//Wirelmage/Getty Images

In 2015, Volkswagen admitted that it had deceived both customers and government about the emissions and fuel efficiency of 11 million diesel cars worldwide. The cheating began in 2008 when engineers installed software to beat emissions tests (Ewings, 2015). Why does the fraud illustrate conflict theory?

motherhood penalty (also called *motherhood wage penalty* or *mommy penalty*) a pay gap between women who are and aren't mothers.

to raise the industry's minimum wage to $15.00 an hour, and 63 percent of small-business owners find qualified new employees through word-of-mouth referrals (Berfield, 2013; Jacobe, 2013; Phillips, 2014). Thus, many Americans are using several strategies to succeed economically.

11-4c Feminist Theories

Feminist scholars agree with conflict theorists that there's widespread workplace inequality. They emphasize, however, that gender is a critical factor in explaining the inequity.

KEY CHARACTERISTICS

Nationally, women and men are equally satisfied with job security and relations with coworkers. Pay is the primary exception, and women are much more dissatisfied than men (Saad, 2014). Such data aren't surprising because at all income levels and in all occupations, women—especially Latinas and black women—earn less than men (see *Figure 11.4*).

Employment rates increased after the recession, but the gains for women have been largely in low-paying jobs, particularly waitressing, in-home health care, food preparation, and housekeeping. From 2009 to 2012, 60 percent of the increase in women's employment, versus 20 percent for men, was in jobs that pay less than $10.10 an hour (National Women's Law Center, 2014). Both historically and currently, as you saw in Chapters 6 and 9, women have lower earnings than men because of occupational sex segregation, gender pay gaps, glass ceilings, and glass escalators.

Many women's earnings suffer from a **motherhood penalty** (also called *motherhood wage penalty* or *mommy penalty*), a pay gap between women who are and aren't mothers. On average, women without children make 90 cents to a man's dollar, mothers make 73 cents to a man's dollar, and single mothers make only 60 cents to a man's dollar (Rowe-Finkbeiner, 2012; Entmacher et al., 2014).

After declining for several decades, the share of stay-at-home mothers rose from 23 percent in 1999 to 29 percent in 2012. There are many reasons for this increase, but low wages and motherhood penalties are at the top of the list. Many mothers in two-income households have dropped out of the labor force because child care costs rose more than 70 percent from 1985 to 2011 (Cohn et al., 2014).

CRITICAL EVALUATION

Critics note that men don't always dominate women in the workplace. There are many situations where female supervisors have power and authority over women and men, and top female managers aren't always concerned about gender wage gaps or support qualified women's promotions (Huffman, 2013). Second, many feminist scholars maintain that capitalism exploits women by crowding them into lower paying occupations. According to critics, however, there's considerable economic gender inequality in socialist, communist, and mixed economies in both industrialized and developing nations (Cudd and Holstrom, 2010).

Third, some people question whether the economy reinforces sex discrimination or simply reflects cultural sexism. For example, 51 percent of Americans, including women, say that children are better off if the mother is at home and doesn't work; only 8 percent say the same about stay-at-home dads (Cohn et al., 2014). Thus, patriarchal values may be a more critical factor than discrimination in explaining many women's economic inequality.

11-4d Symbolic Interaction

Symbolic interactionists rely on micro-level approaches to explain the day-to-day meaning of work. They're especially interested in how work shapes people's self-identity.

KEY CHARACTERISTICS

Symbolic interaction has provided numerous insights on how people define and experience work. Regardless of income level, 63 percent of U.S. workers who are engaged in their jobs, and even 42 percent of those who aren't engaged, say they would continue to work in their current job if they won a $10 million lottery. All of them might quit, of course, but such responses show that work provides many people with "a source of identity, purpose, and satisfaction that money alone may not replace" (Harter and Agrawal, 2013).

Interactionists also study how people are socialized into their jobs and the informal rules that shape behavior. In medicine, surgeons teach their interns and residents that it's normal to make some mistakes, but that carelessness and continued errors are unprofessional (Bosk, 1979). Assembly-line workers also control coworkers who overproduce or underproduce by punishing or rewarding them (see Chapter 6).

CRITICAL EVALUATION

The most common criticism is that symbolic interaction, although providing in-depth analyses, sacrifices scope. Studies of small work groups can tell us much about people's interaction. We don't know, however, if the findings are applicable to large companies or different parts of the country because the research is based on small and nonrepresentative samples.

Another weakness is that interactionism neglects macro-level social forces that affect people's work and

choices. In Baltimore, for example, manufacturing employment dropped by 38 percent between 2000 and 2012 when large companies like General Motors and Westinghouse shut down plants. A community college was eager to help laid-off steelworkers retool their skills, but "any workers who tried to enroll met a tangle of overlapping and sometimes conflicting requirements from the college and the county, state, and federal governments." Others didn't sign up for the programs after learning that, even with new credentials, they would probably get only $14 an hour jobs, quite a drop from $28 an hour jobs in the past (Mangan, 2013: A24). Thus, organizational obstacles can thwart individual choices and efforts.

11-5 GLOBAL POLITICAL SYSTEMS

Every society has a **government**, a formal organization that has the authority to make and enforce laws. Governments are expected to maintain order, provide social services, regulate the economy, establish educational systems, create armed forces to discourage (real or imagined) attacks by other countries, and ensure their residents' safety.

Worldwide, governments vary from democratic to totalitarian, but there are also authoritarian governments and monarchies. Let's begin by looking at democracy.

11-5a Democracy

A **democracy** is a political system in which, ideally, citizens have control over the state and its actions. Democracies are based on several principles:

▶ Individuals participate in governmental decisions, and select leaders who are responsive to the wishes of the majority of the people.

▶ Suffrage (the right to vote) is universal, and elections are frequent, free, fair, and secret.

▶ The government recognizes individual rights, such as freedom of speech (including dissent), press, and assembly, and the right to organize political parties whose members compete for public office.

▶ The "rule of law" requires everyone to obey the law and to be held accountable if they violate it.

government a formal organization that has the authority to make and enforce laws.

democracy a political system in which, ideally, citizens have control over the state and its actions.

FIGURE 11.5 POLITICAL FREEDOM AROUND THE WORLD, 2015

Free	89 countries (46%)	
Partly free	55 countries (28%)	
Not free	51 countries (26%)	

This map shows the degree of political freedom, measured by political rights and civil liberties, in 195 countries around the world. *Political rights* **include free and fair elections, competitive parties, and no discrimination against minorities.** *Civil liberties* **include freedom of speech and the press; freedom to practice one's religion; and the freedom to discuss political issues without fear of physical violence or intimidation by the government.**

Source: Freedom House, 2015.

▶ A system of terror that relies on secret police and the military to intimidate people into conformity and to punish dissenters.

▶ Total control by the government over other institutions, including the military, education, family, religion, economy, media, and all cultural activities, including the arts and sports.

Contemporary examples of the most repressive totalitarian governments include Central African Republic, North Korea, Saudi Arabia, Somalia, Sudan, Syria, and Uzbekistan. Within these countries and territories, state control over daily life is pervasive and wide-ranging, independent organizations and political opposition are banned or suppressed, and those who criticize the government are imprisoned, tortured, or killed (Makinen, 2014; Freedom House, 2015).

Worldwide, 36 percent of the world's population (2.6 billion people) live in countries with repressive governments that deny basic political and civil rights. People in sub-Saharan and North Africa, the Middle East, China, and Russia are the least likely to have political freedom (see *Figure 11.5*). Despite some countries' attempts to establish democracies, democracy declined around the world in 2015 for the ninth consecutive year (Freedom House, 2015).

11-5b Totalitarianism and Dictatorships

At the opposite end of the continuum from democracy is **totalitarianism**, a political system in which the government controls almost every aspect of people's lives. Totalitarianism has several distinctive characteristics (Taylor, 1993; Tormey, 1995; Arendt, 2004):

▶ A pervasive ideology that legitimizes state control and instructs people how to act in their public and private lives.

▶ A single political party controlled by one person, a *dictator*—a supreme, sometimes idolized leader—who stays in office indefinitely.

totalitarianism the government controls almost every aspect of people's lives.

authoritarianism the state controls the lives of citizens but permits some degree of individual freedom.

11-5c Authoritarianism and Monarchies

Many nations have some version of democracy or totalitarianism. However, a number of countries are characterized by **authoritarianism**, a political system in which the state controls the lives of citizens but permits some degree of individual freedom. Authoritarian governments,

Vladimir Putin has served as Russia's Prime Minister and/or President since 1999. In 2014, he occupied and annexed the Ukraine's Crimean peninsula, an important military base and port on the Black Sea. Because Putin once described the Soviet Union's collapse as the twentieth century's worst disaster, some political analysts believe that seizing Crimea is just a first step toward usurping other countries that became independent in the early 2000s ("The End of the Beginning?" 2014; Wynnyckj, 2014).

ALEXEY DRUZHININ/AFP/Getty Images

In 1995, at age 3, Oyo Nyimba Kabamba Iguru became the world's youngest monarch when he was crowned as king of the Ugandan Kingdom of Toro. At age 18, he officially took control of the kingdom.

PETER BUSOMOKE/Getty Images

like Russia, may permit elections, often controlled, but prohibit free speech, punish those who publicly question state policies, and imprison people without giving them a trial (Whitson, 2013; Ford, 2014; Kirişci, 2014; Ruth, 2014).

A **monarchy**, the oldest type of authoritarian regime, is a political system in which power is allocated solely on the basis of heredity and passes from generation to generation. A member of a royal family, usually a king or queen, reigns over a kingdom. A monarch's power and authority are legitimized by tradition and religion. There are more than 40 monarchies around the world. In some countries—especially in the Middle East and parts of Africa—monarchs have absolute control. Those in Norway, Denmark, Belgium, Britain, Japan, and many other countries have little political power because they're limited by democratic constitutions and serve primarily ceremonial roles (Glauber, 2011).

We see, then, that political systems vary worldwide. Despite the variation, political leaders wield considerable power and authority.

11-6 POLITICS, POWER, AND AUTHORITY

Power and authority are important concepts in sociological analyses of politics. The use and misuse of power and authority affect the quality of our everyday lives.

11-6a Power

Power is the ability of a person or group to influence others, even if they resist. For example, a government can quash a peaceful protest or, conversely, protect people's right to protest. *Legitimate power* comes from having a role, position, or title that people accept as legal and appropriate (e.g., Congress has the right to pass laws, the police have the right to enforce laws). *Coercive power* relies on force or the threat of force to impose one's will on others (e.g., a dictator who imprisons or executes protestors).

Because power is relational, it's expressed downward and upward. With downward power, state senators and representatives affect their constituents' lives because of the laws they pass on taxes, the environment, and other issues. In upward power, voters can protest, refuse to follow laws, and vote politicians out of office.

Whether legitimate or coercive, power—especially political power—is about controlling others. Because people may revolt against sheer force, many governments depend on authority to establish order, shape people's attitudes, and control their behavior.

11-6b Authority

Authority, the legitimate use of power, has three characteristics. First, people *consent* to authority because they believe that their obedience is for the greater good (e.g., following traffic rules). Second, people see the authority as *legitimate*—valid, justifiable, and necessary (e.g., paying taxes to support public education, increase national security, and remove trash and snow). Third, people accept authority because it's *institutionalized*, accepted by a large number of people, in organizations (e.g., police departments and government agencies).

Max Weber (1925/1978) described three ideal types of legitimate authority: traditional, charismatic, and rational-legal (see *Table 11.4*). Remember that ideal types are models that describe the basic characteristics of any phenomenon (see Chapters 1 and 5). In reality, these types of authority often overlap.

TRADITIONAL AUTHORITY

Traditional authority is power based on customs that justify the ruler's position. The source of power is personal because the ruler inherits authority on the basis of long-standing customs, traditions, or religious beliefs. Traditional authority is most common in nonindustrialized societies where power resides in kinship groups, tribes, and clans. In many African, Middle Eastern, and Asian countries today, power is passed down among men within a family line (Tétreault, 2001; see also Chapter 8).

---

monarchy power is allocated solely on the basis of heredity and passes from generation to generation.

power the ability of a person or group to influence others, even if they resist.

authority the legitimate use of power.

traditional authority power based on customs that justify the ruler's position.

TABLE 11.4 WEBER'S THREE TYPES OF AUTHORITY

Type of Authority	Description	Source of Power	Examples
Traditional	Power is based on customs, traditions, and/or religious beliefs.	Personal	Medieval kings and queens, emperors, tribal chiefs
Charismatic	Power is based on exceptional personal abilities or a calling.	Personal	Adolf Hitler, Gandhi, Martin Luther King, Jr.
Rational-legal	Power is based on the rules and laws that are inherent in an elected or appointed office.	Formal	U.S. presidents, members of Congress, state officials, police, judges

CHARISMATIC AUTHORITY

Charismatic authority is power based on exceptional individual abilities and characteristics that inspire devotion, trust, and obedience. Like traditional authority, charismatic authority is personal and reflects extraordinary deeds or even a belief that a leader has been chosen by God, but the leaders don't pass their power down to their offspring.

Charismatic leaders can inspire loyalty and passion whether they're heroes or tyrants. Examples of the latter include historical figures Adolf Hitler, Napoleon, Ayatollah Khomeini, and Fidel Castro. These and other dictators have been spellbinding orators who radiated magnetism, dynamism, and tremendous self-confidence, and promised to improve a nation's future (Taylor, 1993).

RATIONAL-LEGAL AUTHORITY

Rational-legal authority is power based on the belief that laws and appointed or elected political leaders are legitimate. Unlike traditional and charismatic authority, rational-legal authority comes from rules and regulations that pertain to an office rather than to a person. For example, anyone running for mayor must have specific qualifications (e.g., U.S. citizenship). When a new mayor (or governor or other politician) is elected, the rules don't change because power is vested in the office rather than the person currently holding the office.

MIXED AUTHORITY FORMS

People may enjoy more than one type of authority. Some historians believe that presidents Abraham Lincoln, Theodore Roosevelt, John F. Kennedy, and Ronald

First Lady Michelle Obama is an example of a charismatic political leader. She works to end childhoood obesity, helps military families, and visits other countries as a good will ambassador.

Reagan enjoyed charismatic appeal beyond their rational-legal authority. In some countries, including Japan and England, emperors, kings, and queens have traditional authority and perform symbolic state functions (e.g., attending ceremonies). Both of these countries also have parliaments that exercise legal-rational authority in determining laws and policies.

11-7 POLITICS AND POWER IN U.S. SOCIETY

Politics plays a critical role in our everyday lives. Three important components of the political system are political parties, special-interest groups, and voters.

11-7a Political Parties

A **political party** is an organization that tries to influence and control government by recruiting, nominating, and electing its members to public office. Political

---

charismatic authority power based on exceptional individual abilities and characteristics that inspire devotion, trust, and obedience.

rational-legal authority power based on the belief that laws and appointed or elected political leaders are legitimate.

political party an organization that tries to influence and control government by recruiting, nominating, and electing its members to public office.

TABLE 11.5 HOW DO DEMOCRATS AND REPUBLICANS DIFFER?

Issue	Many Democrats Believe That...	Many Republicans Believe That...
Economy	The government should raise taxes on the wealthy and corporations, and expand programs for the poor.	The government should lower taxes on the wealthy and corporations to encourage investment and economic growth; poor people are too dependent on the government.
Abortion	Women should have the right to choose abortion in practically all cases; the government should pay for abortions for low-income women.	The unborn should be protected in all cases; there should be no public money for abortions.
Gay Rights	Gay men and lesbians should have the right to marry and to adopt children. They should be fully accepted in the military and other institutions.	Marriage and adoption should be limited to heterosexuals. Inclusion of gay men and lesbians in the armed forces creates interpersonal problems.
Education	The government should strengthen public schools by raising salaries for teachers and decreasing classroom sizes. It shouldn't use tax dollars to help students attend private schools.	The government should increase state and local control of schools. It should use tax dollars to fund students to attend private schools of their choice if their local school is underperforming.

Sources: Doherty et al., 2014; Johnson, 2014; Wilke and Newport, 2014; Dugan, 2015; Saad, 2015.

parties provide a way for citizens to shape public policy at the local, state, and national levels.

FUNCTIONS OF U.S. POLITICAL PARTIES

Political parties engage in a wide variety of activities—everything from mailing flyers and calling voters to drafting laws if a candidate is elected. Parties perform a number of vital political functions, especially recruiting candidates for public office, organizing elections, and if elected, running the government.

Ideally, that's how political parties *should* work. In reality, parties perform some of these functions more effectively than others. For example, parties typically concentrate on winning elections rather than making promised changes once candidates are in office.

THE TWO-PARTY SYSTEM

Many democracies around the world have a number of major political parties: 3 in Canada, 5 in Germany, 9 in Italy and Holland, 34 in Israel, and more than 1,000 in India. Thus, compared with many other countries, the two-party system of Democrats and Republicans in the United States is uncommon. Political parties base their activities on an *ideology*, a set of ideas that constitute a person's or group's beliefs, goals, expectations, and actions. *Table 11.5* offers some examples of major ideological differences between Democrats and Republicans.

Until the late 1980s, about equal percentages of Americans identified themselves as either Republicans or Democrats. Because of dissatisfaction with both parties, however, in late 2013, 22 percent identified themselves as Republicans, 29 percent as Democrats, and a record 46 percent as independent, up from 30 percent in 1991. A large majority (60 percent) of Americans say that we need a third major party that would do a better job of representing the people (McCarthy, 2015). Because there's no national Independent Party, however, independents usually vote for Democrats, Republicans, or don't vote at all.

11-7b Special-Interest Groups

Some people are active in political parties. Others join a **special-interest group** (also called an *interest group*), people who seek or receive benefits or special treatment. Whereas political parties include diverse individuals, special-interest groups are usually made up of people who are similar in social class and political objectives. There are thousands of U.S. special-interest groups that include everything from agribusiness to unions. Special-interest groups use many tactics to influence the government, but the most effective are campaign contributions and lobbying.

CAMPAIGN CONTRIBUTIONS

Money is usually the most important factor in winning an election. Money doesn't guarantee a victory, but candidates with little financing usually have little public visibility. Among all presidential and congressional candidates, the spending doubled from $3.1 billion in the 2000 election to $6.3 billion in the 2014 election (Center for Responsive Politics, 2014; Montanaro et al., 2014).

Contributions come from individuals and national, state, and local party committees, but the most powerful are **political action committees (PACs)**, special-interest groups that raise money to elect one or more

special-interest group (also called an *interest group*) a group of people that seeks or receives benefits or special treatment.

political action committee (PAC) a special-interest group that raises money to elect one or more candidates to public office.

In 5 to 4 decisions in 2010 and 2014, the Supreme Court overturned previous limits on federal campaign donations. Now, individuals, corporations, unions, political parties, and PACs can contribute almost as much as they want (see Fischer, 2015).

candidates to public office. PACs have mushroomed— from 608 in 1974 to almost 15,700 in 2014 (35 percent of them representing corporate interests), and have received billions of dollars.

Some PACs, like the International Brotherhood of Electrical Workers, support primarily Democrats, whereas the National Rifle Association contributes only to Republicans. A number of PACs, however—including the Credit Union National Association and the National Association of Realtors—contribute to both parties to cover their bases regardless of who's elected (Center for Responsive Politics, 2014; Federal Election Commission, 2014). PACs provide access to politicians and obligate lawmakers to pass legislation that a PAC wants. In contrast with small business firms with few resources, national and transnational corporations can contribute huge amounts of money to PACs, increasing the corporations' influence.

LOBBYISTS

A **lobbyist** is someone hired by a special-interest group to influence legislation on the group's behalf. At the federal level, the number of lobbyists rose from 10,400 in 1998 to almost 11,000 in 2015, an average of 21 lobbyists per congressional member. Some of the best-paid lobbyists include former members of Congress, former congressional aides, and lawmakers' relatives (Drutman and Furnas, 2014; "Lobbying Database," 2015).

In 2014, lobbyists spent $3.2 billion to send members of Congress and their family members on expensive vacations, and, most importantly, made generous

lobbyist someone hired by a special-interest group to influence legislation on the group's behalf.

campaign contributions. In return, corporations and other organizations received favorable legislation, such as Congress passing lax gun and banking laws, approving corporate mergers, and not investigating monopolies (Schweizer, 2013; Hamburger and Gold, 2014).

Lobbying also generates *earmarks* (known commonly as "pork")—funding requests by Congress to provide federal money to companies, projects, groups, and organizations in their district—that don't require external competitive bidding. Despite calls to ban earmarks, in 2012 taxpayers paid $3.3 billion for pork spending. Some examples of recent pork barrel projects include $200 million for corporations to promote their products overseas, $4.9 million for 11 states to determine how to use wood, and a $350 million tower in Mississippi to test rocket engines that NASA didn't want, didn't need, and hasn't used but costs $840,000 annually to maintain (Kennedy and Gelber, 2013; Salant, 2014).

11-7c Who Votes, Who Doesn't, and Why

In 2012, 62 percent of Americans voted in the presidential election, down slightly from almost 64 percent in 2008. These were larger than usual percentages, but lower than the numbers of those who voted in presidential elections during the 1950s and 1960s (File and Crissey, 2010; File, 2013).

Of 110 countries around the world that hold democratic presidential elections, 69 have higher voter turnout rates than the United States; in 17 countries, at least 80 percent of the eligible population votes (Friedman, 2012; International Institute for Democracy and Electoral Assistance, 2014).

Why are U.S. voting rates so low? Who votes, who doesn't, and why reflect demographic characteristics, attitudes about politics, and situational and structural factors.

DEMOGRAPHIC FACTORS

Many demographic factors (e.g., marital status, geographic region, religion) affect registration and voting. The most important, however, are sex, age, social class, and race and ethnicity.

Sex. In every presidential election since 1996, women—and across all racial/ethnic groups—have voted at higher rates than men. In the 2012 election, 64 percent of women and 60 percent of men voted. The biggest gender voting gap was among African Americans—black women voted at higher rates than black men by almost 9 percentage points (File, 2013). The reasons for these differences are unclear, but may reflect higher turnouts by women who believe that social services, abortion rights, and health care coverage are threatened (Jones, 2012).

TABLE 11.6 VOTER RATES IN 2012 PRESIDENTIAL ELECTION, BY SELECTED CHARACTERISTICS

A. Age

18–24	41%
25–44	57
45–64	68
65–74	74
75 and older	70

B. Education

High school, no diploma	38%
High school graduate	53%
Some college or associate degree	64%
Bachelor's degree	75%
Advanced degree	81%

C. Annual Family Income

Under $20,000	50%
$20,000 to $29,999	56%
$30,000 to $39,999	58%
$40,000 to $49,999	63%
$50,000 to $74,999	68%
$75,000 to $99,999	74%
$100,000 to $149,999	77%
$150,000 and over	90%

Source: Based on U.S. Census Bureau, Voting and Registration, 2013, Tables 5 and 7.

Age. Historically, and across all elections, voting rates increase with age. In 2012, voting rates were much higher among older than younger people (see *Table 11.6A*). Why the age difference? A key factor is registration. Only 55 percent of young people (ages 18–24) are registered to vote compared with 79 percent of those age 65 and older (U.S. Census Bureau, Voting and Registration, 2013).

Young people have lower registration rates because they're more geographically mobile than older people and are less likely to reregister after a move. Moreover, they often feel uninformed about politics, believe that elections aren't relevant in their lives, and may also be preoccupied with major life events such as going to college and finding jobs (Schachter, 2004; Child Trends Data Bank, 2013). In effect, young people's low voting rates means that they have little effect on political processes.

Social Class. Social class has a significant impact on voting behavior. At each successive level of educational attainment, the voting rate increases, and those with advanced degrees are the most likely to vote (see *Table 11.6B*). People with higher educational levels are usually more informed about and interested in the political process and more

likely to believe that their vote counts (File and Crissey, 2010; "The Politics of Financial Insecurity," 2015).

Voting rates also increase with income levels (see *Table 11.6C*). People with higher incomes are more likely to be employed and to have assets (e.g., houses, stock) and, therefore, are more likely to vote to protect or increase their resources. In contrast, low-income people usually have few assets and may be too disillusioned with the political system to vote ("Who Votes . . . ," 2006; File and Crissey, 2010).

Race and Ethnicity. The 2012 presidential election was the first time since at least 1996, the earliest year with comparable data, that African Americans voted at a higher rate than whites (see *Figure 11.6*). Here again, a key to voter turnout is registration. Across all racial-ethnic groups—including both native-born and naturalized citizens—the higher the voter registration, the higher the voting rate (File and Crissey, 2010; File, 2015).

In 2012, blacks and whites had a higher voter registration rate (74 percent) than Latinos (59 percent) and Asian Americans (56 percent), but why did black voters outpace whites? The increase in black voting, especially since 2008, stems from enthusiasm for Barack Obama's presidential candidacy, high levels of voter mobilization, and a decline of the white population. In both 2008 and 2012, however, much of the surge in turnout rates was driven by blacks ages 45 and older, particularly women, who voted in record numbers (File, 2013).

In contrast to blacks, Latinos and Asian Americans who are eligible to vote don't do so, and their turnout fell between 2008 and 2012 (see *Figure 11.6*). About 75 percent of both groups were born in the United States, but have lower voter registration rates than blacks and whites. Latinos have lower education and income levels

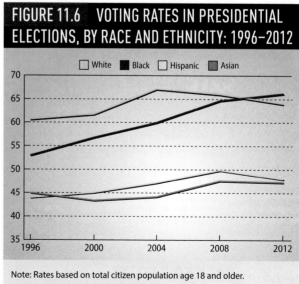

FIGURE 11.6 VOTING RATES IN PRESIDENTIAL ELECTIONS, BY RACE AND ETHNICITY: 1996–2012

Note: Rates based on total citizen population age 18 and older.

Source: Based on File, 2013, Figure 1.

than most other groups, both of which are associated with lower voter turnout rates. Low Asian American voter turnout is puzzling: They have the highest education and income levels (see Chapters 8 and 10), but even among the college-educated, voter turnout has lagged behind whites, blacks, and Latinos (Jones, 2013; Krogstad, 2014).

ATTITUDES

Millions of Americans have a low opinion of the government and Congress. In 2014, 75 percent of Americans said there was widespread corruption in government, up from 66 percent in 2009. When asked to rate 22 occupations according to honesty and ethics, those at the bottom of the list were state officeholders (14 percent), car salespeople (9 percent), members of Congress (8 percent), and lobbyists (6 percent) (Swift, 2013; "75% in U.S. See…," 2015).

Such distrust may help explain why many people don't vote: They believe that their lives won't improve regardless of who's elected. On the other hand, few political incumbents are voted out of office election after election. Why? Only 17 percent of registered voters say that members of Congress should be re-elected, but almost half believe that *their* congressional representatives are doing a good job (Dugan and Hoffman, 2014). Perhaps, then, "the government we have is the government we deserve" (Cillizza, 2014).

How Would *You* Describe Congress?

In 2010, the Pew Research Center asked Americans to use one word that described Congress. The responses were put into a "word cloud." The larger the word, the more times it was used. What one word would *you* use to describe Congress?

Source: Word cloud graphic, created using http://wordle.net, from “ Congress in a Wordle”, Mar. 22, 2010, The Pew Research Center For the People & the Press, a project of the Pew Research Center.

---

pluralism a political system in which power is distributed among a variety of competing groups in a society.

SITUATIONAL AND STRUCTURAL FACTORS

Situational and structural factors also affect voting. Some of the high turnout rates in other countries are partly due to *compulsory voting*: The government imposes fines for not voting, may cut off benefits such as public assistance, or requires evidence of voting to get a passport or a driver's license. Such sanctions send the message that voting isn't a privilege but a civic responsibility. In many European and other countries, voters are registered automatically when they pay taxes or receive public services, elections are held on weekends, and people vote electronically (Seward, 2012; International Institute for Democracy and Electoral Assistance, 2014).

Instead of making voting easier, 34 states (particularly those controlled by Republican legislatures) have passed or are considering laws that will make voter registration more difficult. Some of the new rules require government-issued photo identification (e.g., a driver's license or passport) and a birth certificate or proof of citizenship. Some states have reduced early voting, limited the time polls are open, done away with same-day voter registration, and moved polling locations outside of areas with large racial-ethnic populations. Proponents contend that the new laws will crack down on voter fraud in future elections; opponents argue that the new regulations will stifle voter turnout, especially among students, the poor, and minorities, who tend to vote for Democrats (Whiteaker et al., 2014; Yaccino and Alvarez, 2014; Eckholm, 2015).

Because many Americans don't vote, who has the most power? And do government leaders represent the average citizen? Such questions have generated considerable debate among sociologists and other social scientists.

11-8 SOCIOLOGICAL PERSPECTIVES ON POLITICS AND POWER

According to symbolic interaction, power is socially constructed through interactions with others; people learn to be loyal to a political system—whether it's a democracy or a monarchy—and to show respect for its symbols and leaders. For the most part, however, sociologists rely on functionalist, conflict, and feminist theories to explain political power. *Table 11.7* summarizes these perspectives.

11-8a Functionalism: A Pluralist Model

For functionalists, the people rule through **pluralism**, a political system in which power is distributed among a variety of competing groups in a society (Riesman, 1953; Polsby, 1959; Dahl, 1961).

TABLE 11.7 SOCIOLOGICAL EXPLANATIONS OF POLITICAL POWER

	Functionalism: A Pluralist Model	Conflict Theory: A Power Elite Model	Feminist Theories: A Patriarchal Model
Who has political power?	The people	Rich upper-class people—especially those at top levels in business, government, and the military	White men in Western countries; most men in traditional societies
How is power distributed?	Very broadly	Very narrowly	Very narrowly
What is the source of political power?	Citizens' participation	Wealthy people in government, business corporations, the military, and the media	Being white, male, and very rich
Does one group dominate politics?	No	Yes	Yes
Do political leaders represent the average person?	Yes, the leaders speak for a majority of the people.	No, the leaders are most concerned with keeping or increasing their personal wealth and power.	No, the leaders are rarely women who have decision-making power.

KEY CHARACTERISTICS

In the pluralist model, individuals have little direct power over political decision making but can influence government policies through special interest groups—unions, professional organizations, and so on (see *Figure 11.7*). The various groups rarely join ranks because they concentrate on single issues such as health care, pollution, or education. This focus on different issues fragments groups, but also results in a broad representation of interests and a distribution of power.

Because there are a number of single-issue groups, functionalists maintain, there are multiple leaderships (Dye and Ziegler, 2003). As a result, many leaders, not just a few, can shape decisions that represent numerous groups and issues. Pluralists note that people also have power outside of interest groups: They can vote, run for office, contact lawmakers, and put specific issues on a ballot. Therefore, there are continuous checks and balances as individuals and groups vie for power and try to influence laws and policies.

CRITICAL EVALUATION

Does pluralism work as democratically as functionalists maintain? Critics argue that interest groups have unequal resources. The poor and disadvantaged rarely have the skills and educational backgrounds to organize or promote their interests. In contrast, wealthy individuals and organizations can influence government through political contributions and personal connections.

FIGURE 11.7 PLURALIST AND POWER ELITE PERSPECTIVES OF POLITICAL POWER

PLURALIST MODEL

Power is dispersed among multiple groups that influence the government. For example,

- government employees
- victims' rights groups
- labor unions
- banks and other financial institutions
- realtors and home builders
- teachers
- environmental groups
- women's rights groups

POWER ELITE MODEL

Power is concentrated in a very small group of people who make all the key decisions. For example,

Top level (1 percent)—CEOs of large corporations, high-ranking lawmakers in the executive branch, and top military leaders

Middle level (8 percent)—most members of Congress, lobbyists, entrepreneurs of small businesses, leaders of labor unions and other interest groups (in law, education, medicine, etc.), and influential media commentators

The masses (91 percent)—people who are unorganized and exploited and either don't know or don't care about what's going on in government

Critics also maintain that pluralists aren't realistic about the emergence of multiple leaderships because the affluent control the government. For example:

▶ The median net worth of a member of Congress was $1.03 million in 2013, compared with $56,355 for the average American household.

▶ More than half of the members of Congress are millionaires, and five are worth $100 million or more.

▶ All of the Supreme Court Justices and top executive branch members are also multimillionaires (Rodrigue and Reeves, 2014; Choma, 2015).

Because of such wealth inequality, according to an owner of a one-man glass business, most politicians are totally disconnected from the lives of ordinary Americans. What would he do if he were president? "I'd fire everyone in the House and the Senate, and put working class people in who actually know what it's like to be out here" (Geller, 2014).

11-8b Conflict Theory: A Power Elite Model

Conflict theorists contend that the United States is ruled not by pluralism but a **power elite**, a small group of influential people who make the nation's major political decisions. Sociologist C. Wright Mills (1956) coined the term *power elite* to describe a pyramid of power that he believed characterized American democracy.

KEY CHARACTERISTICS

According to Mills, the power elite is made up of three small but dominant groups of people at the top level who run the country: political leaders, corporate heads, and military chiefs (see *Figure 11.7*). Practically all of the members of these groups are white, Anglo-Saxon, Protestant men who form an inner circle of power. Those at the middle power level include members of Congress, lobbyists, influential media commentators, and others. The bottom level, and the largest and least powerful group, the masses, is composed of everyone else.

The power elite tolerates the masses—including their elections and laws—but in the end simply does what it wants. In 2013, almost 80 percent of Americans wanted limits on the amount of money that congressional candidates could spend on their campaigns (Saad, 2013). The following year, the Supreme Court did just the opposite by striking down laws that limited contributions to political candidates, parties, and PACs.

---

power elite a small group of influential people who make the nation's major political decisions.

Contemporary conflict theorists maintain, like Mills, that the United States isn't a democracy because of the close ties and "revolving door" between the power elite in politics, business, and the military. For example, fewer than 31,390 people—less than 0.01 percent of the nation's population—contributed more than $1.6 billion to political campaigns in 2012, and nobody was elected to Congress without the economic elite's money. In return, politicians pass favorable legislation such as $83 billion in taxpayer-financed subsidies to the country's banks, award earmarks and lucrative military and other contracts, and may later serve as corporate lobbyists (Gilson, 2014; Kristof, 2014; Winship, 2015).

CRITICAL EVALUATION

Functionalists accuse conflict theorists of exaggerating the power elite's importance. Functionalists point out, for example, that "the masses" aren't just puppets but are aware of the power elite's influence. As a result, they vote, mobilize, support particular interest groups, and protest current political and corporate policies.

Also, according to critics, the power elite perspective assumes, incorrectly, that the wealthy people at the top have the same interests and goals. Because the Democratic and Republican parties and their top officials endorse very different agendas, the power elite are rarely unified.

Moreover, critics maintain, the wealthy can't always buy political offices. In 2010, for example, Meg Whitman (the founder of eBay) spent almost $142 million of her personal fortune in her bid for California governor. She lost to Jerry Brown, who contributed only $24,000 of his own money to his $4 million campaign funds (Crowley, 2010; Palmeri and Green, 2010).

11-8c Feminist Theories: A Patriarchal Model

As you saw in Chapter 9, relatively few women, especially minority women, serve in Congress and at the highest government levels. In a recent national poll, 62 percent of men and 69 percent of women said that the country would be better off if there were more women in political office (Nelson, 2013). Why, then, are there so few female officeholders?

KEY CHARACTERISTICS

Feminist theorists maintain that women are generally shut out of the most important political positions because, in a patriarchal society, men rule. A "gender gap in political ambition" begins early in the socialization process. Parents often encourage their sons, not their daughters, to think about politics as a career. As a result, men are more likely than women to take government or political science

In 2012, at age 63, Elizabeth Warren (left), a Massachusetts Democrat, ran for the U.S. Senate. The *Boston Herald* repeatedly referred to her as "granny," "Liz," and "Lizzy." She's now called Madame Senator. Some of the *Huffington Post*'s feminist writers said that Michele Bachmann (right)—a candidate for the Republican nomination in the 2012 presidential election—might have won "if she had let down her necklines" (Seltzer, 2012). Do male candidates ever experience similar sexism?

classes and to join political clubs in college. In adulthood, men are more likely than women to be encouraged to run for a political office. Thus, despite comparable educational and occupational backgrounds, men are almost 60 percent more likely than women to view themselves as "very qualified" to run for office (Lawless and Fox, 2013).

A larger number of female officeholders alone doesn't guarantee greater gender equality. Some Republican women in Congress have opposed equal pay for equal work laws because, they maintain, well-qualified women are already paid the same as men, that such legislation would increase the number of civil lawsuits, and would hurt businesses (Lowery, 2014; McDonough, 2014).

Unlike their male counterparts, female political candidates encounter sexist insults from TV talk show hosts (particularly Fox News) and male candidates, sexist debate questions from both male and female moderators, and reporters' obsession with the women's fashion choices—all of which diminish the women's credibility. And, according to 40 percent of Americans, women are held to higher standards than men when they seek the highest levels of either business or politics (Seltzer, 2012; Parker and Horowitz, 2015).

CRITICAL EVALUATION

Critics fault feminist explanations for several reasons. First, there are powerful women in the most influential congressional committees (Kiefer, 2015), and some of the world's political elite are women (e.g., Angela Merkel, Chancellor of Germany; Dilma Rousseff, President of Brazil; Geun-hye Park, President of South Korea; Christine Lagarde, Managing Director of the International Monetary Fund; and Janet Yellen, Chair of the U.S. Federal Reserve).

Second, women in politics can be just as political as men. Presidential candidate Hillary Clinton, for example, has earned millions of dollars from speeches to companies and associations with federal lobbyists and generous donors (Elliott, 2015; Gold et al., 2015).

One might also question many feminists' contention that patriarchy is the root of political power differences between women and men. Some of the most patriarchal societies—those in parts of Africa, Latin America, and Arab countries, for example—have more women in high-ranking political positions than does the United States, which professes gender equality. Women have the fewest rights in societies where male political leaders are religious zealots who don't believe in gender equality (see Jones, 2014). Thus, religion may be more important than patriarchy in quashing women's political leadership.

STUDY TOOLS 11

READY TO STUDY? IN THE BOOK, YOU CAN:

☐ Check your understanding of what you've read with the Test Your Learning Questions provided on the chapter review card at the back of the book.

☐ Rip out the chapter review card for a handy summary of the chapter and key terms.

ONLINE AT CENGAGEBRAIN.COM YOU CAN:

☐ Collect StudyBits while you read and study the chapter.

☐ Quiz yourself on key concepts.

☐ Prepare for tests with chapter flash cards as well as flash cards that you create.

☐ Watch videos to explore sociology further.

12 | Families and Aging

LEARNING OUTCOMES

After studying this chapter, you will be able to…

12-1 Describe how families are similar and different in the United States and worldwide.

12-2 Describe how and explain why U.S. families are changing.

12-3 Describe, illustrate, and explain why intimate partner violence, child maltreatment, and elder abuse occur.

12-4 Describe, illustrate, and explain how the U.S. older population is changing, and its impact on our society.

12-5 Compare and evaluate the theoretical explanations of families and aging.

After you finish this chapter go to **PAGE 244** for **STUDY TOOLS.**

Cohabitation, marriage, divorce, and grandparents raising grandchildren—these are just some of the family characteristics that continue to change. This chapter examines the connections between families and our aging society. Before considering what sociologists mean by *family* and *aging*, take the short quiz on page 226 to see how much you already know about these topics.

12-1 WHAT IS A FAMILY?

Ask five of your friends to define *family*. Their definitions will probably differ not only from each other's but also from yours. For our purposes, a **family** is an intimate group consisting of two or more people who (1) have a committed relationship, (2) care for one another and any children, and (3) share close emotional ties and functions. This definition includes households (e.g., foster families, same-sex couples) whose members aren't related by birth, marriage, or adoption.

What do you think?

I would undergo medical treatments to live to age 120 or longer.

1	2	3	4	5	6	7

strongly agree   strongly disagree

Contemporary households are complex; family structures vary across cultures and have changed over time. In some societies, a family includes uncles, aunts, and other relatives. In others, only parents and their children are viewed as a family.

family an intimate group consisting of two or more people who (1) have a committed relationship, (2) care for one another and any children, and (3) share close emotional ties and functions.

12-1a How Families Are Similar Worldwide

The family, a social institution, exists in some form in all societies. Worldwide, families are similar in fulfilling some functions that have persisted over time.

FAMILY FUNCTIONS

Families vary considerably in the United States and globally but fulfill five important functions that ensure a society's survival (Parsons and Bales, 1955):

▶ **Sexual activity.** Every society has norms regarding who may engage in sexual relations, with whom, and under what circumstances. In the United States sexual intercourse is banned with someone younger than 18 (or 16 in some states), but some societies permit marriage with girls as young as 8 (see Chapter 9).

   One of the oldest rules that regulates sexual behavior is the **incest taboo**, cultural norms and laws that forbid sexual intercourse between close blood relatives (e.g., brother and sister, father and daughter, uncle and niece).

▶ **Procreation and socialization.** Procreation is an essential family function because it replenishes a country's population. Through socialization, children acquire language; absorb the accumulated knowledge, attitudes, beliefs, and values of their culture; and learn the social and interpersonal skills they need to function effectively in society (see Chapters 3 and 4).

▶ **Economic security.** Families provide food, shelter, clothing, and other material resources for their members.

▶ **Emotional support.** Families supply the nurturance, love, and emotional sustenance that people need to be happy, healthy, and secure. Our friends may come and go, but our family is usually our emotional anchor.

▶ **Social class placement.** Social class affects all aspects of family life. Initially, our social position is based on our parents' social class, but we can move up or down the social hierarchy in adulthood (see Chapter 8).

Some sociologists include *recreation* as a basic U.S. family function. This function isn't critical for survival, but since the 1950s many parents have spent much more time with their children on leisure activities (e.g., visiting amusements parks, playing video games together) that bolster interpersonal bonds.

> **incest taboo** cultural norms and laws that forbid sexual intercourse between close blood relatives.

MARRIAGE

Marriage, a socially approved mating relationship that people expect to be stable and enduring, is also universal. Countries vary in their specific norms and laws dictating who can marry whom and at what age, but marriage everywhere is an important rite of passage that marks adulthood and its related responsibilities, especially providing for a family.

ENDOGAMY AND EXOGAMY

All societies have rules, formal or informal, about the "right" marriage partner. **Endogamy** (often used interchangeably with *homogamy*) is a cultural practice of marrying within one's group. The partners are similar in religion (e.g., Catholics marrying Catholics), race or

---

marriage a socially approved mating relationship that people expect to be stable and enduring.

endogamy (often used interchangeably with *homogamy*) cultural practice of marrying within one's group.

exogamy (often used interchangeably with *heterogamy*) cultural practice of marrying outside one's group.

nuclear family a family form composed of married parents and their biological or adopted children.

---

ethnicity (e.g., blacks marrying blacks), social class (e.g., college-educated marrying college-educated), and/or age (e.g., young people marrying young people). Across the Arab world and in some African nations, marrying first or second cousins is not only commonplace but desirable because such unions reinforce kinship ties and increase a family's resources (Bobroff-Hajal, 2006).

Exogamy (often used interchangeably with *heterogamy*) is a cultural practice of marrying outside one's group, such as not marrying one's relatives. In the United States, 25 states prohibit marriage between first cousins, but violations are rarely prosecuted (National Conference of State Legislatures, 2014). U.S. exogamy is growing. For example, interracial marriages rose from 3 percent of all marriages in 1980 to 10 percent by 2010, but the spouses are usually homogamous (similar) in social class and age (Wang, 2012). Thus, whom we marry isn't entirely a matter of choice but is limited by cultural mate selection norms and values.

12-1b How Families Differ Worldwide

Families also differ around the world. Some variations affect the family's structure, whereas others regulate where people reside and who has the most household power and authority.

NUCLEAR AND EXTENDED FAMILIES

In Western societies, the typical family form is the **nuclear family** composed of married parents and their biological or adopted children. In much of the world, however,

© Mike Baldwin / Cornered

"Empty-nesters. They're hoping to sell before the flock tries to move back in."

Mike Baldwin/CartoonStock.com

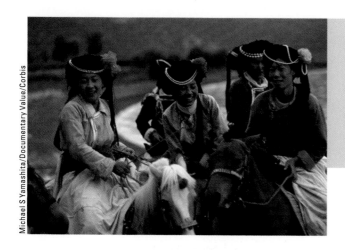

The Mosuo in China's Himalayas is an example of a matriarchal society. For the majority of Mosuo, a family is a household consisting of a woman, her children, and the daughters' offspring. A man will join a lover for the night and then return to his mother's or grandmother's house in the morning. Any offspring resulting from these unions belong to the female; she and her relatives raise the children (Barnes, 2006).

the most common family form is the **extended family**, composed of parents, children, and other kin (e.g., uncles and aunts, nieces and nephews, cousins, grandparents).

As the number of single-parent families increases in industrialized countries, extended families are more common. By helping out with household tasks and child care, other adult members make it easier for a single parent to work outside the home. Many Americans assume that the nuclear family is the most common arrangement, but such families have declined—from 40 percent of all U.S. households in 1970 to 20 percent in 2012 (Vespa et al., 2013).

RESIDENCE PATTERNS

In a **patrilocal residence pattern**, newly married couples live with the husband's family. In a **matrilocal residence pattern**, they live with the wife's family. In a **neolocal residence pattern**, the couple sets up their own residence.

Around the world, the most common residence pattern is patrilocal. In industrialized societies, married couples are typically neolocal. Since the early 1990s, however, the tendency for young married adults to live with the parents of either the wife or the husband—or sometimes with the grandparents of one of the partners—has increased. Such "doubled-up" U.S. households have always existed, but escalated during the 2007–2009 recession for economic reasons (Parker, 2012).

As a result, a recent phenomenon is the **boomerang generation**, young adults who move back into their parents' home after living independently for a while or never leave home in the first place. In 2014, for example, and five years after the 2007–2009 recession ended, 14 percent of adults between the ages of 24 and 34 were living at home with their parents (Jones, 2014). Parents try to launch their children into the adult world, but like boomerangs, some keep coming back.

AUTHORITY AND POWER

Residence patterns often reflect who has authority and power in the family. In a **matriarchal family system**, the oldest females (usually grandmothers and mothers) control cultural, political, and economic resources and, consequently, have power over males. Some American Indian tribes were matriarchal, and in some African countries, the oldest women have considerable authority and influence. For the most part, however, matriarchal societies are rare.

Worldwide, a more typical pattern is a **patriarchal family system**, in which the oldest males (grandfathers, fathers, and uncles) control cultural, political, and economic resources and, consequently, have power over females. In some patriarchal societies, women have few rights within or outside the family; they may not be permitted to work outside the home or attend college. In other patriarchal societies, women may have considerable decision-making power in the home but few legal or political rights, such as getting a divorce or running for political office (see Chapter 9).

In an **egalitarian family system**, both partners share power and authority fairly equally. Many Americans

extended family a family form composed of parents, children, and other kin.

patrilocal residence pattern newly married couples live with the husband's family.

matrilocal residence pattern newly married couples live with the wife's family.

neolocal residence pattern a newly married couple sets up its own residence.

boomerang generation young adults who move back into their parents' home or never leave it in the first place.

matriarchal family system the oldest females control cultural, political, and economic resources and, consequently, have power over males.

patriarchal family system the oldest males control cultural, political, and economic resources and, consequently, have power over females.

egalitarian family both partners share power and authority fairly equally.

think they have egalitarian families, but our families tend to be patriarchal. For example, employed women shoulder almost twice as much housework and child care as men, and are more likely than men to provide caregiving to aging family members (Cynkar and Mendes, 2011; see also Chapter 9).

COURTSHIP AND MATE SELECTION

Sociologists often describe U.S. dating as a **marriage market**, a courtship process in which prospective spouses compare the assets and liabilities of eligible partners, and choose the best available mate. Marriage markets don't sound very romantic, but open dating fulfills several important functions: recreation and companionship; a socially acceptable way of pursuing love; opportunities for sexual intimacy and experimentation; and finding a spouse (Benokraitis, 2015).

Many societies, in contrast, discourage open dating and have **arranged marriages** in which parents or relatives choose the children's spouses. An arranged marriage is a family rather than an individual decision that increases solidarity between families and preserves endogamy. Children may have veto power, but they believe that if partners are compatible, love will result. In some of India's urban areas, arranged marriages also rely on nontraditional methods (e.g., online dating services) to find prospective spouses (Cullen and Masters, 2008; see also Chapter 9).

MONOGAMY AND POLYGAMY

In **monogamy**, one person is married exclusively to another person. Where divorce and remarriage rates are high, as in the United States, people engage in **serial monogamy**. That is, they marry several people but one at a time—they marry, divorce, remarry, divorce, and so on.

Polygamy, in which a man or woman has two or more spouses, is subdivided into *polygyny*—one man married to two or more women—and *polyandry*—one woman is married to two or more men. Nearly 1,000 cultures around the world allow some form of polygamy, either officially or unofficially (Epstein, 2008).

Although rare, there are pockets of polyandry in some remote and isolated parts of India, and among the Pimbwe in western Tanzania, Africa. Polyandry serves several functions: The family is more likely to survive in harsh environments if there's more than one husband to provide food, and if one husband dies, the others care for the widow (Borgerhoff Mulder, 2009; Polgreen, 2010).

In contrast to polyandry, polygyny is common in many societies, especially in some regions of Africa, South America, and the Middle East. Men benefit from polygyny in several ways: They can have many legal sexual partners, more chances to have male heirs, a bigger income if some of the wives are employed, and high social status because they can support multiple wives and children (Al-Jassem, 2011; Nossiter, 2011).

Western and industrialized societies forbid polygamy, but there are pockets of isolated polygynous groups in the United States, Europe, and Canada. The Church of Jesus Christ of Latter-Day Saints (Mormons) banned polygamy in 1890 and excommunicates members who follow such beliefs. Still, males of the Fundamentalist Church of Jesus Christ of Latter Day Saints (FLDS), a polygynous sect that broke away off from the mainstream Mormon Church more than a century ago, head an estimated 300,000 families in Texas, Arizona, Utah, and Canada. Wives who have escaped from these groups have reported forced marriage between men in their 60s and girls as young as 10 years old, sexual abuse, and incest (Janofsky, 2003).

India's Ziona Chana has the world's largest polygynous family—39 wives, 94 children, and 33 grandchildren (so far). They all live together in a 100-room mansion ("After 39 Wives...," 2013).

marriage market prospective spouses compare the assets and liabilities of eligible partners and choose the best available mate.

arranged marriage parents or relatives choose the children's spouses.

monogamy one person is married exclusively to another person.

serial monogamy individuals marry several people, but one at a time.

polygamy a man or woman has two or more spouses.

12-2 HOW U.S. FAMILIES ARE CHANGING

American families have changed considerably since the 1950s. Some of the most important changes are related to marriage, divorce, cohabitation, nonmarital childbearing, and single-parent families.

12-2a Marriage and Divorce

The United States has one of the highest marriage and divorce rates in the world. By age 65, 95 percent of Americans have been married at least once; over a lifetime, about half of first marriages end in divorce (Vespa et al., 2013). Despite the high divorce rate, most people aren't disillusioned about marriage. Indeed, nearly 85 percent who divorce remarry, half of them within 4 years, and 8 percent have been married three times or more (Kreider and Ellis, 2011; Livingston, 2014).

TRENDS

U.S. marriage and divorce rates rose steadily during the twentieth century, but have declined since 1990 (see *Figure 12.1*). In 1960, only 9 percent of U.S. adults age 25 and older had never been married. By 2012, 20 percent had never been married—an historic high. Marriage rates are expected to decrease further because the number of newlyweds has been falling since 2012 (Cohn et al., 2011; Wang and Parker, 2014).

Divorce rates are high, but *lower* today than they were between 1980 and 2009 (see *Figure 12.1*). On average, first marriages last about 9 years, but many marriages are lasting longer than they were a decade ago. In 2009, 55 percent of couples had been married for at least 15 years, 35 percent had reached their twenty-fifth anniversary, and 6 percent had been married 50 years or longer. These percentages are several points higher than they were in 1996 (Kreider and Ellis, 2011; Aughinbaugh et al., 2013).

WHY ARE MARRIAGE AND DIVORCE RATES FALLING?

There are several *macro-level reasons* for declining marriage and divorce rates. First, U.S. values have been changing. In 2010, 39 percent of Americans said that marriage is becoming obsolete, up from 28 percent in 1978 (Cohn et al., 2011; Fry, 2012). In 2013, 64 percent of Americans, down from 76 percent in 2006, said that people should marry, and only 44 percent believed that having children is a "very important" reason to marry. Only 30 percent of Americans think that premarital sex is immoral, open marriage markets provide many opportunities for nonmarital sex, and nonmarital births are now socially acceptable (Cohn, 2013; Poushter, 2014). Thus, the traditional reasons for marriage have waned.

Second, the economy affects marriage and divorce rates. Economic depressions, recessions, and unemployment tend to delay marriage, especially for men. Because of the recent recession, 20 percent of 18- to 34-year-olds have postponed marriage. Moreover, when incomes plummet and people are insecure about their jobs, unhappy married couples tend to stay together: They can't afford to divorce and risk the possibility of not being able to maintain separate households (Cohn, 2012; Martin et al., 2014).

Demographic variables also affect marriage and divorce rates. The median age at first marriage is 29 for men and almost 27 for women, up from 23 for men and 21 for women in 1970 ("Marital Status," 2015).

About 64 percent of Americans with college degrees are married, compared with 47 percent of those with a high school diploma or less, who are more likely to cohabit than marry. And, increasingly, the highly educated are marrying each other. In effect, there's a growing "marriage divide" between those with and without college degrees (Cohn et al., 2011; Fry, 2012; Greenwood et al., 2014).

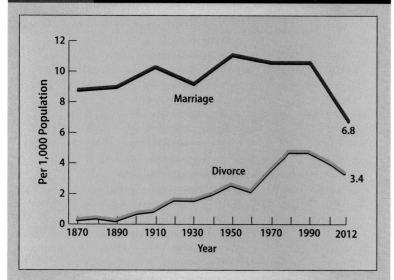

FIGURE 12.1 U.S. MARRIAGE AND DIVORCE RATES, 1870–2012

Sources: Based on Plateris 1973, Table 1; U.S. Census Bureau 2012, Table 78; and CDC/NCHS National Vital Statistics System, 2015.

Americans with a bachelor's degree or higher also have lower divorce rates than those without college degrees. By age 46, 30 percent of those with a college degree or higher are divorced compared with 59 percent with less than a high school diploma (Aughinbaugh et al., 2013). College graduates have lower divorce rates not because they're smarter but because going to college postpones marriage. As a result, better-educated couples are often more mature and capable of dealing with personal crises. They also have higher incomes and health care benefits, both of which lessen marital stress over financial problems (Kreider and Ellis, 2011).

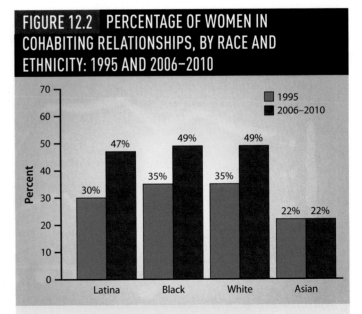

Rubberball/Mike Kemp/Jupiter Images

There are also *micro-level (individual) reasons* for falling marriage and divorce rates. About 84 percent of unmarried Americans say that love is a "very important" reason to marry. As a result, a third are still waiting for their "ideal mate" or "one true love" (Wang and Parker, 2014).

The most common micro-level reasons for divorce include infidelity, communication and financial problems, substance and spousal abuse, premarital doubts, continuous disagreements about how to raise and discipline children, and expecting to change a partner after marriage (see Benokraitis, 2015, for a discussion of these studies). Because many people are delaying marriage, they're usually more mature in handling the challenges of married life, which decreases the likelihood of divorce (Glenn et al., 2010).

12-2b Cohabitation

Cohabitation is an arrangement in which two unrelated people aren't married but live together and are in a sexual relationship (shacking up, in plain English). Because it's based on emotional rather than legal ties, "cohabitation is a distinct family form, neither singlehood nor marriage" (Brown, 2005: 33).

TRENDS

Married couples comprise 49 percent of all households, a sharp decline from 78 percent in 1950. Some of the decline was due to falling marriage rates and those who haven't remarried yet, but most of the decrease was due to cohabitation. The number of unmarried

heterosexual couples surged from 430,000 in 1960 to almost 8.1 million in 2013. This number rises by at least another 605,000 if we include same-sex unmarried partners. However, only about 7 to 9 percent of the population is cohabiting at any given time, as in 2013 (Vespa et al., 2013).

By age 30, 74 percent of U.S. women aged 15–44 have cohabited compared with 62 percent in 1995 (Copen et al., 2013). Between 2005 and 2010, Asian women were the least likely to cohabit, and the only group that didn't experience an increase in cohabitation (see *Figure 12.2*). Asian women have low cohabitation rates because marriage is the norm and premarital sex and cohabitation are stigmatized (see Chapters 9 and 10). In contrast, many Hispanics come from countries in Latin and Central America that have much higher cohabitation rates than those in the United States (Oropesa and Landale, 2004).

Women's cohabitation rates have increased, but the likelihood of cohabiting decreases as their educational levels increase. On average, college-educated women cohabit for the shortest periods (17 months), and are more likely than those without college degrees to transition to marriage (Copen et al., 2012, 2013).

WHY HAS COHABITATION INCREASED?

On a *micro level*, some people drift gradually into **dating cohabitation**, a living arrangement in which a

---

cohabitation two unrelated people aren't married but live together and are in a sexual relationship.

dating cohabitation a couple that spends a great deal of time together decides to move in together.

FIGURE 12.2   PERCENTAGE OF WOMEN IN COHABITING RELATIONSHIPS, BY RACE AND ETHNICITY: 1995 AND 2006–2010

	1995	2006–2010
Latina	30%	47%
Black	35%	49%
White	35%	49%
Asian	22%	22%

Note: These data are for never-married women aged 22–44.

Sources: Based on Cohn et al., 2011; "Marital Status," 2013; U.S. Census Bureau, Current Population Survey..., 2013.

Do you think that women or men benefit more from cohabitation? Why?

couple that spends a great deal of time together decides to move in together. Dating cohabitation is essentially an alternative to singlehood because the decision may be based on a combination of reasons (e.g., convenience, finances, companionship, and sexual accessibility), but there's no long-term commitment. In this type of cohabitation, and especially among young adults, there's considerable *serial cohabitation*, living with different sexual partners over time. Even if there's an unplanned pregnancy, the man, especially, may decide to move on to another cohabiting arrangement (Manning and Smock, 2005; Wartik, 2005).

For many people, premarital cohabitation is a step between dating and marriage. In **premarital cohabitation**, the couple lives together before getting married. They may or may not be engaged but plan to marry. Such "almost-married" cohabitation may be

especially attractive to partners who wonder if they can deal successfully with problems that arise from differences in personalities, interests, finances, ethnicity, religion, or other issues.

On a *macro level*, and across all ethnic groups, 3 out of 4 people who cohabit say that they're delaying marriage because "Everything's there except money." Many low-income black women and Latinas don't want to marry because they believe that their live-in partners will be poor providers, unemployed, unfaithful, irresponsible fathers, or immature even though "he's the love of my life" (Xie et al., 2003; Lichter et al., 2006).

Those with high income and education levels have little to gain from cohabitation. Unlike people in lower socioeconomic groups, they're more likely to have jobs, to afford their own housing, and, consequently, to have more options in living independently even though they're involved romantically (Fry and Cohn, 2011; Sassler and Miller, 2011).

Does cohabitation lead to better marriages? Recent studies show that women who cohabit are no more likely to divorce than those who didn't cohabit *if* there's a commitment (definite plans to marry) and the cohabitation is short. However, marital success also depends on the cohabitors' age, socioeconomic status, commitment, and attitudes towards marriage (Reinhold, 2010; Manning and Cohen, 2012). Cohabitation, like any other relationship, has benefits and costs (see *Table 12.1*).

premarital cohabitation a couple lives together before getting married.

TABLE 12.1 SOME BENEFITS AND COSTS OF COHABITATION

Benefits	Costs
• Couples can pool their resources instead of paying for separate housing, utilities, and so on. They can also have the emotional security of an intimate relationship but maintain their independence by spending time with their friends and family members separately (McRae, 1999; Fry and Cohn, 2011).	• U.S. laws don't specify a cohabitant's rights and responsibilities. For example, there's no automatic inheritance if a partner dies without a will, and it's more difficult to collect child support from a cohabiting partner than an ex-spouse (Silverman, 2003; Grall, 2013).
• Couples who postpone marriage have a lower likelihood of divorce because being older is one of the best predictors of a stable marriage (Copen et al., 2012).	• Compared with married couples, cohabitants have a poorer quality of relationship and lower levels of happiness and satisfaction (Sassler et al., 2012; Wiik et al., 2012).
• Couples find out how much they really care about each other when they have to cope with unpleasant realities (e.g., a partner who doesn't pay bills or rarely showers).	• Cohabitation dilutes intergenerational ties. Compared with their married peers, the longer people live together, the less likely they are to give or receive help from their parents, and to be involved in extended family activities (Eggebeen, 2005).
• Children in cohabiting households reap economic advantages by living with two adult earners instead of a single mother (Cabrera et al., 2008).	• Because cohabiting parents are more than twice as likely as married parents to break up, children's academic, emotional, behavioral, and financial problems often increase (Fomby and Estacion, 2011; Rackin and Gibson-Davis, 2012).

12-2c Nonmarital Childbearing

More than 9 in 10 Americans aged 45 or older either have children (86 percent) or wish they had (7 percent). Such positive attitudes toward childbearing are virtually the same as in 1990, but since then U.S. birth rates have dropped by 11 percent (Martin et al., 2013; Newport and Wilke, 2013). A notable exception is nonmarital childbearing, which has increased significantly since 1970. Nearly half of American men ages 15 to 44 report that at least one of their children was born outside of marriage, and 31 percent say that all of their children were born outside of marriage (Livingston and Parker, 2011).

TRENDS

In 1950, only 3 percent of all U.S. births were to unmarried women. In 2013, there were more than 1.6 million, accounting for 41 percent of all U.S. births. Thus, 4 in 10 American babies are now born outside of marriage, a new record (Curtin et al., 2014).

Births to unmarried women vary widely across racial-ethnic groups. White women have more nonmarital babies than do other groups. Proportionately, however, nonmarital birthrates are highest for black women and lowest for Asian American women (see *Figure 12.3*).

As in marriage and cohabitation, nonmarital births vary by social class. Among college graduates, only 9 percent of recent births were outside of marriage compared with 57 percent for high-school dropouts (see *Table 12.2A*). The percentage of nonmarital births decreases as household income increases (see *Table 12.2B*). Consequently,

TABLE 12.2 BIRTHS TO UNMARRIED MOTHERS, BY EDUCATION AND HOUSEHOLD INCOME

A. Educational Attainment	
Less than high school	57%
High school graduate	49%
Some college	40%
Bachelor's degree or more	9%

B. Annual Household Income	
Less than $10,000	69%
$10,000 to $24,999	55%
$25,000 to $49,999	42%
$50,000 to $99,999	26%
$100,000 to $199,999	17%
$200,000 and above	9%

Source: Based on Shattuck and Kreider, 2013, Table 2.

the poorest and least-educated women are the most likely to have babies outside of marriage, decreasing the children's chances of moving up the social class ladder (Sawhill and Venator, 2014).

Unmarried teenage births peaked in the early 1990s, have declined steadily since then, and are now at their lowest level since 1968. Teenagers account for 15 percent of all nonmarital births compared with 74 percent for women ages 20 to 34. Thus, the number of births to unmarried women ages 20 to 34 is almost five times greater than for teenagers. Still, 86 percent of teenage births are nonmarital, compared with 60 percent for women in their 20s, and 14 percent for those ages 30 to 34. Moreover, 20 percent of unmarried teenagers ages 15 to 19 have a repeat birth: 86 percent have a second child, and 15 percent have 3 to 6 children (Gavin et al., 2013; Shattuck and Kreider, 2013).

WHY HAS NONMARITAL CHILDBEARING INCREASED?

On a *micro (individual) level*, many teens and young adults use contraception inconsistently or incorrectly because they think they or their partners are sterile, they think

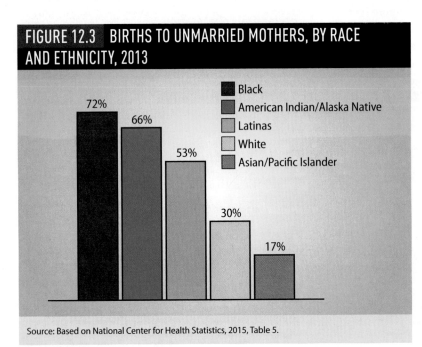

FIGURE 12.3 BIRTHS TO UNMARRIED MOTHERS, BY RACE AND ETHNICITY, 2013

- Black — 72%
- American Indian/Alaska Native — 66%
- Latinas — 53%
- White — 30%
- Asian/Pacific Islander — 17%

Source: Based on National Center for Health Statistics, 2015, Table 5.

THE NUMBER OF BIRTHS TO UNMARRIED WOMEN AGES 20 TO 34 IS ALMOST FIVE TIMES GREATER THAN FOR TEENAGERS.

A recent study found that the reality TV show *16 and Pregnant* reduced U.S. teen births by almost 6 percent in the 18 months following its release (Kearney and Levine, 2014). Should middle schools require adolescents to watch such programs?

not taking the pill a few times "doesn't really matter," or the males don't want to use condoms (Frohwirth et al., 2013). About 67 percent of all unmarried or cohabiting women report having unintended or unwanted pregnancies, but may not get abortions because of personal or religious values (Wildsmith et al., 2011; Curtin et al., 2014).

Demographic variables also affect nonmarital childbearing. Social class is a key factor in nonmarital births (see *Table 12.2*). Educated women are more likely than less-educated women to postpone parenthood until marriage, to be more aware of family planning, and to have more decision-making power in their relationships, such as insisting that men use condoms (Livingston and Cohn, 2013).

Gender, race, and ethnicity intersect, however. Among college-educated women, 41 percent of black women, compared with 22 percent of Latinas and only 9 percent of white women, choose single motherhood. Nonmarital childbearing is a "rational choice" for black women who don't want to marry less educated black men or across racial lines, believe that they may never marry, or that marriage will come too late for them to bear children (Keels, 2014).

Macro-level variables increase nonmarital birthrates in several ways. Until the 1970s, births outside of marriage were virtually unheard of and often kept as secret as possible. Because U.S. attitudes have changed, nonmarital childbearing has become normal. In South Korea and several other industrialized countries, in contrast, nonmarital birth rates are much lower because unmarried mothers are stigmatized (Borowiec, 2014).

About 40 percent of U.S. female teens ages 15 to 17 have had sex, but 80 percent of them hadn't received any formal sex education before the first time they had sex. Publicly funded family planning services that provide counseling and contraceptives, particularly for teenagers and young adults, reduce unintended nonmarital pregnancies and births. Many states, however, have decreased funding for sex education and family planning organizations (Cox et al., 2014; see also Chapter 9).

In what some researchers describe as a pattern of "negative assimilation," 41 percent of second-generation Latinas and Asian mothers are unmarried, compared with 23 percent of recent immigrant women ("Second-Generation Americans," 2013). Thus, many second-generation women are espousing U.S. values that births outside of marriage are acceptable.

12-2d Single-Parent Households

Most children spend the majority of their childhood living with two parents, but single-parent families are increasingly common. Women head most one-parent families.

TRENDS

In 2014, 64 percent of children under age 18 lived with two married parents, down from 77 percent in 1980 (Federal Interagency Forum on Child and Family Statistics, 2015). Asian American children are the most likely to grow up in two-parent homes (see *Figure 12.4*). Until 1980, married couple families were the norm in African American families. Since then, black children have been more likely than children in other racial-ethnic groups to grow up with only one parent, usually the mother. About 65 percent of Latino children live in two-parent families, down from 78 percent in 1970 (Lugaila, 1998).

Nationally, of the nearly 650,000 same-sex households, 19 percent are raising children. Among those living alone or with a partner or spouse, 48 percent of LGBT women and 20 percent of LGBT men are raising a child under age 18. Between 2000 and 2010, the number of same-sex couples with adopted children increased from 9 to 20 percent

FIGURE 12.4 WHERE U.S. CHILDREN LIVE, BY RACE AND ETHNICITY, 2014

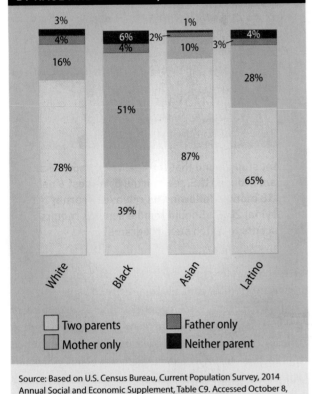

White: 3%, 4%, 16%, 78%

Black: 6%, 4%, 51%, 39%

Asian: 1%, 2%, 10%, 87%

Latino: 4%, 3%, 28%, 65%

- ☐ Two parents
- ☐ Mother only
- ☐ Father only
- ■ Neither parent

Source: Based on U.S. Census Bureau, Current Population Survey, 2014 Annual Social and Economic Supplement, Table C9. Accessed October 8, 2015 (census.gov).

(Gates, 2013). Thus, large numbers of lesbians and gay men—single, partnered, or married—are parents.

WHY HAVE FEMALE-HEADED HOUSEHOLDS INCREASED?

The reasons for the growing number of female-headed households are similar to those for the increase in nonmarital childbearing. On a *micro level*, having sex at an early age, and not using contraception at all, incorrectly, or sporadically results in unintended or unwanted pregnancies, especially among teenagers. Among unmarried parents, 63 percent have multiple partners. In these relationships, most of the fathers invest little time and money in their children because the fathers are young, have low education levels, or are incarcerated. The fathers don't live with their children, know little about them, and move from one sexual relationship to another (Scommegna, 2011; Dodson and Luttrell, 2011).

Demographic variables, especially social class, also affect the rise of female-headed households. College-educated people, as you've seen, tend to postpone marriage, marry rather than cohabit, and delay

fictive kin nonrelatives who are accepted as part of a family.

parenthood—factors that decrease the likelihood of divorce, single-parent households, and, consequently, children's experiencing poverty (Redd et al., 2011).

Not all female-headed households are poor, of course. Among mother-only families, 17 percent of the women have at least a college degree, and 38 percent own their own homes (Vespa et al., 2013). Instead of waiting for "Mr. Right," women with economic resources are deciding to raise children on their own.

In many African American and Latino families, adult males (e.g., sons, brothers, uncles, grandfathers) often provide emotional and financial support to their female kin who head households (Sudarkasa, 2007). Many black families also welcome **fictive kin**, nonrelatives who they accept as part of the family. Fictive kin have strong bonds with biological family members and provide important services (e.g., caring for children when young mothers are employed or negligent) (Billingsley, 1992; Dilworth-Anderson et al., 1993). A recent variation of fictive kin includes single mothers—many of whom are white, unmarried, college-educated women—who turn to one another for companionship and child care help (Bazelon, 2009).

At the *macro level*, values and views about single parenthood have shifted. The share of Americans who view the growing trend of single mothers as a "big problem" decreased from 71 percent in 2007 to 64 percent in 2013 (Wang et al., 2013). The economy also affects the number of female-headed households. As in the case of cohabitation and nonmarital childbearing, low-income women often drift into parenthood with low-income or

Karen Struthers/Shutterstock.com

Is the absent black father a stereotype? A recent national study of white, black, and Latino men found that, among fathers who don't live with their children, black fathers were the *most* likely to parent their children from birth to age 18. The parenting included, and on a daily basis, feeding, bathing, dressing, playing with and reading to children, taking them to and from activities, and helping them with or checking homework (Jones and Mosher, 2013).

unemployed men. Seeking financial security and stability, women gain little or nothing by marrying the babies' fathers or other low-income men.

12-3 FAMILY CONFLICT AND VIOLENCE

Conflict is a normal part of family life, but violence is *not* normal. Over a lifetime, we're much more likely to be assaulted or killed by a family member or current/former spouse, boyfriend, or girlfriend than by a stranger (Truman and Langton, 2014).

12-3a Intimate Partner Violence

Intimate partner violence (IPV) is abuse that occurs between people in a close relationship. The term *intimate partner* refers to current and former spouses, couples who live together, and current and former boyfriends or girlfriends.

TRENDS

IPV is pervasive in U.S. society. Nationwide, 36 percent of females and 29 percent of males have been victims of IPV at some time in their lives (Black et al., 2011). Nearly half of all adults have experienced psychological aggression by an intimate partner (Copen et al., 2013), but IPV impacts all types of people (see *Figure 12.5*).

IPV resulted in 1,517 homicides in 2010; 89 percent of the victims were females. When victims survive IPV assaults, 40 percent of females compared with 5 percent of males suffer a serious physical injury (e.g., gunshot or knife wounds, internal injuries, broken bones) (Catalano, 2013). Women are also much more likely than men to experience IPV over a lifetime regardless of age, race or ethnicity, or social class (Black et al., 2011; Truman and Morgan, 2014).

From 1994 to 2012, the IPV rate declined 67 percent—from 9.8 per 1,000 persons age 12 or older in 1994 to 3.2 in 2012. During this period, 82 percent of the victims were females. In 2012, there were nearly 1 million intimate partner victimizations, but this number is conservative. An estimated 56 percent of victims don't report IPV to the police because they're ashamed, believe that no one can help, or fear reprisal (Truman and Morgan, 2014).

WHY DOES INTIMATE PARTNER VIOLENCE OCCUR?

The reasons for IPV are complex and often the product of multiple individual, demographic, and societal factors. On a *micro level*, violence escalates if one or both partners engage in heavy alcohol or drug use, if they have more children than they can afford (which intensifies financial problems),

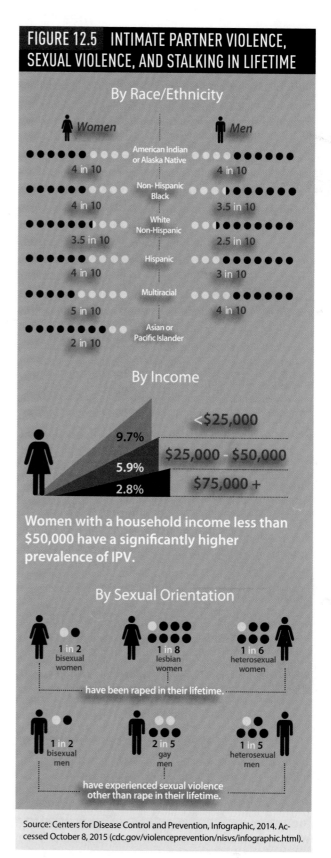

FIGURE 12.5 INTIMATE PARTNER VIOLENCE, SEXUAL VIOLENCE, AND STALKING IN LIFETIME

By Race/Ethnicity

Women / Men

American Indian or Alaska Native — 4 in 10 / 4 in 10
Non-Hispanic Black — 4 in 10 / 3.5 in 10
White Non-Hispanic — 3.5 in 10 / 2.5 in 10
Hispanic — 4 in 10 / 3 in 10
Multiracial — 5 in 10 / 4 in 10
Asian or Pacific Islander — 2 in 10

By Income

<$25,000 — 9.7%
$25,000 - $50,000 — 5.9%
$75,000 + — 2.8%

Women with a household income less than $50,000 have a significantly higher prevalence of IPV.

By Sexual Orientation

1 in 2 bisexual women / 1 in 8 lesbian women / 1 in 6 heterosexual women
have been raped in their lifetime.

1 in 2 bisexual men / 2 in 5 gay men / 1 in 5 heterosexual men
have experienced sexual violence other than rape in their lifetime.

Source: Centers for Disease Control and Prevention, Infographic, 2014. Accessed October 8, 2015 (cdc.gov/violenceprevention/nisvs/infographic.html).

intimate partner violence (IPV) abuse that occurs between people in a close relationship.

or if either partner has been raised in a violent household, has low self-esteem, and has a controlling personality. The most common disagreements that can escalate into abuse and violence have four sources: gender role expectations (who does what housework); money (saving and spending); children (especially discipline); and infidelity, both personally and online (Spivak et al., 2014; Benokraitis, 2015).

Regarding *demographic variables*, IPV often begins at a young age. Among those who experienced rape, physical violence, and/or stalking by an intimate partner the first time, 22 percent of females and 15 percent of males are between 11 and 17 years old; 47 percent of females and 39 percent of males are between 18 and 24 years old (Black et al., 2011). IPV cuts across all social classes, but women living in households with an annual income of less than $7,500 are nearly seven times more likely to be abused than those living in households with an annual income of $75,000 or more (Thompson et al., 2006).

Macro-level variables, particularly unemployment and poverty, increase the likelihood of financial stress and violence. The absence of legal or social sanctions against IPV and a scarcity of shelters discourage victims from trying to escape abuse (Matjasko et al., 2013). A study of girls aged 11 to 16 who had experienced sexual assault concluded that girls (and later women) often didn't report the incidents because such violence had been "normalized in their communities." The girls were ashamed and feared retribution, but also believed that men "can't help it," perceived everyday harassment and abuse as "normal male behavior," and assumed that male authority figures (e.g., police officers, judges) would accuse the girls of overreacting (Hlavka, 2014).

12-3b Child Maltreatment

Child maltreatment (also called *child abuse*) includes a broad range of behaviors that place a child at serious risk or result in serious harm, including physical and sexual abuse, neglect, and emotional mistreatment. The victims often experience several types of maltreatment.

TRENDS

U.S. child maltreatment rates have decreased since 2007. Because only a fraction of the cases are reported, however, federal agencies assume that millions of children experience abuse and neglect on a daily basis. In 2013, an estimated 680,000 children were victims of child maltreatment, representing almost 9 per 1,000 U.S. children, and nearly 1,500

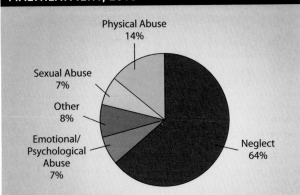

FIGURE 12.6 TYPES OF CHILD MALTREATMENT, 2013

Note: "Neglect" includes medical neglect (almost 2 percent of these cases). "Other" includes categories that some states report, such as a parent's drug/alcohol abuse.

Source: Based on U.S. Department of Health and Human Services, 2015, Table 3-9.

died of abuse and neglect (U.S. Department of Health and Human Services, 2015).

Neglect is the most common type of abuse (see *Figure 12.6*). From birth to age 17, girls (51 percent) are slightly more likely than boys (49 percent) to be maltreated. The most vulnerable children, those younger than 3 years old, account for 27 percent of all child victims; another 20 percent are ages 3 to 5. Victimization rates per 1,000 children vary considerably by race and ethnicity: 15 for African Americans, 13 for American Indian/Alaska Natives, 11 for multiracial, 8 for Latinos, 8 (each) for whites and Pacific Islanders, and under 2 for Asians (U.S. Department of Health and Human Services, 2015).

Almost 81 percent of the perpetrators are one or both parents; in 37 percent of the cases, the abusers are mothers alone. An additional 7 percent of assailants are relatives, and 4 percent are the parents' intimate partners, usually boyfriends (U.S. Department of Health and Human Services, 2013).

WHY DOES CHILD MALTREATMENT OCCUR?

On a *micro level*, child mistreatment is most common in single-parent and stepparent households, those with parental substance abuse and mental illness, if there's adult or sibling violence, and if the child has emotional, developmental, or physical disabilities (Finkelhor et al., 2011). *Demographic variables* that increase the likelihood of child abuse include low socioeconomic status, a parent's young age, and parental conflict before, during, and after a hostile divorce (Truman and Smith, 2012; White and Lauritsen, 2012). On a *macro level*, economic hardship, unemployment, and poverty increase the

child maltreatment (also called *child abuse*) a broad range of behaviors that place a child at serious risk or result in serious harm, including physical and sexual abuse, neglect, and emotional mistreatment.

The movie *Precious* (2009) received many awards, but some critics contended that the film stereotyped black female teenagers—especially those in low-income households—as obese, illiterate, and living with abusive mothers (Amusa, 2010). If you've seen the movie, do you agree with such criticism or not? Why?

likelihood of stress that leads to child abuse, including infant deaths due to severe head traumas (Eckenrode et al., 2014; National Academy of Sciences, 2014).

Almost 26 percent of all children live in homes where parents or other adults engage in violence (Hamby et al., 2011). Whether children are targets of abuse or see it, a growing body of research has linked childhood experience of violence with lifelong developmental problems, including depression, delinquency, suicide, alcoholism, low academic achievement, unemployment, and medical problems in adulthood (Hibbard et al., 2012; Sacks et al., 2014).

12-3c Elder Abuse and Neglect

Elder abuse (sometimes called *elder mistreatment*) is any knowing, intentional, or negligent act by a caregiver or other person that causes harm to people age 65 or older. This term includes physical, psychological, and sexual abuse; neglect; isolation from family and friends; deprivation of basic necessities such as food and heat; not providing needed medications; and financial exploitation.

TRENDS

A national study of people ages 60 and older found that almost 12 percent had experienced at least one of the following types of mistreatment: emotional (4.6 percent), physical (1.6 percent), sexual (0.6 percent), financial (5.2 percent), and neglect (6 percent) (Acierno et al., 2010). The actual rates are probably much higher because only an estimated 20 percent of all elder abuse and neglect is reported (Administration on Aging, 2013).

Who are the victims? About 83 percent are white; the average age is 76; 76 percent are women; 84 percent live in their own homes; 86 percent have a chronic disease or other health condition; 57 percent are married

or cohabiting; 53 percent haven't graduated from high school; 50 percent suffer from dementia, Alzheimer's, or other mental illness; 46 percent feel socially isolated; and the average combined household income is less than $35,000 a year (Acierno et al., 2010; Jackson and Hafemeister, 2011). These data suggest that the most likely victims of elder abuse are those who are the most vulnerable physically, mentally, socially, and financially.

Who are the perpetrators? Most are adult children, spouses or cohabiting partners, or other family members. Less than a third are acquaintances, neighbors, or nonfamily service providers. The average age of the abuser is 45; 77 percent are white; 61 percent are males; 82 percent have a high school diploma or less; 50 percent abuse alcohol and/or other drugs; 46 percent have a criminal record; 42 percent are financially dependent on the elder; 37 percent live with the elder; 29 percent are chronically unemployed; and 25 percent have mental health problems (Jackson and Hafemeister, 2011).

WHY DO ELDER ABUSE AND NEGLECT OCCUR?

On a *micro level*, abuse of alcohol and other drugs is more than twice as likely among family caregivers who abuse elders as among those who don't. Both victims and offenders often report a childhood history of witnessing or experiencing family violence, poor family relationships in the past and currently, and communication problems. In addition, older people with cognitive impairment—due to dementia (deteriorated mental condition) after a stroke, or the onset of Alzheimer's disease—are abused at higher rates than those without such disabilities (Heisler, 2012; National Center on Elder Abuse, 2012).

On a *macro level*, a shared residence is a major risk factor for elder mistreatment because the caregiver(s) may depend on the older person for housing, whereas the elder is dependent on the caregiver(s) for physical help. These situations compound the likelihood of everyday

elder abuse (sometimes called *elder mistreatment*) any knowing, intentional, or negligent act by a caregiver or other person that causes harm to people age 65 or older.

tensions and conflict. Financial stress may also increase the risk of experiencing or inflicting abuse. Unlike low-income families, those in the middle class aren't eligible for admission to public facilities, yet few can pay for the in-home nursing care, high-quality nursing homes, and other services that upper-class families can afford (Acierno et al., 2010; Jackson and Hafemeister, 2011).

12-4 OUR AGING SOCIETY

One of our friends, 55, was shocked and insulted when he ordered a meal at a fast-food restaurant, and the young worker called out "one senior meal!" The *Baltimore Sun* (2012) recently reported on the death of an "elderly" man in a house fire. He was 62.

12-4a When Is "Old"?

What images come to mind when you hear the word *old*? In a recent national survey, people in their 40s said that a person is old at age 63, those in their 70s said age 75, and a 90-year-old woman said that a woman isn't old "until she hits 95" (*AARP Magazine*, 2014: 40). So, how old is "old"?

It depends on whom you ask because "old" is a social construction. In some African nations, where people rarely live past 50 because of diseases and civil wars, 40 is old. In industrialized societies, where the average person lives to age 78, 40 is considered young. Still, regardless of how we feel physically, society usually defines old in chronological age. In the United States, for instance, people are deemed old at age 65, 66, or 67 because they can retire and become eligible for Medicare and Social Security benefits.

Many older Americans are vigorous and productive, but others experience physical and mental limitations as they age. There are significant differences between the *young-old* (65-74 years old), the *old-old* (75-84 years old), and the *oldest-old* (85 years and older) in their ability to live independently, to work, and in their health. Generally, a 65-year-old woman is much less likely than an 85-year-old woman to need caregiving from family and friends.

12-4b Life Expectancy

Life expectancy is the average expected number of years of life remaining at a given age. American children

Itsuo Inouye/AP Images

Between the ages of 65 and 91, avid mountaineer Hulda Crooks scaled Mount Whitney (the highest mountain in the continental United States) 23 times. She died at the age of 101 in 1997.

born in 2010 have a life expectancy of almost 79 years (compared with 71 in 1970 and 47 in 1900).

Life expectancy varies by sex, social class, and race-ethnicity. Historically, currently, and across all racial-ethnic groups, women live longer than men, and Latinas have the longest lifespans (see *Figure 12.7*).

The higher rates of cigarette smoking, heavy drinking, gun use, employment in hazardous occupations, and risk-taking in recreation and driving are responsible for many males' shorter lifespans. The life expectancy gender gap has narrowed since the 1990s, however, because there's been an increase in women's smoking, use of alcohol and other drugs, obesity (which increases the risk of hypertension and heart disease), and stresses due to multiple roles (e.g., juggling employment while caring for children and older family members) ("Catching Up," 2013; see also Chapter 14).

Generally, higher socioeconomic levels increase life expectancy. People with a college degree and higher are more likely than less educated adults to be employed, to be financially secure, and to have employment-related health insurance. They're also more likely to exercise, to not smoke, to drink alcohol in moderation, and to maintain a healthy body weight. An exception is Latinos, who

life expectancy the average expected number of years of life remaining at a given age.

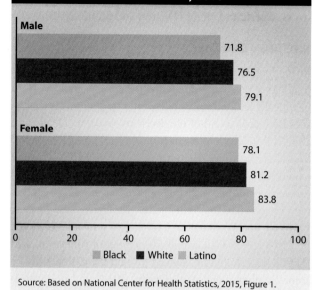

FIGURE 12.7 U.S. LIFE EXPECTANCY AT BIRTH, BY SEX AND RACE-ETHNICITY, 2013

Male
71.8
76.5
79.1

Female
78.1
81.2
83.8

Black ■ White □ Latino

Source: Based on National Center for Health Statistics, 2015, Figure 1.

have among the lowest education levels and household incomes, but the highest life expectancy rates.

Researchers hypothesize that Latinos have longer life spans both because only the healthiest migrate to the United States and because many foreign-born Latinos return to their country of origin when their health deteriorates. Also, recent immigrants, compared with whites, are less likely to smoke and more likely to have strong support networks during illness (Pollard and Scommegna, 2013; Scommegna, 2013).

12-4c How Our Graying Nation Is Changing

The number of Americans age 65 and older is booming, increasing, and becoming more ethnically and racially diverse. For example,

▶ More than 43 million (14 percent of the total population) are 65 or older, a dramatic increase from just 4 percent in 1900, and will comprise more than 20 percent of U.S. residents by 2030.

▶ One of the fastest growing groups is the oldest-old, whose numbers increased from 100,000 in 1900 to almost 6 million in 2012. By 2030, this group will comprise almost 3 percent of the population.

▶ By 2030, 28 percent of the 65-and-older population will be minority, up from 20 percent in 2010; 23 percent of the 85-and-older population will be minority, up from 15 percent in 2010 (Federal

Interagency Forum on Aging-Related Statistics, 2012; "U.S. Census Bureau Projections…," 2012; Ortman et al., 2014).

Our aging population has three important (and interrelated) implications. First, as you'll see shortly, health care and caregiving needs, services, and costs will surge. Second, the number of older Americans is increasing, whereas the proportion of young people is decreasing (see *Figure 12.8*). As a result, many adult children—even in their sixties—will be caring for aging parents, grandparents, and other older relatives for more years than in the past.

Disability rates among older Americans declined during the 1980s and 1990s, but have recently increased, particularly among **baby boomers** (people born between 1946 and 1964) and those 10 years younger than baby boomers. The higher disability rates might be reflecting better medical diagnoses, but baby boomers are more likely than the previous generation to be obese, have diabetes, or high blood pressure (Scommegna, 2013). Thus, the boomer generation might be sicker and for more years than their predecessors who had lower life expectancies.

Third, as the number of older people increases, so does the **old-age dependency ratio**, the number of people age 65 and older who aren't in the labor force

baby boomers people born between 1946 and 1964.

old-age dependency ratio the number of people age 65 and older who aren't in the labor force relative to the number of working-age adults ages 18 to 64.

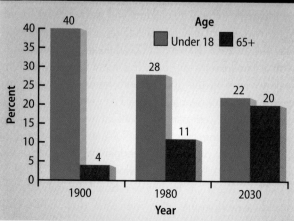

FIGURE 12.8 THE YOUNG AND THE OLD IN THE UNITED STATES, 1900–2030

Sources: Based on U.S. Senate Special Committee on Aging et al., 1991: 9, and Ortman et al. 2014, Table 2.

relative to the number of working-age adults ages 18 to 64. In effect, this ratio shows the burden on the working-age population to support those 65 and older who aren't employed. The old-age dependency ratio has risen. By 2020, only 3 working-age people will be supporting each older person, compared with 14 workers per older person in 1900 (Jacobsen et al., 2011; Ortman et al., 2014). Thus, most people will have to pay much higher federal and state taxes to sustain our ever-growing older population.

12-4d Some Current Aging Issues

The *sandwich generation* is composed of midlife women and men who care for aging parents, are raising a child under age 18, or supporting a grown child. About 71 percent who do so are ages 40 to 59 (Parker and Patten, 2013). The sandwich generation experiences considerable stress because older people usually need assistance at a time when their adult children's and grandchildren's lives are complicated and demanding. Three specific outcomes that the sandwich generation is most likely to experience are the rise of multigenerational households, right-to-die issues, and competition for scarce resources.

MULTIGENERATIONAL HOUSEHOLDS

The share of the U.S. population living in multigenerational households rose from 12 to 16 percent between 1980 and 2010. Asians (28 percent) are more likely to live in multigenerational households than blacks (26 percent), Latinos (25 percent), and whites (14 percent) (Taylor et al., 2013; Vespa et al., 2013; see also Chapter 4 on multigenerational households).

There are many reasons for several generations living under one roof. Graying boomers with small retirement savings may move in with their adult children and grandchildren to avoid poverty. Adult children may move back home with their parents (often with children and girlfriends/boyfriends in tow) because of unemployment, low wages, or home foreclosures. In 2012, 10 percent (7.1 million) of children lived with a grandparent—up from 5 percent in 1990—and the majority lived in the grandparent's home. Extended family living provides intergenerational support, including child care for employed parents. Cultural values, particularly among recent immigrants, reflect long-standing practices of caring for aging parents and grandparents (Ellis, 2013; Fry and Passel, 2014; "National Grandparents Day...," 2014).

The effects of multigenerational households are mixed. Grandparents who live with their adult children and grandchildren may provide crucial economic and child care support. On the other hand, if the multigenerational household is struggling financially, if the grandparents are also raising their own children, and if there's mother–grandmother conflict, grandparents experience emotional problems, including unhappiness, stress, worry, and anger (Deaton and Stone, 2013).

THE RIGHT TO DIE

Living longer and with more years of chronic diseases and disabilities raises questions about later-life choices. Both historically and currently, one of the most controversial issues is regarding end-of-life decisions. Oregon's Death with Dignity Act, which took effect in 1997, authorized prescriptions for lethal doses when two doctors agreed that a patient would die within six months and freely chose *physician-assisted suicide* (also called *physician-aid-in-dying*). Modeling Oregon's law, four other states (Montana, New Mexico, Vermont, and Washington) have passed similar legislation.

In a recent Gallup poll, 70 percent of Americans said that doctors should be allowed to "end the patient's life by some painless means" (McCarthy, 2014), but doing so is illegal in most of the country. Advocates argue that terminally ill people who face a long and painful death and huge medical costs should have the legal right to die with dignity on their own terms instead of lingering in a nursing home, hospital, or hospice. Opponents maintain that actively ending a life, regardless of a person's frailty or suffering, is a moral violation, and that patients might be pushed to die early for the caregiver's convenience (Eckholm, 2014).

Medical care for patients in the last year of life accounts for more than 25 percent of annual Medicare costs (Carr, 2012). Is this a good investment of our resources?

COMPETITION FOR SCARCE RESOURCES

When Congress passed the Social Security Act in 1935, life expectancy was just below 62 years compared with almost 79 today. In the years ahead, the increasing numbers of older Americans will put a significant strain on the nation's health care services and retirement income programs.

The size of the U.S. older population (13 percent) is about half that of the population age 18 and younger (26 percent). Nonetheless, the federal government spends 31 percent on the elderly, primarily through Social Security and Medicare, compared with 10 percent on children (Isaacs et al., 2012; Matthews and Schmit, 2014). The older people get, the higher the medical costs. For example,

▸ A study of 1.8 million Medicare recipients who died in 2008 found that a third had surgery in the last year of life. Nearly 1 in 5 had surgery in the last month of life, and nearly 1 in 10 had surgery in the last week of life. Among those undergoing end-of-life surgery, almost 60 percent were 80 and older (Kwok et al., 2011).

▸ About 60 percent of prostate cancer is diagnosed in men aged 65 or older. It costs $93,000 per patient to treat advanced prostate cancer; the treatment prolongs life by an average of four months (Beil, 2012).

Younger generations can't count on federally financed health care and retirement benefits in the future. Medicare funds are supposed to last until 2026, and Social Security through 2033 (The Board of Trustees..., 2013), but there are no guarantees. As baby boomers age, the costs of these programs will mushroom. Because boomers have higher educational levels than previous generations, they're likely to demand more and more expensive health care services (see Chapter 14).

Despite rising health care costs, 20 percent of Americans want to live to age 100—about 21 years longer than the current average U.S. life expectancy. Moreover, 38 percent support medical treatments that would allow them to live to at least age 120 (Lipka and Stencel, 2013; Lugo et al., 2013). Who'll pay for such radical life extensions?

12-5 SOCIOLOGICAL EXPLANATIONS OF FAMILY AND AGING

The four sociological perspectives are useful in understanding families and aging. *Table 12.3* summarizes the key points of these theories.

12-5a Functionalism

We began this chapter by looking at the vital functions that families perform—such as procreation and economic security—that promote societal stability and individual well-being. Functionalists recognize that families differ in structure (e.g., nuclear vs. extended), but believe that their similar functions ensure a society's continuity.

STABILITY AND ACTIVITY

For many functionalists, marriage, followed by procreation, is critical in promoting social order and cohesion. Parenthood is a crucial social role. One of its tasks, socialization, is essential to maintaining any culture. Problems arise, for instance, when parents can't or don't provide their children with the necessary financial and emotional support or when they divorce.

TABLE 12.3	SOCIOLOGICAL PERSPECTIVES ON FAMILIES AND AGING	
Theoretical Perspective	**Level of Analysis**	**Key Points**
Functionalist	Macro	Families are important in maintaining societal stability and meeting family members' needs.
		Older people who are active and engaged are more satisfied with life.
Conflict	Macro	Families promote social inequality because of social class differences.
		Many corporations view older workers as disposable.
Feminist	Macro and micro	Families both mirror and perpetuate patriarchy and gender inequality.
		Women have an unequal burden in caring for children as well as older family members and relatives.
Symbolic Interactionist	Micro	Families construct their everyday lives through interaction and subjective interpretations of family roles.
		Many older family members adapt to aging and often maintain previous activities.

Retirement benefits society because it provides younger people jobs. Even in retirement, however, **activity theory** proposes that many older people remain engaged in numerous roles and activities, including work. Moreover, those who are active adjust better to aging and are more satisfied with their lives (see Atchley and Barusch, 2004, for a summary of some of this research).

CRITICAL EVALUATION

Functionalism has several weaknesses. First, it's questionable whether some family functions are as universal or necessary as functionalists claim. Procreation often occurs outside of marriage, and the government has assumed some of the family's functions, including caring for some children (as in foster homes) and the aged (through Medicare, for example) (Lundberg and Pollak, 2007).

Second, is activity theory as representative of older people as some functionalists claim? Many people continue to work even in their 80s, but usually not by choice. They can't afford to retire, even though they're in poor health and unhappy in their low-income jobs. As health deteriorates, many older people become less active and more isolated (Lee, 2009).

Third, 10 percent of the huge wave of baby boomers say that they'll never retire, primarily because they've saved very little and carry too much debt (Harter and Agrawal, 2014). This means that there isn't always an orderly progression to retirement and opening up job slots to the younger generation.

12-5b Conflict Theory

Conflict theorists agree that families serve important functions, but point out that some groups benefit more than others. Families are sources of social inequality that mirror the larger society, and the inequities persist from birth to old age (Cruikshank, 2009).

INEQUALITY, SOCIAL CLASS, AND POWER

For conflict theorists, families in high-income brackets have the greatest share of capital, including wealth that they can pass down to the next generation. An inheritance reduces the likelihood that all families can compete for resources that include education, decent housing, and health care (see Chapter 8).

Economic inequality affects almost all aspects of family life. Child abuse and neglect are highest in U.S.

The Zimmers may be the world's oldest rock band. This British group, composed of about 50 people, has a combined age of about 3,000 years.

counties with the greatest gaps between rich and poor. Moreover, the more distance between these social classes, the less likely that the poor will receive needed services and support that reduce child maltreatment rates (Eckenrode et al., 2014).

Unequal access to resources continues into old age. Most employers insist that they value older workers' loyalty, work ethic, reliability, and experience, but many are less likely to hire or retain older people because they're usually more expensive than younger workers. Because many large companies have cut their pension plans, numerous older workers must work long after they expected to retire (Kromer and Howard, 2013; Winerip, 2013).

CRITICAL EVALUATION

Conflict theory is limited for several reasons. First, conflict theorists tend to overlook the fact that many families, especially those in the middle and upper classes, fund public assistance programs such as Medicaid (a social health care program for low-income individuals and families). They also pay billions each year to cover the costs of unintended pregnancies and the resulting infant care expenses for poor women (Sonfield and Kost, 2013).

Second, conflict theory links family inequality to capitalism and social class, but there's also considerable family inequality in countries that aren't capitalist (see Chapters 8 and 11). Thus, social class affects but doesn't determine whether a child will succeed economically.

Third, many older people don't have to struggle for resources. As you saw earlier, the U.S. government's spending on children has declined but increased for people age 65 and older. Older people, regardless of

---

activity theory proposes that many older people remain engaged in numerous roles and activities, including work.

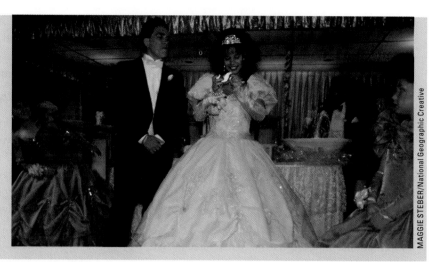

In most Latino communities, the *quinceañera* (pronounced "keen-say-ah-NYAIR-ah") is an important coming-of-age ritual that celebrates a girl's entrance into adulthood on her fifteenth birthday. It's an elaborate and dignified religious and social event that reaffirms strong ties with family, relatives, and friends. According to some critics, however, the budget—which often starts at $10,000—should be saved for college instead of "blown on a party" ("The Power of a Party," 2013: 29).

social class, have generous health care coverage through Medicare or Medicaid. They vote in large numbers, have influential lobbyists in Congress, and flood lawmakers with hundreds of thousands of letters and email messages when there's a threat of cutting Medicare or Social Security payments (Johnson, 2011; see also Chapter 11).

12-5c Feminist Theories

Feminist scholars agree with conflict theorists that there's considerable inequality between low-income and wealthy families in accessing necessary resources. However, feminist theorists emphasize the inequality of gender roles in families, especially in patriarchal societies (including the United States).

GENDER ROLES AND PATRIARCHY

For feminist scholars, families both mirror and perpetuate patriarchy and gender inequality. In most countries, males pass laws about property and inheritance rights, marriage and divorce, and many other regulations that give men authority over women. The United States is similar to most other nations in that men (particularly lawmakers and Catholic Church officials) control women's decisions about reproduction and access to abortion that, in turn, increases unwanted pregnancies and non-marital birth rates.

Male aggression against women and children is common in patriarchal societies in which men hold most of the power, status, and privilege. Females, on the other hand, are marginalized and socialized to accept male domination. In many cases, legal, political, and religious institutions don't take violence against women and children seriously (Lindsey, 2005). Recently, for instance, a judge in Florida "sentenced" a man charged with domestic battery to take his wife, the victim, to Red Lobster for

dinner and then bowling (McEwan, 2012). Thus, some judges still view domestic violence charges as frivolous.

Feminist sociologists have documented the ways that sexism, racism, and classism intersect to oppress women and children. The poorest older adults are most likely to be minority women. Out of 190 nations, the United States, Papua New Guinea, and Swaziland are the only countries that have no national paid parental leave policy. As a result, caregivers, who are predominantly women, must often leave their jobs or work only part-time to care for children and aging parents (Walsh, 2011; Fox et al., 2013).

CRITICAL EVALUATION

Feminist explanations of families and aging have several limitations. First, according to critics, feminist scholars overstate women's oppression. For example, a record 40 percent of all households with children under the age of 18 include mothers who are either the sole or primary source of family income, up from just 11 percent in 1960 (Wang et al., 2013).

A second criticism is that feminist theories tend to gloss over data which show that intimate partner violence is often mutual. Several dozen studies have found that women initiate from 30 to 73 percent of violent incidents (Straus, 2011, 2014). You'll recall that mothers mistreat their children, and almost 40 percent of the perpetrators of elder abuse are women (Jackson and Hafemeister, 2011).

Third, the poorest older adults are most likely to be minority women, but social class and marital status are important factors. Older married women and those in higher socioeconomic levels are less likely to experience poverty in later life than those who are single (never married, divorced, or widowed) and from lower socioeconomic levels (Hokayem and Heggeness, 2014; Hunter, 2014). Thus, patriarchy alone doesn't determine family gender roles, even in later life.

12-5d Symbolic Interaction

For symbolic interactionists, people create subjective meanings of what a family is and does. Thus, people learn, through interaction with others, how to act as a parent, a grandparent, a teenager, a stepchild, and so on throughout the life course.

LEARNING FAMILY AND AGING ROLES

Throughout the socialization process, family members establish trust and build emotional bonds. Between 90 and 97 percent of Americans believe that it's important for *both* mothers and fathers to provide discipline and emotional support, and to teach values ("The New American Father," 2013).

Interactionists often use exchange theory to explain mate selection and family roles. The fundamental premise of **exchange theory** is that people seek through their social interactions to maximize their rewards and minimize their costs. In mate selection, people trade their resources (e.g., wealth, good looks, youth, and/or status) for more, better, or different assets. People may stay in unhappy marriages and other intimate relationships because the rewards seem equal to the costs. Many women tolerate abuse because they fear loneliness or losing the economic benefits that a man provides (Khaw and Hardesty, 2009; see also *Table 12.1* on the benefits and costs of cohabitation).

A well-known psychologist who interviewed more than 200 couples over a 20-year period found that the difference between lasting marriages and those that split up was a "magic ratio" of 5 to 1. That is, if there are five positive interactions between partners for every negative one, the marriage is likely to be stable over time (Gottman, 1994). Thus, we can improve our family relationships by learning to interact in positive ways.

Stereotypes about older people are deeply rooted in U.S. society. About 84 percent of Americans age 60 and older report incidents of **ageism**, or discrimination against older people, including insulting jokes, disrespect, patronizing behavior, and assumptions about being frail or unhealthy (Roscigno, 2010). **Continuity theory** posits that older adults can substitute satisfying new roles for

those they've lost (Atchley and Barusch, 2004). For example, a retired music teacher can join a local chorus or orchestra. Thus, developing new roles may lessen some of the emotional distress due to ageism.

CRITICAL EVALUATION

A common criticism is that symbolic interaction, a micro-level perspective, doesn't address macro-level constraints. For example, U.S. nonmarital teen births have fallen primarily because of structural factors (e.g., greater access to family planning services, expanded educational opportunities for disadvantaged young women) rather than teen attitudes about childbearing (Kearney and Kost, 2014).

Second, exchange theory is limited because people don't always calculate the potential costs and rewards of every decision. In the case of women who care for older family members, genuine love, concern, and a sense of obligation override cost–benefit decisions, even when the person receiving care is abusive (Jackson and Hafemeister, 2013).

Third, continuity theory ignores societal obstacles that discourage older people from engaging in satisfying new roles. After retirement, many must get by on low, fixed incomes and can't afford recreational activities that they enjoyed in the past. Also, older people who experience illnesses because of low-quality health care over a lifetime may become too sick to participate in family and community activities or pursue hobbies.

exchange theory posits that people seek through their social interactions to maximize their rewards and minimize their costs.

ageism discrimination against older people.

continuity theory posits that older adults can substitute satisfying new roles for those they've lost.

STUDY TOOLS 12

READY TO STUDY? IN THE BOOK, YOU CAN:

☐ Check your understanding of what you've read with the Test Your Learning Questions provided on the chapter review card at the back of the book.

☐ Rip out the chapter review card for a handy summary of the chapter and key terms.

ONLINE AT CENGAGEBRAIN.COM YOU CAN:

☐ Collect StudyBits while you read and study the chapter.

☐ Quiz yourself on key concepts.

☐ Prepare for tests with chapter flash cards as well as flash cards that you create.

☐ Watch videos to explore sociology further.

© Anelina/Shutterstock.com

SOC
ONLINE

PREPARE FOR TESTS ON THE STUDYBOARD!

CORRECT

INCORRECT

INCORRECT

INCORRECT

Personalize Quizzes from Your StudyBits

Take Practice Quizzes by Chapter

CHAPTER QUIZZES

Chapter 1

Chapter 2

Chapter 3

Chapter 4

4LTR PRESS

Access SOC ONLINE at www.cengagebrain.com

13 | Education and Religion

LEARNING OUTCOMES

After studying this chapter, you will be able to...

13-1 Differentiate between education and schooling.

13-2 Compare and evaluate sociological perspectives on education.

13-3 Identify, describe, and illustrate some of the contemporary issues in U.S. education.

13-4 Explain how religion, religiosity, and spirituality differ.

13-5 Describe and illustrate religious organization and some of the major world religions.

13-6 Explain how and why U.S. religion is diverse, and whether or not secularization is increasing.

13-7 Compare and evaluate sociological perspectives on religion.

After you finish this chapter go to **PAGE 270** for **STUDY TOOLS.**

Education and religion, two important social institutions, teach values, shape attitudes, maintain traditions, bring people together, control behavior, and grapple with social change. The second half of this chapter examines U.S. and global religions. Let's begin with education.

13-1 EDUCATION AND SOCIETY

Education transmits attitudes, knowledge, beliefs, values, norms, and skills to a society's members. Education can be formal or informal and can occur in a variety of settings. **Schooling**, a narrower term, is formal training and instruction provided in a classroom setting.

U.S. education and schooling have undergone four significant changes since the beginning of the twentieth century: Universal education has expanded, community

> **What do you think?**
>
> I'm optimistic about getting a good job after graduating from college.
>
> 1 2 3 4 5 6 7
> strongly agree   strongly disagree

education a social institution that transmits attitudes, knowledge, beliefs, values, norms, and skills.

schooling formal training and instruction provided in a classroom setting.

colleges have flourished, public higher education has burgeoned, and student diversity has increased as more women and racial-ethnic groups enrolled at colleges and universities. As a result of these and other changes, a record number of Americans have completed high school (90 percent) and obtained a bachelor's or higher degree (34 percent) (Kena et al., 2014).

13-2 SOCIOLOGICAL PERSPECTIVES ON EDUCATION

Sociologists agree that education and schooling are important, but offer different explanations of their purpose and outcomes. *Table 13.1* summarizes the four major perspectives.

13-2a Functionalism: What Are the Benefits of Education?

For functionalists, education contributes to society's stability, solidarity, and well-being, and provides people with an opportunity for upward mobility. In their analyses, functionalists distinguish between manifest and latent functions.

MANIFEST FUNCTIONS OF EDUCATION

Some of the functions of education are *manifest*; that is, they're open, intended, and visible. For example:

▶ Schools are *socialization agencies* that teach children how to get along with others and prepare them for adult economic roles (Durkheim, 1898/1956; Parsons, 1959).

▶ Education *transmits knowledge and culture*. Schools teach skills like reading, writing, and counting; they also instill cultural values that encourage competition, achievement, and democracy (see Chapter 3).

▶ Similar values increase *cultural integration*, the social bonds that people have with each other and with the community at large, and *societal cohesion*.

▶ Education promotes *cultural innovation*. Faculty at research universities receive billions of dollars every year to develop computer technology, treatments for diseases, and programs to address social problems.

▶ Education *benefits taxpayers* because more highly educated people tend to pay higher taxes, are less likely to rely on public assistance programs, and lead

TABLE 13.1 SOCIOLOGICAL EXPLANATIONS OF EDUCATION

Theoretical Perspective	Level of Analysis	Key Points
Functionalist	Macro	Contributes to society's stability, solidarity, and cohesion and provides opportunities for upward mobility
Conflict	Macro	Reproduces and reinforces inequality and maintains a rigid social class structure
Feminist	Macro and micro	Produces inequality based on gender
Symbolic Interactionist	Micro	Teaches roles and values through everyday face-to-face interaction and behavior

healthier lifestyles, reducing state and federal health care costs (Baum et al., 2013).

Many Americans agree with functionalists that a college degree is "very important": 70 percent (up from only 36 percent in 1979) believe that a college education increases job prospects; 74 percent say that a college degree leads to a better quality of life; and in a survey of incoming first-year college students at four-year colleges and universities, 88 percent said that the most important reason for going to college is to get a good job (Newport and Busteed, 2013; Schneider, 2013; Calderon and Sorenson, 2014). Among college graduates, 74 percent said their education helped them grow intellectually (Parker et al., 2011).

Besides expanding a person's intellectual horizons, education increases earnings. On average, college graduates earn 60 to 80 percent more per year than high school graduates. Even academically marginal students who manage to get a four-year degree have earnings that are as high, on average, as those of their peers with higher grades (Oreopoulos and Petronijevic, 2013).

Over a lifetime, education affects earnings five times more than other demographic factors like sex, race, ethnicity, and age (Schramm et al., 2013; Daly and Bengali, 2014). By age 64, people with a bachelor's degree earn six times more than those with only a high school diploma (Selingo, 2013). Not surprisingly, lifetime earnings are even higher for those with advanced degrees (see *Figure 13.1*).

Regardless of race, ethnicity, family background, and marital status, people with a bachelor's degree live longer, report better physical and emotional health, and are better able to handle stress because of their economic and social position. They're also more likely than non-college graduates to hold jobs that offer a greater sense of accomplishment, more autonomy, more opportunities for creativity, and more social interaction with coworkers (Rheault and McGeeney, 2011; Oreopoulos and Petronijevic, 2013).

LATENT FUNCTIONS OF EDUCATION

Education also has *latent functions*—hidden, unstated, and sometimes unintended consequences. For example:

▶ Schools *provide child care*, particularly after-school programs, for the growing number of single-parent and two-income families.

FIGURE 13.1 EDUCATION PAYS OFF

Lifetime earnings for full-time, year-round workers (in millions of dollars)

Education Level	Earnings
Doctorate degree	$3.5
Professional degree	$4.1
Master's degree	$2.8
Bachelor's degree	$2.4
Associate's degree	$1.8
Some college	$1.6
High school graduate	$1.3
Not high school graduate	$1.1

Source: Based on Julian, 2012, Table 1.

- High schools and colleges are *matchmaking institutions* that bring together unmarried people.

- Education *decreases job competition*; the more time that young adults spend in school, the longer the jobs of older workers are safe.

- Educational institutions *create social networks* that can lead to jobs or business opportunities.

- Education is *good for business.* Thousands of companies offer services that tutor and test students and produce textbooks and related materials.

CRITICAL EVALUATION

Do functionalists exaggerate education's benefits? About 35 percent of the world's billionaires don't have a bachelor's degree, and some even dropped out of high school; 22 members of the 2013–2014 Congress have only a high school diploma (Manning, 2013; Wealth-X and UBS, 2014). Moreover, some of the highest paying jobs (e.g., radiation therapists, dental hygienists, commercial pilots, registered nurses) that will be in high demand in the coming decades require only a two-year degree or vocational training (Ogunro, 2012).

Among those with a bachelor's degree, 41 percent say that their work doesn't require a college degree (Saad, 2013). In addition, what people study affects their financial payoff. The lifetime earnings of people with bachelor's degrees vary dramatically. Earnings vary by occupation even with the same major field of study. Someone with a degree in the social sciences who works in management can earn $3.4 million versus $1.9 million with a job in education (Julian, 2012; Carnevale and Cheah, 2013). Thus, the combination of what people study in college *and* the careers they pursue after graduation can make a big difference in lifetime earnings.

Also, according to some critics, functionalists tend to gloss over education's dysfunctions, including rising college costs and student loan debt, topics we'll consider shortly. Conflict theorists, especially, argue that educational institutions produce and reproduce inequality.

13-2b Conflict Theory: Does Education Perpetuate Social Inequality?

From preschool to graduate school, conflict theorists maintain, education creates and perpetuates social inequality based on social class, race, and ethnicity. Schools are also gatekeepers that control and maintain the status quo.

SOCIAL CLASS, RACE/ETHNICITY, AND ACHIEVEMENT GAPs

An **achievement gap** is the difference in academic performance that shows up in grades, standardized test scores, and college completion rates. College enrollments have surged since 1990, especially among minority groups, but who graduates? Asian Americans have made the largest gains (see *Figure 13.2*), but there's considerable variation across subgroups. Of the 48 groups in the Asian population, 87 percent of Asian Indians have at least a bachelor's degree compared with only 21 percent of Samoans (Teranishi, 2011). Such variations, as with other minority subgroups, are due to many factors, including English language proficiency, but the best predictor of educational attainment is social class.

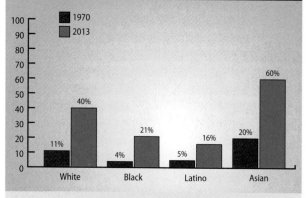

FIGURE 13.2 ATTAINMENT OF BACHELOR'S DEGREE OR HIGHER, BY RACE AND ETHNICITY, 1970 AND 2013

Note: 1970 data aren't available for American Indians/Alaska Natives and Pacific Islanders. In 2013, 17 percent and 25 percent, respectively, had a bachelor's or higher degree.

Sources: Based on U.S. Census Bureau, 2012, Table 229; Kena et al., 2014, Figure 3.

achievement gap the difference in academic performance that shows up in grades, standardized test scores, and college completion rates.

Blend Images/Ariel Skelley/Getty Images

Fifty years ago, the largest achievement gap was between blacks and whites; today it's between social classes. The gap between children from high- and low-income families has grown by about 40 percent since the 1960s (Reardon, 2011).

CUMULATIVE REASONS FOR ACHIEVEMENT GAPS

When the 2007–2009 recession began, a number of schools experienced federal, state, and local budget cuts. Five years after the recession, class sizes at many public schools increased, but staff positions—including teachers, reading and math specialists, and guidance counselors—continued to decrease. Wealthier communities, in contrast, increased property taxes to accommodate higher enrollments (Rich, 2013).

Many of the nation's schools now have a high level of "double segregation" because students are increasingly separated not only by race but also by income. Nationally, 40 percent of black students attend a high-poverty public school compared with only 6 percent of white students. Educational problems linked to poverty *and* racial segregation include less-experienced and less-qualified teachers, high teacher turnover, inadequate facilities (e.g., computers, science labs), outdated textbooks, high dropout rates, and school buildings that are falling apart (Jordan, 2014; Reich, 2014).

Poverty and inferior schooling are significant academic roadblocks for low-income students. They're less likely to attend high schools that offer high-level courses in mathematics and science. This can mean the difference between passing or failing required courses during the first few years of college. Even when top students from low-income public high schools enter selective colleges,

they often struggle to maintain passing grades. ("Selective" schools are those that admit a relatively low percentage of applicants.) Many college youths from low- and middle-income families must also work part-time or full-time and, consequently, have more stress and less time for academic work (Ross et al., 2012; Black et al., 2014).

EDUCATION AND SOCIAL CONTROL

Functionalists see education as an avenue for upward mobility. Conflict theorists maintain that education restricts mobility because of a hidden curriculum, credentialism, and privilege.

Hidden Curriculum. Every school has a formal curriculum that includes reading, writing, and learning other skills. Schools also have a **hidden curriculum**, practices that transmit nonacademic knowledge, values, attitudes, norms, and beliefs that legitimate "economic inequality and the staffing of unequal work roles" (Bowles and Gintis, 1977: 108).

Schools in low-income and working-class neighborhoods tend to stress obedience, following directions, and punctuality so that students can fill low-paid jobs (e.g., restaurants, nursing homes, hospitals) that require these characteristics (Kozol, 2005). Schools in middle-class neighborhoods emphasize proper behavior and appearance, cooperation, conforming to rules, and deference to authority because many of these students will go to college and work in bureaucracies that require such attributes (Hedges, 2011). In contrast, selective schools encourage leadership, creativity, independence, and people skills—all prized characteristics in elite circles (Persell and Cookson, 1985; see also Chapter 8). In effect, then, the hidden curriculum reproduces the existing class structure and provides workers for jobs and occupations in the stratification hierarchy.

Credentialism. Have you noticed that many faculty, doctors', lawyers', and dentists' offices are usually wall-papered with framed degrees? Such tangible symbols of people's achievements reflect **credentialism**, an emphasis on certificates or degrees to show that people have certain skills, educational attainment levels, or job qualifications.

hidden curriculum school practices that transmit nonacademic knowledge, values, attitudes, norms, and beliefs.

credentialism an emphasis on certificates or degrees to show that people have certain skills, educational attainment levels, or job qualifications.

Functionalists maintain that credentialism rewards people for their accomplishments, sorts out those who are the most qualified for jobs, and stimulates upward social mobility. Conflict theorists contend, however, that for many positions, people can gain skills on the job with a few weeks of training or succeed because of ability or other factors.

Because of a large supply of high school graduates, employers can demand higher levels of education even though some jobs (e.g., retail sales, law enforcement) don't require a college degree for competent performance, a process called *credential inflation*. As more people obtain a college degree, its value diminishes, and students from low-income families, who are the least likely to have access to a college education, fall further behind (Bollag, 2007).

Privilege. According to one observer, "We're the only rich nation to spend less educating poor kids than we do educating kids from wealthy families" (Reich, 2014). Many selective colleges claim that they're committed to admitting talented low-income and minority students. In fact, in 2010–2011, colleges and universities awarded $5.3 billion in financial aid to students from high-income families. Among students with a GPA of 3.5 or higher on a 4.0 scale, white students received more than three times as much in merit-based scholarships as minority students

(Baum and Payea, 2011; Kantrowitz, 2011). Among students with similar high school GPAs and SAT scores, since 1995 the proportion of high-income students receiving scholarships from colleges, the federal government, or the states has increased but has fallen for low-income students (Burd, 2013; Carey, 2013; Mettler, 2014).

Another privileged group is *legacies*, the children of alumni who have "reserved seats" regardless of their accomplishments or ability. For example, President George W. Bush—who had mediocre high school grades and standardized test scores—was admitted as a legacy at Yale University, which his father and grandfather had attended (Golden, 2006). Legacies also include students with inferior academic records whose parents, including celebrities, make million-dollar donations. Among selective colleges and universities, almost 75 percent use legacies, which account for up to 30 percent of some universities' student body (Massey, 2007; Hurwitz, 2011).

CRITICAL EVALUATION

Conflict theories have several weaknesses. First, from 1992 to 2012, racial-ethnic diversity increased about 56 percent at community colleges and four-year public and private colleges (Zweifler, 2013). Attaining a bachelor's degree or higher has also risen for minorities (see *Figure 13.2* on page 249). Such data show that racial-ethnic inequality isn't as fixed as many conflict theorists assert.

Second, a hidden curriculum doesn't necessarily determine job placement because many students aren't passive recipients of educational systems. Below the bachelor's degree level, for example, nearly 25 million U.S. adults have acquired professional certificates and licenses in wide-ranging fields (e.g., nursing, computer and information sciences, business management). These and other "alternative credentials" have labor market value that increases employment and earnings (Ewert and Kominski, 2014). Conflict theorists may denounce credentialism and credential inflation, but employers often view credentials as useful initial screening tools that signal job candidates' discipline and drive.

Third, do conflict theorists overstate the importance of social class achievement gaps? A study tracked more than 5,200 Asian American and white students from kindergarten through high school and found that the former, regardless of social class, performed better than their white counterparts. The researchers attributed the Asian American students' greater success to cultural beliefs that achievement is learned rather than innate, parental pressures to succeed, and, especially among recent immigrants, ethnic community resources such as private tutoring and vital information about navigating the education system (Hsin and Xie 2014; see also Chapter 10).

Universal Pictures/Fotos International/Getty Images

How important is graduating from a prestigious college or university? Nationally, 84 percent of business leaders say that the amount of knowledge a job candidate has in a particular field is "very important." Only 9 percent say that where a person attended school is very important (Calderon and Sidhu, 2014). Steven Spielberg, the well-known movie producer and director, graduated from California State University at Long Beach after being rejected by the more prestigious University of Southern California and University of California at Los Angeles film schools.

13-2c Feminist Theories: Is There a Gender Gap in Education?

Since 1982, women have been graduating from college at higher rates than men. Some view this trend as an indicator of greater gender equality; others are alarmed about the supposed "male crisis" in and "feminization" of higher education (Mullen, 2012; see also Chapter 9). Has women's progress come at the cost of men? Or are there still gender gaps, particularly in higher education?

WHO'S GETTING DEGREES

Almost equal numbers of girls and boys are high school graduates. Women earn more associate's, bachelor's, and master's degrees than men, but their percentage of professional and doctoral degrees drops considerably (see *Figure 13.3*).

Across all racial-ethnic groups, students from affluent families are the most likely to earn a bachelor's degree, but men are more likely to do so than women (Mullen, 2012). In 2012, Asian females (50 percent) and Asian males (54 percent) had the highest rates of attaining a bachelor's degree or higher, followed by white males (36 percent) and white females (34 percent). In contrast, black women and Latinas had higher rates than their male counterparts (Snyder and Dillow, 2013; Kena et al., 2014). Such data contradict the description of higher education as feminized, but why do women achieve more degrees than men up to the master's level?

In both high school and college, women spend more time than men studying, earn better grades, hold more leadership positions, and are more involved in student clubs and community volunteer work. Women (81 percent) are also more likely than men (67 percent) to value higher education (Goldin et al., 2006). This may be one reason why women are less likely to drop out of college and enter the labor force. Regardless of the reasons for women's success, many colleges have been giving men preferential treatment in admissions to avoid large gender imbalances in their student bodies (Gewertz, 2009; Kahlenberg, 2010; Dwyer et al., 2013).

MAJORS AND CAREERS

A major gender gap is women's underrepresentation in the high-paying fields of science, technology, engineering, and mathematics (STEM). In elementary, middle, and high school, girls and boys take math and science courses in roughly equal numbers, and about the same numbers leave high school planning to pursue STEM majors in college. After a few years, however, men outnumber women in nearly every STEM field, and in some—mathematics, engineering, and computer science—women earn only 10 to 20 percent of the bachelor's degrees. Their numbers decline further at the graduate level and yet again in the workplace (Hill et al., 2010; Landivar, 2013; Snyder and Dillow, 2013).

Why is there a gradual attrition? In college, females are initially as persistent as men in a STEM major and earn

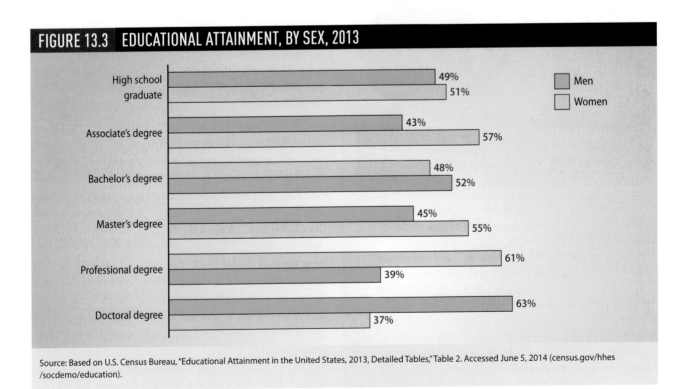

FIGURE 13.3 EDUCATIONAL ATTAINMENT, BY SEX, 2013

High school graduate — Men 49%, Women 51%
Associate's degree — Men 43%, Women 57%
Bachelor's degree — Men 48%, Women 52%
Master's degree — Men 45%, Women 55%
Professional degree — Women 61%, Men 39%
Doctoral degree — Women 63%, Men 37%

Source: Based on U.S. Census Bureau, "Educational Attainment in the United States, 2013, Detailed Tables," Table 2. Accessed June 5, 2014 (census.gov/hhes /socdemo/education).

higher grades. However, they're less satisfied than men with the core courses and more likely to doubt their ability to succeed in a male-dominated discipline. As a result, women's self-confidence falters and they change majors (Shapiro and Williams, 2012; Jagacinski, 2013). The exit from a STEM major is also associated with having few female faculty role models, and some science professors' beliefs that female students won't benefit from mentoring because they're less competent than men (Moss-Racusin et al., 2012; Williams and Ceci, 2012).

CRITICAL EVALUATION

A common criticism is that feminist scholars address women's education barriers but not their choices. Socialization, gender stereotypes, and teachers' expectations affect our behavior (see Chapters 4 and 9), but it's still not clear why many women choose fields of study that they know are on the lower end of the pay scale (e.g., health, education).

Some fault feminist theorists, who support gender equality, for being more interested in women's than men's gender gaps, especially at the associate's, bachelor's, and master's levels (Baenninger, 2011). There's also the question of why feminist scholars devote little attention to issues such as why boys are more disruptive in elementary school, which leads to suspensions and decreases the chance of attending college by at least 16 percent for each suspension; whether the general public's image of men as underachievers affects men's motivation; and

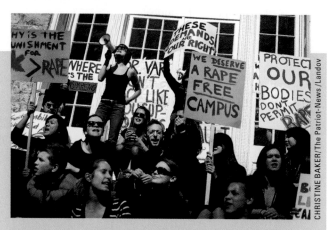

Over a five-year college career, nearly 20 percent of female students experience rape or other sexual assault on campus. About 6 percent of college men rape or attempt to rape women. Of this group, 63 percent are repeat offenders, committing an average of 6 rapes (Lisak and Miller, 2002; Krebs et al., 2007). In mid-2014, the U.S. Department of Education announced that it would investigate 55 universities for possible violations in handling sexual assault accusations.

why parents are more likely to pay for their daughters' (40 percent) than their sons' (29 percent) college education (Bertrand and Pan, 2011; Wang and Parker, 2011).

13-2d Symbolic Interaction: How Do Social Contexts Affect Education?

None of us is born a student or a teacher. Instead, these roles, like others, are socially constructed (see Chapters 3–5). For symbolic interactionists, education is an active *process* that includes students, teachers, peers, and parents and involves tracking, labeling, and student engagement.

TRACKING

Beginning in kindergarten, practically all schools sort students by aptitude. Such sorting results in **tracking** (also called *streaming* or *ability grouping*), assigning students to specific educational programs and classes on the basis of test scores, previous grades, or perceived ability.

Some educators believe that tracking is beneficial because students learn better in groups with others like themselves, and it allows teachers to develop curricula for students with similar ability. Many interactionists maintain, however, that tracking creates and reinforces inequality.

▶ High-track students take classes that involve critical thinking, problem solving, and creativity that high-status occupations require. Low-track students take classes that are limited to simple skills (e.g., punctuality and conformity) that usually characterize lower status jobs.

▶ High-track students have more homework, better quality instruction, and more enthusiastic teachers. One result is that high-track students are more likely to see themselves as "bright," whereas low-track students see themselves as "dumb" or "slow."

▶ The effects of tracking are usually cumulative and lasting. Teachers tend to have low expectations for low-track students, who therefore fall further behind every year in reading, mathematics, and interaction skills (Oakes, 1985; Hanushek and Woessman, 2005).

Middle school and high school become even more stratified as high-track students are sorted into gifted, honors, and advanced courses. In college, students continue to be tracked and sorted into honors programs and accelerated undergraduate courses.

tracking (also called *streaming* or *ability grouping*) assigning students to specific educational programs and classes on the basis of test scores, previous grades, or perceived ability.

LABELING

Tracking often leads to labeling, a serious problem because "there's a widespread culture of disbelief in the learning capacities of many of our children, especially children of color and the economically disadvantaged" (Howard, 2003: 83). Labeling, in turn, can result in a *self-fulfilling prophecy*. That is, students live up or down to teachers' expectations and evaluations that are influenced by a student's social class, skin color, hygiene, accent, and test scores (see Chapter 5).

Parents affect their children's academic outcomes through conscious or unintentional labeling. Only half of parents with annual incomes of less than $25,000 expect their child to attain a four-year college degree compared with 88 percent of parents with incomes over $75,000. Because higher-income parents expect their children to graduate from college, they talk about what's going on in the classroom, provide out-of-school learning opportunities, foster positive attitudes toward school, and applaud academic achievement. As a result, the children, regardless of innate abilities, internalize personal goals that include earning a college degree (Kagan, 2011; Child Trends Data Bank, 2012).

iStockphoto.com/mevans

STUDENT ENGAGEMENT

Many schools assess performance not only through tests but also in terms of *student engagement*, how involved students are in their own learning. A team of researchers who spent thousands of hours in more than 2,500 elementary school classrooms concluded that the typical U.S. child has only a 1 in 14 chance of being in a school that encourages her or his engagement. Because of standardized tests, teachers in public elementary schools spend most of the day on basic reading and math drills and little time on problem solving, reasoning, science, and social studies. As a result, many classrooms are often "dull and bleak places" where kids don't get much teacher feedback or face-to-face interaction with their peers (Pianta et al., 2007).

U.S. high school students aren't as engaged in their education as they could be: 26 percent admit that they usually don't do their homework, and 10 percent drop out. The students who are most likely to be disengaged are from low-income families, are African American or Latino, don't live with both biological parents, attend financially strapped urban public high schools that are typically overcrowded and understaffed, and have fewer resources such as computers and even textbooks (Finn, 2006; Planty et al., 2008).

In college, the typical undergraduate spends more time partying, socializing, and working for pay than studying. On average, even full-time college students study only

14 hours a week compared with 24 hours a week in 1961 (Babcock and Marks, 2011). This is well below the 24 to 30 hours faculty members say students should be spending on class preparation if they're taking three courses.

According to one college instructor, "Education is the only business in which the clients want the least for their money" (Perlmutter, 2001: B1). However, some analysts also blame professors for students' not studying. Many faculty have watered down their required readings, say nothing when students don't prepare for classes or do assignments, and often give easy exams to avoid complaints about low grades. Only 55 percent of first-year college students and 61 percent of seniors say their courses "challenged them to do their best work" (National Survey of Student Engagement, 2013: 9). When faculty dilute their course requirements and tests, some students don't study because they're bored by courses that don't stimulate them (Benton, 2011; Glenn, 2011).

CRITICAL EVALUATION

One weakness is that interactionists gloss over the power of individuals to change the course of events, including overcoming the effects of labeling. For example, a number of low-income parents, despite language barriers among recent immigrants, contact school personnel on a regular basis. Others go to college open houses to speak to college representatives, with their children translating the questions and answers (Tornatzky et al., 2002).

Another limitation is that interactionists, because of their micro-level analysis, neglect the macro-level structural constraints that are built into society. Black and Latino male students enter community colleges with higher aspirations than their white peers, but white males are six times more likely to graduate in 3 years with a certificate or degree. A major reason for the difference is that minority students tend to enter college with weaker academic skills because many have attended underfunded and understaffed high schools (Center for Community College Student Engagement, 2014; see also Rios, 2012).

13-3 SOME CURRENT ISSUES IN U.S. EDUCATION

In an open-ended question about "the most important problem facing this country today," only 5 percent of Americans named education (Riffkin, 2014). A dysfunctional government and weak economy ranked the highest, but the U.S. education system has problems at all levels, some more controversial than others.

13-3a Elementary Schools

Advanced economies, including ours, rely not on physical labor but on "cognitive labor" that requires formal analytical abilities, written communications, and specific technical knowledge—skills that people acquire and cultivate beginning in preschool (Autor, 2014). Three of the ongoing issues regarding elementary education are low test scores, achievement gaps, and effective teaching.

LOW TEST SCORES AND ACHIEVEMENT GAPS

Children who read proficiently by the end of the third grade are more likely to do well in other subjects, including mathematics, and to graduate from high school. Between 1992 and 2013, the average reading and mathematics scores of fourth- and eighth-grade students increased, but there are considerable achievement gaps by race and ethnicity. Asian and white children have the highest proficiency rates in both reading and math, but only about a third of U.S. students are proficient in both subjects (see *Table 13.2*).

Social class explains some of the variations. In 2012, 75 percent of 3- to 5-year-olds whose parents had either a graduate or professional degree, compared with 60 percent of those whose parents had only a high school diploma, were enrolled in kindergarten, preschool, and nursery school programs that provided educational experiences (Kena et al., 2014). In 2013, 80 percent of fourth graders in low-income families were below proficiency in reading compared with 49 percent of higher-income children (National Center for Education Statistics, 2013).

Social class alone doesn't explain student performance, however. Homework also affects test scores. Many elementary and middle school children aren't getting or doing their homework, but parents and the public often blame teachers for the students' poor academic performance (Loveless, 2014).

EFFECTIVE TEACHING

A multi-year project concluded that the most accurate way to evaluate elementary school teachers is to use a three-pronged approach that includes student test scores, multiple classroom observations, and student evaluations (Cantrell and Kane, 2013), but 40 states use only student test scores to evaluate teachers. Much research shows that test scores are invalid and unreliable measures of teacher effectiveness. Other factors affect test scores, including a teacher's quality of undergraduate education, years of experience, being assigned to low-track or high-track classrooms, whether or not students study, and school funding (Kalogrides et al., 2013; Amrein-Beardsley, 2014; Polikoff and Porter, 2014).

13-3b High Schools

High schools, like elementary schools, must grapple with low achievement scores and effective teaching. In addition, the SAT is becoming an increasingly controversial issue.

INTERNATIONAL TEST SCORES

Among 65 countries, U.S. 15-year-olds rank 30th in math, 23rd in science, and 20th in reading. These rankings have slipped since 2009, and many other countries' best students outperform ours. Of 32 nations assessed on computer-based mathematics literacy, the United States comes in only 13th (Banchero, 2013; Kelly et al., 2013; OECD, 2013).

The United States is one of the wealthiest countries in the world; spends more per student from the ages of 6 to 15 than 30 other industrialized countries; ranks above average in the share of high-SES high school students; and has fewer non-English speaking students than other Western countries, including Canada (Carnoy and Rothstein, 2013; OECD, 2013). Why, then, are U.S. students doing so poorly? A major difference between the United States and other developed nations is teacher preparation and quality.

TABLE 13.2	READING AND MATHEMATICS PROFICIENCY, BY RACE AND ETHNICITY, 2013					
	National Average	African American	American Indian	Latino	White	Asian and Pacific Islander
Fourth graders who scored at or above proficient in reading	34%	17%	22%	19%	45%	51%
Eighth graders who scored at or above proficient in math	34%	14%	21%	21%	44%	60%

Note: "Proficient" indicates the ability to handle challenging subject matter.

Source: Based on Annie E. Casey Foundation, 2014, Table 1.

About 7 percent of Americans ages 16 to 24 are high school dropouts, down from 12 percent in 1990. The dropout rate is slightly higher for males than females, and highest among foreign-born Latinos and youth whose families are in the bottom 25 percent of all family incomes (Kena et al., 2014).

Hola Images/Getty Images

TEACHER PREPARATION AND QUALITY

American teachers spend, on average, 20 to 50 percent more hours teaching in class than do their Australian, Canadian, and Finnish counterparts, all of whose students outrank Americans in international tests (Lawrence, 2013). A combination of structural factors helps explain why, despite their individual efforts, many U.S. teachers aren't as effective as they could be.

First, the top-performing countries accept only the top applicants for education programs. Finland, Singapore, and South Korea recruit the top 5 to 30 percent of high school graduates, but accept only 10 percent who have high standardized test scores in science, math, and reading (Auguste et al., 2010; Sawchuck, 2012). Several recent reports described U.S. teacher preparation programs as "an industry of mediocrity" that has low or no academic entry standards, easy coursework, and gives out many easy A's (Greenberg et al., 2013; Putnam et al., 2014). In top-performing countries, 100 percent of teachers are in the top third of their college graduating classes. In the United States, 47 percent of kindergarten through twelfth-grade teachers come from the bottom third. "In other words, we hire lots of our lowest performers to teach, and then we scream when our kids don't excel" (Cloud, 2010: 48).

Second, compared with teachers in top-performing countries, a higher number of U.S. teachers don't have a bachelor's degree or higher in the subjects they teach. For example, almost 30 percent of public high school teachers who teach math didn't major in the field, and 54 percent who teach chemistry didn't major in the subject (Hill, 2011).

Third, the top-performing countries offer teachers competitive salaries. In South Korea and Singapore, teachers on average earn more than lawyers and engineers. In these countries—Finland and many others—teachers are highly regarded and enjoy the same prestige as physicians

and other high-status professions and have lifelong careers, resulting in very low teacher attrition. In contrast, 40 to 50 percent of new U.S. teachers leave the profession within their first 5 years on the job. High turnover rates are especially harmful to students in low-performing schools: Replacing a teacher—especially a good one—is costly, disruptive, and demoralizing for both students and parents (OECD, 2011; Ronfeldt et al., 2011; Kafka et al., 2014).

THE GROWING CONTROVERSY OVER SATs, ACTs, AND AP COURSES

In 2013, about 3.5 million high school students took either the SAT or the ACT, standardized college entrance examinations. The administrators/owners of both exams say that the scores can indicate where students are falling behind in college readiness, measure students' capability to do college-level work, and even predict college success (College Board, 2013; ACT, Inc., 2014).

A growing number of critics, including college officials, contend that the tests are little more than gatekeeping tools that exclude lower socioeconomic students from higher education, don't predict college success, and that students who submit test scores and those who don't perform similarly in college. As a result, 850 four-year colleges don't require applicants to submit SAT or ACT scores (FairTest, 2013; Botstein, 2014; Fraire, 2014; Hiss and Franks, 2014).

Advanced Placement (AP) offers high school students college-level curricula and exams. About 78 percent of degree-granting colleges and universities give course credit to students who get high scores on the exams (Berrett, 2014). In 2013, 30 percent of high school students took at least one AP exam. Of those who took the tests, 27 percent were low income, and only 48 percent of this group scored high enough to get college credit (College Board, 2014).

Like the ACT and SAT, AP is becoming more controversial as critics point to a number of problems. First, AP courses are rarely as demanding as college courses and don't offer the same content. Second, to increase a high school's prestige, superintendents and principals may pressure teachers to offer AP courses and students to take them. Both groups may be unprepared to do so, and failure can make students and teachers feel inferior. Finally, only about 45 percent of the nation's public high schools offer AP courses, and those that do so are typically in high-income and predominantly white neighborhoods (Jackson, 2012; Simon, 2013; Berrett, 2014).

13-3c Colleges and Universities

Public postsecondary institutions often feel embattled by many constituencies and external pressures. Many colleges and universities are grappling with affirmative action issues, grade inflation and cheating, low graduation rates, and rising student loan debt.

AFFIRMATIVE ACTION

In higher education, *affirmative action* refers to admission policies that provide equal access for groups, particularly women and minorities, that have been historically excluded or underrepresented. The enrollment rates of these groups has increased steadily since the 1970s, but using race and ethnicity in college admission decisions has been, and continues to be, a hotly debated topic (see "The Affirmative Action Debate in Higher Education").

In 2006, Michigan voters approved an amendment to the state's constitution to ban using race as a factor in deciding who's admitted to the state's public universities. In 2014, the U.S. Supreme Court (*Schuette v. Coalition to Defend Affirmation Action*) ruled that Michigan's ban was constitutional. The court's decision legitimated similar laws in seven other states (Schmidt, 2014). Affirmative action opponents praised the decision for reinforcing **meritocracy**, a system that rewards people because of their individual accomplishments. We say things like "work hard and you'll get ahead" and "pull yourself up by your bootstraps," but is that what happens? As you've seen, the best predictor of who goes to college is not ability but income.

GRADE INFLATION AND CHEATING

Grade inflation is widespread across all education levels, and some children start cheating on tests as early as the third grade. The consequences of grade inflation and cheating in postsecondary education are more serious, however, because someone who cheated her or his way through college or graduate school may be your dentist, lawyer, doctor, or tax accountant.

Grade Inflation. Has an A replaced the C of the 1970s? In 1971, only 26 percent of first-year college students

WOULD YOU WANT A MILITARY OFFICER, A FINANCIAL PLANNER, A DENTIST, OR A DOCTOR WHO CHEATED HER OR HIS WAY THROUGH COLLEGE OR GRADUATE SCHOOL?

expected to earn at least a B average in college, compared with 52 percent in 1995, and a whopping 70 percent in 2010 (Pryor, 2011), and even though students are studying less than in the past.

Many faculty give high grades because it decreases student complaints, involves less time and thought in grading exams and papers, and reduces the chances of students' challenging a grade. Some faculty believe that they can get favorable course evaluations from students by handing out high grades, and others accept students' view of high grades as a reward for simply showing up in class. Inflating grades decreases student attrition and satisfies administrators, especially when state legislators base funding on graduation rates (Bartlett and Wasley, 2008; Gillespie, 2014).

Cheating. Despite grade inflation, cheating is common in college and graduate school. Recently, New York

meritocracy a system that rewards people because of their individual accomplishments.

The Affirmative Action Debate in Higher Education

Supporters argue that affirmative action programs...

▶ give disadvantaged and underrepresented students a needed boost in overcoming past and current discrimination

▶ have doubled or tripled the number of minority applications, resulting in student populations that are more representative of the surrounding community

▶ have increased minorities' upward social and economic mobility

▶ promote diversity and encourage people to work together effectively in a multicultural society

Opponents argue that affirmative action programs...

▶ are discriminatory because they're based on race rather than academic achievement

▶ benefit primarily middle- and upper-class minorities

▶ are condescending because they imply that minorities can't succeed without preferential treatment

▶ admit students who often can't live up to an institution's academic standards, resulting in high failure rates

What else would you add to either side of the debate?

Sources: Messerli, 2012; Krishnamurthy and Edlin, 2014.

Keith Brofsky/Getty Images

authorities charged 20 students from an affluent Long Island community with paying impersonators up to $3,600 per SAT to increase their chances of getting into a selective school (Winter, 2011). Between 50 and 75 percent of college students admit having cheated at least once on tests and assignments. Many studies have uncovered cheating at the Air Force Academy and in many graduate programs, including business administration and dentistry (McCabe et al., 2006; Frosch, 2007; Wasley, 2007; Barrett, 2014).

Cheating is prevalent because most students aren't caught. If caught, the sanctions are usually mild. In 2013, for example, Harvard University found that 125 undergraduates cheated on a government take-home exam. More than half had to withdraw for a few semesters, some received probation, and others weren't punished at all (Levitz, 2013). Despite widespread grade inflation and cheating, millions of students don't graduate from college.

GRADUATION RATES

The United States has one of the lowest college graduation rates in the industrialized world. In 2012, only 59 percent of full-time students at four-year colleges completed a bachelor's degree within 6 years, and 31 percent at 2-year colleges earned a certificate or associate's degree within 3 years (OECD, 2013; Kena et al., 2014).

Once on campus, 36 percent must take one or more remedial courses and, as you saw earlier, many students admit that they don't study very much. Dropout rates are higher for males than females, part-time than full-time students, and students from lower-income families. At four-year colleges, the more selective the institution, the higher the graduation rate: 33 percent at colleges with open admissions policies, 60 percent at colleges that accept 50 percent or more of applicants, and 86 percent at colleges that accept only 25 percent of applicants (Kena et al., 2014).

Family and work commitments can reduce the likelihood of completing college. Fully three-quarters of undergraduates are managing some combination of family responsibilities, jobs, and commuting to class (Princiotta et al., 2014). Students also drop out because they get sick, accept attractive job offers, or decide that college isn't for them, but cost is a major reason for attrition.

STUDENT LOAN DEBT

Student loan debt has almost tripled since 2004, is now more than $1 trillion, and is second only to mortgage debt (Lee, 2013). About 71 percent of students who graduated from four-year colleges in 2012 had an average debt of almost $30,000, up from $9,500 in 1993. Those at private nonprofit colleges and for-profit colleges (e.g., University of Phoenix) owed much more—$32,300 and almost $40,000, respectively (Reed and Cochrane, 2013; Institute for College Access & Success, 2014).

There's no simple answer for the rising student loan debt because much depends on factors like an institution's endowment, in-state and out-of-state tuition prices, and the growing number and costs of full-time administrators (*Chronicle of Higher Education*, 2014). However, a major reason for the rise is that, since 2000, public spending on higher education has dropped 30 percent even though enrollment at public colleges has jumped 34 percent (Kena et al., 2014). As state funding has decreased, tuition has increased, and so has borrowing to cover tuition, fees, and, for many students, the costs for room and board (see *Figure 13.4*).

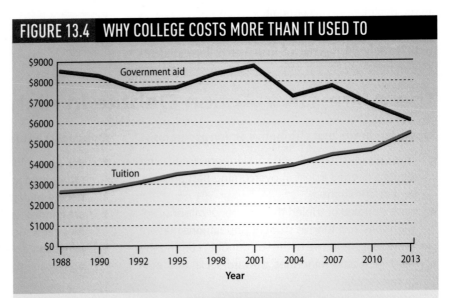

FIGURE 13.4 WHY COLLEGE COSTS MORE THAN IT USED TO

Note: These are public higher-education revenues per student, in 2013 dollars.

Source: Based on State Higher Education Executive Officers, 2014, Figure 3.

YOU BET THEY'LL BE ON THE FINAL.

cartoonistgroup.com
©2013 Grimmy, Inc. grimmy.com
Distributed by King Features Inc. 2/14

Mike Peters/Cartoonist Group

13-4 RELIGION AND SOCIETY

Like education, religion has a profound effect on society. Some form of religion exists in all societies and cultures. Throughout history, religion has been a central part of the human experience, affecting the family, economy, and other social institutions. Why is religion important to many people? And does it always benefit society? You'll see that the four sociological perspectives answer these and other questions differently.

13-4a What Is Religion?

Religion is a social institution that involves shared beliefs, values, and practices related to the supernatural. It unites believers into a community, but customs and practices differ across cultures and groups. For example, being Catholic requires confessing one's sins to a priest, whereas silent confessions directly to God are more common among Protestants, Jews, Muslims, and other religious groups.

Émile Durkheim (1961) distinguished between the sacred and the profane. **Sacred** refers to anything that people see as awe-inspiring, supernatural, holy, and not part of the physical world. In contrast, **profane** refers to the ordinary and everyday elements of life that aren't related to religion. Sociologists differentiate religion from religiosity and spirituality.

13-4b Religion, Religiosity, and Spirituality

Religion is a belief system, but religious expression can vary. When sociologists examine **religiosity**, the ways people demonstrate their religious beliefs, they find that religion and religiosity differ. For example, 58 percent of Americans say that religion is "very important" in their lives, but only 37 percent attend worship services once a week or more (Lugo et al., 2012).

Spirituality is a personal quest to feel connected to a reality greater than oneself. About 37 percent of Americans describe themselves as "spiritual but not religious." Within this group, however, 44 percent pray at least once a day, and 55 percent are "absolutely certain" that there's a God (Lugo et al., 2012). Thus, religious people are spiritual, but spiritual people aren't necessarily religious.

13-5 RELIGIOUS ORGANIZATION AND MAJOR WORLD RELIGIONS

People manifest their religious beliefs most commonly through organized groups, including cults, sects, denominations, and churches. The major world religions differ in their membership and beliefs.

13-5a Cults (New Religious Movements)

A **cult** is a religious group that is devoted to beliefs and practices that are outside of those accepted in

---

religion a social institution that involves shared beliefs, values, and practices related to the supernatural.

sacred anything that people see as awe-inspiring, supernatural, holy, and not part of the natural world.

profane the ordinary and everyday elements of life that aren't related to religion.

religiosity the ways people demonstrate their religious beliefs.

cult a religious group that is devoted to beliefs and practices that are outside of those accepted in mainstream society.

Hare Krishna is the popular name for the International Society of Krishna Consciousness (ISKCON), a new religious movement based in Hinduism. Established in the United States in 1965, its practices include an austere and simple life, vegetarianism, abstinence from drugs and alcohol, chanting, and evangelism.

mainstream society. Many sociologists use **new religious movement (NRM)** rather than *cult* because the media have used the latter term in derogatory ways to describe any unfamiliar, new, or seemingly bizarre religious group (Roberts, 2004).

NRMs usually organize around a **charismatic leader** (like Jesus) whom followers see as having exceptional or superhuman powers and qualities (see Chapter 11). Some NRMs have become established religions. The early Christians were a renegade group that broke away from Judaism, and Islam, the world's second largest religion, began as a cult around Muhammad. Most contemporary cults are fragmentary, loosely organized, and temporary, but others (e.g., the Church of Scientology) have developed into lasting and highly bureaucratic international organizations.

new religious movement (NRM) term used instead of *cult* by most sociologists.

charismatic leader someone that followers see as having exceptional or superhuman powers and qualities.

sect a religious group that has broken away from an established religion.

denomination a subgroup within a religion that shares its name and traditions and is generally on good terms with the main group.

church a large established religious group that has strong ties to mainstream society.

13-5b Sects

A **sect** is a religious group that has broken away from an established religion. People who begin sects are usually dissatisfied members who believe that the parent religion has become too secular and has abandoned key original doctrines. Like cults, some sects are small and disappear after a time, whereas others become established and persist. Examples of sects that have persisted include the Amish, the Jewish Hassidim, Jehovah's Witnesses, Quakers, and Seventh-Day Adventists (Bainbridge, 1997). Some sects develop into denominations.

13-5c Denominations

A **denomination** is a subgroup within a religion that shares its name and traditions and is generally on good terms with the main group. Denominations can form slowly or develop rapidly, depending on factors like geography, immigration, and a country's birth rate. Some scholars describe a denomination as somewhere between a sect and a church. Like sects, denominations have a professional ministry. Unlike sects, denominations view other religious groups as valid and don't make claims that only they possess the truth (Hamilton, 2001).

Denominations typically accommodate themselves to the larger society instead of trying to dominate or change it. As a result, people may belong to the same denomination as did their grandparents or even great-grandparents. Denominations exist in all religions, including Christianity, Judaism, and Islam. In the United States, the many Protestant denominations include Episcopalians, Baptists, Lutherans, Methodists, and Evangelicals.

13-5d Churches

A **church** is a large established religious group that has strong ties to mainstream society. Because leadership is attached to an office rather than a specific leader, new generations of believers replace previous ones, and members follow tradition or authority rather than a charismatic leader. As in a denomination, people are usually born into a church, but may later decide to leave it.

Churches (e.g., Roman Catholic Church, Anglican Church in North America) are typically bureaucratically organized, have formal worship services and trained clergy, and often maintain some degree of control over political or educational institutions (see Chapter 11). Because churches are an integral part of the social order, they often become dependent on, rather than critical of, the ruling classes (Hamilton, 1995).

13-5e Some Major World Religions

Worldwide, the largest religious group is Christians, followed by Muslims. If the world's population is represented as an imaginary village of 100 people, it has about:

▶ 32 Christians

▶ 23 Muslims

▶ 16 Unaffiliated (people who may or may not believe in God but don't identify with any particular religious group)

▶ 15 Hindus

▶ 7 Buddhists

▶ 6 Folk religionists (includes followers of African traditional religions, Chinese folk traditions, Native American religions, and Australian aboriginal religions)

▶ 1 Other (includes Jews, Baha'is, Sikhs, Jains, Shintoists, and many others) (based on Pew Research Center, 2014)

Note that no religious group comes close to being a global majority, that the third largest group is religiously unaffiliated, and that non-Christians outnumber Christians 2 to 1. Five religious groups in particular have had a worldwide impact on economic, political, and social issues. *Table 13.3* provides a brief overview of these groups.

TABLE 13.3 CHARACTERISTICS OF FIVE MAJOR WORLD RELIGIONS

Religion		Date of Origin	Founder	Number of Followers
Christianity	iStockphoto.com/Onfokus	0 C.E.	Jesus Christ	2.2 billion
	Jesus, the son of God, sacrificed his life to redeem humankind. Those who follow Christ's teachings will enter the Kingdom of Heaven. Sinners who don't repent will burn in hell for eternity.			
Islam	iStockphoto.com/kickstand	600 C.E.	Muhammad	1.6 billion
	God is creator of the universe, omnipotent, omniscient, just, forgiving, and merciful. Those who sincerely repent and submit (the literal meaning of Islam) to God will attain salvation; the wicked will burn in hell.			
Hinduism	iStockphoto.com/Ferenc Cegledi	Between 4000 and 1500 B.C.E.	No specific founder	887 million
	Life in all its forms is an aspect of the divine. The aim of every Hindu is to use pure acts, thoughts, and devotion to escape a cycle of birth and rebirth (samsara) determined by the purity or impurity of past deeds (karma).			
Buddhism	iStockphoto.com/Pablo Salgado Barrientos	525 B.C.E.	Siddhartha Gautama	386 million
	Life is misery and decay with no ultimate reality. Meditation and good deeds will end the cycle of endless birth and rebirth, and the person will achieve nirvana, a state of liberation and bliss.			
Judaism	iStockphoto.com/Howard Sandler	2000 B.C.E.	Abraham	15 million
	God is the creator and the absolute ruler of the universe. By obeying the divine law God gave them, Jews bear special witness to God's mercy and justice.			

Note: C.E. (Common Era) is the nondenominational abbreviation for A.D. (Anno Domini, Latin for "In the year of our Lord") and B.C.E. (Before the Common Era) is the nondenominational abbreviation for B.C. (Before Christ).

Sources: Based on a number of sources including the Center for the Study of Global Christianity, 2007; "Religions of the World…," 2007; and "Global Christianity…," 2011.

13-6 RELIGION IN THE UNITED STATES

Among U.S. adults, 90 percent believe in God, 8 percent don't, and 2 percent don't know or don't care (Lugo et al., 2013). For sociologists, religiosity is a better measure of being religious than simply asking people whether they believe in God (or a universal spirit) and which religion they follow. Religiosity includes a number of variables, but the most common are religious belief, affiliation, and participation.

13-6a Religious Belief

Some 58 percent of Americans say that religion is very important in their lives, down from 75 percent in 1952 (Newport, 2013). Not surprisingly, religion isn't important for *agnostics* (those who say that it's impossible to know whether there's a God), *atheists* (those who believe that there's no God), or others who are skeptics. Nonetheless, about 25 percent of Americans, including some atheists and agnostics, embrace the tenets of some Eastern religions or elements of New Age spirituality that include reincarnation, meditation, astrology, and the evil eye (casting curses and evil spells) (Lugo et al., 2012).

13-6b Religious Affiliation

Since 2007, the greatest declines have occurred among evangelical and mainstream white Protestants. In contrast, 20 percent of Americans say that they have no religious preference or affiliation (see *Table 13.4*), up from only 8 percent in 1990 (Kosmin and Keysar, 2009). "Nones," a popular term for the religiously unaffiliated, include Asians, young people, males, political independents, those living in the Pacific and New England regions, college graduates, and people in higher-income brackets (Lugo et al., 2012; Saad, 2013; Newport, 2014).

About half of U.S. adults—especially those who were raised as Catholics or Protestants—have changed their religious affiliation since childhood. Some have done so more than once because of intermarriage or finding a religion they liked better. The most important reasons for becoming unaffiliated include not believing the teachings, seeing many religious people as hypocritical or judgmental, losing respect for religious leaders who focus on power and money, and negative teachings or treatment of lesbians and gays (Pew Forum on Religion & Public Life, 2009; Cooperman et al., 2014; Taylor et al., 2014).

13-6c Religious Participation

Turning to the third measure of religiosity, religious participation, among Americans who say that religion is

TABLE 13.4 RELIGIOUS AFFILIATION IN THE UNITED STATES, 2012

		Percentage Change Since 2007
CHRISTIAN	73%	−5%
Protestant	48	−5
Catholic	22	−1
Mormon	2	−
OTHER FAITH	7	+2
UNAFFILIATED	20	+4.3
Atheist	2.4	+0.8
Agnostic	3.3	+1.2
Nothing in particular	14	+2.3

Source: Based on Lugo et al., 2012.

important to them, 27 percent seldom or never attend religious services. Thus, many people are more likely to believe in a religion than to practice it by attending services regularly. Mormons (81 percent), Protestants (64 percent), Muslims (64 percent), and Catholics (60 percent) have higher attendance rates at religious services than do Jews (34 percent) or non-Christians (Pew Forum on Religion and Public Life, 2008; Newport, 2012).

13-6d Some Characteristics of Religious Participants

Americans differ in their beliefs and affiliations. Religious participation also varies by sex, age, race and ethnicity, and social class.

SEX

Across all age and faith groups in 145 countries, women tend to be more religious than men in believing in God, praying, attending services, and saying that religion is very important in their lives. In the United States, atheists and agnostics are much more likely to be male (64 percent) than female (36 percent) (Taylor et al., 2009; Lugo et al., 2012).

It may be that women are expected to be more pious and spiritual because, especially as nurturers, they transmit religious values to their children (see Chapter 9). Women may turn to religion because they have a harder life than most men. As you saw in previous chapters, women worldwide are more likely than men to experience assaults and poverty. Also, women, on average, live longer than men, and religiosity increases with age (Newport, 2012).

AGE

Generally, Americans age 65 and older are more likely than younger people to describe themselves as religious, to say that religion is very important in their lives, and to attend services at least weekly. These age-related differences may reflect several factors: Older Americans grew up decades ago when church attendance was higher, they seek spiritual comfort as elderly friends and relatives die, they want to lessen a sense of isolation or loneliness, and they're preparing for death (Taylor et al., 2009).

"Millennials" (those born after 1980) are less religious than their elders: They're less likely to pray, to regularly attend worship services, or to identify themselves with a religious group. For example, 32 percent of adults under 30 have no religious affiliation compared with just 9 percent who are 65 and older. Also, young adults are much more likely to be unaffiliated than previous generations were at a similar stage in their lives (Lugo et al., 2012; Martínez and Lipka, 2014).

RACE AND ETHNICITY

Blacks are more religious than any other U.S. racial or ethnic group. They're the most likely to report a religious affiliation, to belong to a church, to pray every day, and to attend religious services every week. The vast majority (78 percent) are Protestant; within this group, 40 percent are Baptist (Sahgal and Smith, 2009; Newport, 2012).

Latinos' religious landscape has been shifting. About 55 percent are Catholic (down from 67 percent in 2010), 22 percent are Protestant and primarily evangelical rather than mainline, and 18 percent are unaffiliated (up from 10 percent in 2010). Some Catholic and evangelical Protestant churches have been especially successful in attracting recent Latino immigrants because the churches offer services in Spanish, the ceremonies are expressive rather than formal, and the clergy are more responsive to their members' social and economic needs (Dias, 2013; Cooperman et al., 2014; "Select-a-Faith," 2014).

Asian Americans may be the most diverse religious group in America. Nationally, 42 percent are Christian, 26 percent are unaffiliated, 24 percent are Hindu or Buddhist, and 7 percent are Muslim, Sikh, or some other religion. There's also considerable variation across subgroups. For example, 65 percent of Filipinos are Catholic, 52 percent of Chinese Americans are unaffiliated, and 61 percent of Korean Americans are Protestant (Funk et al., 2012).

SOCIAL CLASS

People with less education are generally more religious than those with higher educational levels. Because education and income are highly correlated, as education increases, the importance of religion generally decreases (see *Figure 13.5*). Among the "nones," for instance, 38 percent of atheists and agnostics have an annual family income of at least $75,000 compared with 29 percent of those who have a religious affiliation (Lugo et al., 2012).

13-6e Secularization: Is Religion Declining or Thriving?

Industrialized nations have been experiencing **secularization**, a process in which religion loses its social and cultural influence. Some social scientists, including sociologists, maintain that secularization is increasing; others contend that religiosity is flourishing.

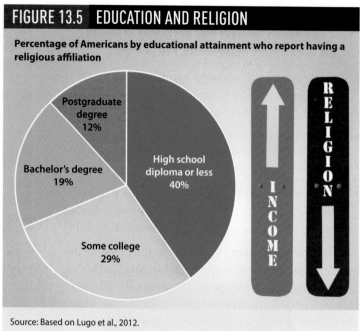

FIGURE 13.5 EDUCATION AND RELIGION

Percentage of Americans by educational attainment who report having a religious affiliation

- Postgraduate degree 12%
- Bachelor's degree 19%
- High school diploma or less 40%
- Some college 29%

Source: Based on Lugo et al., 2012.

IS SECULARIZATION INCREASING?

You've seen that the number of "nones" is growing, attendance at religious services has dropped, and that fewer Americans say that religion is "very important" in their lives. There's other evidence that secularization is rising in the United States and globally. For example:

▶ 45 percent of Americans have a "great deal" of confidence in the church and organized religion, down from 68 percent in 1975 (Riffkin, 2014)

▶ 70 percent of Americans attended religious services on Christmas Eve or Day when they were children; only 54 percent do so in adulthood (Cooperman et al., 2013)

▶ 68 percent of people worldwide describe themselves as religious, down from 77 percent in 2005 (WIN-Gallup International, 2013)

There's also evidence of greater "forced secularization" worldwide. About 64 percent of the world's population—up from 58 percent in 2007—lives in countries with high government restrictions on religion. Some of the laws and policies ban particular faiths, prohibit conversions, or limit preaching. Other laws and practices allow religion-related conflict or terrorism, mob violence against a religious minority, and harassment over religious attire (Grim and Cooperman, 2014; "Trouble in Trappes," 2014).

secularization a process in which religion loses its social and cultural influence.

IS RELIGION THRIVING?

Many sociologists contend that the extent of secularization has been greatly exaggerated. For example:

▶ 61 percent of Americans say that it's important for members of Congress to have strong religious beliefs ("Growing Number of Americans...," 2010)

▶ 83 percent support Christmas displays on government property (Pew Research Center for the People & the Press, 2010)

▶ 93 to 99 percent of people in the Middle East, Asia/Pacific, Latin America, and Africa say that believing in God is essential to morality (Simmons et al., 2014)

Fundamentalism, the belief in the literal meaning of a sacred text (e.g., Christian Bible, Muslim Qur'an, Jewish Torah), is widespread in practically all countries except Europe. About 75 percent of Americans say that everything or almost everything in the Bible should be taken literally because it's the word of God (Saad, 2014).

The U.S. Constitution established a separation of church and state. Over the years, that separation has become fuzzier. Several southern states have passed laws that allow public schools and government buildings to display traditional Christmas scenes and symbols; many public elementary and high schools allow prayer before class, permit students to leave classes for religious instruction, support religious student clubs that faculty sponsor, and offer after-school evangelical

Pascal Deloche/Terra/Corbis

fundamentalism the belief in the literal meaning of a sacred text.

programs. A growing number of public colleges and universities has approved building privately funded "religious dorms" (Bruinius, 2013; Campo-Flores, 2013; Lawrence, 2013).

One of the most divisive issues has been over which explanations for human origins should be taught in public schools. For scientists, it's "a fact" that human beings (and other creatures) evolved from earlier animals through a process known as "natural selection" over several million years (National Science Board, 2014). In contrast, *creationism*, based on a fundamentalist interpretation of the Bible, argues that God created humans in their present form about 10,000 years ago. About 42 percent of Americans espouse creationism. Because of such beliefs, Republicans in conservative states (e.g., Louisiana, Missouri, Tennessee, Oklahoma) have introduced bills for public schools to teach creationism in science classes ("Faith and Reason," 2014; Newport, 2014).

AP Images/Harry Cabluck

BRENDAN SMIALOWSKI/Getty Images

In 2005, the U.S. Supreme Court ruled that a 6-foot-high monument containing the Ten Commandments outside the Texas state capitol was constitutional (left), even though it's on government property. In 2014, the court ruled that prayer in civic life, including legislative bodies, is not only constitutional, but can use explicit Christian or other religious language (right). Do such decisions illustrate greater secularization or religiosity?

iStockphoto.com/PaulMaguire

"WE ESTABLISH NO RELIGION IN THIS COUNTRY. WE COMMAND NO WORSHIP. WE MANDATE NO BELIEF, NOR WILL WE EVER. CHURCH AND STATE ARE AND MUST REMAIN SEPARATE." RONALD REAGAN, REPUBLICAN, 40TH U.S. PRESIDENT (1911–2004).

Another indicator that religion is booming in the United States and elsewhere is the prevalence of **civil religion** (sometimes called *secular religion*), integrating religious beliefs into secular life. The Pledge of Allegiance was written in 1892; the phrase "under God" was added in 1954. Other examples of civil religion include the phrase "In God We Trust" on U.S. currency, prayer in public schools, a legally mandated National Day of Prayer held on the first Thursday of May, and many local jurisdictions' closing public schools, public libraries, and government offices on Good Friday, a Christian holy day.

13-7 SOCIOLOGICAL PERSPECTIVES ON RELIGION

You've seen that religion plays an important role in many people's lives. Why? How does society affect religion? And how does religion affect society? *Table 13.5* summarizes the major sociological perspectives on religion.

13-7a Functionalism: Religion Benefits Society

Durkheim (1961: 13, 43) described religion as a "permanent aspect of humanity" whose "essential task is to maintain, in a positive manner, the normal course of life." Religion can also be dysfunctional, but first let's consider its benefits.

RELIGION IS A SOCIAL GLUE

Religion fulfills a variety of important functions at individual, community, and societal levels. All of the following, according to functionalists, contribute to a society's survival, stability, and solidarity:

▸ *Belonging and identity.* Communal worship and rituals increase social contacts, develop a sense of acceptance and identity, and reinforce people's feeling of belonging to a group. During Christmas, 86 percent of Americans, including "nones," attend holiday activities with family or friends (Cooperman et al., 2013).

▸ *Purpose and emotional comfort.* Religion provides meaning in life and offers hope for the future. Religiosity is highest among the world's poorest countries because religion helps people cope with daily struggles to survive (Crabtree, 2010; WIN-Gallup International, 2013).

▸ *Well-being.* Very religious Americans report being healthier and experiencing less depression and worry than those who aren't religious (Newport et al., 2012; Newport and Himelfarb, 2013).

▸ *Social service.* Religious groups raise millions of dollars for the victims of natural disasters, distribute food and clothing, and find shelter for displaced families. About 88 percent of Americans, including 78 percent of "nones," agree that churches and other religious organizations are important because they help the poor and sick (Lugo et al., 2012).

▸ *Social control.* Because many religious people fear punishment in the afterlife, they try to follow societal rules. Doing so suppresses deviant and antisocial behavior and promotes societal cohesion (Durkheim, 1961; Shariff and Aknin, 2014).

RELIGION PROMOTES SOCIAL CHANGE

Religion can also spearhead social change. Mohandas Gandhi (1869–1948), a spiritual leader in India, worked for his country's independence from Great Britain through nonviolence and peaceful negotiations. In the

civil religion (sometimes called *secular religion*) integrating religious beliefs into secular life.

TABLE 13.5 SOCIOLOGICAL EXPLANATIONS OF RELIGION

Theoretical Perspective	Level of Analysis	Key Points
Functionalist	Macro	Religion benefits society by providing a sense of belonging, identity, meaning, emotional comfort, and social control over deviant behavior.
Conflict	Macro	Religion promotes and legitimates social inequality, condones strife and violence between groups, and justifies oppression of poor people.
Feminist	Macro and micro	Religion subordinates women, excludes them from decision-making positions, and legitimizes patriarchal control of society.
Symbolic Interactionist	Micro	Religion provides meaning and sustenance in everyday life through symbols, rituals, and beliefs, and binds people together in a physical and spiritual community.

United States, religious leaders, especially the Reverend Martin Luther King, Jr., were at the forefront of the civil rights movement during the late 1960s.

In 2011, after decades of debate, the Presbyterian Church, one of the nation's largest Protestant denominations, approved ordaining people with same-sex partners as ministers, elders, and deacons. Numerous Lutheran churches have also welcomed gay pastors. In 2014, and despite many of its congregations' threats of leaving, the Presbyterian Church voted to allow same-sex marriages. Other religious groups (e.g., United Church of Christ, Episcopal Church, Quakers, Reform and Conservative Judaism) have taken similar steps (Goodstein, 2014).

Max Weber (1920/1958) asserted that religion sparks economic development. His study of Calvinism, a Christian sect that arose in Europe during the sixteenth century, led to his coining the term **Protestant ethic**, which he described as a belief that hard work, diligence, self-denial, frugality, and economic success would lead to salvation in the afterlife. According to Weber, the harder the early Calvinists worked and saved, the more likely they were to accumulate money, to become successful, and to drive the growth of capitalism.

Was Weber right about the relationship between the Protestant ethic and the rise of capitalism? The data are mixed. Some studies show that people are diligent not because of religious beliefs but simply because amassing savings provides resources when disaster strikes (as when droughts wipe out crops). On the other hand, a study of 59 industrialized and developing countries found that religious beliefs that encourage hard work and thrift spur economic growth (Cohen, 2002; Barro and McCleary, 2003).

IS RELIGION DYSFUNCTIONAL?

Functionalists emphasize its benefits, but also recognize that religion can be dysfunctional when it harms individuals, communities, and societies. Religious intolerance can spark conflict between groups (including vandalizing churches, mosques, and synagogues) and attacks on religious minorities. The United States has a long history, beginning with the Pilgrims in 1620, of people discriminating against other religious groups even though they themselves sought religious freedom (Davis, 2010).

Recently, a scathing United Nations report accused the Roman Catholic Church and the Vatican of "systematically" protecting predator priests for at least five decades, allowing tens of thousands of children to be abused, humiliating victims' families into

JOHN GRESS/Reuters/Landov

Megachurches are Christian congregations that have a regular weekly attendance of more than 2,000 and tend to be evangelical. They represent only 0.5 percent of all U.S. churches, but their number has almost tripled (to more than 1,600) since 2000 (Bird and Thumma, 2011). From a functionalist perspective, why do you think megachurches are so popular?

Protestant ethic a belief that hard work, diligence, self-denial, frugality, and economic success will lead to salvation in the afterlife.

silence, and still not removing perpetrators from their posts (United Nations Committee on the Rights of the Child, 2014).

CRITICAL EVALUATION

One weakness is that functionalists, by emphasizing religion's benefits, imply that religion is indispensable to leading a good life. People who aren't religious do many good things for society, and some religious people commit heinous acts. Primarily because of priests' sexual abuse of young children, Americans' rating of the clergy's honesty and ethics fell from 67 percent in the mid-1980s to 47 percent in 2013 (Swift, 2013).

A second limitation is functionalists' exaggerating religion's positive effect on physical and emotional health. Jews, who are one of the least religious U.S. groups, report the highest rates of well-being, and "nones," including atheists, report higher rates of well-being than do religious groups (Newport et al., 2012; Leurent et al., 2014).

A third, and major, criticism is that functionalists gloss over numerous dysfunctional aspects of religion that generate wars, terrorism, and genocide. Conflict theory addresses these issues.

13-7b Conflict Theory: Religion Promotes Social Inequality

From a conflict perspective, religion promotes and reinforces social inequality. Throughout history and currently, religion has created discord and divisiveness within groups, between groups, and in the larger society.

"THE OPIUM OF THE PEOPLE"

Much conflict theory reflects the work of Karl Marx (1845/1972), who described religion as "the sigh of the oppressed creature" and "the opium of the people" because it encouraged passivity and acceptance of social and economic inequality. Marx viewed religion as a form of **false consciousness**, an acceptance of a system of beliefs that prevents people from protesting oppression. Contemporary conflict theorists don't view religion as an opiate, but they agree with Marx that religion often justifies violence and promotes social inequality.

RELIGION JUSTIFIES INTOLERANCE AND VIOLENCE

For thousands of years, many governments and religious leaders have condoned or perpetrated widespread violence in the name of religion. Religion promotes conflict when religious groups differentiate between "we" and "they" ("We're right and they're wrong," "Our God is 'the *real* God'"). Such self-righteousness condones aggression, oppression, and brutality within and across societies. For example:

- The "eternal war" between Iraq's Sunnis and Shiites, two major Islamic sects, dates back more than 1,300 years, and there are long-standing conflicts and wars between Muslims and Jews in Israel and Palestine (Crowley, 2014).

- Because of widespread violence against Christians, they now make up only 5 percent of the Middle East's population, down from 20 percent in 1914 (Baker, 2014).

- In nearly 75 percent of the world's countries, religious groups experience harassment and intimidation, including physical or verbal assaults, desecration of holy sites, and discrimination in employment, education, and housing (Theodorou and Henne, 2014).

- In several of Asia's Buddhist-majority nations, monks have incited bigotry and violence—mostly against Muslims (Beech, 2013).

RELIGION PROMOTES SOCIAL INEQUALITY

Conflict theorists see religion as a tool dominant groups use to control society, protect their own interests, and derail social change. U.S. Roman Catholic bishops have a long history of helping the poor and fostering social justice. Most recently, however, they have opposed comprehensive health care because it would fund abortions, have publicly chastised pro-choice Catholic politicians, and have campaigned against same-sex marriage—all of which fuel social inequality and conflict (Doyle, 2011).

Conflict theorists also point out that private evangelical schools, colleges, and universities receive massive state and federal financial aid that diverts much-needed tax revenues from public schools and colleges. Many of the schools' textbooks describe homosexuals and abortion rights supporters as "evil," and are hostile toward other religions, including nonevangelical Protestants, Jews, and Catholics (Berkowitz, 2011; Tabachnick, 2011).

CRITICAL EVALUATION

Functionalists may overemphasize consensus and harmony, but conflict theorists, according to some

false consciousness an acceptance of a system of beliefs that prevents people from protesting oppression.

"Maybe You should tell them that You and Allah are the same person."

critics, often ignore religion's role in challenging homophobia and uniting people. Nationally, for instance, the share of American religious congregations allowing an openly gay or lesbian couple to become full-fledged members grew from 37 percent in 2006 to 48 percent in 2012 (Chaves and Anderson, 2014).

A second criticism is that conflict theorists attach too much credence to Marx's concept of false consciousness. Instead of passively submitting to religious oppression, many people revolt, complain, or resist. The EEOC, for instance, has filed lawsuits against a trucking company that requires its Muslim drivers to deliver alcohol, which Islam prohibits. Others have sued employers, sometimes successfully, who denied requests to not work on holy days, and whose secular dress codes don't allow head scarves, turbans, or long, shaggy beards (Trottman, 2013). Whether religious discrimination is real or imagined, many Americans protest, demonstrate, and seek legal remedies.

13-7c Feminist Theories: Religion Subordinates and Excludes Women

Feminist theorists agree with conflict theorists that religion can foster violence and inequality. They go further, however, by criticizing organized religions as sexist, patriarchal, and shutting women out of leadership positions.

SEXISM, PATRIARCHY, AND SUBORDINATION

"Religion is probably the most important force in continuing the oppression of women worldwide," a feminist blogger writes, and "religiosity and sexism go hand in hand" (Marcotte, 2014). From a feminist perspective, most religions are patriarchal: They emphasize men's experiences and a male point of view and see women as subordinate to men. According to the apostle Paul, "Wives should submit to their husbands in everything" (*Ephesians* 5: 24). The idea that Eve was created out of Adam's rib is often used to justify men's domination of women. Many religious teachings and institutions continue to propagate such beliefs (Ferguson, 2014).

In Orthodox Judaism, a man's daily prayers include this line: "Blessed art thou, O Lord, our God, King of the Universe, that I was not born a woman." The Qur'an tells Muslims that men are in charge of women and that women should obey men. Almost all contemporary religions worship a male deity, and none of the major world religions treat women and men equally (Gross, 1996; Jeffreys, 2011).

Feminist scholars offer alternatives to patriarchal interpretations of Scripture. Some contend that Jesus was a feminist. For example, he defended women against men, encouraged women's intellectual pursuits when it wasn't the norm, and there are numerous passages in the Bible about women spreading Jesus' teachings (Gross, 1996; Ferguson, 2014).

Muslim feminists, similarly, note that men have interpreted sacred Islamic texts to ensure male dominance and control. Women were among some of Muhammad's earliest converts, and the Qur'an has numerous passages that establish women's equal rights in inheritance and family roles (Menissi, 1991, 1996). Thus, women's subordination in many Islamic societies isn't due to religious tenets but to men's interpretations of Scripture to maintain their power and privilege (Smith, 1994).

In the Middle East and North Africa, 35 percent of the countries have religious police who, among other things, enforce strict segregation of the sexes and punish women who are perceived as behaving or dressing improperly. Worldwide, harassing women over "indecent" attire increased from 7 percent of countries in 2007 to 25 percent in 2011 (Stencel and Grim, 2013; Theodorou, 2014). Thus, it's women's fault if men have lustful thoughts or assault women.

EXCLUSION OF WOMEN FROM LEADERSHIP POSITIONS

In 2011, women earned 32 percent of theology degrees, a dramatic increase from only 2 percent in 1970. Still, women make up only 21 percent of the nation's clergy, lead only 8 percent of congregations, and earn 20 percent less than their male counterparts (Chaves et al., 2009; Snyder and Dillow, 2013; BLS Reports, 2014).

Roman Catholicism, Orthodox Judaism, and the Mormon Church don't allow women to be ordained because "women should serve and not lead." U.S. Catholic women contribute $6 billion a year during Sunday Masses, but "the presence of women anywhere within the institutional power structure is virtually nil" (Miller, 2010: 39; Gibson, 2012). Mormon women can be missionaries and provide "welfare and compassionate service," but are excluded from the priesthood and the all-male

In 2014, the Mormon Church excommunicated Kate Kelly (left), leader of the Ordain Women group. According to some U.S. Catholic bishops, the ideas of Sister Simone Campbell (right), and other nuns who advocate ordaining women, border on heresy. Both women are human rights lawyers.

central leadership that governs the church (Kantor and Goodstein, 2014).

Some Protestant denominations justify women's exclusion from leadership positions based on biblical passages such as "I permit no woman to teach or have authority over men; she is to keep silent" (I *Timothy* 2: 11–12). Many Protestant groups—including Southern Baptists and evangelical born-again Christians—interpret this and similar passages to mean that women should never, under any circumstances, instruct men, within or outside of religious institutions. Even in liberal Protestant congregations, female clergy tend to be relegated to specialized ministries with responsibilities for music, youth, or Bible studies (Banerjee, 2006; Richie, 2013).

What happens when religious women openly protest their exclusion? In 2013, Pope Benedict XVI accused the Leadership Conference of Women Religious, which represents 80 percent of U.S. nuns, of promoting "radical feminism" for challenging the Catholic Church's policies of not ordaining women, and set up re-education programs "to bring the nuns back into line." His successor, Pope Francis, recently praised American nuns for their service and sacrifices, but said nothing about ordaining women (Carballo, 2014).

CRITICAL EVALUATION

Feminist perspectives on religion have three weaknesses. First, some feminist Muslim scholars have criticized Western feminists for misreading Islamic and other sacred scriptures, and reducing practically all discussions of gender to the *hijab* (a veil or scarf that Muslim women wear) instead of focusing on justice for both women and men in marriage, employment, and other areas (Fakhraie, 2009; Daneshpour, 2013).

Second, religious women aren't as oppressed as many feminist scholars claim. For instance, a Muslim won a lawsuit against Abercrombie & Fitch after she was fired for wearing a head scarf instead of conforming to the company's dress code (Ballhaus, 2013). Even in the most conservative religions, women don't blindly submit to religious dogma. Instead, they challenge existing doctrines, stop volunteering at church, don't attend weekly services, and establish their own parishes (Padgett, 2010; Miller, 2012).

Third, some feminist scholars overlook the progress and variation of women's leadership in organized religion. In the U.S. Presbyterian Church, for instance, 36 percent of the active ministers are female, up from only 3 percent in 1977 (Hartford Institute for Religion Research, 2012; Presbyterian Mission Agency, 2013). Between 2002 and 2012, the number of female clergy in the Church of England rose from 7 to 21 percent and, in 2014, its lawmaking body cast a final vote to ordain women bishops (Bennhold, 2014).

13-7d Symbolic Interaction: Religion Is Socially Constructed

For symbolic interactionists, religion isn't innate but socially constructed. As a result, people can learn and interpret the same religion differently across cultures and over time. Symbols, rituals, and beliefs are three of the most common means of learning and internalizing religion.

SYMBOLS

A *symbol* is anything that stands for or represents something else to which people attach meaning (see Chapter 3). Many religious symbols are objects (a cross, a steeple, a Bible), but also include behaviors (kneeling or bowing one's head), words ("Holy Father," "Allah," "the Prophet"), and physical appearance (e.g., wearing head scarves, skull caps, turbans, clerical collars).

Religious symbols, like all symbols, are short-hand communication tools. Some interactionists define a religion as "a system of symbols" because it's a community that's unified by its symbols (Berger and Luckmann, 1966; Geertz, 1966). Religious symbols can also be divisive (e.g., a Ten Commandments monument on government property that alienates non-Christians and atheists).

RITUALS

A *ritual* (sometimes called a *rite*) is a formal and repeated behavior that unites people (see Chapter 3). Religious rituals, like secular ones, strengthen a participant's self-identity (Reiss, 2004). Religious rites of passage—*bat mitzvah* for girls and *bar mitzvah* for boys in the Jewish community, and first communion and confirmation for Catholic children—reinforce the individual's sense of belonging to a particular religious group. A group's rituals symbolize its spiritual beliefs and include a wide range of practices—praying, chanting, fasting, singing, dancing, and offering sacrifices.

All religions have rituals like birth and marriage that mark significant life events (see Chapters 3, 5, and 12). Death rituals are probably the most elaborate and sacred worldwide. They vary across religious groups and societies, but all of them comfort the living and show respect for the dead.

BELIEFS

Rituals and symbols come from *beliefs*, convictions about what people think is true. Religious beliefs can be passive (believing in God but never attending services) or active (participating in rituals and ceremonies). Beliefs bind people together into a spiritual community.

One of the strongest beliefs worldwide is that prayer is important. Islam requires prayer five times a day. In the United States, 55 percent of U.S. adults say that they pray every day, and for a variety of reasons, including feeling close to God, as well as requesting better health, more money, and cures for sick pets. Even 21 percent of the "nones" pray at least once a day (Wicker, 2009; Cooperman et al., 2014). Prayer offers psychological and spiritual benefits such as comfort and a sense of unity among those who pray together, but depending on God can also diminish people's motivation to actively shape their own lives (Schieman, 2010).

CRITICAL EVALUATION

A common criticism is that interactionists' focus on micro-level behavior ignores the ways that religion promotes social inequality at the macro level. Conflict theorists and feminist scholars, especially, maintain that people often use religion to justify violence and women's subordination.

Some critics also wonder if interactionists paint too rosy a picture of religion even on a micro level because people's beliefs can wreak considerable havoc. The Islamic zealots who targeted the World Trade Center on 9/11 didn't see themselves as "crazed terrorists" but as "true believers": They were carrying out "God's will" by imposing their religious beliefs on others or destroying their "religious enemies" (Juergensmeyer, 2003).

STUDY TOOLS 13

READY TO STUDY? IN THE BOOK, YOU CAN:

☐ Check your understanding of what you've read with the Test Your Learning Questions provided on the chapter review card at the back of the book.

☐ Rip out the chapter review card for a handy summary of the chapter and key terms.

ONLINE AT CENGAGEBRAIN.COM YOU CAN:

☐ Collect StudyBits while you read and study the chapter.

☐ Quiz yourself on key concepts.

☐ Prepare for tests with SOC5 Flash Cards as well as those you create.

☐ Watch videos to explore sociology further.

© Anelina/Shutterstock.com

SOC
ONLINE
REVIEW FLASHCARDS
ANYTIME, ANYWHERE!

Create Flashcards from Your StudyBits

Review Key Term Flashcards Already Loaded on the StudyBoard

4LTR
PRESS

Access SOC ONLINE at www.cengagebrain.com

14 | Health and Medicine

LEARNING OUTCOMES

After studying this chapter, you will be able to...

14-1 Describe social epidemiology and explain why there are global health gaps.

14-2 Explain how and why environmental, demographic, and lifestyle factors affect U.S. health and illness. Illustrate your explanation with specific examples.

14-3 Compare U.S. with other countries' health care systems in terms of access, costs, and outcomes.

14-4 Compare and evaluate the theoretical explanations of health and medicine.

After you finish this chapter go to **PAGE 291** for **STUDY TOOLS.**

Most of us would probably agree with American poet Ralph Waldo Emerson, who wrote, in 1860, "Health is the first wealth." **Health** is the state of physical, mental, and social well-being. We can usually determine physical health by using objective measures like weight and blood pressure. Mental and social well-being are more difficult to gauge because they rely on subjective definitions that change over time.

14-1 GLOBAL HEALTH AND ILLNESS

How does health differ around the world? And why? Social epidemiology addresses both questions.

14-1a Social Epidemiology

All people experience *disease*, a disorder that impairs a person's normal physical and/or mental condition. **Social epidemiology** examines how societal factors affect the distribution of disease within a population. Why, for example, do people live much longer in some countries than in others?

In answering this and other questions, epidemiologists look at two factors. One is *incidence*, the number of new cases of a health problem that occur in a

What do you think?

The government should be responsible for making sure that all Americans have health care.

| 1 | 2 | 3 | 4 | 5 | 6 | 7 |

strongly agree   strongly disagree

given population during a given time period (e.g., in 2012, there were 1.7 million new diabetes cases among Americans aged 20 or older). The other measure is *prevalence*, the total number of cases (extent) of an illness or health problem within a population or at a particular point in time (e.g., in 2012, 12 percent of the U.S. population aged 20 or older had diabetes) (*National Diabetes Statistics...*, 2014). Epidemiological studies show large health differences across countries.

14-1b Global Health Disparities

The World Bank classifies countries into four broad economic groups based on gross national income (GNI) per capita. *Figure 14.1* shows the number and location of these countries and compares them on life expectancy, infant mortality, and per capita health expenditures. These three variables are among the most frequently used to measure health status. They indicate a population's living standards, people's average socioeconomic status, and the quality and financial resources that a country devotes to **health care**, the prevention, treatment, and management of illness.

The 80 high-income countries make up only 13 percent of the world's population, but have the greatest access to clean water, sanitation, food, and health services. In contrast, chronic malnutrition is the most common killer of children under 5 years old in lower-income

countries. These death rates increased from 37 percent in 1990 to 44 percent in 2013. More than a third of the world's population lacks basic sanitation, almost half are at risk of dying of malaria, and 44 percent of people in low- and lower-middle-income countries can't afford medicine to treat infections and diseases (World Health Organization, 2015).

High income populations live longer, have low infant mortality rates, and spend more on health (see *Figure 14.1*), but experience "diseases of wealth" (e.g., diabetes, heart disease, various cancers). Compared with 16 other high-income countries, the United States has a large and growing "health disadvantage" and ranks last or near-last in nine key health areas including drug-related deaths, chronic lung and heart disease, obesity, and the prevalence of HIV and AIDS (Woolf and Aron, 2013).

health the state of physical, mental, and social well-being.

social epidemiology examines how societal factors affect the distribution of disease within a population.

health care the prevention, management, and treatment of illness.

FIGURE 14.1 WORLDWIDE HEALTH GAPS

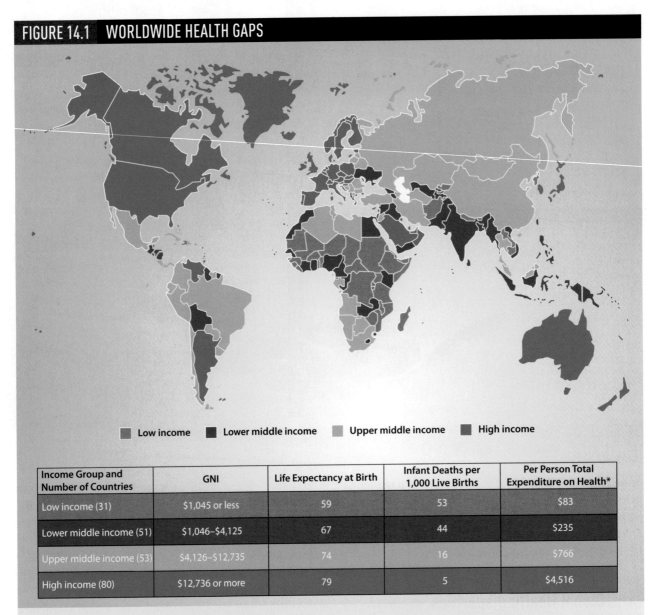

Income Group and Number of Countries	GNI	Life Expectancy at Birth	Infant Deaths per 1,000 Live Births	Per Person Total Expenditure on Health*
Low income (31)	$1,045 or less	59	53	$83
Lower middle income (51)	$1,046–$4,125	67	44	$235
Upper middle income (53)	$4,126–$12,735	74	16	$766
High income (80)	$12,736 or more	79	5	$4,516

*These figures include private, employer, and government expenditures in U.S. dollars.

Sources: Based on World Bank, 2015; World Health Organization, 2015, Tables 1 and 7.

14-2 HEALTH AND ILLNESS IN THE UNITED STATES

Why do Americans have a health disadvantage, including high disability rates? Compared with other high-income nations, non-medical social determinants contribute to large health disparities in the United States.

disability any physical or mental impairment that limits a person's ability to perform a basic life activity.

14-2a Understanding Disability

A **disability** is any physical or mental impairment, temporary or permanent, that limits a person's ability to perform a basic life activity. These activities include walking; doing errands alone; bathing, dressing, or feeding oneself; and working.

Almost a third of Americans aged 18 and over have a disability. Age is a major predictor of disability, affecting 59 percent of adults age 65 and older compared with 25 percent of those ages 18 to 64. Disability also increases with age—from 26 percent for people aged 65 to 74 to 73 percent for those age 85 and older. Besides age, disability rates are higher for women (35 percent) than men

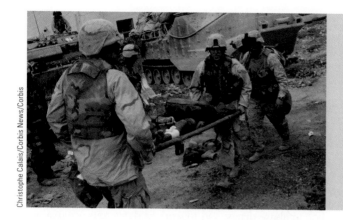

Christophe Calais/Corbis News/Corbis

Advances in battlefield medicine and body armor mean than many soldiers are surviving injuries, but have shattered bones, brain damage, and amputated limbs. About 20 percent of the soldiers who served in Afghanistan and Iraq have posttraumatic stress disorder (PTSD) or other mental health problems. Nearly half of the 1.6 million soldiers who served in these countries receive disability benefits, and have submitted more disability claims than ever—over 1 million between 2010 and 2013 alone ("The Waiting Wounded," 2013; Thompson, 2014).

(26 percent), for people living in poverty (41 percent versus 23 percent who aren't poor), and are highest among American Indians (36 percent) (He and Larsen, 2014; National Center for Health Statistics, 2015).

The number of Americans experiencing one or more disabilities rose from 61 million in 1997 to almost 75 million in 2013 (National Center for Health Statistics, 2015). Why the increase? First, as you'll see shortly, behavior previously considered normal has been medicalized. Second, medical advances help many people survive diseases, car accidents, and wars, but they may have lifelong health ailments. Third, as the proportion of the population age 65 and older increases, so does the proportion living with disabilities.

14-2b Social Determinants of Health and Illness

No single variable explains well-being. Stem cells replace older cells as they die off. About 65 percent of cancers are the result of "bad luck," the product of random and spontaneous genetic changes that occur when stem cells make mistakes copying healthy DNA. This means that 35 percent of cancers are preventable (Tomasetti and Vogelstein, 2015).

Whether it's cancer or other diseases, genes affect all of us. Nonetheless, environmental, demographic, and lifestyle factors—all of which are interrelated—have a significant impact on health and illness.

ENVIRONMENTAL FACTORS

Many environmental hazards affect our bodies. For example:

▶ There's a strong association between early-life exposure to air pollution and autism, schizophrenia, and learning disabilities (Roberts et al., 2013; Allen et al., 2014).

▶ Many toxic chemicals in food and everyday products (e.g., dishwashing detergents, shampoos, shaving products, and makeup) have been linked to asthma,

some types of cancer, and children's behavioral disorders (Sathyanarayana et al., 2013; Blake, 2014; Grandjean and Landrigan, 2014).

▶ Norovirus—a very contagious virus—is the leading cause of disease outbreaks from contaminated food. About 21 million Americans get sick from norovirus each year, 71,000 are hospitalized, and about 800 die. Outbreaks on cruise ships account for only 1 percent of all norovirus; 70 percent of cases are due to food workers who are infected, don't wash their hands, or go to work when sick (Hall et al., 2014).

Access to health care can prolong life, but an estimated 440,000 Americans die each year because of preventable medical errors (e.g., doctors operating on the wrong patient or body part, patient infections). Medication errors—misdiagnosing an illness, prescribing the wrong drugs, or giving patients drugs that interact dangerously—injure about 1.3 million Americans and kill about 7,000 every year. Moreover, thousands of patients a year leave operating rooms with surgical sponges in their bodies that cause infections, lifetime digestive problems, removal of intestines, and even death (Brown and McGann, 2011; Eisler, 2013; James, 2014).

A growing problem is antibiotic resistance, which kills an estimated 23,000 Americans each year. Bacteria have become more resistant because of the overuse of antibiotics: People take them for illnesses for which antibiotics aren't effective, and farmers feed them to animals to promote growth. No new antibiotics have been discovered since 1987, largely because pharmaceutical companies reap much higher profits by developing very expensive and specialized medicines for small populations with chronic conditions, including multiple sclerosis, cancer, HIV, and hepatitis C ("The Drugs Don't Work," 2014: 54; Tozzi, 2014).

DEMOGRAPHIC FACTORS

Religion, family size, marital status, urban-rural residence, and other variables affect health. Four of the most important demographic factors are age, gender, social class, and race and ethnicity.

Age. Not surprisingly, age is the single best predictor of illness and death. Death rates drop sharply shortly after birth, begin to rise at about age 45, and escalate, particularly after age 78.

From birth through elementary school, children have health problems over which they have no control. Nearly 1 in 100 U.S. babies are born with *fetal alcohol spectrum disorders*, a range of permanent birth defects caused by a mother drinking alcohol during pregnancy. The disorders include intellectual infirmities, speech and language delays, poor social skills, mental retardation, and physical abnormalities like congenital heart defects (National Organization on Fetal Alcohol Syndrome, 2012; Weinhold, 2012).

Many adolescents have health problems because of their own choices: 21 percent of twelfth graders take drugs without a doctor's prescription, 28 percent smoke, 47 percent drink alcohol, and 60 percent text or email while driving (Kann et al., 2014). Such risky behavior may be due to peer pressure, to the fact that some parts of the teen brain are still developing, and because of adolescents' becoming more independent as they break away from parental supervision (see Chapter 4).

Health problems emerge and increase during one's late thirties because of genes, lifestyle choices (which we'll address shortly), and because physical decline is normal and inevitable as we age. No matter how well tuned we keep our bodies, the parts start wearing down: Reflexes slow, hearing and eyesight dim, and stamina and muscle strength decrease.

After age 65, chronic rather than acute diseases comprise the majority of health problems, including disability. **Chronic diseases** (e.g., asthma and high blood pressure) are long-term or lifelong illnesses that develop gradually or are present from birth. In contrast, **acute diseases** (e.g., chicken pox) are illnesses that strike suddenly and often disappear rapidly but can cause

incapacitation and sometimes death. Chronic diseases increase as people age. *Dementia*, a loss of mental abilities, most commonly occurs during one's 70s. The most debilitating form of dementia is **Alzheimer's disease**, a progressive, degenerative disorder that attacks the brain and impairs memory, thinking, and behavior.

Gender. The gender gap in life expectancy has decreased since 1975, but women, on average, still live longer than men (see Chapter 12). Men of all racial-ethnic groups are two to three times more likely than women to die in motor vehicle crashes, to be victims of homicide, to smoke and drink alcohol, to abuse drugs, and to work in dangerous occupations (e.g., construction, law enforcement) (National Center for Health Statistics, 2015). Women are also more likely than men to have a regular physician, to see a doctor when something's wrong, and to keep appointments (Levine, 2014).

Depression is the single most common mental health problem among adolescents, usually begins at about age 14, and affects 29 percent of U.S. high school students. Girls (36 percent) are more likely than boys (22 percent) to report being depressed, but the reasons are similar: substance abuse, bullying (either as a victim or perpetrator), sexual or physical abuse, parental divorce, and a family history of mental disorders. In adulthood, and across all ages, women are more likely than men to be taking antidepressant medications (Murphey et al., 2013; National Center for Health Statistics, 2015).

Depression may lead to suicide. Among high school students, females are twice as likely as males to attempt suicide. Nationally, however, men are about four times more likely than women to die by suicide, and the highest

chronic diseases long-term or lifelong illnesses that develop gradually or are present from birth.

acute diseases illnesses that strike suddenly and often disappear rapidly but can cause incapacitation and sometimes death.

Alzheimer's disease a progressive, degenerative disorder that attacks the brain and impairs memory, thinking, and behavior.

PhotoPlus Magazine/Getty Images
Thomas Trutschel/Getty Images
UniversalImagesGroup/Getty Images

Three reasons why women live longer than men.

rates are among white men aged 75 and older. Suicide rates have risen since 2007, particularly for young females aged 10 to 14. The reasons for the upturns are unclear, but may be due to the increased availability of firearms and media coverage of incidents that escalate the risk of "suicide contagion" (National Center for Health Statistics, 2015; Sullivan et al., 2015).

Social Class. Social class has a strong effect on health. As early as age 45, the lower the family income, the greater the likelihood that adults will experience two or more chronic diseases. Living in poor neighborhoods increases the likelihood of stress (which affects diet, smoking, and alcohol/drug usage), a sedentary lifestyle because of limited recreational facilities, and less access to health care (Beckles and Truman, 2013; Hummer and Hernandez, 2013).

Policymakers have tried to eliminate *food deserts*, geographic areas where access to fresh, healthy, and affordable food is limited or nonexistent because grocery stores are too far away. Social class disparities persist even when consumers have access to the same food, however. Even in the same store, less educated households purchase less nutritious food, primarily because of preferences for unhealthy food and higher prices of healthier food (Handbury et al., 2015). Thus, access to nourishing food doesn't necessarily create healthier eaters.

People in lower social classes are also more likely to have dangerous jobs. In 2014, nearly 4,700 U.S. workers died due to occupation-related injuries. More than 3 million experienced workplace injuries or illnesses caused, among other things, by falls, overexertion, and accidents using equipment ("Employer-Reported...," 2014; "National Census of Fatal...," 2015).

Race and Ethnicity. A good measure of a population's health is its *infant mortality rate*, the number of babies under age 1 who die per 1,000 live births in a given year. Infants born to black women are more than twice as likely to die before age 1 as those born to white, Asian, and Hispanic women (see *Figure 14.2*). The higher black infant mortality rate may be due to a combination of factors: More babies born before 24 weeks who don't survive, less access to quality health care, and black mothers' greater likelihood of having some conditions (high blood pressure, diabetes, obesity) that are associated with negative birth outcomes (Chen et al., 2014; MacDorman et al., 2014).

You've seen in previous chapters that blacks are disproportionately poor, which raises the likelihood of living in neighborhoods with high crime rates, and not receiving quality health care. Moreover, encountering racism in day-to-day life intensifies stress, anxiety, and fears that are

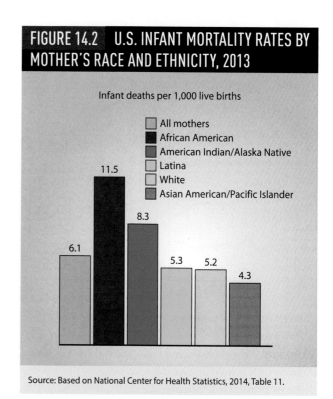

FIGURE 14.2 U.S. INFANT MORTALITY RATES BY MOTHER'S RACE AND ETHNICITY, 2013

Infant deaths per 1,000 live births

- All mothers
- African American
- American Indian/Alaska Native
- Latina
- White
- Asian American/Pacific Islander

All mothers: 6.1
African American: 11.5
American Indian/Alaska Native: 8.3
Latina: 5.3
White: 5.2
Asian American/Pacific Islander: 4.3

Source: Based on National Center for Health Statistics, 2014, Table 11.

associated with mental health problems, and higher premature death rates due to hypertension and other diseases.

You'll recall that Latinos, despite their generally low education levels and household incomes, have the highest life expectancy rates (see *Table 12.3* and the related text). Some researchers speculate that this "Hispanic health paradox" may be due to several reasons: (1) a greater likelihood of being undercounted than other groups because of their unauthorized status, therefore leading to a false appearance of greater longevity; (2) those who migrate to the United States tend to be healthier than those who don't migrate; and (3) compared with other racial/ethnic groups, Latinos are the least likely to engage in unhealthy behavior and to die from illicit drugs and prescription drug abuse (Frieden, 2011; Riosmena et al., 2013).

The longer Latinos live in the United States, however, the higher their rates of heart disease, high blood pressure, some types of cancer, and diabetes. Acculturation may result in improved access to health care and preventive services, but also to adopting unhealthy behaviors (e.g., smoking, excessive alcohol consumption), less physical activity, and unhealthier diets (Dominguez et al., 2015; "Hispanic Health," 2015; Riosmena et al., 2015).

LIFESTYLE FACTORS

Our lifestyle choices can improve or impair our health. The top three preventable lifestyle health hazards, in order of priority, are smoking, obesity, and substance abuse. Sexually transmitted diseases also cause infections and illness.

FIGURE 14.3 U.S. CIGARETTE SMOKING, BY SEX AND AGE, 2002–2013

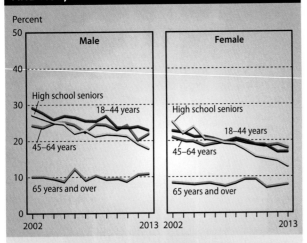

Source: National Center for Health Statistics, 2015, Figure 8.

Smoking. Worldwide and in the United States, tobacco use, primarily cigarette smoking, is the leading cause of preventable disease, disability, and death. U.S. smoking declined from 42 percent in 1965 to 18 percent in 2013 among adults, and from 36 percent in 1997 to 16 percent in 2013 among high school students (Agaku et al., 2014; Kann et al., 2014). Smoking continues to be higher among males than females across all age groups, and hasn't changed much for the oldest population (see *Figure 14.3*).

There are now more former than current smokers, but tobacco use is responsible for about 1 in 5 deaths

From 2013 to 2014, the number of middle and high school students using electronic cigarettes more than tripled, rising from 780,000 to 2.5 million students (Arrazola et al., 2015). To attract adolescents, the e-cigarette industry uses celebrities, cartoon characters, and "kid-friendly" flavors like chocolate, gummy bears, and mint candy (Zimmerman, 2014).

annually, and, on average, smokers die about 10 years earlier than nonsmokers. Because tobacco harms nearly every human organ, smoking is linked to cancer, heart disease, stroke, Type 2 diabetes, macular degeneration, erectile dysfunction, lung diseases (including emphysema and bronchitis), and birth defects (U.S. Department of Health and Human Services, 2014).

Prevention is difficult because the tobacco industry spends almost $10 billion a year to market its products, half of all movies for children under 13 contain scenes of tobacco use, half of the states continue to allow smoking in public places, and images and messages normalize tobacco use in magazines, on the Internet, and at retail stores frequented by youth. During 2014, states collected more than $25 billion from tobacco taxes and legal settlements, but spent only 2 percent of the money on tobacco control programs (Agaku et al., 2014).

Public health experts are debating whether electronic cigarettes, or e-cigarettes, are safer than tobacco cigarettes. E-cigarettes deliver nicotine in a liquid smoke-free vapor (often called "vaping"), and don't contain the same carcinogens and other toxic chemicals as regular cigarettes. One of the earliest studies found that e-cigarettes were "modestly effective" in helping smokers quit tobacco products, but the results were similar to wearing a nicotine patch (Bullen et al., 2013). Since then, researchers have found that adolescents who use e-cigarettes are less likely to smoke tobacco cigarettes as well. Among adolescents who have never smoked, however, those who try e-cigarettes are likely to start smoking conventional cigarettes within a year (Arrazola et al., 2015; Leventhal et al., 2015).

Obesity. Obesity is the second leading and preventable cause of disease, disability, and death. As you saw in Chapter 6, childhood and adult obesity has increased since the early 1970s. Preschoolers who are overweight or obese are 5 times more likely than normal-weight children to be overweight or obese as adults, increasing their risk for heart disease, high blood pressure, strokes, diabetes, osteoporosis, and several types of cancer. Nonetheless, almost 79 percent of parents of obese preschoolers describe their children as "about the right weight" (Cunningham et al., 2014; Duncan et al., 2015).

In 1990, not a single state had an obesity rate above 15 percent. Now *all* states have obesity rates above 20 percent and 22 states have obesity rates equal to or greater than 30 percent ("Obesity Prevalence Maps," 2015). Some researchers predict that, at the current rate, 51 percent of American adults will be obese in 2030 (see *Figure 14.4*), and 9 percent of that group will be severely obese (Finkelstein et al., 2012).

FIGURE 14.4 | U.S. ADULTS KEEP GETTING FATTER

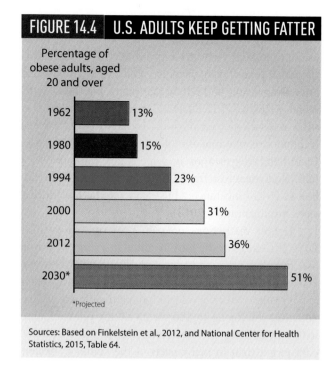

Percentage of obese adults, aged 20 and over

- 1962 — 13%
- 1980 — 15%
- 1994 — 23%
- 2000 — 31%
- 2012 — 36%
- 2030* — 51%

*Projected

Sources: Based on Finkelstein et al., 2012, and National Center for Health Statistics, 2015, Table 64.

Why is obesity more prevalent than in the past? People are getting fatter because of poor eating habits, large food portions, and not exercising, but structural factors also affect obesity rates. Healthy food is less expensive than during the 1950s, more widely available than ever before, and easy to prepare. However, a smaller share of income now buys many more calories, including sugar-sweetened beverages and prepared food that are high in unhealthy carbohydrates or fat. Thus, since 1970, the average American's consumption of calories has risen by about 20 percent, fueling an "obesity epidemic" (Sturm and An, 2014).

Low-fat diets rarely reduce weight because much of our food contains sugar that's added during preparation, processing, or at the table. Products with added sugar include those with packaging statements like "made with whole grain," "excellent source of calcium," "fat-free," "good source of Vitamin D," or "100% juice." However, a small container of Greek yogurt can have three or more teaspoons of added sugar.

Substance Abuse. The nation's third leading lifestyle-related and preventable cause of death is **substance abuse**, a harmful overindulgence in or dependence on a drug or other chemical. Nearly 85,000 Americans die each year because of excessive drinking, and more than 4 million visit emergency rooms for alcohol-related problems ("Alcohol and Public Health," 2014).

Excessive alcohol use includes *binge drinking*, having 4 or more drinks on a single occasion for women or 5 or more drinks on a single occasion for men, generally within about 2 hours. *Heavy drinking* involves consuming 15 or more drinks per week for men, and 8 or more

substance abuse a harmful overindulgence in or dependence on a drug or other chemical.

How Much Sugar Are You Really Eating?

The American Heart Association recommends limiting our daily dose of added sugar to 24 grams (about 6 teaspoons) for women and 36 grams (about 9 teaspoons) for men.

Vitamin-enhanced water (20 ounces)	Blueberry muffin	Prepared coleslaw (1 cup)	Thin Mints (10)	Slice of frozen pizza	Fast-food chicken sandwich
32 g of sugar	22 g of sugar	23 g of sugar	26 g of sugar	6 g of sugar	16 g of sugar

TABLE 14.1 SOME CONSEQUENCES OF ABUSIVE COLLEGE DRINKING

Researchers estimate that, each year, among college students aged 18 to 24...

1,825	Die from alcohol-related unintentional injuries, including motor vehicle crashes
599,000	Are unintentionally injured while drunk
696,000	Are assaulted by another student who's been drinking
97,000	Are victims of alcohol-related sexual assault or date rape
400,000	Have unprotected sex while intoxicated, and more than 100,000 were too drunk to remember if they consented to having sex
3.4 million	Drive while drunk
25%	Have academic problems because of their drinking, including missing classes, doing poorly on exams or papers, earning lower grades overall, and flunking out of college

Sources: Based on National Institute on Alcohol Abuse and Alcoholism, 2013, 2014.

drinks for women. Some 23 percent of Americans are binge drinkers and 6 percent are heavy alcohol users. Among young people ages 12 to 20, drinking fell from 29 percent in 2002 to 23 percent in 2014. The reasons for the downturn are unclear, but may be due to adolescents' belief that alcohol is more risky than smoking marijuana (Center for Behavioral Health Statistics and Quality, 2015; Substance Abuse and Mental Health Services Administration, 2015). *Table 14.1* summarizes some of the consequences of excessive college drinking on students, their families, and the larger community.

Excessive alcohol use results in myriad short- and long-term health risks that include injuries, violence, risky sexual behaviors, miscarriage and stillbirth, physical and mental birth defects, alcohol poisoning, unemployment, psychiatric problems, heart disease, several types of cancer, and liver disease. For teenagers and young adults under age 25, alcohol consumption can irreversibly damage a part of the brain that involves memory, learning, and social interaction (Crowe et al., 2012).

Illicit (illegal) *drugs* include marijuana/hashish, cocaine and crack, heroin, hallucinogens (e.g., LSD, PCP), inhalants, and any prescription-type psychotherapeutic drug, including stimulants and sedatives, used nonmedically. More than 10 percent of Americans (27 million people) age 12 and older use illegal drugs. Illicit drug use is more common among men than women, among people aged 18 to 25, and least common among Asian Americans (National Center for Health Statistics, 2015).

Since 2012, Alaska, Colorado, Oregon, and Washington have legalized marijuana for recreational use, and 23 states (so far) and the District of Columbia allow marijuana for medical use. Federal laws still forbid possessing, using, and selling marijuana, but the Department of Justice isn't challenging state laws that enact strict regulations to control its use and sale. Marijuana is the most commonly used illicit drug (see *Figure 14.5*). Even before legalization in some states, the number of marijuana users rose from 14.5 million in 2007 to 19 million in 2012 (Substance Abuse and Mental Health Services Administration, 2013).

Tighter restrictions on pain prescriptions such as Oxycontin and Vicodin have resulted in a surge of heroin use, doubling among persons age 12 and older between 2007 and 2014 (Center for Behavioral Health Statistics and Quality, 2015). Prescription pain pills containing narcotics sell for up to $80 for an 80 mg pill compared with only $9 for a dose of heroin. Heroin has become so easy to get that, in many parts of the country, "dealers deliver to the suburbs and run specials to attract young, professional, upper-income customers." According to the substance abuse director at a North Carolina medical center, "Our heroin patients come from the five best neighborhoods" (Leger, 2013).

Most illicit drugs have the same health risks as excessive alcohol use but can also have immediate life-threatening consequences, including death. In addition, illicit drug users, compared with non-users, have higher rates of mental illness, suicidal thought and behavior, major depressive episodes, and impaired ability to function at school, home, or work (Hedegaard et al., 2014; Jones et al., 2015).

Sexually Transmitted Diseases. There are nearly 19 million new sexually transmitted disease (STD) cases in the United States every year. People infected with

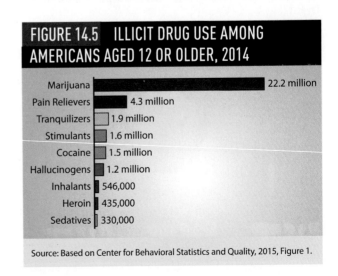

FIGURE 14.5 ILLICIT DRUG USE AMONG AMERICANS AGED 12 OR OLDER, 2014

Marijuana	22.2 million
Pain Relievers	4.3 million
Tranquilizers	1.9 million
Stimulants	1.6 million
Cocaine	1.5 million
Hallucinogens	1.2 million
Inhalants	546,000
Heroin	435,000
Sedatives	330,000

Source: Based on Center for Behavioral Statistics and Quality, 2015, Figure 1.

STDs are at least two to five times more likely than unin-fected individuals to contract HIV, the virus that causes AIDS. More than 1.2 million Americans are living with HIV, including 14 percent who aren't aware of their in-fection. Men who have sex with men (MSM)—whether straight, gay, or bisexual—represent only 4 percent of the male population but 78 percent of new HIV infections. Another high-risk group is people who inject drugs. HIV-infected people live much longer than in the past, but almost 14,000 Americans died of AIDS in 2012 (CDC Fact Sheet, 2014; "HIV in the United States...," 2015).

People aged 55 and older account for 26 percent of the 1.2 million living with HIV. Of the older Americans with new infections, 23 percent are women. Some are injection drug users, but most don't insist that men use condoms because they no longer worry about getting pregnant. The women aren't aware that their partners are MSM, inject drugs, or have been recently tested pos-itive for a STD, including HIV (Benokraitis, 2015; "HIV Among People...," 2015; Spiller et al., 2015).

We've looked at some of the reasons why some peo-ple are healthier than others. Who gets health care? And who pays for it?

14-3 HEALTH CARE: UNITED STATES AND GLOBAL

Health care encompasses a number of components. One of the most important is **medicine**, a social institution that deals with illness, injury, and other health problems (see also Chapter 6 on institutions).

14-3a U.S. Health Care Coverage

In 2010, Congress passed the Patient Protection and Afford-able Care Act, also called the Affordable Care Act (ACA) or "Obamacare." The ACA's goal was to give more Americans age 64 and under access to affordable, quality health insur-ance, and to reduce the growth in health care spending.

Has the ACA worked? The uninsured rate de-creased from 18 percent in 2013 (before the ACA took effect) to 10 percent in 2014, and continues to drop. The ACA produced the most significant gains in coverage for people living below or just above the poverty line (see Chapter 8) and Latinos—the groups most likely to lack insurance (Levy, 2015; Smith and Medalia, 2015).

Who are the 33 million uninsured? People who live in states where Medicaid doesn't cover the "nearly

poor," are undocumented immigrants, haven't enrolled in Medicaid, or have incomes above the federal poverty level but can't afford the programs offered by employ-ers or private insurers. The largest group of the unin-sured—about 45 percent—foregoes coverage by choice for a variety of reasons: Paying for care out-of-pocket as needed is less expensive than health insurance, they rely on free or low-cost clinics, go without care or medi-cine, and/or use emergency rooms (Radnofsky, 2015; Robert Wood Johnson Foundation, 2015).

14-3b Who Pays for Health Care?

About 64 percent of Americans have health coverage through private insurance, primarily employer-based programs that cover from some to all costs for work-ers and their family members. Another 36 percent have government health insurance funded by taxpayers (see *Figure 14.6*).

Employer-based health insurance coverage has de-teriorated. Some employers with 50 or more full-time workers are considering stopping health insurance be-cause the ACA's penalties will be less expensive than the insurance costs. Many large companies (Target, Home Depot, Walgreens) continue to maximize their record profits by ending health plans for part-time workers and increasing the premiums of full-time employees. Since 1999, workers at both large and small firms have been paying higher premiums, deductibles, copayments, and

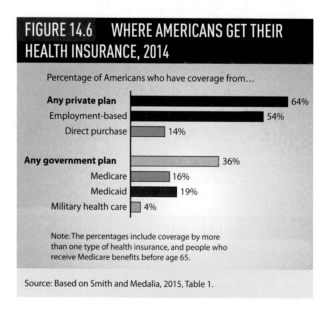

FIGURE 14.6 WHERE AMERICANS GET THEIR HEALTH INSURANCE, 2014

Percentage of Americans who have coverage from...

Any private plan	64%
Employment-based	54%
Direct purchase	14%
Any government plan	36%
Medicare	16%
Medicaid	19%
Military health care	4%

Note: The percentages include coverage by more than one type of health insurance, and people who receive Medicare benefits before age 65.

Source: Based on Smith and Medalia, 2015, Table 1.

medicine a social institution that deals with illness, injury, and other health problems.

other out-of-pocket costs, whereas wages, including those of middle-class families, have stagnated (Kellerman and Auerbach, 2013; Jilani, 2014; Eibner, 2015).

Most people pay a variety of taxes to states and the federal government to support programs like Medicare and Medicaid. Medicare pays most of the medical costs of Americans age 65 and over, regardless of income. Thus, even billionaires are eligible for Medicare. Medicaid, another government program, provides medical care, regardless of age, for people living below the poverty level (see Chapter 12).

14-3c The United States Compared with Other Countries

The United States spends more on health care than any other nation in the world—more, in fact, than the next 10 highest-spending developed countries combined. U.S. health care costs have risen steadily—per person, nationally, and as a percentage of the gross domestic product. We spend more on health care than 34 other high-income countries, but rank last among the top 11 in quality of care, access to care, efficiency, and health outcomes (Davis et al., 2014; OECD, 2014).

By 2022, health care expenses will consume $1 of every $5 in the economy (see *Table 14.2*). Why, then, are many Americans worse off than their counterparts in other high-income countries? One reason is that from a third to almost half of U.S. health care spending is wasteful (Redberg, 2012; Newhouse et al., 2013).

In Canada, Germany, Great Britain, France, Japan, and Sweden, the government picks up most of the health care bill. In these and other countries, patients and health care workers don't have to submit bills to several insurance providers, resulting in considerable administrative savings. The government, nonprofit organizations, or large groups (cities and industries) have considerable power in keeping down drug costs and setting prices for health care providers and services. There are also private hospitals, but most are nonprofit (OECD, 2014).

In the United States, "hospitals are the most expensive part of the world's most expensive health system." Even nonprofit hospitals are businesses: Up to 26 percent of their revenues are profits, executives can earn almost $6 million a year, and they sell as many services as possible at the highest price (Brill, 2013; "Prescription for Change," 2013).

With its high prescription prices, the United States spends far more per capita on medicines than other developed countries. Unlike other countries where the government sets a national wholesale price for each drug, the United States lets pharmaceutical companies set the prices. Congressional lawmakers—heavily lobbied by the pharmaceutical industry, the American Hospital Association, and other groups—have forbidden Medicare, the nation's largest medical insurer and the world's largest buyer of prescription drugs, to negotiate drug prices or the cost of health products. Thus, a hospital can charge $24.00 for each niacin (Vitamin B3) pill that drugstores sell for about a nickel apiece, and $1.50 for an acetaminophen tablet (which relieves pain and fever) that costs $1.49 for a box of 100 at Amazon—a 10,000 percent markup (Brill, 2013). Recently, a medication for hepatitis C, a liver disease, jumped from $36,000 to $84,000 for a 12-week course of treatment. Taxpayers, who pay for most of Medicare's and Medicaid's prescription drugs, will bear the costs (Cohn, 2015; Ornstein, 2015).

Medical technology has also increased health care costs. There are 32 magnetic resonance imaging (MRI) machines per 1 million people in the United States compared with only 6 per 1 million in the United Kingdom. U.S. doctors do 71 percent more CT scans than do doctors in Germany. Technology can save lives, but U.S. doctors, compared with those in many other countries, are more likely to adopt new and expensive machines that are highly profitable for both physicians and hospitals. The more machines there are, the more doctors use them, increasing a patient's, insurer's, and nation's medical costs (Brill, 2013, 2015). *Figure 14.7* provides more examples of why U.S. health care costs are so high. Because of the

TABLE 14.2 THE INCREASING COST OF U.S. HEALTH CARE, 1980–2022

	1980	1990	2000	2010	2022
Average cost per person	$1,100	$2,864	$4,878	$8,402	$14,664
National health expenditure	$256 billion	$724 billion	$1.4 trillion	$2.6 trillion	$5 trillion
Percent of gross domestic product (GDP)	9%	13%	14%	18%	20%

Notes: The "average cost per person" includes medical care, supplies, drugs, and health insurance. All numbers are in current dollars; those for 2022 are projected.

Sources: Based on Centers for Medicare & Medicaid Services, 2010, Table 1, and 2014, Table 1.

FIGURE 14.7 SHOULD YOU TRAVEL ABROAD FOR MEDICAL CARE?

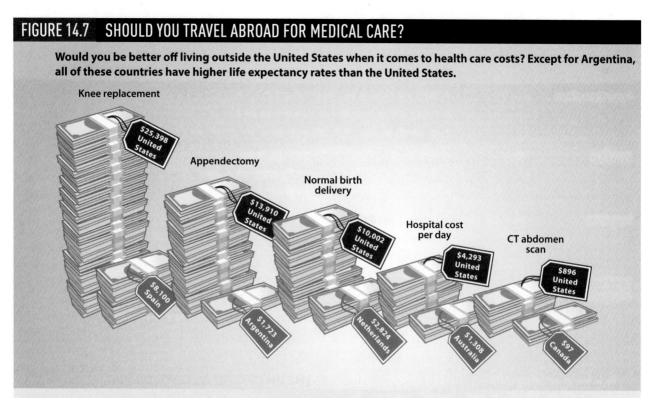

Would you be better off living outside the United States when it comes to health care costs? Except for Argentina, all of these countries have higher life expectancy rates than the United States.

Knee replacement

$25,398 United States

$8,100 Spain

Appendectomy

$13,910 United States

$1,723 Argentina

Normal birth delivery

$10,002 United States

$2,824 Netherlands

Hospital cost per day

$4,293 United States

$1,308 Australia

CT abdomen scan

$896 United States

$97 Canada

Source: Based on International Federation of Health Plans, 2014. See also Howard, 2014, for a comparison of surgery costs in the United States and other countries.

high costs, each year up to 750,000 Americans engage in *medical tourism*, traveling to another country for medical care ("Medical Tourism," 2015).

 14-4 SOCIOLOGICAL PERSPECTIVES ON HEALTH AND MEDICINE

Sociologists focus on different aspects of health, health care, and medicine. *Table 14.3* summarizes these perspectives.

14-4a Functionalism: Good Health and Medicine Benefit Society

For functionalists, good health and medicine are critical for a society's survival and stability. Thus, countries try to develop a medical care system that, ideally, benefits the entire population regardless of age, sex, race, ethnicity, social class, or other characteristics.

HEALTH CARE AND OTHER INSTITUTIONS

Health care systems are connected to many institutions. Despite ongoing debates about the ACA and the

government's requiring people to buy health insurance, millions of individuals and families are now eligible for government assistance to pay for coverage, and can get free preventive health services (e.g., diabetes tests, colonoscopies, routine vaccinations).

The government is deeply involved in health care in other ways. It funds much of the scientific research at universities as well as federal agencies like the Centers for Disease Control and Prevention and the National Institutes of Health, both of which deal with prevention, treatment, and health care policy. At the national and state levels, numerous government agencies are responsible for passing and enforcing regulations regarding new drugs, medical procedures, and access to medical care.

Health care affects the economy because it's the nation's largest employer. "Demand for health services will rise as Obamacare expands insurance and Americans grow older, fatter, and sicker" ("The Health Paradox," 2013: 27). Our huge health care spending threatens funding for other public programs, including education and transportation. Functionalists would point out, however, that the trillions spent on health care have generated millions of jobs at hospitals, drug companies, insurance companies, nursing homes, and information-technology firms (see Chapter 11).

TABLE 14.3 SOCIOLOGICAL PERSPECTIVES ON HEALTH AND MEDICINE

Perspective	Level of Analysis	Key Points
Functionalism	Macro	• Health and medicine are critical in ensuring a society's survival and are closely linked to other institutions. • Illness is dysfunctional because it prevents people from performing expected roles. • Sick people are expected to seek professional help and get well.
Conflict	Macro	• There are gross inequities in the health care system. • The medical establishment is a powerful social control agent. • A drive for profit ignores people's health needs.
Feminist	Macro and micro	• Women are less likely than men to receive high-quality health care. • Gender stratification in medicine and the health care industry reduces women's earnings. • Men control women's health.
Symbolic Interaction	Micro	• Illness and disease are socially constructed. • Labeling people as ill increases their likelihood of being stigmatized. • Medicalization has increased the power of medical associations, parents, and mental health advocates and the profits of pharmaceutical companies.

Besides generating jobs, health care systems shape corporate policies which, in turn, benefit families. Recently, for instance, the pharmacy chain CVS banned sales of all tobacco products, Target is promoting employees' wellness and adding more organic and natural food to grocery aisles, and some restaurants (e.g., McDonald's, Wendy's, Red Lobster) have taken sodas off their children's menus and reduced sodium and fat in kids' meals (D'Innocenzio, 2015; Martin, 2015).

THE SICK ROLE

Talcott Parsons (1951), an influential sociologist, introduced the concept of the **sick role**, a social role that excuses people from normal obligations because of illness. In Parsons' model, sick people aren't responsible for their condition and, therefore, have legitimate reasons for not performing their usual social roles ("I missed the exam because I had the flu").

Parsons emphasized that the sick role is temporary, and that individuals must seek medical help to hasten their recovery. Otherwise, people will view them as hypochondriacs, slouchers, and malingerers who aren't living up to their responsibilities. Thus, from a functionalist perspective, the sick role is legitimate if it's short-lived, but dysfunctional if people feign illness to shirk their duties long-term in the family, the workplace, or other groups.

sick role a social role that excuses people from normal obligations because of illness.

THE PHYSICIAN AS GATEKEEPER

From a functionalist perspective, physicians play a key gatekeeping role in limiting the sick role. They verify a person's condition as sick and provide an excuse for temporarily not performing necessary roles. They also designate a patient as "recovered" and ready to meet societal role expectations again. Doctors' specialized knowledge gives them considerable authority in defining health and

Iakov Filimonov/Shutterstock.com

How sick is too sick to miss work or class? Are menstrual cramps enough? What about a severe cold? A hangover? And when does being sick become dysfunctional—after a day, a week, a month?

illness that's unmatched by other health care providers, such as nurses and pharmacists.

CRITICAL EVALUATION

Functionalist theories are limited for several reasons. First, health care policies can be divisive. For example, 56 percent of Americans say that health care coverage isn't the government's responsibility (Wilke, 2013). Moreover, most of the job growth in the health care sector has been in low-paying occupations (hospital orderlies, nursing home assistants, lab technicians) (Ross et al., 2014). Thus, many people aren't benefiting from the booming health care industry.

Second, people often view those with chronic illnesses like arthritis and diabetes as healthy enough to perform their expected roles regardless of how they sick they feel. Despite a life-threatening illness, however, many people can't assume even a short-lived sick role. They don't have the resources to see doctors or may lose a portion of their earnings if they miss workdays. Almost one in five Americans with health insurance has put off treatment for a serious condition because of high out-of-pocket costs (Brown, 2013).

Third, because physicians are powerful gatekeepers, many prosper by prolonging sick roles. The more times that doctors see patients or visit them at the hospital, and the more tests that doctors order, the higher their salaries. Conflict theorists, especially, maintain that much of the health care industry is dysfunctional because access to health care varies considerably, and only the wealthy don't have to worry about receiving and paying for the best available medical care.

14-4b Conflict Theory: Health Care and Medicine Don't Benefit Everyone

For conflict theorists, medicine and the health care industry benefit some groups much more than others. In contrast to functionalists, conflict theorists argue that the medical system reinforces social inequality, exerts social control to maintain the status quo, and is often driven by a profit motive rather than a concern for people's well-being.

SOCIAL CONTROL OF MEDICINE AND HEALTH CARE

From a conflict perspective, those at the top of the medical hierarchy have considerable control. Physicians, for instance, have almost absolute power in diagnosing an illness, providing treatment, and deciding on medical procedures.

The American Medical Association (AMA) and the health care industry can maintain the status quo because of the **medical-industrial complex**, a network of business enterprises that influences medicine and health care. The medical-industrial complex includes many groups—doctors, nurses, hospitals, lawyers, nursing homes and hospices, insurance companies, drug manufacturers, accountants, banks, and real estate and construction businesses.

Who benefits from the medical-industrial complex? All of those involved, especially hospitals and drug companies. Big hospitals have been buying up previously independent cancer clinics and doctors' offices. Doing so gives them the power to raise prices for private insurers and government programs, particularly Medicare. As a result, they can charge higher rates, sometimes three times as much, for cancer drugs and outpatient visits and services (Bach, 2015; Bai and Anderson, 2015).

The percentage of Americans taking five or more prescription drugs increased from 4 percent in 1994 to 20 percent in 2012, and consumption of narcotic painkillers, or opioids, jumped 300 percent between 1999 and 2010. The prices for dozens of drugs that treat everything from blood pressure to multiple sclerosis have doubled since 2007, and the prices of some generic drugs have soared by 1,000 percent or more (Institute for Clinical and Economic Review, 2015; Jaret, 2015; National Center for Health Statistics, 2015).

Direct-to-consumer prescription drug marketing ("Ask your doctor about…"), which began in the 1980s, increases a drug's sales. People who see drug advertising are more likely to ask their doctors for a particular brand-name medication. Doctors may feel pressure to accommodate their patients' requests, which, in turn, increases a drug's price and sales (Timmermans and Oh, 2010; Niederdeppe et al., 2013).

HEALTH, PROFIT, AND WASTE

From a conflict perspective, health care is big business. In 2013, the average compensation of an health insurance CEO was almost $14 million (a 19 percent increase over the previous year) compared with $35,000 for the average hospital worker. Large hospitals are crowded with "senior VPs, VPs of this, that, and the other" whose salaries are two to three times higher than those of surgeons. A Wisconsin surgeon discovered that a brief

---

medical-industrial complex a network of business enterprises that influences medicine and health care.

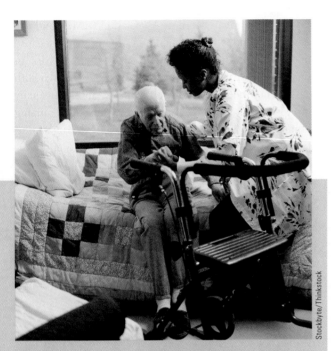

End-of-life care is a billion dollar industry. Compared with their nonprofit counterparts, for-profit hospices spend less on nursing per patient, provide fewer services, and push out patients whose care becomes expensive. Medicare provides about 90 percent of funding for hospices and nursing homes, but rarely punishes either for violating health and safety rules (Hallman and Shifflett, 2014; Whoriskey and Keating, 2014).

services that have the same health outcomes as lower-priced alternatives, preventable hospital injuries and readmissions that cost Medicare almost $5 billion a year, unnecessary tests and treatments to guard against liability in malpractice lawsuits, and electronic medical records that result in needlessly complex and time-consuming work for physicians and their staff (Eisler and Hansen, 2013; Office of Inspector General, 2013; Krauthammer, 2015; Pear, 2015).

CRITICAL EVALUATION

The most common criticism is that conflict theorists often ignore the contributions of medical and health care systems. Without them, people would suffer more, die at a younger age, and have a lower quality of life. Second, not all medical providers are motivated by profit. For example, doctors in family practices earn, on average, only about half as much ($208,000) as surgeons (Bureau of Labor Statistics, 2014–2015).

Third, doctors, medical scientists, and some medical associations (e.g., American Academy of Family Physicians) have been among the most vocal critics of unneeded surgery, tests, and procedures. Examples include cardiac stress tests, antibiotic prescriptions for sinusitis that usually resolves itself within a few weeks, and recommending vitamins and other supplements that don't benefit health. Others have urged ending expensive end-of-life care that simply prolongs the dying process. Most recently, nearly 120 of the nation's leading oncologists have publicly protested drug companies' out of control greed which is bankrupting cancer patients (Kellerman, 2012; Guallar et al., 2013; Agnvall, 2014; Tefferi et al., 2015).

Finally, most people aren't simply victims of a malicious medical-industrial complex. Many make unhealthy lifestyle choices (unprotected sex, drug abuse), and complain rather than change their behavior when employers impose health insurance penalties for smoking and obesity. Others join forces with drug companies to lobby the FDA to decrease current restrictions on potent narcotic painkillers (Meier and Lipton, 2013; Brody, 2015).

14-4c Feminist Theories: Health Care and Medicine Benefit Men More Than Women

Both feminist and conflict theorists emphasize the connection between health and inequality. Feminist scholars go further by addressing the health costs of being a woman, gender stratification in medicine and health care, and men's control over women's choices.

outpatient appendectomy he had performed for a fee of $1,700 generated over $12,000 in hospital bills. Thus, "the biggest bucks are earned not through the delivery of care, but from overseeing the business of medicine" (Healthcare-NOW! 2014; Rosenthal, 2014: SR4).

Most physicians are concerned about quality care, but Medicare is a source of wealth for many. In 2012, 2 percent of doctors received 23 percent of all Medicare fees, including a Florida ophthalmologist who took in more than $26 million to treat fewer than 900 patients. In the same year, Medicare also paid $7 million to at least eight doctors whose medical licenses had been revoked or suspended (Terhune et al., 2014; Tozzi et al., 2014).

About 37 percent of the Medicaid and Medicare spending is wasteful. Between 10 and 20 percent of all surgeries are unnecessary, and more than 2,200 doctors nationwide are suspected of writing elderly and disabled patients prescriptions, particularly opioids, that are expensive, addictive, and medically unnecessary. Other wasteful spending includes the following: expensive

Is Being Fashionable Hazardous to Women's Health?

Squeezing into skinny jeans and tight pants can cause nerve compression, numbness, and digestive problems. Narrow-toed shoes and stilettos wreak havoc: blisters, bunions, corns, calluses, hammertoes, nerve damage, back and neck pain, painful inflammation, stress fractures, and ankle sprains. Still, according to a podiatrist, "Women will wear their high-heeled shoes until their feet are bloody stumps" (Ianzito, 2013: 3; Gleiber, 2014).

Many cosmetics contain chemicals that can trigger skin problems (e.g., itching, rashes, acne). Others—contaminated with bacteria, yeasts, or molds—can lead to a range of problems from simple rashes to serious infections that can cause swelling and breathing difficulties. A recent study that tested 32 commonly sold lipsticks and lip glosses found that they contained lead, cadmium, chromium, aluminum, and five other metals—some at potentially toxic levels that are linked to stomach tumors. The more often and heavily that women use lipstick, the greater the health risks (Liu et al., 2013).

Except for some hair color additives, the FDA doesn't require companies to test products for safety or list toxic ingredients. Japan, Canada, and the European Union have banned more than 500 cosmetic products containing toxic ingredients and that are sold in the United States (Rano and Houlihan, 2012). Skin Deep (ewg.org/skindeep) ranks the safety of a range of cosmetic products.

maron/Age Fotostock

THE HEALTH COSTS OF BEING A WOMAN

Women, compared with men, have greater health care needs and risks, especially during their reproductive years. On average, 28 percent of women aged 15 to 44 with private health insurance and 39 percent with Medicaid have filled a prescription for an opioid. Taking these medications early in pregnancy, often before women know they're pregnant, can increase the risk for birth defects and preterm births (Ailes et al., 2015).

Almost 33 percent of women who give birth in the United States have a cesarean section (C-section), up from only 5 percent in 1970. The main reason for the increase is hospital administrators' and maternity clinicians' fear of lawsuits if a vaginal birth has a negative outcome (Morris, 2014). Because C-sections are major surgical procedures, the risks include infection, blood clots, and injury to other organs, particularly the bladder.

U.S. maternal mortality rates have increased. American women are now more than three times as likely to die from pregnancy-related complications as their counterparts in Britain, the Czech Republic, Germany, or Japan. The upturn is probably due to increasingly common health conditions (e.g., obesity, hypertension, diabetes) that make delivery more dangerous ("Maternal Mortality," 2015).

After following nearly 90,000 women aged 40 to 59 for 25 years, the researchers concluded that women who got mammograms every year for five years were no less likely to die of breast cancer than those who didn't, and recommended cutting back on the screenings (Miller et al., 2014). Mammograms can save women's lives, but also have health risks. If a tumor is benign and there's no history of breast cancer, patients could be subjected to unnecessary procedures—surgery, chemotherapy, and radiation—that do more harm than good given their complications (Ong and Mandl, 2015). Nonetheless, many major organizations, including the American Cancer Society, continue to recommend annual screenings beginning at age 40.

On average, women live longer than men but are more likely than men, particularly at later ages, to suffer from stress, obesity, hypertension, and chronic illnesses. When men report pain, they're often taken more seriously than women. Doctors, for instance, are more likely to prescribe painkillers for men, but sedatives for women (Chemaly, 2015).

For decades, heart disease "has been studied in men, for men, and by men." Even though heart disease is the leading cause of women's death, and kills more women than men, only 35 percent of participants in heart-related studies are women. Such research gaps "leave women's health to chance" (Johnson et al., 2014; Merz, 2015: A16).

GENDER STRATIFICATION IN HEALTH CARE OCCUPATIONS

Economists expect job growth in health care occupations to surge until at least 2022 because of a graying population and corporations' inability to offshore most medical services (Kolet and Chandra, 2012; see also Chapter 11 on offshoring). Many of these jobs, however, including home health aides and dental assistants (not dental hygienists), are low-paying and dominated by women.

In health care jobs, men consistently earn more than women (see *Table 14.4*). Among physical therapists, for example, males earn almost $77,000 a year compared with $68,000 for females even when both have the same level of education, years of experience, and work responsibilities (see also Chapters 8 and 11).

Feminist scholars attribute much of the gender wage gap to male doctors' gatekeeping. Because many registered nurses, pharmacists, and physical therapists (occupations that have more women than men) now receive doctoral degrees, they want to use the honorific title of "doctor." Doing so would win more respect from patients, help women land top administrative jobs that pay more, provide more autonomy in treating patients, and bring higher fees from health insurers. Physicians oppose the idea for several reasons: They want to maintain their prestige in the health care industry; they treat

patients first, whereas nurses and others play only secondary roles; and they have considerably more education and training than others to diagnose and treat illness and disease (Harris, 2011).

MEN'S CONTROL OF WOMEN'S HEALTH

Feminist scholars maintain that in patriarchal societies, including ours, men control many aspects of women's health. Catholic bishops, all of whom are men, condemn contraceptives as sinful, even though 98 percent of U.S. Catholic women have used birth control (Jones and Dreweke, 2011).

Most recently, the Supreme Court ruled, 5-4 (*Burwell v. Hobby Lobby Stores*), that privately held for-profit corporations don't have to pay for their employees' contraceptive coverage, as the ACA requires, if doing so conflicts with the owners' personal religious beliefs. In effect, a religious objection can trump a federal law, and businesses can shift insurance costs to taxpayers or private insurers. According to some Johns Hopkins doctors, "The Supreme Court decision has started us down a dangerous path on which the religious beliefs of a third party enter the examination room and interfere with the doctor-patient relationship" (Singal et al., 2014: 17; see also Rosenfeld, 2014). All of the justices who ruled against women's contraceptive coverage were males.

In many states, conservative lawmakers have passed laws to restrict or block women's ability to get abortions (see Chapter 9). They aren't passing similar laws to limit the availability of Viagra and other erectile dysfunction medications for men because "Viagra is a wonderful drug" (Beadle, 2012). The Hobby Lobby chain doesn't have to pay for IUDs and emergency contraceptive pills, but covers Viagra and other erectile dysfunction drugs.

CRITICAL EVALUATION

Critics have questioned feminist theories for three reasons. First, feminist scholars sometimes gloss over the fact that social class, rather than gender, has a big effect on people's health and receiving health care services (Kindig and Cheng, 2013; Montez and Zajacova, 2013).

Second, it's not entirely clear that men can be blamed for gatekeeping because 52 percent of

TABLE 14.4 WOMEN EARN LESS THAN MEN IN HEALTH CARE OCCUPATIONS, 2014

	Men	Women	Percentage of Women in This Occupation
All health care practitioners	**$65,312**	**$51,116**	**74**
Pharmacists	$113,152	$98,904	52
Physicians and surgeons	$104,104	$64,792	37
Physical therapists	$76,856	$67,964	61
Registered nurses	$61,880	$55,952	89
Licensed practical and vocational nurses	$45,136	$38,324	89
Clinical laboratory technologists and technicians	$45,656	$42,536	73

Note: The data are median annual earnings of full-time workers.

Source: Based on U.S. Bureau of Labor Statistics, 2015.

Nurse practitioners (NPs), 90 percent of whom are women, complete a master's or doctoral program that includes diagnosing and treating patients, managing acute and chronic illnesses, and prescribing medications. Because of a projected shortage of doctors, NPs are pressing 34 states to relax laws requiring doctors to oversee their work. Many doctors' groups, led by the AMA, are fighting the NPs, arguing that less supervision would jeopardize patient safety. Several national studies have found no evidence that NPs provide lower-quality care than physicians (Beck, 2013; Pettypiece, 2013). How might you explain the AMA's resistance from a feminist perspective?

14-4d Symbolic Interaction: The Social Construction of Health and Illness

Symbolic interactionists focus on how we define and construct views about health, illness, and medicine and then implement these definitions in everyday life. Social constructions include labeling, stigmatizing behavior, and medicalizing attitudes and behaviors as normal or sick.

THE SOCIAL CONSTRUCTION OF ILLNESS

Medical models assume that illness is anything that deviates from normal biological functioning, and that an illness has specific features that a doctor can recognize. In contrast, interactionists view illness and medicine as social constructions that can change over time. In 1956, the AMA declared that alcoholism is a treatable disease that's due to a genetic predisposition rather than a person's lacking moral character or self-discipline. And, in 2013, the AMA officially designated obesity as a disease—rather than a lifestyle habit—that requires research funding and medical intervention, including surgery and drugs (Newberry, 2013; Pollack, 2013).

With or without medical intervention, individuals construct and manage their illnesses differently. Some people's social worlds shrink when they become immersed in the day-to-day aspects of managing a chronic illness like rheumatoid arthritis: They become increasingly cut off from everyday routines if they're unable to work, spend less time with family and friends, or can't move about freely. Others create new self-identities by describing themselves as survivors (of breast cancer, for instance), exchange information about treatment options, and participate in local or national fundraising events (Conrad and Barker, 2010).

Social class also affects social constructions of illness. As educational level rises, people report poorer health, presumably because they're more aware of medical knowledge and evaluate their own physical well-being more critically (Schnittker, 2009).

THE MEDICALIZATION OF HEALTH

In 1952, the American Psychiatric Association (APA) published the first edition of the *Diagnostic and Statistical Manual of Mental Disorders* (DSM), which has become the global "bible of mental illness." The DSM relies on subjective definitions and criteria, but is influential in labeling (or unlabeling) mental illness. The number of disorders increased from 106 in 1952 to nearly 400 in 2013. Some of the new mental illnesses include "binge eating" (frequent overeating), "bereavement" (mourning

physicians and surgeons are women. Also, perhaps the gatekeeping by physicians (men or women) is justified because a doctoral degree in nursing, pharmacy, or physical therapy requires considerably less schooling and training than getting a Doctor of Medicine (M.D.) degree (Harris, 2011).

Third, and like conflict theorists, feminist scholars rarely address lifestyle choices that affect people's health. Among adult women, for example, 16 percent smoke, 36 percent are obese, and many use illicit drugs (National Center for Health Statistics, 2015).

the loss of a loved one), "caffeine intoxication" (drinking too much coffee), "disruptive mood dysregulation disorder" (children's temper tantrums), and "hoarding" (accumulating too much stuff) (American Psychiatric Association, 2013).

The DSM's ever-changing diagnoses and labels are an example of **medicalization**, a process that defines a nonmedical condition or behavior as an illness, disorder, or disease that requires medical treatment. The APA once classified homosexuality as a psychological disorder, but dropped sexual orientation from its roster of mental illnesses in 1973. The DSM affects patients, doctors, insurers, pharmaceutical companies, and taxpayers because psychiatrists and other physicians use it to bill insurance companies. DSM-based diagnoses also determine whether people get special services at school, qualify for disability benefits, are stigmatized, and even whether they can adopt children.

Medicalization is a lucrative business. Pharmaceutical corporations have reaped enormous profits because everyday normal anxieties, discomforts, and stresses (frustration with traffic, boredom with routine housekeeping chores, feeling sad or personally insecure) can be "fixed" by popping a Prozac or other pill (Herzberg, 2009). The more behaviors that the DSM defines as mental illness, the more likely psychiatrists are to increase their number of patients. Mental rights advocates and parents also benefit from medicalization, as when children who are diagnosed with ADHD get insurance coverage or special treatment, including more attention from teachers and health specialists (Conrad and Barker, 2010; Rochman, 2012).

According to critics, the DSM encourages misdiagnosis, overdiagnosis, the medicalization of normal behavior, and the prescription of a large number of unnecessary drugs, especially in the case of ADHD. In 2014, more than 10,000 2- to 3-year-olds (that's right, 2- to 3-year-olds!), particularly those on Medicaid, were being medicated for ADHD. About 11 percent of children aged 4 to 17 have been diagnosed with ADHD. Between

Monkey Business Images/Shutterstock.com

2003 and 2012, ADHD rates in this age group climbed by 42 percent, and more than two-thirds of those with a current diagnosis were receiving prescriptions for stimulants (e.g., Ritalin, Adderall). ADHD diagnoses among high-school students are particularly high—10 percent for girls and 19 percent for boys. Such numbers alarm many physicians because ADHD medications have many side effects that include delayed growth, sleep problems, decreased appetite, headaches, stomachaches, moodiness, and irritability (Rettew, 2014; Schwarz, 2014; Visser et al., 2014).

ADHD is diagnosed about 25 times more often in the United States than in the United Kingdom. Attitudes vary by country, but many European parents, teachers, and doctors are reluctant to use medication for what they consider routine childhood behavioral problems. Even though some Europeans believe that ADHD is a disease, "parents are loath to get their child labeled" (Kelley, 2013: 26). Thus, America's high ADHD rates and drug sales illustrate cultural differences in the social construction of health and illness.

MEDICALIZATION, LABELING, AND STIGMA

From an interactionist perspective, conditions and behaviors that are diagnosed and labeled as illness or disease change over time, differ among groups, and vary across countries. Medicalization and labeling stigmatize some illnesses and diseases more than others. Male impotence, which commonly increases as men age and was stigmatized, is now called "erectile dysfunction," and treated with drugs, regardless of age, to enhance a man's sexual experience. Ads and commercials ("Cialis is ready when you are") have flooded magazines, newspapers, and online sites.

Women's sexuality is also being medicalized. The FDA has recently approved the sale of flibanserin, an experimental drug, to increase postmenopausal women's sex drive. In clinical trials, only 10 percent of women said that the pill increased their sexual desire. Flibanserin has dangerous side effects, like fainting, because it interacts with other medications and alcohol. Women's sex drives fluctuate throughout life, diminish due to bad relationships or stress, and there's nothing wrong with low sexual desire (Nagoski, 2015; Weisman, 2015; see also Chapter 9). Pharmaceutical companies, however,

medicalization a process that defines a nonmedical condition or behavior as an illness, disorder, or disease that requires medical treatment.

will reap huge profits by convincing women that their sexuality is abnormal.

CRITICAL EVALUATION

Symbolic interaction theories are limited for several reasons. The most common criticism is that they don't address structural factors such as laws and government policies. One reason for Europe's low ADHD rates is that the governments prohibit direct-to-consumer drug ads (Kelley, 2013). After Quebec expanded insurance coverage for prescription medications in 1997, there was a significant increase in Ritalin use compared with the rest of Canada, which didn't provide similar insurance coverage (Currie et al., 2014).

Second, millions of people successfully resist medicalization. Many deaf people have ignored recommendations to get cochlear implants (devices that can increase hearing). They don't see deafness as a medical disability but as a social reality that helps them form a community and identify with other deaf people (Conrad and Barker, 2010). A "fat pride" movement argues that obesity's health risks are exaggerated and focuses, instead, on changing discrimination against overweight people (see Chapter 16).

Third, interactionists emphasize that health and illness are social constructions, but there are many serious health problems whether or not they're medicalized. People die from unknown diseases, for example. Others suffer from "contested" illnesses, including chronic fatigue syndrome, that many physicians question because the symptoms are difficult to diagnose and treat (Institute of Medicine, 2015).

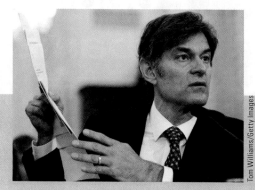

How do authority figures affect people's health perceptions and behavior? Dr. Mehmet Oz is a highly respected and published cardiothoracic surgeon, and host of the popular *Dr. Oz Show*. Products he endorses on his show "are almost guaranteed to fly off the shelves" (Weathers, 2014). Members of a Senate committee recently reprimanded Dr. Oz, however, for giving millions of people medical advice that's false and even dangerous (e.g., some diet pills being a "magic weight-loss cure").

STUDY TOOLS 14

READY TO STUDY? IN THE BOOK, YOU CAN:

☐ Check your understanding of what you've read with the Test Your Learning Questions provided on the chapter review card at the back of the book.

☐ Rip out the chapter review card for a handy summary of the chapter and key terms.

ONLINE AT CENGAGEBRAIN.COM YOU CAN:

☐ Collect StudyBits while you read and study the chapter.

☐ Quiz yourself on key concepts.

☐ Prepare for tests with SOC5 Flash Cards as well as those you create.

☐ Watch videos to explore sociology further.

15 Population, Urbanization, and the Environment

LEARNING OUTCOMES

After studying this chapter, you will be able to...

15-1 Explain how and why populations change, and evaluate the population growth theories.

15-2 Describe how and explain why global and U.S. cities are changing, describe the consequences of urbanization, and compare and evaluate the theoretical explanations of urbanization.

15-3 Describe and illustrate the major environmental issues, and discuss whether sustainable development is achievable.

After you finish this chapter go to **PAGE 313** for **STUDY TOOLS**.

In 2015, the world hit a population milestone of 7.3 billion people, and is projected to increase to 9.6 billion by 2050. Our planet has a record number of inhabitants, and many are living longer than ever before. This chapter examines how such changes affect the world's population, urbanization, and environment. Let's begin with population.

15-1 POPULATION CHANGES

Global population has grown rapidly since 1800, and more than tripled since 1900 (see *Figure 15.1*). By 2100, the world population will climb to 11.2 billion, and nearly all of the growth will occur in eight developing African countries (United Nations Population Division, 2015).

The world adds about 400,000 people each day, but population changes vary across countries. By 2050, India, with more than 1.7 billion people, will surpass China as the world's largest country, Nigeria will replace the United States as the third largest country, and three African nations will be among the most populous (see *Table 15.1*).

What do you think?

Individuals can do little to prevent global warming.

| 1 | 2 | 3 | 4 | 5 | 6 | 7 |

strongly agree   strongly disagree

Information about these and other changes comes from **demography**, the scientific study of human populations. A **population** is a group of people

FIGURE 15.1 WORLD POPULATION GROWTH THROUGH HISTORY

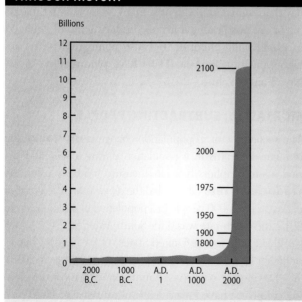

Billions

Source: Adapted from Population Reference Bureau, 2008, "World Population Projections to 2100," accessed at www.prb.org.

who share a geographic territory. A territory can be as small as a town or as vast as the planet, depending on a researcher's focus. Demographers analyze

populations in terms of size, composition, distribution, and change. They also study personally relevant topics, including your probability of getting married or divorced, the kind of job you'll probably have, how many times you'll move, and how long you'll probably live (McFalls, 2007).

15-1a Why Populations Change

Populations are never static. Their growth or decline involves three key factors: fertility (how many people are born), mortality (how many die), and migration (how many move from one area to another).

FERTILITY: ADDING NEW PEOPLE

The study of population changes begins with **fertility**, the number of babies born during a specified period in a particular society. Demographers use several fertility

demography the scientific study of human populations.

population a group of people who share a geographic territory.

fertility the number of babies born during a specified period in a particular society.

TABLE 15.1 WORLD'S TEN MOST POPULOUS COUNTRIES, 2015 AND 2050

	2015			2050	
	Country	Population (In Millions)		Country	Population (In Millions)
1	China	1,376	1 ↑	India	1,705
2	India	1,311	2 ↓	China	1,348
3	United States	322	3 ↑	Nigeria	399
4	Indonesia	258	4 ↓	United States	389
5	Brazil	208	5 ↓	Indonesia	321
6	Pakistan	189	6	Pakistan	310
7	Nigeria	182	7 ↓	Brazil	238
8	Bangladesh	161	8	Bangladesh	202
9	Russia	143	9	Democratic Republic of Congo	195
10	Mexico	127	10	Ethiopia	188

Source: Based on United Nations Population Division, 2015, Table S.3.

measures depending on the level of specificity needed. One of the most commonly used measures is the **crude birth rate** (also called the *birth rate*), the number of live births for every 1,000 people in a population in a given year. "Crude" implies that the rate is a general measure of a society's childbearing. It's based on the total population rather than more specific measures (e.g., a woman's age or marital status). In 2015, the crude birth rate was 20 worldwide, 36 for Africa, 13 for the United States, and 11 for Europe (Kaneda and Bietsch, 2015).

Another standard measure, the **total fertility rate (TFR)**, is the average number of children born to a woman during her lifetime. In the early 1900s, U.S. women had an average TFR of 3.5 compared with 1.8 in 2014. Worldwide, TFRs are much higher in the least developed nations (4.3) than in developed nations (1.7), ranging from a low of under 1.4 in some Asian and European countries to 6.0 and higher in 9 African nations (Kaneda and Bietsch, 2015).

crude birth rate (also called the *birth rate*) the number of live births per 1,000 people in a population in a given year.

total fertility rate (TFR) the average number of children born to a woman during her lifetime.

mortality the number of deaths in a population during a specified period.

crude death rate (also called the *death rate*) the number of deaths per 1,000 people in a population in a given year.

To maintain a stable population, a woman must have an average of two children, the *replacement rate* for herself and her partner. TFRs above 2.1 indicate that a country's population is increasing and getting younger. Rates below 2.1 mean that a country's population is decreasing and getting older. The older a country's population, the more difficult it is to support people who are living into their eighties and longer and to provide resources for the young (United Nations Population Division, 2015; see also Chapter 12 on the old-age dependency ratio).

In nearly all countries, fertility tends to fall as people grow richer and women are better educated. Some nations, particularly those in East Asia, worry that their "ultra low" TFRs of 1.4 or less aren't at replacement levels. In Tokyo, Seoul, Bangkok, and other Asian cities, female college graduates outnumber males and female labor force participation is high. According to cultural norms, however, women should give up work after the birth of a child. And, employed or not, women do at least three more hours of housework a day than men. Consequently, many women aren't marrying or are postponing marriage, reducing their likelihood of ever having children. Because nonmarital births are taboo and rare in Asia, TFRs have plummeted ("Asia's New Family Values," 2015).

MORTALITY: SUBTRACTING PEOPLE

The second factor in population change is **mortality**, the number of deaths in a population during a specified period. Demographers typically measure mortality using the **crude death rate** (also called the *death rate*), the number of deaths per 1,000 people in a population in a given year. In 2015, the death rate was 6 for South America, 7 for Asia, 8 worldwide and for the United States, 11 for Europe, but 14 or higher for 11 African nations (Kaneda and Bietsch, 2015).

Developed countries, compared with most developing countries, have lower death rates, but also large proportions of people age 65 and older who are no longer employed but require costly medical services. Thus, a better measure of a population's health is the *infant mortality rate*, the number of deaths of infants younger than 1 year per 1,000 live births. Generally, as the standard of living improves—access to clean water, adequate sanitation,

Africa's Niger and South Sudan have the world's highest TFRs, 7.6 and 6.9, respectively. The reasons for the high fertility include poverty and women's limited access to education and birth control. In contrast, some European towns are facing "slow death" due to low fertility and immigration rates. Germany is trying to boost its shrinking TFR (1.4 compared with 2.5 in the 1960s) by offering families tax breaks and free day care for all children age 1 and older (Daley and Kulish, 2013; Rossi and Jucca, 2014).

and medical care—the infant mortality rate decreases. In 2015, the infant mortality rate was 5 in developed countries, 62 in the least developed countries, and 90 or higher in 6 African nations (Kaneda and Bietsch, 2015).

Lower infant mortality greatly raises *life expectancy*, the average number of years that people who were born at about the same time can expect to live. Worldwide, the average life expectancy is 71—69 for males and 73 for females. Again, however, there are considerable variations across countries—from a high of 84 in Hong Kong and 83 in Japan to a low of 44 in several African nations. The United States, with a life expectancy of 79, ranks below at least 25 other industrialized countries and only slightly higher than less developed countries such as Cuba and Uruguay (Kaneda and Bietsch, 2015). In almost every society, people with a higher socioeconomic status live longer and healthier lives because their occupations are physically safe, and their resources include clean drinking water, sanitation, and medical services.

MIGRATION: ADDING AND SUBTRACTING PEOPLE

The third demographic factor in population change is **migration**, the movement of people into or out of a specific geographic area. There are two types of migration: international and internal. *International migration*, movement to another country, includes *emigrants* (people who move out of a country) and *immigrants* (people who move into a country).

Migrants are people who choose to leave their country to seek a better life elsewhere. *Refugees* (sometimes called *displaced persons*) are people who are forced to flee their country—often with no warning—to escape war, persecution, torture, or death. An international migrant can have several of these characteristics. For example, my family members and I were refugees who fled Lithuania (on a horse and wagon) from Communism during World War II, lived in displaced persons camps in Germany, and then immigrated to the United States.

International Migration. The number of international migrants reached an all-time high of 232 million in 2013, and is expected to double by 2050. The largest flows have moved from one developing country to another (as from Indonesia to Saudi Arabia) or from a developing to an industrialized country (as from Mexico to the United States). Also, most emigrants move to a neighboring country (from Mexico to the United States, for example, rather than from Mexico to Canada). The United States has 20 percent of the world's migrants, more than any other country in the world, and is the only industrialized nation where one-quarter of migrants are unauthorized (Martin, 2013).

With an estimated 1 in every 122 people on earth now displaced, there are currently more refugee exiles worldwide than at any time in recorded history. By the end of 2014, nearly 60 million people had been driven from their homes because of conflict, violence, or human rights violations. More than half of all current global refugees come from just three countries—Syria, Afghanistan, and Somalia (United Nations High Commissioner for Refugees, 2015). The most common reason for other

migration the movement of people into or out of a specific geographic area.

international migration is economic, but also includes fleeing bad schools, high crime rates, and natural disasters. Thus, the reasons for international migration today are similar to those of European immigrants who came to the United States during the twentieth century.

Internal Migration. International migration has increased, but 92 percent of adults worldwide who change residences do so because of *internal migration*, movement within a country. Over a five-year period, 24 percent of U.S. adults move within the country, similar to rates reported in other advanced economies, including New Zealand, Finland, and Norway (Esipova et al., 2013).

Between 2013 and 2014, 36 million people living in the United States (12 percent of the population) moved to a different residence. Why do people move? The most common reasons (48 percent) are housing-related (e.g., "Wanted new or better home/apartment," eviction), 30 percent are family-related (e.g., change in marital status), and 20 percent are job-related (e.g., new job, job transfer, to look for work) (Ihrke, 2014; "U.S. Mover Rate...," 2015).

Social class and race/ethnicity also affect internal migration. In 2014, 84,000 people left Puerto Rico for the U.S. mainland, a 38 percent increase from 2010. There are now more Puerto Ricans living in the mainland (5.1 million) than on the island (3.5 million). The top reasons for moving stateside are economic (42 percent) and family-related (38 percent), but many also cite crumbling schools and pervasive crime, particularly drug trafficking (Alvarez, 2014; Krogstad, 2015).

After Hurricanes Katrina and Rita devastated much of the U.S. Gulf coast in 2005, 75 percent of New Orleans' population—primarily low-income blacks—moved to other states; they didn't have home insurance policies that covered the cost of the damages. Ten years later, blacks make up 60 percent of the city's population compared with 67 percent before Katrina (Jonsson, 2015).

15-1b Population Composition and Structure

Demographers study age and gender to understand a population's composition and structure. Two of the most common measures are *sex ratios* and *population pyramids*.

SEX RATIOS

A **sex ratio** is the proportion of men to women in a population. A sex ratio of 100 means that there are equal numbers of men and women; a ratio of 95 means that there are 95 men for every 100 women (fewer males than females). Worldwide, without human intervention, 102

So far, more than 11 million Syrians (about half of the population) have fled the civil war that began in 2011 or have been internally displaced from their homes. "Many have been on the run for a year or more...moving from village to village, up to as many as 20 times, before they finally made it across an international border" (Bengali, 2014: 10). The death of Syrian refugee Aylan Kurdi, 3, sparked an outburst of compassion worldwide.

to 107 boys are born for every 100 girls. In the United States, the sex ratio is 105 at birth, but skewed in many countries: 115 in Curaçao, 114 in Armenia, 112 in India and Vietnam, 111 in China and Albania, and as high as 138 in some of these countries' rural areas (Gilles and Feldman-Jacobs, 2012; *CIA World Factbook*, 2014).

These sex ratio imbalances are attributed primarily to *female infanticide* (sometimes called *gendercide*)—the intentional killing of baby girls. In many Asian countries, including China and India, there's a preference for boys because males are expected to carry on the family name, care for elderly parents, inherit property, and play a central role in family rituals. Consequently, and particularly in rural areas, hundreds of thousands of female infants die every year because of neglect, abandonment, and starvation. Others are aborted after ultrasound scanners reveal the child's sex ("Sex-Selective Abortion...," 2013).

China is a good example of some of the unintended negative consequences of lopsided sex ratios. In 1979, China passed a one-child policy to combat poverty and overpopulation. The policy succeeded in limiting population growth because of at least 600 million abortions and sterilizations. The country's TFR fell from 6 in the late 1960s to 1.5 by 2010, but well below the 2.1 replacement rate to maintain a constant population (Eberstadt, 2013; "Family Planning...," 2014).

Because of the unbalanced sex ratio for the last 30 years, by 2020 China will have about 35 million more young men than women. That means a large number of men won't be able to find wives—a problem that has already increased the illegal trafficking of women from poorer neighboring countries, including Cambodia, Myanmar, and Vietnam—and there'll be millions of aging bachelors (Tsai, 2012; Larson, 2014).

sex ratio the proportion of men to women in a population.

To reduce the burden of elder care, the government eased the policy in 2013 and in 2015 by allowing married couples to have two children. The urban, educated middle class—the group that China's leaders want to see increase its family size—hasn't shown much interest in doing so because of high costs of living and expenses in raising a second child. Thus, China's working-age population is expected to start declining rapidly in 2015, resulting in fewer workers to support a growing older population (Burkitt, 2013; "Reforming the One-Child Policy…," 2013).

POPULATION PYRAMIDS

A **population pyramid** is a graphic depiction of a population's age and sex distribution at a given point in time. As *Figure 15.2* shows, Mexico is a young country: Much of its population is under age 45 (which also means that many women are in their childbearing years), and there are relatively few people 65 years and older. In contrast, Italy is an old country, and the United States is somewhere in the middle.

The shape of the pyramid (a triangle for Mexico, a rectangle for the United States, and a diamond for Italy) has future implications for young and old countries. Italy has a relatively small number of women ages 15 to 44 (in their reproductive years) and a bulge of people ages 45 to 79. This suggests that there'll be a scarcity of workers to support an aging population and a greater need for social services for older people than for children and adolescents (Kochhar and Oates, 2014; see also Chapter 12). Thus, population pyramids give us a snapshot of a country's demographic profile and indicate some of the problems that countries are likely to face in the future.

15-1c Population Growth: A Ticking Bomb?

Some of the world's largest countries, many of them in the developing world, will grow even more by 2050 (see *Table 15.1*). So, has population growth gotten out of hand? There are many views on this question, but two of the most influential have been Malthusian theory (which argues that the world can't sustain its unprecedented population surge), and demographic transition theory (which maintains that population growth is slowing).

MALTHUSIAN THEORY

For many demographers, population growth is a ticking bomb. They subscribe to **Malthusian theory**, which maintains that the population is growing faster than the

FIGURE 15.2 POPULATION PYRAMID PROJECTIONS, 2025

Source: U.S. Census Bureau, International Database, www.census.gov/ipc/www/idb/pyramids.html. Accessed January 20, 2007.

food supply needed to sustain it. This theory is named after Thomas Malthus (1766–1834), an English economist, clergyman, and college professor who argued that humans are multiplying faster than the earth's ability to produce sufficient food.

According to Malthus (1798/1965), population grows at a *geometric rate* (2, 4, 8, and so on), whereas the food supply grows at an *arithmetic rate* (1, 2, 3, 4, and so on). That is, two parents can have 4 children and

population pyramid a graphic depiction of a population's age and sex distribution at a given point in time.

Malthusian theory maintains that the population is growing faster than the food supply needed to sustain it.

16 grandchildren within 50 years. The available number of acres of land, farm animals, and other sources of food can increase in that time period, but certainly not quadruple. In effect, then, the food supply will not keep up with population growth. Because there are millions of parents, the results could be catastrophic, with masses of people living in poverty or dying of starvation.

Malthus posited that two types of checks affect population size. *Positive checks* (famine, disease, war) limit reproduction, raise the death rate, and lower the overall population. *Preventive checks* (contraception, postponing marriage, abortion, and extramarital sex) also limit reproduction by reducing birth rates and, consequently, ensure a higher standard of living for all (Malthus, 1872/1991).

Malthusian theory has had a lasting influence. *Neo-Malthusians* (or New Malthusians) agree that the population is exploding beyond food supplies. The world's population reached its first billion in 1800. In the 200 years that followed, the world added another 5 billion people (see *Figure 15.1*). As a result of this growth, according to some influential neo-Malthusians, the earth has become a "dying planet"—a world with insufficient food and a rapidly expanding population that pollutes the environment (Ehrlich, 1971; Ehrlich and Ehrlich, 2008).

An estimated one in eight people in the world suffer from chronic hunger that prevents them from working and having a normal life. Resources and technical knowledge are available to increase food production by 70 percent by 2050, but poverty and environmental problems plague the countries that need food the most (FAO et al., 2013).

DEMOGRAPHIC TRANSITION THEORY

Some demographers are more optimistic than neo-Malthusians. **Demographic transition theory** maintains that population growth is kept in check and stabilizes as countries experience economic and technological development,

which, in turn, affects birth and death rates. According to this theory, population growth changes as societies undergo industrialization, modernization, technological progress, and urbanization. During these processes, a nation goes through four stages (see *Figure 15.3*), from high birth and death rates to low birth and death rates. Put simply, the greater the human development, the lower the fertility and death rates.

▶ **Stage 1: Preindustrial society.** In this initial stage, there's little population growth. The birth rate is high because people rarely use birth control: They want as many children as possible to provide unpaid agricultural labor and support parents in old age, but a high death rate offsets the high birth rate. No country exists in this current stage of the demographic transition.

▶ **Stage 2: Early industrial society.** There's significant population growth because the birth rate is higher than the death rate. The birth rate may even increase over what it was in Stage 1 because mothers and their children enjoy improved health care. Couples may still have large numbers of children because they fear that many of them will die, but the death rate declines because of better sanitation, better nutrition, and medical advances (e.g.,

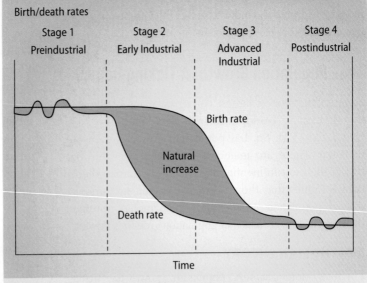

FIGURE 15.3 THE CLASSICAL DEMOGRAPHIC TRANSITION MODEL

Source: U.S. Census Bureau, International Database, www.census.gov/ipc/www/idb /pyramids.html. Accessed January 20, 2007.

demographic transition theory maintains that population growth is kept in check and stabilizes as countries experience economic and technological development.

A number of countries, particularly in Europe and parts of Asia, have below-replacement fertility rates. For some policy analysts, the recent influx of refugees and migrants is an opportunity to shore up labor forces with young workers whose taxes will support graying and dwindling native-born populations (OECD/European Union, 2015; Portes, 2015). Do you agree?

immunizations and antibiotics). Most of the world's poorest countries—including sub-Saharan Africa, Afghanistan, and Yemen—are currently in Stage 2.

▶ **Stage 3: Advanced industrial society.** As the infant mortality rate declines, parents have fewer children. Effective birth control reduces family size. The decrease in child care responsibilities, in turn, enables women to work outside the home. Brazil, Mexico, and some Asian countries are currently in Stage 3.

▶ **Stage 4: Postindustrial society.** In this stage, the demographic transition is complete, and the society has low birth and death rates. Women tend to be well educated and to have full-time jobs or careers. If there's little immigration, the population may even decrease because the birth rate is low. This is the case today in Canada, Japan, Singapore, Hong Kong, Australia, New Zealand, the United States, and Europe.

CRITICAL EVALUATION

The dire predictions of Malthus and his successors that global population growth would lead to worldwide famine, disease, and poverty haven't come true. Still, 1.2 billion people (22 percent of the world's population) live in abject poverty, subsisting on less than $1.25 a day (United Nations Development Program, 2014; Malik, 2014; see also Chapter 8).

Despite neo-Malthusians' fears, global fertility is half of what it was in 1972. The population of some industrialized countries is declining because people aren't having enough babies to replace themselves. These countries are experiencing **zero population growth (ZPG)**, a stable population level that occurs when each woman has no more than two children.

Fearing that there won't be enough young workers to pay for social security and the rising cost of health care for aging populations, some low-birth nations with fertility rates below ZPG are paying women to have more children. Russia gives mothers with one child $12,500 for each additional baby; Japan has expanded its day care facilities and offers families a monthly allowance of $145 per child younger than 15; Germany and France have liberal parental leave; South Korea turns off its office lights once a month so people have more time to go home and multiply and offers other incentives, including cash gifts, for staff who produce more than two babies; China is encouraging married couples to have two children; and Singapore, besides giving couples almost $5,000 for having a child, lets families with babies go to the front of the queue for high quality, affordable public housing (Hales, 2014; Nechepurenko, 2014; "Pro-natalism," 2015).

Some neo-Malthusians maintain, however, that it's irresponsible for *any* country to encourage higher fertility rates. They worry about the consequences of adding 3 billion more inhabitants to the planet in less than 50 years, especially for many developing countries "with desperate economic outlooks" (Sachs, 2005; Shorto, 2008).

One result of population growth is urban growth. Cities attract people because of jobs and cultural activities, but urbanization has also created numerous problems.

15-2 URBANIZATION

If you've flown over the United States, you've probably noticed that people tend to cluster in and around cities. After sunset, some areas glow with lights, whereas others are engulfed in darkness. The average person, in the United States and worldwide, is more likely to live in a city than a rural area, and this trend is rising.

A *city* is a geographic area where a large number of people live relatively permanently and make a living primarily through nonagricultural activities. **Urbanization**, which increases the size of cities, is the movement of people from rural to urban areas. Most of this discussion focuses on U.S. cities, but let's begin with a brief look at global urbanization.

zero population growth (ZPG) a stable population level that occurs when each woman has no more than two children.

urbanization people's movement from rural to urban areas.

During the Industrial Revolution, many cities, like this one in Hamburg, Germany, constructed canals that provided an inexpensive means of transportation and distributing goods.

TABLE 15.2 URBANIZATION AROUND THE WORLD

Percentage of People Living in Urban Areas

	1950	2014	2050 (Projected)
World	**29**	**54**	**66**
Africa	14	40	56
Asia	18	48	64
Latin America and the Caribbean	41	80	86
North America	64	81	87
Europe	51	73	82
Oceania	62	71	74

Source: Based on United Nations Department of Economic and Social Affairs, 2012, Table 2, and United Nations, 2014, Table 1.

15-2a Urbanization: A Global View

In 2008, for the first time in history, a majority of the world's population lived in urban areas. By 2050, 66 percent of the world's population is projected to be urban (United Nations, 2014).

ORIGIN AND GROWTH OF CITIES

Cities are one of the most striking features of modern life, but they've existed for centuries. About 7,000 years ago, for example, people built small cities in the Middle East and Latin America to protect themselves from attackers and to increase trade. By 1800, 56 cities in Western Europe had a population of 40,000 or more (Chandler and Fox, 1974; De Long and Shleifer, 1992).

Before the Industrial Revolution, which began in the late eighteenth century, urban settlements in Europe, India, and China developed largely because people figured out how to use natural resources (e.g., mining coal and transporting water for irrigation and consumption). The Industrial Revolution spurred ever-increasing numbers of people to move to cities in search of jobs, schooling, and improved living conditions. As a result, the urban population surged—from 3 percent of the world's population in 1800 to 14 percent in 1900 (Sjoberg, 1960; Mumford, 1961). As industrialization advanced, urbanization increased.

WORLD URBANIZATION TRENDS

Between 1920 and 2014, the world's urban population increased from 270 million to 3.9 billion, and is expected to rise to 6.3 billion by 2050. Africa and Asia are home to nearly 90 percent of the world's rural population, but both are urbanizing faster than other regions (see *Table 15.2*). Between 2014 and 2050, just three countries—India, China, and Nigeria—are expected to account for 37 percent of the world's urban population growth (United Nations, 2014).

Many of the world's largest cities are **megacities**, metropolitan areas with at least 10 million inhabitants. In 1950, the world had only two megacities: Tokyo (11.3 million) and New York-Newark (12.3 million). By 2030, there'll be 41 megacities. Only two of them will be in the United States—New York-Newark (19.9 million), and Los Angeles-Long Beach-Santa Ana (13.3 million)—and both will be much smaller than Tokyo (37.2 million), Delhi (36.1 million), Shanghai (30.8 million), and other Asian cities. Thus, the number of megacities is growing, and they're much bigger than in the past (United Nations, 2014).

Should the explosive growth of cities and megacities concern us? Generally, cities provide jobs, offer better health services, and have more educational opportunities, but not everyone benefits from such advantages. The urban poor are often crowded into slums where children are less likely to be enrolled in school, sanitation is inadequate, and the economic gap between the haves and have-nots is widening (Laneri, 2011).

15-2b Urbanization in the United States

Like many other countries, the United States is becoming more urban. How, specifically, has the urban landscape changed? And what are some of the consequences?

megacities metropolitan areas with at least 10 million inhabitants.

iStockphoto.com/naveen0301

Because of high population growth, many of India's streets are chaotic: "Cars, trucks, buses, motorcycles, taxis, rickshaws, cows, donkeys, and dogs jostle for every inch of the roadway as horns blare and brakes squeal. Drivers run red lights and jam their vehicles into available space, ignoring pedestrians" (Hamm, 2007: 49–50).

HOW URBAN AMERICA IS CHANGING

During the Industrial Revolution, millions of Americans in agricultural areas migrated to cities to find jobs. As a result, between 1900 and 2000, the urban population surged from 39 to 79 percent (Riche, 2000; U.S. Census Bureau, 2012).

Suburbanization. As cities grew more crowded, dirtier, and noisier, urban growth sparked *suburbanization*, people's moving to communities just outside a city. In 1920, about 15 percent of Americans lived in the suburbs compared with more than half today (Palen, 2014).

During the 1950s, two-thirds of urban dwellers moved to suburbs. The federal government, fearing a return to the economic depression of the 1930s, underwrote the construction of much new housing in the suburbs. The general public obtained low-interest mortgages, veterans were offered the added incentive of being able to purchase a home with a $1 down payment, and massive highway construction programs enabled commuting by car. As a result, suburbs mushroomed (Rothman, 1978).

Suburban life is appealing. Some of the benefits include more privacy and space, one's own yard and garages, better schools, safer streets, and, if there are jobs nearby, shorter commutes. Suburban elites have sprawling "McMansions," considerable acreage, and gated communities.

Especially since 1980, however, much of the social and physical separation between cities and suburbs has blurred. More Latinos, Asians, and blacks now live in the suburbs than in the city, representing up to 35 percent of the suburban population in more than a third of the nation's largest metropolitan areas. One in three poor Americans now live in suburbs and crime rates, although lower than in cities, have increased (Kneebone and Berube, 2013; Frey, 2014). Suburbs are also attracting fewer young adults, especially those with young children. This means that many communities no longer have the taxes that support schools, services for older people, and public recreational facilities (El Nasser and Overberg, 2011). For these and other reasons, as you'll see shortly, many big cities are growing faster than suburbs.

Edge Cities and Exurbs. Originally, most suburbs were bedroom communities for commuters with jobs in the city. Over the last few decades, suburbanization has generated **edge cities**, business centers that are within or close to suburban residential areas and include offices, schools, shopping, entertainment, malls, hotels, and medical facilities.

People have also created **exurbs**, areas of new development beyond the suburbs that are more rural but on the fringe of urbanized areas. About 6 percent of Americans live in exurbs. The average exurbanite is white, a middle-income earner, married with children, a "super commuter" (one who travels two or more hours a day for work), and owns a large house outside of an expensive metropolitan suburb (Berube et al., 2006; Lalasz, 2006).

Metropolitan Statistical Areas. Together, suburbs, edge cities, and exurbs form **metropolitan statistical areas** (MSAs), also called *metro areas*, that consist of a central city of at least 50,000 people and the urban areas linked to it. The vast majority of Americans (84 percent) live in 366 MSAs. With a population of nearly 20 million, New York-Newark-Jersey City is the largest MSA; Carson City, Nevada, with a population of 55,000, is the smallest. The fastest-growing metro areas have been in the South and West (e.g., Dallas-Fort Worth, Atlanta, Los Angeles), but six have been in the Midwest (e.g., Bismarck, North Dakota, and Casper, Wyoming), where mining, quarrying, and oil and gas extraction fueled job opportunities (Wilson et al., 2012; U.S. Census Bureau Newsroom, 2014).

Since 2011, and for the first time in nearly a hundred years, metro areas—particularly those with more than 1 million people—have grown faster than suburbs (Frey, 2014). Two groups—young professionals and Baby Boomers—have driven the trend in metro living. Because many young professionals are postponing

edge cities business centers that are within or close to suburban residential areas.

exurbs areas of new development beyond the suburbs that are more rural but on the fringe of urbanized areas.

metropolitan statistical area (MSA, also called *metro area*) a central city of at least 50,000 people and urban areas linked to it.

Micro-apartments, sometimes smaller than college dorm rooms, are cropping up in many U.S. metro areas to accommodate the growing number of single professionals, students, and older people. The rents are high, however. In San Francisco and other metro areas, renting a 17-by-17-foot apartment can cost up to $2,500 a month (Elsen, 2015).

marriage and parenthood (see Chapter 12), they don't have to worry about the quality of schools. Moreover, many work in cities, want to reduce commute times, and can't afford the down payment for a house in the suburbs. A number of Baby Boomers, particularly those who are retiring, are moving to metro areas for a variety of reasons (e.g., better public transportation, greater access to cultural events, social services for older people, no longer being healthy enough to maintain a suburban house) (Kneebone and Berube, 2013; Westcott, 2014).

SOME CONSEQUENCES OF URBANIZATION

Cities offer many benefits, including a vast array of culturally diverse restaurants, shops, and activities, but urbanization also creates problems. Some of the drawbacks include urban sprawl, increased traffic congestion, a scarcity of affordable housing, and residential segregation.

Urban Sprawl. Urban sprawl—the rapid, unplanned, and uncontrolled spread of development into regions adjacent to cities—is widespread. According to some estimates, urbanization in the Southeast will increase by up to 190 percent by 2060. Development on that scale will

urban sprawl the rapid, unplanned, and uncontrolled spread of development into regions adjacent to cities.

gentrification a process in which upper-middle-class and affluent people buy and renovate houses and stores in downtown urban neighborhoods.

result in losing 15 percent of agricultural land, 12 percent of grasslands, and 10 percent of forests (Terando et al., 2014).

Urban sprawl has created rapid *job sprawl*, which occurs when companies move jobs from metropolitan areas to suburbs. The more distant the suburb, the less likely low-income people are to hear about employment opportunities through informal networks, to afford houses in these areas, and to have transportation to the jobs (Raphael and Stoll, 2010).

Some large metro areas (e.g., Washington, D.C., New York, Boston, Atlanta, Miami, and Denver) are implementing "walkable urbanism" to curb urban sprawl. That is, the developments are neighborhoods where everyday destinations (e.g., work, school, restaurants) are concentrated and within walking distances. For the most part, however, most urban development is built around driving rather than walking (Leinberger and Lynch, 2014).

Traffic Congestion. Generally, the only way to get around in urban sprawl areas is by automobile. This means that most suburban households face the costs of buying, fueling, insuring, and maintaining multiple cars. The U.S. population has increased by 23 percent over the last 25 years, but total highway miles have increased by only 5 percent, resulting in greater traffic congestion within and outside cities. More than 10 million Americans (8 percent of all people who don't work at home) now travel 60 minutes or longer one-way, a proportion that has increased by 95 percent since 1990. Traffic snarls and long commutes increase air pollution and stress and decrease the time that people have for family and leisure activities (McKenzie, 2013).

Lack of Affordable Housing. In some cases, the poor are pushed out by **gentrification**, a process in which upper-middle-class and affluent people buy and renovate houses and stores in downtown urban neighborhoods. Governments in many older cities encourage gentrification to increase dwindling populations, to revitalize urban areas, and to augment tax revenues. Gentrification can benefit an entire community because it breaks up concentrated poverty pockets, the new residents may demand improvements in schools and crime control, and an influx of retail stores can generate new jobs. Rent increases, however, have displaced many low-income residents and small businesses. Also, some black longtime residents have complained that aggressive policing tactics, like stopping and frisking people on the street, increase when wealthier people move into a neighborhood (Buntin, 2015; Jonsson, 2015).

Residential Segregation. Since 1970, racial residential segregation has decreased. Among metropolitan areas with a population of 500,000 or more, the least segregated are in the South and West, and the most segregated

Angela Hampton/Bubbles Photolibrary/Alamy

Gentrification improves old city neighborhoods and increases property values, but also displaces low-income residents.

FIGURE 15.4 HOUSEHOLDS THAT LIVE MAINLY AMONG THEMSELVES, 1980 AND 2010

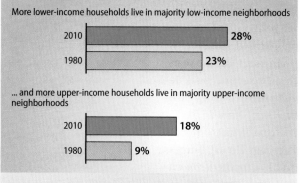

More lower-income households live in majority low-income neighborhoods

2010	28%
1980	23%

... and more upper-income households live in majority upper-income neighborhoods

2010	18%
1980	9%

Note: Lower-income households had an annual income of less than $34,000 (in 2010 dollars); upper-income households had an annual income of $104,000 and above.

Source: Based on Fry and Taylor, 2012: 1.

are mainly in the Northeast and Midwest. There may be a "continued easing of the color line" because racial segregation has waned even in long-standing "hypersegregated" cities like Chicago and New York. The nation's 20 most multiethnic metropolitan regions are now "global neighborhoods" that have substantial numbers of whites, blacks, Latinos, and Asians (Scommegna, 2011; Iceland et al., 2013: 119).

Whereas racial residential segregation has decreased, social class residential segregation has increased (see *Figure 15.4*). The share of families living in middle-income neighborhoods dropped from 65 to 42 percent between 1970 and 2009. The residential segregation gap between affluent and all other social classes is more pronounced than in the past, and the most residentially segregated are low-income blacks and Latinos (Fry and Taylor, 2012; Bischoff and Reardon, 2013).

Residential segregation along income lines has increased primarily because of the long-term rise in income inequality (see Chapters 8 and 11). The geographic isolation of affluent neighborhoods has led to their residents'

lower investments in resources and services, including public schools and parks, that would benefit more people. Some wealthy neighborhoods have even fought off state laws that mandate constructing affordable housing for low- and middle-income families. According to the high-income homeowners, building less expensive houses would raise crime rates, lower property values, and "There's plenty of affordable housing *in neighboring communities*" (McCabe, 2014: 39, emphasis added).

15-2c Sociological Explanations of Urbanization

How and why do cities change? In answering this and other questions, functionalists underscore urban development, conflict theorists emphasize the impact of capitalism and big business, feminist scholars focus on women's safety and space, and symbolic interactionists examine the quality of city life (*Table 15.3* summarizes these perspectives).

TABLE 15.3 SOCIOLOGICAL EXPLANATIONS OF URBANIZATION

Perspective	Level of Analysis	Key Points
Functionalist	Macro	Cities serve many important social and economic functions, but urbanization can also be dysfunctional.
Conflict	Macro	Driven by greed and profit, large corporations, banks, developers, and other capitalist groups shape cities' growth or decline.
Feminist	Macro and micro	Whether they live in cities or suburbs, women generally experience fewer choices and more constraints than men.
Symbolic Interactionist	Micro	City residents differ in their types of interaction, lifestyles, and perceptions of urban life.

FUNCTIONALISM: HOW AND WHY CITIES CHANGE

Urban ecology studies the relationships between people and urban environments. Over the years, social scientists have revised urban ecology theories (see *Figure 15.5*).

Sociologists Robert Park and Ernest Burgess (1921) proposed *concentric zone theory* to explain the distribution of social groups within urban areas. According to this model, a city grows outward from a central point in a series of rings. The innermost ring, the central business district, is surrounded by a zone of transition, which contains industry and poor-quality housing. The third and fourth rings have housing for the working and middle classes. The outermost ring is occupied by people who live in the suburbs and commute daily to work in the central business district.

Economist Homer Hoyt's (1939) *sector theory* refined concentric zone theory. He proposed that cities develop in sectors (instead of rings) that radiate from the central business district depending on various economic and social activities. Some sectors are predominantly industrial, some contain stores and offices, and others, generally farther away from the central business district, are middle- and upper-class residential areas.

Geographers Chauncey Harris and Edward Ullman (1945) developed another influential model, *multiple-nuclei theory*, which proposed that a city contains more than one center around which activities revolve. For example, a "minicenter" often includes an outlying business district with stores and offices that are accessible to middle- and upper-class residential neighborhoods, whereas airports typically attract hotels and warehouses.

As cities grew after World War II, these models no longer described urban spaces. Thus, Harris (1997)

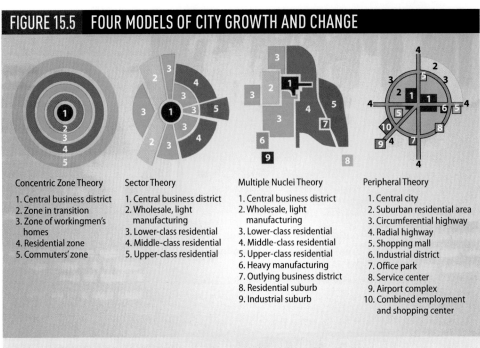

FIGURE 15.5   FOUR MODELS OF CITY GROWTH AND CHANGE

Concentric Zone Theory
1. Central business district
2. Zone in transition
3. Zone of workingmen's homes
4. Residential zone
5. Commuters' zone

Sector Theory
1. Central business district
2. Wholesale, light manufacturing
3. Lower-class residential
4. Middle-class residential
5. Upper-class residential

Multiple Nuclei Theory
1. Central business district
2. Wholesale, light manufacturing
3. Lower-class residential
4. Middle-class residential
5. Upper-class residential
6. Heavy manufacturing
7. Outlying business district
8. Residential suburb
9. Industrial suburb

Peripheral Theory
1. Central city
2. Suburban residential area
3. Circumferential highway
4. Radial highway
5. Shopping mall
6. Industrial district
7. Office park
8. Service center
9. Airport complex
10. Combined employment and shopping center

Source: Based on Park and Burgess, 1921; Hoyt, 1939; Harris and Ullman, 1945; and Harris, 1997.

proposed a *peripheral* theory of urban growth, which emphasized the development of suburbs around a city but away from its center. As suburbs and edge cities burgeon, highways that link the city's central business district to outlying areas and beltways that loop around the city provide relatively easy access to airports, the downtown, and surrounding areas.

Functionalists also examine urban dysfunctions, including overcrowding, poverty, deviant behavior, and environmental destruction. From 2000 to 2010, for instance, the number of vacant housing units—many of them in older industrial cities like Baltimore and Detroit—increased by 4.5 million, or 44 percent. Abandoned buildings can increase crime rates and a neighborhood's social disorganization (Williams, 2013).

CONFLICT THEORY: THE IMPACT OF CAPITALISM AND BIG BUSINESS

For functionalists, people's choices shape urban changes. In contrast, **new urban sociology**, a perspective heavily influenced by conflict theory, views urban changes as largely the result of decisions made by powerful capitalists and high-income groups. That is, economic and political factors favorable to the rich, not ordinary citizens, determine urban growth or decline. When a local government wants to rejuvenate parts of the inner city, it typically offers tax breaks, changes zoning laws, and allows real estate, construction, and banking industries to seek profits with little regard for the needs of low-income households or

---

urban ecology studies the relationships between people and urban environments.

new urban sociology view that urban changes are largely the result of decisions made by powerful capitalists and high-income groups.

the homeless (Feagin and Parker, 1990; Macionis and Parrillo, 2007).

For conflict theorists, urban space is a commodity that's bought and sold for profit. It's not the average American, they argue, but bankers, developers, politicians, and influential businesspeople who determine how to use urban space. Because increasing a property's value is a higher priority than community needs, poor and low-income people are crowded into dilapidated neighborhoods (Logan and Molotch, 1987; Gottdiener and Hutchison, 2000).

Some analysts are especially critical of the tech sector for being "bad urbanists." To illustrate, in 2011 Twitter received generous tax incentives to move its new headquarters into one of San Francisco's poorest neighborhoods. City officials expected that the company's presence would revitalize the community. Instead, Twitter employees eat all their meals in the dining area, rarely leave the building to shop because small businesses can't afford the area's skyrocketing rents, and the neighborhood's evictions increased 38 percent by 2013 because longtime residents couldn't pay the soaring rent increases (Arieff, 2013; Butler, 2014; Steinmetz, 2014).

FEMINIST THEORIES: GENDER ROLES, SPACE, AND SAFETY

Feminist theories emphasize gender-related constraints. Whether they live in cities or suburbs, women generally experience more problems than men because living spaces are usually designed by men who have tended to ignore women's changing roles.

Many women fear the city, especially urban public spaces like streets, parks, and public transportation

Slums are one of the most dysfunctional by-products of urbanization. Megacities—particularly those in India, Egypt, Pakistan, Kenya, and Mexico—have the world's largest slums.

(Domosh and Seager, 2001). They see these places as risky for their physical safety, despite the fact that most violence against women occurs at home. A few cities provide public transportation (e.g., minivans that operate seven nights a week) to prevent crimes against women, usually minority women, who must travel to work after 8:00 p.m. and return home before dawn. For the most part, however, such services are rare (Hayden, 2002).

In the suburbs, both women's and men's physical mobility is limited because of the scarcity of public transportation systems. Consequently, many suburban households have two or more cars, but women living in the suburbs usually experience greater problems than men. For example, there may be fewer job opportunities (especially if mothers want to be at home when children return from school), and women often spend much of their time maintaining a single-family home, decreasing their time for educational or leisure activities (Cichocki, 1981; Hayden, 2002).

SYMBOLIC INTERACTION THEORY: HOW PEOPLE EXPERIENCE CITY LIFE

Symbolic interactionists are most interested in the impact of urban life on city residents. In a classic essay, sociologist Louis Wirth (1938: 14) described the city as a place where "our physical contacts are close, but our social contacts are distant."

Wirth defined the city as a large, dense, and socially and culturally diverse area. These characteristics produce *urbanism*, a way of life that differs from that of rural dwellers. Wirth saw urbanites as more tolerant than residents of small towns or rural areas of a variety of lifestyles, religious practices, and attitudes.

He also believed that urbanism has negative consequences, including alienation, friction because of physical congestion, impersonal relationships, and a disintegration of kinship and friendship ties. Some studies have supported Wirth's theory of urbanism (see Guterman, 1969), but others have challenged his views. In a recent national study, people living in large metropolitan areas scored higher than those living in small towns and rural areas on physical and emotional health, access to basic necessities, and being satisfied with life (Witters, 2010).

CRITICAL EVALUATION

Functionalists tend to overlook urbanization's negative political and economic impact, especially when profit and greed guide urban planning. Conflict theory seems to assume that residents are helpless victims as developers and corporations raze low-income houses. In fact, environmental groups have had considerable success in pushing through legislation to maintain and even increase open public spaces and build energy-saving homes in low-income neighborhoods (Moore, 2008).

Feminist sociologists have made important contributions through studies of the everyday lives of low-income women, especially in central cities (see Chapter 11), but urbanization has received much less attention. Urbanites are more diverse than some symbolic interactionists claim. People living in cities aren't necessarily more self-centered or isolated than those in small towns or rural areas. Instead, many have close family bonds, friends, and satisfying relationships with coworkers (Crothers, 1979; Wilson, 1993). Also, symbolic interaction doesn't show how political, educational, religious, and economic factors shape people's experiences of city life (Hutter, 2007).

You've seen that the world's population is growing rapidly and becoming more urbanized. Both population growth and urbanization are taxing the planet's limited resources.

15-3 THE ENVIRONMENT

Consider the following:

▶ Each year, Americans throw away 31 to 40 percent of their food—50 percent more than in the 1970s—that ends up in landfills (Gunders, 2012; Neff et al., 2015).

▶ The average American throws away 5 pounds of solid waste every day. As waste decomposes in landfills, it gives off many gases, including methane, which contributes to global warming (Powell et al., 2015).

▶ Up to 3.5 million Americans get sick every year from polluted beach water that causes a wide range of diseases, including ear-nose-eye infections, hepatitis, skin rashes, and respiratory illnesses (Dorfman and Haren, 2014).

▶ The United States uses 1.2 billion pounds of pesticides a year, but only 0.01 percent reaches the intended target. The other 99.99 percent contaminates the food, air, and water (Hsu et al., 2014).

Such environmental problems threaten our **ecosystem**, a community of living and non-living organisms that share a physical environment. Plants, animals, and humans depend on each other for survival. Because the ecosystem is interconnected worldwide, what happens in one country affects others. Water, air pollution, and climate change are three interrelated factors that are endangering the global ecosystem.

ecosystem a community of living and non-living organisms that share a physical environment.

Among developing countries, 25 percent of girls aren't in school compared with 17 percent of boys. One reason for this difference is that girls are more likely than boys to be responsible for collecting the family's water, making it difficult for them to attend school (Global WASH Fast Facts, 2013).

15-3a Water

An expanding world population, extreme weather patterns, and industrial pollution are jeopardizing already limited water supplies. More than 768 million people worldwide don't have clean water, and 1.8 billion drink contaminated water. Contaminated water, inadequate availability of water, and lack of access to sanitation together contribute to 88 percent of deaths from diarrheal diseases. In some developing countries, families spend up to 25 percent of their income to purchase water, and many women and children spend up to 6 hours a day carrying it home (World Health Organization and UNICEF, 2014; WWAP, 2014).

AVAILABILITY AND CONSUMPTION

Worldwide, water scarcity is our biggest problem. Just 3 percent of the earth's water is fresh, and two-thirds of it is locked up in the ground, glaciers, and ice caps. That leaves about 1 percent of the earth's water for the world's more than 7 billion people. The world's demand for water has tripled over the last half century. Water scarcity, which affects at least 40 percent of people around the world, is projected to increase. More than half of Earth's 37 largest aquifers (underground layers of rock that store water) are being depleted "at alarming rates" to keep pace with demands from agriculture, growing

How Much Water It Takes to Make...

In 1950, Americans used 150 billion gallons of water every day compared with 400 billion gallons today. It takes water to make everything. For example, it takes:

Jacob Kearns/Shutterstock.com

2,600 gallons to make a pair of blue jeans

Christopher Elwell/Shutterstock.com

634 gallons to make an average hamburger

magicoven/Shutterstock.com

53 gallons to produce a cup of to-go latte coffee

Stocksnapper/Shutterstock.com

713 gallons to make a cotton shirt

Evgeny Karandaev/Shutterstock.com

1.5 gallons to produce an average 18 ounces of bottled water

Sources: Based on Connell, 2011; Postel, 2012.

---

populations, and industries like mining (WWAP, 2014; Richey et al., 2015; United Nations, 2015).

Some refer to water as "blue gold" because it's becoming one of the earth's most precious and scarce commodities. Water shortages—rather than oil or diamonds—are behind conflicts and even wars in a number of countries. According to a past Secretary-General of the United Nations, "Too often, where we need water, we find guns..." (World Water Assessment Program, 2009: 20; Reynolds, 2012).

Industrialized nations not only have greater access to clean water than the developing world, they use more and pay less for it. The average person in the United States uses about 163 gallons of water per day compared with 40 gallons in Germany, 30 in Denmark, 23 in China, and less than 3 in Mozambique. On average, Americans pay only $.48 per gallon compared with $1.65 a gallon in Denmark. In the developing world, people typically pay five times as much as Europeans (Maxwell, 2012). Thus, in many countries, clean water is a luxury rather than a basic human right.

THREATS TO WATER SUPPLIES

Precipitation (in the form of rain, snow, sleet, or hail) is the ecosystem's main source of water. Clean water has been depleted for many reasons, including pollution, privatization, and waste.

Pollution. Every year, more than 860 billion gallons of sewage, pesticides, fertilizers, automotive chemicals, and trash spoil the country's freshwater. Industrial sites and farm fertilizers are two of the major pollution sources. As a result, 20 percent of the nation's lakes and 55 percent of its rivers and streams can't provide drinking water or support healthy aquatic life (EPA, 2013, 2014).

In 2014, a chemical spill in West Virginia left 300,000 people without access to running water and unable to use their faucets to drink, cook, or bathe for five days. The chemical industry maintains that such spills are rare, but they occur every day. In 2013, for instance, 76 U.S. companies had almost 3,900 spills. From 2001 to 2010, there were 992 oil and gas fluid spills in three Colorado counties alone (Chemi, 2014; Walsh, 2014).

Few corporate polluters are fined, and many are exempt from major environmental laws, including the 1974 Safe Water Drinking Act. Consider hydraulic fracturing, or *fracking*, a process that injects water, sand, and chemicals at high pressure to extract gas and oil from rock that lies deep underground. Many U.S. lawmakers—some of them with deep ties to industry that finances political campaigns—endorse fracking to increase energy supplies (Blake, 2014).

Fracking has risks because gases can escape into drinking water. Some of the 750 chemicals injected into the ground contaminate water, increasing the risks of cancer, infertility, and other health problems (Kassotis et al., 2014; EPA, 2015). Fracking has also increased the number and magnitude of earth tremors and earthquakes from Colorado to the Atlantic coast. Cities and counties nationwide have passed 430 measures to control or ban fracking, but companies typically prevail despite opposition (Shauk and Olson, 2014; Skoumal et al., 2015).

Privatization. Water is a big business because of *privatization*, transferring some or all of the assets or operations of public systems into private hands. Perrier, Evian, Coca-Cola, PepsiCo—and particularly the French giants Vivendi and Suez—have been buying the rights to extract water in the United States and other countries at will from aquifers, then bottling and selling it around the world.

About 70 percent of Americans say they drink bottled water. In 2014, bottled water companies spent more than $84 million on advertising to compete with each other and to convince consumers that bottled water is healthier and safer than soda and tap water. Bottling water is lucrative for corporations, but there are many environmental drawbacks. It depletes local water supplies, whether the water comes from municipal sources (40 percent) or local springs (60 percent). Moreover, about 86 percent of the empty plastic bottles in the United States clog landfills instead of being recycled (Food and Water Watch, 2013; Lohan, 2015).

Waste. Most water problems are due to human mismanagement and waste, not nature. Of all available water worldwide, agriculture consumes about 70 percent, industry uses 20 percent, and 10 percent is residential (WWAP, 2014). In agriculture, many irrigation systems are inefficient, farmers and agribusiness (large agriculture companies) often grow water-hungry crops like cotton and sugarcane in arid areas, and pesticide and chemical fertilizer runoff from fields pollutes streams, rivers, and lakes.

Some Americans are trying to conserve water, but they're a minority. For example, nearly 75 percent of residential water use in California—which has experienced four years of severe droughts—goes to outdoor purposes, mostly landscaping. Affluent California neighborhoods use five times more water per person than the statewide average. After Governor Brown called for a 25 percent reduction in water use, consumption in a wealthy community went *up* by 9 percent. The average Palm Springs golf course uses the same amount of water in one day that a family of four does in 5 years (Pickert, 2014; Kuznia, 2015; Lurie, 2015).

In the United States, a significant water pipe bursts, on average, every 2 minutes somewhere in the country,

losing 7 billion gallons of water a day. Most of the nation's pipes, especially in cities along the eastern seaboard, are nearly 200 years old, and some are made of wood. A water pipe that leaks or bursts wastes huge amounts of drinking water, damages streets and homes, and seeps dangerous pollutants into drinking water. Despite such hazards, replacing a city's substandard pipes has decreased. From 1890 to 1920, for instance, Chicago replaced aging water and sewage pipes at a rate of about 75 miles a year. That number has declined steadily over the years, dropping to 30 miles a year in 2003 (Duhigg, 2010; Pearlstine, 2013).

15-3b Air Pollution, Global Warming, and Climate Change

Global warming is a serious environmental problem. Let's begin with air pollution, the major cause of climate change and global warming.

AIR POLLUTION: SOME SOURCES AND CAUSES

There are many reasons for air pollution, but four are among the most common. First, a major source of air pollution is the burning of *fossil fuels*, substances obtained from the earth, including coal, petroleum, and natural gas. The exhaust gases of cars, trucks, and buses contain poisons—sulfur dioxide, nitrogen oxide, carbon dioxide (CO_2), and carbon monoxide. Power plants that produce electricity by burning coal or oil account for nearly 40 percent of U.S. carbon dioxide emissions.

Second, manufacturing plants that produce consumer goods pour pollutants into the air. Formaldehyde-based vapors that can lead to cancer and respiratory problems are emitted by many household and personal care products: pressed wood (often used in furniture), plastic grocery bags, waxed paper, latex paints, detergents, nail polish, cosmetics, and shampoos ("Formaldehyde," 2012; see also Chapter 14).

Third, winds blow contaminants in the air across borders and oceans. Air pollution from Asia, particularly China, has affected the air quality in the Sequoia and Kings Canyon national parks in California. Air pollution originating in Europe has been tracked to Asia, the Arctic, and even rural areas in the western United States where there's little industry or automobile traffic (Lamb, 2009; Cooper et al., 2010).

A fourth reason for air pollution is that government policies and enforcement vary from one administration

iStockphoto.com/gmutlu

China is the planet's biggest air polluter. Sixteen of the world's 20 most polluted cities are in China. The smog is so thick on some days that visibility drops to a few yards, and cities close schools, airports, and highways (Denyer, 2013; Cendrowski, 2014).

to another. Between 2002 and 2006, Justice Department lawsuits against polluters declined by 70 percent, and criminal and civil fines for polluting decreased by more than half compared with the period from 1996 to 2000. The Bush administration blocked the efforts of 18 states to cut emissions from cars and trucks because it believed that tougher regulations would hurt the U.S. economy (Environmental Integrity Project, 2007; "No Action on Greenhouse Gases," 2008).

President Obama promised to decrease air pollution, but environmental groups have accused his administration of being too soft on the fossil-fuel industry by not limiting their carbon dioxide emissions. The rules also exclude Indian reservations—which have some of the most polluting coal-fired power plants in the country—from meeting state goals to reduce carbon dioxide emissions by 2030 (Eilperin and Bernstein, 2014; Magill, 2014).

In early August, 2015, the Obama administration finalized the Clean Power Plan, which scientists and environmentalists hailed as the nation's historically strongest legislation to decrease carbon dioxide. A few weeks later, he gave Royal Dutch Shell final approval to resume oil and gas drilling in the Arctic, one of the world's most ecologically vulnerable regions, and home to polar bears and other endangered species (Loki, 2015). Shell abandoned the project six weeks later because of high costs.

GLOBAL WARMING AND THE GREENHOUSE EFFECT

Global warming is an increase in the average temperature of earth's atmosphere. The warming is due to several factors. Some are natural, such as changes in solar radiation, the earth's orbit, and the frequency or intensity of volcanic activity. At least 95 percent of global warming is due to human activities, however, and not natural climate swings (Intergovernmental Panel on Climate Change, 2014).

Global warming begins with the **greenhouse effect**, the heating of earth's atmosphere because of the presence of certain atmospheric gases. When the sun's heat enters the atmosphere, some of it is absorbed by earth's surface, and some of it is reflected back to space. Greenhouse gases in the atmosphere trap some of this heat. Heat is necessary to support life, but when greenhouse gases increase, earth becomes warmer than it would be otherwise, endangering public health and the welfare of current and future generations (U.S. Environmental Protection Agency, 2009).

Air pollutants, primarily CO_2, ignite the greenhouse effect. Between 2000 and 2010, greenhouse-gas emissions grew at 2.2 percent a year—almost twice as fast as in the previous 30 years. During this decade, 90 percent of the world's CO_2 emissions came from burning fossil fuels, primarily from coal-fired power plants. The volume of carbon dioxide hit a record in 2013, the largest year-to-year increase since 1984, when reliable global record keeping began (Intergovernmental Panel on Climate Change, 2014; World Meteorological Organization, 2014).

China is now emitting almost twice as much CO_2 as the next-biggest polluter, the United States. At current rates, it will produce 500 billion tons of carbon dioxide between 1990 and 2050—as much as the whole world produced between the start of the Industrial Revolution and 1970. China is the world's top coal producer, consumer, and importer, accounting for about half of global coal usage. It burns nearly as much coal every year (more than 4 billion tons) as the rest of the world combined, an important factor in its carbon dioxide emissions ("China's Environment...," 2014; U.S. Energy Information Administration, 2014).

In the United States, about 80 percent of all greenhouse gases come from industrial coal-fired power plants like Southern Company and its subsidiaries in Georgia, Alabama, Indiana, Ohio, and West Virginia owned by America Electric Power. At least 10 other states also have

global warming the increase in the average temperature of earth's atmosphere.

greenhouse effect the heating of earth's atmosphere because of the presence of certain atmospheric gases.

Mike Peters/Cartoonist Group

high-polluting power plants (Food and Agriculture Organization of the United Nations, 2011; Cappiello, 2012).

SOME EFFECTS OF CLIMATE CHANGE

Climate change is a change in overall temperatures and weather conditions over time. Global warming probably began thousands of years ago when humans began changing the planet by cutting down and burning forests to grow food (Fischman, 2009). Such *deforestation*, clearing massive amounts of trees, affects global climate changes because forests recycle carbon dioxide into oxygen.

Regardless of when it began, climate change is "unequivocal," has "moved firmly into the present," and is having a profound impact on every ecosystem worldwide. Since record-keeping began in 1880, 2014 was the hottest year on earth, and 9 of the 10 warmest years have occurred since 2000. Between 1750 and 2005, seven countries produced 63 percent of all global warming. The United States, accounting for 20 percent of all climate-changing pollution, was the leading offender, followed by China and Russia (8 percent each). Scientists predict that extreme weather, especially heat waves, will produce heavier rainfall in some regions; more floods, landslides, and uncontained wildfires; stronger hurricanes and tornadoes; and more intense droughts around the world (Matthews et al., 2014; NOAA National Centers . . . , 2015).

Many changes have already occurred because of climate change. For example:

▶ Higher CO_2 levels are turning oceans more acidic, resulting in some shellfish (e.g., clams and crabs) dying or not developing (Spotts, 2009).

▶ Global weather- and climate-related disasters rose from 743 between 1971–1980 to 3,496 between 2001–2010 (World Meteorological Organization, 2014).

▶ Excessive heat in the United States claims more lives each year than floods, lightning, tornadoes, and hurricanes combined (National Oceanic and Atmospheric Administration, 2014).

▶ The world's largest ice sheets in Antarctica and western Greenland are melting at a rate of 120 cubic miles a year, the highest speed in the last 20 years. The melting glaciers could contribute to sea levels rising 2 to 3 feet by 2100 depending on the region (Helm et al., 2014; Intergovernmental Panel on Climate Change, 2014).

▶ Some Alaskan native villages are facing disaster. Climate change has thinned the Arctic ice so much that it's become too dangerous to hunt whales, a traditional means of sustenance, or even to live in the state's coastal communities. As the ice melts, the towns are in "imminent danger" from erosion and being washed to sea during storms (Mooney, 2015).

Such data show that our planet is ailing. Can people slow some of earth's devastation through better environmental policies and practices, particularly by endorsing sustainable development?

15-3c Is Sustainable Development Possible?

Sustainable development refers to economic activities that don't threaten the environment. There are reasons to be both pessimistic and optimistic about achieving sustainable development.

REASONS TO BE PESSIMISTIC

Which country is the greenest? Not the United States. For example:

▶ A study of environmental performance ranked 178 countries on factors like water quality, air pollution,

---

climate change a change in overall temperatures and weather conditions over time.

sustainable development economic activities that don't threaten the environment.

and protecting the ecosystem. Switzerland, Luxembourg, Australia, Singapore, and the Czech Republic were the top five. The United States ranked 33rd, and below a number of developing countries, including Slovakia and Serbia (Hsu et al., 2014).

▶ In a global study of 132 nations, the United States ranked only 69th in ecosystem sustainability (Porter and Stern, 2014).

▶ Among the world's 16 largest economies, the United States ranks only 13th in energy-efficiency policies and programs (e.g., fuel economy standards for vehicles, energy efficiency standards for appliances), and below developing countries like China and India (Young et al., 2014).

Why does the United States rank so low in environmental performance? There are many reasons, but three are especially important. First, many Americans' worries about global warming have waned and they see economic

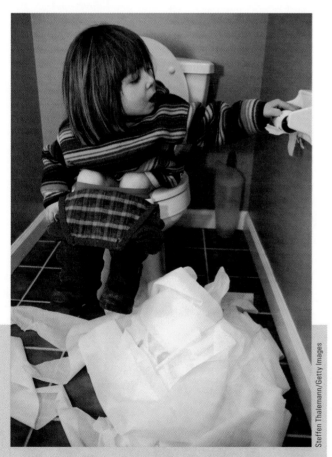

Fluffy and ultra-plush toilet paper is made by chopping down and grinding up the pulp of trees that are decades or even a century old. U.S. corporations have introduced "earth-friendly" toilet paper, but "customers are unwavering in their desire for the softest paper possible" (Fahrenthold, 2009: A1).

TABLE 15.4 IS ECONOMIC GROWTH OR THE ENVIRONMENT MORE IMPORTANT?

Percentage of Americans Who...

	Worry about Global Warming	Believe That Economic Growth Is More Important Than Protecting the Environment
1999	68%	28%
2007	64%	37%
2014	56%	41%

Note: The numbers may not sum to 100 percent because of no responses and rounding.

Sources: Based on Newport, 2014, and Swift, 2014.

growth as more important than protecting the environment (see *Table 15.4*). Only half of Americans believe that climate change is due mostly to human activity, such as burning fossil fuels. Some 25 percent say there's no solid evidence of global warming, up from 11 percent in 2009. Equal numbers—40 percent each—support and oppose fracking (20 percent have no opinion). About 85 percent of Americans say that they recycle, but only 34 percent really do so, and only 29 percent recycle electronic gadgets (EPA, 2014; Funk and Rainie, 2015; Sullivan, 2015; Swift, 2015).

Second, environmental problems are a low priority. Among almost 2,000 U.S. companies, 67 percent of businesspeople say that social and environmental matters are very important but only 10 percent report that their companies are developing sustainability policies (Kiron et al., 2013). Chevron Corporation pulled back from key efforts to produce clean energy because the profits from non-renewable oil and gas sales were much higher.

Some environmentalists are especially dismayed by *greenwashers*, companies and other organizations that pollute the planet while presenting an environmentally responsible public image. In 2000, for instance, BP (British Petroleum) rebranded itself "Beyond Petroleum" and changed its logo to a green-and-yellow sunburst, but the majority of its business produces fossil fuels (Elgin, 2014).

A third reason for America's low environmental rank is lawmakers' often accommodating big business. Oil and utilities industries, which finance many elections, have successfully lobbied Congress to drop laws that would've required them to develop renewable wind and solar energy sources. The National Park Service, a federal agency, recently stopped selling disposable bottles and let visitors refill reusable ones with public drinking water. Under pressure from the powerful

ARE AMERICANS ECO-FRIENDLY ONLY WHEN IT'S CONVENIENT?

bottled water industry, some Senators are trying to pass an amendment that would prevent the agency's banning bottled water sales. Corporations and many small businesses also profit from government subsidies for agriculture and exports, but taxpayers bear the costs of air and water pollution (Saperstein, 2014; Loki, 2015; Rein, 2015; see also Chapter 11 on the close ties between the government and corporations).

Americans say that they want cleaner energy, but most oppose wind farms because the 40-foot windmills would "disturb my view of the landscape" (Dunlap, 2007). Thus, are Americans eco-friendly only when it's convenient?

REASONS TO BE OPTIMISTIC

There's been considerable progress since 1970, when the United States celebrated its first Earth Day. Most cars no longer burn leaded gasoline, ozone-destroying chlorofluorocarbons (CFCs) have been generally phased out, and total emissions of the six major air pollutants declined by 54 percent during the same period as the U.S. population increased by 47 percent. The U.S. recycling rate increased from less than 10 percent in 1980 to over 34 percent in 2012. As a result, landfill trash decreased from 89 to 54 percent during this interval (Sperry, 2008; EPA, 2014).

Many companies have found that being green is good for their profits, image, and environment, and are now more eco-friendly. For example:

▶ Each pair of Levi's new Waste<Less jeans is composed of eight recycled plastic bottles (Berfield, 2012).

▶ Ford, Honda, and Toyota use recycled materials— jeans, sweaters, plastic bottles, and sugar cane by-products—for dashboards, tires, seat covers, and cushions (Fleming, 2014).

▶ SABMiller, the world's second-largest brewer, is working with India's wheat farmers to use water more efficiently to ensure that the company has water to brew beer ("A New Green Wave," 2014).

▶ Coca-Cola is years ahead of schedule in replacing the water that it uses around the world to make its beverages, and Nike is using more synthetic material that's less dependent on weather conditions (Davenport, 2014; Schwartz, 2015).

Economic self-interest rather than a concern for the environment sometimes drives such eco-friendly practices. Some corporations are eager to reduce their reliance on water-intensive products like cotton, for instance. Regardless of motives, enacting environmentally friendly policies promotes sustainability.

Current efficiency standards for appliances, lighting, and other equipment that were implemented during the early 1980s will save the United States the equivalent of two years of energy use ($1.1 trillion by 2035) and slash greenhouse gas emissions. Also, local governments in southern California and some Texas cities are recycling wastewater for drinking (Lowenberger et al., 2012; Preston, 2014).

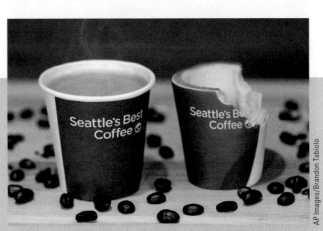

Many individuals and businesses are contributing to sustainable development. Hawaii has some of the nation's greenest cities, admired for their excellent air quality and extensive use of rooftop solar panels. KFC, a restaurant chain, is introducing edible cups at Seattle's Best Coffee, part of Starbucks, to decrease the environmental impact of packaging (Strom, 2015).

In other cases, progress has been mixed. On an individual level, 71 percent of Americans say that the United States "should do whatever it takes to protect the environment." Out of 40 nations, however, Americans rank only 24th in being "very concerned" about global climate change. On a group level, environmentalists support clean energy, but have sued large companies and some state governments that want to build solar farms that might damage habitats for birds and other creatures (Carle, 2015; Pew Research Center, 2015; Young, 2015).

On an organizational level, Walmart says that it recycles 80 percent of the waste in domestic stores, and hopes to achieve zero waste. Some environmentalists contend, however, that Walmart's manufacturing low-quality products increases landfills and accelerates consumer consumption (Clifford, 2012). In 2014, the Supreme Court upheld an EPA regulation requiring Midwestern states with coal-fired power plants to reduce harmful emissions that drift to the East Coast, but also ruled in favor of an electronics manufacturing corporation accused of contaminating a North Carolina community (Savage and Banerjee, 2014; Peeples, 2014). The EPA has imposed stiffer penalties on automakers that overstate their fuel economy, but allows using 82 chemicals, including pesticides that Europe restricts or bans (Hakim, 2015; Kessler, 2015).

STUDY TOOLS 15

READY TO STUDY? IN THE BOOK, YOU CAN:

☐ Check your understanding of what you've read with the Test Your Learning Questions provided on the chapter review card at the back of the book.

☐ Rip out the chapter review card for a handy summary of the chapter and key terms.

ONLINE AT CENGAGEBRAIN.COM YOU CAN:

☐ Collect StudyBits while you read and study the chapter.

☐ Quiz yourself on key concepts.

☐ Prepare for tests with chapter flash cards as well as flash cards that you create.

☐ Watch videos to explore sociology further.

© Anelina/Shutterstock.com

Social Change: Collective Behavior, Social Movements, and Technology

LEARNING OUTCOMES

After studying this chapter, you will be able to...

16-1 Compare, illustrate, and evaluate the major types of collective behavior, their theories, and functions.

16-2 Compare, illustrate, and evaluate the major types of social movements, their theories, and functions.

16-3 Describe and illustrate recent technological advances, their benefits, costs, and ethical controversies.

After you finish this chapter go to **PAGE 331** for **STUDY TOOLS**.

For the most part, as you've seen throughout this textbook, sociologists study behavior and social processes that are relatively institutionalized, routine, stable, and predictable. This chapter examines **social change**, the transformations of societies and social institutions over time. Some collective behavior is short-lived with few long-term changes; others, particularly social movements, can have lasting effects. Technology has also played a critical role in sparking social change. Let's begin with collective behavior.

16-1 COLLECTIVE BEHAVIOR

Do you have boxes crammed with collectibles like Barbie dolls or Star Wars merchandise? A tattoo? Have you ever joined a club or sports team? Texted? Taken a selfie? If so, you've engaged in collective behavior.

social change transformations of societies and social institutions over time.

What do you think?

I sometimes feel overwhelmed in keeping up with texting, e-mail, Twitter, and Facebook.

1	2	3	4	5	6	7

strongly agree   strongly disagree

Allison Shelley/Getty Images

16-1a What Is Collective Behavior?

Collective behavior is the spontaneous and unstructured behavior of a large number of people. Collective behavior encompasses a wide range of actions, including riots, fads, fashion, panic, rumors, and responses to disasters.

Collective behavior has two important characteristics. First, it's an act rather than a state of mind. You may *feel* panic when a tornado threatens your town, but you don't engage in collective behavior until you actually *leave* your home and head for a safer location.

Second, collective behavior varies in its degree of spontaneity and structure. Panic, the least structured form of collective behavior, is typically short-lived. Fads (e.g., diets, "ugly" Christmas sweaters) are more structured. They may last several years, are planned, and may be expensive. Other forms of collective behavior, including pro- and antiabortion groups, become highly institutionalized social movements that include a staff, budget, and lobbying.

16-1b Why Collective Behavior Occurs

Over the years, sociologists have proposed many explanations of collective behavior. Four of the most influential are contagion, convergence, emergent norm, and structural strain theory. Each has strengths and weaknesses.

CONTAGION THEORY

Contagion theory proposes that individuals act emotionally and irrationally due to a crowd's almost hypnotic influence. This perspective is rooted in the work of French scholar Gustave Le Bon (1841–1931), who asserted that people in crowds, particularly mobs and riots, change radically. The anonymity of the crowd, a feeling of power, and an infectious "mob mind" embolden people to abandon personal responsibility and to engage in antisocial, and often violent, behavior (Le Bon, 1896/1968).

American sociologists later refined Le Bon's theory. They emphasized "milling," a process of people moving about, talking to one another, and becoming increasingly excited. Milling produces a "circular reaction" in which the participants become more intense and more unified in attacking a target (Park and Burgess, 1921; Blumer, 1969). Because people feel anonymous in crowds, they

collective behavior the spontaneous and unstructured behavior of a large number of people.

abandon self-control and adopt behavior that they'd reject in other settings (Brown and Goldin, 1973).

CONVERGENCE THEORY

Contagion theory asserts that crowds are highly suggestible and out of control. In contrast, *convergence theory* proposes that crowds consist of like-minded people who deliberately assemble in a place to pursue a common goal. Thus, collective behavior occurs when people who share similar values, beliefs, attitudes, emotions, and goals come together in, or converge on, a certain location. Instead of the crowd affecting individuals, as contagion theory claims, convergence theory argues that individuals influence a crowd. Examples of convergence theory include marches, rallies, and acts of civil disobedience that oppose a government's domestic or foreign policies.

EMERGENT NORM THEORY

Emergent norm theory emphasizes that social norms shape crowd behavior. When people perceive a situation as unusual, they create rules that guide their behavior in rational ways. Those who are the most committed may become the leaders and pacesetters; those who disagree or have a different agenda, like looting, are typically ignored. In effect, then, people aren't spontaneously "infected" with the emotions of others, as contagion theory postulates. Instead, crowd members develop norms as a situation unfolds. Unionized workers may orchestrate a strike against an airline or hospital. If, however, the negotiators can't reach an agreement, the strikers may become more intense and develop new norms that include sit-ins and boycotts (Brown and Goldin, 1973; Turner and Killian, 1987).

In 2014, from Los Angeles to New York City, hundreds of fast-food workers and their supporters were arrested as they escalated a fight for better pay with strikes, rallies, and sit-ins. Which collective behavior theory would you use to explain the protests? Why?

STRUCTURAL STRAIN THEORY

Structural strain theory (also called *value-added theory*) proposes that collective behavior occurs if six conditions are present (Smelser, 1962). These conditions are "value-added" in the sense that each condition leads to the next one, ending in an episode of collective behavior:

1. *Structural conduciveness* refers to the social conditions that foster collective behavior. When channels for expressing a grievance either aren't available or fail, like-minded people may resort to protests.

2. *Structural strain* occurs when societal problems (e.g., discrimination, war) make people angry, frustrated, or interfere with their everyday lives.

3. In the *growth and spread of a generalized belief* stage, people begin to see a situation as a widespread problem instead of their own fault, attribute the problem to a person or group, and believe that something should be done.

4. *Precipitating factors* are incidents or events that trigger action.

5. During the *mobilization for action* stage, leaders emerge.

6. In the *social control* stage, opposing groups prevent, interrupt, or repress those advocating social change. Government officials, the police, community and business leaders, courts, the mass media, and other social control agents—all of whom benefit from the status quo—may ridicule or quash the emerging collective behavior.

CRITICAL EVALUATION

Over time, scholars found little evidence for contagion theory's claim that crowds are typically irrational or out of control. Crowds can sway emotions, but can also be effective in meeting their objectives, and even mobs can be focused and rational (Couch, 1968; McPhail, 1991).

Convergence theories are useful in describing crowds that are deliberate and planned, but have several limitations. They don't explain why some people who share the same perspectives join a protest, rally, or demonstration while others don't, and don't account for the shifts in crowd behavior, no matter how well planned (Berk, 1974; McPhail, 1991).

Emergent norm theory helps explain why many crowds are orderly and focused rather than disorderly and mob-like. In the case of protestors, those who join a crowd quickly learn "the rules" ("Ignore hecklers" or "Don't chat with spectators"). Nonetheless, emergent norm theory doesn't explain which norms emerge, why, and how they differ in crowds. In the case of large crowds—those numbering in the thousands, for instance—how are emergent norms disseminated so quickly and accepted by the participants? It's also not clear why some participants conform to emergent norms while others ignore them and

A day after the CDC confirmed the first U.S. Ebola case in Texas, and emphasized that the risk of infection is low, there were hundreds of social media rumors that the disease had spread to Iowa and other states. Why did the rumors proliferate? Almost 67 percent of Americans believed that Ebola spreads easily (it doesn't), 40 percent expected a major outbreak in the United States (highly unlikely), and 26 percent were personally terrified that they or someone in their immediate family would get the Ebola virus (also highly unlikely) (Harvard School of Public Health, 2014).

Facts *about* **Ebola** in the U.S.

You CAN'T get Ebola through AIR

You CAN'T get Ebola through WATER

You CAN'T get Ebola through FOOD grown or legally purchased in the U.S.

Center for Disease Control

act irrationally (as looting stores and vandalizing property) (Brown and Goldin, 1973; Berk, 1974).

Structural strain theory helps to predict when and where collective behavior might occur, and offers insights on why, at every stage, collective behavior may either fade or escalate. However, the model doesn't explain all forms of collective behavior. In the case of fads and rumors, as you'll see shortly, all six stages don't necessarily occur. A second limitation is that the sequence of stages isn't necessarily the same as Smelser outlined. Moreover, the theory doesn't say how much structural strain must exist to spark collective behavior (Rule, 1988; Locher, 2002).

16-1c Types of Collective Behavior

There are many forms of collective behavior; some are more fleeting or harmful than others (Turner and Killian, 1987). Let's begin with rumors, one of the most common types of collective behavior.

RUMORS

There were widespread rumors that on January 1, 2000, a glitch (Y2K) in operating systems would cause computers around the world to crash, leading to global power outages, banks losing all of their customers' statements, and even airplanes falling from the skies. None of this occurred.

A **rumor** is unfounded information that people spread quickly. Through modern communication technology, a rumor can spread to millions of people within seconds, and incite riots, panic, or widespread anxiety. Because of the Y2K rumor, thousands of people built underground shelters, and millions stocked up on bottled water, canned food, batteries, and medical supplies.

Most rumors (that rock stars Elvis Presley and John Lennon are alive, for example) are harmless. Others can be damaging. After a woman claimed that she found part of a human finger in her cup of Wendy's beef chili, the restaurant's business dropped by half nationally, and rumors

warning people to stop eating fast food altogether spread over the Internet (Richtel and Barrionuevo, 2005). The woman admitted, ultimately, that she had planted the finger to try to get a lucrative settlement. Nonetheless, some customers are still leery about eating at fast-food restaurants.

Rumors are typically false, so why do so many people believe them? First, rumors often deal with an important subject about which—especially during uncertain economic times or natural disasters—people are anxious, insecure, or stressed. This makes people especially suggestible. In Hurricane Katrina's aftermath, the media reported numerous rumors of carjacking, murders, thefts, and rapes, the overwhelming majority of which subsequently proved to be false.

Second, there's often little factual information to counter a rumor, or people distrust the sources of information. During Y2K, the people who stockpiled groceries and so forth didn't believe computer scientists or federal officials who assured the general public that there wouldn't be a calamity. Third, rumors offer entertainment, diversion, and drama in our otherwise mundane daily lives (Campion-Vincent, 2005; Heath, 2005). Gossip and urban legends are two of the most common types of rumor.

Gossip. Gossip is the act of spreading rumors, often negative, about other people's personal lives. Someone once said that "no one gossips about other people's virtues." Because of its tendency to be derogatory, gossip makes us feel superior ("Did you know that Margie just had breast implants? Isn't she pathetic?"). Gossip is also interesting, entertaining, and exposes hypocrisy (Epstein, 2011).

---

rumor unfounded information that people spread quickly.

gossip rumors, often negative, about other people's personal lives.

Sometimes, individuals gossip to control other people's behavior and to reinforce a community's moral standards. Comments about someone's drug abuse or marital infidelity strengthen norms of what's deviant or unacceptable, and bonds us with like-minded people (see Chapter 4). In other cases, people gossip because they resent or envy someone's physical appearance or accomplishments. Besides hurting personal and workplace relationships, gossip often creates fear, resentment, and stress—all of which can increase the likelihood of illness (Sunstein, 2009; Drexler, 2014).

Urban Legends. Another form of rumor is **urban legends** (also called *contemporary legends* and *modern legends*), stories—funny, horrifying, or just odd—that supposedly happened somewhere. Some of the most common and enduring urban legends, but with updated variations, deal with food contamination, like the finger at Wendy's. Others have targeted politicians. We still hear, for example, that President Obama wasn't born in the United States, even after he produced his Hawaii birth certificate.

Health-related urban legends also persist. In 1998, medical researchers showed conclusively that there's no relationship between autism and vaccinations for measles, mumps, and rubella. Despite the scientific evidence, during the 2015 Republican presidential debates, Donald Trump asserted, again, that childhood vaccines can cause autism. Two other presidential candidates—Dr. Ben Carson, a retired neurosurgeon, and Senator Paul Rand, an ophthalmologist—didn't disagree with Trump. Doctors, health organizations, and autism advocacy groups quickly criticized and tried to correct the misinformation, but such "ongoing antivaccine noise

by celebrities" helps explain mumps and measles outbreaks in the United States and Europe (Sifferlin, 2014: 7; Welch, 2015; see also Chapters 2 and 14).

Why do urban legends persist much longer than gossip? First, they reflect contemporary anxieties and fears—about contaminated food, unscrupulous companies, and corrupt and unresponsive governments. Second, urban legends are cautionary tales that warn us to watch out in a dangerous world. There have been tales that sunscreens cause blindness, and that women have died sniffing perfume samples sent to them in the mail. Third, we tend to believe urban legends because we hear them from people we trust, especially family members, coworkers, and friends. Fourth, rejecting misinformation and outright lies is hard work because doing so requires rethinking already-held beliefs. Finally, urban legends—like the one about alligators living in New York City's sewer system or that AriZona tea contains human urine—are fun to tell and "too beguiling to fade away" (Brunvand, 2001; Ellis, 2005; Emery, 2015).

PANIC AND MASS HYSTERIA

In 2003, an indoor fireworks display to kick off a heavy metal concert in West Warwick, Rhode Island, set off a fire that killed 100 people and injured 200 others. As thick black smoke poured through the audience, hundreds of patrons stampeded for the front door (even though there were three other exits), trampling and crushing those who had fallen beneath them.

Most of the deaths in West Warwick weren't due to the fire but to **panic**—a collective flight, often irrational, from a real or perceived danger. The danger seems so overwhelming that people desperately jam an escape route, jump from high buildings, leap from a sinking ship, or sell off their stock. Fear drives panic: "Each person's concern is with his [or her] own safety and personal security, whether the danger is physical, psychological, social, or financial" (Lang and Lang, 1961: 83).

urban legends (also called *contemporary legends* and *modern legends*) rumors about stories that supposedly happened somewhere.

panic a collective flight, often irrational, from a real or perceived danger.

Every year, more than 2 million Muslims make a pilgrimage to Mecca (in Saudi Arabia) to participate in the Hajj, an important religious ritual. In 2015, nearly 2,200 of the pilgrims were crushed to death during a stampede. The suggested causes, still being investigated, include heat (110 degrees Fahrenheit), people rushing to complete the rituals and pushing against each other in opposite directions, confusion among the many first-timers, and panic as people fell and were trampled (Gambrell, 2015; Yan, 2015).

Panic is similar to **mass hysteria**—an intense, fearful, and anxious reaction to a real or imagined threat by large numbers of people. In mid-2009, the World Health Organization issued an alert that H1N1 (the "swine flu") would become a worldwide epidemic and result in numerous deaths. Millions of Americans stood in line for hours, sometimes overnight, because the vaccine was initially in short supply, but the swine flu never materialized on the predicted scale. Unlike panic, which usually subsides quickly, mass hysteria may last longer because warnings—especially about health and food—reinforce our general fears and anxieties about life's dangers.

FASHIONS

As the "Betty Crocker Makeover" on the next page shows, General Mills has been modernizing "Betty" since the 1930s. Doing so may reinforce people's cynical views about corporate manipulation. On the other hand, updating Betty's image and fashion may help explain why General Mills has successfully attracted new generations of consumers over the decades, and why Betty Crocker products are now sold worldwide.

Fashion is a popular way of dressing during a particular time or among a particular group of people. Over the years, black women's hairstyles have changed—from Afros in the late 1960s, to straightened hair during the 1980s, to braids, cornrows, dreadlocks, hair extensions, and coloring more recently. All reflect gender politics, racial solidarity, generational differences, identity, and changing images of beauty (Banks, 2000; Desmond-Harris, 2009).

Fashion involves periodic changes in the popularity of clothes, furniture, music, language usage, books, automobiles, sports, recreational activities, the names parents give their children, and even the dogs that people own. People who want to be fashionable buy clothes with prominent labels, but what's fashionable changes from year to year. Thus, U.S. adults regularly wear only about 20 percent of their wardrobe (Smith, 2013).

Why do fashions, particularly clothes, change fairly quickly? One reason is that designers, manufacturers, and retailers must continuously generate demand for new products and services to maintain a profit. Retailers have created a number of "shopping holidays" to move merchandise, including clothes (e.g., Amazon's "Prime Day," Kohl's senior discount Wednesdays, and many large department stores' "Black Friday" sales the day after Thanksgiving). Second, many people keep up with fashion because they don't want to seem different, out-of-date, or dowdy. Third, shopping for new clothes and other products decreases the boredom of everyday routine. Also, clothes and other merchandise are status symbols that signal being an insider: "Others…will admire me for…making stylish choices" (Best, 2006: 85–86; for classic analyses of fashion and collective behavior, see Veblen, 1899/1953; Packard, 1959; and Bourdieu, 1984).

FADS

A **fad** is a fashion that spreads rapidly and enthusiastically but lasts only a short time (Turner and Killian, 1987; Lofland, 1993). Fads include *products* (e.g., bean bag chairs, Crocs, pet rocks), *activities* (e.g., twerking, step aerobics, grapefruit diets), and *popular personalities and television characters* (e.g., the Lone Ranger during the 1950s, the Kardashians more recently).

Why do fads pop up? A major reason is profit. Because children and adolescents are especially likely to adopt fads, manufacturers create numerous products and activities that they hope will catch on. The products include toys, sportswear, and new cereals. The hottest fads are usually the must-have Christmas toys that children plead for every year. A few months later, the toy may be thrown away or stuffed in the back of a closet.

mass hysteria an intense, fearful, and anxious reaction to a real or imagined threat by large numbers of people.

fashion a popular way of dressing during a particular time or among a particular group of people.

fad a fashion that spreads rapidly and enthusiastically but lasts only a short time.

BETTY CROCKER MAKEOVER

1936

1955

1965

1968

1972

1980

1986

1996

AP Images/General Mills

In 1921, General Mills created Betty Crocker, a fictitious woman, to answer thousands of questions about baking that came in from consumers every year. As fashions and hairstyles changed, so did Betty Crocker's image. The original image of a stern, gray-haired older woman has changed over the years so that, by 1996, she had a darker complexion and wore casual attire. Can you think of other brands that have changed their image over the years to keep up with changing trends?

In the early 1980s, Cabbage Patch Kids dolls were extremely popular. Barely a decade later, "the dolls had fallen victim to the toy industry's perennial challenge: winning a new crop of fans every few years when kids outgrow their toys." To pump up sagging sales and profit, in 2014 the toymaker reinvigorated the dolls by introducing Cabbage Patch Kids wearing new outfits and Twinkle Toes, a popular line of sneakers that sparkle and light up (Townsend, 2014: 27).

Some people dismiss fads as "ridiculous" or "silly," but they serve several functions. In a mass society, where people often feel anonymous, a fad can develop strong in-group feelings and a sense of belonging, especially among people who have similar interests and attitudes. Fads can also be fun, promise to resolve a nagging problem (e.g., being overweight), and help us keep up with technological changes (Marx and McAdam, 1994; Best, 2006).

Most fads are soon forgotten, but some become established. Pez candy dispensers, which originated in 1952 and cost 49 cents, are still inexpensive (under $2.00), are sold in more than 60 countries, and have been continuously updated to include popular television characters like the Simpsons (Paul, 2002). Other fads, like streaking (running around nude in public places), reemerge from time to time ("Streaking," 2005).

DISASTERS

A **disaster** is an unexpected event that causes widespread damage, destruction, distress, and loss. Some disasters are due to *social causes* like war, genocide, terrorist attacks, and civil strife. Some are due to *technological causes*, including oil spills, nuclear accidents, burst dams, building collapses. Others—like floods, landslides, earthquakes, hurricanes, and tsunamis—have environmental origins (Marx and McAdam, 1994).

Disasters often inspire organized behavior rather than chaos. Instead of panicking, most people are rational, cooperative, and altruistic. They often care for family members instead of fleeing, and thousands of volunteers offer financial, medical, and other help.

PUBLICS, PUBLIC OPINION, AND PROPAGANDA

A **public** is a group of people, not necessarily in direct contact with each other, who are interested in a particular issue. A public is different than the general public, which consists of everyone in a society.

There are as many publics as there are issues—abortion, gun control, pollution, health care, and same-sex marriage, to name just a few. Even within one organization or institution, there may be several publics that are concerned about entirely different issues. At a college, students may be most concerned about the cost of tuition, faculty may spend much time discussing instructional technology, and maintenance employees may be most interested in wages.

The interaction within a public is often indirect rather than face-to-face—as through the mass media, social media, blogs, or professional journals. Because publics

disaster an unexpected event that causes widespread damage, destruction, distress, and loss.

public a group of people, not necessarily in direct contact with each other, who are interested in a particular issue.

aren't organized groups with memberships, they're often transitory. Publics expand or contract as people lose or develop interest in an issue. A public may surge during a highly publicized and controversial incident, like removing a patient's life support, but then evaporate quickly. In other cases, publics organize and become enduring social movements (a topic we'll examine shortly).

Some publics express themselves through **public opinion**, widespread attitudes on a particular issue. Public opinion has three components: (1) It's a verbalization rather than an action; (2) about a matter that concerns many people; and (3) involves a controversial issue (Turner and Killian, 1987). Like publics, public opinions wax and wane over time. People's interest in crime and

One of the best-known examples of propaganda in the United States is this Uncle Sam poster, designed in 1917, and used to recruit soldiers for World Wars I and II. Opponents of the Vietnam War used it as an antiwar poster during the 1960s and 1970s.

education drops, for example, when they're more concerned about pressing issues such as jobs and income.

Public opinion can be swayed by **propaganda**, spreading information (or misinformation) to influence people's attitudes or behavior. Propaganda isn't a type of collective behavior, but it affects collective behavior in several important ways. First, it can create attitudes that will arouse collective outbursts (e.g., strikes or riots). Second, propaganda can try to or succeed in preventing collective outbursts, as when corporations convince employees that job losses are due to a weak economy rather than to offshoring (see Chapter 11). Third, propaganda tries to gain followers for a cause, whatever it might be (Smelser, 1962). Propaganda is institutionalized in advertising, political campaign literature, and government policies. It's conveyed in many ways: The mass media, social media, political speeches, religious groups, rumor, and symbols (e.g., flags, bumper stickers).

CROWDS

Much collective behavior is scattered geographically, but crowds are concentrated in a limited physical space. A **crowd** is a temporary gathering of people who share a common interest or participate in a particular event. Regardless of size—whether it's a few dozen people or millions—crowds come together for a specific reason, including a religious leader's death, a concert, or a riot.

Crowds differ in their motives, interests, and emotional level:

▶ A *casual crowd* is a loose collection of people who have little in common except for being in the same place at the same time and participating in a common activity or event. There's little if any interaction, the gathering is temporary, and there's little emotion. Examples include people watching a street performer, spectators at the scene of a fire, and shoppers at a busy mall.

▶ A *conventional crowd* is a group of people that assembles for a specific purpose and follows established norms. Unlike casual crowds, conventional crowds are structured, their members may interact, and they conform to rules that are appropriate for the situation. Examples include people attending religious services, funerals, graduation ceremonies, and parades.

public opinion widespread attitudes on a particular issue.

propaganda spreading information (or misinformation) to influence people's attitudes or behavior.

crowd a temporary gathering of people who share a common interest or participate in a particular event.

"Public mourning" includes strangers coming together to show their respect to the dead. Less than a day after Apple's co-founder Steve Jobs' death in 2011, fans across the globe created memorials in Japan and other countries. Is this kind of collective behavior an example of a conventional, expressive, or acting crowd? Why?

- ▶ An *expressive crowd* is a group of people who show strong emotions toward some object or event. The feelings—which can range from joy to grief—pour out freely as the crowd reacts to a stimulus. Examples include attendees at religious revivals, revelers during Mardi Gras, and enthusiastic fans at a football game.

- ▶ An *acting crowd* is a group of people who have intense emotions and a single-minded purpose. The event may be planned, but acting crowds can also be spontaneous. Examples include people fleeing a burning building, soccer fans storming a field, and college students having a water balloon fight.

- ▶ A *protest crowd* is a group of people who assemble to achieve a specific goal. Protest crowds demonstrate their support of or opposition to an idea or event. Most demonstrations—like antiwar protests, civil rights marches, boycotts, and labor strikes—are usually peaceful. Peaceful protesters can become aggressive, however, resulting in destruction and violence (Blumer, 1946; McPhail and Wohlstein, 1983; Pell, 2011).

One type of crowd can easily change into another. A conventional crowd at a nightclub can turn into an acting crowd if a fire causes people to panic and flee for safety. Any of the five types of crowds—from casual to protest—can become a mob or a riot.

mob a highly emotional and disorderly crowd that uses force, the threat of force, or violence against a specific target.

riot a violent crowd that directs its hostility at a wide and shifting range of targets.

MOBS

A **mob** is a highly emotional and disorderly crowd that uses force, the threat of force, or violence against a specific target. The target can be a person, group, or property. Mobs often arise when people are demanding radical societal changes, like removing a corrupt government official. In other cases, especially when authority breaks down, people who take advantage of a situation may engage in mob behavior. After a 2010 earthquake in Port-au-Prince, Haiti, most of the city's 3 million survivors focused on clearing the streets of debris and pulling bodies out of the rubble. However, dozens of armed men—some wielding machetes, others with sharpened pieces of wood—"dodged from storefront to storefront, battering down doors and hauling away whatever they could carry…" (Romero and Lacey, 2010: A1). After attacking, a mob tends to dissolve quickly.

RIOTS

Compared with mobs, riots usually last longer. A **riot** is a violent crowd that directs its hostility at a wide and shifting range of targets. Unlike mobs, which usually have a specific target, rioters unpredictably attack whomever or whatever gets in their way. Most riots arise out of long-standing anger, frustration, or dissatisfaction that may have smoldered for years or even decades. Some of these long-term tensions are due to discrimination, poverty, unemployment, economic deprivation, or other unaddressed grievances.

There are numerous protests in the United States every year, but race riots have been the most violent and destructive. More than 150 U.S. cities experienced race riots after the assassination of Martin Luther King Jr. in 1968. In 1992, riots broke out in 11 cities after four white police officers were acquitted of beating Rodney King, a black motorist, in Los Angeles. The violence resulted in deaths and considerable property damage because of fires and looting. In 2014 and 2015, riots erupted in Ferguson, Missouri, and Baltimore, Maryland, after the controversial deaths of black men at the hands of the police. As in the past, looting and/or burning local businesses accompanied both of the riots.

Riots are usually expressions of deep-seated hostility, but this isn't always the case. In the 2010 Championship Series, riots broke out after the Los Angeles Lakers beat the Boston Celtics. Such "celebration riots" are due to extreme enthusiasm and excitement rather than anger or frustration, and can lead to "an orgy of gleeful destruction" (Locher, 2002: 95).

Much collective behavior, like a mob or riot, is spontaneous and short-lived. Collective behavior that's structured and enduring brings long-term social changes. Social movements are important vehicles for creating or suppressing societal changes.

16-2 SOCIAL MOVEMENTS

There are hundreds of social movements in the United States alone. Why are they so prevalent? And why do they matter?

16-2a What Is a Social Movement?

A **social movement** is a large and organized group of people who want to promote or resist a particular social change. "Social movements are as American as apple pie. The abolition of slavery, women's right to vote, unions, open admissions to public colleges and student aid, and Head Start are all changes in our society that were won through social movements" (Ewen, 1998: 81–82).

Unlike other forms of collective behavior, social movements are goal-oriented, deliberate, structured, and can have a lasting impact on a society. And in contrast to many other forms of collective behavior (like crowds, mobs, and riots), the people who make up a social movement are dispersed over time and space, and usually have little face-to-face interaction (Turner and Killian, 1987; Lofland, 1996).

Some U.S. social movements, like white supremacists, are relatively small. Others are large and have subgroups that appeal to different segments of the population. The U.S. environmental movement has at least 50 smaller groups, including Earth First!, Greenpeace, the National Audubon Society, the Union of Concerned Scientists, and the Wilderness Society.

16-2b Types of Social Movements

Sociologists generally classify social movements according to their goals (changing some aspect of society or resisting change) and the amount of change they seek (limited or widespread). Some social movements are perceived as more threatening than others because they challenge the existing social order (see *Table 16.1*).

Alternative social movements focus on changing some people's attitudes or behavior in a specific way. They typically emphasize spirituality, self-improvement, or physical well-being. These movements are the least threatening to the status quo because they seek limited change and only for some people. For instance, millions of non-Asian Americans, influenced by Asian religions, have embraced yoga, meditation, and healing practices like acupuncture (Cadge and Bender, 2004).

Redemptive social movements (also called *religious* or *expressive movements*) propose a dramatic change, but only for some people. They're usually based on spiritual or supernatural beliefs, promising to renew

TABLE 16.1 FIVE TYPES OF SOCIAL MOVEMENTS

Movement	Goal	Examples
Alternative	Change some people in a specific way	Alcoholics Anonymous, transcendental meditation
Redemptive	Change some people, but completely	Jehovah's Witnesses, born-again Christians
Reformative	Change everyone, but in specific ways	Gay rights advocates, Mothers Against Drunk Driving (MADD)
Resistance	Preserve status quo by blocking or undoing change	Antiabortion groups, white supremacists
Revolutionary	Change everyone completely	Right-wing militia groups, Communism, Islamic State in Iraq and Syria (ISIS)

people and guarantee some form of salvation or rebirth. Examples include any religious movements that actively seek converts (e.g., Jehovah's Witnesses, some Christian evangelical groups) (see Chapter 13).

Reformative social movements want to change everyone, but only in a specific way. These movements, the most common type in U.S. society, don't want to replace the existing economic, political, or social class arrangements, but to change society in some particular way. Examples include groups that champion the rights of the disabled, gays, crime victims, fat people, and animals.

Resistance social movements (also called *reactionary movements*) try to preserve the status quo by blocking change or undoing change that has already occurred. Resistance movements are often called *countermovements* because they usually form immediately after an earlier movement has succeeded in creating change. For example, antiabortion groups that arose in the United States shortly after the Supreme Court decision in *Roe v. Wade* (1973), which legalized abortion, seek to reverse that decision.

Revolutionary social movements want to completely destroy the existing social order and replace it with a new one. These movements range from utopian groups that withdraw from society and try to create their own to terrorists who use violence and intimidation. Examples of the latter include ISIS in the Middle East and Boko Haram in Africa.

16-2c Why Social Movements Emerge

A social movement is "an answer either to a threat or a hope" (Touraine, 2002: 89), but not everyone who feels threatened or hopeful joins a social movement. Why not? Let's look at four explanations, beginning with the oldest.

social movement a large and organized group of people who want to promote or resist a particular social change.

MASS SOCIETY THEORY

Early on, sociologists believed that the people who formed social movements felt powerless, insignificant, and isolated in modern mass societies, which are impersonal, industrialized, and highly bureaucratized. Thus, according to *mass society theory*, social movements offer a sense of belonging to people who feel alienated and disconnected from others (Kornhauser, 1959).

CRITICAL EVALUATION

Mass society theory may explain why some people form extreme political movements like Fascism and Nazism, but subsequent research has shown that movement organizers are typically not isolated but well-integrated into their families and communities. Also, historically, many political activists in the United States, including those behind the civil rights and women's rights movements during the late 1960s, weren't powerless but came from relatively privileged backgrounds (McAdam and Paulsen, 1994).

RELATIVE DEPRIVATION THEORY

Relative deprivation theory is broader than mass society theory. **Relative deprivation** is a gap between what people have and what they think they should have compared with others in a society. What people think they should have includes money, social status, power, or privilege.

Relative deprivation theorists note two other elements. First, people often feel that they *deserve* better than they have ("I've worked hard all my life"). Second, they believe that they *can't attain their goals through conventional channels* ("I've contacted my senators, but they ignore me"). Thus, shared beliefs combined with unfulfilled expectations can trigger change-oriented social movements (Davies, 1979; Morrison, 1971).

CRITICAL EVALUATION

Relative deprivation theory helps explain why some social movements emerge. Critics point out, however, that there's a certain degree of relative deprivation in all societies, but people don't always react by forming social movements. Relative deprivation theory also doesn't explain why some people join movements even though they don't see themselves as deprived, and don't expect to gain anything personally if the movement succeeds (Gurney and Tierney, 1982; Johnson and Klandermans, 1995).

RESOURCE MOBILIZATION THEORY

It takes more than feeling alienated (mass society theory) or disadvantaged (relative deprivation theory) to sustain

> **relative deprivation** a gap between what people have and what they think they should have compared with others in a society.

a social movement. Instead, according to *resource mobilization theory*, a social movement will succeed if it can put together (or mobilize) an organization and leadership dedicated to advancing its cause (Oberschall, 1995; Gamson, 1990). Other important resources include money, devoted volunteers, paid staff, access to the media, effective communication systems, special technical or legal knowledge and skills, equipment, physical space, alliances with like-minded groups, and lobbyists who influence legislation (see Chapter 11).

Consider U.S. gun-rights and gun-control movements. In 2014, Georgia passed a law that allows people with firearm licenses to carry guns into bars, churches, schools, some airport areas, and government buildings not protected by security guards during business hours. The gun control movement has seen some victories, including strengthening gun-control laws in 15 states and the District of Columbia. Still, the National Rifle Association (NRA), a powerful leader of the gun-rights movement, defeated a federal bill mandating tougher background checks, something 85 percent of Americans favor. The NRA has brushed off gun-control efforts because it has "millions of members…who will give us small amounts of money [that] add up to an annual budget of more than $300 million" ("Gun Laws: A Shot and a Beer," 2014).

CRITICAL EVALUATION

A key contribution of resource mobilization theory is its emphasis on structural factors (like organization and leadership) in explaining why some social movements thrive whereas others shrivel. A major criticism, however, is that resource mobilization theory largely ignores the role of relative deprivation in a social movement's formation. If there aren't large numbers of dissatisfied people to initiate a movement, even plentiful resources won't be able to sustain it (Klandermans, 1984; Buechler, 2000).

Consider gun control. Between 2011 and 2014, a mass shooting occurred every 64 days, but most people's interest in gun control fades shortly after a mass shooting. Perhaps Americans don't start or join a movement, despite the availability of resources, because they assume that there's no way to stop the killings (Beres, 2015; Follman, 2015).

NEW SOCIAL MOVEMENTS THEORY

New social movements theory, which became prominent during the 1970s, emphasizes the linkages among culture, politics, and ideology. Unlike the earlier perspectives, new social movements theory proposes that many recent movements (like those that work for peace and environmental protection) promote the rights and welfare of *all* people rather than specific groups in particular countries (Laraña et al., 1994; Melucci, 1995). Thus, new social movements

IF THERE AREN'T LARGE NUMBERS OF DISSATISFIED PEOPLE TO INITIATE A MOVEMENT, EVEN PLENTIFUL RESOURCES WON'T BE ABLE TO SUSTAIN IT.

theory is especially interested in "the struggle to liberate the voices of the dispossessed" (Schehr, 1997: 6).

For these theorists, recent social movements differ from older ones in several ways. They attract a disproportionate number of people who are well-educated and relatively affluent, who represent a wide array of professions (e.g., educators, scientists, actors, businesspeople, political leaders), and who share a broad goal—improving the quality of life for all people around the world. Moreover, recent social movements pursue goals or advance values that may not personally benefit its members, such as eradicating measles or tuberculosis in developing countries (Obach, 2004).

CRITICAL EVALUATION

Unlike earlier perspectives, new social movements theory helps us understand collective behavior that crosses international boundaries. According to some critics, however, neither this perspective nor the groups that it examines are novel. Some social movements (like feminism and environmentalism) have been around for a long time and still focus on the same basic issues, like women's second-class citizenship and population growth. In addition, some scholars point out that educated middle-class or wealthy activists were as common in the old social movements as in more recent ones (Rose, 1997; Buechler, 2000; Sutton, 2000).

Critics also contend that new social movements theory often overstates people's altruistic motivations. Many people who join environmental groups do so for reasons referred to as NIMBY (not in my backyard). That is, they're concerned about their own community's environment, but show little interest in ecological threats to people elsewhere.

These four theories, despite their limitations, offer insights about social movements because "no single theory is sufficient to explain the complexities of any social movement" (Blanchard, 1994: 8). The next question is why some social movements flourish and others collapse.

16-2d The Stages of Social Movements

Most social movements are short-lived. Some never really get off the ground; others meet their goals and disband. Social movements generally go through four stages: emergence, organization, institutionalization, and decline (see *Figure 16.1*) (Tilly, 1978).

EMERGENCE

During *emergence*, the first stage of a social movement, a number of people are distressed about some condition and want to change it. One or more individuals, serving as agitators or prophets, emerge as leaders. They verbalize the feelings of the discontented, crystallize the issues, and push for action. If leaders don't get much support, the movement may die.

If, on the other hand, the discontent resonates among growing numbers of people, public awareness increases and the movement attracts like-minded people. In India, women's groups have had some success in combatting pervasive sexual violence because growing numbers of both women and men are protesting rape and sexual assaults. Some Indian cities have passed new laws that include sanctions for police who refuse to register rape allegations, but many offenders still aren't charged or prosecuted (Lahiri and Rana, 2013).

ORGANIZATION

Once people's consciousness has been raised, the second stage is *organization*. The most active members form alliances, seek media coverage, develop strategies and

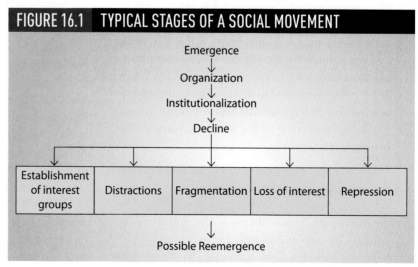

FIGURE 16.1 TYPICAL STAGES OF A SOCIAL MOVEMENT

Emergence → Organization → Institutionalization → Decline → { Establishment of interest groups | Distractions | Fragmentation | Loss of interest | Repression } → Possible Reemergence tactics, recruit members, and acquire the necessary resources. The organization establishes a division of labor in which leaders make policy decisions and followers perform necessary tasks like preparing mass mailings, developing websites, and responding to phone calls and emails. At this stage, the movement may develop chapters at local, regional, national, and international levels.

The fast-food workers of the strikes that began in 2012 have demanded a pay raise to $15.00 per hour and the right to unionize. According to some analysts, the protesters may be spearheading a broader "living wage movement" for several reasons: They have been organized by Fast Food Forward (a powerful labor group in New York City), have been financed by the Service Employees International Union, and have received considerable (and supportive) media coverage (Weissmann, 2014). Nonetheless, and despite predictions, the strikes haven't attracted millions of unhappy retail and other low-income earners, and a living wage movement hasn't become institutionalized.

INSTITUTIONALIZATION

As a movement grows, it becomes *institutionalized* and more bureaucratic: The number of staff positions increases, members draw up bylaws, the organization may hire outsiders (writers, attorneys, and lobbyists) to handle some of the necessary tasks, and the leaders may spend more of their time on speaking tours, in media interviews, and at national or international meetings. As the social movement grows and becomes more bureaucratic and self-sufficient, the original leaders may move on to better paying and more influential positions in government or the private sector.

DECLINE

Almost all social movements end sooner or later. This *decline*, the last stage, may take several forms:

▶ If a social movement is successful, it can become an *interest group* and a part of society's fabric. A small antismoking movement that began in the mid-1970s now has the enthusiastic support of numerous prestigious organizations, including the American Cancer Society, the American Heart Association, the American Medical Association, the World Health Organization, and governing bodies within and outside the United States (Wolfson, 2001).

▶ Those involved in a social movement may become *distracted* because the group loses sight of its original goals, their enthusiasm diminishes, or both. In the mid-1960s, when Ralph Nader initially condemned the automobile industry for car safety defects, he gained a large following. As Nader and

Food movements have burgeoned worldwide, especially in urban centers. Many U.S. cities have institutionalized food carts and trucks by passing or revising regulations that support food vending. Doing so encourages neighborhood entrepreneurs to use available city spaces, attracts "foodies," and offers residents and tourists a variety of ethnic, organic, and gourmet "culinary experiences" (Hanser and Hyde, 2014: 47).

his consumer rights groups expanded their focus to include environmental issues and corporate crimes, many of the initial followers lost interest.

▶ A social movement may experience *fragmentation* because the participants disagree about goals, strategies, or tactics. The environmental movement encompasses numerous groups that focus on different issues—air quality, marine life, land use, and global warming. Participants may also drift away because of time constraints, strategy disagreements, health problems, or similar reasons (see Blee, 2015).

▶ Social movements may decline because of *repression*. Many autocratic governments quash dissent. A government can crush an emerging social movement by arresting protestors and imprisoning or even executing leaders (see Chapter 11).

Social movements can wane for a combination of reasons. In 2011, the Occupy Wall Street (OWS) movement protested socioeconomic inequality, corporate greed, and corporate power over the U.S. government. The group's slogan, "We are the 99%," referred to the income inequality between the top 1 percent and the rest of the population. Despite similar demonstrations in 70 major U.S. cities and over 600 communities, and 900 cities worldwide, the movement faded away. Why? OWS raised awareness of the injustice of this nation's inequality, but ultimately failed because it lacked leadership, clear objectives, solutions, organization, and a long-term commitment (Madrick, 2013; Sandbu, 2013).

One Teenager + Technology = Social Change

In 2012, Sarah Kavanagh— a high school sophomore in Hattiesburg, Mississippi—started a Change.org petition to get PepsiCo to remove brominated vegetable oil (BVO) from its Gatorade products. BVO contains a chemical compound that keeps sports and citrus-flavored drinks more uniform by preventing oily ingredients from separating. It's also used as a flame retardant on upholstered furniture and children's products. Sarah gathered more than 200,000 signatures, and PepsiCo eliminated BVO in early 2013. She then created another successful petition that resulted in Coca-Cola's removing BVO from all its beverages by the end of 2014 (Strom, 2014; Velasco, 2014).

JAMES EDWARD BATES/The New York Times/Redux

A social movement that declines can sometimes experience a resurgence. For example, there have been several waves of the women's rights movement in the United States since the mid-nineteenth century. The first wave ensured women's right to vote in 1920; the second wave expanded employment and educational rights during the 1960s and 1970s; and the third wave, during the 1990s, focused on economic and other inequalities experienced by women of different social classes, sexual orientations, and nationalities. Despite considerable backlash over the past hundred years, American feminism has appealed to diverse groups, agitated in a multitude of spheres ranging from athletics to religion, and has used a variety of strategies to press for social change (Cobble et al., 2014).

16-2e Why Social Movements Matter

On an *individual level*, many of us enjoy a variety of rights as workers, consumers, voters, and victims—rights that we owe to highly dedicated people who were determined to change inequitable laws and customs.

On an *institutional level*, social movements can change general practices. Shopping for healthy food is much easier today than it was before the 1990s, when veggie burgers, tofu, and nutritional labeling were practically nonexistent. Now, mainstream grocery stores have large organic food sections, a variety of fruits and vegetables, and breads and cereals made with whole grains, nuts, and less salt, sugar, and chemical additives. The "farm-to-table" movement, with its emphasis on locally grown produce and meat, has benefited people's health and regional economies. In effect, then, vegetarian and consumer groups have changed the way many farms operate and loosened corporate control of food products (O'Connor, 2015; Kowitt, 2015).

On a *societal level*, social movements have had a major impact in the United States and globally. Most of the world's great religions began as protest movements (see Chapter 13). Also, democratic forms of government in the United States and other countries grew out of the activities of revolutionary groups that sought greater political and economic freedom (Giugni et al., 1999; della Porta and Diani, 1999).

You've seen that social movements can generate or resist change. However, "what all of these social movements have in common is the desire by ordinary citizens to have a say in the operation of their society" (Locher, 2002: 246). Technology is another important source of societal change.

16-3 TECHNOLOGY AND SOCIAL CHANGE

This brief "history" of how to cure an earache has circulated on the Internet:

- 2000 B.C.—Here, eat this root.
- 1000 A.D.—That root is heathen, say this prayer.
- 1850 A.D.—That prayer is superstition, drink this potion.
- 1940 A.D.—That potion is snake oil, swallow this pill.
- 1985 A.D.—That pill is ineffective, take this antibiotic.
- 2000 A.D.—That antibiotic is artificial. Here, eat this root.

As this anecdote suggests, despite technological progress over the centuries, some of the old remedies are enjoying renewed popularity (peoplespharmacy.com, for instance, offers home treatments for everything from arthritis to whooping cough). Still, technological advances have brought enormous social changes, including benefits, costs, and ethical concerns.

16-3a Some Recent Technological Advances

Technology, the application of scientific knowledge for practical purposes, is vital to human life: "For good or ill, [technologies] are woven inextricably into the fabric of our lives, from birth to death, at home, in school, in paid work" (MacKenzie and Wajcman, 1999: 3).

In the next 50 years, technology is likely to change our lives more dramatically than ever before. Several companies are working on "smart" pills to improve mental ability and restore brains that have been impaired by disease or injury. Our houses may be built with sensors that automatically test for anthrax, environmental contaminants, allergens, and radioactivity, and with devices that can defend against chemical and biological agents (Rubin, 2004; Murphy, 2005). Whether such predictions will become reality is anyone's guess, but seven companies—including General Motors, Google, Tesla, and Nissan—plan to sell driverless cars by 2020. In the meantime, technological advances are changing our lives. Let's look briefly at a few of the most influential.

COMPUTER TECHNOLOGY

In 1887, English mathematician Charles Babbage designed the first programmable computer. Since then, computers have gone through seven generations of evolution. *Artificial intelligence (AI)*, a branch of computer science, is developing theories and computer systems to perform tasks that normally require human intelligence (e.g., speech recognition, decision making, and understanding). AI advances have spurred the development of *robots*, machines that are programmed to perform humanlike functions. Industrial robots, known as "collaborative robots" or "cobots," already work alongside people in doing tasks like stacking heavy packages. By 2025, according to some experts, AI and robots will affect nearly every aspect of daily life (Smith et al., 2014; Swan, 2014).

Robots have performed repetitive, dirty, or dangerous human tasks, including toxic waste cleanup, mining, minefield sweeping, underwater and space exploration, and locating bombs. Based on the most up-to-date information, Watson, IBM's supercomputer, can answer thousands of questions that doctors and nurses might ask about the symptoms and treatment of diabetes and other diseases. It can also answer lawyers' questions on

technology the application of scientific knowledge for practical purposes.

Some European nursing homes are using Zora, a robot, to teach their residents exercises and to chat with children in their classrooms. One woman, who hadn't spoken in four months, surprised everyone when she blurted out "I'm well" after Zora asked how she was doing, and then carried on a brief conversation (McFarland, 2015).

bankruptcy and other topics, complete with citations from legislation or case law. In time, robots may be able to discuss stock market investment strategies, give you advice about a personal problem, read a book to you in any desired language or in a voice of either sex, cook and serve your meals, and be emotionally savvy companions who cheer you up when you're unhappy ("Artificial Intelligence," 2015; Kharif, 2015).

BIOTECHNOLOGY

Biotechnology is a broad term that applies to all practical uses of living organisms. It covers anything from the use of microorganisms (e.g., yeast) to ferment beer to *genetic engineering*, sophisticated techniques that can change the makeup of cells and move genes across species to produce new organisms.

In agriculture, biotechnology produces genetically modified (GM) crops like corn, soybeans, and cotton. Between 1996 and 2012, the total earth's land cultivated with GM crops rose from 4.2 to 395 million acres. U.S. farmers continue to be the chief users of GM crops, producing 43 percent of the world's total (Atici, 2014). Biotech crops increase production and make agriculture less costly because new varieties of plants can tolerate herbicides and resist uncontrollable diseases that destroy millions of tons of food each year. Critics contend, however, that GM crops may not be safe, may harm soil quality, and may disrupt ecosystems (Nestle, 2012; Caplan, 2013).

In medicine, biotechnology includes genetic testing/screening that can determine a child's biological parents, a person's ancestry, inherited disorders, and a person's chance of developing or passing on a genetic disease. You'll see shortly, however, that genetic screening also raises worrisome ethical questions.

NANOTECHNOLOGY

Another promising innovation is *nanotechnology*, the ability to build objects one atom or molecule at a time. The key characteristic of these objects is tiny size. Nanotechnology is based on structures measured in nanometers, a unit of measurement equal to 1 billionth of a meter, or 1/80,000th the width of a human hair.

In medicine, researchers foresee the day when thousands of nanotubes will be packed into a hairlike capsule the size of a splinter and painlessly implanted under the skin. The nanotubes would continuously monitor blood sugar, cholesterol, and hormone levels, and destroy tumors without also frying adjacent healthy cells, one of the negative effects of current chemotherapy and radiation treatments for cancer (Weiss, 2004, 2005).

Scientists are also working on nanoparticles that hunt down and kill tumor cells, and nanosensors—microscopic robots with legs, propellers, and cameras—that would live in the bloodstream and send messages to smartphones about signs of infection, an impending heart attack, or other issues that are early warnings for disease or death. Swallowing little pills that contain nanosensors may become a routine part of doctors' office visits in the future (Alkhatib et al., 2014; Cha, 2014; "Hunting as a Pack," 2014).

These and other technological innovations promise longer and healthier lives in the future, but what about the present? What are some of the current benefits and costs of technological changes?

16-3b Some Benefits and Costs of Recent Technologies

The ever-accelerating pace of change means that millennials (people born after 2000) are using technologies that didn't exist just a decade ago. Indeed, much of this "iGeneration" views even those in their 20s as outdated in their tech skills (Rosen, 2010). Technology may be creating greater generation gaps than in the past, but 59 percent of Americans believe that technological advancements will improve people's lives (Smith, 2014). Large numbers also say that it'd be almost impossible to give up some of their hi-tech gadgets (see *Figure 16.2*).

Is the average American too upbeat about new technology? A computer scientist has noted that every technology has a "dark side" because practically every benefit also has a cost (Lohr, 2011: A1). Here are a few contemporary examples.

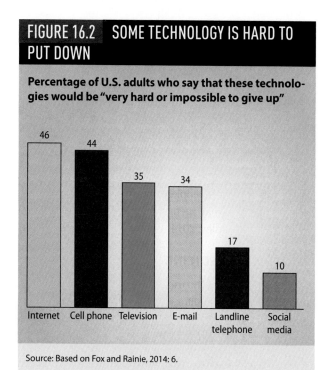

FIGURE 16.2 SOME TECHNOLOGY IS HARD TO PUT DOWN

Percentage of U.S. adults who say that these technologies would be "very hard or impossible to give up"

Internet: 46
Cell phone: 44
Television: 35
E-mail: 34
Landline telephone: 17
Social media: 10

Source: Based on Fox and Rainie, 2014: 6.

COLLABORATIVE CONSUMPTION

▶ *Benefits*: Technology has spurred "collaborative consumption" (also called a "sharing economy" and "access economy") that enables people to earn and save money by sharing goods and services. Airbnb matches budget travelers and people with a spare room or other lodging for rent, RelayRides allows people to rent your car, and taxi-like services (e.g., Lyft, Uber, Sidecar) bring drivers and passengers together.

▶ *Costs*: The sharing economy has created conflict: Airbnb hosts have sometimes ignored zoning laws, don't always pay city or state taxes, and neighbors have complained about noise and a steady stream of strangers. Taxicab companies have protested that ridesharing has decreased their business because the drivers have avoided all regulations, inspections, fees, and insurance requirements (Goodale, 2015; Stein, 2015).

3-D PRINTING

▶ *Benefits*: 3-D printing—a process for making a physical object from a three-dimensional digital model— is revolutionizing manufacturing. Because the process can reduce manufacturing costs by 25 percent or more, the CEO of an aluminum corporation describes 3-D printing as "the beginning of...a second Industrial Revolution" (Geier, 2014: 76). Among other things, 3-D printing produces medical implants (such as knee replacements), synthetic heart valves, prosthetics that are easy to use, models of buildings, and even cars that are safe and fuel efficient. Desktop 3-D printers produce a wide range of objects like toys, clocks, ceramics, and wrenches.

▶ *Costs*: As with any new technology, 3-D printers have negative impacts: They consume 50 to 100 times more electrical energy than traditional molds, increase usage of plastics that aren't recycled, and emit toxic particles when used in the home that can pose health risks. Moreover, people can print weapons, including plastic guns, that metal detectors or x-ray scanners can't spot (Bodley, 2014; Fung, 2014; Gilpin, 2014).

HEALTH AND MEDICINE

▶ *Benefits*: People who use health-related mobile gadgets and apps can monitor their blood pressure, calories, and physical fitness. The number of robotic procedures increases by 30 percent each year. Robotic surgery reduces physician fatigue because she/he doesn't have to stand over the patient for hours, and a robot's "hands" can reach into tight spots and move in ways that human hands can't. In addition, minimally invasive robotic surgeries usually result in less blood loss and faster recoveries because there's a smaller incision to heal.

▶ *Costs*: Health-related apps encourage "iPochondria" among people who are anxious about their health which, in turn, increases doctors' workloads and insurance costs. Robotic surgery is expensive, the outcomes are generally similar to traditional operations, there are no national training standards for robotic surgery, one company dominates the market, and adverse events inceased 34 percent between 2011 and 2012 alone (Langreth, 2013; Howard, 2013/2014; "M-Health...," 2014).

MONITORING SYSTEMS

▶ *Benefits*: New technology allows older people to live at home rather than in an institutional setting. Sensors can track people's medications and falls. Adult children can monitor an aging parent or relative remotely via cellular connections, seeing if they go outside or get home safely after short drives.

▶ *Costs*: Medicaid pays for some aging-at-home technology, but Medicare doesn't. Even if families can afford the technology, some older people feel intimidated by even one-button devices and don't use them (Abrahms, 2014; Tsukayama, 2014).

Some policy analysts fear that technology will produce more jobs for robots than for humans, even in occupations like teaching, sales, nursing, and stock trading.

Driverless cars could displace millions of truck drivers, bus drivers, and others who drive for a living. Others contend that computers will never replace jobs that require "people skills" like teamwork, creativity, managing diverse employees, leadership, and decision-making (Colvin, 2014; "Rise of the Robots," 2014).

There's considerably more consensus that almost every technological advance reduces privacy and increases data breaches. China is allegedly building massive databases of millions of Americans' personal and employment information, sometimes including fingerprints, by hacking government agencies and health care companies. Government auditors and other investigators have warned the White House for several decades about cybersecurity weaknesses, but the warnings have usually been ignored (Davis, 2015; Nakashima, 2015).

Many "data mining" companies routinely collect information as people click from site to site. Much of this web tracking is done anonymously, but a new crop of "snooper" sites is making it easier than ever before for anyone with Internet access to assemble and sell personal information, including your name, Social Security number, address, what you buy, whom you love, and which sites you've visited.

The data come from a variety of sources, including public records on campaign contributions, property sales, and court cases; networking sites where people provide information about themselves, their jobs, relatives, and friends; cell phone companies; and hospitals that share their medical records with state health agencies, which in turn sell the information to private data-mining companies (Leber, 2013; Robertson, 2013; Suich, 2014).

Google compiles enough data to build comprehensive portfolios of most users—who they are, where they go, and what they do. To create a new stream of profit, Facebook sells marketers detailed information about its users. Marketers, in turn, can flood millions of websites and mobile apps with targeted ads. At many department stores, cameras have

**"Dear Andy: How have you been?
Your mother and I are fine. We miss you.
Please sign off your computer and come
downstairs for something to eat. Love, Dad."**

become so sophisticated that companies can analyze what shoppers are looking at, and even their mood (Efrati, 2013; Goel, 2014; "Marketing in the Digital Age," 2015).

Only 1 to 9 percent of Americans are "very confident" that the government agencies or businesses that collect data about them will keep the information private and secure. However, millions provide information—including gender, age, and income—for as little as a "$1 off" online coupon or a $5.00 to $10.00 prepaid gift card, and agree to be tracked over GPS, Wi-Fi, and cellular networks. The companies then sell the data to storeowners, online retailers, and app developers (Clifford and Hardy, 2013; Madden and Rainie, 2015).

16-3c Some Ethical Issues

DNA testing has given millions of people information about their genetic predispositions for diseases like cystic fibrosis, cancer, and Huntington's disease (an incurable neurological disorder). Having such information has helped doctors and patients make better health care decisions, but the expanding use of genetic testing has an unforeseen consequence: More people who don't show any symptoms are being told they have genes for potentially fatal diseases that may develop "next month, next year, when you are 60 years old, or never" (Marcus, 2013: D1).

Those affected become "patients-in-waiting" who undergo continuous screening and worry about their condition. Asymptomatic patients with an inherited risk for a genetic heart muscle disease may have a defibrillator implanted, which involves regular maintenance, possible equipment failures, infections, and other complications. Some doctors believe that such preventive measures for patients-in-waiting are unethical because they're unnecessary, risky, and people live in limbo not knowing whether or when a disease will develop (Marcus, 2013).

Stem cell research is also controversial. A *stem cell* is a building block of the human body that can replicate indefinitely and thus serve as a continuous source of new cells. These self-regenerating cells are found in embryos and umbilical cords, but also in parts of adult bodies, including the brain, blood, heart, skin, bone marrow, intestines, and other organs. Fetal stem cells are more valuable in research because they can produce any cell type of the body, whereas adult stem cells are generally limited to the cell types of their tissue of origin like a liver or kidney (National Institutes of Health, 2006; McDaniels and Cohn, 2015).

Opponents of stem cell research argue that all embryos should be protected. Proponents maintain that hundreds of thousands of embryos that fertility clinics dispose of every year are a wasted resource: They could be used to treat heart disease, leukemia and other cancers, diabetes, Parkinson's disease, and numerous other health problems. In 2013, scientists for the first time created stem cells by cloning embryos. The experiments' success rekindled debate among bioethicists who have renounced "reproductive cloning" as unethical (Healy, 2013).

There's also concern about parents' and scientists' right to create "designer babies" by choosing genes for a child's physical attributes, as eye and hair color, height, complexion, and even athleticism. Some people believe it's unethical to bioengineer children because better-off parents could use the technology to give their kids a competitive edge. On the other hand, 26 percent of U.S. adults believe that prospective parents should be able to alter their children's DNA to produce smarter, healthier, or more athletic offspring (Naik, 2013; Smith, 2014).

Another ethical dilemma is that biotechnological advances, like other technological developments, are most readily available to higher-income people. Screening expectant mothers for fetal abnormalities like Down syndrome is becoming more standard, but not all insurers cover the $1,200 to $2,700 costs of more sophisticated screening tests for high-risk women (Lewis, 2014). Insurance also rarely covers pediatric prosthetics, such as up to $40,000 for a hand that'll have to be replaced about every two years as a child grows (Cohn, 2014). Thus, technological advances are fraught with both promise and pitfalls.

STUDY TOOLS 16

READY TO STUDY? IN THE BOOK, YOU CAN:

☐ Check your understanding of what you've read with the Test Your Learning Questions provided on the chapter review card at the back of the book.

☐ Rip out the chapter review card for a handy summary of the chapter and key terms.

ONLINE AT CENGAGEBRAIN.COM YOU CAN:

☐ Collect StudyBits while you read and study the chapter.

☐ Quiz yourself on key concepts.

☐ Prepare for tests with chapter flash cards as well as flash cards that you create.

☐ Watch videos to explore sociology further.

© Anelina/Shutterstock.com

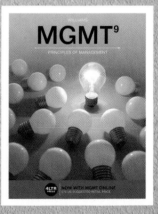

REFERENCES

The references that are new to this edition are printed in red.

"2012 Survey of Business Owners." 2015. American Factfinder, Table SB1200CSA01. Accessed October 3, 2015 (factfinder.census.gov).

"2013 Indian Gaming Industry Report." 2015. 500 Nations. Accessed October 4, 2015 (500nations.com).

"75% in U.S. See Widespread Government Corruption." 2015. Gallup, September 19. Accessed October 7, 2015 (gallup.com).

"75%—A Nation of Flag Wavers." 2011. Pew Research Center. Accessed July 15, 2011 (www.pewresearch.org).

AAA Foundation for Traffic Safety. 2015. "2014 Traffic Safety Culture Index." January. Accessed July 21, 2015 (aaafoundation.org).

AARP Magazine. 2014. "You're Old, I'm Not." February 3, 40–42.

AAUW. 2013. *The Simple Truth About the Gender Pay Gap, Fall 2013 Edition.* Washington, DC: American Association of University Women.

Abrahms, Sally. 2014. "Is This the End of the Nursing Home?" *AARP Bulletin,* March, 20, 22.

Abudabbeh, Nuha. 1996. "Arab Families." Pp. 333–346 in *Ethnicity and Family Therapy,* 2nd edition, edited by Monica McGoldrick, Joe Giordano, and John K. Pearce. New York: Guilford Press.

Academy of Medical Royal Colleges. 2011. "Induced Abortion and Mental Health: A Systematic Review of the Mental Health Outcomes of Induced Abortion, Including Their Prevalence and Associated Factors." National Collaborating Centre for Mental Health, December. Accessed February 9, 2013 (www.aomrc.org.uk).

Acierno, Ron, Melba Hernandez, Amanda B. Amstadter, Heidi S. Resnick, Kenneth Steve, Wendy Muzzy, and Dean G. Kilpatrick. 2010. "Prevalence and Correlates of Emotional, Physical, Sexual, and Financial Abuse and Potential Neglect in the United States: The National Elder Mistreatment Study." *American Journal of Public Health* 100 (February): 292–297.

Acosta, Yesenia D., Luke J. Larsen, and Elizabeth M. Grieco. 2014. "Noncitizens Under Age 35: 2010–2012." U.S. Census Bureau, February. Accessed March 22, 2014 (www.census.gov).

ACT, Inc. 2014. "Improving College and Career Readiness and Success for Everyone." Accessed June 21, 2014 (www.act.org).

Adams, Bert N., and R. A. Sydie. 2001. *Sociological Theory.* Thousand Oaks, CA: Pine Forge Press.

Adamson, David M. 2010. "The Influence of Personal, Family, and School Factors on Early Adolescent Substance Use." Rand Health. Accessed August 20, 2011 (www.rand.org).

Adkins, Amy. 2015. "Majority of U.S. Employees Not Engaged Despite Gains in 2014." Gallup, January 28. Accessed October 7, 2015 (gallup.com).

Adkins, Amy. 2015. "Only 35% of U.S. Managers Are Engaged in Their Jobs." Gallup, April 2. Accessed August 22, 2015 (gallup.com).

Administration on Aging. 2013. "Protect Seniors in the Year of Elder Abuse Prevention." Accessed November 30, 2014 (www.aoa.gov).

"Adultery in New England: Love Free or Die." 2014. *The Economist,* April 19, 24.

"After 39 Wives, Mizo Man Wants to Marry Again." 2013. Matters India, April 27. Accessed May 16, 2014 (www.mattersindia.com).

Agaku, Israel T., et al. 2014. "Tobacco Product Use Among Adults—United States, 2012–2013." *MMWR* 63 (June 27): 542–547.

Agence France-Presse. 2013. "Bill Gates' 'Hand in Pocket' Draws Criticism in South Korea." April 23. Accessed November 10, 2013 (www.rawstory.com).

Agnvall, Elizabeth. 2014. "10 Tests to Avoid." *AARP Bulletin,* March, 12, 14, 16.

Aguila, Raul. 2014. "What We Hate About Holiday Travel." *Consumer Reports,* November, 7.

Aguilar, Leslie, and Linda Stokes. 1995. *Multicultural Customer Service: Providing Outstanding Service Across Cultures.* New York: McGraw Hill.

Ailes, Elizabeth C., et al. 2015. "Opioid Prescription Claims among Women of Reproductive Age—United States, 2008–2012." *MMWR* 64 (2): 37–41.

Aizer, Anna, and Joseph J. Doyle, Jr. 2013. "Juvenile Incarceration, Human Capital and Future Crime: Evidence from Randomly-Assigned Judges." National Bureau of Economic Research, June. Accessed October 19, 2014 (www.nber.org).

Akechi, Hironori, Atsushi Senju, Helen Uibo, Yukiko Kikuchi, Toshikazu Hasegawa, and Jari K. Hietanen. 2013. "Attention to Eye Contact in the West and East: Autonomic Responses and Evaluative Ratings." *PLoS One* 8 (3): e59312. Accessed December 5, 2013 (www.plosone.org).

Akers, Ronald L. 1997. *Criminological Theories: Introduction and Evaluation,* 2nd edition. Los Angeles: Roxbury.

"Alcohol and Public Health." 2014. Centers for Disease Control and Prevention, March 14. Accessed July 30, 2014 (www.cdc.gov).

Alexander, Karl L., Doris Entwisle, and Linda Olson. 2014. *The Long Shadow: Family Background, Disadvantaged Urban Youth, and the Transition to Adulthood.* New York: Russell Sage Foundation.

Al-Jassem, Diana. 2011. "Women in Polygamous Marriages Suffering Psychological Torture." ArabNews, March 8. Accessed September 12, 2012 (www.arabnews.com).

Alkhatib, Hasan, et al. 2014. "IEEE CS 2022 Report." IEEE Computer Society, February. Accessed October 8, 2014 (www.computer.org).

Allen, Joshua, et al. 2014. "Early Postnatal Exposure to Ultrafine Particulate Matter Air Pollution: Persistent Ventriculomegaly, Neurochemical Disruption, and Glial Activation Preferentially in Male Mice." *Environmental Health Perspectives,* June 5, advance publication. Accessed July 30, 2014 (www.ehponline.org).

Allport, Gordon W. 1954. *The Nature of Prejudice.* Reading, MA: Addison-Wesley.

Al-Mahmood, Syed Z. 2013. "Bangladesh to Raise Workers' Pay." *Wall Street Journal,* May 13, B4.

Alman, Ashley. 2014. "One Percenter Convicted of Raping Infant Child Dodges Jail Because He 'Will Not Fare Well.'" *Huffington Post,* March 30. Accessed March 31, 2014 (www.huffingtonpost.com).

Alsever, Jennifer. 2014. "Immigrants: America's Job Creators." *Fortune,* June 16, 56.

Alsever, Jennifer. 2014. "Prison Labor's New Frontier." *Fortune,* June 16, 60.

Altonji, Joseph G., Sarah Cattan, and Iain Ware. "Identifying Sibling Influence on Teenage Substance Use." National Bureau of Economic Research, October. Accessed July 24, 2011 (www.nber.org).

Alvaredo, Facundo, Anthony B. Atkinson, Thomas Piketty, and Emmanuel Saez. 2013. "The Top 1 Percent in International and Historical Perspective." *Journal of Economic Perspectives* 27 (Summer): 3–20.

Alvarez, Lizette. 2014. "Economy and Crime Spur New Puerto Rican Exodus." *New York Times,* February 9, A1.

American Association of Suicidology. 2009. "Elderly Suicide Fact Sheet." June 23. Accessed December 17, 2009 (www.suicidology.org).

"American Indian and Alaska Native Heritage Month: November 2014." 2014. U.S. Census Bureau News, November 12. Accessed November 22, 2014 (www.census.gov).

American Psychiatric Association. 2013. *Diagnostic and Statistical Manual of Mental Disorders,* 5th edition. Arlington, VA: American Psychiatric Association.

American Psychological Association. 2014. "About Transgender People, Gender Identity, and Gender Expression." Accessed March 2, 2015 (apa.org).

American Psychological Association. 2014. "Are Teens Adopting Adults' Stress Habits?" February 11. Accessed April 12, 2014 (www.stressinamerica.org).

American Society for Aesthetic Plastic Surgery. 2015. "Cosmetic Surgery National Data Bank Statistics: 2014." Accessed August 18, 2015 (surgery.org).

American Sociological Association. 1999. *Code of Ethics and Policies and Procedures of the ASA Committee on Professional Ethics.* Accessed January 10, 2010 (www.asanet.org).

"American Time Use Survey—2014 Results." 2015. BLS News Release, June 24. Accessed September 29, 2015 (bls.gov).

Amnesty International. 2015. "Death Penalty 2015: The Good and the Bad." July 23. Accessed September 2, 2015 (www.amnesty.org).

Amrein-Beardsley, Audrey. 2014. "Recommended Readings on VAMs." VAMboozled. Accessed June 21, 2014 (www.vamboozled.com).

Amusa, Malena. 2010. "'Precious' Pushes Past Controversy to Oscar Night." Women's eNews, March 5. Accessed March 8, 2010 (www.womensenew.org).

Andersen, Margaret L., and Patricia Hill Collins. 2010. "Why Race, Class, and Gender Still Matter." Pp. 1–16 in *Race, Class, and Gender: An Anthology,"* 7th edition, edited by Margaret L. Andersen and Patricia Hill Collins. Belmont, CA: Wadsworth.

Anderson, James F., and Laronistine Dyson. 2002. *Criminological Theories: Understanding Crime in America.* Lanham, MD: University Press of America.

Anderson, Monica, and Andrew Perrin. 2015. "15% of Americans Don't Use the Internet. Who Are They?" Pew Research Center, July 28. Accessed August 14, 2015 (pewresearch.org).

Anderson, Monica. 2015. "6 Facts About Americans and Their Smartphones." Pew Research Center, April 1. Accessed August 14, 2015 (pewresearch.org).

Anderson, Monica. 2015. "A Rising Share of the U.S. Black Population is Foreign Born." Pew Research Center, April 9. Accessed October 4, 2015 (pewresearch.org).

Andrews, Edmund L. 2015. "Stanford Study Finds Blacks and Hispanics Typically Need Higher Incomes Than Whites to Live in Affluent Neighborhoods." Stanford Report, June 25. Accessed October 4, 2015 (news.stanford .edu).

Andrzejewski, Adam. 2014. "The Federal Transfer Report." Open the Books, March. Accessed April 12, 2014 (www.openthebooks.com).

Anestis, Michael D., and Joye C. Anestis. 2015. "Suicide Rates and State Laws Regulating Access and Exposure to Handguns." *American Journal of Public Health* 105 (October): 2049–58.

"A New Green Wave." 2014. *The Economist,* August 30, 61.

Annie E. Casey Foundation. 2014. "Race for Results: Building a Path to Opportunity for All Children." Accessed June 21, 2014 (www .aecf.org).

Annie E. Casey Foundation. 2015. *2015 Kids Count Data Book.* Baltimore, Maryland.

Arab American Institute Foundation. 2012. "Quick Facts About Arab Americans." Accessed March 22, 2014 (www.aaiusa.org).

Arendt, Hannah. 2004. *The Origins of Totalitarianism.* New York: Schocken.

Arieff, Allison. 2013. "What Tech Hasn't Learned from Urban Planning." *New York Times,* December 13. Accessed September 19, 2014 (www.nytimes.com).

Ariès, Phillippe. 1962. *Centuries of Childhood.* New York: Vintage.

Armstrong, Elizabeth A., Laura Hamilton, and Paula England. 2010. "Is Hooking Up Bad for Young Women?" *Contexts* 9 (Summer): 22–27.

Armstrong, Elizabeth A., Paula England, and Alison C. K. Fogarty. 2012. "Accounting for Women's Orgasm and Sexual Enjoyment in College Hookups and Relationships." *American Sociological Review* 77 (June): 435–462.

Arrazola, René A., et al. 2015. "Tobacco Use Among Middle and High School Students—United States, 2011–2014." *MMWR* 64 (No. 14): 381–385.

"Artificial Intelligence." 2015. *The Economist,* October 3, 81–82.

ASA Research Department. 2013. "Recruitment and Retention of Sociology Majors." *ASA Footnotes* 41 (January): 1, 4.

Asch, Solomon. 1952. *Social Psychology.* Englewood Cliffs, NJ: Prentice-Hall.

Ashburn, Elyse. 2007. "A Race to Rescue Native Tongues." *Chronicle of Higher Education,* September 28, B15.

"The Ashley Madison Hack." 2015. *The Economist,* August 22, 14.

Asi, Maryam, and Daniel Beaulieu. 2013. "Arab Households in the United States: 2006–2010." U.S. Census Bureau, May.

Accessed March 22, 2014 (www.census .gov).

"Asia's New Family Values." 2015. *The Economist,* August 22, 36.

"Asian/Pacific American Heritage Month: May 2015." 2015. *U.S. Census Bureau News,* April 29. Accessed September 22, 2015 (census.gov).

"Asians Fastest-Growing Race or Ethnic Group in 2012, Census Bureau Reports." 2013. Newsroom, U.S. Census Bureau, June 13. Accessed March 22, 2014 (www.census .gov).

"A Smarter War on Poverty." 2014. *Bloomberg Businessweek,* January 13–19, 8.

Association of Certified Fraud Examiners. 2012. "2012 Report to the Nations on Occupational Fraud and Abuse." Accessed January 12, 2014 (www.acfe.com).

AT&T Newsroom. 2015. "Smartphone Use While Driving Grows Beyond Texting to Social Media, Web Surfing, Selfies, Video Chatting." May 19. Accessed July 21, 2015 (att.com).

Atchley, Robert C., and Amanda S. Barusch. 2004. *Social Forces and Aging: An Introduction to Social Gerontology,* 10th edition. Belmont, CA: Wadsworth.

Atici, Cemal. 2014. "Climate Change: Genetically Modified Organisms (GMO) as a Mitigating Measure?" Food and Agriculture Organization of the United Nations, May 6. Accessed October 8, 2014 (www.fao.org).

Attinasi, John J. 1994. "Racism, Language Variety, and Urban U.S. Minorities: Issues in Bilingualism and Bidialectalism." Pp. 319–347 in *Race,* edited by Steven Gregory and Roger Sanjek. New Brunswick, NJ: Rutgers University Press.

Aughinbaugh, Alison, Omar Robles, and Hugette Sun. 2013. "Marriage and Divorce: Patterns by Gender, Race, and Educational Attainment." *Monthly Labor Review* (October): 1–19.

Auguste, Byron, Paul Kihn, and Matt Miller. 2010. "Closing the Talent Gap: Attracting and Retaining Top-Third Graduates to Careers in Teaching." McKinsey & Company, September. Accessed March 20, 2012 (www .mckinsey.com).

Aunola, Kaisa, and Jari-Erik Nurmi. 2005. "The Role of Parenting Styles in Children's Problem Behavior." *Child Development* 76 (November/December): 1144–1159.

Auster, Carol J., and Claire S. Mansbach. 2012. "The Gender Marketing of Toys: An Analysis of Color and Type of Toy on the Disney Store Website." *Sex Roles* 67 (October): 375–388.

Austin, Algernon. 2013. "Native Americans and Jobs." Economic Policy Institute, December 17. Accessed March 22, 2014 (www.epi.org).

Autor, David H. 2014. "Skills, Education, and the Rise of Earnings Inequality Among the 'Other 99 Percent.'" *Science* 344 (May): 843–851.

Auxier, Richard. 2010. "Congress in a Wordle." Pew Research Center, March 22. Accessed March 26, 2010 (www.pewresearch.org).

Avellar, Sarah, and Pamela Smock. 2005. "The Economic Consequences of the Dissolution of Cohabiting Unions." *Journal of Marriage and Family* 67 (May): 315–327.

Axtell, Roger E., Tami Briggs, Margaret Corcoran, and Mary Beth Lamb. 1997. *Do's and Taboos Around the World for Women in Business.* New York: John Wiley & Sons.

Babbie, Earl. 2013. *Social Research Counts.* Belmont, CA: Cengage.

Babcock, Philip, and Mindy Marks. 2011. "The Falling Time Cost of College: Evidence

from Half a Century of Time Use Data." *The Review of Economics and Statistics* 83 (May): 468–478.

Bach, Peter B. 2015. "Seeking a Cure for Drug-Price Insanity." *Fortune,* September 15, 63–64.

Baenninger, MaryAnn. 2011. "For Women on Campuses, Access Doesn't Equal Success." *Chronicle of Higher Education,* October 7, A26.

Bagri, Neha Thirani, and Nida Najar. 2015. "Chicken: It's What's for Dinner in an Indian State, Even Among Caged Carnivores." *New York Times,* March 30, A4.

Bahadur, Nina. 2013. "'Men Taking Up Too Much Space on the Train Raises Some Interesting Questions About Being Male in Public." *Huffington Post,* September 13. Accessed December 4, 2013 (www.huffingtonpost .com).

Bahadur, Nina. 2014. "'Misconception' Reveals the Dark, Misleading World of Crisis Pregnancy Centers." *Huffington Post,* September 18. Accessed September 29, 2015 (huffingtonpost .com).

Bai, Ge, and Gerard F. Anderson. 2015. "Extreme Markup: The Fifty US Hospitals with the Highest Charge-to-Cost Ratios." *Health Affairs* 34 (6): 922–928.

Bainbridge, William S. 1997. *The Sociology of Religious Movements.* New York: Routledge.

Baker, Aryn. 2014. "Unholy Choices." *Time,* March 21, 36–41.

Baker, Aryn. 2015. "Out in Africa." *Time,* August 17, 34–39.

Bales, Robert F. 1950. *Interaction Process Analysis.* Reading, MA: Addison-Wesley.

Ballhaus, Rebecca. 2013. "Muslim Worker Wins Bias Case." *Wall Street Journal,* September 10, A3.

Baltimore Sun. 2012. "Elderly Man Dies in Fire at Single-Family Home." December 9, 5.

Banchero, Stephanie. 2013. "Students Slip in Global Tests." *Wall Street Journal,* December 3, A8.

Bandura, Albert, and Richard H. Walters. 1963. *Social Learning and Personality Development.* New York: Holt, Rinehart & Winston.

Bandy, Tawana. 2012. "What Works for Male Children and Adolescents: Lessons from Experimental Evaluations of Programs and Interventions." Child Trends Research Brief, August. Accessed September 22, 2012 (www .childtrends.org).

Banerjee, Neela. 2006. "Clergywomen Find Hard Path to Bigger Pulpit." *New York Times,* August 26, A1, A12.

Banfield, Edward C. 1974. *The Unheavenly City Revisited.* Boston: Little, Brown.

Banks, Duren, and Tracey Kyckelhahn. 2011. "Characteristics of Suspected Human Trafficking Incidents, 2008–2010." Office of Justice Programs, April. Accessed October 25, 2011 (www.ojp.gov).

Banks, Ingrid. 2000. *Hair Matters: Beauty, Power, and Black Women's Consciousness.* New York: New York University Press.

Barash, David P. 2012. "The Evolutionary Mystery of Homosexuality." *Chronicle Review,* November 21, B4–B5.

Barker, James B. 1993. "Tightening the Iron Cage: Concertive Control in Self-Managing Teams." *Administrative Science Quarterly* 38 (September): 408–437.

Barker, Megan M. 2011. "Manufacturing Employment Hard Hit During the 2007–09 Recession." *Monthly Labor Review* 134 (April): 28–33.

Barnes, Cynthia. 2006. "China's 'Kingdom of Women.'" *Slate*, November 17. Accessed June 12, 2007 (www.slate.com).

Barnes, Robert. 2011. "Limits on Video Games Rejected." *Washington Post*, June 28, A1.

Barnes, Taylor. 2012. "Watch Your Tongue: Prejudiced Comments Illegal in Brazil." *Christian Science Monitor*, December 4. Accessed November 10, 2013 (www.csmonitor.com).

Barreto, Michelle, Micheal K. Ryan, and Manuela T. Schmitt, eds. 2009. *The Glass Ceiling in the 21st Century: Understanding Barriers to Gender Equality*. Washington, DC: American Psychological Association.

Barrett, Paul M. 2014. "Bad Sports." *Bloomberg Businessweek*, March 3–9, 51–55.

Barro, Robert J., and Rachel M. McCleary. 2003. "Religion and Economic Growth Across Countries." *American Sociological Review* 68 (October): 760–781.

Barry, Ellen, and Mansi Choksi. 2013. "Gang Rape in India, Routine and Invisible." *New York Times*, October 27, A1.

Barry, Ellen. 2013. "Policing Village Moral Codes as Women Stream to India's Cities." *New York Times*, October 20, A6.

Barstow, David, and Suhasini Raj. 2015. "Caste Quotas in India Come Under Attack." *New York Times*, August 31, A1, A3.

Bartlett, Thomas, and Paula Wasley. 2008. "Just Say 'A': Grade Inflation Undergoes Reality Check." *Chronicle of Higher Education*, September 5. Accessed June 21, 2014 (www.chronicle.com).

Bartlett, Tom. 2011. "Caffeine Is Definitely Good/Bad for You." *Chronicle of Higher Education*, January 20. Accessed January 25, 2011 (www.chronicle.com/blogs).

Barton, Allen H. 1980. "A Diagnosis of Bureaucratic Maladies." Pp. 27–36 in *Making Bureaucracies Work*, edited by Carol H. Weiss and Allen H. Barton. Beverly Hills, CA: Sage.

Bass, Frank, and Dakin Campbell. 2013. "Poor Neighborhoods See Branches Disappear." *Bloomberg Businessweek*, May 13–19, 48–49.

Bates, Nancy, and Theresa J. DeMaio. 2013. "Measuring Same-Sex Relationships." *Contexts* 12 (Winter): 66–69.

Baum, Sandy, and Kathleen Payea. 2011. "Trends in Student Aid 2011." College Board Advocacy & Policy Center. Accessed April 12, 2012 (www.collegeboard.org).

Baum, Sandy, Jennifer Ma, and Kathleen Payea. 2013. "Education Pays 2013: The Benefits of Higher Education for Individuals and Society." College Board. Accessed June 21, 2014 (www.collegeboard.org).

Baumrind, Diana. 1968. "Authoritarian versus Authoritative Parental Control." *Adolescence* 3 (11): 255–272.

Baumrind, Diana. 1989. "Rearing Competent Children." Pp. 349–378 in *Child Development Today and Tomorrow*, edited by William Damon. San Francisco: Jossey-Bass.

Bax, Pauline. 2013. "In Ghana, Death Has Become Big Business." *Bloomburg Businessweek*, August 29–September 1, 23–26.

Bazelon, Emily. 2009. "2 Kids + 0 Husbands = Family." *New York Times Magazine*, February 1, 30.

Beadle, Amanda Peterson. 2012. "Minnesota Senator Thinks Abortion Pill Is Wrong, While Viagra Is a 'Wonderful Drug.'" Alternet, May 3. Accessed May 24 (www.alternet.org).

Beard, Henry, and Christopher Cerf. 2015. *Spinglish: The Definitive Dictionary of Deliberately Deceptive Language*. New York: Blue Rider Press.

Beck, Melinda. 2013. "Nurse Practitioners Seek Right to Treat Patients on Their Own." *Wall Street Journal*, August 15, A3.

Becker, Howard S. 1963. *Outsiders: Studies in the Sociology of Deviance*. New York: Free Press.

Becker, Howard, and Ruth H. Useem. 1942. "Sociological Analysis of the Dyad." *American Sociological Review* 7 (February): 3–26.

Beckles, Gloria L., and Benedict I. Truman. 2013. "Education and Income—United States, 2009 and 2011." *MMWR* 62 (November 22): 9–19.

Beckwith, Ryan Teague. 2015. "Transcript: Read the Full Text of the Second Republican Debate." *Time*, September 18. Accessed September 29, 2015 (time.com).

Beech, Hannah. 2013. "The Face of Buddhist Terror." *Time*, July1, 42–50.

Beech, Hannah. 2013. "Why China Needs More Children." *Time*, December 2, 36–49.

Begala, Paul. 2011. "Who You Calling Lazy?" *Newsweek*, December 15, 8.

Begley, Sharon. 2010. "Sins of the Grandfathers." *Newsweek*, November 8, 48–50.

Beil, Laura. 2012. "How Much Would You Pay for Three More Months of Life?" *Newsweek*, September 3, 40–44.

Belkin, Douglas, and Caroline Porter. 2012. "Web Profiles Haunt Students." *Wall Street Journal*, October 4, A3.

Belknap, Joanne. 2007. *The Invisible Woman: Gender, Crime, and Justice*, 3rd edition. Belmont, CA: Wadsworth Press.

Bell, Claes. 2015. "Budgets Can Crumble in Times of Trouble." *Bankrate*, January 7. Accessed August 26, 2015 (bankrate.com).

Bendavid, Naftali. 2013. "Countries Expand Recognition for Alternative 'Intersex' Gender." *Wall Street Journal*, October 31, A9.

Bengali, Shashank. 2014. "Toll of Refugees Fleeing Syrian Strife Tops 3 Million, U.N. Says." *Baltimore Sun*, August 30, 10.

Bennett, Jessica, and Jesse Ellison. 2010. "'I Don't': The Case Against Marriage." *Newsweek*, June 21, 42–45.

Bennett, Jessica. 2009. "Tales of a Modern Diva." *Newsweek*, April 6, 42–43.

Bennhold, Katrin. 2014. "Major Gain for Women in Church of England." *New York Times*, November 18, A4.

Benokraitis, Nijole V. 1997. *Subtle Sexism: Current Practices and Prospects for Change*. Thousand Oaks, CA: Sage.

Benokraitis, Nijole V. 2015. *Marriages & Families: Changes, Choices, and Constraints*, 8th edition. Upper Saddle River, NJ: Prentice Hall.

Benson, Michael L. 2002. *Crime and the Life Course*. Los Angeles: Roxbury.

Benton, Thomas H. 2011. "A Perfect Storm in Undergraduate Education." *Chronicle of Higher Education*, February 25, A43, A45. Part 2 in April 8, A45–A46.

Beres, Damon. 2015. "Interest in 'Gun Control' Is Fading, Google Data Shows." *Huffington Post*, October 12. Accessed October 25, 2015 (huffingtonpost.com).

Berfield, Susan. 2012. "Levi's Has a New Color for Blue Jeans: Green." *Bloomberg Businessweek*, October 22–28, 26–28.

Berfield, Susan. 2013. "Fast-Food Workers of the World, Unite!" *Bloomberg Businessweek*, December 9–15, 20–22.

Berger, Peter L., and Thomas Luckmann. 1966. *The Social Construction of Reality: A Treatise in the Sociology of Knowledge*. New York: Doubleday.

Berk, Richard A. 1974. *Collective Behavior*. Dubuque, IA: Wm. C. Brown Company.

Berkos, Kristen M., Terre H. Allen, Patricia Kearney, and Timothy G. Plax. 2001. "When Norms Are Violated: Imagined Interactions as Processing and Coping Mechanisms." *Communication Monographs* 68 (September): 289–300.

Berkowitz, Bill. 2011. "Why Is Jerry Falwell's Evangelical University Getting Filthy Rich Off Your Tax Money?" AlterNet, June 29. Accessed June 29, 2011 (www.alternet.org).

Berlin, Overton B., and Paul Kay. 1969. *Basic Color Terms*. Berkeley: University of California Press.

Berrett, Dan. 2014. "Some Elite Colleges Reject AP Credits in Favor of New Core Classes." *Chronicle of Higher Education*, February 21, A13.

Bersani, Bianca E. 2014. "A Game of Catch-Up? The Offending Experience of Second-Generation Immigrants." *Crime & Delinquency* 60 (February): 60–84.

Bertrand, Marianne, and Jessica Pan. 2011. "The Trouble with Boys: Social Influences and the Gender Gap in Disruptive Behavior." National Bureau of Economic Research, October. Accessed April 14, 2012 (www.nber.org).

Berube, Alan, and Natalie Holmes. 2015. "Some Cities Are Still More Unequal Than Others—An Update." Brookings Institution, March 17. Accessed September 31, 2015 (brookings .edu).

Berube, Alan, Audrey Sinter, Jill H. Wilson, and William H. Frey. 2006. "Finding Exurbia: America's Fast-Growing Communities at the Metropolitan Fringe." Brookings Institution. Accessed February 2, 2008 (www.brookings .edu).

Best, Joel. 2006. *Flavor of the Month: Why Smart People Fall for Fads*. Berkeley: University of California Press.

Betcher, R. William, and William S. Pollack. 1993. *In a Time of Fallen Heroes: The Re-Creation of Masculinity*. New York: Atheneum.

Bettinger, Eric P., Bridget T. Long, Philip Oreopoulos, and Lisa Sanbonmatsu. 2012. "The Role of Application Assistance and Information in College Decisions: Results from the H&R Block FAFSA Experiment." *Quarterly Journal of Economics* 127 (August): 1205–1242.

Bharadwaj, Prashant, Mallesh M. Pai, and Agne Suziedelyte. 2015. "Mental Health Stigma." National Bureau of Economic Research, June. Accessed July 22, 2015 (nber.org).

Bhardwaj, Deepika. 2014. "Delhi Takes Forward 'Kiss of Love' Protest Amid Police Barricades and Assaults." *Times of India*, November 10. Accessed September 27, 2015 (timesofindia .indiatimes.com).

Bialik, Carl. 2011. "Irreconcilable Claim: Facebook Causes 1 in 5 Divorces." *Wall Street Journal*, March 12. Accessed November 8, 2011 (www.online.wsj.com).

"Biggest Transnational Companies." 2012. *The Economist*, July 10. Accessed April 12, 2014 (www.economist.com).

Billingsley, Andrew. 1992. *Climbing Jacob's Ladder: The Enduring Legacy of African-American Families*. New York: Simon & Schuster.

Bird, Chloe E., et al. 2010. "Neighborhood Socioeconomic Status and Biological 'Wear & Tear' in a Nationally Representative Sample of U.S. Adults." *Journal of Epidemiology and Community Health* 64 (October): 860–865.

Bird, Warren, and Scott Thumma. 2011. "A New Decade of Megachurches: 2011 Profile of Large Attendance Churches in the United States." Leadership and Network. Accessed June 21, 2014 (www.leadnet.org).

Bischoff, Kendra, and Sean F. Reardon. 2013. "Residential Segregation by Income, 1970–2009." Russell Sage Foundation, October 16. Accessed September 19, 2014 (www.russellsage.org).

Bitterman, Amy, Rebecca Goldring, and Lucinda Gray. 2013. "Characteristics of Public and Private Elementary and Secondary School Principals in the United States: Results from the 2011–12 Schools and Staffing Survey." National Center for Education Statistics, August. Accessed November 17, 2014 (www.nces.ed.gov).

Bivens, Josh, and Lawrence Mishel. 2015. "Understanding the Historic Divergence Between Productivity and a Typical Worker's Pay." Economic Policy Institute, September 2. Accessed October 7, 2015 (epi.org).

"Black (African-American) History Month: February 2015." 2015. U.S. Census Bureau News, January 21. Accessed October 3, 2015 (census.gov).

Black, M. C., et al. 2011. "National Intimate Partner and Sexual Violence Survey: 2010 Summary Report." Centers for Disease Control and Prevention. Accessed November 2013 (cdc.gov).

Black, Sandra E., Jane A. Lincove, Jenna Cullinane, and Rachel Veron. 2014. "Can You Leave High School Behind?" National Bureau of Economic Research, January. Accessed June 21, 2014 (www.nber.org).

Black, Thomas. 2010. "More Car Jobs Shift to Mexico." Bloomberg Businessweek, June 28–July 4, 10–11.

Blair, J. Pete, and Katherine W. Schweit. 2014. "A Study of Active Shooter Incidents in the United States Between 2000 and 2013." Federal Bureau of Investigation, September 16. Accessed September 2, 2015 (fbi.gov).

Blake, Mariah. 2014. "Are Any Plastics Safe?" Mother Jones, March/April, 19–25, 60–61.

Blake, Mariah. 2014. "The Chevron Communiques: A Trove of Secret Cables Shows How Hillary Clinton's State Department Sold Fracking to the World." Mother Jones, September/October, 51–54, 72.

Blanchard, Dallas A. 1994. The Anti-Abortion Movement and the Rise of the Religious Right: From Polite to Fiery Protest. New York: Twayne Publishers.

Blau, Peter M. 1986. Exchange and Power in Social Life, revised edition. New Brunswick, NJ: Transaction.

Blau, Peter M., and Marshall W. Meyer. 1987. Bureaucracy in Modern Society, 3rd edition. New York: Random House.

Blay, Zeba. 2015. "How Feminist TV Became the New Normal." Huffington Post, June 18. Accessed September 29, 2015 (huffington-post.com).

Blee, Kathleen. 2015. "How Grassroots Groups Lose Political Imagination." Contexts 14 (1): 32–37.

Bloomberg News. 2014. "Madoff Scam Recovery Hits $10 Billion, Almost 60% of Lost Money, at a Cost of $1 Billion." November 22. Accessed December 9, 2014 (www.investmentnews.com).

BLS News Release. 2014. "Usual Weekly Earnings of Wage and Salary Workers: Third Quarter 2014." Bureau of Labor Statistics, October 24. Accessed December 2, 2014 (www.bls.gov/cps).

BLS News Release. 2015. "The Employment Situation—September 2015." Bureau of Labor Statistics, October 2. Accessed October 26, 2015 (bls.gov/cps).

BLS News Release. 2015. "Volunteering in the United States – 2014." Bureau of Labor Statistics, USDL-15-0280, February 25. Accessed August 19, 2015 (bls.gov/cps).

BLS Reports. 2013. "A Profile of the Working Poor, 2011." Bureau of Labor Statistics, April. Accessed February 6, 2014 (www.bls.gov).

BLS Reports. 2014. "Women in the Labor Force: A Databook." Bureau of Labor Statistics, May. Accessed June 21, 2014 (www.bls.gov).

BLS Reports. 2015. "A Profile of the Working Poor, 2013." U.S. Bureau of Labor Statistics, July. Accessed September 11, 2015 (bls.gov).

Blumer, Herbert. 1946. "Collective Behavior." Pp. 65–121 in New Outline of the Principles of Sociology, edited by Alfred M. Lee. New York: Barnes & Noble.

Blumer, Herbert. 1969. Symbolic Interactionism: Perspective and Method. Englewood Cliffs, NJ: Prentice Hall.

Board of Trustees, The. 2013. The 2013 Annual Report of the Board of Trustees of the Federal Old-Age and Survivors Insurance and Federal Disability Insurance Trust Funds. Accessed July 28, 2013 (www.socialsecurity.gov).

Bobroff-Hajal, Anne. 2006. "Why Cousin Marriage Matters in Iraq." Christian Science Monitor, December 26, 9.

Bodley, Michael. 2014. "Custom-Made Knees Using 3-D Printing." Baltimore Sun, July 20, 10.

Bogage, Joseph. 2015. "Why Companies Are Speaking Up About Gay Marriage." Washington Post, June 26. Accessed September 30, 2015 (washingtonpost.com).

Bogle, Kathleen A. 2008. Hooking Up: Sex, Dating, and Relationships on Campus. New York: New York University Press.

Bohannon, Paul, ed. 1971. Divorce and After. New York: Doubleday.

Bollag, Burton. 2007. "Credential Creep." Chronicle of Higher Education, June 22, A10–A12.

Bonacich, Edna. 1972. "A Theory of Ethnic Antagonism: The Split Labor Market." American Sociological Review 37 (October): 547–559.

Bonner, Robert. 2007. "Research—Male Elementary Teachers: Myths and Realities." Men Teach. Accessed June 12, 2008 (www.menteach.org).

Booth, Robert. 2014. "Why the Facebook Experiment Toying with Your Emotions Is So Terrifying." Alternet, June 29 (accessed August 8, 2014).

Borgerhoff Mulder, Monique. 2009. "Serial Monogamy as Polygyny or Polyandry?" Human Nature 20 (Summer): 130–150.

Borowiec, Steven. 2014. "Adoptees Try to Lift Stigma." Christian Science Monitor Weekly, April 7, 16–17.

Bosk, Charles. 1979. Forgive and Remember: Managing Medical Failure. Chicago: University of Chicago Press.

Boslaugh, Sarah. 2007. Secondary Data Sources for Public Health: A Practical Guide. New York: Cambridge University Press.

"Boss Sells His Company, Shares Proceeds." 1999. Baltimore Sun, September 12, 24A.

Botstein, Leon. 2014. "The SAT Is Part Hoax, Part Fraud." Time, March 24, 17.

Bourdieu, Pierre. 1984. Distinction: A Social Critique of the Judgement of Taste. Translated by Richard Nice. Cambridge, MA: Harvard University Press.

Bourdieu, Pierre. 1986. "The Forms of Capital." Pp. 241–258 in Handbook for Theory and Research for the Sociology of Education, edited by J. G. Richardson. Westport, CT: Greenwood.

Bowen, Brent D., and Dean E. Headley. 2015. "Airline Quality Rating 2015." ERAU Scholarly Commons, April 13. Accessed August 22, 2015 (commons.erau.edu/aqrr).

Bowles, Samuel, and Herbert Gintis. 1977. Schooling in Capitalist America: Educational Reform and the Contradictions of Economic Life. New York: Basic Books.

Boyle, Matthew. 2013. "Yes, Real Men Drink Beer and Use Skin Moisturizer." Bloomberg Businessweek, October 7–13, 32–33.

Bradshaw, York W., and Michael Wallace. 1996. Global Inequalities. Thousand Oaks, CA: Pine Forge Press.

Braga, Anthony A. 2003. "Systematic Review of the Effects of Hot Spots Policing on Crime." Unpublished paper. Accessed May 14, 2005 (www.campbellcollaboration.org).

Braitman, Keli A., Neil K. Chaudhary, and Anne T. McCart. 2011. "Effect of Passenger Presence on Older Drivers' Risk of Fatal Crash Involvement." Insurance Institute for Highway Safety, March. Accessed July 4, 2011 (www.iihs.org).

Brescoll, Victoria L., and Eric L. Uhlmann. 2008. "Can an Angry Woman Get Ahead? Status Conferral, Gender, and Expression of Emotion in the Workplace." Psychological Science 19 (March): 268–275.

Brill, Steven. 2013. "Bitter Pill: How Outrageous Pricing and Egregious Profits Are Destroying Our Health Care." Time, March 4, 14–55.

Brill, Steven. 2015. "What I Learned from My $190,000 Surgery." Time, January 19, 34–43.

"Broad Public Support for Legal Status for Undocumented Immigrants." 2015. Pew Research Center, June 4. Accessed October 4, 2015 (pewresearch.org).

Brody, Jane E. 2015. "Limit Children's Screen Time, and Your Own" New York Times, July 14, D7.

Brody, Jane E. 2015. "Screens Separate the Obsessed From Life." New York Times, July 7, D7.

Brody, Jane E. 2015. "The More We Learn, the More We Ignore." New York Times, October 13, D7.

"Broke in the 'Burbs." 2013. The Economist, July 20, 30.

Brooks, Arthur C. 2007. "I Love My Work." The American (online), September–October. Accessed September 30, 2007 (www.theamericanmag.com).

Brown, Alyssa. 2013. "Costs Still Keep 30% of Americans from Getting Treatment." Gallup, December 9. Accessed July 26, 2014 (www.gallup.com).

Brown, Archie. 2010. "Signposts: Why Did Communism End When It Did?" History Today, March. Accessed April 12, 2014 (www.historytoday.com).

Brown, DeNeen L. 2009. "The High Cost of Poverty: Why the Poor Pay More." Washington Post, May 18, C1.

Brown, Michael, and Amy Goldin. 1973. Collective Behavior: A Review and Reinterpretation of the Literature. Pacific Palisades, CA: Goodyear.

Brown, Pamela A., and Elizabeth McGann. 2011. "Medication Error Prevention: A Shared Responsibility." Medscape Medical News, June 14. Accessed April 24, 2012 (www.medscape.com).

Brown, Susan I. 2005. "How Cohabitation Is Reshaping American Families." Contexts 4 (Summer): 33–37.

Browne, Andrea N. 2015. "3 Things That Cost Women More." Baltimore Sun, June 7, 5.

Bruinius, Harry. 2013. "Whose Holidays Are They?" Christian Science Monitor, December 16, 21–23.

Brummelman, Eddie, Sander Thomaes, Stefanie A. Nelemans, Bram Orobio de Castro, Geertjan Overbeek, and Brad J. Bushman. 2015. "Origins of Narcissism in Children." *Proceedings of the National Academy of Sciences* 112 (March 25): 3659–3662.

Brunvand, Jan H. 2001. *The Truth Never Stands in the Way of a Good Story.* Urbana and Chicago: University of Illinois Press.

Bryant, Karl. 2014. "Teaching the Nature-Nurture Debate." *Contexts* 13 (Fall): 22–23.

Buchheit, Paul. 2015. "The Corporate Debt to Society: $10,000 Per Household, Per Year." AlterNet, February 22. Accessed August 26, 2015 (alternet.org).

Buechler, Steven M. 2000. *Social Movements in Advanced Capitalism: The Political Economy and Cultural Construction of Social Activism.* New York: Oxford University Press.

Bullen, Christopher, et al. 2013. "Electronic Cigarettes for Smoking Cessation: A Randomized Controlled Trial." *The Lancet* 382 (November 16): 1629–1637.

Bullock, Karen. 2005. "Grandfathers and the Impact of Raising Grandchildren." *Journal of Sociology and Social Welfare* 32 (March): 43–59.

Buntin, John. 2015. "The Myth of Gentrification." *Slate,* January 14. Accessed October 23, 2015 (slate.com).

Burd, Stephen. 2013. "Undermining Pell: How Colleges Compete for Wealthy Students and Leave the Low-Income Behind." New America Foundation, May. Accessed June 21, 2014 (www.newamerica.org).

Burd-Sharps, Sarah, and Rebecca Rasch. 2015. "Impact of the US Housing Crisis on the Racial Wealth Gap Across Generations." Social Science Research Council, June. Accessed October 4, 2015 (ssrc.org).

Bureau of Labor Statistics. 2014. "Charting the Labor Market: Data from the Current Population Survey (CPS)." April 4. Accessed April 12, 2014 (www.bls.gov).

Bureau of Labor Statistics. 2014–2015. *Occupational Outlook Handbook, 2014–15 Edition.* "Surgeons and Physicians." Accessed July 26, 2014 (www.bls.gov).

Burkitt, Laurie. 2013. "China Eases Limits on Births as Labor Shortage Looms." *Wall Street Journal,* November 16–17, A1, A8.

Bushman, Brad J., Patrick E. Jamieson, Llana Weitz, and Daniel Romer. 2013. "Gun Violence Trends in Movies." *Pediatrics* 32 (5): 1014–18.

Butler, Kiera. 2014. "Google's Magic Bus." *Mother Jones,* May–June, 70–71.

Butterfield, Fox. 2002. "Father Steals Best: Crime in an American Family." *New York Times* (August 15): 1A.

Cabrera, Natasha J., Jay Fagan, and Danielle Farrie. 2008. "Explaining the Long Reach of Fathers' Prenatal Involvement on Later Paternal Engagement." *Journal of Marriage and Family* 70 (December): 1094–1107.

Cadge, Wendy, and Courtney Bender. 2004. "Yoga and Rebirth in America: Asian Religions Are Here to Stay." *Contexts* 3 (Winter): 45–51.

Calderon, Valerie J., and Preety Sidhu. 2014. "Business Leaders Say Knowledge Trumps College Pedigree." Gallup, February 25. Accessed June 21, 2014 (www.gallup.com).

Calderon, Valerie J., and Susan Sorenson. 2014. "Americans Say College Degree Leads to a Better Life." Gallup, April 7. Accessed June 21, 2014 (www.gallup.com).

Camarota, Steven A. 2015. "Welfare Use by Immigrant and Native Households." Center for Immigration Studies, September. Accessed October 4, 2015 (cis.org).

Cameron, Deborah. 2007. *The Myth of Mars and Venus.* New York: Oxford University Press.

Campion-Vincent, Véronique. 2005. "From Evil Others to Evil Elites: A Dominant Pattern in Conspiracy Theories Today." Pp. 103–122 in *Rumor Mills: The Social Impact of Rumor and Legend,* edited by Gary Alan Fine, Veronique Campion-Vincent, and Chip Heath. New Brunswick, NJ: Transaction Publishers.

Campo-Flores, Arian. 2013. "Religious Dorms Sprout Up." *Wall Street Journal,* September 3, A3.

Cano, Regina Garcia. 2015. "S.D. Tribe to Open Nation's 1st Pot Resort." *Baltimore Sun,* September 30, 10.

Cantrell, Steven, and Thomas J. Kane. 2013. "Ensuring Fair and Reliable Measures of Effective Teaching." MET Project, Policy and Practice Brief. Accessed June 21, 2014 (www.metproject.org).

Cantril, Hadley, Hazel Gaudet, and Herta Herzog. 1952. *The Invasion from Mars.* Princeton, NJ: Princeton University Press.

Capaccio, Tony, and Jonathan D. Salant. 2014. "Military: The Pentagon Loves to Pay Top Dollar." *Bloomberg Businessweek, July* 21–27, 26–27.

"Capital Punishment." 2015. *The Economist,* July 4, 50–51.

Caplan, Arthur L. 2013. "Genetically Modified Food: Good, Bad, Ugly." *Chronicle Review,* September 13, B4–B5.

Cappiello, Dina. 2012. "EPA: Power Plant Are Main Global Warming Culprits." *USA Today,* January 11. Accessed January 12, 2012 (www.usatoday.com).

Carballo, José Rodríguez. 2014. "Final Report on the Apostolic Visitation of Women Religious in the United States of America." Holy See Press Office, December 16. Accessed December 28, 2014 (press.vatican.va).

Card, David, and Laura Giuliano. 2011. "Peer Effects and Multiple Equilibria in the Risky Behavior of Friends." National Bureau of Economic Research, May. Accessed July 24, 2011 (www.nber.org).

Carey, Anne R., and Paul Trap. 2014. "How Much Are 'Mom Jobs' Worth?" *USA Today,* May 5–9, 1A.

Carey, Benedict. 2010. "Revising Book on Disorders of the Mind." *New York Times,* February 10, 1.

Carey, Kevin. 2013. "Too Much 'Merit Aid' Requires No Merit." *Chronicle of Higher Education,* February 22, A20–A21.

Carl, Traci. 2002. "Amid Latte, Mocha Craze, Coffee Growers Go Hungry in Paradise." Accessed September 19, 2002 (www.yahoo.com).

Carle, Jill. 2015. "Climate Change Seen as Top Global Threat." Pew Research Center, July 14. Accessed October 23, 2015 (pewresearch.org).

Carnagey, Nicholas L., and Craig A. Anderson. 2005. "The Effects of Reward and Punishment in Violent Video Games on Aggressive Affect, Cognition, and Behavior." *Psychological Science* 16 (November): 882–889.

Carnevale, Anthony P., and Ban Cheah. 2013. "Hard Times: College Majors, Unemployment, and Earnings." Georgetown Center on Education and the Workforce, May. Accessed June 24, 2014 (www.cew.georgetown.edu).

Carney, Ginny. 1997. "Native American Loanwords in American English." *Wicazo SA Review,* 12 (Spring): 189–203

Carnoy, Martin, and Richard Rothstein. 2013. "What Do International Tests Really Show About U.S. Student Performance?" Economic Policy Institute, January 28. Accessed June 21, 2014 (www.epi.org).

Carr, Deborah. 2012. "The Social Stratification of Older Adults' Preparations for End-of-Life Health Care." *Journal of Health and Social Behavior* 53 (3): 297–312.

Carrell, Scott E., Mark Hoekstra, and James E. West. 2010. "Is Poor Fitness Contagious? Evidence from Randomly Assigned Friends." National Bureau of Economic Research, November. Accessed July 24, 2011 (www.nber.org).

Carrell, Scott E., Mark Hoekstra, and James E. West. 2015. "The Impact of Intergroup Contact on Racial Attitudes and Revealed Preferences." National Bureau of Economic Research, February. Accessed May 1, 2015 (nber.org).

Carson, E. Ann. 2014. "Prisoners in 2013." Bureau of Justice Statistics, September. Accessed November 6, 2014 (www.bjs.gov).

Caston, Richard J. 1998. *Life in a Business-Oriented Society: A Sociological Perspective.* Boston: Allyn & Bacon.

Catalano, Shannan. 2013. "Intimate Partner Violence: Attributes of Victimization, 1993–2011." Bureau of Justice Statistics, November. Accessed November 25, 2014 (www.bjs.gov).

Catalyst. 2014. "Women CEOs of the Fortune 1000." January 15. Accessed February 25, 2014 (www.catalyst.org).

Catalyst. 2015. "2014 Catalyst Census: Women Board Directors." January 13. Accessed September 29, 2015 (catalyst.org).

"Catching Up." 2013. *The Economist,* January 12, 67–68.

CDC Fact Sheet. 2014. "Reported STDs in the United States: 2012 National Data for Chlamydia, Gonorrhea, and Syphilis." Centers for Disease Control and Prevention, January. Accessed July 26, 2014 (www.cdc.gov).

CDC/NCHS National Vital Statistics System. 2015. "National Marriage and Divorce Rate Trends." February 19. Accessed October 8, 2015 (cdc.gov).

Cellini, Stephanie R., Signe-Mary McKernan, and Caroline Ratcliffe. 2008. "The Dynamics of Poverty in the United States: A Review of Data, Methods, and Findings." *Journal of Policy Analysis and Management* 27 (Summer): 577–605.

Cendrowski, Scott. 2014. "Business Created China's Pollution Problem." *Fortune,* April 28, 90–93.

Center for American Women and Politics. 2015. "Women in Elective Office 2015." Accessed September 29, 2015 (cawp.rutgers.edu).

Center for Behavioral Health Statistics and Quality. 2015. Behavioral Health Trends in the United States: Results from the 2014 National Survey on Drug Use and Health. Accessed October 16, 2015 (samhsa.gov/data).

Center for Community College Student Engagement. 2014. *Aspirations to Achievement: Men of Color and Community Colleges.* Austin, TX: The University of Texas at Austin, Program in Higher Education Leadership.

Center for Responsive Politics. 2014. "The Money Behind the Elections." Accessed April 12, 2014 (www.opensecrets.org).

Center for Responsive Politics. 2014. "Top PACs." Accessed April 12, 2014 (www.opensecrets.org).

Center for Strategic and International Studies. 2014. "Net Losses: Estimating the Global

Cost of Cybercrime: Economic Impact of Cybercrime II." Intel Security, June. Accessed September 2, 2015 (csis.org).

Center for the Study of Global Christianity. 2007. "Global Table 5: Status of Global Mission, Presence and Activities, AD 1800–2005." Gordon-Conwell Theological Seminary. Accessed August 24, 2007 (www.gcts.edu).

Centers for Medicare & Medicaid Services. 2010. "NHE Tables for Selected Calendar 1960–2010." April 11. Accessed April 24, 2010 (www.cms.gov).

Centers for Medicare & Medicaid Services. 2014. "National Health Expenditure Projections 2012–2022." May 6. Accessed July 26, 2014 (www.cms.gov).

Cha, Ariana Eunjung. 2014. "'Smart Pills' with Chips, Cameras and Robotic Parts Raise Legal, Ethical Questions." *Washington Post*, May 24. Accessed October 8, 2014 (www .washingtonpost.com).

Chambliss, William J. 1969. *Crime and the Legal Process*. New York: McGraw-Hill.

Chambliss, William J., and Robert B. Seidman. 1982. *Law, Order, and Power*, 2nd edition. Reading, MA: Addison-Wesley.

Chandler, Tertius, and Gerald Fox. 1974. *3000 Years of Urban Growth*. New York: Academic Press.

Chandra, Anjani, William D. Mosher, and Casey Copen. 2011. "Sexual Behavior, Sexual Attraction, and Sexual Identity in the United States: Data from the 2006–2008 National Survey of Family Growth." *National Health Statistics Reports*, No. 36, March 3. Accessed March 2, 2014 (www.cdc.gov /nchs).

Chasin, Barbara H. 2004. *Inequality & Violence in the United States: Casualties of Capitalism*, 2nd edition. Amherst, NY: Humanity Books.

Chaves, Mark, and Shawna L. Anderson. 2014. "Changing American Congregations: Findings from the Third Wave of the National Congregations Study." *Journal for the Scientific Study of Religion* 53 (December): 676–689.

Chaves, Mark, Shawna Anderson, and Jason Byassee. 2009. "American Congregations at the Beginning of the 21st Century." National Congregation Study. Accessed June 21, 2014 (www.soc.duke.edu/natcong).

Chemaly, Soraya. 2015. "The Shocking Ways Sexism Affects Women's Health Every Day." AlterNet, June 25. Accessed October 17, 2015 (alternet.org).

Chemi, Eric. 2014. "One Year: 3,885 Spills." *Bloomberg Businessweek*, February 3–9, 55.

Chen, Alice, Emily Oster, and Heidi Williams. 2014. "Why Is Infant Mortality Higher in the US Than in Europe?" National Bureau of Economic Research, September. Accessed December 5, 2014 (www.nber.org).

Chen, Frances S., Julia A. Minson, Maren Schöne1, and Markus Heinrichs. 2013. "In the Eye of the Beholder: Eye Contact Increases Resistance to Persuasion." *Psychological Science* 24 (November): 2254–2261.

Chen, Juan, and David T. Takeuchi. 2011. "Intermarriage, Ethnic Identity, and Perceived Social Standing among Asian Women in the United States." *Journal of Marriage and Family* 73 (August): 876–888.

Chen, Michelle. 2012. "Health Care System Leaves Patients Frustrated—Nurses Work for a Solution." AlterNet, June 4. Accessed June 5, 2012 (www.alternet.org).

Chesler, Phyllis. 2006. "The Failure of Feminism." *Chronicle of Higher Education*, February 26, B12.

Chesney-Lind, Meda, and Lisa Pasko. 2004. *The Female Offender: Girls, Women, and Crime*, 2nd edition. Thousand Oaks, CA: Sage.

Chesney-Lind, Meda, and Nikki Jones, eds. 2010. *Fighting for Girls: New Perspectives on Gender and Violence*. Albany: State University of New York Press.

Chetty, Raj, Nathaniel Hendren, and Lawrence F. Katz. 2015. "The Effects of Exposure to Better Neighborhoods on Children: New Evidence from the Moving to Opportunity Experiment." National Bureau of Economic Research, May. Accessed September 16, 2015 (nber.org).

Chetty, Raj, Nathaniel Hendren, Patrick Kline, Emmanuel Saez, and Nicholas Turner. 2014. "Is the United States Still a Land of Opportunity? Recent Trends in Intergenerational Mobility." National Bureau of Economic Research, January. Accessed February 6, 2014 (www.nber.org).

"Child Brides in West Africa: Girls Fight Back." 2014. *The Economist*, August 23, 42.

"Child Maltreatment." 2014. Centers for Disease Control and Prevention. Accessed October 26, 2014 (www.cdc.gov).

Child Trends Data Bank. 2012. "Parental Expectations for Their Children's Academic Attainment: Indicators on Children and Youth." Accessed June 23, 2014 (www .childtrends.org).

Child Trends Data Bank. 2013. "Youth Voting." June. Accessed April 20, 2014 (www .childtrendsdatabank.org).

Children's Bureau. 2013. "Trends in Foster Care and Adoption: FY2002–FY2012." U.S. Department of Health and Human Services. Accessed November 18, 2013 (www.acf.hhs .gov).

Children's Defense Fund. 2014. "The State of America's Children 2014." Accessed February 14, 2014 (www.childrensdefense.org).

"China's Environment: A Small Breath of Fresh Air." 2014. *The Economist*, February 8, 14, 16.

Choma, Russ. 2015. "One Member of Congress = American Households: Lawmakers' Personal Finances Far from Average." Open Secrets, January 12. Accessed October 7, 2015 (opensecrets.org).

Christakis, Erika. 2012. "The Overwhelming Maleness of Mass Homicide." *Time*, July 24. Accessed November 20, 2013 (www.time .com).

Christina, Greta. 2011. "Wealthy, Handsome, Strong, Packing Endless Hard-Ons: The Impossible Ideals Men Are Expected to Meet." Independent Media Institute, June 20. Accessed June 21, 2011 (www.alternet .org).

Chronicle of Higher Education. 2014. "Data Point." April 18, A23.

Churchill, Ward. 1997. *A Little Matter of Genocide: Holocaust and Denial in the Americas, 1492 to the Present*. San Francisco: City Lights Books.

CIA World Factbook. 2014. "Sex Ratio." March 6. Accessed September 19, 2014 (www.cia.gov).

Cichocki, Mary K. 1981. "Women's Travel Patterns in a Suburban Development." Pp. 151–163 in *New Space for Women*, edited by Gerda R. Wekerle, Rebecca Peterson, and David Morley. Boulder, CO: Westview Press.

Cillizza, Chris. 2014. "Our C-Minus Government (and Why We Deserve It)." *Washington Post*, October 15. Accessed November 23, 2014 (www.brookings.edu).

Clayson, Dennis E., and Debra A. Haley. 2011. "Are Students Telling Us the Truth? A

Critical Look at the Student Evaluation of Teaching." *Marketing Education Review* 21 (Summer): 101–112.

Clement, Scott. 2011. "Workplace Harassment Drawing Wide Concern." *Washington Post*, November 16. Accessed February 25, 2012 (www.washingtonpost.com).

Clemente, Frank, and Marc Auerbach. 2015. "The Walmart Web: How the World's Biggest Corporation Secretly Uses Tax Havens to Dodge Taxes." Americans for Tax Fairness, June. Accessed September 11, 2015 (americansfortaxfairness.org).

Clements, David D. 2012. *Corporations Are Not People: Why They Have More Rights Than You Do and What You Can Do About It*. San Francisco: Berrett-Koehler.

Clemmons, Nakia S., Paul A. Gastanaduy, Amy P. Fiebelkorn, Susan B. Redd, and Gregory S. Wallace. 2015. "Measles—United States, January 4–April 2, 2015." *MMWR* 64 (April 17): 373–6.

Clifford, Stephanie, and Quentin Hardy. 2013. "Attention, Shoppers: Store Is Tracking Your Cell." *New York Times*, July 15, A1.

Clifford, Stephanie. 2012. "Unexpected Ally Helps Wal-Mart Cut Waste." *New York Times*, April 13, B1.

Cloke, Kenneth, and Joan Goldsmith. 2002. *End of Management and the Rise of Organizational Democracy*. San Francisco: Jossey-Bass.

Cloud, John. 2010. "How to Recruit Better Teachers." *Time*, September 20, 47–52.

"Clueless." 2014. *The Economist*, June 28, 26.

Cobble, Dorothy Sue, Linda Gordon, and Astrid Henry. 2014. "What 'Lean In' Leaves Out." *Chronicle Review*, September 16, B4–B5.

Cohen, Jere. 2002. *Protestantism and Capitalism: The Mechanisms of Influence*. New York: Aldine de Gruyter.

Cohen, Patricia. 2015. "Public-Sector Jobs Vanish, and Blacks Take Blow." *New York Times*, May 25, A1.

Cohen, Patricia. 2015. "The New Shop Class." *New York Times*, March 11, Bi, B4.

Cohen, Patricia. 2015. "When Company is Fined, Taxpayers Often Share Bill." *New York Times*, February 4, B1, B9.

Cohen, Stefanie. 2012. "Why Women Writers Still Take Men's Names." *Wall Street Journal*, December 7, D9.

Cohn, D'Vera, Eileen Patten, and Mark Hugo. 2014. "Puerto Rican Population Declines on Island, Grows on U.S. Mainland." Pew Research Center, August 11. Accessed September 19, 2014 (www.pewresearch.org).

Cohn, D'Vera, Gretchen Livingston, and Wendy Wang. 2014. "After Decades of Decline, a Rise in Stay-at-Home Mothers." Pew Research Center, April 8. Accessed April 12, 2014 (www.pewresearch.org).

Cohn, D'Vera, Jeffrey S. Passel, Wendy Wang, and Gretchen Livingston. 2011. "Barely Half of U.S. Adults Are Married—A Record Low." Pew Research Center, December 14. Accessed May 18, 2014 (www.pewsocialtrends.org).

Cohn, D'Vera, Paul Taylor, Mark H. Lopez, Catherine A. Gallagher, Kim Parker, and Kevin T. Maass. 2013. "Gun Homicide Rate Down 49% Since 1993 Peak; Public Unaware." Pew Research Center, May 7. Accessed October 12, 2013 (www .pewresearch.org).

Cohn, D'Vera. 2011. "India Census Offers Three Gender Options." Pew Research Center, February 7. Accessed October 25, 2011 (www.pewresearch.org).

Cohn, D'Vera. 2012. "Divorce and the Great Recession." Pew Research Center, May 2.

Accessed May 22, 2014 (www.pewsocialtrends .org).

Cohn, D'Vera. 2013. "Love and Marriage." Pew Research Social & Demographic Trends, February 13. Accessed May 16, 2014 (www .pewsocialtrends.org).

Cohn, Meredith. 2014. "Kids Are Outfitted with New Hands Made on 3-D Printers." *Baltimore Sun*, September 29, 1, 11.

Cohn, Meredith. 2015. "Hepatitis C Drugs Can Cure—If Patients Can Afford Them." *Baltimore Sun*, August 18, 1, 16.

Colapinto, John. 1997. "The True Story of John/ Joan." *Rolling Stone* (December 11): 54–73, 92–97.

Colapinto, John. 2001. *As Nature Made Him: The Boy Who Was Raised as a Girl.* New York: Harper Perennial.

Colapinto, John. 2004. "What Were the Real Reasons Behind David Reimer's Suicide?" *Slate*, June 3. Accessed April 24, 2008 (www .slate.com).

College Board. 2013. "2013 College-Bound Seniors: Total Group Profile Report." Accessed June 21, 2014 (www.collegeboard.org).

College Board. 2014. "The 10th Annual AP Report to the Nation." February 11. Accessed June 21, 2014 (apreport.collegeboard.com).

Collins, Jason, and Franz Lidz. 2013. "The Gay Athlete." *Sports Illustrated*, May 6, 34–41.

Colvin, Geoff. 2012. "The Art of the Self-Managing Team." *Fortune*, December 3, 22–23.

Colvin, George. 2014. "In the Future, Will There Be Any Work Left for People to Do?" *Fortune*, June 16, 193–202.

Conley, Dalton. 1999. *Being Black, Living in the Red: Race, Wealth, and Social Policy in America.* Berkeley: University of California Press.

Conley, Dalton. 2004. *The Pecking Order: Which Siblings Succeed and Why.* New York: Pantheon.

Connell, Dave. 2011. "30 Days of H20: Final Days." The Nature Conservancy, February 22. Accessed June 6, 2012 (blog.nature.org).

Conrad, Peter, and Kristin K. Barker. 2010. "The Social Construction of Illness: Key Insights and Policy Implications." *Journal of Health and Social Behavior* 51 (Suppl.): S67–S79.

Considine, Austin. 2011. "For Asian-American Stars, Many Web Fans." *New York Times*, July 31, ST6.

Consumer Reports. 2014. "When You Need Medical Care, Stat." August, 13.

ConsumerReports.org. 2009. "6 Top Reasons for Not Having Sex." February. Accessed March 4, 2014 (www.consumerreports.org).

Cook, Thomas D., and Donald T. Campbell. 1979. *Quasi-Experimentation: Design and Analysis Issues for Field Settings.* Chicago: Rand McNally.

Cooky, Cheryl, Michael A. Messner, and Michela Musto. 2015. " 'It's Dude Time!': A Quarter Century of Excluding Women's Sports in Televised News and Highlight Shows." *Communication & Sport* 3 (September): 261–287.

Cooley, Charles Horton. 1909/1983. *Social Organization: A Study of the Larger Mind.* New Brunswick, NJ: Transaction Books.

Cooper, David. 2015. "Given the Economy's Growth, the Federal Minimum Wage Could Be Significantly Higher." Economic Policy Institute, April 14. Accessed October 7, 2015 (epi.org).

Cooper, O. R., et al. 2010. "Increasing Springtime Ozone Mixing Ratios in the Free Troposphere over Western North America." *Nature* 463 (January 21): 344–348.

Cooperman, Alan, Gregory Smith, and Besheer Mohamed. 2013. "Celebrating Christmas and the Holidays, Then and Now." Pew Research Center, December 18. Accessed June 21, 2014 (www.pewresearch.org).

Cooperman, Alan, Mark H. Lopez, Cary Funk, and Jessica H. Martínez. 2014. "The Shifting Religious Identity of Latinos in the United States: Nearly One-in-Four Latinos Are Former Catholics." Pew Research Center, May 7. Accessed June 21, 2014 (www .pewresearch.org).

Copen, Casey E., Anjani Chandra, and Gladys Martinez. 2012. "Prevalence and Timing of Oral Sex with Opposite-Sex Partners Among Females and Males Aged 15–24 Years: United States, 2007–2010." *National Health Statistics Reports*, No. 56, August 16. Accessed March 2, 2014 (www.cdc.gov/nchs).

Copen, Casey E., Kimberly Daniels, and William D. Mosher. 2013. "First Premarital Cohabitation in the United States: 2006–2010 National Survey of Family Growth." *National Health Statistics Reports*, April 4. Accessed May 16, 2014 (www.cdc.gov/nchs).

Copen, Casey E., Kimberly Daniels, Jonathan Vespa, and William D. Mosher. 2012. "First Marriages in the United States: Data from the 2006–2010 National Survey of Family Growth." *National Health Statistics Reports*, March 22. Accessed May 16, 2014 (www.cdc .gov/nchs).

Corak, Miles. 2013. "Income Inequality, Equality of Opportunity, and Intergenerational Mobility." *Journal of Economic Perspectives* 27 (Summer): 79–102.

Corporation for National & Community Service. 2015. "The State of Volunteering in America 2014." Accessed August 19, 2015 (volunteeringinamerica.gov).

Coser, Lewis. 1956. *The Functions of Social Conflict.* New York: Free Press.

Couch, Carl J. 1968. "Collective Behavior: An Examination of Some Stereotypes." *Social Problems* 15 (Winter): 310–322.

Council for Global Equity. 2014. "The Facts on LGBT Rights in Russia." Accessed March 2, 2014 (www.globalequity.org).

Covenant Eyes, Inc. 2013. "Pornography Statistics." Accessed March 2, 2014 (www.covenanteyes .com).

Cox, Shanna, Karen Pazol, Lee Warner, Lisa Romero, Alison Spitz, Lorrie Gavin, and Wanda Barfield. 2014. "Vital Signs: Births to Teens Aged 15–17 Years—United States, 1991–2012." *MMWR* 63 (April 11): 312–317.

Crabtree, Steve. 2010. "Religiosity Highest in World's Poorest Nations." Gallup, June 3. Accessed August 31, 2012 (gallup.com).

Craig, Lyn. 2015. "How Mothers and Fathers Allocate Time to Children: Trends, Resources, and Policy Context." *Family Focus*, Spring, F9–F10.

Cressey, Donald R. 1953. *Other People's Money: A Study in the Social Psychology of Embezzlement.* Glencoe, IL: Free Press.

Crothers, Charles. 1979. "On the Myth of Rural Tranquility: Comment on Webb and Collette." *American Journal of Sociology* 84 (May): 1441–1445.

Crowe, Ann H., Tracy G. Mullins, Kimberley A. Cobb, and Nathan C. Lowe. 2012. "Effects and Consequences of Underage Drinking." *Juvenile Justice Bulletin*, Office of Juvenile Justice and Delinquency Prevention, September. Accessed July 30 (www.ojjdp .gov).

Crowley, Michael. 2010. "What $120 Million Buys." *Time*, October 11, 35–37.

Crowley, Michael. 2014. "Iraq's Eternal War." *Time*, June 30, 28–34.

Cruikshank, Margaret. 2009. *Learning to Be Old: Gender, Culture, and Aging.* Lanham, MD: Rowman & Littlefield.

Cudd, Ann E., and Nancy Holstrom. 2010. *Capitalism, For and Against: A Feminist Debate.* New York: Cambridge University Press.

Culbert, Samuel A. 2011. "Why Your Boss Is Wrong About You." *New York Times*, March 3, A25.

Culbert, Samuel A., and Lawrence Rout. 2010. *Get Rid of the Performance Review!: How Companies Can Stop Intimidating, Start Managing—and Focus on What Really Matters.* New York: Business Plus.

Cullen, Lisa T., and Coco Masters. 2008. "We Just Clicked." *Time*, January 28, 86–89.

Cumberworth, Erin. 2010. "Homeboy Industries." *Pathways*, Summer, 22–23.

Cunningham, Solveig A., Michael R. Kramer, and K. M. Venkat Narayan. 2014. "Incidence of Childhood Obesity in the United States." *New England Journal of Medicine* 370 (5): 403–411.

Currie, E. 1985. *Confronting Crime: An American Challenge.* New York: Pantheon.

Currie, Janet, Mark Stable, and Lauren E. Jones. 2014. "Do Stimulant Medications Improve Educational and Behavioral Outcomes for Children with ADHD?" National Bureau of Economic Research, June. Accessed July 26, 2014 (www.nber.org).

Curtin, Sally C., Stephanie J. Ventura, and Gladys M. Martinez. 2014. "Recent Declines in Nonmarital Childbearing in the United States." National Center for Health Statistics, August. Accessed November 30, 2014 (www.nchs.gov).

Curtiss, Susan. 1977. *Genie.* New York: Academic Press.

Cynkar, Peter, and Elizabeth Mendes. 2011. "More Than One in Six American Workers Also Act as Caregivers." Gallup, July 26. Accessed March 10, 2012 (www.gallup.com).

D'Innocenzio, Anne. 2015. "Target Stores Aim to Instill Healthier Habits." *Baltimore Sun*, September 17, 10.

D'Onofrio, Brian M., and Benjamin B. Lahey. 2010. "Biosocial Influences on the Family: A Decade Review." *Journal of Marriage and Family* 72 (June): 762–782.

Dahl, Robert A. 1961. *Who Governs? Democracy and Power in an American City.* New Haven. CT: Yale University Press.

Dahlin, E. 2011. "There's No 'I' in Innovation." *Contexts* 10 (Fall): 22–27.

Daley, Suzanne, and Nicholas Kulish. 2013. "Germany Fights Population Drop." *New York Times*, August 14, A1.

Daly, Mary C., and Leila Bengali. 2014. "Is It Still Worth Going to College?" Federal Reserve Bank of San Francisco, May 5. Accessed June 21, 2014 (www.frbsf.org).

Daneshpour, Manijeh. 2013. "A Mini-Narrative About My Praxis as a Muslim Feminist." *Family Focus* 57.4 (Winter): F1–F3.

Daniel, Mary-Alice. 2014. "What White People Need to Learn." AlterNet, February 7. Accessed March 22, 2014 (www.alternet.org).

Dank, Meredith, et al. 2014. "Estimating the Size and Structure of the Underground Commercial Sex Economy in Eight Major US Cities." Urban Institute, March. Accessed November 20, 2014 (www.urban.org).

Dao, James. 2013. "In Debate over Military Sexual Assault, Men Are Overlooked Victims." *New York Times*, June 24, A12.

Davenport, Coral. 2014. "Industry Awakens to Threat of Climate Change." *New York Times*, January 24, A1.

Davey, Monica. 2014. "Immigrants Seen a Way to Refill Detroit Ranks." *New York Times*, January 24, A12.

Davidson, Paul. 2011. "Season of Part-Time Jobs Kicks Off with Holidays." *USA Today*, November 25–27, A1–A2.

Davies, James C. 1979. "The J-Curve of Rising and Declining Satisfaction as a Cause of Revolution and Rebellion." Pp. 413–436 in *Violence in America: Historical and Comparative Perspectives*, edited by Hugh D. Graham and Ted R. Gurr. Beverly Hills, CA: Sage.

Davis, Julie Hirschfeld. 2015. "Hacking Exposed 21 Million in U.S., Government Says." *New York Times*, July 10, A1, A3.

Davis, Julie Hirschfeld. 2015. "What to Wear? First Ladies Are Steeped in Protocol." *New York Times*, January 30, A15.

Davis, Karen, Kristof Stremikis, David Squires, and Cathy Schoen. 2014. "Mirror, Mirror on the Wall: How the Performance of the U.S. Health Care System Compares Internationally." The Commonwealth Fund, June. Accessed July 26, 2014 (www.commonwealthfund.org).

Davis, Kenneth C. 2010. "America's True History of Religious Tolerance." *Smithsonian*, October. Accessed September 28, 2010 (www.smithsonian.com).

Davis, Kingsley, and Wilbert E. Moore. 1945. "Some Principles of Stratification." *American Sociological Review* 10 (April): 242–249.

De La Haye, Kayla, Garry Robins, Philip Mohr, and Carlene Wilson. 2013. "Adolescents' Intake of Junk Food." *Journal of Research on Adolescence* 23 (September): 524–536.

de las Casas, Bartolome. 1992. *The Devastation of the Indies: A Brief Account*. Baltimore, MD: Johns Hopkins University Press.

De Long, J. Bradford, and Andrei Shleifer. 1992. "Princes and Merchants: European City Growth Before the Industrial Revolution." National Bureau of Economic Research, December. Accessed January 10, 2008 (www.nper.org).

de Pommereau, Isabelle. 2013. "A Lift for Europe's Women?" *Christian Science Monitor Weekly*, August 5,16.

de Vogue, Ariane, and Jeremy Diamond. 2015. "Supreme Court Rules in Favor of Same-Sex Marriage Nationwide." CNN, June 27. Accessed September 29, 2015 (cnn.com).

"Deadly Intolerance." 2014. *The Economist*, March 1, 42.

Death Penalty Information Center. 2015. "Facts about the Death Penalty." September 2. Accessed September 3, 2015 (deathpenaltyinfo.org).

Deaton, Angus, and Arthur A. Stone. 2013. "Grandpa and the Snapper: The Wellbeing of the Elderly Who Live with Children." National Bureau of Economic Research, June. Accessed July 10, 2013 (www.nber.org).

Deegan, Mary Jo. 1986. *Jane Addams and the Men of the Chicago School, 1892–1918*. New Brunswick, NJ: Transaction Books.

Deggans, Eric. 2012. "For Asians and Latinos, Stereotypes Persist in Sitcoms." National Public Radio, February 24. Accessed March 22, 2014 (www.npr.org).

DeGraw, David. 2014. "How the Ultra-Rich .01% Have Sucked Up Even More of America's Wealth Than You Think." AlterNet, August 22. Accessed November 10, 2014 (www.alternet.org).

DeKeseredy, Walter S. 2011. *Violence Against Women: Myths, Facts, Controversies*. Toronto: University of Toronto Press.

della Porta, Donatella, and Mario Diani. 1999. *Social Movements: An Introduction*. Malden, MA: Blackwell.

DeNavas-Walt, Carmen, and Bernadette D. Proctor. 2015. "Income and Poverty in the United States: 2014." U.S. Census Bureau, September. Accessed September 16, 2015 (census.gov).

Dennis, Brady. 2013. "PG-13 Movies Match R Rating, Study Says." *Washington Post*, November 11. Accessed November 18, 2013 (www.washingtonpost.com).

Denny, C. H., J. Tsai, R. L. Floyd, and P. P. Green. 2009. "Alcohol Use Among Pregnant and Nonpregnant Women of Childbearing Age—United States, 1991–2005." *MMWR* 58 (May 22): 529–532.

Denyer, Simon. 2013. "Season of Smog Drifting over China." *Baltimore Sun*, October 23, 13.

DeParle, Jason. 2012. "Harder for Americans to Rise from Lower Rungs." *New York Times*, January 5, A1.

Department of Defense. 2013. "Annual Report on Sexual Assault in the Military: Fiscal Year 2012." Department of Defense Sexual Assault Prevention and Response, May. Accessed January 12, 2014 (www.sapr.mil).

Department of Defense. 2015. "Department of Defense Annual Report on Sexual Assault in the Military: Fiscal Year 2014." Accessed September 29, 2015 (sapr.mil).

Deprez, Esmé E., and William Selway. 2014. "Guns Allowed All Over—Except Near Schools." *Bloomberg Businessweek*, May 19–25, 27–28.

DeSilver, Drew. 2014. "5 Facts About Economic Inequality." Pew Research Center, January 7. Accessed February 14, 2014 (www.pewresearch.org).

DeSilver, Drew. 2014. "Q/A: What the *New York Times*' Polling Decision Means." Pew Research Center, July 28. Accessed October 26, 2014 (www.pewresearch.org).

DeSilver, Drew. 2015. "Businesses Owned by Women, Minorities Lag in Revenue Share." Pew Research Center, September 1. Accessed October 4, 2015 (pewresearch.org).

Desmond-Harris, Jenee. 2009. "Why Michelle's Hair Matters." *Time*, September 7, 55–57.

Deveny, Kathleen. 2008. "They're No Baby Einsteins." *Newsweek*, January 14, 61.

Dewan, Shaila, and Robert Gebeloff. 2012. "Among the Wealthiest One Percent, Many Variations." *New York Times*, January 15, A1.

Dholia, Esha. 2014. "Too Many Tropes: Top 3 Stereotypes of South Asians in the Media." *Videshi Magazine*, March 13. Accessed March 22, 2014 (www.videshimagazine.com).

Diamond, Milton, and H. Keith Sigmundson. 1997. "Sex Reassignment at Birth: Long-Term Review and Clinical Implications." *Archives of Pediatrics & Adolescent Medicine* 15 (March): 298–304.

Dias, Elizabeth. 2013. "¡Evangelicos!" *Time*, March 15, 20–28.

Dickey, Jack. 2015. "Save Our Vacation." *Time*, June 1, 44–49.

Dilworth-Anderson, Peggye, Linda M. Burton, and William L. Turner. 1993. "The Importance of Values in the Study of Culturally Diverse Families." *Family Relations* 42 (July): 238–242.

DiSalvo, David. 2010. "Are Social Networks Messing with Your Head?" *Scientific American Mind* 20 (January/February): 48–55.

"DNA Reveals White Supremacist's Black Heritage." 2013. *Baltimore Sun*, November 14, 8.

Dockterman, Eliana. 2013. "The Digital Parent Trap. Should Your Kids Avoid Tech—or Embrace It?" *Time*, August 19, 54.

Dodson, Lisa, and Wendy Luttrell. 2011. "Families Facing Untenable Choices." *Contexts* 10 (Winter): 38–42.

Doeringer, Peter B., and Michael J. Piore. 1971. *Internal Labor Markets and Manpower Analysis*. Lexington, MA: Heath-Lexington Books.

"Dog-lovers v. Dog-eaters: Pet Food." 2015. *The Economist*, June 20, 46.

Doherty, Carroll, and Alec Tyson. 2015. "In Debate Over Legalizing Marijuana, Disagreement over Drug's Dangers." Pew Research Center, April 14. Accessed August 23, 2015 (pewresearch.org).

Doherty, Carroll, and Alec Tyson. 2015. "Most Say Government Policies Since Recession Have Done Little to Help Middle Class, Poor." Pew Research Center, March 4. Accessed August 26, 2015 (pewresearch.org).

Doherty, Carroll, and Jocelyn Kiley. 2015. "Support for Same-Sex Marriage at Record High, but Key Segments Remain Opposed." Pew Research Center, June 8. Accessed September 29, 2015 (pewresearch.org).

Doherty, Carroll, Juliana M. Horowitz, and Michael Dimock. 2014. "Most See Inequality Growing, but Partisans Differ Over Solutions." Pew Research Center, January 23. Accessed April 12, 2014 (www.pewresearch.org).

Doherty, Carroll. 2015. "Continue Bipartisan Support for Expanded Background Checks on Gun Sales." Pew Research Center, August 13. Accessed August 23, 2015 (pewresearch.org).

Dolnick, Sam. 2011. "Dance, Laugh, Drink, Save the Date: It's a Ghanaian Funeral." *New York Times*, April 12, A1.

Domhoff, G. William. 2013. "Interlocking Directorates in the Corporate Community." Who Rules America, October. Accessed April 12, 2014 (whorulesamerica.net).

Dominguez, Kenneth, et al. 2015. "Vital Signs: Leading Causes of Death, Prevalence of Diseases and Risk Factors, and Use of Health Services among Hispanics in the United States—2009–2013" *MMWR* 64 (17): 469–478.

Domosh, Mona, and Joni Seager. 2001. *Feminist Geographers Make Sense of the World*. New York: Guilford Press.

Donadio, Rachel. 2013. "When Italians Chat, Hands and Fingers Do the Talking." *New York Times*, July 1, A6.

Dorfman, Mark, and Angela Haren. 2014. "Testing the Waters." Natural Resources Defense Council, June. Accessed September 19, 2014 (nrdc.org).

Douglas, Susan J., and Meredith W. Michaels. 2004. *The Mommy Myth: The Idealization of Motherhood and How It Has Undermined Women*. New York: Free Press.

Dovidio, John F. 2009. "Racial Bias, Unspoken but Heard." *Science* 326 (December 18): 1641–1642.

Downs, Edward, and Stacy L. Smith. 2010. "Keeping Abreast of Hypersexuality: A Video Game Character Content Analysis." *Sex Roles* 62 (June): 721–733.

Doyle, Francis X. 2011. "Bishops' Missed Opportunity." *Baltimore Sun*, November 14, 15.

Drake, Bruce. 2013. "Americans See Growing Gap Between Rich and Poor." Pew Research Center, December 5. Accessed February 14, 2014 (www.pewresearch.org).

Drexler, Peggy. 2012. "The New Face of Infidelity." *Wall Street Journal*, October 21, C3.

Drexler, Peggy. 2013. "The Delicate Protocol of Hugging." *Wall Street Journal*, September 14–15, C3.

Drexler, Peggy. 2014. "Why We Love to Gossip." *Huffington Post*, November 14. Accessed October 25, 2015 (huffingtonpost.com).

"The Drugs Don't Work." 2014. *The Economist*, May 3, 54.

Drutman, Lee, and Alexander Furnas. 2014. "K Street Pays Top Dollar for Revolving Door Talent." Sunlight Foundation, January 21. Accessed April 20, 2014 (sunlightfoundation.com).

Du Bois, W. E. B. 1986. *The Souls of Black Folk.* New York: Library of America.

Dugan, Andrew, and Brad Hoffman. 2014. "Record Low Say Own Representative Deserves Re-Election." Gallup, January 24. Accessed April 20, 2014 (www.gallup.com).

Dugan, Andrew, and Frank Newport. 2013. "In U.S., Fewer Believe 'Plenty of Opportunity' to Get Ahead." Gallup, November 30. Accessed February 6, 2014 (www.gallup.com).

Dugan, Andrew. 2013. "In U.S., Support for Complete Smoking Ban Increases to 22%." Gallup, July 29. Accessed January 12, 2014 (www.gallup.com).

Dugan, Andrew. 2013. "More Say Crime Is Serious Problem in U.S. Than Locally." Gallup, November 1. Accessed January 12, 2014 (www.gallup.com).

Dugan, Andrew. 2014. "More Americans Worse Off Financially Than a Year Ago." Gallup, January 15. Accessed April 12, 2014 (www.gallup.com).

Dugan, Andrew. 2015. "Once Taboo, Some Behaviors Now More Acceptable in U.S." Gallup, June 1. Accessed July 26, 2015 (gallup.com).

Dugan, Andrew. 2015. "Trust Differs Most by Ideology for Church, Police, Presidency." Gallup, June 30. Accessed October 7, 2015 (gallup.com).

Dugan, Andrew. 2015. "U.S. Support for Increased Immigration Up to 25%." Gallup, August 10. Accessed October 4, 2015 (gallup.com).

Duggan, Maeve, Amanda Lenhart, Cliff Lampe, and Nicole B. Ellison. 2015. "Parents and Social Media." Pew Research Center, July 16. Accessed August 14, 2015 (pewresearch.org).

Duggan, Maeve, Nicole B. Ellison, Cliff Lampe, Amanda Lenhart, and Mary Madden. 2015. "Social Media Update 2014." Pew Research Center, July 28. Accessed January 9, 2015 (pewresearch.org).

Duggan, Maeve. 2014. "Online Harassment." Pew Research Center, October 22. Accessed August 14, 2015 (pewresearch.org).

Duhigg, Charles. 2010. "Saving U.S. Water and Sewer Systems Would Be Costly." *New York Times*, March 14, A1.

Dumaine, Brian. 2015. "Can a [Billionaire] Hedge Fund Manager Fix Income Inequality?" *Fortune*, September 1, 87–91.

Duncan, D. T., A. R. Hansen, W. Wang, F. Yan, and J. Zhang. 2015. "*Change* in Misperception of Child's Body Weight among Parents of American Preschool Children." *Childhood Obesity*, National Center for Biotechnology Information, May. Accessed October 15, 2015 (ncbi.nlm.nih.gov).

Dunifon, Rachel, and Lori Kowaleski-Jones. 2007. "The Influence of Grandparents in Single-Mother Families." *Journal of Marriage and Family* 69 (May): 465–481.

Dunlap, Riley E, 2007. "The State of Environmentalism in the U.S." Gallup, April 19. Accessed March 27, 2007 (www.gallup.com).

Durden, Tyler. 2011. "Here Are the 29 Public Companies with More Cash Than the US Treasury." Zero Hedge, July 15. Accessed January 9, 2012 (www.zerohedge.com).

Durkheim, Émile. 1893/1964. *The Division of Labor in Society.* New York: Free Press.

Durkheim, Émile. 1897/1951. *Suicide: A Study in Sociology,* translated by John A. Spaulding and George Simpson, edited by George Simpson. New York: Free Press.

Durkheim, Émile. 1898/1956. *Education and Sociology,* translated by Sherwood D. Fox. Glencoe, IL: Free Press.

Durkheim, Émile. 1961. *The Elementary Forms of the Religious Life.* New York: Collier Books.

Durose, Matthew R., Alexia D. Cooper, and Howard N. Snyder. 2014. "Recidivism of Prisoners Released in 30 States in 2005: Patterns from 2005 to 2010." Bureau of Justice Statistics, April. Accessed September 2, 2015 (bjs.gov).

Dwyer, Rachel E., Randy Hodson, and Laura McCloud. 2013. "Gender, Debt, and Dropping Out of College." *Gender & Society* 27 (February): 30–55.

Dye, Thomas R., and Harmon Ziegler. 2003. *The Irony of Democracy: An Uncommon Introduction to American Politics,* 12th edition. Belmont, CA: Wadsworth.

Eagan, Kevin, Jennifer B. Lozano, Sylvia Hurtado, and Matthew H. Case. 2014. *The American Freshman: National Norms Fall 2013.* Los Angeles: Higher Education Research Institute, UCLA.

Eaton, Danice K., et al. 2012. "Youth Risk Behavior Surveillance—United States, 2011." *MMWR* 61 (June 8): 1–162.

Eavis, Peter. 2015. "Judge's Ruling Against 2 Banks Points to Wide Fraud in '08 Crash." *New York Times*, May 12, A1, B2.

Eberstadt, Nicholas. 2009. "Poor Statistics." *Forbes*, March 2, 26.

Eberstadt, Nicholas. 2013. "China's Coming One-Child Crisis." *Wall Street Journal*, November 27, A15.

Eckenrode, John, Elliott G. Smith, Margaret E. McCarthy, and Michael Dineen. 2014. "Income Inequality and Child Maltreatment in the United States." *Pediatrics* 33 (March): 454–461.

Eckhoff, Vickery. 2015. "Exposing America's Billionaire Welfare Ranchers." AlterNet, March 24. Accessed August 26, 2015 (alternet.org).

Eckholm, Erik. 2014. " 'Aid in Dying' Movement Takes Hold in Some States." *New York Times*, February 8, A1.

Eckholm, Erik. 2015. "Texas ID Law Ruled to Breach Voters' Rights." *New York Times*, August 6, A1, A18.

Eckstein, Rick, Rebecca Schoenike, and Kevin Delaney. 1995. "The Voice of Sociology: Obstacles to Teaching and Learning the Sociological Imagination." *Teaching Sociology* 23 (October): 353–363.

Edin, Kathryn J., and H. Luke Shaefer. 2015. *$2.00 a Day: Living on Almost Nothing in America.* New York: Houghton Mifflin Harcourt.

Efrati, Amir. 2013. "Google's Data-Trove Dance." *Wall Street Journal*, July 31, B1, B4.

Eggebeen, David J. 2005. "Cohabitation and Exchanges of Support." *Social Forces* 83 (May): 1097–1110.

Egley, Arlen, and James C. Howell. 2013. "Highlights of the 2011 National Youth Gang Survey." Office of Juvenile Justice and Delinquency Prevention, September. Accessed January 12, 2014 (www.ojjdp.gov).

Ehrlich, Paul R., and Anne H. Ehrlich. 2008. *The Dominant Animal: Human Evolution and the Environment.* Washington, DC: Island Press.

Ehrlich, Paul. 1971. *The Population Bomb*, 2nd edition. San Francisco: Freeman.

Eibner, Christine. 2015. "How Much Are You Paying for Health Care?" Rand Corporation, January 15. Accessed October 16, 2015 (rand.org).

Eilperin, Juliet, and Lenny Bernstein. 2014. "Environmental Groups Say Obama Needs to Address Climate Change More Aggressively." *Washington Post*, January 16. Accessed September 19, 2014 (www.washingtonpost.com).

Eisenberg, Abne M., and Ralph R. Smith, Jr. 1971. *Nonverbal Communication.* Indianapolis: Bobbs-Merrill.

Eisler, Peter, and Barbara Hansen. 2013. "Doctors Perform Thousands of Unnecessary Surgeries." *USA Today*, June 20. Accessed July 26, 2014 (www.usatoday.com).

Eisler, Peter. 2013. "What Surgeons Leave Behind Costs Some Patients Dearly." *USA Today*, March 8. Accessed July 26, 2014 (www.usatoday.com).

Ekman, Paul, and Wallace V. Friesen. 1984. *Unmasking the Face: A Guide to Recognizing Emotions from Facial Clues.* Palo Alto, CA: Consulting Psychologists Press.

El Nasser, Haya, and Paul Overberg. 2011. "In Many Neighborhoods, Kids Are Only a Memory." *USA Today*, June 3, 1A, 6A.

Elgin, Ben. 2014. "An Oil Giant Dims the Lights on Green Power." *Bloomberg Businessweek*, June 2–8, 21–22.

Eliot, Lise. 2012. *Pink Brain, Blue Brain: How Small Differences Grow into Troublesome Gaps—And What We Can Do About It.* Oxford, England: Oneworld Publications.

Ellingwood, Ken. 2009. "Not Being on Time a High Art in Mexico." *Los Angeles Times*, September 12. Accessed September 22, 2009 (www.latimes.com).

Elliott, Philip. 2015. "Ready for Speaking Fees." *Time*, June 1, 12.

Ellis, Bill. 2005. "Legend/AntiLegend: Humor as an Integral Part of the Contemporary Legend Process." Pp. 123–140 in *Rumor Mills: The Social Impact of Rumor and Legend*, edited by Gary Alan Fine, Veronique Campion-Vincent, and Chip Heath. New Brunswick, NJ: Transaction Publishers.

Ellis, Renee R., and Tavia Simmons. 2014. "Coresident Grandparents and Their Grandchildren: 2012." *U.S. Census Bureau*, October. Accessed October 25, 2014 (www.census.gov).

Ellis, Renee. 2013. "Changes in Coresidence of Grandparents and Grandchildren." *Family Focus* (Summer): F13, F15, F17.

Ellison, Jesse. 2011. "The 2011 Global Women's Progress Report." *Newsweek*, September 26, 27–33.

Elsen, Tracy. 2015. "The Smallest Apartment for Rent in SF Is Just 233 Square Feet." Curbed, July 17. Accessed October 23, 2015 (sf.curbed.com).

Emery, David. 2015. "Does AriZona Tea Contain Human Urine?" Urban Legends, July 25. Accessed October 25, 2015 (urbanlegends.about.com).

Emmons, William R., and Bryan J. Noeth. 2015. "Race, Ethnicity, and Wealth." Federal Reserve Bank of St. Louis, February. Accessed September 16, 2015 (stlouisfed.org).

"Employer-Reported Workplace Injuries and Illnesses—2013." 2014. Bureau of Labor Statistics, December 4. Accessed October 16, 2015 (bls.gov).

"The End of the Beginning." 2014. *The Economist*, March 8, 22–24.

"Ending the Shame." 2013. *The Economist*, September 14, 16, 18.

England, Paula, and Reuben J. Thomas. 2009. "The Decline of the Date and the Rise of the College Hook Up." Pp. 141–152 in *Family in Transition*, 15th edition, edited by Arlene S. Skolnick and Jerome H. Skolnick. Boston: Pearson Higher Education.

England, Paula. 2005. "Emerging Theories of Care Work." *Annual Review of Sociology* 31(August): 381–399.

Entmacher, Joan, Lauren Frohlich, Katherine G. Robbins, Emily Martin, and Liz Watson. 2014. "Underpaid & Overloaded: Women in Low-Wage Jobs." National Women's Law Center. Accessed November 23, 2014 (www .nwlc.org).

Environmental Integrity Project. 2007. "Paying Less to Pollute: Environmental Enforcement Under the Bush Administration." Accessed February 12, 2008 (www.environmentalintegrity .org).

EPA. 2013. "The National Rivers and Streams Assessment 2008–2009: A Collaborative Survey." March. Accessed September 19, 2014 (epa.gov/aquaticsurveys).

EPA. 2014. "Municipal Solid Waste Generation, Recycling, and Disposal in the United States: Facts and Figures for 2012." February. Accessed September 19, 2014 (epa.gov /wastes).

EPA. 2014. "Nutrient Pollution." August 26. Accessed September 19, 2014 (epa.gov /nutrientpollution).

EPA. 2015. "Assessment of the Potential Impacts of Hydraulic Fracturing for Oil and Gas on Drinking Water Resources. Executive Summary." June. Accessed October 23, 2015 (epa.gov).

Epstein, Joseph. 2011. *Gossip: The Untrivial Pursuit*. New York: Houghton Mifflin Harcourt.

Epstein, Robert. 2008. "Same-Sex Marriage Is Too Limiting." *Los Angeles Times*, December 4. Accessed December 6, 2008 (www.latimes .com).

Equal Employment Opportunity Commission. 2011. "Sexual Harassment Charges EEOC & FEPAs Combined: FY 1997–FY 2011." Accessed March 2, 2014 (www.eeoc.gov).

Equal Employment Opportunity Commission. 2015. "Charges Alleging Sexual Harassment: FY 2010–FY 2014." Accessed September 2, 2015 (eeoc.gov).

Erikson, Kai T. 1966. *Wayward Puritans: A Study in the Sociology of Deviance*. New York: John Wiley & Sons.

Ernst & Young. 2013. "Navigating Today's Complex Business Risks: Europe, Middle East, India and Africa Fraud Survey 2013." Accessed January 12, 2014 (www.ey.com).

Esipova, Neli, Anita Pugliese, and Julie Ray. 2013. "381 Million Adults Worldwide Migrate Within Countries." Gallup, May 15. Accessed September 19, 2014 (www.gallup.com).

Esping-Andersen, Gosta. 1990. *The Three Worlds of Welfare Capitalism*. Cambridge, United Kingdom: Polity Press.

Esterl, Mike. 2013. "The Natural Evolution of Food Labels." *Wall Street Journal*, November 6, B1–B2.

Etzioni, Amitai. 1975. *A Comparative Analysis of Complex Organizations*. New York: Free Press.

Evans, Becky. 2013. "Bill Gates 'Disrespects' South Korea's Female President by Shaking Hands While Keeping Other in His Pocket." MailOnline, April 23. Accessed November 10, 2013 (dailymail.co.uk).

Evans, Harold, Gail Buckland, and David Lefer. 2006. *They Made America: From the Steam Engine to the Search Engine: Two Centuries of Innovators*. New York: Little, Brown.

Ewen, Lynda Ann. 1998. *Social Stratification and Power in America: A View from Below*. Six Hills, NY: General Hall.

Ewert, Stephanie, and Robert Kominski. 2014. "Measuring Alternative Educational Credentials: 2012." U.S. Census Bureau, January. Accessed June 21, 2014 (www.census.gov).

Ewing, Jack. 2015. "Volkswagen Engine-Rigging Scheme Said to Have Begun in 2008." *New York Times*, October 5, B1, B10.

"Extramarital Affairs." 2014. Pew Research Global Attitudes Project. Accessed April 18, 2014 (www.pewglobal.org).

Fahrenthold, David A. 2009. "Environmentalists Seek to Wipe Out Plush Toilet Paper." *Washington Post*, September 24, A1.

FairTest. 2013. "SAT Score Averages Stagnate as Number of Test-Takers Declines: Persistent Race, Gender Gaps Show Failure of Test-Driven Schooling." National Center for Fair & Open Testing, September 26.

"Faith and Reason." 2014. *The Economist*, February 22, 28.

Fakhraie, Fatemeh. 2009. "Feminists Don't Understand Muslim Women." Double X, May 20. Accessed April 26, 2010 (www .doublex.com).

Falah, Ghazi-Wald, and Caroline Nagel, eds. 2005. *Geographies of Muslim Women: Gender, Religion, and Space*. New York: Guilford Press.

"Family Planning: One-Child Proclivity." 2014. *The Economist*, July 19, 40.

Fang, Ferric C., R. Grant Steen, and Arturo Casadevall. 2012. "Misconduct Accounts for the Majority of Retracted Scientific Publications." *Proceedings of the National Academy of Sciences* 109 (42): 17028–17033.

FAO, IFAD, and WFP. 2013. *The State of Food Insecurity in the World 2013: The Multiple Dimensions of Food Security*. Rome: FAO.

Faris, Robert, and Diane Felmlee. 2011. "Status Struggles: Network Centrality and Gender Segregation in Same- and Cross-Gender Aggression." *American Sociological Review* 76 (February): 48–73.

Farley, Melissa. 2001. "Prostitution: The Business of Sexual Exploitation." Pp. 879–891 in *Encyclopedia of Women and Gender*, volume 2, edited by Judith Worrell. New York: Academic Press.

Farley, Sally D., Amie M. Ashcraft, Mark F. Stasson, and Rebecca L. Nusbaum. 2010. "Nonverbal Reactions to Conversational Interruption: A Test of Complementarity Theory and the Status/Gender Parallel." *Journal of Nonverbal Behavior* 34 (December): 193–206.

"Father's Day: June 21, 2015." 2015. U.S. Census Bureau News. Accessed September 29, 2015 (census.org).

Feagin, Joe R. 2001. "Social Justice and Sociology: Agendas for the Twenty-First Century." *American Sociological Review* 66 (February): 1–20.

Feagin, Joe R., and Melvin P. Sikes. 1994. *Living with Racism: The Black Middle-Class Experience*. Boston: Beacon Press.

Feagin, Joe R., and Robert Parker. 1990. *Building American Cities: The Urban Real Estate Game*, 2nd edition. Englewood Cliffs, NJ: Prentice Hall.

Federal Bureau of Investigation. 2003. 2012. "Financial Crimes Report to the Public." Accessed February 14, 2014 (www.fbi.org).

Federal Bureau of Investigation. 2014. *Crime in the United States, 2013*. Fall. Accessed September 2, 2015 (fbi.gov).

Federal Bureau of Investigation. 2015. "2014 Internet Crime Report." Accessed August 22, 2015 (www.ic3.gov).

Federal Election Commission. 2014. "PAC Count—1974 to Present." January. Accessed April 12, 2014 (www.fec.gov).

Federal Interagency Forum on Child and Family Statistics. 2013. *America's Children: Key National Indicators of Well-Being, 2013*. Washington, DC: U.S. Government Printing Office.

Federal Interagency Forum on Child and Family Statistics. 2015. *America's Children: Key National Indicators of Well-Being, 2015*. Washington, DC: U.S. Government Printing Office.

Feldman-Jacobs, Charlotte, and Donna Clifton. 2014. "Female Genital Mutilation/Cutting: Data and Trends, Update 2014." Population Reference Bureau. Accessed March 2, 2014 (www.prb.org).

Ferdman, Roberto A. 2015. "France Is Making it Illegal for Supermarkets to Throw Away Edible Food." *Washington Post*, May 22. Accessed July 26, 2015 (washingtonpost.com).

Ferguson, Christopher J., and Cheryl K. Olson. 2013. "Video Game Violence Use Among 'Vulnerable' Populations: The Impact of Violent Games on Delinquency and Bullying Among Children with Clinically Elevated Depression or Attention Deficit Symptoms." *Journal of Youth and Adolescence* 42 (August), published online.

Ferguson, Emily. 2014. "11 Reasons Christianity Needs Feminism." *Huffington Post*, May 21. Accessed June 21, 2014 (www.huffingtonpost .com).

Fernald, Anne, Virginia A. Marchman, and Adriana Weisleder. 2013. "SES Differences in Language Processing Skill and Vocabulary are Evident at 18 Months." *Developmental Science* 16 (March): 234–248.

Ferree, Myra M. 2005. "It's Time to Mainstream Research on Gender." *Chronicle Review*, August 12, B10.

Fields, Jason, and Lynne. M. Casper. 2001. "America's Families and Living Arrangements: 2000." U.S. Census Bureau, Current Population Reports. Accessed February 12, 2003 (www.census.gov).

File, Thom, and Camille Ryan. 2014. "Computer and Internet Use in the United States: 2013." U.S. Census Bureau, American Community Survey Reports, November. Accessed August 22, 2015 (census.gov).

File, Thom, and Sarah Crissey. 2010. "Voting and Registration in the Election of November 2008: Population Characteristics." U.S. Census Bureau, Current Population Reports. Accessed May 15, 2010 (www.census.gov).

File, Thom. 2013. "The Diversifying Electorate— Voting Rates by Race and Hispanic Origin in 2012 (and Other Recent Elections)." U.S. Census Bureau, May. Accessed April 20, 2014 (census.gov).

File, Thom. 2015. "Who Votes? Congressional Elections and the American Electorate: 1978–2014." U.S. Census Bureau, July. Accessed October 7, 2015 (census.gov).

"Financial Crime." 2015. *The Economist*, May 23, 61.

Fingerman, Karen L., Yen-Pi Cheng, Eric D. Wesselmann, Steven Zarit, Frank Furstenberg, and Kira S. Birditt. 2012. "Helicopter Parents and Landing Pad Kids: Intense Parental Support of Grown Children." *Journal of Marriage and Family* 74 (August): 880–896.

Finkelhor, David, Heather Turner, Sherry Hamby, and Richard Ormrod. 2011. "Polyvictimization: Children's Exposure to Multiple Types of Violence, Crime, and Abuse." Office of Justice Programs, October. Accessed March 7, 2012 (www.ojp.usdoj .gov).

Finkelstein, Eric A., et al. 2012. "Obesity and Severe Obesity Forecasts Through 2030." *American Journal of Preventive Medicine* 42 (6). Accessed May 14, 2012 (www.ajponline .org).

Finn, Jeremy D. 2006. *The Adult Lives of At-Risk Students: The Roles of Attainment and Engagement in High School.* Washington, DC: U.S. Department of Education.

Fischman, Josh. 2009. "Global Warming Before Smokestacks." *Chronicle Review,* November 6, B11–B12.

Flavin, Jeanne. 2001. "Feminism for the Mainstream Criminologist: An Invitation." *Journal of Criminal Justice* 29 (July/August): 271–285.

Fleegler, Eric W., et al. 2013. "Firearm Legislation and Firearm-Related Fatalities in the United State." *JAMA Internal Medicine,* March 6. Accessed January 15, 2014 (www.archinte .jamanetwork.com).

Fleishman, Jeffrey, and Amro Hassan. 2009. "Gadget to Help Women Feign Virginity Angers Many in Egypt." *Los Angeles Times,* October 7. Accessed October 28, 2009 (www .latimes.com).

Fleming, Charles. 2014. "Ford Greenlights Recycling." *Baltimore Sun,* July 5, 8.

Fleming, John. 2015. "Millennials Most Trusting on Safety of Personal Information." Gallup, May 11. Accessed August 14, 2015 (gallup.com).

Fletcher, Douglass Scott, and Ian M. Taplin. 2002. *Understanding Organizational Evolution: Its Impact on Management and Performance.* Westport, CT: Quorum Books.

Fletcher, Michael A., and Steven Mufson. 2014. "Why Did GM Take So Long to Respond to Deadly Defect? Corporate Culture May Hold Answer." *Washington Post,* March 30. Accessed April 12, 2014 (www.washingtonpost.com).

Florida, Richard. 2011. "The Geography of Hate." *The Atlantic,* May 11. Accessed August 27, 2015 (theatlantic.com).

Flower, Shawn M. 2010. "Gender-Responsive Strategies for Women Offenders." National Institute of Corrections, November. Accessed August 20, 2011 (www.nicic.gov).

Fogg, Piper. 2005. "Don't Stand So Close to Me." *Chronicle of Higher Education,* April 29, A10–A12.

Follman, Mark. 2015. "Trigger Warnings." *Mother Jones,* November/December, 22–29.

Fomby, Paula, and Angela Estacion. 2011. "Cohabitation and Children's Externalizing Behavior in Low-Income Latino Families." *Journal of Marriage and Family* 73 (February): 46–66.

Food and Agriculture Organization of the United Nations. 2011. "The State of the World's Land and Water Resources for Food and Agriculture, Managing Systems at Risk: Summary Report." Accessed June 20, 2012 (www.fao.org).

Food and Water Watch. 2013. "Take Back the Tap: Bottled Water Wastes Resources and Money." July. Accessed September 19, 2014 (www .foodandwaterwatch.org).

"Food Companies Propose Cutting Back on Junk Food Marketing Aimed at Children." 2011. *Washington Post,* July 14. Accessed July 20, (www.washingtonpost.com).

Ford, Peter. 2011. "Why Japan Will Rebound." *Christian Science Monitor,* March 28, 16–29.

Ford, Peter. 2014. "Did China Really Soften?" *Christian Science Monitor Weekly,* January 13, 11–12.

"Formaldehyde." 2012. Environmental Protection Agency. Accessed September 20, 2014 (www .epa.gov).

Foroohar, Rana. 2011. "Whatever Happened to Upward Mobility?" *Time,* November 14, 27–34.

Fowler, Tom. 2013. "Former Enron CEO's Sentence Cut to 14 Years." *Wall Street Journal,* June 22–23, A3.

Fox, Justin. 2013. "Which Is Worse: Airline Monopolies or Airline Competition?" Harvard Business Review Blog Network, August 15. Accessed April 12, 2014 (www .blogs.hbr.org).

Fox, Susannah, and Lee Rainie. 2014. "The Web at 25 in the U.S." Pew Research Center, February 27. Accessed October 8, 2014 (www.pewresearch.org).

Fox, Susannah, Maeve Duggan, and Kristen Purcell. 2013. "Family Caregivers Are Wired for Health." Pew Research Center, June 20. Accessed July 10, 2013 (www.pewinternet .org).

Fraire, John. 2014. "Why Your College Should Dump the SAT." *Chronicle of Higher Education,* May 2, A44.

Freedom House. 2015. "Freedom in the World 2015." Accessed October 12, 2015 (freedom-house.org).

Freese, Jeremy. 2008. "Genetics and the Social Science Explanation of Individual Outcomes." *American Journal of Sociology* 114 (Suppl.): S1–S35.

Fremstad, Shawn. 2012. "The Poverty Rate Is Higher Than the Federal Government Says It Is." Center for Economic and Policy Research, September 21. Accessed February 6, 2014 (www.cepr.net).

Frey, Carl B., and Michael A. Osborne. 2013. "The Future of Employment: How Susceptible Are Jobs to Computerization?" University of Oxford, September 17. Accessed April 12, 2014 (www.oxfordmartin.ox.ac.uk).

Frey, William H. 2014. "New Projections Point to a Majority Nation in 2044." Brookings, December 12. Accessed October 4, 2015 (brookings.edu).

Frey, William H. 2014. "The Suburbs: Not Just for White People Anymore." *New Republic,* November 14. Accessed October 23, 2015 (newrepublic.com).

Frey, William H. 2014. "Will This Be the Decade of Big City Growth?" Brookings Institution, May 23. Accessed September 19, 2014 (brookings.edu).

Frieden, Thomas R. 2011. "Foreword." Centers for Disease Control and Prevention. *MMWR* 60 (Suppl): 1–2.

Friedman, Howard S. 2012. "American Voter Turnout Lower Than Other Wealthy Countries." *Huffington Post,* July 10. Accessed April 20, 2014 (www .huffingtonpost.com).

Friedman, Jaclyn. 2011. *What You Really Really Want: The Smart Girl's Shame-Free Guide to Sex and Safety.* Berkeley, CA: Seal Press.

Friedrichs, David O. 2009. *Trusted Criminals: White Collar Crime in Contemporary Society,* 4th edition. Belmont, CA: Wadsworth Publishing.

Frohlich, Thomas C., Michael B. Sauter, and Sam Stebbins. 2015. "10 States with the Most Hate Groups." July 9. Accessed August 25, 2015 (24/7wallst.com).

Frohwirth, Lon, Ann M. Moore, and Renata Maniaci. 2013. "Perceptions of Susceptibility to Pregnancy Among Women Obtaining Abortions." *Social Science & Medicine* 99 (December): 18–26.

"From Alpha to Omega." 2015. *The Economist,* August 15, 61.

Frosch, Dan. 2007. "18 Air Force Cadets Exit over Cheating." *New York Times,* May 2, 18.

Frueh, Sara. 2014. "Career Choices of Female Engineers." The National Academies Press. Accessed November 17, 2014 (www.nap .edu).

Fry, Richard, and D'Vera Cohn. 2011. "Living Together: The Economics of Cohabitation." Pew Research Center, June 27. Accessed March 1, 2012 (www.pewsocialtrends.org).

Fry, Richard, and Jeffrey S. Passel. 2014. "In Post-Recession Era, Young Adults Drive Continuing Rise in Multi-Generational Living." Pew Research Center, July 17. Accessed July 25, 2014 (www.pewresearch .org).

Fry, Richard, and Kim Parker. 2015. "More Millennials Living With Family Despite Improved Job Market." July 29. Accessed August 6, 2015 (pewresearch.org).

Fry, Richard, and Paul Taylor. 2012. "The Rise of Residential Segregation by Income." Pew Research Center, Social & Demographic Trends, August 1. Accessed September 19, 2014 (www.pewsocialtrends.org).

Fry, Richard. 2012. "No Reversal in Decline of Marriage." Pew Research Center, November 20. Accessed February 9, 2013 (www .pewsocialtrends.org).

Fry, Richard. 2013. "Young Adults After the Recession: Fewer Homes, Fewer Cars, Less Debt." Pew Research Center, February 21. Accessed November 20, 2013 (www .pewresearch.org).

Fryer, Roland G., Jr., Devah Pager, and Jorg L. Spenkuch. 2011. "Racial Disparities in Job Finding and Offered Wages." National Bureau of Economic Research, September. Accessed December 7, 2011 (www.nber .org).

Fuentes-Nieva, Ricardo, and Nick Galasso. 2014. *Working for the Few.* Oxfam International, January. Accessed February 10, 2014 (www .oxfam.org).

Fung, Brian. 2014. "So, This Exists: A Working Car Has Been 3D-Printed Out of Carbon Fiber Plastic." *Washington Post,* September 19. Accessed October 8, 2014 (www .washingtonpost.com).

Funk, Cary, and Lee Raine. 2015. "Americans, Politics, and Science Issues." Pew Research Center, July 1. Accessed October 23, 2015 (pewresearch.org).

Funk, Cary, and Lee Rainie. 2015. "Public and Scientists' Views on Science and Society." Pew Research Center, January 29. Accessed July 22, 2015 (pewresearch.org).

Funk, Cary, Luis Lugo, and Alan Cooperman. 2012. "Asian Americans: A Mosaic of Faiths." Pew Research Center, Pew Forum on Religion & Public Life, July 19. Accessed June 21, 2014 (www.pewforum.org).

Gallup Editors. 2013. "Most Americans Practice Charitable Giving, Volunteerism." Gallup, December. Accessed December 19, 2013 (www.gallup.com).

Gallup. 2013. "State of the American Workplace: Employee Engagement Insight for U.S. Business Leaders." Accessed April 12, 2014 (www.gallup.com).

Gallup. 2014. "U.S. Employment (Weekly)." Accessed February 7, 2014 (www.gallup .com).

Galperin, Andrew, et al. 2013. "Sexual Regret: Evidence for Evolved Sex Differences." *Archives of Sexual Behavior* 42 (October): 1145–1161.

Galston, William A. 2013. "The Eroding American Middle Class." *Wall Street Journal*, November 13, A15.

Gambrell, Jon. 2015. "Stampede at Hajj in Saudi Arabia Reportedly Killed at Least 2,111." *Huffington Post*, October 20. Accessed October 25, 2015 (huffingtonpost.com).

Gamson, William. 1990. *The Strategy of Social Protest*, 2nd edition. Belmont, CA: Wadsworth.

Gamson, William. 1999. *Popular Culture & High Culture: An Analysis and Evaluation of Taste*. Revised and updated edition. New York: Basic Books.

Gamson, William. 2005. "Wishes for the Discipline's Future." *Chronicle Review*, August 12, B9.

Gans, Herbert J. 1971. "The Uses of Poverty: The Poor Pay All." *Social Policy*, July/August, 78–81.

Gardner, Marilyn. 2008. "Happiness Is a Warm 'Thank You.'" *Christian Science Monitor*, January 28, 13, 16.

Garfinkel, Harold. 1956. "Conditions of Successful Degradation Ceremonies." *American Journal of Sociology* 61 (March): 420–424.

Garfinkel, Harold. 1967. *Studies in Ethnomethodology* Englewood Cliffs, NJ: Prentice-Hall.

Gates, Gary J. 2011. "How Many People Are Lesbian, Gay, Bisexual, and Transgender?" Williams Institute, April. Accessed March 2, 2014 (www.williamsinstitute.law.ucla.edu).

Gates, Gary J. 2013. "LGBT Parenting in the United States." Williams Institute, February. Accessed May 16, 2014 (www.williamsinstitute.law.ucla.edu).

Gavin, Lorrie, et al. 2013. "Vital Signs: Repeat Births among Teens—United States, 2007–2010." *MMWR* 62 (April 5): 249–255.

Gavrilos, Dina. 2006. "U.S. News Magazine Coverage of Latinos: 2006 Report." National Association of Hispanic Journalists, June. Accessed April 2, 2007 (www.nahj.org).

Geertz, Clifford. 1966. "Religion as a Cultural System." Pp. 1–46 in *Anthropological Approaches to the Study of Religion*, edited by Michael Banton. London: Tavistock.

Geier, Ben. 2014. "Using 3-D Printing to Make Jet Engines." *Fortune*, December 1, 76.

Geller, Adam. 2014. "This Is How Americans Are Grappling with Income Inequality." *Huffington Post*, February 2. Accessed February 4, 2014 (www.huffingtonpost.com).

"Gender Equality Universally Embraced, but Inequalities Acknowledged." 2010. Pew Research Center, Global Attitudes Project, July 1. Accessed November 1, 2011 (www.pewglobal.org).

Geoghegan, Tom. 2013. "Why Do So Many Americans Live in Mobile Homes?" *BBC News Magazine*, September 23. Accessed January 12, 2014 (www.bbc.co.uk).

Gerth, H. H., and C. Wright Mills, eds. 1946. *Max Weber: Essays in Sociology*. New York: Oxford University Press.

Gewertz, Catherine. 2009. "Do Men Deserve a Break in College Admissions?" *Education Week* online, December 9. Accessed December 13, 2010 (www.blogs.edweek.org).

Ghandnoosh, Nazgol. 2015. "Black Lives Matter: Eliminating Racial Inequity in the Criminal Justice System." The Sentencing Project. Accessed September 2, 2015 (www.sentencingproject.org).

Gibson, David. 2012. "Vatican Orders Crackdown on American Nuns." *USA Today*, April 18. Accessed April 20, 2012 (www.usatoday.com).

Gibson, Megan. 2012. "Celebrities Offering Scientific 'Facts'? Just Say No." *Time*, January 3. Accessed January 24, 2012 (www.newsfeed.time.com).

Gilbert, Dennis. 2011. *The American Class Structure in an Age of Growing Inequality*, 8th edition. Thousand Oaks, CA: Pine Forge Press.

Gilles, Kate, and Charlotte Feldman-Jacobs. 2012. "When Technology and Tradition Collide: From Gender Bias to Sex Selection." Population Reference Bureau, September. Accessed September 19, 2014 (www.prb.org).

Gillespie, Michael Allen. 2014. "Grade Degradation." *Chronicle Review*, January 24, B2.

Gilpin, Lindsey. 2014. "The Dark Side of 3D Printing: 10 Things to Watch." TechRepublic, March 5. Accessed October 8, 2014 (www.techrepublic.com).

Gilson, Dave. 2014. "Don't Tread on Me." *Mother Jones*, January/February, 28–33.

Gimpelson, Vladimir, and Daniel Treisman. 2015. "Misperceiving Inequality." National Bureau of Economic Research, May. Accessed September 16, 2015 (nber.org).

Giugni, Marco, Doug McAdam, and Charles Tilley, eds. 1999. *How Social Movements Matter*. Minneapolis: University of Minnesota Press.

GLAAD. 2013. "Where We Are on TV." Accessed March 2, 2014 (www.glaad.org).

Glassner, Barry. 2010. "Still Fearful After All These Years." *Chronicle Review*, January 22, B11–B12.

Glauber, Bill. 2011. "Do Monarchies Still Matter?" *Christian Science Monitor*, February 21, 27–31.

Glaze, Lauren E., and Danielle Kaeble. 2014. "Correctional Populations in the United States, 2013." Bureau of Justice Statistics, December. Accessed September 2, 2015 (bjs.gov).

Gleiber, Michael A. 2014. "The High Price Our Bodies Pay for High Fashion." *Huffington Post*, October 26. Accessed December 5, 2014 (www.huffingtonpost.com).

Glenn, David. 2011. "For Business Majors, Easy Does It." *Chronicle of Higher Education*, April 22, A1, A3–A5.

Glenn, Norval D., Jeremy Uecker, and Robert W. B. Love, Jr. 2010. "Later First Marriage and Marital Success." *Social Science Research* 39 (September): 787–800.

Glionna, John M. 2009. "South Korean Kids Get a Taste of Boot Camp." *Los Angeles Times*, August 22, A1.

"Global Christianity: A Report on the Size and Distribution of the World's Christian Population." 2011. Pew Research Center, December 1. Accessed April 8, 2012 (www.pewresearch.org).

"Global Views on Morality." 2014. Pew Research Center. Accessed September 29, 2015 (www.pewglobal.org).

Global WASH Fast Facts. 2013. "Information on Water, Sanitation, and Hygiene." Centers for Disease Control and Prevention, November 8. Accessed September 19, 2014 (www.cdc.gov).

Glynn, Sarah Jane, and Audrey Powers. 2012. "The Top 10 Facts About the Wage Gap." Center for American Progress, April 16. Accessed November 17, 2014 (www.americanprogress.org).

Goble, Priscilla, Carol L. Martin, Laura D. Hanish, and Richard A. Fabes. 2012." Children's Gender-Typed Activity Choices Across Preschool Social Contexts." *Sex Roles* 67 (October): 435–451.

Godofsky, Jessica, Cliff Zukin, and Carl Van Horn. 2011. "Unfulfilled Expectations: Recent College Graduates Struggle in a Troubled Economy." *Work Trends*, May. Accessed June 5, 2011 (www.heldrich.rutgers.edu).

Goel, Vindu. 2014. "With New Ad Platform, Facebook Opens Gates to Its Vault of User Data." *New York Times*, September 29, B7.

Goffman, Erving. 1959. *The Presentation of Self in Everyday Life*. New York: Doubleday Anchor Books.

Goffman, Erving. 1961. *Asylums: Essays on the Social Situation of Mental Patients and Other Inmates*. Garden City, NY: Anchor Books.

Goffman, Erving. 1963. *Stigma: Notes on the Management of Spoiled Identity*. Englewood Cliffs, NJ: Prentice-Hall.

Goffman, Erving. 1967. *Interaction Ritual: Essays on Face-to-Face Behavior*. New York: Anchor Books.

Goffman, Erving. 1969. *Strategic Interaction*. Philadelphia: University of Pennsylvania Press.

Gold, Matea, Rosalind S. Helderman, and Anu Narayanswamy. 2015. "Hillary Clinton Made $3.2 Million From Tech Sector. Now She's Hitting It Up for Campaign Dollars." *Washington Post*, May 18. Accessed May 19, 2015 (washingpost.com).

Goldberg, Michelle. 2015. "Feminist Writers Are So Besieged by Online Abuse That Some Have Begun to Retire." *Washington Post*, February 20. Accessed September 30, 2015 (washingtonpost.com).

Golden, Daniel. 2006. *The Price of Admission: How America's Ruling Class Buys Its Way into Elite Colleges—and Who Gets Left Outside the Gates*. New York: Crown.

Goldin, Claudia, Lawrence F. Katz, and Ilyana Kuziemko. 2006. "The Homecoming of American College Women: The Reversal of the College Gender Gap." National Bureau of Economic Research. Accessed July 25, 2006 (www.nber.org).

Goldin, Claudia. 2014. "A Grand Gender Convergence: Its Last Chapter." *American Economic Review*, 104 (4): 1091–1119.

Goldring, Rebecca, Lucinda Gray, and Amy Bitterman. 2013. "Characteristics of Public and Private Elementary and Secondary School Teachers in the United States: Results From the 2011–12 Schools and Staffing Survey." National Center for Education Statistics, August. Accessed November 17, 2014 (www.nces.ed.gov).

Goodale, Gloria. 2015. "'Sharing Economy': Helpful or Selfish?" *Christian Science Monitor Weekly*, March 30, 21–23.

Goode, Erica. 2013. "U.S. Prison Populations Decline, Reflecting New Approach to Crime." *New York Times*, July 26, A11.

Goodnough, Abby. 2010. "Making It Clear That a Clear Parking Space Isn't." *New York Times*, December 28, A10.

Goodstein, Laurie. 2014. "Presbyterians Vote to Allow Same-Sex Marriages." *New York Times*, June 20, A11.

Gopnik, Alison, Andrew N. Meltzoff, and Patricia K. Kuhl. 2001. *The Scientist in the Crib: What Early Learning Tells Us About the Mind*. New York: Perennial.

Gottdiener, Mark, and Ray Hutchison. 2000. *The New Urban Sociology*, 2nd edition. New York: McGraw-Hill.

Gottlieb, Mark. 2011. "Attention to Duty." *Christian Science Monitor*, June 6, 44.

Gottman, John M. 1994. *What Predicts Divorce? The Relationships Between Marital Processes and Marital Outcome*. Hillsdale, NJ: Lawrence Erlbaum Associates.

Goudreau, Jenna. 2012. "A New Obstacle for Professional Women: The Glass Escalator." *Forbes*, May 21. Accessed December 18, 2013 (www.forbes.com).

Gouldner, Alvin W. 1962. "Anti-Minotaur: The Myth of a Value-Free Sociology." *Social Problems* 9 (Winter): 199–212.

"Government Efficiency and Effectiveness." 2015. GAO Highlights, GAO-15-440T, March 4. Accessed August 22, 2015 (gao.gov).

"Government Resolves to Start Making Sense Under New Law Forbidding Federal Gibberish." 2011. *Washington Post*, May 20. Accessed May 21, 2011 (www.washingtonpost.com).

Governors Highway Safety Association. 2014. "Distracted Driving Laws." Accessed January 12, 2014 (www.ghsa.org).

Goyette, Braden. 2013. "Oprah Talks to Swiss Newspaper About Alleged Racist Encounter." *Huffington Post*, August 14. Accessed March 22, 2014 (www.huffingtonpost.com).

Graham, Carol. 2015. "The High Costs of Being Poor in America: Stress, Pain, and Worry." Brookings Institution, February 19. Accessed September 31, 2015 (brookings.edu).

Graif, Corina, and Robert J. Sampson. 2009. "Spatial Heterogeneity in the Effects of Immigration and Diversity on Neighborhood Homicide Rates." *Homicide Studies* 13 (August): 242–260.

Grall, Timothy. 2013. "Custodial Mothers and Fathers and Their Child Support: 2011." U.S. Census Bureau, October. Accessed May 16, 2014 (www.census.gov).

Grandjean, Philippe, and Philip J. Landrigan. 2014. "Neurobehavioral Effects of Developmental Toxicity." *The Lancet Neurology* 13 (March): 330–338.

Granville, Kevin. 2015. "9 Recent Cyberattacks Against Big Business." *New York Times*, February 5. Accessed August 22, 2015 (newyorktimes.com).

Grauerholz, Liz, and Sharon Bouma-Holtrop. 2003. "Exploring Critical Sociological Thinking." *Teaching Sociology* 31 (October): 485–496.

Graves, Joseph L., Jr. 2001. *The Emperor's New Clothes: Biological Theories of Race at the Millennium*. New Brunswick, NJ: Rutgers University Press.

Gray, Paul S., John B. Williamson, David R. Karp, and John R. Dalphin. 2007. *The Research Imagination: An Introduction to Qualitative and Quantitative Methods*. New York: Cambridge University Press.

Grazian, David. 2008. *On the Make: The Hustle of Urban Nightlife*. Chicago: University of Chicago Press.

Greeley, Brendan. 2013. "How Inequality Became a Household Word." *Bloomberg Businessweek*, December 16–22, 16–17.

Green, Jeff. 2011. "The Silencing of Sexual Harassment." *Bloomberg Businessweek*, November 21–28.

Greenberg, Julie, Arthur McKee, and Kate Walsh. 2013. "Teacher Prep Review: A Review of the Nation's Teacher Preparation Programs 2013." National Council on Teacher Quality. Accessed June 21, 2014 (www.nctq.org).

Greenfield, Rebecca. 2015. "You Take the Vacation. Your Boss Pays the Bill." *Bloomberg Businessweek*, May 18–24, 22–23.

Greenwood, Jeremy, Nezih Guner, Georgi Kocharkov, and Cesar Santos. 2014. "Marry Your Like: Assortative Mating and Income Inequality." National Bureau of Economic Research, January. Accessed February 6, 2014 (nber.org).

Grim, Brian J., and Alan Cooperman. 2014. "Religious Hostilities Reach Six-Year High." Pew Research Center, January 14. Accessed June 21, 2014 (www.pewresearch.org).

Grind, Kirsten. 2013. "Mother, Can You Spare a Room?" *Wall Street Journal*, May 4–5, B1, B10.

Gross, Rita M. 1996. *Feminism and Religion: An Introduction*. Boston: Beacon Press.

Grossman, David. 2011. "Fliers Come to Fisticuffs Over Reclined Seat." *USA Today*, June 6. Accessed June 7, 2011 (www.usatoday.com).

Groves, Robert, and Frank Vitrano. 2011. "The U.S. Decennial Census and the American Community Survey: Looking Back and Looking Ahead." Population Reference Bureau, April. Accessed November 9, 2011 (www.prb.org).

"Growing Number of Americans Say Obama Is a Muslim." 2010. Pew Research Center, August 19. Accessed April 8, 2012 (www.pewresearch.org).

Guallar, Eliseo, Saverio Stranges, Cynthia Mulrow, Lawrence J. Appel, and Edgar R. Miller III. 2013. "Enough Is Enough: Stop Wasting Money on Vitamin and Mineral Supplements." *Annals of Internal Medicine* 159 (12): 850–851.

Guarino, Mark. 2013. "Auto Jobs Are Back, Too, but at Lower Wages." *Christian Science Monitor Weekly*, May 13, 22–23.

Guerra, Nancy G., Carly B. Dierkhising, and Pedro R. Payne. 2013. "How Should We Identify and Intervene With Youth at Risk of Joining Gangs? A Developmental Approach for Children Ages 0–12." Pp. 63–73 in *Changing Course: Preventing Gang Membership*, edited by Thomas R. Simon, Nancy M. Witter, and Reshma R. Mahendra. National Institute of Justice and Centers for Disease Control and Prevention. Accessed December 18, 2013 (www.nij.gov).

"Gun Laws: A Shot and a Beer." 2014. *The Economist*, May 3, 28.

Gunders, Dana. 2012. "Wasted: How America Is Losing Up to 40 Percent of Its Food from Farm to Fork to Landfill." Natural Resources Defense Council, August. Accessed November 10, 2013 (www.nrdc.org).

Gurkoff, Joe, and Anna Ranieri. 2012. *How Can I Help You? What You Can (and Can't) Do to Counsel a Friend, Colleague or Family Member with a Problem*. North Charleston, SC: CreateSpace Independent Publishing Platform.

Gurney, Joan M., and Kathleen T. Tierney. 1982. "Relative Deprivation and Social Movements: A Critical Look at Twenty Years of Theory and Research." *Sociological Quarterly* 23 (Winter): 33–47.

Guterman, Stanley S. 1969. "In Defense of Wirth's 'Urbanism as a Way of Life.'" *American Journal of Sociology* 74 (March): 492–499.

Guttmacher Institute. 2013. "U.S. Women Who Have Abortions. Infographics. Accessed February 11, 2012 (www.guttmacher.org).

Guttmacher Institute. 2014. "Facts on Induced Abortion in the United States." February. Accessed March 2, 2014 (guttmacher.org).

Guttmacher Institute. 2015. "Laws Affecting Reproductive Health and Rights: State Trends at Midyear, 2015." Accessed September 29, 2015 (guttmacher.org).

Guynn, Jessica. 2015. "Facebook's Diversity Still Lagging." *USA Today*, June 26, B1.

Haas, Ann P., Phillip L. Rodgers, and Jody L. Herman. 2014. "Suicide Attempts Among Transgender and Gender Non-Conforming Adults." Williams Institute, January. Accessed March 2, 2014 (www.williamsinstitute.law.ucla.edu).

Haas, Steven A., and David R. Schaefer. 2014. "With a Little Help from My Friends? Asymmetrical Social Influence on Adolescent Smoking Initiation and Cessation." *Journal of Health and Social Behavior* 55 (June): 126–143.

Hacker, Jacob S. 2002. *The Divided Welfare State: The Battle over Public and Private Social Benefits in the United States*. New York: Cambridge University Press.

Hacker, Jacob S., and Paul Pierson. 2010. *Winner-Take-All Politics: How Washington Made the Rich Richer and Turned Its Back on the Middle Class*. New York: Simon and Schuster.

Hagan, Frank E. 2008. *Introduction to Criminology: Theories, Methods, and Criminal Behavior*, 6th edition. Thousand Oaks, CA: Sage.

Hagerty, James R. 2013. "Help Wanted. A Lot of It." *Wall Street Journal*, June 11, R7.

Hakim, Danny. 2015. "Banned Abroad: Differences in Rules for Pesticides Point Up U.S.-Europe Trade Bumps." *New York Times*, February 24, B1, B4.

Hales, Emily. 2014. "Incentives for Having Kids: Solution for Falling Birth Rates?" *Deseret News*, June 26. Accessed September 21 (www.newsok.com).

"Half of Young Children in the U.S. Are Read to at Least Once a Day, Census Bureau Reports." 2011. U.S. Census Bureau, August 11. Accessed November 18, 2013 (www.census.gov/newsroom).

Hall, Aron J., Mary E. Wikswo, Kimberly Pringle, L. Hannah Gould, and Umesh D. Parashar. 2014. "Vital Signs: Foodborne Norovirus Outbreaks—United States, 2009–2012." *MMWR* 63 (22): 491–495.

Hall, Edward T. 1959. *The Silent Language*. New York: Doubleday.

Hall, Edward T. 1966. *The Hidden Dimension*. Garden City, NY: Doubleday.

Hallman, Ben, and Shane Shifflett. 2014. "When Hospices Mistreat the Dying, They Almost Never Get Punished." *Huffington Post*, December 30. Accessed October 16, 2015 (projects.huffingtonpost.com).

Halperin, David M. 2012. "How to Be Gay." *Chronicle Review*, September 7, B13–B17.

Halsey, Ashley, III. 2010. "Study: Older People Are Driving More, Having Fewer Accidents." *Washington Post*, June 22, A10.

Halsey, Ashley, III. 2011. "Seat in Lap Leads to Scuffle in the Air." *Washington Post*, June 1, A1.

Hamburger, Tom, and Matea Gold. 2014. "Google, Once Disdainful of Lobbying, Now a Master of Washington Influence." *Washington Post*, April 12. Accessed April 20, 2014 (www.washingtonpost.com).

Hamby, Sherry, and David Finkelhor. 2001. "Choosing and Using Child Victimization Questionnaires." *OJJDP Juvenile Justice Bulletin*. Washington, DC: U.S. Department of Justice.

Hamby, Sherry, David Finkelhor, Heather Turner, and Richard Ormrod. 2011. "Children's Exposure to Intimate Partner Violence and Other Family Violence." Office of Justice Programs, October. Accessed May 20, 2014 (www.ojp.usdoj.gov).

Hamilton, Malcolm B. 1995. *The Sociology of Religion: Theoretical and Comparative Perspectives*. New York: Routledge.

Hamilton, Malcolm B. 2001. *The Sociology of Religion*, 2nd edition. New York: Routledge.

Hamm, Steve. 2007. "The Trouble with India." *Business Week*, March 19, 49–58.

Hammadi, Saad. 2013. "Making Asia's Factories Safer." *Christian Science Monitor Weekly*, November 11, 16.

Hamre, Bridget K., and Robert C. Pianta. 2001. "Early Teacher–Child Relationships and the Trajectory of Children's School Outcomes Through Eighth Grade." *Child Development* 72 (March/April): 625–638.

Handbury, Jessie, Ilya Rahkovsky, and Molly Schnell. 2015. "What Drives Nutritional Disparities? Retail Access and Food Purchases Across the Socioeconomic Spectrum." National Bureau of Economic Research, April. Accessed October 16, 2015 (nber.org).

Hanes, Stephanie. 2010. "Texting Bans Ineffective?" *Christian Science Monitor*, October 11, 20.

Hanes, Stephanie. 2011. "Pretty in Pink?" *Christian Science Monitor*, September 26, 26–31.

Haney, Craig, Curtis Banks, and Philip Zimbardo. 1973. "Interpersonal Dynamics in a Simulated Prison." *International Journal of Criminology and Psychology* 1: 69–97.

Hansen, Lawrence A. 2010. "Noxious Groupthink." *Chronicle of Higher Education*, November 12, B6–B8.

Hanser, Amy, and Zachary Hyde. 2014. "Foodies Remaking Cities." *Contexts* 13 (Summer): 44–49.

Hanushek, Eric A., and Ludger Woessman. 2005. "Does Educational Tracking Affect Performance and Inequality? Differences-in-Differences Evidence Across Countries." National Bureau of Economic Research, February. Accessed July 25, 2007 (www.nber.org).

Hardoon, Deborah. 2015. "Wealth: Having It All and Wanting More." Oxfam Issue Briefing, January. Accessed September 13, 2014 (oxfam.org).

Hargrove, Thomas. 2011. "Disappearing Jobs Illustrate U.S. Manufacturing Revolution." Knox News, November 13. Accessed January 9, 2012 (www.knoxnews.com).

Harlow, Harry E., and Margaret K. Harlow. 1962. "Social Deprivation in Monkeys." *Scientific American* 206 (November): 137–146.

Harnish, Verne. 2014. "5 Ways to Tone Your Operations." *Fortune*, October 27, 46.

Harrell, Erika, and Lynn Langton. 2013. "Victims of Identity Theft, 2012." U.S. Department of Justice, Office of Justice Programs, December. Accessed January 12, 2014 (www.ojp.usdoj.gov).

Harrell, Erika, Lynn Langton, Marcus Berzofsky, Lance Couzens, and Hope Smiley-McDonald. 2014. "Household Poverty and Nonfatal Violent Victimization, 2008–2012." Bureau of Justice Statistics, November. Accessed September 11, 2015 (bjs.gov).

Harris, Chauncey D. 1997. "'The Nature of Cities' and Urban Geography in the Last Century." *Urban Geography* 18 (1): 15–35.

Harris, Chauncey D., and Edward L. Ullman. 1945. "The Nature of Cities." *Annals* 242: 7–17.

Harris, Gardiner. 2011. "When the Nurse Wants to Be Called 'Doctor.'" *New York Times*, June 6, A23.

Harris, Gardiner. 2013. "India's New Focus on Rape Shows Only the Surface of Women's Perils." *New York Times*, January 12. Accessed March 4, 2014 (www.nytimes.com).

Harris, Jennifer L., et al. 2013. "Fast Food FACTS 2013: Measuring Progress in Nutrition and Marketing to Children and Teens." Yale Rudd Center for Food Policy & Obesity. Accessed November 10, 2013 (www.fastfoodmarketing.org).

Harter, Jim, and Sangeeta Agrawal. 2013. "'Engaged' Workers Would Keep Jobs upon Winning Lottery." Gallup, December 20. Accessed April 12, 2014 (www.gallup.com).

Harter, Jim, and Sangeeta Agrawal. 2014. "Many Baby Boomers Reluctant to Retire." Gallup, January 20. Accessed May 16, 2014 (www.gallup.com).

Hartford Institute for Religion Research. 2012. "Fast Facts About American Religion." Accessed July 1, 2014 (www.hirr.hartsem.edu).

Harvard School of Public Health. 2014. "Poll Finds Many in U.S. Lack Knowledge About Ebola and Its Transmission." August 21. Accessed October 8, 2014 (www.hsph.harvard.edu).

Harvey, Tom. 2015. "Citing High Level of Fraud in Utah, Lawmakers Pass White Collar Crime Registry." *The Salt Lake Tribune*, March 11. Accessed August 31, 2015 (sltrib.com).

Haskins, Ron. 2008. "Immigration: Wages, Education, and Mobility." Pp. 81–90 in *Getting Ahead or Losing Ground: Economic Mobility in America*, edited by Julia B. Isaacs, Isabel V. Sawhill, and Ron Haskins. Washington DC: Brookings Institution.

Haskins, Ron. 2015. "$2 a Day: A More Complete Picture." Brookings Institution, September 14. Accessed September 31, 2015 (brookings.edu).

Hausmann, Ricardo, Laura D. Tyson, Yasmina Bekhouche, and Saadia Zahidi. 2014. "The Global Gender Gap Report 2014." World Economic Forum. Accessed December 2, 2014 (www.weforum.org).

Hawai'i Free Press. 2012. "Census Bureau: Hawaii 7th Highest Poverty Rate in US." November 15. Accessed February 14, 2014 (www.hawaiifreepress.com).

Hayden, Dolores. 2002. *Redesigning the American Dream: Gender, Housing, and Family Life*. New York: W. W. Norton

He, Wan, and Luke J. Larsen. 2014. "Older Americans with a Disability: 2008–2012." U.S. Census Bureau, December. Accessed December 5, 2014 (www.census.gov).

"The Health Paradox." 2013. *The Economist*, May 11, 27–28.

Health Resources and Services Administration. 2010. "Assuring Access to Essential Health Care." Accessed July 15, 2011 (www.plainlanguage.gov).

"Health-Care Fraud: The $272 Billion Swindle." 2014. *The Economist*, May 31, 26–27.

Healthcare-NOW! 2014. "Health Insurance CEO Pay Skyrockets in 2013." May 5. Accessed July 26, 2014 (www.healthcare-now.org).

Healy, Melissa. 2013. "1st Stem Cells Made by Cloning Embryos." *Baltimore Sun*, May 16, 8.

Heath, Chip. 2005. "Introduction." Pp. 81–85 in *Rumor Mills: The Social Impact of Rumor and Legend*, edited by Gary Alan Fine, Veronique Campion-Vincent, and Chip Heath. New Brunswick, NJ: Transaction Publishers.

Heath, Jennifer. 2008. *The Veil: Women Writers on Its History, Lore, and Politics*. Berkeley: University of California Press.

Hechter, Michael, and Karl-Dieter Opp. 2001. "Introduction." Pp. xi–xx in *Social Norms*, edited by Michael Hechter and Karl-Dieter Opp. New York: Russell Sage Foundation.

Hedegaard, Holly, and Li-Hui Chen. 2014. "Rates of Drug Deaths Involving Heroin, by Selected Age and Racial/Ethnic Groups—United States, 2002 and 2011." *MMWR* 63 (July 11): 595.

Hedges, Chris. 2011. "Our Public Schools Are Churning Out Drones for the Corporate State." AlterNet, April 11. Accessed April 14, 2011 (www.alternet.org).

Heffer, Simon. 2010. "America Is the Acceptable Face of Cultural Imperialism." *Telegraph Media Group*, July 24. Accessed November 10, 2013 (telegraph.co.uk).

Hegewisch, Ariane, and Heidi Hartmann. 2015. "The Gender Wage Gap: 2014." Institute for Women's Policy Research, September. Accessed September 29, 2015 (iwpr.org).

Hegewisch, Ariane, Emily Ellis, and Heidi Hartmann. 2015. "The Gender Wage Gap: 2014, Earnings by Race and Ethnicity." Institute for Women's Policy Research, March. Accessed October 8, 2015 (iwpr.org).

Heider, Eleanor R., and Donald C. Olivier. 1972. "The Structure of the Color Space in Naming and Memory for Two Languages." *Cognitive Psychology* 3: 337–354.

Heilbroner, R. L., and L. C. Thurow. 1998. *Economics Explained: Everything You Need to Know About How the Economy Works and Where It's Going*. New York: Touchstone.

Heim, Joe. 2015. "Indian Americans Dominate the National Spelling Bee. Why Should They Take Abuse on Social Media for It?" *Washington Post*, May 25. Accessed October 4, 2015 (washingtonpost.com).

Heiman, Julia R., J. Scott Long, Shawna N. Smith, William A. Fisher, Michael S. Sand, and Raymond C. Rosen. 2011. "Sexual Satisfaction and Relationship Happiness in Midlife and Older Couples in Five Countries." *Archives of Sexual Behavior* 40 (4): 741–753.

Heimer, Karen, Stacy Wittrock, and Halime Ünal. 2006. "The Crimes of Poverty: Economic Marginalization and the Gender Gap in Crime." Pp. 115–126 in *Gender and Crime: Patterns of Victimization and Offending*, edited by Karen Heimer and Candace Kruttschnitt. New York: New York University Press.

Heisler, Candace. 2012. "Elder Abuse." Office for Victims of Crime. Accessed May 16, 2014 (www.ovc.ncjrs.gov).

Helliwell, John, Richard Layard, and Jeffrey Sachs (eds). 2013. *World Happiness Report 2013*. United Nations. Accessed February 6, 2014 (www.unsdsn.org).

Helm, V., A. Humbert, and H. Miller. 2014. "Elevation and Elevation Change of Greenland and Antarctica Derived from CryoSat-2." *The Cryosphere* (August): 1539–1559.

Henry, Meghan, Alvaro Cortes, Azim Shivji, and Katherine Buck. 2014. "The 2014 Annual Homeless Assessment Report (AHAR) to Congress." U.S. Department of Housing and Urban Development, October. Accessed September 16, 2015 (www.hudexchange.info).

Herzberg, David. 2009. *Happy Pills in America: From Miltown to Prozac*. Baltimore, MD: Johns Hopkins University Press.

Hetherington, E. Mavis, Ross D. Parke, and Virginia Otis Locke. 2006. *Child Psychology: A Contemporary Viewpoint*, 6th edition. Boston: McGraw-Hill.

Hibbard, Roberta, Jane Barlow, and Harriet MacMillan. 2012. "Psychological Maltreatment." American Academy of Pediatrics. Accessed May 4, 2014 (www.pediatrics.aappublications.org).

Hicks, Josh. 2015. "HUD Paid $37 Million a Month in Subsidies to Ineligible Households." *Washington Post*, February 19. Accessed August 23, 2015 (washingtonpost.com).

Hightower, Jim. 2015. "CEOs Call for Wage Increases for Workers! What's the Catch?" AlterNet, August 26. Accessed August 26, 2015 (alternet.org).

Hilbert, Richard A. 1992. *The Classical Roots of Ethnomethodology: Durkheim, Weber, and Garfinkel*. Chapel Hill: University of North Carolina Press.

Hill, Catey. 2015. "5 Times It's More Expensive to Be a Woman." Market Watch, April 14. Accessed September 29, 2015 (marketwatch.com).

Hill, Catherine, Christianne Corbett, and Andresse St. Rose. 2010. "Why So Few? Women in Science, Technology, Engineering, and Mathematics." American Association of University Women. Accessed October 25, 2011 (www.aauw.org).

Hill, Jason G. 2011. "Education and Certification Qualifications of Departmentalized Public High School-Level Teachers of Core Subjects: Evidence from the 2007–08 Schools and Staffing Survey." National Center for Education Statistics, May. Accessed June 21, 2014 (www.nces.org).

Hillaker, B. D., H. E. Brophy-Herb, F. A. Villarruel, and B. E. Hass. 2008. "The Contributions of Parenting to Social Competencies and Positive Values in Middle School Youth: Positive Family Communication, Maintaining Standards, and Supportive Family Relationships." Family Relations 57 (December): 591–601.

Hillin, Taryn. 2014. "The Divorce Mistakes You Don't Even Know You're Making." Huffington Post, March 18. Accessed March 18, 2014 (www.huffingtonpost.com).

Hinkle, Stephen, and John Schopler. 1986. "Bias in the Evaluation of In-Group and Out-Group Performance." Pp. 196–212 in Psychology of Everyday Intergroup Relations, 2nd edition, edited by Stephen Worchel and William G. Austin. Chicago: Nelson-Hall.

Hirsch, Jerry. 2014. "Recall Fever Spreading Fast." Baltimore Sun, May 17, 10.

Hirschi, Travis. 1969. Causes of Delinquency. Berkeley: University of California Press.

"Hispanic Health." 2015. Centers for Disease Control and Prevention, May. Accessed October 16, 2015 (cdc.gov/vitalsigns).

"Hispanic Heritage Month 2015." 2015. U.S. Census Bureau News, September 14. Accessed September 22, 2015 (census.gov).

Hiss, William C., and Valerie W. Franks. 2014. "Defining Promise: Optional Standardized Testing Policies in American College and University Admissions." National Association for College Admission Counseling, February 5. Accessed June 21, 2014 (www.nacacnet.org).

"HIV Among People Aged 50 and Older." 2015. Centers for Disease Control and Prevention, March. Accessed October 15, 2015 (cdc.gov).

"HIV in the United States: At a Glance." 2015. Centers for Disease Control and Prevention, July 1. Accessed August 26, 2015 (cdc.gov).

Hlavka, Heather R. 2014. "Normalizing Sexual Violence: Young Women Account for Harassment and Abuse." Gender & Society 20 (June): 1–22.

Hoecker-Drysdale, Susan. 1992. Harriet Martineau: First Woman Sociologist. Providence, RI: Berg.

Hoffman, Piper. 2014. "7 Countries That Still Kill 'Witches.'" Care2 Causes, October 21. Accessed October 24, 2014 (www.care2.com).

Hokayem, Charles, and Misty L. Heggeness. 2014. "Living in Near Poverty in the United States: 1966–2012." U.S. Census Bureau, May. Accessed May 16, 2014 (www.census.gov).

Holloway, Kali. 2015. "8 Myths That Fuel the Assault on Abortion Rights." AlterNet, July 25. Accessed September 29, 2015 (alternet.org).

Holmes, Elizabeth. 2013. "Grooming Tips from Giant Men." Wall Street Journal, November 27, D1.

Homans, George. 1974. Social Behavior: Its Elementary Forms, revised edition. New York: Harcourt Brace Jovanovich.

Hondagneu-Sotelo, Pierrette. 2001. Domestica: Immigrant Workers Cleaning and Caring in the Shadows of Affluence. Berkeley: University of California Press.

Hooke, Alexander E. 2014. "Speak Not, Lest Ye Be Judged." Baltimore Sun, May 27, 13.

Horowitz, Juliana M., Erin Carriere-Kretschmer, and Jacob Poushter. 2010. "Gender Equality Universally Embraced, but Inequalities Acknowledged." Pew Research Center, July 1. Accessed March 2, 2014 (www.pewresearch.org).

Hotz, V. Joseph, and Juan Pantano. 2013. "Strategic Parenting, Birth Order and School Performance." National Bureau of Economic Research, October. Accessed October 26, 2014 (www.nber.org).

Howard, Beth. 2013/2014. "Meet Dr. Robot." AARP Magazine, December/January, 18, 23.

Howard, Beth. 2014. "Should You Have Surgery Abroad?" AARP Magazine, October/November, 26–30.

Howard, Jeff. 2003. "Still at Risk: The Causes and Costs of Failure to Educate Poor and Minority Children for the Twenty-First Century." Pp. 81–97 in A Nation Reformed? American Education 20 Years After A Nation at Risk, edited by David T. Gordon and Patricia A. Graham. Cambridge, MA: Harvard Education Press.

Hoyt, Homer. 1939. The Structure and Growth of Residential Neighborhoods in American Cities. Washington, DC: Federal Housing Administration.

Hsin, Amy, and Yu Xie. 2014. "Explaining Asian Americans' Academic Advantage Over Whites." Proceedings of the National Academy of Sciences, Early Edition, May 5. Accessed June 21, 2014 (www.pnas.org).

Hsu, Angel, et al. 2014. "The 2014 Environmental Performance Index: Full Report and Analysis." Yale Center for Environmental Law & Policy. Accessed September 19, 2014 (www.epi.yale.edu).

Hubbard, Ruth. 1990. The Politics of Women's Biology. New Brunswick, NJ: Rutgers University Press.

Huff, Rodney, Christian Desilets, and John Kane. 2010. "The 2010 National Public Survey on White Collar Crime." National White Collar Crime Center. Accessed January 12, 2014 (www.nw3c.org).

Huffington Post. 2011. "New American Bible Changes Some Words, Including 'Holocaust.'" March 3. Accessed July 16, 2011 (www.huffingtonpost.com).

Huffman, Matt L. 2013. "Organizations, Managers, and Wage Inequality." Sex Roles 68 (February): 216–222.

Hughes, Everett C. 1945. "Dilemmas and Contradictions of Status." American Journal of Sociology 50: 353–359.

Huizinga, David, Shari Miller, and the Conduct Problems Prevention Research Group. 2013. "Developmental Sequences of Girls' Delinquent Behavior." Office of Juvenile Justice and Delinquency Prevention, December. Accessed January 12, 2014 (www.ojp.usdoj.gov).

"Human Rights Violations." 2015. Unite to End Violence Against Women. Accessed September 26, 2015 (un.org).

Hummer, Robert A., and Elaine M. Hernandez. 2013. "The Effect of Educational Attainment on Adult Mortality in the United States." Population Bulletin 68 (June): 1–16.

Hunt, Darnell, and Ana-Christina Ramón. 2015. "2015 Hollywood Diversity Report: Flipping the Script." Ralph J. Bunche Center for African American Studies at UCLA, February. Accessed October 4, 2015 (bunchecenter.ucla.edu).

Hunter, Lori M. 2014. "Unmarried Baby Boomers Face Disadvantages as They Grow Older." Population Reference Bureau, February. Accessed May 16, 2014 (www.prb.org).

"Hunting as a Pack." 2014. Economist Technology Quarterly, December 6, 6.

Hurley, Lawrence. 2015. "Supreme Court Upholds Broad Housing Discrimination Claims." Reuters, June 25. Accessed October 4, 2015 (reuters.com).

Hurwitz, Michael. 2011. "The Impact of Legacy Status on Undergraduate Admissions at Elite Colleges and Universities." Economics of Education Review 30 (June): 480–492.

Hutter, Mark. 2007. Experiencing Cities. Boston: Allyn & Bacon.

Ianzito, Christina. 2013. "Killer Heels." Baltimore Sun, August 1, 3.

Iceland, John, Gregory Sharp, and Jeffrey M. Timberlake. 2013. "Sun Belt Rising: Regional Population Change and the Decline in Black Residential Segregation, 1970–2009." Demography 50 (February): 97–123.

Ihrke, David. 2014. "Reason for Moving: 2012 to 2013." U.S. Census Bureau, June. Accessed September 19, 2014 (www.census.gov).

Independent Television Service. 2003. "Race the Power of an Illusion: What Is This Thing Called Race?" Accessed March 22, 2014 (www.itvs.org).

Independent Television Service. 2014. "Economic Empowerment." Public Broadcasting Service. Accessed February 10, 2014 (www.pbs.org/independentlens).

Ingraham, Christopher. 2015. "The Ugly Truth About Hate Crimes—in 5 Charts and Maps." Washington Post, June 18. Accessed August 27, 2015 (washingtonpost.com).

Institute for Clinical and Economic Review. 2015. "PCSK9 Inhibitors for Treatment of High Cholesterol: Effectiveness, Value, and Value-Based Price Benchmarks. Draft Report." October 8. Accessed October 16, 2015 (icer-review.org).

Institute for College Access & Success. 2014. "Quick Facts About Student Debt." March. Accessed June 21, 2014 (www.ticas.org).

Institute of Medicine. 2015. Beyond Myalgic Encephalomyelitis/Chronic Fatigue Syndrome: Redefining an Illness. Washington, DC: National Academies Press.

Insurance Institute for Highway Safety. 2010. "Q&A: Older Drivers." Accessed October 28, 2011 (www.iihs.org).

Insurance Institute for Highway Safety. 2013. "Status Report." 45 (September 28). Accessed October 18, 2013 (www.iihs.org).

Intergovernmental Panel on Climate Change. 2014. "Climate Change 2014: Impacts, Adaptation and Vulnerability; Summary to Policymakers." Accessed September 19, 2014 (www.ipcc.ch).

International Federation of Health Plans. 2014. "2013 Comparative Price Report: Variation in Medical and Hospice Prices by Country." March 14. Accessed July 26, 2014 (www.ifhp.com).

International Institute for Democracy and Electoral Assistance. 2014. "VAP Turnout." Accessed April 20, 2014 (www.idea.int).

Inter-Parliamentary Union. 2053. "Women in National Parliaments." Accessed September 29, 2014 (ipu.org).

Ioannidis, John. 2005. "Contradicted and Initially Stronger Effects in Highly Cited Clinical Research." JAMA 294 (July 13): 218–228.

Irving, Shelley K., and Tracy A. Loveless. 2015. "Dynamics of Economic Well-Being: Participation in Government Programs,

2009–2012: Who Gets Assistance?" U.S. Census Bureau, May. Accessed September 11, 2015 (census.gov).

Isaacs, Julia, Katherine Toran, Heather Hahn, Karina Fortuny, and C. Eugene Steurle. 2012. "Kids' Share 2012: Report on Federal Expenditures on Children Through 2011." Urban Institute, July. Accessed May 16, 2014 (www.urban.org).

Ivry, Bob, Bradley Keoun, and Phil Kuntz. 2011. "Secret Fed Loans Gave Banks $13 Billion Undisclosed to Congress." *Bloomberg Businessweek,* November 27. Accessed August 26, 2015 (bloomberg.com).

Jackson, C. Kirabo. 2012. "Do College-Prep Programs Improve Long-Term Outcomes?" National Bureau of Economic Research, February. Accessed April 14, 2012 (www.nber.org).

Jackson, D. D. 1998. "'This Hole in Our Heart': Urban Indian Identity and the Power of Silence." *American Indian Culture and Research Journal* 22 (4): 227–254.

Jackson, Shelly L., and Thomas L. Hafemeister. 2011. "Financial Abuse of Elderly People vs. Other Forms of Elder Abuse: Assessing Their Dynamics, Risk Factors, and Society's Response." U.S. Department of Justice, February. Accessed March 5, 2012 (www.ncjrs.gov).

Jackson, Shelly L., and Thomas L. Hafemeister. 2013. "Understanding Elder Abuse." National Institute of Justice, June. Accessed May 16, 2014 (www.nij.gov).

Jacobe, Dennis. 2013. "U.S. Small Businesses Struggle to Find Qualified Employees." Gallup, February 15. Accessed April 12, 2014 (www.gallup.com).

Jacobs, Jerry A. 2005. "Multiple Methods in *ASR.*" *Footnotes* 33 (December): 1, 4.

Jacobsen, Linda A., Mary Kent, Marlene Lee, and Mark Mather. 2011. "America's Aging Population." *Population Bulletin* 66 (February): 1–16.

Jacoby, Sanford M. 1997. *Modern Manors: Welfare Capitalism Since the New Deal.* NJ: Princeton University Press.

Jacoby, Susan. 2005. "Sex in America." *AARP* (July/August): 57–62, 114.

Jagacinski, Carolyn M. 2013. "Women Engineering Students: Competence Perceptions and Achievement Goals in the Freshman Engineering Course." *Sex Roles* 69 (December): 644–657.

James, John T. 2013. "A New, Evidence-Based Estimate of Patient Harms Associated with Hospital Care." *Journal of Patient Safety* 9 (September): 122–128.

Jandt, Fred E. 2001. *Intercultural Communication: An Introduction.* Thousand Oaks, CA: Sage.

Janis, Irving L. 1972. *Victims of Groupthink: A Psychological Study of Foreign-Policy Decisions and Fiascoes.* Boston: Houghton Mifflin.

Janis, Irving L.1982. *Groupthink: A Psychological Study of Policy Decisions and Fiascoes.* Boston: Houghton Mifflin Company.

Janofsky, Michael. 2003. "Young Brides Stir New Outcry on Utah Polygamy." *New York Times,* February 28, 1.

Jaret, Peter. 2015. "Prices Spike for Some Generics." *AARP Bulletin,* July-August, 8, 10.

Jargon, Julie. 2013. "Not-So-Fast Food." *Wall Street Journal,* November 15, B2.

Jeffreys, Sheila. 2011. *Man's Dominion: The Rise of Religion and the Eclipse of Women's Rights.* New York: Routledge.

Jernigan, David. 2010. "Alcohol Marketing and Youth: Why It's a Problem and What You Can Do." Center for Alcohol Marketing and Youth, December 14. Accessed April 12, 2012 (www.camy.org).

Jewkes, Rachel, Emma Fulu, Tim Roselli, and Claudia Garcia-Moreno. 2013. "Prevalence of and Factors Associated with Non-Partner Rape Perpetration: Findings from the UN Multi-Country Cross-Sectional Study on Men and Violence in Asia and the Pacific." *Lancet Global Health* 1 (October): e208–218.

Jilani, Zaid. 2011. "Profits at Largest 500 Corporations Grew by 81 Percent in 2010." Think Progress, May 5. Accessed May 7, 2011 (www.alternet.org).

Jilani, Zaid. 2014. "As Walmart's Sales Increase, It Decides to Cut Healthcare for 30,000 Employees." AlterNet, October 8. Accessed December 5, 2014 (www.alternet.org).

Johnson, Dave. 2014. "8 Phony GOP Solutions for Poverty That Will Only Bring More Economic Pain." AlterNet, January 16. Accessed April 12, 2014 (www.alternet.org).

Johnson, Dave. 2014. "8 Ways Being Poor Is Wildly Expensive in America." AlterNet, January 29. Accessed February 14, 2014 (www.alternet.org).

Johnson, Hank, and Bert Klandermans, eds. 1995. *Social Movements and Culture.* Minneapolis: University of Minnesota Press.

Johnson, Ileana. 2010. "Why Has Communism Failed?" *Orthodoxy Today,* October 30. Accessed April 12, 2014 (www.orthodoxytoday.org).

Johnson, Kevin, and H. Darr Beiser. 2013. "Aging Prisoners' Costs Put Systems Nationwide in a Bind." *USA Today,* July 11. Accessed January 15, 2014 (www.usatoday.com).

Johnson, Kirk. 2011. "Between Young and Old, a Political Collision." *New York Times,* June 4, A10.

Johnson, Paula A., Therese Fitzgerald, Alina Salganicoff, Susan F. Wood, and Jill M. Goldstein. 2014. "Sex-Specific Medical Research: Why Women's Health Can't Wait." Brigham and Women's Hospital. Accessed October 17, 2014 (brighamandwomens.org).

Johnston, Pamela. 2005. "Dressing the Part." *Chronicle of Higher Education,* August 10. Accessed August 11, 2005 (www.chronicle.com).

Jones, Ann. 2013. *How the Wounded Return from America's Wars—The Untold Story.* Chicago, IL: Haymarket Books.

Jones, Jeff, and Lydia Saad. 2015. "Gallup Poll Social Series: Values and Beliefs." Gallup, May 6–10. Accessed August 28, 2015 (gallup.com).

Jones, Jeffrey M. 2012. "Gender Gap in 2012 Vote Is Largest in Gallup's History." Gallup, November 9. Accessed April 20, 2014 (www.gallup.com).

Jones, Jeffrey M. 2013. "In U.S., Voter Registration Lags Among Hispanics and Asians." Gallup, November 9. Accessed April 20, 2014 (www.gallup.com).

Jones, Jeffrey M. 2013. "U.S. Blacks, Hispanics Have No Preferences on Group Labels." Gallup, July 26. Accessed March 22, 2014 (www.gallup.com).

Jones, Jeffrey M. 2014. "In U.S., 14% of Those Aged 24 to 34 Are Living with Parents." Gallup, February 13. Accessed May 22, 2014 (www.gallup.com).

Jones, Jeffrey M. 2014. "Reports of Alcohol-Related Family Trouble Remain Up in U.S." Gallup, July 29. Accessed November 4, 2014 (www.gallup.com).

Jones, Jeffrey M. 2015. "Confidence in U.S. Institutions Still Below Historical Norms." Gallup, June 15. Accessed August 22, 2015 (gallup.com).

Jones, Jeffrey M. 2015. "In U.S., 65% Favor Path to Citizenship for Illegal Immigrants." Gallup, August 12. Accessed October 4, 2015 (gallup.com).

Jones, Jeffrey M. 2015. "In U.S., Confidence in Police Lowest in 22 Years." Gallup, June 19. Accessed August 22, 2015 (gallup.com).

Jones, Jeffrey M. 2015. "Majority in U.S. Now Say Gays and Lesbians Born, Not Made." Gallup, May 20. Accessed September 29, 2015 (gallup.com).

Jones, Jo, and William D. Mosher. 2013. "Fathers' Involvement with Their Children: United States, 2006–2010." *National Health Statistics Reports,* December 20. Accessed May 16, 2014 (www.cdc.gov/nchs).

Jones, Nicholas A., and Jungmiwha Bullock. 2012. "The Two or More Races Population: 2012." U.S. Census Bureau, September. Accessed March 22, 2014 (www.census.gov).

Jones, R. K., L. Frohwirth, and A. M. Moore. 2013. "More Than Poverty: Disruptive Events Among Women Having Abortions in the USA." *Journal of Family Planning and Reproductive Health Care* 39 (January): 36–43.

Jones, Rachel K., and Jenna Jerman. 2014. "Abortion Incidence and Service Availability in the United States, 2011." *Perspectives on Sexual and Reproductive Health* 46 (1): 3–14.

Jones, Rachel K., and Joerg Dreweke. 2011. "Countering Conventional Wisdom: New Evidence on Religion and Contraceptive Use." Guttmacher Institute, April. Accessed April 24, 2012 (www.guttmacher.org).

Jones, Sophia. 2014. "Turkish Activists Say Their Country Is Sliding Backward on Women's Rights." *Huffington Post,* April 18. Accessed April 20, 2014 (www.huffingtonpost.com).

Jonsson, Jan O., David B. Grusky, Reinhard Pollak, and Matthew Di Carlo. 2009. "Recent Trends in Social Mobility in the United States: A New Approach to Modeling Trend in Big Class, Gradational, and Microclass Reproduction." Stanford Center for the Study of Poverty and Inequality, September. Accessed September 22, 2011 (www.inequality.com).

Jonsson, Patrik. 2007. "In US Justice, How Much Bias?" *Christian Science Monitor,* September 21, 1, 10.

Jonsson, Patrik. 2015. "Reborn on the Bayou." *Christian Science Monitor Weekly,* August 17 & 24, 27–32.

Jonsson, Patrik. 2015. "The Downside of Urban Renewal." *Christian Science Monitor Weekly,* May 11, 10.

Jordan, Reed. 2013. "Poverty's Toll on Mental Health." Metrotrends, November 25. Accessed February 6, 2014 (www.blogs.metrotrends.org).

Jordan, William. 2013. "Americans Divided Over Blackface Halloween Make Up." YouGov, November 1. Accessed March 18, 2014 (www.today.yougov.com).

Joshi, Suchi P., Jochen Peter, and Patti M. Valkenburg. 2011. "Scripts of Sexual Desire and Danger in US and Dutch Teen Girl Magazines: A Cross-National Content Analysis." *Sex Roles* 64 (April): 463–474.

Juergensmeyer, Mark. 2003. "Thinking Globally About Religion." Pp. 3–13 in *Global Religions: An Introduction,* edited by Mark Juergensmeyer. New York: Oxford University Press.

Julian, Tiffany. 2012. "Work-Life Earnings by Field of Degree and Occupation for People with a Bachelor's Degree: 2011." U.S. Census Bureau, October. Accessed June 21, 2014 (www.census.gov).

June, Audrey Williams. 2015. "Despite Progress, Only 1 in 4 College Presidents Are Women." *Chronicle of Higher Education*, March 20, A6.

Kafka, Stephanie, Jordan Jeromcheck, and Susan Sorenson. 2014. "States in West and Midwest Lead Nation in Teacher Respect." Gallup, April 10. Accessed June 21, 2014 (www.gallup.com).

Kagan, Jerome. 2011. "Want Better Students? Teach Their Parents." *Christian Science Monitor*, February 21, 34.

Kahan, Dan M., Hank Jenkins-Smith, and Donald Braman. 2011. "Cultural Cognition of Scientific Consensus." *Journal of Risk Research* 14 (2): 147–174.

Kahlenberg, Richard D. 2010. "10 Myths About Legacy Preferences in College Admissions." *Chronicle of Higher Education*, October 1, A23, A25.

Kalev, Alexandra, Frank Dobbin, and Erin Kelly. 2006. "Best Practices or Best Guesses? Assessing the Efficacy of Corporate Affirmative Action and Diversity Policies." *American Sociological Review* 71 (August): 589–617.

Kalish, Rachel, and Michael Kimmel. 2011. "Hooking Up." *Australian Feminist Studies* 26 (March): 137–151.

Kalogrides, Demetra, Susanna Loeb, and Tara Beteille. 2013. "Systematic Sorting: Teacher Characteristics and Class Assignments." *Sociology of Education* 86 (April): 103–123.

Kane, Nazneen. 2015. "Stay-at-Home Fatherhoods." *Contexts* 14 (Spring): 74–76.

Kaneda, Toshiko, and Kristin Bietsch. 2015. "2015 World Population Data Sheet." Washington, DC: Population Reference Bureau. Accessed September 13, 2014 (prb.org).

Kang, Cecilia. 2015. "Facebook Doubles Black Hires, but They Only Make Up About 1.5% of Staff." *Washington Post*, July 1. Accessed October 4, 2015 (washingtonpost.com).

Kann, Laura, et al. 2014. "Youth Risk Behavior Surveillance—United States, 2013." *MMWR* 63 (June 13): 1–168.

Kantor, Jodi, and David Streitfeld. 2015. "Amazon's Bruising, Thrilling Workplace." *New York Times*, August 16, A1.

Kantor, Jodi, and Laurie Goodstein. 2014. "From Mormon Women, a Flood of Requests and Questions on Their Role in the Church." *New York Times*, March 7, A2.

Kantrowitz, Mark. 2011. "The Distribution of Grants and Scholarships by Race." September 2. Accessed March 20, 2012 (www.finaid.org).

Kaplan, David A. 2010. "The Best Company to Work For." *Fortune*, February 8, 57–72.

Kapp, Diana. 2013. "Can New Building Toys for Girls Improve Math and Science Skills?" *Wall Street Journal*, April 17, D1, D3.

Kassotis, Christopher D., Donald E. Tillitt, J. Wade Davis, Annette M. Hormann, and Susan C. Nagel. 2014. "Estrogen and Androgen Receptor Activities of Hydraulic Fracturing Chemicals and Surface and Ground Water in a Drilling-Dense Region." *Endocrinology* 155 (March): 897–907.

Kasumovic, Michael M., and Jeffrey H. Kuznekoff. 2015. "Insights Into Sexism: Male Status and Performance Moderates Female-Directed Hostile and Amicable Behavior." *PLoS One* 10 (7): e0131613. doi:10.1371/journal.pone.0131613.

Katz, Jackson. 2006. *The Macho Paradox: Why Some Men Hurt Women and How All Men Can Help*. Naperville, IL: Sourcebooks.

Katzenbach, Jon. 2003. *Why Pride Matters More Than Money: The Power of the World's Greatest Motivational Force*. New York: Crown Business.

Kaufman, Alexander C. 2015. "Why Marriage Equality Is Great For the Economy." *Huffington Post*, June 26. Accessed September 29, 2015 (huffingtonpost.com).

Kay, Aaron C., Martin V. Day, Mark P. Zanna, and A. David Nussbaum. 2013. "The Insidious (and Ironic) Effects of Positive Stereotypes." *Journal of Experimental Social Psychology* 49 (March): 287–291.

Kearney, Melissa S., and Phillip B. Levine. 2014. "Media Influences on Social Outcomes: The Impact of MTV's *16 and Pregnant* on Teen Childbearing." Brookings Institution, January. Accessed May 16, 2014 (www.brookings.edu).

Kearney, Melissa S., and Phillip B. Levine. 2014. "Teen Births Are Falling: What's Going On?" Brookings Institution, March. Accessed May 16, 2014 (www.brookings.edu).

Kearney, Melissa S., Benjamin H. Harris, Elisa Jácome, and Lucie Parker. 2014. "Ten Economic Facts About Crime and Incarceration in the United States." The Hamilton Project, May. Accessed November 4, 2014 (www.hamiltonproject.org).

Keels, Micere. 2014. "Choosing Single Motherhood." *Contexts* 13 (Spring): 70–72.

Keeter, Scott, and Carroll Doherty. 2015. "The Politics of Financial Insecurity." Pew Research Center, January 8. Accessed August 26, 2015 (pewresearch.org).

Keeter, Scott, Kyley McGeeney, Ruth Igielnik, Andrew Mercer, and Nancy Mathiowetz. 2015. "From Telephone to the Web: The Challenge of Mode of Interview Effects in Public Opinion Polls." Pew Research Center, May 13. Accessed July 22, 2015 (pewresearch.org).

Keeter, Scott. 2009. "New Tricks for Old—and New—Dogs: Challenges and Opportunities Facing Communications Research." Pew Research Center Publications, March 3. Accessed December 31, 2009 (www.pewresearch.org).

Keeter, Scott. 2010. "Ask the Expert." Pew Research Center, December 29. Accessed December 30, 2010 (www.pewresearch.org).

Kellermann, Arthur L. 2012. "Should We Ration End-of-Life Care?" The RAND Blog, October 30. Accessed July 26, 2014 (www.rand.org).

Kellermann, Arthur R., and David I. Auerbach. 2013. "Health Care Cost Growth Is Hurting Middle-Class Families." RAND Health Affairs Blog, January 8. Accessed July 26, 2014 (www.rand.org).

Kelley, Trista. 2013. "The Big Bucks in Keeping Kids Focused." *Bloomberg Businessweek*, October 14–20, 25–27.

Kelly, Dana, Holly Xie, Christine Winquist Nord, Frank Jenkins, Jessica Ying Chan, and David Kastberg. 2013. "Performance of U.S. 15-Year-Old Students in Mathematics, Science, and Reading Literacy in an International Context." National Center for Education Statistics, December. Accessed June 21, 2014 (www.nces.org).

Kemp, Alice Abel. 1994. *Women's Work: Degraded and Devalued*. Englewood Cliffs, NJ: Prentice Hall.

Kena, Grace, Susan Aud, Frank Johnson, Xiaolei Wang, Jijun Zhang, Amy Rathbun, Sidney Wilkinson-Flicker, and Paul Kristapovich. 2014. "The Condition of Education 2014." National Center for Education Statistics, May. Accessed June 21, 2014 (www.nces.ed.gov).

Kendall, Diana. 2002. *The Power of Good Deeds: Privileged Women and the Social Reproduction of the Upper Class*. Lanham, MD: Rowman & Littlefield.

Kennedy, Allison, Katherine LaVail, Glen Nowak, Michelle Basker, and Sarah Landry. 2011. "Confidence About Vaccines in the United States: Understanding Parents' Perceptions." *Health Affairs* 30 (June): 1151–1159.

Kennedy, Sean, and Luke Gelber. 2012. "2012 Congressional Pig Book Summary." Citizens Against Government Waste. Accessed April 20, 2014 (www.cagw.org).

Kenny, Charles. 2015. "Immigrants to the Rescue." *Bloomberg Businessweek*, August 3–9, 10–11.

Kessler, Aaron M. 2015. "E.P.A Issues Stiffer Rules on Vehicle Fuel Ratings." *New York Times*, February 24, B1, B8.

Kettner, Peter M., Robert M. Moroney, and Lawrence L. Martin. 1999. *Designing and Managing Programs: An Effectiveness-Based Approach*, 2nd edition. Thousand Oaks, CA: Sage.

Khadaroo, Stacy Teicher. 2015. "Keeping Kids from Crime." *Christian Science Monitor Weekly*, February 16, 27–32.

Khan, Shamus R. 2012. *Privilege: The Making of an Adolescent Elite at St. Paul's School*. Princeton, NJ: Princeton University Press.

Kharif, Olga. 2015. "A Technology that Reveals Your Feelings." *Bloomberg Businessweek*, August 10–23, 38–39.

Khaw, Lyndal B. L., and Jennifer L. Hardesty. 2009. "Leaving an Abusive Partner: Exploring Boundary Ambiguity Using the Stages of Change Model." *Journal of Family Theory & Review* 1 (March): 38–53.

Kidd, Dustin. 2007. "Harry Potter and the Functions of Popular Culture." *The Journal of Popular Culture* 40 (February): 69–89.

Kiefer, Francine. 2015. "Behind Landmark Sex-Trafficking Bills, 20 Women Senators." *Christian Science Monitor Weekly*, March 30, 14.

Kiel, Paul, and Dan Nguyen. 2015. "Bailout Tracker." ProPublica, September 14. Accessed September 15, 2015 (projects.propublica.org).

Killerman, Samuel. 2014. "30+ Examples of Middle-to-Upper Class Privilege." It's Pronounced Metrosexual, August 10. Accessed October 4, 2015 (itspronouncedmetrosexual.com).

Kilmann, R. H. 2011. *Quantum Organizations: A New Paradigm for Achieving Organizational Success and Personal Meaning*. Newport Coast, CA: Kilmann Diagnostics.

Kim, Jerry W., Bruce Kogut, and Jae-Suk Yang. 2015. "Executive Compensation, Fat Cats, and Best Athletes." *American Sociological Review* 80 (April): 299–328.

Kindig, David L., and Erika R. Cheng. 2013. "Even as Mortality Fell in Most US Counties, Female Mortality Nonetheless Rose in 42.8 Percent of Counties from 1992 to 2006." *Health Affairs* 32 (March): 451–458.

Kindy, Kimberly. 2013. "Tribes' Fight with Government Shakes Contractors." *Baltimore Sun*, December 29, 8.

King, Colbert I. 2015. "The Key Reason Why Racism Remains Alive and Well in America." *Washington Post*, June 26. Accessed October 4, 2015 (washingtonpost.com).

Kinsey Institute. 2011. "Continuum of Human Sexuality." Accessed November 24, 2012 (www.indiana.edu/~kinsey).

Kinsey, Alfred C., Wardell B. Pomeroy, and Clyde E. Martin. 1948. *Sexual Behavior in the Human Male*. Philadelphia: Saunders.

Kirisci, Kemal. 2014. "Turkey, the Twitter Ban, and Upcoming Local Elections." Brookings

Institution, March 26. Accessed April 12, 2014 (www.brookings.edu).

Kirkegaard, Jacob Funk. 2015. "The True Levels of Government and Social Expenditures in Advanced Economies." Peterson Institute for International Economics, March. Accessed September 11, 2015 (piie.org).

Kirkland, Allegra. 2014. "There Goes the Neighborhood: 6 Billionaires Who Moved Somewhere and Ruined It." AlterNet, September 26. Accessed November 10, 2014 (www.alternet.org).

Kirkpatrick, David D. 2010. *The Facebook Effect: The Inside Story of the Company That Is Connecting the World.* New York: Simon & Schuster.

Kiron, David, Nina Kruschwitz, Holger Rubel, Martin Reeves, and Sonja-Katrin Fuisz-Kehrbach. 2013. "Sustainability's Next Frontier: Walking the Talk on the Sustainability Issues That Matter Most." *MIT Sloan Management Review* 55 (December): 1–28.

Kivisto, Peter, and Dan Pittman. 2001. "Goffman's Dramaturgical Sociology: Personal Sales and Service in a Commodified World." Pp. 311–334 in *Illuminating Social Life: Classical and Contemporary Theory Revisited*, 2nd edition. Thousand Oaks, CA: Pine Forge Press.

Klandermans, Bert. 1984. "Mobilization and Participation: Social Psychological Explanations of Resource Mobilization Theory." *American Sociological Review* 49 (October): 583–600.

Klein, Allison, and Josh White. 2011. "Technology, Police Tactics Are Reining in Car Thefts." *Washington Post*, July 24, A1.

Klein, Ezra. 2011. "Do We Still Need Unions? Yes: Why They're Worth Fighting For." *Newsweek*, March 7, 18.

Klos, Diana Mitsu. 2013. "The Status of Women in the U.S. Media 3013." Women's Media Center. Accessed March 2, 2014 (www.womensmediacenter.com).

Kluger, Jeffrey. 2014. "Why Mass Killers Are Always Male." *Time*, May 25. Accessed August 6, 2015 (time.com).

Kneebone, Elizabeth, and Alan Berube. 2013. *Confronting Suburban Poverty in America.* Washington, DC: Brookings Institution.

Kneebone, Elizabeth, and Alan Berube. 2014. "Does the Suburbanization of Poverty Mean the War on Poverty Failed?" Brookings, January 8. Accessed February 6, 2014 (www.brookings.edu).

Koba, Mark. 2013. "Workers Unite! (So You Can Become Capitalists)." CNBC, December 13. Accessed December 19, 2013 (www.nbcnews.com).

Kochhar, Rakesh, and Rich Morin. 2014. "Despite Recovery, Fewer Americans Identify as Middle Class." Pew Research Center, January 27. Accessed February 6, 2014 (www.pewresearch.org).

Kochhar, Rakesh, and Russ Oates. 2014. "Attitudes About Aging: A Global Perspective." Pew Research Center, January 30. Accessed September 19, 2014 (www.pewresearch.org).

Kochhar, Rakesh. 2007. "1995–2005: Foreign-Born Latinos Make Progress on Wages." Pew Hispanic Center, August 21. Accessed October 10, 2007 (www.pewhispanic.org).

Koenig, David, and Scott Mayerowitz. 2015. "1 or 2 Airlines Controlling Many Airports." *Baltimore Sun*, July 15, 14.

Kohut, Andrew. 2015. "Are Americans Ready for Obama's 'Middle Class' Populism?" Pew Research Center, February 19. Accessed August 26, 2015 (pewresearch.org).

Kolet, Ilan, and Shobhana Chandra. 2012. "Health Care's Jobs Boom." *Bloomberg Businessweek*, February 6–12, 20–21.

Kornhauser, William. 1959. *The Politics of Mass Society.* New York: Free Press.

Kosmin, Barry A., and Ariela Keysar. 2009. *American Religious Identification Survey [ARIS 2008].* Summary Report, March. Accessed April 25, 2010 (www.americanreligionsurvey-aris.org).

Kosova, Weston, and Pat Wingert. 2009. "Crazy Talk." *Newsweek*, June 8, 54–62.

Kowitt, Beth. 2015. "The War on Big Food." *Fortune*, June 1, 61–70.

Kozol, Jonathan. 2005. *The Shame of the Nation: The Restoration of Apartheid Schooling in America.* New York: Crown.

Krache, D. 2008. "How to Ground a 'Helicopter Parent.'" CNN, August 19. Accessed July 10, 2009 (www.cnn.com).

Kramer, Adam D. I., Jamie E. Guillory, and Jeffrey T. Hancock. 2014. "Experimental Evidence of Massive-Scale Emotional Contagion through Social Networks." *Proceedings of the National Academy of Sciences* 111 (24): 8788–8790.

Kramer, Laura, and Ann Beutel. 2015. *The Sociology of Gender: A Brief Introduction*, 4th edition. New York: Oxford University Press.

Krantz, Matt. 2015. "11 Big Profitable Companies Pay No Tax." America's Markets, March 4. Accessed September 11, 2015 (americasmarkets.usatoday.org).

Krantz, Matt. 2015. "9 CEOs Paid 800 Times More Than Their Workers." *USA Today*, August 5. Accessed October 7, 2015 (usatoday.com).

Kraska, Peter B. 2004. *Theorizing Criminal Justice: Eight Essential Orientations.* Long Grove, IL: Waveland Press.

Krasnova, Hanna, Helena Wenninger, Thomas Widjaja, and Peter Buxmann. 2013. "Envy on Facebook: A Hidden Threat to Users' Life Satisfaction?" Unpublished paper presented at the 11th International Conference on Wirtschaftinformatik, February 27–March 1, Leipzig, Germany. Accessed December 5, 2013 (www.aisel.aisnet.org).

Krauthammer, Charles. 2015. "Why Doctors Quit." *Washington Post*, May 28. Accessed October 16, 2015 (washingtonpost.com).

Kreager, Derek A., and Jeremy Staff. 2009. "The Sexual Double Standard and Adolescent Peer Acceptance." *Social Psychology Quarterly* 72 (June): 143.

Krebs, Christopher P., Christine H. Lindquist, Tara D. Warner, Bonnie S. Fisher, and Sandra L. Martin. 2007. "The Campus Sexual Assault (CSA) Study." U.S. Department of Justice, December. Accessed June 21, 2014 (www.doj.gov).

Kreeger, Karen Y. 2002. "Sex-Based Differences Continue to Mount." *The Scientist* 16 (February 18). Accessed November 20, 2013 (www.the-scientist.com).

Kreeger, Karen Y. 2002. "X and Y Chromosomes Concern More Than Reproduction." *The Scientist* 16 (February 4). Accessed November 20, 2013 (www.the-scientist.com).

Kreider, Rose M., and Renee Ellis. 2011. "Number, Timing, and Duration of Marriages and Divorces: 2009." U.S. Census Bureau, Current Population Reports, May. Accessed March 10, 2012 (www.census.gov).

Krishnamurthy, Prasad, and Aaron Edlin. 2014. "Affirmative Action and Stereotypes in Higher Education Admissions." National Bureau of Economic Research, October. Accessed December 5, 2014 (www.nber.org).

Kristof, Nicholas. 2014. "A Nation of Takers?" *New York Times*, March 26, A31.

Krogstad, Jens Manuel, and Mark Hugo Lopez. 2014. "Hispanic Nativity Shift: U.S. Births Drive Population Growth as Immigration Stalls." Pew Research Center, April 29. Accessed November 25, 2014 (www.pewresearch.org).

Krogstad, Jens Manuel. 2014. "Asian American Voter Turnout Lags Behind Other Groups; Some Non-Voters Say They're 'Too Busy.'" Pew Research Center, April 9. Accessed April 20, 2014 (www.pewresearch.org).

Krogstad, Jens Manuel. 2014. "Census Bureau Explores New Middle East/North Africa Ethnic Category." Pew Research Center, March 24. Accessed October 4, 2015 (pewresearch.org).

Krogstad, Jens Manuel. 2015. "Puerto Ricans Leave in Record Numbers for Mainland U.S." Pew Research Center, October 14. Accessed October 23, 2015 (pewresearch.org).

Kroll, Luisa, and Kerry A. Dolan. 2013. "Forbes Billionaires." *Forbes*, March 25, 85–144, 164–174, 190–200.

Kromer, Braedyn, and David Howard. 2013. "Labor Force Participation and Work Status of People 65 Years and Older." U.S. Census Bureau, January. Accessed May 16, 2014 (www.census.gov).

Kross, Ethan, et al. 2013. "Facebook Use Predicts Declines in Subjective Well-Being in Young Adults." *PLoS One* 8 (8): e69841. Accessed December 5, 2013 (www.plosone.org).

Krueger, Alan B., Judd Cramer, and David Cho. 2014. "Are the Long-Term Unemployed on the Margins of the Labor Market?" Paper presented at the Brookings Panel on Economic Activity, March 20–21. Accessed April 12, 2014 (www.brookings.edu).

Kuhn, Manford H. and Thomas S. McPartland. 1954. "An Empirical Investigation of Self-Attitudes." *American Sociological Review* 19 (1): 68–76.

Kulczycki, Andrei, and Arun P. Lobo. 2002. "Patterns, Determinants, and Implications of Intermarriage Among Arab Americans." *Journal of Marriage and the Family* 64 (February): 202–210.

Kunkel Dale L., Jessica S. Castonguay, and Christine R. Filer. 2015. "Evaluating Industry Self-Regulation of Food Marketing to Children." *American Journal of Preventive Medicine* 49 (August): 181–187.

Küntzle, Julia, and Paul Blondé. 2013. "The Unhappy Fate of Ghanaian Witches." VICE United States, October 22. Accessed February 14, 2014 (www.vice.com).

Kurzman, Charles, and David Schanzer. 2015. "The Other Terror Threat." *New York Times*, June 16, A27.

Kusenbach, Margarethe. 2009. "Salvaging Decency: Mobile Home Residents' Strategies of Managing the Stigma of 'Trailer' Living." *Qualitative Sociology* 32 (December): 399–428.

Kuznia, Rob. 2015. "Rich Californians Balk at Limits: 'We're Not All Equal When It Comes to Water.'" *Washington Post*, June 13. Accessed October 23, 2015 (washingtonpost.com).

Kwok, Alvin C., et al. 2011. "The Intensity and Variation of Surgical Care at the End of Life: A Retrospective Cohort Study." *The Lancet*, October 6. Accessed April 15, 2012 (www.lancet.com).

Kyckelhahn, Tracey. 2013. "State Corrections Expenditures, FY 1982–2010." Bureau of Justice Statistics, December 11. Accessed January 15, 2014 (www.bjs.gov).

Kymlicka, Will. 2012. *Multiculturalism: Success, Failure, and the Future.* Washington, DC: Migration Policy Institute.

La France, Betty H., David D. Henningsen, Aubrey Oates, and Christina M. Shaw. 2009. "Social-Sexual Interactions? Meta-Analyses of Sex Differences in Perceptions of Flirtatiousness, Seductiveness, and Promiscuousness." *Communication Monographs* 76 (September): 263–285.

La Puma, John. 2014. "Don't Ask Your Doctor About 'Low T.'" *New York Times,* February 4, A21.

"Labor Day 2015: Sept. 1. 7." 2015. U.S. Census Bureau News, July 23. Accessed September 12, 2015 (census.gov).

Lahiri, Tripti, and Preetika Rama. 2013. "In India, New Anger at Police Response to Rapes." *Wall Street Journal,* May 1, A13.

Lakoff, Robin. T. 1990. *Talking Power: The Politics of Language.* New York: Basic Books.

Lalasz, Robert. 2006. "Americans Flocking to Outer Suburbs in Record Numbers." Population Reference Bureau. Accessed January 24, 2008 (www.prg.org).

Lamb, Gregory M. 2009. "National Parks Face New Threats." *Christian Science Monitor,* September 27, 36–37.

Lamb, Sharon, and Lyn Mikel Brown. 2007. *Packaging Girlhood: Rescuing Our Daughters from Marketers' Schemes.* New York: St. Martin's Press.

Landivar, Liana C. 2013. "Disparities in STEM Employment by Sex, Race, and Hispanic Origin." U.S. Census Bureau, September. Accessed June 21, 2014 (www.census.gov).

Landsberger, Henry A. 1958. *Hawthorne Revisited.* Ithaca, NY: Cornell University Press.

Laneri, Raquel. 2011. "Slumdog Millions." *Forbes,* May 9, 100–101.

Lang, James M. 2013. "How Orwell and Twitter Revitalized My Course." *Chronicle of Higher Education,* November 1, A35–A37.

Lang, Kurt, and Gladys Engel Lang. 1961. *Collective Dynamics.* New York: Thomas Y. Crowell.

Langreth, Robert. 2013. "Do Robot Surgeons Do No Harm?" *Bloomberg Businessweek,* March 14, 17–18.

Lankford, Adam. 2015. "Mass Shooters, Firearms, and Social Strains: A Global Analysis of an Exceptionally American Problem." Paper presented at the American Sociological Association, 110th Annual Meeting, August 23, Chicago, Illinois.

Lannutti, Pamela J., Melanie Laliker, and Jerold L. Hale. 2001. "Violations of Expectations and Social-Sexual Communication in Student/ Professor Interactions." *Communication Education* 50 (January): 69–82.

Laraña, Enrique, Hank Johnston, and Joseph R. Gusfield, eds. 1994. *New Social Movements: From Ideology to Identity.* Philadelphia: Temple University Press.

Larence, Eileen. 2011. "Private Sector and Law Enforcement Collaborate to Deter and Investigate Theft." U.S. Government Accountability Office, GAO-11-675, June. Accessed January 12, 2014 (www.gao.gov).

LaRossa, Ralph, and Donald C. Reitzes. 1993. "Symbolic Interactionism and Family Studies." Pp. 135–163 in *Sourcebook of Family Theories and Methods: A Contextual Approach,* edited by Pauline G. Boss, William J. Doherty, Ralph LaRossa, Walter R. Schumm, and Suzanne K. Steinmetz. New York: Plenum Press.

Larson, Christina. 2014. "China Needs Girls. More Are on the Way." *Bloomberg Businessweek,* August 4–10, 13–14.

Lauerman, John. 2013. "Colleges Slow to Investigate Assaults." *Baltimore Sun,* June 20, 10.

Laughlin, Lynda. 2014. "A Child's Day: Living Arrangements, Nativity, and Family Transitions: 2011 (Selected Indicators of Child Well-Being)." U.S. Census Bureau, December. Accessed August 6, 2015 (census.gov).

Lavy, Victor, and Edith Sand. 2015. "On the Origins of Gender Human Capital Gaps: Short and Long Term Consequences of Teachers' Stereotypical Biases." National Bureau of Economic Research, January. Accessed September 16, 2015 (nber.org).

Lawless, Jennifer L., and Richard L. Fox. 2013. "Girls Just Wanna Not Run: The Gender Gap in Young Americans' Political Ambition." Women & Politics Institute, March. Accessed April 21, 2014 (www.american.edu).

Lawn, Richard. 2013. *Experiencing Jazz,* 2nd edition. New York: Routledge.

Lawrence, Lee. 2013. "Was God Expelled?" *Christian Science Monitor,* June 17, 26–32.

Lawrence, Lee. 2015. "The Overbooked Generation." *Christian Science Monitor Weekly,* May 11, 26–32.

Lazear, Edward P., Kathryn L. Shaw, and Christopher T. Stanton. 2013. "The Value of Bosses." National Bureau of Economic Research, June. Accessed December 10, 2013 (www.nber.org).

Leaper, Campbell, and Melanie M. Ayres. 2007. "A Meta-Analytic Review of Gender Variations in Adults' Language Use: Talkativeness, Affiliative Speech, and Assertive Speech." *Personality and Social Psychology Review* 11 (November): 328–363.

Le Bon, Gustave. 1896/1968. *The Crowd: A Study of the Popular Mind,* 2nd edition. Dunwoody, GA: Norman S. Berg.

Leber, Jessica. 2013. "How Wireless Carriers Are Monetizing Your Movements." *MIT Technology Review,* April 12. Accessed October 8, 2014 (www.technologyreview.com).

Ledger, Kate. 2009. "Sociology and the Gene." *Contexts* 8 (Summer): 16–20.

Lee, David M., James Nazroo, Daryl B. O'Connor, Margaret Blake, and Neil Pendleton. 2015. "Sexual Health and Well-being Among Older Men and Women in England: Findings from the English Longitudinal Study of Ageing." *Archives of Sexual Behavior,* January 27, 1–12. Accessed September 29, 2015 (link.springer.com).

Lee, Donghoon. 2013. "Household Debt and Credit: Student Debt." Federal Reserve Bank of New York, February 28. Accessed June 21, 2014 (ny.frb.org).

Lee, Jennifer, and Min Zhou. 2015. *The Asian American Achievement Paradox.* New York, NY: Russell Sage Foundation.

Lee, Marlene. 2009. "Aging, Family Structure, and Health." Population Reference Bureau, October. Accessed April 10, 2010 (www.prb.org).

Legatum Prosperity Index. 2013. "The Prosperity Index 2013." Accessed February 6, 2014 (www.prosperity.com).

Leger, Donna L. 2013. "OxyContin a Gateway to Heroin for Upper-Income Addicts." *USA Today,* April 25. Accessed July 28, 2014 (www.usatoday.com).

Leinberger, Christopher B., and Patrick Lynch. 2014. "Foot Traffic Ahead: Ranking Walkable Urbanism in America's Largest Metros." Smart Growth America. Accessed September 19, 2014 (www.smartgrowthAmerica.org).

Lemert, Edwin M. 1951. *Social Pathology: A Systematic Approach to the Theory of Sociopathic Behavior.* New York: McGraw-Hill.

Lemert, Edwin M. 1967. *Human Deviance, Social Problems and Social Control.* Englewood Cliffs, NJ: Prentice Hall.

Lencioni, Patrick. 2002. *The Five Dysfunctions of a Team: A Leadership Fable.* San Francisco: Jossey-Bass.

Lengermann, Patricia Madoo, and Jill Niebrugge-Brantley. 1992. "Contemporary Feminist Theory." Pp. 308–357 in *Contemporary Sociological Theory,* 3rd edition, edited by George Ritzer. New York: McGraw-Hill.

Lenhart, Amanda, Aaron Smith, Monica Anderson, Maeve Duggan, and Andrew Perrin. 2015. "Teens, Technology & Friendships." Pew Research Center, August 6. Accessed August 14, 2015 (pewresearch.org).

Leonardsen, Dag. 2004. *Japan as a Low-Crime Nation.* New York: Palgrave Macmillan.

Leung, Maxwell. 2013. "Jeremy Lin's Model Minority Problem." *Contexts* 12 (Summer): 53–56.

Leurent, B., et al. 2013. "Spiritual and Religious Beliefs as Risk Factors for the Onset of Major Depression: An International Cohort Study." *Psychological Medicine* 43 (10): 2109–2120.

LeVay, Simon. 2011. *Gay, Straight, and the Reason Why: The Science of Sexual Orientation.* New York: Oxford University Press.

Leventhal, Adam M., et al. 2015. "Association of Electronic Cigarette Use with Initiation of Combustible Tobacco Product Smoking in Early Adolescence." *JAMA* 314 (7): 700–707.

Levin, Diane E., and Jean Kilbourne. 2009. *So Sexy So Soon: The New Sexualized Childhood and What Parents Can Do to Protect Their Kids.* New York: Ballantine Books.

Levin, Ezra, Jeremie Greer, and Ida Rademacher. 2014. "From Upside Down to Right-Side Up." Corporation for Enterprise Development. Accessed September 11, 2015 (cfed.org).

The Levin Institute. 2013. "Pop Culture." State University of New York. Accessed November 10, 2013 (www.globalization101.org).

Levine, Hallie. 2014. "5 Reasons Women Live Longer Than Men." *Health,* October 13. Accessed October 16, 2015 (news.health.com).

Levine, Lawrence W. 1998. *Highbrow/Lowbrow: The Emergence of Cultural Hierarchy in America.* Cambridge, MA: Harvard University Press.

Levine, Linda. 2012. "An Analysis of the Distribution of Wealth Across Households, 1989–2010." Congressional Research Service, July 17. Accessed February 6, 2014 (www.crs.gov).

Levitz, Jennifer. 2013. "Harvard Punishes Dozens of Students for Cheating." *Wall Street Journal,* February 3, A2.

Levy, Jenna. 2015. "In U.S., Uninsured Rate Dips to 11.9% in First Quarter." Gallup, April 13. Accessed October 16, 2015 (gallup.com).

Lewin, Kurt, Ronald R. Lippit, and Ralph K. White. 1939. "Patterns of Aggressive Behavior in Experimentally Created 'Social Climates.'" *Journal of Social Psychology* 10 (2): 271–301.

Lewin, Tamar. 2013. "New Milestone Emerges: Baby's First iPhone App." *New York Times,* October 26, A17.

Lewis, David Levering. 1993. *W. E. B. Du Bois: Biography of a Race, 1868–1919.* New York: Henry Holt.

Lewis, Oscar. 1966. "The Culture of Poverty." *Scientific American* 115 (October): 19–25.

Lewis, Ricki. 2014. "Noninvasive Prenatal DNA Test Okay for Low-Risk Pregnancies." Medscape, February 27. Accessed October 8, 2014 (www.medscape.com).

Lichter, Daniel T., Zhenchao Qian, and Leanna M. Mellott. 2006. "Marriage or Dissolution? Union Transitions Among Poor Cohabiting Women." *Demography* 43 (May): 223–240.

"Life Size Barbie Shows Young Girls the Dangers of Unrealistic Body Expectations." 2011. Diets in Review, May 3. Accessed October 25, 2011 (www.dietsinreview.com).

Lilly, J. Robert, Francis T. Cullen, and Richard A. Ball. 1995. Criminological Theory: Context and Consequences, 2nd edition. Thousand Oaks, CA: Sage.

Lindner, Andrew M. 2012. "An Old Tool with New Promise." Contexts 11 (Winter): 70–72.

Lindsey, Linda L. 2005. Gender Roles: A Sociological Perspective, 4th edition. Upper Saddle River, NJ: Prentice Hall.

Lino, Mark. 2014. "Expenditures on Children by Families, 2013." U.S. Department of Agriculture, Center for Nutrition Policy and Promotion. Accessed November 18, 2014 (www.cnpp.usda.gov).

Linton, Ralph. 1936. The Study of Man. New York: Appleton-Century-Crofts.

Linton, Ralph. 1964. The Study of Man: An Introduction. New York: Appleton-Century-Crofts.

Lipka, Michael, and Sandra Stencel. 2013. "Racial and Ethnic Groups View 'Radical Life Extension' Differently." Pew Research Center, August 6. Accessed May 16, 2014 (www.pewresearch.org).

Lipman, Joanne. 2015. "Let's Expose the Gender Pay Gap." New York Times, August 13, A19.

Lisak, David, and Paul M. Miller. 2002. "Repeat Rape and Multiple Offending Among Undetected Rapists." Violence and Victims 17 (1): 73–84.

Liu, Sa, S. Katherine Hammond, and Ann Rojas-Cheatham. 2013. "Concentrations and Potential Health Risks of Metals in Lip Products." Environmental Health Perspectives 121 (June): 705–710.

Livingston, Gretchen, and Anna Brown. 2014. "Birth Rate for Unmarried Women Declining for First Time in Decades." Pew Research Center, August 11. Accessed August 14, 2014 (www.pewresearch.org).

Livingston, Gretchen, and D'Vera Cohn. 2013. "Long-Term Trend Accelerates Since Recession: Record Share of New Mothers Are College Educated." Pew Research Center, May 10. Brookings Institution, March. Accessed May 16, 2014 (pewresearch.org).

Livingston, Gretchen, and Kim Parker. 2011. "A Tale of Two Fathers." Pew Social & Demographic Trends, June 15. Accessed May 20, 2014 (pewsocialtrends.org).

Llana, Sara M., and Sibylla Brodzinsky. 2012. "Latin America's Silent Scourge." Christian Science Monitor, November 26, 18–20.

"Lobbying Database." 2015. Center for Responsive Politics. Accessed October 12, 2015 (opensecrets.org).

Locher, David A. 2002. Collective Behavior. Upper Saddle River, NJ: Prentice Hall.

Lodge, Amy C., and Debra Umberson. 2012. "All Shook Up: Sexuality of Mid- To Later Life Married Couples." Journal of Marriage and Family 74 (June): 428–443.

Lofland, John. 1993. "Collective Behavior: The Elementary Forms." Pp. 70–75 in Collective Behavior and Social Movements, edited by Russell L. Curtis, Jr. and Benigno E. Aguirre. Boston: Allyn & Bacon.

Lofland, John. 1996. Social Movement Organizations: Guide to Research on Insurgent Realities. New York: Aldine De Gruyter.

Lofquist, Daphne, Terry Lugaila, Martin O'Connell, and Sarah Feliz. 2012. "Households and Families: 2010." Census Briefs, April. Accessed May 15, 2012 (www.census.gov).

Logan, John, and Harvey Molotch. 1987. Urban Fortunes: The Political Economy of Place. Berkeley: University of California Press.

Lohan, Tara. 2015. "Why America's Deadly Love Affair with Bottles Water Has to Stop." AlterNet, September 17. Accessed October 23, 2015 (alternet.org).

Lohr, Steve. 2011. "Computers That See You and Keep Watch Over You." New York Times, January 1, A1.

Loki, Reynard. 2015. "Dangerous and Radical: Republican Senators Vote to Sell Off America's National Forests." AlterNet, May 5. Accessed October 23, 2015 (alternet.org).

Loki, Reynard. 2015. "Environmentalists Blast Obama's Decision to Let Shell Drill in Arctic." AlterNet, August 23. Accessed October 23, 2015 (alternet.org).

Lopez, Mark. H., and Ana Gonzalez-Barrera. 2013. "What Is the Future of Spanish in the United States?" Pew Research Center, September 5, 2013. Accessed November 10, 2013 (www.pewresearch.org).

Lorber, Judith, and Lisa Jean Moore. 2007. Gendered Bodies: Feminist Perspectives. Los Angeles: Roxbury.

Loumarr, Ess. 2015. "How Much Weight Can You Lose by Not Drinking Soda?" Live Strong, January 28. Accessed August 23, 2015 (livestrong.com).

Loveless, Tom. 2014. "The 2014 Brown Center Report on American Education: How Well Are American Students Learning?" Brown Center on Education Policy, March. Accessed June 21, 2014 (www.brookings.edu).

Lowenberger, Amanda, Joanna Mauer, Andrew deLaski, Marianne DiMascio, Jennifer Amann, and Steven Nadel. 2012. "The Efficiency Boom: Cashing in on the Savings from Appliance Standards." American Council for an Energy-Efficient Economy, and Appliance Standards Awareness Project, March. Accessed June 22, 2012 (www.aceee.org).

Lowery, Wesley. 2014. "Senate Republicans Unanimously Reject Equal Pay Bill." Washington Post, April 9. Accessed April 21, 2014 (www.washingtonpost.com).

Lowrey, Annie. 2013. "Top 10% Took Home Half of U.S. Income in 2012." New York Times, September 11, B4.

Lublin, Joann S. 2013. "Tyco's Former CEO Set for 2014 Parole." Wall Street Journal, December 4, B2.

Luckey, John R. 2008. "CRS Report for Congress: The United States Flag: Federal Law Relating to Display and Associated Questions." Congressional Research Service, April 14. Accessed July 20, 2011 (www.senate.gov).

Lugaila, Terry. A. 1998. "Marital Status and Living Arrangements: March 1998 (Update)." U.S. Census Bureau, Current Population Reports. Accessed August 8, 2000 (www.census.gov).

Lugo, Luis, Alan Cooperman, and Cary Funk. 2013. "Living to 120 and Beyond: Americans' Views on Aging, Medical Advances, and Radical Life Extension." Pew Research Center, August 6. Accessed May 16, 2014 (www.pewresearch.org).

Lugo, Luis, Alan Cooperman, and Gregory A. Smith. 2013. "A Portrait of Jewish Americans." Pew Research Center, October 1. Accessed June 21, 2014 (www.pewresearch.org).

Lugo, Luis, Alan Cooperman, Cary Funk, and Gregory A. Smith. 2012. 'Nones' on the Rise: One-in-Five Adults Have No Religious Affiliation." Pew Forum on Religion & Public Life, October 9. Accessed June 21, 2014 (www.pewforum.org).

Luling, Todd Van. 2013. "Hey Politicians, It's the 21st Century. Time to Stop Being Sexist Idiots." Huffington Post, October 8. Accessed March 2, 2014 (www.huffingtonpost.com).

Lundberg, Shelly, and Robert A. Pollak. 2007. "The American Family and Family Economics." Journal of Economic Perspectives 21 (Spring): 3–26.

Lurie, Julia. 2015. "We're So Hosed: Where Does California's Dwindling Water Supply Go?" Mother Jones, July/August, 62–63.

Lutz, William. 1989. Doublespeak. New York: Harper & Row.

Lynas, Mark. 2015. "Even in 2015, the Public Doesn't Trust Scientists." Washington Post, January 30. Accessed July 22, 2015 (washingtonpost.com).

Lynch, Eleanor W., and Marci J. Hanson, eds. 1999. Developing Cross-Cultural Competence: A Guide for Working with Children and Their Families, 2nd edition. Baltimore, MD: Paul H. Brookes.

Lynn, David B. 1969. Parental and Sex Role Identification: A Theoretical Formulation. Berkeley, CA: McCutchen.

Lynn, M., and M. Todoroff. 1995. "Women's Work and Family Lives." Pp. 244–271 in Feminist Issues: Race, Class, and Sexuality, edited by Nancy Mandell. Scarborough, Ontario: Prentice Hall Canada.

Maccoby, Eleanor E., and John A. Martin. 1983. "Socialization in the Context of the Family: Parent-Child Interaction." Pp. 1–101 in Socialization, Personality, and Social Development: Vol. 4. Handbook of Child Psychology, edited by E. Mavis Hetherington. New York: Wiley.

MacDorman, Marian F., T. J. Mathews, Ashna D. Mohangoo, and Jennifer Zeitlin. 2014. "International Comparisons of Infant Mortality and Related Factors: United States and Europe, 2010." National Vital Statistics Reports 63 (5): 1–7.

Mach, Annie l., and Ada S. Cornell. 2013. "Federal Employees Health Benefits Program (FEHBP): Available Health Insurance Options." Congressional Research Service, November, 13. Accessed November 17, 2014 (www.crs.gov).

Macionis, John J., and Vincent N. Parrillo. 2007. Cities and Urban Life, 4th edition. Upper Saddle River, NJ: Prentice Hall.

MacKenzie, Donald, and Judy Wajcman, eds. 1999. The Social Shaping of Technology, 2nd edition. Philadelphia: Open University Press.

Madden, Mary, Amanda Lenhart, Sandra Cortesi, Urs Gasser, Maeve Duggan, and Aaron Smith. 2013. "Teens, Social Media, and Privacy." Pew Research Center, May 21. Accessed December 4, 2013 (www.pewinternet.org).

Madden, Mary, and Lee Rainie. 2015. "Americans' Attitudes About Privacy, Security and Surveillance." Pew Research Center, May 20. Accessed August 23, 2015 (pewinternet.org).

Madigan, Nick. 2009. "Seniors Increasingly Targeted by Con Artists; Police 'Struggling to Keep Up.'" Baltimore Sun, January 14, 3.

Madrick, Jeff. 2013. "The Anti-Economist: The Fall and Rise of Occupy Wall Street." Harper's Magazine, March, 9–11.

Magill, Bobby. 2014. "Clean Power Plan Exempts Major CO_2 Emitters." Climate Central, June 17. Accessed September 19, 2014 (www.climatecontrol.org).

Magnier, Mark. 2009. "In Northern India, Village Elders Order 'Honor Killings.'" Los Angeles Times, September 26. Accessed September 28, 2009 (www.latimes.com).

Maines, David R. 2001. *The Faultline of Consciousness: A View of Interactionism in Sociology.* New York: Aldine De Gruyter.

Major, Brenda, Mark Appelbaum, Linda Beckman, Mary Ann Dutton, Nancy Felipe Russo, and Carolyn West. 2008. "Report of the APA Task Force on Mental Health and Abortion." American Psychological Association, August 13. Accessed September 5, 2008 (www.apa .org).

Major, William. 2011. "Thoreau's Cellphone Experiment." *Chronicle of Higher Education,* January 21, A33, A35.

Makinen, Julie. 2013. "Law on Elder Visits in Effect." *Baltimore Sun,* August 6, 6.

Makinen, Julie. 2014. "Report Shows 'Brutal' N. Korea." *Baltimore Sun,* February 18, 5.

Makino, Catherine. 2009. "Rape Victim Presses Case of Police Abuse in Japan." Women's eNews, May 5. Accessed May 7, 2009 (womensenews.org).

Malik, Khalid. 2014. "Human Development Report 2014." United Nations Development Program. Accessed September 20, 2014 (hdr.undp.org).

Malone, Noreen. 2015. " 'I'm No Longer Afraid': 35 Women Tell Their Stories About Being Assaulted by Bill Cosby, and the Culture That Wouldn't Listen." *New York Magazine,* July 26. Accessed September 29, 2015 (nymag .org).

Malthus, Thomas Robert. 1798/1965. *An Essay on Population.* New York: Augustus Kelley.

Malthus, Thomas Robert. 1872/1991. *An Essay on the Principle of Population,* 7th edition. London: Reeves & Turner.

Mangan, Katherine. 2013. "Comanche Nation College Tries to Rescue a Lost Tribal Language." *Chronicle of Higher Education,* June 14, A18–A19.

Mangan, Katherine. 2013. "Life After Steel." *Chronicle of Higher Education,* April 26, A33–A35.

Manning, Jennifer E. 2013. "Membership of the 113th Congress: A Profile." Congressional Research Service, August 2013. Accessed June 21, 2014 (www.crs.gov).

Manning, Wendy D., and Jessica A. Cohen. 2012. "Premarital Cohabitation and Marital Dissolution: An Examination of Recent Marriages." *Journal of Marriage and Family* 74 (April): 377–387.

Manning, Wendy D., and Pamela J. Smock. 2005. "Measuring and Modeling Cohabitation: New Perspectives from Qualitative Data." *Journal of Marriage and Family* 67 (November): 989–1002.

"Manufacturing Day: Oct. 4, 2013." 2013. U.S. Census Bureau News, October 1. Accessed April 12, 2014 (www.census.gov).

Manyika, James, Michael Chui, Jacques Bughin, Richard Dobbs, Peter Bisson, and Alex Marrs. 2013. "Disruptive Technologies: Advances That Will Transform Life, Business, and the Global Economy." McKinsey Global Institute, May. Accessed April 12, 2014 (www.mckinsey.com).

Marcotte, Amanda. 2014. "Is Religion Inherently Oppressive?" AlterNet, April 9. Accessed June 21, 2014 (www.alternet.org).

Marcus, Amy Dockser. 2013. "Genetic Testing Leaves More Patients Living in Limbo." *Wall Street Journal,* November 19, D1–D2.

Marikar, Sheila. 2015. "The Company Boosts Growth by Treating Its Employees Like Royalty." *Baltimore Sun,* June 7, 4.

"The Marines Need a Few Good Men." 2015. *Bloomberg Businessweek,* September 28–October 4, 10.

"Marital Status." 2013. "2012 American Community Survey 1-Year Estimates." American FactFinder, Table S1201. Accessed May 20, 2014 (census.gov).

"Marital Status." 2015. U.S. Census Bureau, Table MS-2. Accessed October 8, 2015 (census.gov).

"Marketing in the Digital Age." 2015. *The Economist,* August 29, 51–52.

Markey, Richard M., Charlotte N. Markey, and Juliana E. French. 2014. "Violent Video Games and Real-World Violence: Rhetoric Versus Data." *Psychology of Popular Media Culture,* August 18. APA PsycNet. Accessed August 6, 2015 (psycnet.apa.org).

Marquand, Robert. 2011. "Europe Rejects Multiculturalism." *Christian Science Monitor,* February 21, 14.

"Marriage in India." 2015. *The Economist,* September 5, 43–44.

Martens, Richard. 2015. "In Small-Town America, Open Arms." *Christian Science Monitor Weekly,* March 23, 21–23.

Martin, Andrew. 2015. "Making Washington Fall in Love with Pizza Again." *Bloomberg Businessweek,* March 9–15, 35–36.

Martin, Joyce A., Brady E. Hamilton, Michelle J. K. Osterman, Sally C. Curtin, and T. J. Mathews. 2013. "Births: Final Data for 2012." *National Vital Statistics Reports* 62 (December 30): 1–87.

Martin, Molly A. 2008. "The Intergenerational Correlation in Weight: How Genetic Resemblance Reveals the Social Role of Families." *American Journal of Sociology* 114 (Suppl.): S67–S105.

Martin, Philip. 2013. "The Global Challenge of Managing Migration." *Population Bulletin* 68 (November): 1–16.

Martin, Steven, Nan Astone, and Elizabeth Peters. 2014. "Fewer Marriages, More Divergence: Marriage Projections for Millennials to Age 40." Urban Institute, April 29. Accessed July 25, 2014 (www.urban.org).

Martinez Gladys, Casey E. Copen, and Joyce C. Abma. 2011. "Teenagers in the United States: Sexual Activity, Contraceptive Use, and Childbearing, 2006–2010 National Survey of Family Growth." *Vital and Health Statistics,* 2011, No. 31. October. Accessed December 14, 2012 (www.cdc.gov/nchs).

Martínez, Jessica, and Michael Lipka. 2014. "Hispanic Millennials Are Less Religious Than Older U.S. Hispanics." Pew Research Center, May 8. Accessed June 21, 2014 (www.pewresearch.org).

Maruschak, Laura M., and Thomas P. Boncjar. 2013. "Probation and Parole in the United States, 2012." Bureau of Justice Statistics, December. Accessed January 15, 2014 (www.bjs.gov).

Marx, Gary T., and Douglas McAdam. 1994. *Collective Behavior and Social Movements: Process and Structure.* Upper Saddle River, NJ: Prentice Hall.

Marx, Karl. 1844/1964. *Economic and Philosophic Manuscripts of 1844.* New York: International Publishers.

Marx, Karl. 1845/1972. "The German Ideology." Pp. 110–164 in *The Marx-Engels Reader,* edited by Robert C. Tucker. New York: W. W. Norton.

Marx, Karl. 1867/1967. *Capital,* edited by Friedrich Engels. New York: International Publishers.

Marx, Karl. 1934. *The Class Struggles in France.* New York: International Publishers.

Marx, Karl. 1964. *Karl Marx: Selected Writings in Sociology and Social Philosophy,* translated by T. B. Bottomore. New York: McGraw-Hill.

Masci, David, and Seth Motel. 2015. "5 Facts About Same-Sex Marriage." Pew Research Center, June 26. Accessed September 29, 2015 (pewresearch.org).

Mason, Josephine, and Alastair Sharp. 2015. "Hacker's Ashley Madison Data Dump Threatens Marriages, Reputations." Reuters, August 19. Accessed August 23, 2015 (reuters.com).

Massey, Douglas S. 2007. *Categorically Unequal: The American Stratification System.* New York: Russell Sage Foundation.

"Maternal Mortality." 2015. *The Economist,* July 18, 28.

Mather, Mark, and Beth Jarosz. 2014. "The Demography of Inequality in the United States." *Population Bulletin* 69 (November): 1–15.

Mather, Mark, and Charlotte Feldman-Jacobs. 2015. "Women and Girls at Risk of Female Genital Mutilation/Cutting in the United States." Population Reference Bureau, February. Accessed September 29, 2015 (prb.org).

Mather, Mark, Keven Pollard, and Linda A. Jacobsen. 2011. *First Results from the 2010 Census.* Washington, DC: Population Reference Bureau.

Matjasko, Jennifer L., Phyllis H. Niolon, and Linda A. Valle. 2013. "The Role of Economic Factors and Economic Support in Preventing and Escaping from Intimate Partner Violence." *Journal of Policy Analysis and Management* 32 (October): 122–141.

Matthews, H. Damon, Tanya L. Graham, Serge Keverian, Cassandra Lamontagne, Donny Seto, and Trevor J Smith. 2014. "National Contributions to Observed Global Warming." *Environmental Research Letters* 9 (January), doi:10.1088/1748-9326/9/1/014010. Accessed October 22, 2015 (iopscience.iop.org).

Matthews, Hannah, and Stephanie Schmit. 2014. "Child Care Assistance Spending and Participation in 2012." CLASP, February. Accessed May 16, 2014 (www.clasp.org).

Maxwell, Steve. 2012. "Water Is *Still* Cheap: Demonstrating the True Value of Water." *American Water Works Association* 104 (May): 31–36.

Mayo, Elton. 1945. *The Problems of an Industrial Civilization.* Cambridge, MA: Harvard University Press.

Mayo, Michael. 2012. "Home for the Holidays: Convicted Killers Goodman, LeVin." *Sun Sentinel,* December 19. Accessed January 12, 2014 (www.articles.sun-sentinel.com).

McAdam, Doug, and Ronnelle Paulsen. 1994. "Specifying the Relationship Between Social Ties and Activism." *American Journal of Sociology* 99 (November): 640–667.

McCabe, Brian J. 2014. "When Property Values Rule." *Contexts* 13 (Winter): 38–43.

McCabe, Donald L., Kenneth D. Butterfield, and Linda K. Trevino. 2006. "Academic Dishonesty in Graduate Business Programs: Prevalence, Causes, and Proposed Action." *Academy of Management Learning Education* 5 (September): 294–305.

McCabe, Janice, Karin L. Brewster, and Kathryn H. Tillman. 2011. "Patterns and Correlates of Same-Sex Sexual Activity Among U.S. Teenagers and Young Adults." *Perspectives on Sexual and Reproductive Health* 43 (September): 15–21.

McCarthy Justin. 2014. "Seven in 10 Americans Back Euthanasia." Gallup, June 18. Accessed June 21, 2014 (www.gallup.com).

McCarthy, Julie. 2015. "An Angry Young Man Leads Protests Against India's Affirmative Action." NPR, August 31. Accessed September 11, 2015 (npr.org).

McCarthy, Justin. 2015. "Americans' Satisfaction with Job Aspects Up from 2005." Gallup, August 28. Accessed October 7, 2015 (gallup.com).

McCarthy, Justin. 2015. "Confidence in U.S. Branches of Government Remains Low." Gallup, June 15. Accessed October 7, 2015 (gallup.com).

McCarthy, Justin. 2015. "Majority in U.S. Maintain Need for Third Major Party." Gallup, September 25. Accessed October 7, 2015 (gallup.com).

McCarthy, Justin. 2015. "U.S. Support for Gay Marriage Stable After High Court Ruling." Gallup, July 17. Accessed July 22, 2015 (gallup.com).

McCoy, Charles, and Roscoe Scarborough. 2015. "Why Do We Watch Trashy TV?" *Washington Post*, May 27. Accessed July 26, 2015 (washingtonpost.com).

McCoy, Kevin. 2013. "Ex-Enron CEO Skilling's Resentenced to 14 Years." *USA Today*, June 21. Accessed February 14, 2014 (www.usatoday.com).

McCoy, Kevin. 2015. "Report: IRS Gave $18.8M to Firms with Tax Debts." *USA Today*, June 25.

McDaniels, Andrea, K., and Meredith Cohn. 2015. "Fetal Tissue Can Be Vital, Researchers Say." *Baltimore Sun*, August 16, 1, 20.

McDermott, Rose, James H. Fowler, and Nicholas A. Christakis. 2013. "Breaking Up Is Hard to Do, Unless Everyone Else Is Doing It Too: Social Network Effects on Divorce in a Longitudinal Sample." *Social Forces* 92 (2): 491–519.

McDonald, Kim A. 1999. "Studies of Women's Health Produce a Wealth of Knowledge on the Biology of Gender Differences." *Chronicle of Higher Education*, June 25, A19, A22.

McDonough, Katie. 2014. "The GOP's Other War on Women: 5 Gender Battlegrounds Beyond Abortion and Contraception." AlterNet, March 10. Accessed April 21, 2014 (www.alternet.org).

McEwan, Melissa. 2012. "Unbelievable: Man Beats Wife, Judge Orders Him to Take Her Out to Red Lobster and the Bowling Alley." AlterNet, February 9. Accessed February 10, 2012 (www.alternet.org).

McFadden, Robert D., and Angela Macropoulos. 2008. "Wal-Mart Employee Trampled to Death." *New York Times*, November 29, A16.

McFalls, Joseph A., Jr. 2007. "Population: A Lively Introduction, 5th edition." *Population Bulletin* 62 (March): 1–31.

McFarland, Mac. 2015. "How a Robot Ended Up Teaching Exercise Classes in a Dutch Retirement Home." *Washington Post*, May 27. Accessed October 25, 2015 (washingtonpost.com).

McGregor, Jena, and Steve Hamm. 2008. "Managing the Workforce." *BusinessWeek*, January 28, 34–43.

McGregor, Jena. 2015. "Google's Diversity Numbers Changed Little in Past Year." *Washington Post*, June 1. Accessed October 4, 2015 (washingtonpost.com).

McHale, Susan M. 2001. "Free-Time Activities in Middle Childhood: Links with Adjustment in Early Adolescence." *Child Development* 76 (November/December): 1764–1778.

McHale, Susan M., Kimberly A. Updegraff, and Shawn D. Whiteman. 2012. "Sibling Relationships and Influences in Childhood and Adolescence." *Journal of Marriage and Family* 74 (October): 913–930.

McIntosh, Peggy. 1995. "White Privilege and Male Privilege: A Personal Account of Coming to See Correspondences Through Work in Women's Studies." Pp. 76–87 in *Race, Class, and Gender: An Anthology*, 2nd edition, edited by Margaret L. Andersen and Patricia Hill Collins. Belmont, CA: Wadsworth.

McKenzie, Brian. 2013. "Out-of-State and Long Commutes: 2011." U.S. Census Bureau, February. Accessed September 19, 2014 (www.census.gov).

McKernan, Signe-Mary, Caroline Ratcliffe, Eugene Steuerle, and Sisi Zhang. 2013. "Less Than Equal: Racial Disparities in Wealth Accumulation." Urban Institute, April. Accessed February 6, 2014 (www.urban.org).

McKinnon, Mark. 2011. "Do We Still Need Unions? No: Let's End a Privileged Class." *Newsweek*, March 7, 19.

McNeil, David G. 2010. "U.S. Apologizes for Syphilis Program in Guatemala." *New York Times*, October 1, A1.

McNeill, David. 2012. "Japanese Fraud Case Highlights Weaknesses in Scientific Publishing." *Chronicle of Higher Education*, October 12, A16–A17.

McPhail, Clark, and Ronald T. Wohlstein. 1983. "Individual and Collective Behavior Within Gatherings, Demonstrations, and Riots." *Annual Review of Sociology* 9 (August): 579–600.

McPhail, Clark. 1991. *The Myth of the Madding Crowd*. New York: deGruyter.

McRae, Susan. 1999. "Cohabitation or Marriage?" Pp. 172–190 in *The Sociology of the Family*, edited by Graham Allan. Malden, MA: Blackwell Publishers.

Mead, George Herbert. 1934. *Mind, Self, and Society*. Chicago: University of Chicago Press.

Mead, George Herbert. 1964. *On Social Psychology*. Chicago: University of Chicago Press.

Mecia, Tony. 2015. "Financial Infidelity Poll: 6% Hid Bank Account from Spouse or Partner." Credit Cards, January 20. Accessed September 29, 2015 (creditcards.com).

"Media Coverage of Hispanics." 2009. Pew Hispanic Center Project for Excellence in Journalism, December 7. Accessed March 12, 2010 (pewhispanic.org).

"Medical Tourism." 2015. Centers for Disease Control and Prevention, February 23. Accessed October 17, 2015 (cdc.gov).

Mehl, Matthias R., Simine Vazire, Nairan Ramfrez-Esparza, Richard B. Slatcher, and James W. Pennebaker. 2007. "Are Women Really More Talkative Than Men?" Science 317 (July): 82.

Meier, Barry, and Eric Lipton. 2013. "F.D.A. Shift on Painkillers Was Years in the Making." *New York Times*, October 28, A1.

Meitiv, Danielle. 2015. "When Letting Your Kids Out of Your Sight Becomes a Crime." *Washington Post*, February 13. Accessed August 6, 2015 (washingtonpost.com).

Melton, Glennon. 2013. "5 Reasons Social Media Is Dangerous for Me." *Huffington Post*, October 1. Accessed December 4, 2013 (www.huffingtonpost.com).

Melucci, Alberto. 1995. "The New Social Movements Revisited: Reflections on a Sociological Misunderstanding." Pp. 107–119 in *Social Movements and Social Classes: The Future of Collective Action*, edited by Louis Maheu. Thousand Oaks, CA: Sage.

Mencher, Steve. 2013. "Visit Your Parents or Be Sued: Is That Un-American?" AARP, July 8. Accessed November 2, 2013 (www.aarp.org).

Mendelsohn, Oliver, and Maria Vicziany. 1998. *The Untouchables, Subordination, Poverty and the State in Modern India*. New York: Cambridge University Press.

Meng, Liu. 2011. "Chinese College Drops Plan to Discourage Kissing and Other 'Uncivilized' Behavior' on Campus." *Global Times*, April 22. Accessed October 25, 2011 (www.china.globaltimes.cn).

Menissi, Fatima. 1991. *The Veil and the Male Elite: A Feminist Interpretation of Women's Rights in Islam*. Reading, MA: Addison-Wesley.

Menissi, Fatima. 1996. *Women's Rebellion and Islamic Memory*. Atlantic Highlands, NJ: Zed Books.

Merryman, Ashley. 2013. "Losing Is Good for You." *New York Times*, September 25, A29.

Merton, Robert K. 1938. "Social Structure and Anomie." *American Sociological Review* 3 (December): 672–682.

Merton, Robert K. 1948/1996. "The Self-Fulfilling Prophecy." Pp. 183–201 in *Robert K. Merton: On Social Structure and Science*, edited by Piotr Sztompka. Chicago: University of Chicago Press.

Merton, Robert K. 1949. "Discrimination and the American Creed." Pp. 99–126 in *Discrimination and National Welfare*, edited by Robert M. MacIver. New York: Harper.

Merton, Robert K. 1968. *Social Theory and Social Structure*. New York: Free Press.

Merton, Robert K., and Alice K. Rossi. 1950. "Contributions to the Theory of Reference Group Behavior." Pp. 40–105 in *Continuities in Social Research*, edited by Robert K. Merton and Paul L. Lazarsfeld. New York: Free Press.

Merz, C. Noel Bairey. 2015. "Improving Gender Equity in Medical Research (Letter to the Editor)." *New York Times*, July 27, A16.

Messerli, Joe. 2012. "Should Affirmative Action Policies, Which Give Preferential Treatment Based on Minority Status, Be Eliminated?" BalancedPolitics.org, January 7. Accessed June 21, 2014 (www.balancedpolitics.org).

Messing, Jill T., Jacquelyn Campbell, Janet Sullivan Wilson, Sheryll Brown, and Beverly Patchell. 2014. "Police Departments' Use of the Lethality Assessment Program: A Quasi-Experimental Evaluation." National Institute of Justice, March 31. Accessed September 2, 2015 (www.nij.gov).

Meston, Cindy M., and David M. Buss. 2007. "Why Humans Have Sex." *Archives of Sexual Behavior* 36 (August): 477–507.

Meteyer, Karen B., and Maureen Perry-Jenkins. 2009. "Dyadic Parenting and Children's Externalizing Symptoms." *Family Relations* 58 (July): 289–302.

Mettler, Suzanne. 2014. "Equalizers No More: Politics Thwart Colleges' Role in Upward Mobility." *Chronicle Review*, March 7, B7–B11.

"M-Health: Health and Appiness." 2014. *The Economist*, February 1, 56–57.

Michels, Robert. 1911/1949. *Political Parties*. Glencoe, IL: Free Press.

Milan, Lynn. 2012. "Characteristics of Doctoral Scientists and Engineers in the United States: 2008." National Science Foundation, December. Accessed May 15, 2013 (www.nsf.gov).

Milgram, Stanley. 1963. "Behavioral Study of Obedience." *Journal of Abnormal and Social Psychology* 67 (4): 371–378.

Milgram, Stanley. 1965. "Some Conditions of Obedience and Disobedience to Authority." *Human Relations* 18 (February): 57–76.

"Millennials Outnumber Baby Boomers and Are Far More Diverse, Census Bureau Reports." 2015. U.S. Census Bureau, June 25. Accessed October 4, 2015 (census.gov).

Miller, Anthony B., Claus Wall, Cornelia J. Barnes, Ping Sun, Teresa To, and Steven A. Narod. 2014. "Twenty-Five Year Follow-Up for Breast Cancer Incidence and Mortality of the

Canadian National Breast Screening Study: Randomised Screening Trial." *BMJ* 348 (February): 1–5.

Miller, Claire C. 2013. "Angry over U.S. Surveillance, Tech Giants Bolster Defenses." *New York Times*, November 1, A1.

Miller, Claire C., Somini Sengupta. 2013. "Selling Secrets of Phone Users to Advertisers." *New York Times*, October 6, A1.

Miller, D. W. 2001. "DARE Reinvents Itself—With Help from Its Social-Scientist Critics." *Chronicle of Higher Education*, October 16, A12–A14.

Miller, Elizabeth, et al. 2015. "A School Health Center Intervention for Abusive Adolescent Relationships: A Cluster RCT." *Pediatrics* 135 (January): 76–85.

Miller, Lisa. 2010. "A Woman's Place Is in the Church." *Newsweek*, April 12, 34–41.

Miller, Lisa. 2012. "Feminism's Final Frontier? Religion." *Washington Post*, March 8. Accessed March 9, 2012 (www.washingtonpost.com).

Mills, C. Wright. 1956. *The Power Elite*. New York: Oxford University Press.

Mills, C. Wright. 1959. *The Sociological Imagination*. New York: Oxford University Press.

Mills, Theodore M. 1958. "Some Hypotheses on Small Groups from Simmel." *American Journal of Sociology* 63 (May): 642–650.

Mirabella, Lorraine. 2015. "More Foreign Labor OK'd." *Baltimore Sun*, June 6, 1, 15.

Mischel, Walter. 1966. "A Social Learning View of Sex Differences." Pp. 57–81 in *The Development of Sex Differences*, edited by Eleanor E. Maccoby. Stanford, CA: Stanford University Press.

Mishel, Lawrence, and Alyssa Davis. 2015. "Top CEOs Make 300 Times More Than Typical Workers." Economic Policy Institute, June 21. Accessed October 7, 2015 (epi.org).

Mishel, Lawrence, Josh Bivens, Elise Gould, and Heidi Shierholz. 2012. *The State of Working America*, 12th edition. Ithaca, NY: Cornell University Press.

Mitnik, Pablo A., and David B. Grusky. 2015. "Economic Mobility in the United States." The Pew Charitable Trusts, July. Accessed September 11, 2015 (pewtrusts.org).

"Modern Immigration Wave Brings 59 Million to U.S., Driving Population Growth and Change Through 2065." 2015. Pew Research Center, September 28. Accessed October 4, 2015 (pewresearch.org).

Molin, Anna. 2012. "In Sweden, Playtime Goes Gender-Neutral for Holidays." *Wall Street Journal*, November 29, D1–D2.

Money, John, and Anke A. Ehrhardt. 1972. *Man & Woman, Boy & Girl: The Differentiation and Dimorphism of Gender Identity from Conception to Maturity*. Baltimore, MD: Johns Hopkins University Press.

Montanaro, Domenico, Rachel Wellford, and Simone Pathe. 2014. "Money Is Pretty Good Predictor of Who Will Win Elections." Public Broadcasting Service, November 11. Accessed November 23, 2014 (www.pbs.org).

Montez, Jennifer Karas, and Anna Zajacova. 2013. "Explaining the Widening Education Gap in Mortality Among U.S. White Women." *Journal of Health and Social Behavior* 54 (June): 165–181.

Monto, Martin and Anna Carey. 2014. "A New Standard of Sexual Behavior? Are Claims Associated with the 'Hookup Culture' Supported by General Social Survey Data?" *Journal of Sex Research* 51(6): 605–615.

Mooney, Chris. 2015. "The Remote Alaskan Village That Needs to Be Relocated Due to Climate Change." *Washington Post*, February 24. Accessed October 23, 2015 (washingtonpost.com).

Moore, Elizabeth S. 2006. "It's Child's Play: Advergaming and the Online Marketing of Food to Children." Henry J. Kaiser Family Foundation, July. Accessed April 12, 2008 (www.kff.org).

Moore, Kathleen. 2008. "Low-Income Homes Green—and Affordable." *Daily Gazette*, July 1. Accessed September 21, 2008 (www.dailygazette.com).

Moore, Latetia V., and Frances E. Thompson. 2015. "Adults Meeting Fruit and Vegetable Intake Recommendations—United States, 2013." *MMWR* 64 (26): 719–725.

Moore, Peter. 2015. "A Third of Young Americans Say They Aren't 100% Heterosexual." YouGov, August 20. Accessed September 29, 2015 (today.yougov.com).

Morello, Carol, and Scott Clement. 2014. "'Happy Days' No More: Middle-Class Families Squeezed as Expenses Soar, Wages Stall." *Washington Post*, April 26. Accessed March 27, 2014 (www.washingtonpost.com).

Morello, Carol. 2010. "An Unexpected Result for Some Census Takers: The Wrath of Irate Americans." *Washington Post*, June 20, A4.

Morello, Carol. 2014. "Census to Change the Way It Counts Gay Married Couples." *Washington Post*, May 26. Accessed October 23, 2014 (www.washingtonpost.com).

Morgan, Rachel E., and Britney J. Mason. 2014. "Crimes Against the Elderly, 2003–2013." Bureau of Justice Statistics, November. Accessed September 2, 2015 (bjs.gov).

Morgenson, Gretchen. 2004. "No Wonder C.E.O.'s Love Those Mergers." *New York Times*, July 18, C1.

Morin, Rich, and Seth Motel. 2012. "A Third of Americans Now Say They Are in the Lower Class." Pew Social & Demographic Trends, September 10. Accessed February 6, 2014 (www.pewsocialtrends.org).

Morin, Richard. 2009. "What Divides America?" Pew Research Center, September 24. Accessed March 2, 2010 (www.pewresearch.org).

Morris, Desmond. 1994. *Bodytalk: The Meaning of Human Gestures*. New York: Crown Trade Paperbacks.

Morris, Theresa. 2014. "C-Section Epidemic." *Contexts* 13 (Winter): 70–72.

Morrison, Denton E. 1971. "Some Notes Toward Theory on Relative Deprivation, Social Movements, and Social Change." *American Behavioral Scientist* 14 (May–June): 675–690.

Moss-Racusin, Corrine A. John F. Dovidio, Victoria L. Brescoli, Mark, J. Graham, and Jo Handelsman. 2012. "Science Faculty's Subtle Gender Biases Favor Male Students." *Proceedings of the National Academy of Sciences* 109 (October 9): 16474–16479.

Motel, Seth, and Eileen Patten. 2013. "Statistical Portrait of the Foreign-Born Population in the United States, 2011." Pew Research Hispanic Trends Project, January 29. Accessed March 22, 2014 (www.pewhispanic.org).

Motivans, Mark. 2013. "Federal Justice Statistics 2010—Statistical Tables." Bureau of Justice Statistics, December. Accessed January 12, 2014 (www.bjs.gov).

Moyer, Imogene L. 2001. *Criminological Theories: Traditional and Nontraditional Voices and Themes*. Thousand Oaks, CA: Sage.

Moyer, Imogene L. 2003. "Jane Addams: Pioneer in Criminology." *Women & Criminal Justice* 14 (3/4): 1–14.

Mozzafarian, Dariush, and David S. Ludwig. 2015. "Stop Fearing Fat." *New York Times*, July 9, A25.

Mudallal, Zainab. 2015. "Nearly Half of Americans Didn't Take a Vacation Day in 2014." Quartz, January 5. Accessed October 7, 2015 (qz.com).

Mullen, Ann. 2012. "The Not-So-Pink Ivory Tower." *Contexts* 4 (Fall): 34–38.

Mumford, Lewis. 1961. *The City in History: Its Origins, Transformations, and Its Prospects*. New York: Harcourt, Brace.

Munsell, Christina R., Jennifer L. Harris, Vishnudas Sarda, and Marlene B. Schwartz. 2015. "Parents' Beliefs About the Healthfulness of Sugary Drink Options: Opportunities to Address Misperceptions." *Public Health Nutrition*, March 11, 1-9 published online ahead of print. Accessed October 25, 2015 (http://dx.doi.org/10.1017/S1368980015000397).

Murdock, George P. 1940. "The Cross-Cultural Survey." *American Sociological Review*, 5: 361–370.

Murphey, David, Megan Barry, and Brigitte Vaughn. 2013. "Mental Health Disorders." Child Trends, January. Accessed July 26, 2014 (www.childtrends.org).

Murphy, Cait. 2005. "Fast-Forward to the Future." *Fortune*, September 19, 271.

Murphy, Caryle. 2009. "Behind the Veil." *Christian Science Monitor*, December 13, 12–17.

Murphy, John. 2004. "S. Africa's New Goal: Economic Equality." *Baltimore Sun*, April 27, 1A, 4A.

Murphy, Kevin. 2013. "Hallmark Responds to 'Gay' Ornament Controversy with Regret." *Huffington Post*, November 1. Accessed November 1, 2013 (www.huffingtonpost.com).

Murray, Charles. 1984. *Losing Ground: American Social Policy, 1950–1980*. New York: Basic Books.

Murray, Charles. 2012. *Coming Apart: The State of White America, 1960–2010*. New York: Crown Forum.

Muskal, Michael. 2015. "Still Wedded to Beliefs, Ky. Clerk Freed from Jail." *Baltimore Sun*, September 9, 5.

Mutzabaugh, Ben. 2011. "Reports: Thai Airline Recruits 'Third-Sex' Attendants." *USA Today*, January 29. Accessed October 25, 2011 (www.usatoday.com).

Myers, John P. 2007. *Dominant–Minority Relations in America: Convergence in the New World*, 2nd edition. Boston: Allyn & Bacon.

Nagoski, Emily. 2015. "Nothing Is Wrong with Your Sex Drive." *New York Times*, January 27, A21.

Naik, Gautam. 2013. "New Advance Toward 'Designer Babies.'" *Wall Street Journal*, October 4, A3.

Naili, Hajer. 2011. "Study Details Sex-Traffic in Post-Saddam Iraq." Women's e-News, November 9. Accessed February 25, 2012 (www.womensenews.org).

Nakashima, Ellen. 2015. "With a Series of Major Hacks, China Builds a Database on Americans." *Washington Post*, June 5. Accessed October 25, 2015 (washingtonpost.com).

Nash, Elizabeth, Rachel Benson Gold, Andrea Rowan, Gwendolyn Rathbun, and Yana Vierboom. 2014. "Laws Affecting Reproductive Health and Rights: 2013 State Policy Review." Guttmacher Institute. Accessed March 2, 2014 (www.guttmacher.org).

National Alliance to End Homelessness. 2015. "The State of Homelessness in America 2015." Accessed September 11, 2015 (endhomelessness.org).

"National Census of Fatal Occupational Injuries in 2014 (Preliminary Results)." 2015. Bureau of Labor Statistics, September 17. Accessed October 16, 2015 (bls.gov).

National Center for Education Statistics. 2013. "The Nation's Report Card: A First Look: 2013 Mathematics and Reading." U.S. Department of Education, November. Accessed June 21, 2014 (www.nationsreprtcard .gov).

National Center for Health Statistics. 2014. "Health, United States, 2013." May. Accessed July 26, 2014 (www.cdc.gov/nchs).

National Center for Health Statistics. 2015. "Health, United States, 2014." May. Accessed October 8, 2015 (cdc.gov/nchs).

National Center on Elder Abuse. 2012. "Statistics/ Data." Accessed May 16, 2014 (www.ncea .aoa.gov).

National Conference of State Legislatures. 2014. "State Laws Regarding Marriages Between First Cousins." Accessed May 16, 2014 (www .ncsl.org).

National Diabetes Statistics Report: Estimates of Diabetes and Its Burden in the United States. 2014. Centers for Disease Control and Prevention. Atlanta, GA: U.S. Department of Health and Human Services.

"National Grandparents Day 2014: Sept. 7." 2014. U.S. Census Bureau, July 8. Accessed October 20, 2014 (www.census.gov).

National Institute of Mental Health. 2011. "The Teen Brain: Still Under Construction." Accessed November 20, 2013 (www.nimh .nih.gov).

National Institute on Alcohol Abuse and Alcoholism. 2013. "College Drinking." July. Accessed July 26, 2014 (www.niaaa.nih.gov).

National Institute on Alcohol Abuse and Alcoholism. 2014. "Alcohol Facts and Statistics." May. Accessed July 30, 2014 (www.niaaa.nih.gov).

National Institute on Drug Abuse. 2012. "Principles of Drug Addiction Treatment: A Research-Based Guide." Accessed January 15, 2014 (www.drugabuse.gov).

National Institutes of Health. 2006. "Stem Cell Information." Accessed August 6, 2008 (www .nih.gov).

National Oceanic and Atmospheric Administration. 2014. "Heat Wave: A Major Summer Killer." NOAA Watch. Accessed September 19, 2014 (www.noaawatch.gov).

National Organization on Fetal Alcohol Syndrome. 2012. "FASD: What Everyone Should Know." Accessed July 25, 2014 (www.nofas.org).

National Public Radio. 2010. "Black Male Privilege?" Interview transcript, March 4. Accessed March 10, 2010 (www.npr.org).

National Retail Federation. 2013. "Retailers Estimate Holiday Return Fraud Will Cost Them $3.4 Billion, According to NRG Survey." December 6. Accessed January 15, 2014 (www.nrf.com).

National Science Board. 2014. *Science and Engineering Indicators 2014.* Arlington VA: National Science Foundation.

National Survey of Student Engagement. 2013. "A Fresh Look at Student Engagement— Annual Results 2013." Bloomington, IN: Indiana University Center for Postsecondary Research. Accessed June 21, 2014 (www.nsse .iub.edu).

National White Collar Crime Center. 2013. "Cyberstalking." May. Accessed December 4, 2013 (www.nw3c.org).

National Women's Law Center. 2014. "Women's Unemployment Rises Despite Job Gains, Growth Concentrated in Low-Wage Sectors, NWLC Analysis Shows." April 4. Accessed April 12, 2014 (www.nwlc.org).

Nechepurenko, Ivan. 2014. "Russia Reverses Birth Decline—But for How Long?" *The Moscow Times,* June 22. Accessed September 21 (www.themoscowtimes.com).

Neelakantan, Shailaja. 2006. "In India, Conservatives Want Women Under Wraps." *Chronicle of Higher Education,* May 26, A47–A48.

Neff, Roni A., Marie L. Spiker, and Patricia L. Truant. 2015. "Wasted Food: U.S. Consumers' Reported Awareness, Attitudes, and Behaviors." *PLoS One* 10 (6): doi:10.1371/journal.pone.0127881. Accessed October 22, 2015 (journals.plos.org).

Neider, Linda L., and Chester A. Schriesheim, eds. 2005. *Understanding Teams.* Greenwich, CT: Information Age.

Nelson, Colleen M. 2013. "Poll: Most Women See Bias in the Workplace." *Wall Street Journal,* April 12, A4.

Nestle, Marion. 2012. "Genetically Modified Food Myths and Truths—A Critical Review of the Science." AlterNet, June 21. Accessed June 26, 2012 (www.alternet.org).

Neuhaus, Jessamyn. 2010. "Marge Simpson, Blue-Haired Housewife: Defining Domesticity on *The Simpsons.*" *The Journal of Popular Culture* 43 (August): 761–781.

"The New American Father." 2013. Pew Research Center, June 14. Accessed May 16, 2014 (www.pewresearch.org).

Newberry, Sydne. 2013. "What's in a Name? Calling Obesity a Disease Could Help Improve Chronic Disease Outcomes." The RAND Blog, October 16. Accessed July 26, 2014 (www.rand.org).

Newhouse, Joseph P., Alan M. Garber, Robin P. Graham, Margaret A. McCoy, Michelle Mancher, and Ashna Kibria, eds. 2013. *Variation in Health Care Spending: Target Decision Making, Not Geography.* Washington, DC: The National Academies Press.

Newkirk, Margaret, and Gigi Douban. 2012. "Legal Immigrants Wanted for Dirty Jobs." *Bloomberg Businessweek,* October 8–14, 34, 36.

Newman, Andrew Adam. 2015. "When 'His' Becomes 'Theirs.'" *New York Times,* March 2, D1, D5.

Newman, John. 2014. "Coaches, Not Presidents, Top Public-College Pay List." *Chronicle of Higher Education,* May 16. Accessed June 14, 2014 (www.chronicle.com).

Newport, Frank, and Brandon Busteed. 2013. "Americans Still See College Education as Very Important." Gallup, December 17. Accessed June 21, 2014 (www.gallup.com).

Newport, Frank, and Igor Himelfarb. 2013. "In U.S., Strong Link Between Church Attendance, Smoking." Gallup, August 5. Accessed June 21, 2014 (www.gallup.com).

Newport, Frank, and Joy Wilke. 2013. "Desire for Children Still Norm in U.S." Gallup, September 25. Accessed May 16, 2014 (www .gallup.com).

Newport, Frank, Dan Witters, and Sangeeta Agrawal. 2012. "In U.S., Very Religious Americans Have Higher Wellbeing Across All Faiths." Gallup, December 1. Accessed April 8, 2012 (www.gallup.com).

Newport, Frank, Jeffrey M. Jones, and Lydia Saad. 2014. "State of the Union: The Public Weighs In on 10 Key Issues." Gallup, January 31. Accessed March 22, 2014 (www.gallup.com).

Newport, Frank. 2011. "Americans Prefer Boys to Girls, Just as They Did in 1941." Gallup, June 23. Accessed October 25, 2011 (www .gallup.com).

Newport, Frank. 2012. "Americans Like Having a Rich Class, as They Did 22 Years Ago." Gallup. May 11. Accessed February 14, 2014 (www.gallup.com).

Newport, Frank. 2012. "Seven in 10 Are Very or Moderately Religious." Gallup, December 4. Accessed June 21, 2014 (www.gallup.com).

Newport, Frank. 2013. "In U.S., 87% Approve of Black-White Marriage, vs. 4% in 1958." Gallup, July 25. Accessed March 22, 2014 (www.gallup.com).

Newport, Frank. 2013. "In U.S., Four in 10 Report Attending Church in Last Week." Gallup, December 24. Accessed June 21, 2014 (www .gallup.com).

Newport, Frank. 2014. "In U.S., 42% Believe Creationist View of Human Origins." Gallup, June 4. Accessed June 21, 2014 (www.gallup .com).

Newport, Frank. 2014. "Mississippi Most Religious State, Vermont Least Religious." Gallup, February 3. Accessed June 21, 2014 (www .gallup.com).

Newport, Frank. 2015. "A Tale of Two Supreme Court Decisions." Gallup, July 2. Accessed September 29, 2015 (gallup.com).

Newport, Frank. 2015. "Americans Greatly Overestimate Percent Gay, Lesbian in U.S." Gallup, May 21. Accessed September 29, 2015 (gallup.com).

Newport, Frank. 2015. "Fewer Americans Identify as Middle Class in Recent Years." Gallup, April 28. Accessed May 2, 2015 (www.gallup .com).

Newport, Frank. 2015. "In U.S., 60% Satisfied with Ability to Get Ahead." Gallup, January 30. Accessed May 2, 2015 (www.gallup.com).

Newport, Frank. 2015. "Most U.S. Smartphone Owners Check Phone at Least Hourly." Gallup, July 9. Accessed August 14, 2015 (gallup.com).

Nichols, James. 2014. "Cecil Chao, Hong Kong Billionaire, Doubles Reward for Any Man Who Can Make Lesbian Daughter Straight." *Huffington Post,* January 24. Accessed March 2, 2014 (www.huffingtonpost.com).

Niederdeppe, Jeff, Sahara Byrne, Rosemary J. Avery, and Jonathan Cantor. 2013. "Direct-to-Consumer Television Advertising Exposure, Diagnosis with High Cholesterol, and Statin Use." *Journal of General Internal Medicine* 28 (July): 886–893.

Nielsen Company, The. 2008. "College Spring Break Study." February 27. Accessed April 10, 2008 (www.alcoholstats.com).

Niquette, Mark, and Richard Rubin. 2014. "States Target Corporate Cash Stashed Overseas." *Bloomberg Businessweek,* April 21–27, 27–28.

Nir, Sarah Maslin. 2015. "The Price of Nice Nails." *New York Times,* May 7. Accessed October 4, 2015 (nytimes.com).

"No Action on Greenhouse Gases." 2008. *Baltimore Sun,* July 12, 2A.

NOAA National Centers for Environmental Information. 2015. "State of the Climate: Global Analysis for July 2015." Accessed October 23, 2015 (ncdc.noaa.gov).

Norris, Tina, Paula L. Vines, and Elizabeth M. Hoeffel. 2012. "The American Indian and Alaska Native Population: 2010." U.S. Census Bureau. 2010 Census Briefs, January. Accessed March 20, 2012 (www.census.gov).

Nossiter, Adam. 2011. "Hinting at an End to a Curb on Polygamy, Interim Libyan Leader Stirs Anger." *New York Times,* October 30, A6.

"Number of Jobs Held, Labor Market Activity, and Earnings Growth Among the Youngest

Baby Boomers: Results from a Longitudinal Survey." 2012. Bureau of Labor Statistics, News Release, July 25. Accessed November 18, 2013 (www.bls.gov).

Nyseth, Hollie, Sarah Shannon, Kia Heise, and Suzy Maves McElrath. 2011. "Embedded Sociologists." *Contexts* 10 (Spring): 44–50.

O'Brien, Jodi, and Peter Kollock. 2001. *The Production of Reality: Essays and Readings on Social Interaction*, 3rd edition. Thousand Oaks, CA: Pine Forge Press.

O'Connor, Anahad. 2014. "New Concern About Testosterone and Heart Risks." *New York Times*, January 30, A13.

O'Connor, Siobhan. 2015. "Farm-to-Fridge Is Heating Up." *Time*, August 24, 28.

O'Donnell, Victoria. 2007. *Television Criticism*. Thousand Oaks, CA: Sage.

O'Keefe, Ed, Matea Gold, and David A. Fahrenthold. 2015. "Bush Tax Forms Show Income of $29 Million Since 2007." *Washington Post*, June 30. Accessed August 31, 2015 (washingtonpost.com).

O'Keefe, Kevin. 2014. "TV's Renaissance for Strong Women Is Happening in a Surprising Place." *The Atlantic*, October 9. Accessed September 29, 2015 (theatlantic.com).

Oakes, Jeannie. 1985. *Keeping Track: How Schools Structure Inequality*. New Haven, CT: Yale University Press.

Obach, Brian K. 2004. *Labor and the Environmental Movement: The Quest for Common Ground*. Cambridge, MA: MIT Press.

Oberschall, Anthony. 1995. *Social Movements: Ideologies, Interests, and Identities*. New Brunswick, NJ: Transaction.

"Obesity Prevalence Maps." 2015. Centers for Disease Control and Prevention, September 11. Accessed October 15, 2015 (cdc.gov).

OECD. 2011. "Building a High-Quality Teaching Profession." Accessed June 21, 2014 (www.oecd.org).

OECD. 2013. "Education at a Glance 2013." Accessed June 21, 2014 (www.oecd.org).

OECD. 2013. "Results from PISA 2012: United States." Accessed June 22, 2014 (www.oecd.org).

OECD. 2014. "Income Inequality Update." June. Accessed November 10, 2014 (www.oecd.org).

OECD. 2014. "Society at a Glance 2014: OECD Social Indicators." Accessed July 26, 2014 (www.oecd.org).

OECD. 2014. *OECD Factbook 2014: Economic, Environmental and Social Statistics*. OECD Publishing. Accessed July 26, 2014 (www.oecd.org).

OECD/European Union. 2015. Indicators of Immigrant Integration 2015: Settling In. Paris: OECD Publishing.

Office of Inspector General. 2013. "Prescribers with Questionable Patterns in Medicare Part D." June. Accessed July 26, 2014 (www.oig.hhs.gov).

Office of the Deputy Chief of Staff for Intelligence. 2006. "Arab Cultural Awareness: 58 Factsheets." U.S. Army Training and Doctrine Command, Ft. Leavenworth, Kansas, January. Accessed February 15, 2006 (www.fas.org).

Office of the Inspector General. 2015. "Qualifying for Disability Benefits in Puerto Rico Based on an Inability to Speak English." Social Security Administration, April. Accessed October 4, 2015 (oig.ssa.gov).

Office of the United Nations High Commissioner for Human Rights. 2010. "The Right to Water. Fact Sheet No. 35." United Nations,

August. Accessed November 10, 2014 (www.un.org).

OfficeTime. 2015. "Top 10 Time Killers." Accessed August 23, 2015 (officetime.net).

Ogburn, William F. 1922. *Social Change with Respect to Culture and Original Nature*. New York: Dell.

Ogunro, Nola. 2012. "10 Great Paying Jobs You Can Get Without a 4 Year College Degree." *Alternet*, September 21. Accessed June 23, 2014 (www.alternet.org).

Ogunwole, Stella U., Malcolm P. Drewery, Jr., and Merarys Rios-Vargas. 2012. "The Population with a Bachelor's Degree or Higher by Race and Hispanic Origin: 2006–2010." U.S. Census Bureau, May. Accessed March 22, 2014 (www.census.gov).

Olinto, Pedro, Kathleen Beegle, Carlos Sobrado, and Hiroki Uematsu. 2013. "The State of the Poor: Where Are the Poor, Where Is Extreme Poverty Harder to Find, and What Is the Current Profile of the World's Poor?" World Bank, October. Accessed February 6, 2014 (www.worldbank.org).

Oliver, Melvin L., and Thomas M. Shapiro. 1995. *Black Wealth/White Wealth: A New Perspective on Racial Inequality*. New York: Routledge.

Olson, Jonathan R. 2010. "Choosing Effective Youth-Focused Prevention Strategies: A Practical Guide for Applied Family Professionals." *Family Relations* 59 (April): 207–220.

Olson, Theodore B. 2010. "The Conservative Case for Gay Marriage." *Newsweek*, January 18, 48–53.

Ong, Mei-Sing, and Kenneth D. Mandl. 2015. "National Expenditure for False-Positive Mammograms and Breast Cancer Overdiagnoses Estimated at $4 Billion a Year." *Health Affairs* 34 (4): 576–583.

Oppel, Richard A., Jr. 2011. "Steady Decline in Major Crime Baffles Experts." *New York Times*, May 4, A17.

Orenstein, Peggy. 2011. "Should the World of Toys Be Gender-Free?" *New York Times*, December 30, A23.

Oreopoulos, Philip, and Uros Petronijevic. 2013. "Making College Worth It: A Review of Research on the Returns to Higher Education." Accessed June 21, 2014 (www.ticas.org).

Ornstein, Charles. 2015. "New Hepatitis C Drugs Are Costing Medicare Billions." *Washington Post*, March 29. Accessed October 16, 2015 (washingtonpost.com).

Oropesa, R. Salvatore, and Nancy S. Landale. 2004. "The Future of Marriage and Hispanics." *Journal of Marriage and Family* 66 (November): 901–920.

Ortman, Jennifer M., Victoria A. Velkoff, and Howard Hogan. 2014. "An Aging Nation: The Older Population in the United States." U.S. Census Bureau, May. Accessed May 16, 2014 (www.census.gov).

Oudekerk, Barbara, Dara Blachman-Demner, and Carrie Mulford. 2014. "Teen Dating Violence: How Peers Can Affect Risk & Protective Factors." National Institute of Justice, November. Accessed August 6, 2015 (nij.gov).

Ovide, Shira, and Rachel Feintzeig. 2013. "Microsoft Abandons Dreaded 'Stack.'" *Wall Street Journal*, November 13, B1, B5.

Owen, Daniela J., Amy M. S. Slep, and Richard E. Heyman. 2012. "The Effect of Praise, Positive Nonverbal Response, Reprimand, and Negative Nonverbal Response on Child Compliance: A Systematic Review." *Clinical*

Child and Family Psychology Review 15 (December): 364–385.

Oxfam. 2013. "The Cost of Inequality: How Wealth and Income Extremes Hurt Us All." January 18. Accessed February 6, 2014 (www.oxfam.org).

Oxfam. 2014. "Even It Up: Time to End Extreme Inequality." October. Accessed November 10, 2014 (www.oxfam.org).

Packard, Vance. 1959. *The Status Seekers*. New York: David McKay.

Padgett, Tim. 2010. "Robes for Women." *Time*, September 27, 53–55.

Palen, J. John. 2014. *The Urban World*, 10th edition. New York: Oxford University Press.

Palermo, Elizabeth. 2015. "What You Need to Know About the New Libido Pill for Women." *Huffington Post*, August 24. Accessed September 29, 2015 (huffingtonpost.com).

Palmeri, Christopher, and Peter S. Green. 2010. "Former Executives Bomb at the Ballot Box." *Bloomberg Businessweek*, November 8–14, 35–36.

Park, Robert, and Ernest Burgess. 1921. *Introduction to the Science of Sociology*. Chicago: University of Chicago Press.

Parker, Kim, and Eileen Patten. 2013. "The Sandwich Generation: Rising Financial Burdens for Middle-Aged Americans." Pew Research Center, January 30. Accessed May 22, 2014 (www.pewsocialtrends.org).

Parker, Kim, and Juliana Menasce Horowitz. 2015. "Family Support in Graying Societies." Pew Research Center, May 21. Accessed July 22, 2015 (pewresearch.org).

Parker, Kim, and Juliana Menasce Horowitz. 2015. "Women and Leadership." Pew Research Center, January 14. Accessed September 29, 2015 (pewresearch.org).

Parker, Kim, and Wendy Wang. 2013. "Modern Parenthood." Pew Research Center, March 14. Accessed March 10, 2014 (www.pewresearch.org).

Parker, Kim, Rich Morin, Juliana Menasce Horowitz, and Mark Hugo Lopez. 2015. "Multiracial in America." Pew Research Center, June 11. Accessed June 13, 2015 (pewresearch.org).

Parker, Kim, Richard Fry, D'Vera Cohn, and Wendy Wang. 2011. "Is College Worth It? College Presidents, Public Assess Value, Quality and Mission of Higher Education." Pew Research Center, May 16. Accessed April 15, 2012 (www.pewsocialtrends.org).

Parker, Kim. 2012. "The Boomerang Generation: Feeling OK About Living with Mom and Dad." Pew Social & Demographic Trends, March 15. Accessed November 20, 2013 (pewsocialtrends.org).

Parker-Pope, Tara. 2011. "Web of Popularity, Achieved by Bullying." *New York Times*, February 14, A1.

Parrott, Scott, and Caroline Titcomb Parrott. 2015. "U.S. Television's 'Mean World' for White Women: The Portrayal of Gender and Race on Fictional Crime Dramas." *Sex Roles* 73 (June): 70–82.

Parsons, Talcott, and Robert F. Bales, eds. 1955. *Family, Socialization, and Interaction Process*. New York: Free Press.

Parsons, Talcott. 1951. *The Social System*. Glencoe, IL: Free Press.

Parsons, Talcott. 1954. *Essays in Sociological Theory*, revised edition. New York: Free Press.

Parsons, Talcott. 1959. "The School Class as a Social System: Some of Its Functions in American Society." *Harvard Educational Review* 29 (Fall): 297–313.

Parsons, Talcott. 1960. *Structure and Process in Modern Societies*. New York: Free Press.

Partnership for Public Service. 2011. "Scores by Effective Leadership." Accessed August 13, 2011 (www.bestplacestowork.org).

Partnership for Public Service. 2013. "Federal Leadership on the Decline." April. Accessed December 19, 2013 (www.ourpublicservice.org).

Passel, Jeffrey S., and D'Vera Cohn. 2008. "U.S. Population Projections: 2005–2050." Pew Research Center, February 11. Accessed November 15, 2011 (www.pewresearch.org).

Passel, Jeffrey S., and D'Vera Cohn. 2015. "Unauthorized Immigrant Population Stable for Half a Decade." Pew Research Center, July 22. Accessed October 4, 2015 (pewresearch.org).

Passel, Jeffrey S., D'Vera Cohn, and Mark Hugo Lopez. 2011. "Hispanics Account for More Than Half of Nation's Growth in Past Decade." Pew Research Center, March 24. Accessed December 7, 2011 (www.pewhispanic.org).

Patchin, Justin W. 2013. "Cyberbullying Research: 2013 Update." Cyberbullying Research Center, November 20. Accessed December 4, 2013 (www.cyberbullying.us).

Paul, Annie Murphy. 2010. *Origins: How the Nine Months Before Birth Shape the Rest of Our Lives*. New York: Free Press.

Paul, Noel. C. 2002. "The Birth of a Would-Be Fad." *Christian Science Monitor*, September 23, 11, 14–16.

Paul, Richard, and Linda Elder. 2007. *The Miniature Guide to Critical Thinking: Concepts and Tools*. Dillon Beach, CA: Foundation for Critical Thinking.

Paulson, Amanda. 2014. "Changes of Address." *Christian Science Monitor Weekly*, May 26, 12–13.

Paulson, Amanda. 2015. "What Does It Mean to Be a Good Parent?" *Christian Science Monitor Weekly*, May 4, 21–23.

Payne, David, and Glenn Somerville. 2015. "Where the Jobs Will Be." *Baltimore Sun*, April 19, 3.

Payne, K. K., and J. Copp. 2013. "Young Adults in the Parental Home and the Great Recession." National Center for Family & Marriage Research. Accessed November 20, 2013 (www.hcfmr.bgsu.edu).

Pazol, Karen, Andreea A. Creanga, Kim D. Burley, and Denise J. Jamieson. 2014. "Abortion Surveillance—United States, 2011." *MMWR* 63 (11): 1–41.

Pear, Robert. 2015. "Tech Rivalries Impede Digital Medical Record Sharing." *New York Times*, May 27, A13.

Pearlstine, Norman. 2013. "Fix This: Water." *Bloomberg Businessweek*, March 25–31, 46–51.

Pedersen, Paul. 1995. *The Five Stages of Culture Shock: Critical Incidents Around the World*. Westport, CT: Greenwood Press.

Pedulla, David S., and Sarah Thébaud. 2015. "Can We Finish the Revolution? Gender, Work-Family Ideals, and Institutional Constraint." *American Sociological Review* 80 (February): 116–139.

Peeples, Lynne. 2014. "Supreme Court Decision Is a Boon for Polluters, Critics Say." *Huffington Post*, June 10. Accessed September 19, 2014 (www.huffingtonpost.com).

Pell, Nicholas. 2011. "Beyond Occupy Wall Street: 11 American Uprisings You've Never Heard of That Changed the World." AlterNet, October 21. Accessed October 22, 2011 (www.alternet.org).

Perlmutter, David D. 2001. "Students Are Blithely Ignorant; Professors Are Bitter." *Chronicle of Higher Education*, July 27, B20.

Perrin, Andrew, and Maeve Duggan. 2015. "Americans' Internet Access: 2000–2015." Pew Research Center June 26. Accessed July 22, 2015 (pewresearch.org).

Perry, Gina. 2013. *Behind the Shock Machine: The Untold Story of the Notorious Milgram Psychology Experiments*. New York: New Press.

Persell, Caroline Hodges, and Peter W. Cookson, Jr. 1985. "Chartering and Bartering: Elite Education and Social Reproduction." *Social Problems* 33 (December): 114–129.

Peter, Tom A. 2012. "Mistreatment of Afghan Women Caused by Far More Than Taliban." *Christian Science Monitor*, January 31. Accessed March 4, 2014 (www.csmonitor.com).

Peters, Mary Ann, and Mark P. Lagon. 2015. "More Women in Office." *Baltimore Sun*, September 17, 17.

Petersen, Andrea. 2012. "Smarter Ways to Discipline Kids." *Wall Street Journal*, December 26, D1, D3.

Peterson, James L., Josefina J. Card, Marvin B. Eisen, and Bonnie Sherman-Williams. 1994. "Evaluating Teenage Pregnancy Prevention and Other Social Programs: Ten Stages of Program Assessment." *Family Planning Perspectives* 26 (May): 116–120, 131.

Peterson, Scott. 2008. "In Iran, Barbie Seen as Cultural Invader." *Christian Science Monitor*, September 15, 4.

Petrosino, Anthony, Carolyn Turpin-Petrosino, Meghan E. Hollis-Peel, and Julia G. Lavenberg. 2013. "Scared Straight and Other Juvenile Awareness Programs for Preventing Juvenile Delinquency: A Systematic Review." Campbell Systematic Reviews. Accessed September 2, 2015 (campbellcollaboration.org).

Pettypiece, Shannon. 2013. "Anything You Can Do, I Can Do Better." *Bloomberg Businessweek*, March 11–17, 27–29.

Pew Center on the States. 2010. "Prison Count 2010." April. Accessed August 24, 2011 (www.pewcenteronthestates.org).

Pew Charitable Trusts. 2015. "The Precarious State of Family Balance Sheets." January. Accessed September 11, 2015 (pewtrusts.org).

Pew Forum on Religion & Public Life. 2008. "U.S. Religious Landscape Survey: Religious Beliefs and Practices: Diverse and Politically Relevant." June. Accessed June 26, 2008 (www.religions.pewforum.org).

Pew Forum on Religion & Public Life. 2009. "Faith in Flux: Changes in Religious Affiliation in the U.S." April. Accessed April 25, 2010 (www.pewforum.org).

Pew Hispanic Center. 2012. "When Labels Don't Fit: Hispanics and Their Views of Identity." Pew Research Center, April 4. Accessed March 22, 2014 (www.pewhispanic.org).

Pew Research Center for the People & the Press. 2010. "83%—Support Christmas Displays in Public." Accessed April 26, 2010 (www.pewresearch.org).

Pew Research Center for the People & the Press. 2014. "Thirteen Years of the Public's Top Priorities." January 27. Accessed March 22, 2014 (www.people-press.org).

Pew Research Center. 2015. "How Americans View the Top Energy and Environmental Issues." January 15. Accessed October 23, 2015 (pewresearch.org).

Pewewardy, Cornel. 1998. "Fluff and Feathers: Treatment of American Indians in the Literature and the Classroom." *Equity & Excellence in Education* 31 (April): 69–76.

Pfeffer, Fabian T., Sheldon Danziger, and Robert F. Schoeni. 2014. "Wealth Levels, Wealth Inequality, and the Great Recession." Russell Sage Foundation, June. Accessed November 10, 2014 (www.russellsage.org).

Pflanz, Mike. 2014. "Briefing: Africa's Stance on Gays." *Christian Science Monitor Weekly*, March 17, 13.

Pflaumer, Alicia. 2011. "Texting Bride Video Goes Viral on the Web." *Christian Science Monitor*, November 1. Accessed November 3, 2011 (www.csmonitor.com).

Phelps, Glenn, and Steve Crabtree. 2013. "More Than One in Five Worldwide Living in Extreme Poverty." Gallup, December 23. Accessed February 6, 2014 (www.gallup.com).

Phelps, Glenn, and Steve Crabtree. 2013. "Worldwide, Median Household Income About $10,000." Gallup, December 16. Accessed February 6, 2014 (www.gallup.com).

Phelps, Glenn, and Steve Crabtree. 2014. "Worldwide, Richest 3% Hold One-Fifth of Collective Income." Gallup, January 3. Accessed February 6, 2014 (www.gallup.com).

Phelps, Timothy M. 2014. "Justice Dept. Policy to Let Tribes Grow, Sell Cannabis." *Baltimore Sun*, December 11, 6.

Phillips, L. Taylor, and Brian S. Lowery. 2015. "The Hard-Knock Life? Whites Claim Hardships in Response to Racial Inequity." *Journal of Experimental Social Psychology* 61 (November): 12–18.

Phillips, Matthew. 2012. "Manufacturing: A Rebound, Not a Renaissance." *Bloomberg Businessweek*, December 17–23, 9–10.

Phillips, Matthew. 2014. "Welders, America Needs You." *Bloomberg Businessweek*, March 24–April 6, 19–21.

Pianta, Robert C., Jay Belsky, Renate Houts, Fred Morrison, and The National Institute of Child Health and Human Development (NICHD) Early Child Care Research Network. 2007. "Teaching: Opportunities to Learn in America's Elementary Classrooms." *Science* 315 (March 30): 1795–1796.

Pickert, Kate. 2014. "Turning Off the Tap." *Time*, August 4, 18–19.

Pierce, Lamar, Daniel Snow, and Andrew McAfee. 2013. "Cleaning House: The Impact of Information Technology Monitoring on Employee Theft and Productivity." Social Science Research Network, August 24. Accessed October 24, 2013 (www.ssrn.com).

Pipkin, Whitney. 2013. "Are Asian Americans Taking over the Internet?" *Asian Fortune*, November 29. Accessed December 4, 2013 (www.asianfortunenews.com).

Pittz, Will. 2005. "Closing the Gap: Solutions to Race-Based Health Disparities." Applied Research Center & Northwest Federation of Community Organizations, June. Accessed April 20, 2007 (www.arc.org).

Planty, M., W. Hussar, T. Snyder, S. Provasnik, G. Kena, R. Dinkes, A. KewalRamani, and J. Kemp. 2008. *The Condition of Education 2008*. Washington, DC: National Center for Education Statistics, U.S. Department of Education.

Plateris, Alexander A. 1973. *100 Years of Marriage and Divorce Statistics: 1867–1967*. Rockville, MD: National Center for Health Statistics.

Polderman, Tinca, J. C., et al. 2015. "Meta-Analysis of the Heritability of Human Traits Based on Fifty Years of Twin Studies." *Nature Genetics* 47 (May): 702–709.

Polgreen, Lydia. 2010. "One Bride for Two Brothers: A Custom Fades in India." *New York Times*, July 16, A4.

"Police Culture." 2015. *The Economist*, April 25, 28–29.

"Policing Philadelphia: Boots on the Street." 2013. *The Economist*, August 24, 33.

Polikoff, Morgan S., and Andrew C. Porter. 2014. "Instructional Alignment as a Measure of Teaching Quality." *Educational Evaluation and Policy Analysis* 36 (December): 399–416.

"The Politics of Financial Insecurity." 2015. Pew Research Center, January 8. Accessed October 7, 2015 (pewresearch.org).

Pollack, Andrew. 2013. "A.M.A. Recognizes Obesity as a Disease." *New York Times*, June 19, B1.

Pollack, Andrew. 2015. "F.D.A. Approves a Drug for Low Libido in Women." *New York Times*, August 19, B1, B5.

Pollard, Kelvin, and Paola Scommegna. 2013. "The Health and Life Expectancy of Older Blacks and Hispanics in the United States." *Today's Research on Aging*, No. 28, June, 1–8.

Pollick, Michael. 2014. "Why Didn't Communism Work in Eastern Europe?" Conjecture Corporation, April 1. Accessed April 12, 2014 (www.wisegeek.org).

Polsby, Nelson W. 1959. "Three Problems in the Analysis of Community Power." *American Sociological Review* 24 (December): 796–803.

Pong, Suet-ling, Lingxin Hao, and Erica Gardner. 2005. "The Roles of Parenting Styles and Social Capital in the School Performance of Immigrant Asian and Hispanic Adolescents." *Social Science Quarterly* 86 (December): 928–950.

Porter, Eduardo. 2013. "America's Sinking Middle Class." *New York Times*, September 19, A1.

Porter, Eduardo. 2014. "A Global Boom, but Only for Some." *New York Times*, March 19, B1.

Porter, Michael E., and Scott Stern. 2014. "Social Progress Index 2014: Executive Summary." Social Progress Imperative. Accessed September 19, 2014 (www.socialprogressimperative.org).

Portes, Jonathan. 2015. "Immigration Is Good for Economic Growth. If Europe Gets It Right, Refugees Can Be Too." *Huffington Post*, September 15. Accessed October 23, 2015 (huffingtonpost.com).

Postel, Sandra. 2012. "Humanity's Growing Impact on the World's Freshwater." AlterNet, February 23. Accessed June 23, 2012 (alternet.org).

Poushter, Jacob. 2014. "Russia's Moral Barometer: Homosexuality Unacceptable, but Drinking, Less So." Pew Research Center, February 6. Accessed March 2, 2014 (www.pewresearch.org).

Poushter, Jacob. 2014. "What's Morally Acceptable? It Depends on Where in the World You Live." Pew Research Center, April 15. Accessed May 16, 2014 (www.pewresearch.org).

Powell, Jon T., Timothy G. Townsend, and Julie B. Zimmerman. 2015. "Estimates of Solid Waste Disposal Rates and Reduction Targets for Landfill Gas Emissions." *Nature Climate Change*, September 21. Accessed October 23, 2015 (nature.com).

"The Power of a Party." 2013. *The Economist*, August 13, 29–30.

Powers, Charles H. 2004. *Making Sense of Social Theory: A Practical Introduction*. Lanham, MD: Rowman & Littlefield.

Presbyterian Mission Agency. 2013. "The Top Most Frequently Asked Questions About the PC (USA)." Accessed July 2, 2014 (www.presbyterianmission.org).

"Prescription for Change." 2013. *The Economist*, June 29, 61–62.

Preston, Darrell. 2014. "Cooling Off with a Nice, Tall Glass of Toilet Water." *Bloomberg Businessweek*, April 28–May 4, 37.

Preston, Julia. 2015. "States Are Divided by the Lines They Draw on Immigration." *New York Times*, March 30, A10, A12.

Price, Barbara Raffel, and Natalie J. Sokoloff, eds. 2004. *The Criminal Justice System and Women: Offenders, Prisoners, Victims, & Workers*, 3rd edition. New York: McGraw-Hill.

Princiotta, Daniel, Laura Lippman, Renee Ryberg, Hannah Schmitz, David Murphey, and Mae Cooper. 2014. "Social Indicators Predicting Postsecondary Success." Child Trends, April 1. Accessed June 23, 2014 (www.childtrends.org).

Project on Government Oversight. 2010. "Letter to NIH on Ghostwriting Academics." November 29. Accessed July 3, 2011 (www.pogo.org).

"Pro-natalism." 2015. *The Economist*, July 25, 47–48.

Pryor, John H. 2011. "The Changing First-Year Student: Challenges for 2011." Higher Education Research Institute at UCLA, January 27. Accessed April 10, 2012 (www.heri.ucla.edu).

Przybylski, Andrew K., and Allison F. Mishkin. 2015. "How the Quantity and Quality of Electronic Gaming Relates to Adolescents' Academic Engagement and Psychosocial Adjustment." *Psychology of Popular Media Culture*, March 2. APA PsycNet. Accessed August 6, 2015 (psycnet.apa.org).

Putnam, Hannah, Julie Greenberg, and Kate Walsh. 2014. "Easy A's and What's Behind Them." National Council on Teacher Quality, November. Accessed December 5, 2014 (www.nctq.org).

Putnam, Robert D. 2015. *Our Kids: The American Dream in Crisis*. New York: Simon & Schuster.

Quinney, Richard. 1980. *Class, State, and Crime*. Boston: Little, Brown.

Rackin, Heather, and Christina M. Gibson-Davis. 2012. "The Role of Pre- and Postconception Relationships for First-Time Parents." *Journal of Marriage and Family* 74 (June): 389–398.

Radesky, Jenny, et al. 2014. "Patterns of Mobile Device Use by Caregivers and Children During Meals in Fast Food Restaurants." *Pediatrics* 133 (4): e843–e849.

Radnofsky, Louise. 2015. "Meet the Health-Law Holdouts." *Wall Street Journal*, June 25, A1, A12.

Rainie, Lee, and Kathryn Zickuhr. 2015. "Americans' Views on Mobile Etiquette." Pew Research Center, August 26. Accessed August 26, 2015 (pewresearch.org).

Rainie, Lee, Cary Funk, and Monica Anderson. 2015. "How Scientists Engage the Public." Pew Research Center, February 15. Accessed July 22, 2015 (pewresearch.org).

Rainie, Lee, Kristen Purcell, and Aaron Smith. 2011. "The Social Side of the Internet." Pew Internet & American Life Project, January 1. Accessed August 10, 2011 (www.pewinternet.org).

Rampell, Catherine. 2011. "Companies Spend on Equipment, Not Workers." *New York Times*, June 10, A1.

Rampell, Catherine. 2014. "The Safety Net Catches the Middle Class More Than the Poor." *Washington Post*, April 7. Accessed March 8, 2014 (www.washingtonpost.com).

Rank, Mark R., and Thomas A. Hirschl. 2015. "The Likelihood of Experiencing Relative Poverty over the Life Course." *PLoS One* 10(7): e0133513. doi:10.1371/journal.pone.0133513. Accessed September 16, 2015 (journals.plos.org).

Rano, Jason, and Jane Houlihan. 2012. "Myths on Cosmetic Safety." Skin Deep Cosmetics Database. Accessed May 10, 2012 (www.ewg.org).

Rape Crisis Center. 2014. "Get the Facts." Accessed February 14, 2014 (www.rccmsc.org).

Raphael, Steven, and Michael A. Stoll. 2010. "Job Sprawl and the Suburbanization of Poverty." Brookings Institute, Metropolitan Policy Program, March. Accessed May 1, 2010 (www.brookings.edu).

Ray, Rebecca, Milla Sanes, and John Schmitt. 2013. "No-Vacation Nation Revisited." Center for Economic and Policy Research, May. Accessed April 12, 2014 (cepr.net).

Rayasam, Renuka. 2007. "Immigrants: The Unsung Heroes of the U.S. Economy." *U.S. News & World Report*, February 26, 58.

Rayman, Noah. 2014. "The World's Top 5 Cybercrime Hotspots." *Time*, August 7. Accessed September 1, 2015 (time.com).

Reaney, Patricia. 2012. "Average Cost of U.S. Weddings Hits $27,021." Reuters, May 23. Accessed June 2, 2012 (www.reuters.com).

Reardon, Sean F. 2011. "The Widening Academic Achievement Gap Between the Rich and the Poor: New Evidence and Possible Explanations." Pp. 91–116 in *Whither Opportunity? Rising Inequality and the Uncertain Life Chances of Low-Income Children*, edited by R. Murnane and G. Duncan. New York: Russell Sage Foundation Press.

Reardon, Sean F., Lindsay Fox, and Joseph Townsend. 2015. "Neighborhood Income Composition by Household Race and Income, 1990–2009." *Annals of the American Academy of Political and Social Science* 660 (July): 78–99.

Redberg, Rita. 2012. "Less Is More." Institute of Medicine, September. Accessed July 26, 2014 (www.iom.edu).

Redd, Zakia, Tahilin Sanchez Karver, David Murphey, Kristin Anderson Moore, and Dylan Knewstub. 2011. "Two Generations in Poverty: Status and Trends among Parents and Children in the United States, 2000–2010." Child Trends Research Brief, November. Accessed January 12, 2012 (www.childtrends.org).

Reddy, Sumathi. 2013. "'I Don't Smoke, Doc,' and Other Patient Lies." *Wall Street Journal*, February 19, D3.

Reed, Matthew, and Debbie Cochrane. 2013. "Student Debt and the Class of 2012." Institute for College Access & Success, December. Accessed June 21, 2014 (www.ticas.org).

Reeves, Jay. 2013. "Alabama Woman's Body Removed from Front Yard Grave." *U.S. News & World Report*, November 15. Accessed August 10, 2014 (www.us.news.com).

Reeves, Richard V., and Joanna Venator. 2015. "Sex, Contraception, or Abortion? Explaining Class Gaps in Unintended Childbearing." Brookings, February. Accessed September 29, 2015 (brookings.edu).

Reeves, Richard V., and Kerry Searle Grannis. 2013. "Five Strong Starts for Social Mobility." Brookings Institution, January 9. Accessed February 14, 2014 (www.brookings.edu).

"Reforming the One-Child Policy." 2013. *The Economist*, March 16, 45–46.

Reger, Jo. 2012. *Everywhere and Nowhere: Contemporary Feminism in the United States.* New York: Oxford University Press.

Regnerus, Mark, and Jeremy Uecker. 2011. *Premarital Sex in America: How Young Americans Meet, Mate, and Think About Marrying.* New York: Oxford University Press.

Reich, Robert. 2014. "The Four Biggest Myths About Income Inequality." *Christian Science Monitor*, May 5. Accessed June 21, 2014 (www.csmonitor.com).

Reich, Robert. 2014. "Why Widening Inequality Is Hobbling Equal Opportunity." AlterNet, February 6. Accessed February 10, 2014 (www.alternet.org).

Reilly, Katie. 2013. "Sesame Street Reaches Out to 2.7 Million American Children with an Incarcerated Parent." Pew Research Center, June 21. Accessed November 18, 2013 (www.pewresearch.org).

Reiman, Jeffrey, and Paul Leighton. 2010. *The Rich Get Richer and the Poor Get Prison: Ideology, Class, and Criminal Justice*, 9th edition. Upper Saddle River, NJ: Prentice Hall.

Rein, Lisa, and Emily Wax-Thibodeaux. 2015. "Veterans Affairs Improperly Spent $6 Billion Annually, Senior Official Says." *Washington Post*, May 14. Accessed August 23, 2015 (washingtonpost.com).

Rein, Lisa. 2011. "Federal Workers Tell Us What Should Be Cut from the Budget." *Washington Post*, April 14, B4.

Rein, Lisa. 2015. "Lockheed Martin Pays $4.7 Million to Settle Charges it Lobbied for Federal Contract with Federal Money." *Washington Post*, August 24. Accessed September 3, 2015 (washingtonpost.com).

Rein, Lisa. 2015. "Park Service to Big Water: No Federal Funding for Bottled Water Bans? We'll Find Our Own Money, Thanks." *Washington Post*, July 30. Accessed October 23, 2015 (washingtonpost.com).

Reinhold, Steffen 2010. "Reassessing the Link Between Premarital Cohabitation and Marital Instability." *Demography* 47 (August): 719–733.

Reiss, Steven. 2004. "The Sixteen Strivings for God." *Zygon* 39 (June): 303–320.

"Religions of the World: Number of Adherents, Names of Houses of Worship. . . ." 2007. Religious Tolerance. Accessed August 24, 2007 (www.religioustolerance.org).

Restaurant Opportunities Centers United, et al. 2014. "The Glass Floor: Sexual Harassment in the Restaurant Industry." New York, NY: Restaurant Opportunities Centers United.

Rettew, David. 2014. "Do You Have a Disorder or Just a Trait?" *Psychology Today*, January/ February, 42–43.

Reynolds, Kelly A. 2012. "The Price of Drinking Water." *Water Conditioning & Purification* 54 (July): 50–52.

Rheault, Magali, and Kyley McGeeney. 2011. "Education Is a Key Predictor of Emotional Health After 65." Gallup, August 19. Accessed April 3, 2012 (www.gallup.com).

Ricciardelli, Rosemary, Kimberley A. Clow, and Philip White. 2010. "Investigating Hegemonic Masculinity: Portrayals of Masculinity in Men's Lifestyle Magazines." *Sex Roles* 63 (March): 64–78.

Rich, Motoko. 2013. "Subtract Teachers, Add Pupils: Math of Today's Jammed Schools." *New York Times*, December 22, A1.

Richburg, Keith B. 2009. "States Seek Less Costly Substitutes for Prison." *Washington Post*, July 13, A1.

Riche, Martha Farnsworth. 2000. "America's Diversity and Growth: Signposts for the 21st Century." *Population Bulletin* 55 (June): 1–41.

Richey, Alexandra S., Brian F. Thomas, Min-Hui Lo, James S. Famiglietti, Sean Swenson, and Matthew Rodell. 2015. "Uncertainty in Global Groundwater Storage Estimates in a Total Groundwater Stress Framework." *Water Resources Research* 51 (July): 5198–5216.

Richie, Christina. 2013. "The Scandal of the (Female) Evangelical Mind." *Chronicle of Higher Education*, June 14, A37–A38.

Richtel, Matt, and Alexei Barrionuevo. 2005. "Wendy's Restaurants." *New York Times*, April 22, A9.

Rideout, Victoria, Ulla G. Foehr, and Donald F. Roberts. 2010. *Generation M2: Media in the Lives of 8- to 18-Year-Olds.* Kaiser Family Foundation, January. Accessed January 30, 2010 (www.kff.org).

Rideout, Victoria. 2013. "Zero to Eight: Children's Media Use in America 2013." Common Sense Media, Fall. Accessed November 18, 2013 (www.commonsensemedia.org).

Riesman, David. 1953. *The Lonely Crowd.* New York: Doubleday.

Riffkin, Rebecca. 2014. "Americans Still Prefer a Male Boss to a Female Boss." Gallup, October 14. Accessed November 17, 2014 (www.gallup.com).

Riffkin, Rebecca. 2014. "Jobs, Government, and Economy Remain Top U.S. Problems." Gallup, May 19. Accessed June 21, 2014 (www.gallup.com).

Riffkin, Rebecca. 2014. "New Record Highs in Moral Acceptability." Gallup, May 30. Accessed August 10, 2014 (www.gallup.com).

"The Right Choices." 2015. *The Economist*, June 20, 23–25.

Riley, Ricky. 2014. "8 TV Shows That Reinforce Negative Stereotypes of Black People." *Atlanta Black Star*, October 29. Accessed October 4, 2015 (atlantablackstar.com).

Rios, Victor. 2012. "Reframing the Achievement Gap." *Contexts* 4 (Fall): 8–10.

Riosmena, Fernando, Elisabeth Root, Jamie Humphrey, Emily Steiner, and Rebecca Stubbs. 2015. "The Waning Hispanic Health Paradox." *Pathways* (Spring): 25–29.

Riosmena, Fernando, Rebecca Wong, and Alberto Palloni. 2013. "Migration Selection, Protection, and Acculturation in Health: A Binational Perspective on Older Adults." *Demography* 50 (June): 1039–1064.

"Rise of the Robots." 2014. *The Economist*, March 29, 13.

Ritzer, George. 1992. *Contemporary Sociological Theory*, 3rd edition. New York: McGraw-Hill.

Ritzer, George. 1996. *The McDonaldization of Society: An Investigation into the Changing Character of Contemporary Social Life.* Thousand Oaks, CA: Pine Forge Press.

Ritzer, George. 2008. *The McDonaldization of Society*, 5th edition. Los Angeles: Pine Forge Press.

Robers, Simone, Jana Kemp, and Jennifer Truman. 2013. "Indicators of School Crime and Safety: 2012." National Center for Education Statistics, June. Accessed December 4, 2013 (www.nces.ed.gov).

Robert Wood Johnson Foundation. 2010. "California Nurse Ratio Law Saves Lives, Improves Nurse Morale, Study Finds." May 16. Accessed June 6, 2012 (www.rwjf.org).

Robert Wood Johnson Foundation. 2015. "Growing Retail Clinic Industry Employs, Empowers Nurse Practitioners." February 20. Accessed August 22, 2015 (rwjf.org).

Robert Wood Johnson Foundation. 2015. "Understanding the Uninsured Now." June. Accessed October 16, 2015 (rwjf.org).

Roberts, Andrea L., et al. 2013. "Perinatal Air Pollutant Exposures and Autism Spectrum Disorder in the Children of Nurses' Health Study II Participants." *Environmental Health Perspectives* 121 (August): 978–984.

Roberts, Keith A. 2004. *Religion in Sociological Perspective*, 4th edition. Belmont, CA: Wadsworth.

Roberts, Yvonne. 2014. "Addicted to Email? The Germans Have an Answer." AlterNet, September 1, 2015. Accessed September 2, 2014.

Robertson, Jordan. 2013. "Your Not-So-Secret Medical History." *Bloomberg Businessweek*, August 12–25, 41–42.

Rochman, Bonnie. 2012. "The End of an Epidemic?" *Time*, February 6, 16.

Rochman, Bonnie. 2013. "Hover No More: Helicopter Parents May Breed Depression and Incompetence in Children." *Time*, February 22. Accessed November 20, 2013 (www.healthland.time.com).

Rodrigue, Edward P., and Richard V. Reeves. 2014. "Horatio Alger Goes to Washington: Representation and Social Mobility." Brookings, October 2. Accessed November 23, 2014 (www.brookings.edu).

Rodrigue, Edward, and Richard V. Reeves. 2015. "Cutting Poverty by Increasing Program Participation." Brookings Institution, June 11. Accessed September 31, 2015 (brookings.edu).

Roethlisberger, F. J., and William J. Dickson. 1939/1942. *Management and the Worker: An Account of a Research Program Conducted at the Western Electric Company, Hawthorne Works, Chicago.* Cambridge, MA: Harvard University Press.

Rohrlich, Justin. 2010. "Why White-Collar Criminals Don't Fear Getting Caught." Minyanville Media, Inc., November 12. Accessed January 12, 2014 (www.minyanville.com).

Roman, Caterina Gouvis, et al. 2012. "Social Networks, Delinquency, and Gang Membership: Using a Neighborhood Framework to Examine the Influence of Network Composition and Structure in a Latino Community." The Urban Institute, February. Accessed January 12, 2014 (www.urban.org).

Romano, Andrew, and Allison Samuels. 2012. "Is Obama Making It Worse?" *Newsweek*, April 6, 40–45.

Romer, Dan. 2011. "After 11 Years of Setting the Record Straight, Stories About Holiday Suicides Still Outnumber Those Debunking the Myth." Annenberg Public Policy Center, December 13. Accessed December 14, 2011 (www.annenbergpublicpolicycenter.org).

Romero, Simon, and Marc Lacey. 2010. "Looting Flares Where Authority Breaks Down." *New York Times*, January 17, A1.

Ronfeldt, Matthew, Hamilton Lankford, Susanna Loeb, and James Wyckoff. 2011. "How Teacher Turnover Harms Student Achievement." National Bureau of Economic Research, June. Accessed April 14, 2012 (www.nber.org).

Roscigno, Vincent J. 2010. "Ageism in the American Workplace." *Contexts* 9 (Winter): 16–21.

Rose, Fred. 1997. "Toward a Class-Cultural Theory of Social Movements: Reinterpreting New Social Movements." *Sociological Forum* 12 (September): 461–494.

Rose, Peter I. 1997. *They and We: Racial and Ethnic Relations in the United States*, 5th edition. New York: McGraw-Hill.

Rosen, Larry D. 2010. *Rewired: Understanding the iGeneration and the Way They Learn.* New York: Palgrave Macmillan.

Rosenberg, Paul. 2014. "10 Things You Might Not Know About Poverty." AlterNet, January 8. Accessed February 6, 2014 (www.alternet.org).

Rosenfeld, Steven. 2014. "The Supreme Court's Radical Right Wing Majority: Waging War on Women and Boosting Corporate Power." AlterNet, June 30. Accessed July 26, 2014 (www.alternet.org).

Rosenthal, Elizabeth. 2014. "Medicine's Top Earners Are Not the M.D.s." *New York Times*, May 18, SR4.

Rosenthal, Robert, and Lenore Jacobsen. 1968. *Pygmalion in the Classroom: Teacher Expectations and Pupils' Intellectual Development*. New York: Holt, Rinehart, and Winston.

Rosin, Hanna. 2012. "Boys on the Side." *The Atlantic*, September. Accessed January 7, 2013 (www.theatlantic.com).

Ross, Martha, Nicole P. Svajlenka, and Jane Williams. 2014. "Part of the Solution: Pre-Baccalaureate Healthcare Workers in a Time of Health System Change." Brookings, July. Accessed December 5, 2014 (www.brookings.edu).

Ross, Terris, Grace Kena, Amy Rathbun, Angelina KewalRamani, Jijun Zhang, Paul Kristapovich, and Eileen Manning. 2012. "Higher Education: Gaps in Access and Persistence Study." National Center for Education Statistics, August. Accessed June 21, 2014 (www.nces.ed.gov).

Rossano, Matt J. 2012. "The Essential Role of Ritual in the Transmission and Reinforcement of Social Norms." *Psychological Bulletin* 138 (May): 529–549.

Rossi, Max, and Lisa Jucca. 2014. "Villages Face 'Slow Death' as Italy's Population Ages." *Baltimore Sun*, March 2, 14.

Rothman, Sheila M. 1978. *Women's Proper Place: A History of Changing Ideals and Practices, 1870 to the Present*. New York: Basic Books.

Rowe-Finkbeiner, Kristin. 2004. *The F-Word: Feminism in Jeopardy, Women, Politics, and the Future*. Emeryville, CA: Seal Press.

Rowe-Finkbeiner, Kristin. 2012. "It's Not a 'Mommy War,' It's a War on Moms." Moms Rising, April 14. Accessed April 12, 2014 (www.momsrising.org).

Rubin, Kenneth, William Bukowski, and Jeffrey G. Parker. 1998. "Peer Interactions, Relationships, and Groups." Pp. 619–700 in *Handbook of Child Psychology: Vol. 3. Social, Emotional, and Personality Development*, edited by William Damon and Nancy Eisenberg. New York: Wiley.

Rubin, Rita. 2004. "'Smart Pills' Make Headway." *USA Today*, July 7, 1D.

Ruetschlin, Catherine, and Dedrick Asante-Muhammad. 2013. "The Challenge of Credit Card Debt for the African American Middle Class." Demos & NAACP, December 4. Accessed March 22, 2014 (www.demos.org).

Rugh, Jacob S., and Douglas S. Massey. 2010. "Racial Segregation and the American Foreclosure Crisis." *American Sociological Review* 75 (October): 629–651.

Rule, James B. 1988. *Theories of Civil Violence*. Berkeley: University of California Press.

Ruth, Jennifer. 2014. "Non-Critical Thinking in China." *Chronicle Review*, February 28, B20.

Rutter, Virginia, and Pepper Schwartz. 2012. *The Gender of Sexuality: Exploring Sexual Possibilities*, 2nd edition. Lanham, MD: Rowman & Littlefield.

Ryan, Camille. 2013. "Language Use in the United States: 2011." U.S. Census Bureau, August. Accessed March 22, 2014 (www.census.gov).

Saad, Lydia. 2001. "Majority Considers Sex Before Marriage Morally Okay." *Gallup Poll Monthly*, No. 428 (May): 46–48.

Saad, Lydia. 2006. "Families of Drug and Alcohol Abusers Pay an Emotional Toll." Gallup News Service, August 25. Accessed August 27, 2006 (www.gallup.com).

Saad, Lydia. 2010. "Nearly 4 in 10 Americans Still Fear Walking Alone at Night." Gallup, November 5. Accessed August 22, 2011 (www.gallup.com).

Saad, Lydia. 2013. "Half in U.S. Support Publicly Financed Federal Campaigns." Gallup, June 24. (Accessed April 16, 2014 (www.gallup.com).

Saad, Lydia. 2013. "In U.S., Rise in Religious 'Nones' Slows in 2012." Gallup, January 10. Accessed June 21, 2014 (www.gallup.com).

Saad, Lydia. 2013. "Majority of U.S. Workers Say Job Doesn't Require a Degree." Gallup, September 9. Accessed June 21, 2014 (www.gallup.com).

Saad, Lydia. 2013. "U.S. Support for Euthanasia Hinges on How It's Described." Gallup, May 29. Accessed May 16, 2014 (www.gallup.com).

Saad, Lydia. 2014. "More Women Than Men in U.S. Workforce Are Irked by Pay." Gallup, October 16. Accessed November 23, 2014 (www.gallup.com).

Saad, Lydia. 2014. "The '40-Hour' Workweek is Actually Longer—by Seven Hours." Gallup, August 29. Accessed November 23, 2014 (www.gallup.com).

Saad, Lydia. 2014. "Three in Four in U.S. Still See the Bible as Word of God." Gallup, June 4. Accessed June 21, 2014 (www.gallup.com).

Saad, Lydia. 2015. "Americans Choose 'Pro-Choice' for First Time in Seven Years." Gallup, May 29. Accessed September 29, 2015 (gallup.com).

Saad, Lydia. 2015. "Americans' Support for Labor Unions Continues to Recover." Gallup, August 17. Accessed October 7, 2015 (gallup.com).

Saad, Lydia. 2015. "Working Women Still Lag Men in Opinion of Workplace." Gallup, September 2. Accessed September 29, 2015 (gallup.com).

Sachs, Jeffrey S. 2005. "Confusion over Population: Growth or Dearth?" *Pop!ulation Press* 11 (Winter/Spring): 17.

Sacks, Vanessa, David Murphey, and Kristin Moore. 2014. "Adverse Childhood Experiences: National and State-Level Prevalence." Child Trends Research Brief, July. Accessed November 25, 2014 (www.childtrends.org).

Saez, Emmanuel, and Gabriel Zucman. 2014. "Wealth Inequality in the United States Since 1913: Evidence from Capitalized Income Tax Data." National Bureau of Economic Research, October. Accessed November 10, 2014 (www.nber.org).

Saez, Emmanuel. 2013. "Striking It Richer: The Evolution of Top Incomes in the United States (Updated with 2012 Preliminary Estimates)." UC Berkeley, September 3. Accessed February 6, 2014 (www.elsa.berkeley.edu).

Sagarin, Edward. 1975. *Deviants and Deviance*. New York: Praeger.

Sahgal, Neha, and Greg Smith. 2009. "A Religious Portrait of African-Americans." Pew Research Center, Pew Forum on Religion & Public Life Project, January 30. Accessed June 21, 2014 (www.pewforum.org).

Sahgal, Neha, and Tim Townsend. 2014. "Four-in-Ten Pakistanis Say Honor Killing of Women Can Be at Least Sometimes Justified." Pew Research Center, May 30. Accessed November 17, 2014 (www.pewresearch.org).

Sahgal, Neha. 2014. "Coke, 'America the Beautiful,' and the Language of Diversity." Pew Research Center, February 3. Accessed November 25, 2014 (www.pewresearch.org).

Salamone, Frank A. 2005. "Jazz and Its Impact on European Classical Music." *The Journal of Popular Culture* 38 (May): 732–743.

Salant, Jonathan D. 2014. "$350M NASA Tower Shows How Congress Protects Pork." *Baltimore Sun*, January 12, 12.

Sanburn, Josh. 2010. "Brief History: Secret Medical Testing." *Time*, October 18, 30.

Sanburn, Josh. 2015. "The Joy of Less." *Time*, March 23, 44–50.

Sánchez, Erika L. 2013. "The Challenge of Defining Muslim Feminism." *Huffington Post*, February 10. Accessed October 12, 2013 (www.huffingtonpost.com).

Sandberg, Sheryl. 2013. "Why I Want Women to Lean In." *Time*, March 18, 44–45.

Sandbu, Martin. 2013. "Talkin' 'Bout a Revolution." *Financial Times*, April 19. Accessed October 8, 2014 (www.ft.com).

Sandstrom, Kent L., Daniel D. Martin, and Gary Alan Fine. 2006. *Symbols, Selves, and Social Reality: A Symbolic Interactionist Approach to Social Psychology and Sociology*, 2nd edition. Los Angeles: Roxbury.

Sapir, Edward. 1929. "The Status of Linguistics as a Science." *Language* 5 (4): 207–214.

Sassler, Sharon, and Amanda J. Miller. 2011. "Class Differences in Cohabitation Processes." *Family Relations* 60 (April): 163–177.

Sassler, Sharon, Fenaba R. Addo, and Daniel T. Lichter. 2012. "The Tempo of Sexual Activity and Later Relationship Quality." *Journal of Marriage and Family* 74 (August): 708–725.

Sathyanarayana, Sheela, et al. 2013. "Unexpected Results in a Randomized Dietary Trial to Reduce Phthalate and Bisphenol A Exposures." *Journal of Exposure Science and Environmental Epidemiology* 23 (February): 378–384.

Saulny, Susan. 2011. "Counting by Race Can Throw Off Some Numbers." *New York Times*, February 10, A1.

Savage, David G., and Neela Banerjee. 2014. "Justices Uphold Midwest Coal Plant Curbs." *Baltimore Sun*, April 30, 14.

Sawchuck, Stephen. 2012. "Teacher Quality, Status Entwined Among Top-Performing Nations." *Education Week* 31 (16): 12–16.

Sawhill, Isabel V., and Joanna Venator. 2014. "Families Adrift: Is Unwed Childbearing the New Norm?" Brookings, October 13. Accessed October 14, 2014 (www.brookings.edu).

Scarlett, W. George, Sophie Naudeau, Dorothy Salonius-Pasternak, and Iris Ponte. 2005. *Children's Play*. Thousand Oaks, CA: Sage.

Schachter, Jason P. 2004. "Geographical Mobility: 2002 to 2003." U.S. Census Bureau, Current Population Reports. Accessed May 15, 2007 (www.census.gov).

Schaeffer, Robert K. 2003. *Understanding Globalization: The Social Consequences of Political, Economic, and Environmental Change*, 2nd edition. Lanham, MD: Rowman & Littlefield.

Scheer, Robert. 2015. *They Know Everything About You: How Data-Collecting Corporations and Snooping Government Agencies Are Destroying Democracy*. New York: Nation Books.

Schehr, Robert C. 1997. *Dynamic Utopia: Establishing Intentional Communities as a New Social Movement*. Westport, CT: Bergin & Garvey.

Schieman, Scott. 2010. "Socioeconomic Status and Beliefs About God's Influence in Everyday Life." *Sociology of Religion* 71 (Spring): 25–51.

Schiesel, Seth. 2011. "Supreme Court Has Ruled; Now Games Have a Duty." *New York Times*, July 29, C1.

Schiffrin, Holly H., Miriam Liss, Haley Miles-McLean, Katherine A. Geary, Mindy J. Erchull, and Taryn Tashner. 2014. "Helping or Hovering? The Effects of Helicopter Parenting on College Students' Well Being." *Journal of Child and Family Studies* 23 (April): 548–557.

Schlesinger, Izchak M. 1991. "The Wax and Wane of Whorfian Views." Pp. 7–44 in *Influence of Language on Culture & Thought*, edited by Robert Cooper and Bernard Spolsky. New York: Mounton de Gruyter.

Schlesinger, Robert. 2011. "Two Takes: Collective Bargaining Rights for Public Sector Unions?" *U.S. News Weekly*, February 25, 15–16.

Schmall, Emily. 2007. "The Cult of Chick-fil-A." *Forbes*, July 23, 80, 83.

Schmidt, Peter. 2008. "2 Studies Raise Questions about Research Based on Student Surveys." *Chronicle of Higher Education*, November 6. Accessed November 9, 2008 (www.chronicle.com).

Schmidt, Peter. 2014. "Supreme Court Exposes Affirmative Action at Colleges to Continued Political Assault." *Chronicle of Higher Education*, May 2, A3–A4.

Schmidt, Peter. 2015. "Racism Is Widespread Among College Students, Researcher Says." *Chronicle of Higher Education*, March 20, A4.

Schnall, Marianne. 2012. "Letting Girls Be Girls—A Global Campaign." Women's Media Center, January 25. Accessed January 2012 (www.womensmediacenter.com).

Schneider, Mark. 2013. "Higher Education Pays: But a Lot More for Some Graduates Than for Others." College Measures. Accessed June 21, 2014 (www.collegemeasures.org).

Schneider, Monica C., and Angela L. Bos. 2014. "Measuring Stereotypes of Female Politicians." *Political Psychology* 25 (April): 245–266.

Schnittker, Jason. 2009. "Mirage of Health in the Era of Biomedicalization: Evaluating Change in the Threshold of Illness, 1972–1996." *Social Forces* 87 (June): 2155–2182.

Schramm, J. B., Chad Aldeman, Andrew Rotherham, Rachael Brown, and Jordan Cross. 2013. "Smart Shoppers: The End of the 'College for All' Debate." College Summit. Accessed June 21, 2014 (www.collegesummit.org).

Schuck, Peter. 2013. "The ObamaCare Website Failure Was Inevitable." *Wall Street Journal*, November 5, A15.

Schultz, Ellen E. 2011. *Retirement Heist: How Companies Plunder and Profit from the Nest Eggs of American Workers*. New York: Penguin.

Schur, Edwin M. 1968. *Law and Society: A Sociological View*. New York: Random House.

Schurman-Kauflin, Deborah. 2000. *The New Predator: Women Who Kill*. New York: Algora.

Schutz, Alfred. 1967. *The Phenomenology of the Social World*. Evanston, IL: Northwestern University Press.

Schwalbe, Michael. 2001. *The Sociologically Examined Life: Pieces of the Conversation*, 2nd edition. Mountain View, CA: Mayfield.

Schwartz, John. 2015. "Coca-Cola Says It's Close to Water Replenishment Goal." *New York Times*, August 26, B3.

Schwartz, Larry. 2014. "8 Mistakes We're Making About Ebola That We Also Made when AIDS Appeared." AlterNet, October 17. Accessed October 18, 2014 (www.alternet.org).

Schwarz, Alan. 2014. "Thousands of Toddlers Are Medicated for A.D.H.D., Report Finds, Raising Worries." *New York Times*, May 17, A11.

Schweizer, Peter. 2013. *Extortion: How Politicians Extract Your Money, Buy Votes, and Line Their Own Pockets*. New York: Houghton Mifflin Harcourt.

Schwyzer, Hugo. 2011. "How Our Sick Culture Makes Girls Think They Have to Be Gorgeous to Be Loved." AlterNet, April 12. Accessed October 29, 2012 (www.alternet.org).

Scommegna, Paola. 2011. "Least Segregated U.S. Metros Concentrated in Fast-Growing South and West." Population Reference Bureau, September 11. Accessed June 23, 2012 (www.prb.org).

Scommegna, Paola. 2011. "U.S. Parents Who Have Children with More Than One Partner." Population Reference Bureau, June. Accessed March 6, 2012 (www.prb.org).

Scommegna, Paola. 2013. "Aging U.S. Baby Boomers Face More Disability." Population Reference Bureau, March. Accessed May 16, 2014 (www.prb.org).

Scommegna, Paola. 2013. "Elderly Immigrants in the United States." *Today's Research on Aging* 29 (October): 1–9.

Scommegna, Paola. 2013. "Exploring the Paradox of U.S. Hispanics' Longer Life Expectancy." Population Reference Bureau, July. Accessed July 26, 2014 (www.prb.org).

Scott, Robert E. 2012. "The China Toll." EPI Briefing Paper 345, Economic Policy Institute, August 23. Accessed April 12, 2014 (www.epi.org).

"Second-Generation Americans." 2013. Pew Research Center, February 7. Accessed March 22, 2014 (www.pewresearch.org).

"Select-a-Faith." 2014. *The Economist*, May 17, 30.

Selingo, Jeffrey J. 2013. "The Diploma's Vanishing Value." *Wall Street Journal*, April 27–28, C3.

Seltzer, Sarah. 2012. "Skinny Minnie: Our Culture's Bizarre Obsession with Stick-Thin Women." AlterNet, October 16. Accessed May 15, 2013 (www.alternet.org).

Seltzer, Sarah. 2012. "The 5 Most Sexist Moments of the Campaign." AlterNet, November 9. Accessed April 21, 2014 (www.alternet.org).

Semuels, Alana. 2010. "Can a Prison Save a Town?" *Los Angeles Times*, May 3. Accessed May 4, 2010 (www.latimes.com).

Sengupta, Somini. 2013. "The Information-Gathering Paradox." *New York Times*, October 27, SR4.

Sengupta, Somini. 2013. "What You Didn't Post, Facebook May Still Know." *New York Times*, March 26, B1.

Sentencing Project, The. 2013. "Report of the Sentencing Project to the United Nations Human Rights Committee." August. Accessed January 15, 2014 (www.sentencingproject.org).

Senter, Mary S., Nicole Van Vooren, and Roberta Spalter-Roth. 2014. "Sociology, Criminology Concentrations, and Criminal Justice: Differences in Reasons for Majoring, Skills, Activities, and Early Outcomes?" American Sociological Association, April. Accessed October 25, 2014 (www.asanet.org).

Seward, Zachary M. 2012. "58 Countries with Better Voter Turnout Than the United States." Quartz, November 6. Accessed April 20, 2014 (www.qz.com).

"Sex-Selective Abortion: Gendercide in the Caucasus." 2013. *The Economist*, September 21, 54–55.

Shamoo, Adil E., and Bonnie Bricker. 2011. "The Threat of Bad Science." *Baltimore Sun*, January 11, 13.

Shanahan, Michael J., Shawn Bauldry, and Jason Freeman. 2010. "Beyond Mendel's Ghost." *Contexts* 3 (Fall): 34–39.

Shapira, Ian. 2014. "In Arizona, a Navajo High School Emerges as a Defender of the Washington Redskins." *Washington Post*, October 26. Accessed October 27, 2014 (www.washingtonpost.com).

Shapiro, Jenessa R., and Amy M. Williams. 2012. "The Role of Stereotype Threats in Undermining Girls' and Women's Performance and Interest in STEM Fields." *Sex Roles* 66 (February): 175–183.

Shariff, Azim F., and Lara B. Aknin. 2014. "The Emotional Toll of Hell: Cross-National and Experimental Evidence for the Negative Well-Being Effects of Hell Beliefs." *PLoS One* 9 (January): e85251.

Sharifzadeh, Virginia-Shirin. 1997. "Families with Middle Eastern Roots." Pp. 441–482 in *Developing Cross-Cultural Competence: A Guide for Working with Children and Families*, edited by Eleanor W. Lynch and Marci J. Hanson. Baltimore, MD: Paul H. Brookes.

Shattuck, Rachel M., and Rose M. Kreider. 2013. "Social and Economic Characteristics of Currently Unmarried Women with a Recent Birth: 2011." U.S. Census Bureau, May. Accessed May 16, 2014 (www.census.gov).

Shauk, Zain, and Bradley Olson. 2014. "Where Frackers Are Friendly Neighbors." *Bloomberg Businessweek*, September 8–14, 37–38.

Shea, Christopher. 2014. "The Liar's 'Tell.'" *The Chronicle Review*, October 17, B6–B9.

Shellenbarger, Sue. 2013. "Finding the Just-Right Level of Self-Esteem for a Child." *Wall Street Journal*, February 27, D1, D3.

Shepherd, Julianne Escobedo. 2011. "Alabama Latinos Flee Immigration Law, Leaving Insufficient Workforce for Tuscaloosa Tornado Cleanup." AlterNet, June 30. Accessed July 2, 2011 (www.alternet.org).

Shepherd, Julianne Escobedo. 2011. "Hard-Partying Rich Boy Kills Two in Hit-and-Run, Buys Himself Out of a Prison Sentence." AlterNet, June 6. Accessed June 10, 2011 (www.alternet.org).

Sheppard, Kate. 2011. "The Hackers and the Hockey Stick." *Mother Jones*, May/June, 33–45.

Sheppard, Kate. 2013. "Scientific Misconceptions: The Junk Science Behind Anti-Abortion Advocates' Wildest Claims." *Mother Jones*, January/February, 14.

Sherwood, Jessica Holden. 2010. *Wealth, Whiteness, and the Matrix of Privilege: The View from the Country Club*. Lanham, MD: Lexington Books.

Shibutani, Tamotsu. 1986. *Social Process: An Introduction to Sociology*. Berkeley: University of California Press.

Shilo, Guy, and Riki Savaya. 2011. "Effects of Family and Friend Support on LGB Youths' Mental Health and Sexual Orientation Milestones." *Family Relations* 60 (July): 318–330.

Shorrocks, Anthony, James B. Davies, Rodrigo Lluberas, Markus Stierli, and Antonios Koutsoukis. 2014. *Global Wealth Report 2014*. Credit Suisse Research Institute, October. Accessed November 14, 2014 (www.credit-suisse.com).

Shorto, Russell. 2008. "No Babies?" *New York Times*, June 29, 34.

Shriver, Maria. 2014. *The Shriver Report: A Woman's Nation Pushes Back from the Brink*. New York: Rosetta Books.

Shteir, Rachel. 2013. "Feminism Fizzles." *The Chronicle Review*, February 1, B6–B9.

Sickmund, Melissa, and Charles Puzzanchera, eds. 2014. *Juvenile Offenders and Victims: 2014 National Report*. Pittsburgh, PA: National Center for Juvenile Justice.

Siegel, Andrea F. 2009. "Dog Killer Gets 3 Years." *Baltimore Sun*, August 21, 10.

Siegman, Aron W., and Stanley Feldstein, eds. 1987. *Nonverbal Behavior and Communication*. Hillsdale, NJ: Lawrence Erlbaum Associates.

Sifferlin, Alexandra. 2014. "Rise of the Mumps: What's Behind the New Cases." *Time*, April 7, 24.

Silverman, Rachel E. 2003. "Provisions Boost Rights of Couples Living Together." *Wall Street Journal*, March 5, D1.

Silverman, Rachel E. 2013. "Tracking Sensors Invade the Workplace." *Wall Street Journal*, March 7, B1, B2.

Simmel, Georg. 1902. "The Number of Members as Determining the Sociological Form of the Group." *American Journal of Sociology* 8 (July): 1–46.

Simmons, Katie, James Bell, Richard Wike, and Alan Cooperman. 2014. "Worldwide, Many See Belief in God as Essential to Morality: Richer Nations Are Exception." Pew Research Center, May 27. Accessed June 21, 2014 (www.pewresearch.org).

Simon, Paula. 2013. "AP Is Not for Everyone." *Baltimore Sun*, August 24, 17.

Singal, Pooja, Adi Rattner, and Meghana Desale. 2014. "Court Ruling Limits a Doctor's Options." *Baltimore Sun*, July 17, 17.

Singer, Natasha. 2013. "Group Criticizes Learning Apps for Babies." *New York Times*, August 8, B8.

Singer, Natasha. 2013. "They Loved Your G.P.A. Then They Saw Your Tweets." *New York Times*, November 10, BU3.

Sjoberg, Gideon. 1960. *The Preindustrial City: Past and Present*. Glencoe, IL: Free Press.

Skoning, Gerald D. 2013. "How Congress Puts Itself Above the Law." *Wall Street Journal*, April 16, A15.

Skoumal, Robert J., Michael R. Brudzinski, and Brian S. Currie. 2015. "Earthquakes Induced by Hydraulic Fracturing in Poland Township, Ohio." *Bulletin of the Seismological Society of America* 105 (February):189–197.

Slater, Dan. 2013. "Darwin Was Wrong About Dating." *New York Times*, January 13, SR1.

Smedley, Audrey. 2007. *Race in North America: Origin and Evolution of a Worldview*, 3rd edition. Boulder, CO: Westview Press.

Smelser, Neil J. 1962. *Theory of Collective Behavior*. New York: Free Press.

Smelser, Neil J. 1988. "Social Structure." Pp. 103–129 in *Handbook of Sociology*, edited by Neil J. Smelser. Newbury Park, CA: Sage.

Smith, Aaron, and Monica Anderson. 2015. "5 Facts About Online Dating." Pew Research Center, April 20. Accessed August 14, 2015 (pewresearch.org).

Smith, Aaron, Janna Anderson, and Lee Rainie. 2014. "AI, Robotics, and the Future of Jobs." Pew Research Center, August 6. Accessed October 8, 2014 (www.pewresearch.org).

Smith, Aaron. "The Best (and Worst) of Mobile Connectivity." Pew Research Center, November 30. Accessed December 4, 2013 (www.pewinternet.org).

Smith, Aaron. 2014. "5 Facts About Online Dating." Pew Research Center, February 13. Accessed March 2, 2014 (www.pewresearch.org).

Smith, Aaron. 2014. "U.S. Views of Technology and the Future: Science in the Next 50 Years."

Pew Research Center, April 17. Accessed October 8, 2014 (www.pewresearch.org).

Smith, Aaron. 2015. "U.S. Smartphone Use in 2015." Pew Research Center, April 1. Accessed August 14, 2015 (pewresearch.org).

Smith, Adam. 1776/1937. *An Inquiry into the Nature and Causes of the Wealth of Nations*. New York: Modern Library.

Smith, Dorothy E. 1987. *The Everyday World as Problematic: A Feminist Sociology*. Toronto: University of Toronto Press.

Smith, Jane I. 1994. "Women in Islam." Pp. 303–325 in *Today's Woman in World Religions*, edited by Arvind Sharma. Albany: State University of New York Press.

Smith, Jessica C., and Carla Medalia. 2015. "Health Insurance Coverage in the United States: 2014." U.S. Census Bureau, Current Population Reports, September. Accessed October 15, 2015 (census.gov).

Smith, N., and A. Leiserowitz. 2013. "American Evangelicals and Global Warming." *Global Environmental Change* 23 (October): 1009–1017.

Smith, Ray A. 2013. "A Closet Filled with Regrets." *Wall Street Journal*, April 18, D1, D4.

Smith, Rebecca. 2014. "The First Baby Has Been Born Following a Womb Transplant." *The Telegraph*, October 3. Accessed October 8, 2014 (telegraph.co.uk).

Smith, Stacy L., and Marc Choueiti. 2011. "Gender Inequality in Cinematic Content? A Look at Females on Screen & Behind-the-Camera in Top-Grossing 2008 Films." Annenberg School for Communication & Journalism, University of Southern California." April 22. Accessed October 25, 2011 (www.annenberg.usc.edu).

Smith, Stacy L., Katherine M. Pieper, Amy Granados, and Marc Choueiti. 2010. "Assessing Gender-Related Portrayals in Top-Grossing G-Rated Films." *Sex Roles* 62 (June): 774–786.

Smith, Stacy L., Marc Choueiti, Katherine Pieper, Traci Gillig, Carmen Lee, and Dylan DeLuca. 2015. "Inequality in 700 Popular Films: Examining Portrayals of Gender, Race, & LGBT Status From 2007 to 2014." USC Annenberg School for Communication and Journalism. Accessed September 29, 2015 (annenberg.usc.edu).

Smith, Tom W., Peter V. Marsden, and Michael Hout. 2011. *General Social Survey, 1972–2010, Cumulative File Codebook*. Ann Arbor, MI: Inter-University Consortium for Political and Social Research. Accessed February 13, 2014 (www.icpsr.umich.edu).

Snell, Tracy L. 2014. "Capital Punishment, 2013—Statistical Tables." Bureau of Justice Statistics, December. Accessed September 2, 2015 (bjs.gov).

Snider, Laureen. 1993. "Regulating Corporate Behavior." Pp. 177–210 in *Understanding Corporate Criminality*, edited by Michael B. Blankenship. London: Garland Press.

Snipp, C. Matthew. 1996. "A Demographic Comeback for American Indians." *Population Today* 24 (November): 4–5.

Snyder, Thomas D., and Sally A Dillow. 2013. *Digest of Education Statistics 2012*. U.S. Department of Education, National Center for Education Statistics, December. Accessed June 21, 2014 (www.nces.ed.gov).

"Social Networking Fact Sheet." 2015. Pew Research Center. Accessed August 14, 2015 (pewresearch.org).

Soguel, Dominique. 2009. "Wage Gap Study Arrives in Time for Equal Pay Day." Women's eNews, April 28. Accessed April 28, 2009 (www.womensenews.org).

Sommers, Dixie, and James C. Franklin. 2012. "Overview of Projections to 2020." *Monthly Labor Review* 135 (January): 3–20.

Sonfield, Adam, and Kathryn Kost. 2013. "Public Costs from Unintended Pregnancies and the Role of Public Insurance Programs in Paying for Pregnancy Infant Care: Estimates for 2008." Guttmacher Institute, October. Accessed May 16, 2014 (www.guttmacher .org).

Sonfield, Adam, and Kathryn Kost. 2015. "Public Costs from Unintended Pregnancies and the Role of Public Insurance Programs in Paying for Pregnancy-Related Care: National Estimates for 2010." Guttmacher Institute, February. Accessed September 29, 2015 (guttmacher.org).

Southern Poverty Law Center. 2015. "Active U.S. Hate Groups." Accessed August 26, 2015 (www.splcenter.org).

Spalter-Roth, Roberta, Nicole Van Vooren, and Mary S. Senter. 2013. "Social Capital for Sociology Majors: Applied Activities and Peer Networks." American Sociological Association. Accessed October 12, 2013 (www.asanet.org).

Speigel, Lee. 2013. "48 Percent of Americans Believe UFOs Could Be ET Visitations." *Huffington Post*, September 11. Accessed September 12, 2013 (www.huffingtonpost .com).

Spencer, Herbert. 1862/1901. *First Principles*. New York: P. F. Collier & Son.

Sperry, Shelley. 2008. "Ozone Defense." *National Geographic* 214 (October): no page given.

Spetz, Joanne, Stephen T. Parente, Robert J. Town, and Dawn Bazarko. 2013. "Scope-of-Practice Laws for Nurse Practitioners Limit Cost Savings That Can Be Achieved in Retail Clinics." *Health Affairs* 32 (November): 1977–1984.

Spiller, Michael W., Dita Broz, Cyprian Wejnert, Lina Nerlander, and Gabriela Paz-Bailey. 2015. "HIV Infection and HIV-Associated Behaviors among Persons Who Inject Drugs—20 Cities, United States, 2012." *MMWR* 64 (10): 270–275.

Spitznagel, Eric. 2013. "Men on the Run." *Bloomberg Businessweek*, March 25–31, 82.

Spivak, Howard R., E. Lynn Jenkins, Kristi VanAudenhove, Debbie Lee, Mim Kelly, and John Iskander. 2014. "CDC Grand Rounds: A Public Health Approach to Prevention of Intimate Partner Violence." *MMWR* 63 (January 17): 38–41.

Spotts, Peter N. 2009. "New Climate Change Signal: Oceans Turning Acidic." *Christian Science Monitor*, December 9. Accessed December 12, 2010 (www.csmonitor.com).

Spradley, J. P., and M. Phillips. 1972. "Culture and Stress: A Quantitative Analysis." *American Anthropologist* 74 (3): 518–529.

Sprigg, Peter. 2011. "Marriage's Public Purpose." *Baltimore Sun*, February 2, 27.

Srivastava, Mehul, and Kartikay Mehrotra. 2013. "India's Second-Class Citizens." *Bloomburg Businessweek*, January 14–20, 10–12.

Srivastava, Mehul. 2013. "In Bangladesh, Outside Inspectors Are Still MIA." *Bloomberg Businessweek*, November 4–10, 28–30.

St. George, Donna. 2015. "Md. Officials: Letting 'Free Range' Kids Walk or Play Alone Is Not Neglect." *Washington Post*, June 11. Accessed August 6, 2015 (washingtonpost.com).

Starr, Christine R., and Gail M. Ferguson. 2012. "Sexy Dolls, Sexy Grade-Schoolers? Media and Maternal Influences on Young Girls' Self-Sexualization." *Sex Roles* 67 (October): 463–476.

State Higher Education Executive Officers (SHEEO). 2014. "State Higher Education Finance FY 2013." Accessed June 21, 2014 (www.sheeo.org).

Statista. 2014. "How Americans Spend Their Online Time." Statista Infographics, April. Accessed August 22, 2015 (statista.com).

Steel, Emily. 2015. "For Cable Network, Crime Certainly Pays." *New York Times*, January 5, B1.

Steffensmeier, Darrell J., Jennifer Schwartz, and Michael Roche. 2013. "Gender and Twenty-First-Century Corporate Crime: Female Involvement and the Gender Gap in Enron-Era Corporate Frauds." *American Sociological Review* 78 (June): 448–476.

Stein, Joel. 2015. "Baby, You Can Drive My Car." *Time*, February 9, 32–40.

Steinberg, Julia R., and Lawrence B. Finer. 2011. "Examining the Association of Abortion History and Current Mental Health: A Reanalysis of the National Comorbidity Survey Using a Common-Risk-Factors Model." *Social Science & Medicine* 72 (1): 72–82.

Steinhauser, Paul. 2014. "CNN Poll: Pathway to Citizenship Trumps Border Security." Cable News Network, February 6. Accessed March 22, 2014 (www.cnn.com).

Steinmetz, Katy. 2014. "Disrupted: The Tech Wealth Transforming San Francisco Has Unemployment Down, Evictions Up and Tensions Flaring." *Time*, February 10, 32–35.

Stencel, Sandra, and Brian J. Grim. 2013. "Headscarf Incident in Sudan Highlights a Global Trend." Pew Research Center, September 18. Accessed June 21, 2014 (www.pewresearch.org).

Stepan-Norris, Judith. 2015. "Social Justice & the Next Upward Surge for Unions." *Contexts* 14 (Spring): 46–51.

Sternbergh, Adam. 2008. "Why White People Like 'Stuff White People Like.'" *New Republic*, March 17. Accessed November 6, 2008 (www.tnr.com).

Steuerle, Eugene, Signe-Mary McKernan, Caroline Ratcliffe, and Sisi Zhang. 2013. "Lost Generations? Wealth Building Among Young Americans." Urban Institute, March. Accessed February 6, 2014 (www.urban.org).

"Steve Jobs: Adopted Child Who Never Met His Biological Father." 2011. *Telegraph*, October 6. Accessed March 22, 2014 (telegraph.co.uk).

Stevenson, Betsey, and Justin Wolfers. 2013. "Subjective Well-Being and Income: Is There Any Evidence of Satiation?" *American Economic Review* 103 (May): 598–604.

Stewart, James B. 2015. "Discipline for Airlines, Pain for Fliers." *New York Times*, June 12, B1, B7.

Stillars, Alan L. 1991. "Behavioral Observation." Pp. 197–218 in *Studying Interpersonal Interaction*, edited by B. M. Montgomery and S. Duck. New York: Guilford Press.

Stilwell, Victoria. 2013. "Body Parts for Sale by Desperate U.S. Workers." *Bloomberg Businessweek*, October 21–27, 22, 24.

Stokes, Bruce. 2013. "The U.S.'s High Income Gap Is Met with Relatively Low Public Concern." Pew Research Center. Accessed February 14, 2014 (www.pewresearch.org).

Stolley, Giordano, and Somchai Taphaneeyapan. 2002. "Stomaching Bugs in Thailand." *Baltimore Sun*, June 20, 2A.

Strasburger, Victor C., and Marjorie J. Hogan. 2013. "Policy Statement: Children, Adolescents, and the Media." *Pediatrics* 132 (5): 958–961.

Straus, Murray A. 2011. "Gender Symmetry and Mutuality in Perpetration of Clinical-Level Partner Violence: Empirical Evidence and Implications for Prevention and Treatment." *Aggression and Violent Behavior* 16 (July–August): 279–288.

Straus, Murray A. 2014. "Addressing Violence by Female Partners Is Vital to Prevent or Stop Violence Against Women: Evidence From the Multisite Batterer Intervention Evaluation." *Violence Against Women* 20 (7): 889–899.

"Streaking." 2005. Wikipedia Encyclopedia. Accessed September 8, 2005 (www.wikipedia.org).

Strean, William B. 2009. "Remembering Instructors: Play, Pain, and Pedagogy." *Qualitative Research in Sport and Exercise* 1 (November): 210–220.

Streib, Jessi. 2011. "Class Reproduction by Four Year Olds." *Qualitative Sociology* 34 (June): 337–352.

Stroebe, Margaret, Maarten van Son, Wolfgang Stroebe, Rolf Kleber, Henk Schut, and Jan van den Bout. 2000. "On the Classification and Diagnosis of Pathological Grief." *Clinical Psychology Review* 20 (January): 57–75.

Strom, Stephanie. 2014. "Coca-Cola to Remove an Ingredient Questioned by Consumers." *New York Times*, May 6, B6.

Strom, Stephanie. 2015. "Kraft Cheese Product Gains Nutritional Seal; Children's Advocates Criticize Move." *New York Times*, March 13, B3.

Strom, Stephanie. 2015. "The Fried Chicken is O.K., but the Cups Are Delicious." *New York Times*, February 26, B3.

"Study Shows Positive Results from Early Head Start Program." 2002. U.S. Department of Health and Human Services, June 3. Accessed July 10, 2002 (www.hhs.gov).

Sturm, Roland, and Ruopeng An. 2014. "Obesity and Economic Environments." *CA: A Cancer Journal for Clinicians* 65 (5): 337–350.

Substance Abuse and Mental Health Services Administration. 2013. *Results from the 2012 National Survey on Drug Use and Health: Summary of National Findings.* Accessed July 26, 2014 (www.store.samhsa.gov).

Substance Abuse and Mental Health Services Administration. 2015. "Behavioral Health Barometer: United States, 2014." Accessed October 15, 2015 (samsa.gov).

Sudarkasa, Niara 2007. "African American Female-Headed Households: Some Neglected Dimensions." In *Black Families*, 4th edition, edited by Harriette Pipes McAdoo, 172–183. Thousand Oaks, CA: Sage.

Suddath, Claire. 2013. "So Hard to Say Goodbye." *Bloomberg Businessweek*, November 4–10, 79–81.

Suddath, Claire. 2013. "You Get a D+ in Teamwork." *Bloomberg Businessweek*, November 11–17, 91.

Suich, Alexander. 2014. "Little Brother." *The Economist*, September 13, 3–16.

Sullivan, Andrew. 2011. "Why Gay Marriage Is Good for Straight America." *Newsweek*, July 25, 12–14.

Sullivan, Erin M., Joseph L Annest, Thomas R. Simon, Feijun Luo, and Linda L. Dahlberg. 2015. "Suicide Trends among Persons Aged 10–24 Years—United States, 1994–2012." *MMWR* 64 (8): 201–225.

Sullivan, Randall. 2015. "One Man's Trash." *Wired*, February, 95–103.

Sumner, William G. 1906. *Folkways*. New York: Ginn.

"Sun Trust: Many Higher Income Households Living Paycheck-to-Paycheck." 2015. Sun Trust Banks, Inc. April 16. Accessed October 7, 2015 (suntrust.com).

Sunstein, Cass R. 2009. *On Rumors: How Falsehoods Spread, Why We Believe Them, What Can Be Done.* New York: Farrar, Straus and Giroux.

"Super Bull Sunday." 2015. Snopes.com, January 30. Accessed July 22, 2015 (snopes.com).

Sutherland, Edwin H. 1949. *White Collar Crime*. New York: Holt, Rinehart, and Winston.

Sutherland, Edwin H., and D. R. Cressey. 1970. *Criminology*, 8th edition. Philadelphia: Lippincott.

Sutton, Philip W. 2000. *Explaining Environmentalism: In Search of a New Social Movement.* Burlington, VT: Ashgate.

Swan, Noelle. 2014. "Rise of the People-Friendly Machines." *Christian Science Monitor Weekly*, October 6, 21–23.

Swift, Art. 2013. "Honesty and Ethics Rating of Clergy Slides to New Low." Gallup, December 16. Accessed April 20, 2014 (www.gallup.com).

Swift, Art. 2015. "Americans Split on Support for Fracking in Oil, Natural Gas." Gallup, March 23. Accessed October 23, 2015 (gallup.com).

Swift, Art. 2015. 2015. "Most Americans Continue to Say Death Penalty Morally OK." Gallup, June 4. Accessed August 22, 2015 (gallup.com).

Szarota, Piotr. 2010. "The Mystery of the European Smile: A Comparison Based on Individual Photographs Provided by Internet Users." *Journal of Nonverbal Behavior* 34 (December): 249–256.

Tabachnick, Rachel. 2011. "The 'Christian' Dogma Pushed by Religious Schools That Are Supported by Your Tax Dollars." AlterNet, May 23. Accessed July 12, 2011 (www.alternet.org).

Tabuchi, Hiroko. 2015. "Target Plans to Raise Pay to at Least $9 an Hour." *New York Times*, March 14, B5.

"Tackling Poverty." 2015. *The Economist*, September 5, 31–32.

Taibbi, Matt. 2014. *The Divide: American Injustice in the Age of the Wealth Gap.* New York: Spiegel & Grau.

Tajfel, Henri. 1982. "Social Psychology of Intergroup Relations." *Annual Review of Psychology* 33 (February): 1–39.

Tannen, Deborah. 1990. *You Just Don't Understand: Women and Men in Conversation.* New York: Ballantine.

Tannenbaum, Frank. 1938. *Crime and the Community*. New York: Columbia University Press.

Tanur, Judith M. 1994. "The Trustworthiness of Survey Research." *Chronicle of Higher Education*, May 25, B1-B3.

Taormina-Weiss, Wendy. 2013. "Psychological and Social Aspects of Disability." Disabled World, June 22. Accessed November 18, 2013 (www.disabled-world.com).

"Tattoos in the Workplace: Ink Blots." 2014. *The Economist*, August 2, 22.

Tavris, Carol. 2013. "How Psychiatry Went Crazy." *Wall Street Journal*, May 18, C5.

Tavris, Carol. 2013. "The Experiments That Still Shock." *Wall Street Journal*, September 6, C1, C7.

Taylor, Carl S., and Pamela R. Smith. 2013. "The Attraction of Gangs: How Can We Reduce It?" Pp. 19–29 in *Changing Course: Preventing Gang Membership*, edited by Thomas R. Simon, Nancy M. Witter, and Reshma R. Mahendra. National Institute of Justice and Centers for Disease Control and Prevention. Accessed December 18, 2013 (www.nij.gov).

Taylor, Frederick W. 1911/1967. *The Principles of Scientific Management.* New York: W. W. Norton & Company.

Taylor, Jay. 1993. *The Rise and Fall of Totalitarianism in the Twentieth Century.* New York: Paragon House.

Taylor, Paul, and Mark H. Lopez. 2013. "Skepticism About the Census Voter Turnout Finding." Pew Research Center, May 15. Accessed August 8, 2014 (www.pewresearch.org).

Taylor, Paul, Carroll Doherty, Rich Morin, and Kim Parker. 2014. "Millennials in Adulthood: Detached from Institutions, Networked with Friends." Pew Research Center, March 7. Accessed June 21, 2014 (www.pewresearch.org).

Taylor, Paul, et al. 2012. "The Lost Decade of the Middle Class." Pew Research Center, August 22. Accessed February 14, 2014 (www.pewsocialtrends.org).

Taylor, Paul, et al. 2013. "A Survey of LGBT Americans: Attitudes, Experiences and Values in Changing Times." Pew Research Center, June 13. Accessed March 2, 2014 (www.pewresearch.org).

Taylor, Paul, et al. 2013. "The Rise of Asian Americans." Pew Research Center, April 4. Accessed May 16, 2014 (www.pewsocialtrends.org).

Taylor, Paul, Rich Morin, Kim Parker, and D'Vera Cohn. 2009. "Growing Old in America: Expectations vs. Reality." Pew Research Center, June 29. Accessed April 30, 2010 (www.pewsocialtrends.org).

Taylor, Susan C. 2003. *Brown Skin: Dr. Susan Taylor's Prescription for Flawless Skin, Hair, and Nails.* New York: HarperCollins.

Tefferi, Andrew. 2015. "In Support of a Patient-Driven Initiative and Petition to Lower the High Price of Cancer Drugs." *Mayo Clinic Proceedings* 90 (8): 996–1000.

Tenbrunsel, Ann, and Jordan Thomas. 2015. "The Street, the Bull and the Crisis: A Survey of the US & UK Financial Services Industry." University of Notre Dame and Lebaton Sucharow, May. Accessed September 2, 2015 (www.secwhistlebloweradvocate.com).

Terando, Adam J., Jennifer Costanza, Curtis Belyea, Robert R. Dunn, Alexa McKerrow, and Jaime A. Collazo. 2014. "The Southern Megalopolis: Using the Past to Predict the Future of Urban Sprawl in the Southeast U.S." *PLoS One* 9 (July): e102261.

Teranishi, Robert. 2011. "Asian Americans and Pacific Islanders: Facts, Not Fiction—Setting the Record Straight." CARE and College Board. Accessed December 7, 2011 (www.nyu.edu/projects/care).

Terhune, Chad, Noam N. Levy, and Doug Smith. 2014. "Medicare Unveils 2012 Payouts: 2% of U.S. Providers Got 23% of Fees." *Baltimore Sun*, April 9, 1, 16.

Tétreault, Mary Ann. 2001. "A State of Two Minds: State Cultures, Women, and Politics in Kuwait." *International Journal of Middle East Studies* 33 (May): 203–220.

Tharoor, Ishaan. 2015. "A Politician in Finland Declared War on Multiculturalism. This Is How His Country Responded." *Washington Post*, July 31. Accessed July 31, 2015 (washingtonpost.com).

Theodorou, Angelina, and Peter Henne. 2014. "How Religious Harassment Varies by Region Across the Globe." Pew Research Center, May 2. Accessed June 21, 2014 (www.pewresearch.org).

Theodorou, Angelina, and Peter Henne. 2014. "Religious Police Found in Nearly One-in-Ten Countries Worldwide." Pew Research Center, March 19. Accessed June 21, 2014 (www.pewresearch.org).

Thibaut, John W., and Harold H. Kelley. 1959. *The Social Psychology of Groups.* New York: Wiley.

Thio, Alex D., Jim D. Taylor, and Martin D. Schwartz. 2012. *Deviant Behavior*, 11th edition. Upper Saddle River, NJ: Pearson.

Thomas, Anita Jones, Jason Daniel Hacker, and Denada Hoxha. 2011. "Gendered Racial Identity of Black Young Women." *Sex Roles* 64 (April): 530–542.

Thomas, W. I., and Dorothy Swaine Thomas. 1928. *The Child in America.* New York: Alfred A. Knopf.

Thompson, Arienne. 2010. "16, Pregnant . . . and Famous: Teen Moms Are Newest Stars." *USA Today*, November 23. Accessed November 24, 2011 (www.usatoday.com).

Thompson, Mark. 2014. "A Troubled Marine's Final Fight." *Time*, February 10, 35–42.

Thompson, Robert S., et al. 2006. "Intimate Partner Violence: Prevalence, Types, and Chronicity in Adult Women." *American Journal of Preventive Medicine* 30 (June): 447–457.

Thorpe-Moscon, Jennifer, and Alixandra Pollock. 2014. "Feeling Different: Being the 'Other' in US Workplaces." Catalyst. Accessed March 22, 2014 (www.catalyst.org).

Tilly, Charles. 1978. *From Mobilization to Revolution.* Reading, MA: Addison-Wesley.

Timberg, Craig, and Ellen Nakashima. 2013. "Amid NSA Spying Revelations, Tech Leaders Call for New Restraints on Agency." *Washington Post*, November 1. Accessed November 10, 2013 (www.washingtonpost.com).

Timberlake, Cotten. 2013. "Don't Even Think About Returning This Dress." *Bloomberg Businessweek*, September 30–October 6, 29–31.

Timmermans, Stefan, and Hyeyoung Oh. 2010. "The Continued Social Transformation of the Medical Profession." *Journal of Health and Social Behavior* 51 (March): S94–S109.

Tobar, Hector. 2009. "Language as a Bridge and an Identity." *Los Angeles Times*, September 22. Accessed September 24, 2009 (www.latimes.com).

Tomasetti, Cristian, and Bert Vogelstein. 2015. "Variation in Cancer Risk Among Tissues Can Be Explained by the Number of Stem Cell Divisions." *Science* 347 (6217): 78–81.

Tormey, Simon. 1995. *Making Sense of Tyranny: Interpretations of Totalitarianism.* New York: Manchester University Press.

Tornatzky, Louis G., Richard Cutler, and Jongho Lee. 2002. "College Knowledge: What Latino Parents Need to Know and Why They Don't Know It." The Tomas Rivera Policy Institute. Accessed January 2, 2005 (www.trpi.org).

Toth, Emily. 2011. "No Girls Aloud." *Chronicle of Higher Education*, April 1, A26.

Touraine, Alain. 2002. "The Importance of Social Movements." *Social Movement Studies* 1 (April): 89–96.

Townsend, Matt. 2014. "The Cabbage Patch Kids Get a Makeover." *Bloomberg Businessweek*, June 9–22, 27–28.

Tozzi, John, David Armstrong, and Caroline Chen. 2014. "Why Medicare Keeps Paying Sketchy Docs." *Bloomberg Businessweek*, May 5–11, 32–34.

Tozzi, John. 2014. "Who Should Pay the Bill for Wonder Drugs?" *Bloomberg Businessweek*, July 10–16, 27–28.

"Transgender Rights." 2015. *The Economist*, June 13, 31.

Triedman, Julie. 2015. "A Few Good Women." *The American Lawyer*, May 28. Accessed September 29, 2015 (americanlawyer.com).

Trottman, Melanie, and Kris Maher. 2013. "Organized Labor Loses Members." *Wall Street Journal*, January 24, A6.

Trottman, Melanie. 2013. "Religious Discrimination Claims on the Rise." *Wall Street Journal*, October 28, B1, B4.

"Trouble in Trappes." 2013. *The Economist*, July 27, 44.

Truman, Jennifer L., and Erica L. Smith. 2012. "Prevalence of Violent Crime Among Households with Children, 1993–2010." Bureau of Justice Statistics, September. Accessed May 16, 2014 (www.bjs.gov).

Truman, Jennifer L., and Lynn Langton. 2014. "Criminal Victimization, 2013." Bureau of Justice Statistics, September. Accessed November 6, 2014 (www.bjs.gov).

Truman, Jennifer L., and Lynn Langton. 2015. "Criminal Victimization, 2014." Bureau of Justice Statistics, August. Accessed September 2, 2015 (bjs.gov).

Truman, Jennifer L., and Rachel E. Morgan. 2014. "Nonfatal Domestic Violence, 2003–2012." Bureau of Justice Statistics, April. Accessed May 16, 2014 (www.bjs.gov).

Truman, Jennifer, Lynn Langton, and Michael Planty. 2013. "Criminal Victimization, 2012." Bureau of Justice Statistics, October. Accessed January 12, 2014 (www.bjs.gov).

Tsai, Tyjen. 2012. "China Has Too Many Bachelors." Population Reference Bureau, January. Accessed September 19, 2014 (www.prb.org).

Tsukayama, Hayley. 2014. "High-Tech Upgrades May Let Aging Boomers Live Independently in Their Own Homes Longer." *Washington Post*, January 20. Accessed October 8, 2014 (www.washingtonpost.com).

Tsukayama, Hayley. 2015. "Tech Companies Are the 'Worst Offenders' When It Comes to Tax Loopholes, Advocate Says." *Washington Post*, August 31. Accessed August 31, 2015 (washingtonpost.com).

Tumin, Melvin M. 1953. "Some Principles of Stratification: A Critical Analysis." *American Sociological Review* 18 (August): 387–393.

Turk, Austin T. 1969. *Criminality and the Legal Order.* Chicago: Rand-McNally.

Turk, Austin T. 1976. "Law as a Weapon in Social Conflict." *Social Problems* 23 (February): 276–291.

Turner, Jennifer. 2013. "A Living Death: Life Without Parole for Nonviolent Offenses." American Civil Liberties Union. Accessed January 15, 2014 (www.aclu.org).

Turner, Margery A., Rob Santos, Diane K. Levy, Doug Wissoker, Claudia Aranda, and Rob Pitingolo. 2013. "Housing Discrimination Against Racial and Ethnic Minorities 2012." The Urban Institute, June. Accessed March 22, 2014 (www.huduser.org).

Turner, Ralph H., and Lewis M. Killian. 1987. *Collective Behavior*, 3rd edition. Englewood Cliffs, NJ: Prentice Hall.

Twenge, Jean M. 2007. *Generation Me: Why Today's Young Americans Are More Confident, Assertive, Entitled—and More Miserable Than Ever Before.* New York: Simon & Schuster.

UN Women. 2014. "Facts and Figures: Ending Violence Against Women." Accessed March 2, 2014 (www.unwomen.org).

UN Women. 2015. "Facts and Figures: Leadership and Political Participation." Accessed September 29, 2015 (unwomen.org).

UNICEF. 2013. "Female Genital Mutilation/Cutting: A Statistical Overview and Exploration of the Dynamics of Change." Unite for Children. Accessed March 2, 2014 (www.childinfo.org).

"Union Members—2014." 2014. Bureau of Labor Statistics, January 23. Accessed January 26, 2015 (bls.gov).

United Human Rights Council. 2004. "History of Genocide." Accessed April 29, 2004 (www.unitedhumanrights.org).

United Nations Committee on the Rights of the Child. 2014. "Concluding Observations on the Second Periodic Report of the Holy See." February 25. Accessed June 21, 2014 (www.ohchr.org).

United Nations Department of Economic and Social Affairs. 2012. "World Urbanization Prospects: The 2011 Revision, Highlights." March. Accessed June 23, 2012 (www.esa.un.org).

United Nations Development Program. 2013. "Humanity Divided: Confronting Inequality in Developing Countries." November. Accessed February 6, 2014 (www.undp.org).

United Nations Development Program. 2014. "Human Development Report 2014." Accessed September 19, 2014 (www.undp.org).

United Nations High Commissioner for Refugees. 2015. "World at War: UNHCR Global Trends—Forced Displacement in 2014." June 18. Accessed October 23, 2015 (unhcr.org).

United Nations Population Division. 2015. "World Population Prospects: The 2015 Revision." Department of Economic and Social Affairs. Accessed October 19, 2015 (un.org).

United Nations. 2014. *World Urbanization Prospects: The 2014 Revision.* Department of Economics and Social Affairs, Population Division. Accessed September 19, 2014 (www.esa.un.org).

United Nations. 2015. "The Millennium Development Goals Report 2015." Accessed September 29, 2015 (un.org).

U.S. Bureau of Labor Statistics. 2014. "Labor Force Participation Rate." Accessed February 13, 2014 (www.bls.gov).

U.S. Bureau of Labor Statistics. 2015. "Household Data, Annual Averages, Table 39." Accessed August 22, 2015 (bls.gov).

U.S. Census Bureau News. 2014. "Unmarried and Single Americans Week Sept. 21–27, 2014." Facts for Features, July 30. Accessed July 28, 2015 (census.gov).

U.S. Census Bureau Newsroom. 2014. "Energy Boom Fuels Rapid Population Growth in Parts of Great Plains; Gulf Coast Also Has High Growth Areas, Says Census Bureau." U.S. Census Bureau, March 27. Accessed September 19, 2014 (www.census.gov).

"U.S. Census Bureau Projections Show a Slower Growing, Older, More Diverse Nation a Half Century from Now." 2012. United States Census Bureau, December 12. Accessed May 16, 2014 (www.census.gov).

U.S. Census Bureau, Current Population Survey. 2013. "Annual Social and Economic Supplements." November. Accessed October 20, 2014 (www.census.gov).

U.S. Census Bureau, Voting and Registration. 2013. "Voting and Registration in the Election of November 2012—Detailed Tables." Census Bureau, May 8. Accessed April 18, 2014 (www.census.gov).

U.S. Census Bureau. 2008. "An Older and More Diverse Nation by Midcentury." August 14. Accessed August 18, 2008 (www.census.gov).

U.S. Census Bureau. 2012. *Statistical Abstract of the United States: 2012,* 131st edition. Washington, DC: Government Printing Office.

U.S. Department of Health & Human Services. 2015. *Child Maltreatment 2013.* Administration on Children, Youth and Families, Children's Bureau. Accessed July 16, 2015 (acf.hhs.gov).

U.S. Department of Health and Human Services. 2013. *Child Maltreatment 2012.* Administration for Children and Families, Children's Bureau. Accessed May 16, 2014 (www.acf.hhs.gov).

U.S. Department of Health and Human Services. 2014. "The Health Consequences of Smoking—50 Years of Progress. A Report of the Surgeon General." Office of the Surgeon General. Accessed July 30, 2014 (www.cdc.gov).

U.S. Department of Justice. 1996. "Policing Drug Hot Spots." National Institute of Justice, January. Accessed May 3, 2005 (www.ncjrs.org).

U.S. Department of Justice. 2011. "U.S. Attorney Announces Drug Endangered Children Task Force." Press Notice, May 31. Accessed July 25, 2011 (www.usdoj.gov).

U.S. Department of State. 2013. "Trafficking in Persons Report: June 2013." Accessed March 2, 2014 (www.state.gov).

U.S. Energy Information Administration. 2014. "China." February 4. Accessed September 19, 2014 (www.eia.gov).

U.S. Environmental Protection Agency. 2009. "EPA's Endangerment Finding." December 7. Accessed May 1, 2010 (www.epa.gov).

U.S. Senate Special Committee on Aging, American Association of Retired Persons, Federal Council on the Aging, and U.S. Administration on Aging. 1991. *Aging America: Trends and Projections, 1991.* Washington, DC: U.S. Department of Health and Human Services.

U.S. Senate. 2004. *Report on the U.S. Intelligence Community's Prewar Intelligence Assessments on Iraq.* Accessed September 13, 2006 (www.gpoaccess.gov).

U.S. Travel Association. 2014. "Working for Free: U.S. Workforce Forfeits $52.4 Billion in Time Off Benefits Annually." October 21. Accessed November 23, 2014 (www.ustravel.org).

Urbina, Ian. 2006. "In Online Mourning, Don't Speak Ill of the Dead." *New York Times,* November 5, 1.

Urrico, Roy. 2015. "The 10 Worst Data Breaches of 2015 (So Far)." *Credit Union Times,* July 7: Accessed August 30 (cutimes.com).

"Vacation Habits Around the World." 2013. Expedia 2013 Vacation Deprivation Study, November 18. Accessed April 12, 2014 (www.viewfinder.expedia.com).

"Valley of the Dudes." 2015. *The Economist,* April 4, 64.

Vallotton, Claire, and Catherine Ayoub. 2011. "Use Your Words: The Role of Language in the Development of Toddlers' Self-Regulation." *Early Childhood Research Quarterly* 26 (2): 169–181.

van Agtmael, Peter. 2013. "Laugh After Death." *Time,* November 18, 44–49.

Van Vooren, Nicole, and Roberta Spalter-Roth. 2010. "Tracking Master's Students Through Programs and into Careers." *Footnotes* (September/October): 10–11.

Veblen, Thorstein. 1899/1953. *The Theory of the Leisure Class.* New York: New American Library.

Velasco, Schuyler. 2014. "Food Additives: More Fuss Than Harm?" *Christian Science Monitor,* May 26, 43.

Venkatesh, Sudhir. 2008. *Gang Leader for a Day: A Rogue Sociologist Takes to the Streets.* New York: Penguin Press.

Vespa, Jonathan, Jamie M. Lewis, and Rose M. Kerner. 2013. "America's Families and Living Arrangements: 2012." U.S. Census Bureau, August. Accessed May 16, 2014 (www.census.gov).

Vicens, A. J. 2015. "Missing in Action." *Mother Jones,* July/August, 13.

Visser, Susanna N., et al. 2014. "Trends in the Parent-Report of Health Care Provider-Diagnosed and Medicated Attention-Deficit/Hyperactivity Disorder: United States, 2003–2011." *Journal of the American Academy of Child & Adolescent Psychiatry* 53 (January): 34–46.

Vitello, Paul. 2006. "The Trouble when Jane Becomes Jack." *New York Times* (August 20): H1, H6.

Vlasic, Bill. 2015. "G.M.'s Ignition Switch Death Toll Hits 100." *New York Times,* May 12, B3.

Vold, George B. 1958. *Theoretical Criminology.* New York: Oxford University Press.

Volz, Matt. 2012. "$3.4 Billion Indian Land Royalty Settlement Upheld." Salon, May 22. Accessed March 22, 2014 (www.salon.com).

Von Drehle, David. 2014. "Breaking the Marriage Bans." *Time,* March 3, 12.

Voosen, Paul. 2014. "As People Shun Pollsters, Researchers Put Online Surveys to the Test." *Chronicle of Higher Education,* September 5, A8, A10.

Voosen, Paul. 2014. "In Backlash over Facebook Research, a Trove of Data Is at Risk." *Chronicle of Higher Education,* July 18, A5.

"Wages Maximus." 2014. *Christian Science Monitor Weekly,* March 10, 33.

"The Waiting Wounded." 2013. *The Economist,* March 23, 33–34.

Waldron, Travis. 2011. "Pastor at Kentucky Church That Banned Interracial Couples Calls for Vote to Reverse Decision." AlterNet, December 3. Accessed December 7, 2011 (www.alternet.org).

Walk Free Foundation. 2014. "The Global Slavery Index 2014." Accessed September 11, 2015 (globalslaveryindex.org).

Wallace, Bruce. 2006. "Japanese Schools to Teach Patriotism." *Los Angeles Times,* December 16. Accessed December 18, 2006 (www.latimes.com).

Wallis, Cara. 2011. "Performing Gender: A Content Analysis of Gender Display in Music Videos." *Sex Roles* 64 (February): 160–172.

Walls, Mark. 2011. "High Court Strikes Down Calif. Law on Violent Video Games." *Education Week,* June 27. Accessed June 28, 2011 (www.edweek.org).

Walmsley, Roy. 2013. "World Prison Population List (Tenth Edition)." International Center for Prison Studies. Accessed January 15, 2014 (www.prisonstudies.org).

Walsh, Bryan. 2014. "After the Spill." *Time,* January 27, 14.

Walsh, Janet. 2011. "Failing Its Families: Lack of Paid Leave and Work-Family Supports in the US." Human Rights Watch. Accessed April 7, 2013 (www.hrw.org).

Wang, Wendy, and Kim Parker. 2011. "Women See Value and Benefits of College; Men Lag on Both Fronts, Survey Finds." Pew Research Center, August 17. Accessed April 15, 2012 (www.pewsocialtrends.org).

Wang, Wendy, and Kim Parker. 2014. "Record Share of Americans Have Never Married: As Values, Economic and Gender Patterns Change." Pew Research Center, September 24. Accessed October 26, 2014 (www.pewresearch.org).

Wang, Wendy, Kim Parker, and Paul Taylor. 2013. "Breadwinner Moms." Pew Research Center, May 29. Accessed May 18, 2014 (www.pewresearch.org).

Wang, Wendy. 2012. "The Rise of Intermarriage." Pew Research Center, Social & Demographic Trends, February 16. Accessed March 22, 2014 (www.pewsocialtrends.org).

Wang, Wendy. 2013. "Parents' Time with Kids More Rewarding Than Paid Work—and More Exhausting." Pew Research Center, October 8. Accessed March 2, 2014 (www.pewresearch.org).

Wang, Y. Claire. 2015. "The Dangerous Silence of Academic Researchers." *Chronicle of Higher Education.* February 27, A48.

Ward, Brian, W., James M. Dahlhamer, Adena M. Galinsky, and Sarah S. Joestl. 2014. "Sexual Orientation and Health Among U.S. Adults: National Health Interview Survey, 2013." *National Health Statistics Reports,* No. 77, July 15. Accessed September 29, 2015 (cdc.gov/nchs).

Warren, Kenneth R. 2012. "NIH Statement on International FASD Awareness Day." National Institutes of Health, September 5. Accessed November 20, 2013 (www.nih.gov).

Wartik, Nancy. 2005. "The Perils of Playing House." *Psychology Today* (July/August): 42–52.

Wasley, Paula. 2007. "46 Students Are Disciplined for Cheating at Indiana University's Dental School." *Chronicle of Higher Education,* May 9. Accessed May 10, 2007 (www.chronicle.com).

Wasley, Paula. 2008. "The Syllabus Becomes a Repository of Legalese." *Chronicle of Higher Education* 54, March 14, A1, A8–A11.

Waters, Mary C., and Marisa Gerstein Pineau, eds. 2015. *The Integration of Immigrants into American Society.* Washington, DC: The National Academies Press.

Wealth-X and UBS. 2014. "Billionaire Census 2014." Accessed November 10, 2014 (www.billionairecensus.com).

Weathers, Cliff. 2014. "4 of the Biggest Quacks Plaguing America with False Claims About Science." AlterNet, July 17. Accessed July 26, 2014 (www.alternet.org).

Weathers, Cliff. 2014. "Research Behind Dr. Oz-Endorsed Diet Supplement Was Bogus, Authors Admit." AlterNet, October 22. Accessed December 5, 2014 (www.alternet.org).

Weathers, Cliff. 2015. "It's 2015 and Congress Is Almost Entirely Made Up of White Christian Men." AlterNet, January 6. Accessed September 29, 2015 (alternet.org).

Weaver, Rosanna Landis. 2015. "The 100 Most Overpaid CEOs." As You Sow. Accessed October 7, 2015 (asyousow.org).

Webb, Alex. 2013. "Germany's Coalition Looks to Gender Quotas." *Bloomberg Businessweek,* December 16–22, 17–18.

Weber, Lynn, Tina Hancock, and Elizabeth Higginbotham. 1997. "Women, Power, and Mental Health." Pp. 380–396 in *Women's Health: Complexities and Differences,* edited by Sheryl B. Ruzek, Virginia L. Olesen, and Adele E. Clarke. Columbus: Ohio State University Press.

Weber, Max. 1920/1958. *The Protestant Ethic and the Spirit of Capitalism,* translated by Talcott Parsons. New York: Charles Scribner's Sons.

Weber, Max. 1925/1947. *The Theory of Social and Economic Organization.* New York: Free Press.

Weber, Max. 1925/1978. *Economy and Society,* edited by Guenther Roth and Claus Wittich. Berkeley: University of California Press.

Weber, Max. 1946. *From Max Weber: Essays in Sociology,* translated and edited by H. H. Gerth and C. Wright Mills. Berkeley: University of California Press.

Weinhold, Bob. 2012. "A Steep Learning Curve: Decoding Epigenetic Influences on Behavior and Mental Health." *Environmental Health Perspectives* 120 (October): A396–A401.

Weintraub, Arlene. 2010. "Break That Hovering Habit Early." *U.S. News & World Report,* September, 42–43.

Weise, Elizabeth, and Tom Vanden Brook. 2015. "More Ashley Madison Files Published." *USA Today,* August 20. Accessed August 21, 2015 (usatoday.com).

Weisman, Carrie. 2015. "Can a Woman Find Sexual Desire in a Pill?" AlterNet, June 15. Accessed October 15, 2015 (alternet.org).

Weiss, Carol H. 1998. *Evaluation: Methods for Studying Programs and Policies,* 2nd edition. Upper Saddle River, NJ: Prentice Hall.

Weiss, Rick. 2004. "Nanomedicine's Promise Is Anything but Tiny." *Washington Post,* January 31, A8.

Weiss, Rick. 2005. "The Power to Divide." *National Geographic* 208 (July): 3–27.

Weissman, Jordan. 2014. "The Fast-Food Strikes Have Been a Stunning Success for Organized Labor." *Slate,* September 7. Accessed October 8, 2014 (www.slate.com).

Welch, Ashley. 2015. "GOP Debate Fact Check: Claims About Vaccines and Autism." CBS News, September 17. Accessed October 25, 2015 (cbsnews.com).

Welch, David. 2011. "For the UAW, a Bargaining Dilemma." *Bloomberg Businessweek,* September 19–25.

Wenneras, Christine, and Agnes Wold. 1997. "Nepotism and Sexism in Peer Review." *Nature* 387 (May 22): 341–343.

Wessel, David. 2015. "The Typical Male U.S. Worker Earned Less in 2014 Than in 1973." Brookings, September 18. Accessed October 7, 2015 (brookings.edu).

West, Candace, and Don H. Zimmerman. 2009. "Accounting for Doing Gender." *Gender & Society* 23 (February): 112–122.

Westcott, Lucy. 2014. "More Americans Moving to Cities, Reversing the Suburban Exodus." The Wire, March 27. Accessed September 19, 2014 (www.thewire.com).

White, Alexander, and Carolyn Witkus. 2013. "How Legal Marijuana May Affect Troubled Families." *Christian Science Monitor Weekly,* March 11, 36.

White, James W. 2005. *Advancing Family Theories.* Thousand Oaks, CA: Sage.

White, Nicole, and Janet L. Lauritsen. 2012. "Violent Crime Against Youth: 1994–2010." Bureau of Justice Statistics, December. Accessed May 4, 2013 (www.bjs.gov).

Whiteaker, Chloe, Laurie Meisler, Yvette Romero, and Karen Weise. 2014. "Voter ID States That Make Voting Super Simple—or Really Hard." *Bloomberg Businessweek,* October 27–November 2, 36.

Whitehead, Jaye C. 2011. "The Wrong Reasons for Same-Sex Marriage." *New York Times,* May 16, A21.

Whitelaw, Kevin. 2000. "But What to Call It?" *U.S. News & World Report,* October 16, 42.

Whiteman, Shawn, D., Susan M. McHale, and Anna Soli. 2011. "Theoretical Perspectives on Sibling Relationships." *Journal of Family Theory & Review* 3 (June): 124–139.

Whitesides, John. 2014. "Anti-Union Group Pledges to Remain on the Offensive." *Baltimore Sun,* February 18, 2.

Whitson, Sarah Leah. 2013. "Restricting Speech Restricts Arab Freedom." *Contexts* 12 (Spring): 20–21.

"Who Votes, Who Doesn't, and Why: Regular Voters, Intermittent Voters, and Those Who Don't." 2006. Pew Research Center for the People & the Press, October 18. Accessed May 3, 2007 (www.peoplepress.org).

Whorf, Benjamin Lee. 1956. *Language, Thought, and Reality.* Cambridge, MA: MIT Press.

Whoriskey, Peter, and Dan Keating. 2014. "Dying and Profits: The Evolution of Hospice." *Washington Post,* December 26. Accessed October 16, 2015 (washingtonpost.com).

Wicker, Christine. 2009. "How Spiritual Are We?" *Parade,* October 4, 4–5.

Wightman, Patrick, Robert Schoeni, and Keith Robinson. 2012. "Familial Financial Assistance to Young Adults." Paper prepared for the annual meetings of the Population Association of America, May 3. Accessed February 13, 2014 (www.npc.umich.edu).

Wiik, Kenneth A., Renske Keizer, and Trude Lappegård. 2012. "Relationship Quality in Marital and Cohabiting Unions Across Europe." *Journal of Marriage and Family* 74 (June): 389–398.

Wilcox, W. Bradford, and Elizabeth Marquardt. 2011. "The State of Our Unions, Marriage in America 2011: When Baby Makes Three." Institute for American Values. Accessed March 2, 2014 (www.stateofourunions.org).

Wilcox, W. Bradford. 2015. "Knot Now: The Benefits of Marrying in Your Mid-to-Late 20s (Including More Sex!)." *Washington Post,* February 20. Accessed September 30, 2015 (washingtonpost.com).

Wildsmith, Elizabeth, Nicole R. Steward-Streng, and Jennifer Manlove. 2011. "Childbearing Outside Marriage: Estimates and Trends in the United States." Child Trends, November. Accessed March 4, 2012 (www.childtrends.org).

Wilke, Joy, and Frank Newport. 2014. "Americans' Satisfaction with Economy Sours Most Since 2001." Gallup, January 16. Accessed April 12, 2014 (www.gallup.com).

Wilke, Joy. 2013. "Majority in U.S. Say Healthcare Not Gov't Responsibility." Gallup, September 18. Accessed July 26, 2014 (www.gallup.com).

Wilkinson, Charles. 2006. *Blood Struggle: The Rise of Modern Indian Nations.* New York: W. W. Norton & Company.

Wilkinson, Lindsey, and Jennifer Pearson. 2013. "High School Religious Context and Reports of Same-Sex Attraction and Sexual Identity in Young Adulthood." *Social Psychology Quarterly* 76 (June): 180–202.

Williams, Claudia, and Barbara Gault. 2014. "Paid Sick Days Access in the United States: Differences by Race/Ethnicity, Occupation, Earnings, and Work Schedule." Institute for Women's Policy Research, March. Accessed April 12, 2014 (www.iwpr.org).

Williams, Frank P., III, and Marilyn D. McShane. 2004. *Criminological Theory,* 4th edition. Upper Saddle River, NJ: Prentice Hall.

Williams, Lauren. 2014. "21 Things You Can't Do While Black." AlterNet, February 12. Accessed March 22, 2014 (www.alternet.org).

Williams, Robin M., Jr. 1970. *American Society: A Sociological Interpretation,* 3rd edition. New York: Knopf.

Williams, Timothy. 2013. "Blighted Cities Prefer Razing to Rebuilding." *New York Times,* November 13, A15.

Williams, Wendy H., and Stephen J. Ceci. 2012. "When Scientists Choose Motherhood." *American Scientist* 100 (March/April). Accessed February 20, 2012 (www.americanscientist.org).

Wilson, Meagan M. 2014. "Hate Crime Victimization, 2004–2012—Statistical Tables." Bureau of Justice Statistics, February. Accessed March 2, 2014 (www.bjs.gov).

Wilson, Steven G., David A. Plane, Paul J. Mackun, Thomas R. Fischetti, and Justyna Goworowska. 2012. "Patterns of Metropolitan and Micropolitan Population Change: 2000 to 2010." U.S. Census Bureau, September. Accessed September 19, 2014 (www.census.gov).

Wilson, Thomas C. 1993. "Urbanism and Kinship Bonds: A Test of Four Generalizations." *Social Forces* 71 (March): 703–712.

Wilson, William Julius. 1996. *When Work Disappears: The World of the New Urban Poor.* New York: Knopf.

Winerip, Michael. 2013. "Pushed Out of a Job Early." *New York Times,* December 6. Accessed May 16, 2014 (www.nytimes.com).

Wing, Nick, and Carly Schwartz. 2014. "Here's the Painful Truth About What It Means to Be 'Working Poor' in America." *Huffington Post,* May 19. Accessed May 19, 2014 (www.huffingtonpost.com).

WIN-Gallup International. 2013. "Global Index of Religiosity and Atheism, 2012." Accessed June 21, 2014 (www.gallup-international.com).

Wingfield, Adia Harvey. 2010. "Are Some Emotions Marked 'Whites Only'? Racialized Feeling Rules in Professional Workplaces." *Social Problems* 57 (May): 258–268.

Winship, Michael. 2015. "Inside the Cushy, Fabulously Compensated Life of an Ex-Congressman." Alternet, April 8. Accessed October 7, 2015 (alternet.org).

Winship, Scott. 2011. "Mobility Impaired." *National Review,* November 7. Accessed February 13, 2014 (www.nationalreview.com).

Winter, Michael. 2011. "13 More Charged in SAT Cheating Scandal in N.Y." *USA Today,* November 23. Accessed November 24, 2011 (www.usatoday.com).

Wirth, Louis. 1938. "Urbanism as a Way of Life." *American Journal of Sociology* 44 (July): 1–24.

Witters, Dan. 2010. "Large Metro Areas Top Small Towns, Rural Areas in Wellbeing." Gallup, May 17. Accessed May 20, 2010 (www.gallup.com).

"Wives Who Earn More Than Their Husbands, 1978–2012." 2014. U.S. Bureau of Labor Statistics, Labor Force Statistics from the Current Population Survey, March 24. Accessed April 9, 2014 (www.bls.gov/cps).

Wolff, Edward N. 2012. "The Asset Price Meltdown and the Wealth of the Middle Class." National Bureau of Economic Research, November. Accessed February 6, 2014 (www.nber.org).

Wolff, Edward N. 2014. "Household Wealth Trends in the United States, 1962–2013: What Happened over the Great Recession?" National Bureau of Economic Research, December. Accessed September 16, 2015 (nber.org).

Wolfson, Mark. 2001. *The Fight Against Big Tobacco: The Movement, the State, and the Public's Health.* New York: Aldine de Gruyter.

Wolven, Jacqueline. 2014. "The 3 Things Middle-Aged Women Need to Stop Doing When Speaking in Public." *Huffington Post,* September 18. Accessed September 18, 2014 (huffingtonpost.com).

Wondergem, Taylor R., and Mihaela Friedlmeier. 2012. "Gender and Ethnic Differences in Smiling: A Yearbook Photographs Analysis from Kindergarten Through 12th Grade." *Sex Roles* 67 (October): 403–411.

Wong, Curtis M. 2013. "Hallmark's 'Holiday Sweater' Keepsake Ornament Omits 'Gay' from 'Deck the Halls' Lyrics." *Huffington Post,* October 28. Accessed November 10, 2013 (www.huffingtonpost.com).

Wood, Graeme. 2012. "Who Are the Millionaires?" *National Review,* December 19. Accessed February 6, 2014 (www.nationalreview.com).

Wood, Julia T. 2015. *Gendered Lives: Communication, Gender, & Culture,* 11th edition. Stamford, CT: Cengage.

Woolf, Steven H., and Laudon Aron. 2013. *U.S. Health in International Perspective: Shorter Lives, Poorer Health.* National Research Council and Institute of Medicine of the National Academies. Accessed July 26, 2014 (www.national-academies.org).

World Bank and Collins. 2013. *Atlas of Global Development: A Visual Guide to the World's Greatest Challenges,* 4th edition. Washington, DC, and Glasgow: World Bank and Collins.

World Bank. 2012. *Gender Equality and Development.* Washington, DC. Accessed September 22, 2011 (www.worldbank.org).

World Bank. 2015. "Country and Lending Groups." Accessed October 14, 2015 (data.worldbank.org).

World Economic Forum. 2013. "Outlook on the Global Agenda 2014." Accessed February 6, 2014 (www.weforum.org).

World Health Organization and UNICEF. 2014. "Progress on Drinking Water and Sanitation." May. Accessed September 19, 2014 (www.unicef.org).

World Health Organization. 2013. "Global and Regional Estimates of Violence Against Women: Prevalence and Health Effects of Intimate Partner Violence and Non-Partner Sexual Violence." Accessed March 2, 2014 (www.who.int).

World Health Organization. 2013. "Global and Regional Estimates of Violence Against Women: Prevalence and Health Effects of Intimate Partner Violence and Non-Partner Sexual Violence." Accessed August 6, 2015 (who.int).

World Health Organization. 2015. *World Health Statistics 2015.* Accessed October 15, 2015 (who.int).

World Meteorological Organization. 2014. "Atlas of Mortality and Economic Losses from Weather, Climate and Water Extremes (1970–2012)." Accessed September 19, 2014 (wmo.int).

World Meteorological Organization. 2014. "The State of Greenhouse Gases in the Atmosphere Based on Global Observations Through 2013." WMO Greenhouse Gas Bulletin, September 9. Accessed September 19, 2014 (www.wmo.int).

World Water Assessment Program. 2009. *The United Nations World Water Development Report 3: Water in a Changing World.* Paris: UNESCO and London: Earthscan.

Worley, Heidi. 2014. "Top 10 Countries Closing Gender Gap." Population Reference Bureau, February. Accessed March 2, 2014 (www.prb.org).

Wozniak, Jesse, and Bob Groves. 2011. "A National Ceremony." *Contexts* 10 (Winter): 12–18.

Wright, Wynne, and Elizabeth Ransom. 2005. "Stratification on the Menu: Using Restaurant Menus to Examine Social Class." *Teaching Sociology* 33 (July): 310–316.

WWAP. 2014. *The United Nations World Water Development Report 2014: Water and Energy.* United Nations World Water Assessment Program. Accessed September 19, 2014 (www.unesco.org).

Wynnyckj, Mychailo. 2014. "Crimea—Putin's Sudetenland." Project Maidan, March 3. Accessed April 12, 2014 (www.projectmaidan.com).

Wysong, Earl, Robert Perrucci, and David Wright. 2014. *The New Class Society: Goodbye American Dream?* 4th edition. Lanham, MD: Rowman & Littlefield.

Xie, Yu, James Raymo, Kimberly Goyette, and Arland Thornton. 2003. "Economic Potential and Entry into Marriage and Cohabitation." *Demography* 40 (May): 351–367.

Yaccino, Steven, and Lizette Alvarez. 2014. "New G.O.P. Bid to Limit Voting in Swing States." *New York Times,* March 30, A1.

Yan, Holly. 2015. "717 People Dead: What Caused the Haji Stampede?" CNN, September 26. Accessed October 25, 2015 (cnn.com).

Yglesias, Matthew. 2013. "The End of Cheap Airfare." Slate, February 19. Accessed April 12, 2014 (www.slate.com).

Yokota, Fumise, and Kimberly M. Thompson. 2000. "Violence in G-rated Animated Films." *JAMA* 283 (May 24/31): 2716–2720.

York, Emily Bronson. 2012. "Form Follows Content at McD's." *Baltimore Sun,* February 2, 14.

You, Danzhen, Phillip Jingxian Wu, and Tessa Wardlaw. 2013. "Levels & Trends in Child Mortality: Report 2013." UN Inter-Agency Group for Child Mortality Estimation. Accessed February 6, 2014 (www.childmortality.org).

Young, Elise. 2015. "A Rough Ride for Renewable Energy." *Bloomberg Businessweek,* May 25-May 31, 28–29.

Young, Rachel, et al. 2014. "The 2014 International Energy Efficiency Scorecard." American Council for an Energy-Efficient Economy. July. Accessed September 19, 2014 (www.aceee.org).

Zahedi, Ashraf. 2008. "Concealing and Revealing Female Hair: Veiling Dynamics in Contemporary Iran." Pp. 250–265 in *The Veil: Women Writers on Its History, Lore, and Politics,* edited by Jennifer Heath. Berkeley: University of California Press.

Zahn, Margaret A., et al. 2010. "Causes and Correlates of Girls' Delinquency." Office of Juvenile Justice and Delinquency Prevention, April. Accessed January 12, 2014 (www.ojjdp.gov).

Zhang, Yuanyuan, Travis L. Dixon, and Kate Conrad. 2010. "Female Body Image as a Function of Themes in Rap Music Videos: A Content Analysis." *Sex Roles* 62 (June): 787–797.

Zimbardo, Philip G., Christina Maslach, and Craig Haney. 2000. "Reflections on the Stanford Prison Experiment: Genesis, Transformations, Consequences." Pp. 193–237 in *Obedience to Authority: Current Perspectives on the Milgram Paradigm,* edited by Thomas Blass. Mahwah, NJ: Lawrence Erlbaum Associates.

Zimbardo, Philip. G. 1975. "Transforming Experimental Research into Advocacy for Social Change." Pp. 33–66 in *Applying Social Psychology: Implications for Research, Practice, and Training,* edited by Morton Deutsch and Harvey A. Hornstein. Hillsdale, NJ: Erlbaum.

Zimmerman, Jonathan. 2014. "Are E-Cigarettes Ensnaring a New Generation of Smokers?" *Christian Science Monitor Weekly,* March 31, 35.

Zong, Jie, and Jeanne Batalova. 2015. "Frequently Requested Statistics on Immigrants and Immigration in the United States." Migration Policy Institute, February 26. Accessed October 4, 2015 (migrationpolicy.org).

Zweifler, Seth. 2013. "Elite Institutions: Far More Diverse Than They Were 20 Years Ago." *Chronicle of Higher Education,* November 1, B16.

Zweigenhaft, Richard L., and G. William Domhoff. 2006. *Diversity in the Power Elite: How It Happened, Why It Matters.* Lanham, MD: Rowman & Littlefield.

NAME INDEX

A

AARP Magazine, 238
AAUW, 166
Abdullah (Crown Prince), 94
Abrahms, Sally, 330
Abudabbe, Nuha, 193
Academy of Medical Royal Colleges, 170
Acierno, Ron, 237–238
ACT, Inc., 256
Adams, Bert N., 8, 13, 16
Adamson, David M., 132
Addams, Jane, 8, 12–13
Adkins, Amy, 109, 113
Administration on Aging, 237
Agaku, Israel T., 278
Agnvall, Elizaeth, 286
Agrawal, Sangeeta, 213, 242
Aguilar, Leslie, 39
Ailes, Elizabeth C., 287
Akechi, Hironori, 93
Aknin, Lara B., 265
Al-Jassem, Diana, 228
Alkhatib, Hasan, 329
Allen, Joshua, 275
Al-Mahmood, Syed Z., 156
Alman, Ashley, 129
Alsever, Jennifer, 137, 182
Altonji, Joseph G., 71
Alvaredo, Facundo, 153
Alvarez, Lizette, 220, 296
American Association of Suicidology, 9
American Psychological Association, 161, 211
American Society for Aesthetic Plastic Surgery, 162
Amnesty International, 137
Amrein-Beardsley, Audrey, 255
Amusa, Malena, 237
An, Ruopeng, 279
Andersen, Margaret L., 176
Anderson, Craig A., 74
Anderson, Gerard F., 285
Anderson, Monica, 98, 189
Anderson, Shawna L., 268
Andrews, Edmund L., 195
Andrzejewski, Adam, 203
Anestis, Joye C., 135
Anestis, Michael D., 135
Annie E. Casey Foundation, 147
Arendt, Hannah, 214

Arieff, Allison, 305
Armstrong, Elizabeth A., 169
Aron, Laudon, 273
Arrazola, René, 278
Asante-Muhammad, Dedrick, 183
ASA Research Department, 7
Asch, Solomon, 104–105
Ashburn, Elyse, 191
Asi, Maryann, 193
Atchley, Robert C., 242, 244
Atici, Cemal, 328
Attinasi, John J., 43
Auerbach, David I., 282
Auerbach, Marc, 155
Aughinbaugh, Alison, 229–230
Auguste, Byron, 256
Aunola, Kaisa, 70
Auster, Carol J., 72
Autor, David H., 255
Ayoub, Catherine, 75

B

Babbage, Charles, 328
Babbie, Earl, 25, 28, 35
Babcock, Philip, 254
Bach, Peter B., 285
Bachmann, Michele, 223
Baenninger, MaryAnn, 253
Bagri, Neha Thirani, 48
Bahadur, Nina, 95, 170
Bai, Ge, 285
Bainbridge, William S., 260
Baker, ARyn, 173, 267
Bales, Robert F., 103–104, 173, 225
Balhaus, Rebecca, 269
Banchero, Stephanie, 255
Bandy, Tawana, 35
Banerjee, Neela, 269, 313
Banfield, Edward C., 147
Banks, Duren, 175
Banks, Ingrid, 319
Barash, David P., 160
Barker, James B., 114
Barker, Kristin K., 289–291
Barker, Megan M., 205
Barnes, Cynthia, 42, 74, 227
Barreto, Michelle, 115
Barrionuevo, Alexei, 317
Barro, Robert J., 266
Barry, Ellen, 58
Barstow, David, 140
Bartlett, Thomas, 257

Barusch, Amanda S., 242, 244
Bates, Nancy, 34
Baum, Sandy, 248, 251
Baumrind, Diana, 70
Bax, Paulin, 50
Bazelon, Emily, 234
Beadle, Amanda Peterson, 288
Beck, Melinda, 289
Becker, Howard S., 85, 103, 133
Beckles, Gloria L., 277
Beech, Hannah, 267
Begley, Sharon, 63
Beil, Laura, 241
Beiser, H. Darr, 134
Belkin, Douglas, 99
Belknap, Joanne, 130–131
Bell, Claes, 144
Bendavid, Naftali, 159
Bender, Courtney, 323
Bengali, Leila, 248
Bengali, Shashank, 296
Bennett, Jessica, 75
Bennhold, Katrin, 269
Benokraitis, Nijole, 21, 71, 78, 98, 174, 228, 230, 236
Benor, Mark, 137
Benson, Michael, 134
Benton, Thomas H., 254
Beres, Damon, 324
Berfield, Susan, 212, 312
Berger, Peter L., 88, 270
Berk, Richard A., 316
Berkowitz, Bill, 267
Berlin, Overton B., 42
Bernstein, Lenny, 309
Berrett, Dan, 256
Bertrand, Marianne, 253
Berube, Alan, 146, 150, 301–302
Best, Joel, 319–320
Betcher, R. William, 173
Bettinger, Eric P., 250
Beutel, Ann, 15
Bharadwaj, Prashant, 29
Bhardwaj, Deepika, 177
Bialik, Carl, 22
Bietsch, Kristin, 152, 175, 294–295
Billingsley, Andrew, 234
Bin Laden, Osama, 126
Bird, Chloe, 145
Bird, Warren, 266
Bischoff, Kendra, 303

Cunningham, Solvig A., 278
Currie, E., 134
Currie, Janet, 291
Curtin, Sally, 233
Curtiss, Susan, 62

D

Dahl, Robert A., 220
Dahlin, E., 54
Daley, Suzanne, 295
Daly, Mary C., 248
Daneshpour, Manijeh, 269
Daniel, Mary-Alice, 180
Dank, Meredith, 175
Davenport, Coral, 312
Davey, Monica, 194
Davidson, Paul, 207
Davies, James C., 324
Davis, Alyssa, 211
Davis, James, 40
Davis, Julie Hirschfeld, 50, 282
Davis, Karen, 330
Davis, Kim, 170
Davis, Kingsley, 154
Davis, Viola, 195
Death Penalty Information Center, 137
Deegan, Mary Jo, 13
Deggans, Eric, 196
DeGraw, David, 146
DeKeseredy, Walter S., 131
De La Haye, Kayla, 72
Della Porta, Donatella, 327
De Long, J., 300
DeMaio, Theresa J., 34
de Mestral, George, 54–55
DeNavas-Walt, Carmen, 146–147, 188–189
Dennis, Brady, 74
Denny, C. H. J., 63
Denyer, Simon, 309
Deparle, Jason, 149
De Pommereau, Isabelle, 172
Deprez, Esmé E., 121
DeSilver, Drew, 29, 157, 190
Desmond-Harris, Jenee, 319
De Vogue, Ariane, 170
Dewan, Shaila, 144
Dholia, Esha, 196
Diamond, Jeremy, 170
Diamond, Milton, 63–64
Diani, Mario, 327
Dias, Elizabeth, 263
Dickey, Jack, 208
Dickson, William J., 110
Dillow, Sally A., 165
Dilworth-Anderson, Peggye, 234
DiSalvo, David, 98
Dockterman, Eliana, 97
Dodson, Lisa, 234

Doeringer, Peter B., 194
Doherty, Carroll, 120, 142, 144, 171, 217
Dolan, Kerry A., 138
Dolnick, Sam, 50
Domhoff, G. William, 143
Dominguez, Kenneth, 277
Domosh, Mona, 305
Donadio, Rachel, 92
Dorfman, Mark, 306
Douban, Gigi, 194
Douglas, Susan J., 87
Dovidio, John F., 196
Downs, Edward, 31
Doyle, Francis X., 267
Drake, Bruce, 157
Drew, Charles R. (Dr.), 189
Drexler, Peggy, 94, 168, 318
Du Bois, W. E. B., 8, 13, 15
Dugan, Andrew, 47, 120, 122, 148, 182, 200, 204, 217, 220
Duggan, Maeve, 29, 97
Duhigg, Charles, 308
Duncan, D. T. A. R., 278
Dunifon, Rachel, 25
Durden, Tyler, 203
Durkheim, Émile, 8–9, 14, 126–127, 247, 259, 265
Dwyer, Rachel E., 252
Dye, Thomas, 221

E

Eagan, Kevin, 45
Eaton, Danice K., 167
Eavis, Peter, 128
Eberstadt, Nicholas, 146
Eckenrode, John, 237, 242
Eckhoff, Vickery, 148
Eckholm, Erik, 220, 240
Eckstein, Rick, 6
Edin, Kathryn J., 152
Edison, Thomas, 54
Efrati, Amir, 331
Ehrlich, Anne H., 298
Ehrlich, Paul R., 298
Eibner, Christine, 282
Eilperin, Juliet, 309
Eisenberg, Abner M., 93
Eisler, Peter, 286
Ekman, Paul, 93
Elder, Linda, 6
Elgin, Ben, 311
Eliot, Lise, 62
Elliott, Philip, 223
Ellis, Bill, 240, 318
Ellis, Renee R., 71, 229–230
Ellison, Jesse, 157
Ellison, Larry, 138
Ellison, Nicole, 97
El Nasser, Haya, 301

Elsen, Tracy, 302
Emerson, Ralph Waldo, 272
Emery, David, 318
Emmons, William R., 150
England, Paula, 130, 169
Entmacher, Joan, 212
Environmental Integrity Project, 309
Environmental Protection Agency (EPA), 307, 309, 311
Epstein, Joseph, 317
Erikson, Erik, 65, 126
Ernst & Young, 129
Esipova, Neil, 296
Esping-Anderson, Gosta, 203
Esterl, Mike, 116
Etzioni, Amitai, 107
Evans, Becky, 46, 54
Ewen, Lynda Ann, 323
Ewing, Jack, 212

F

Fahrenthold, David A., 311
FairTest, 256
Fakhraie, Fatemeh, 269
Fang, Ferric C., 36
Faris, Robert, 22
Farley, Melissa, 91
Feagin, Joe R., 12, 187, 305
Federal Bureau of Investigation, 121–125, 129–131
Federal Election Commission, 218
Federal Interagency Forum on Child and Family Statistics, 233
Feintzeig, Rachel, 112
Feldman-Jacobs, Charlotte, 172, 296
Feldstein, Stanley, 93
Felmlee, 22
Ferdman, Roberto A., 45
Ferguson, Christopher J., 74
Ferguson, Emily, 268
Ferre, Myra, 16
Ferrell, Will, 52
Fey, Tina, 52
Fields, Jason, 77
File, Thom, 96, 218–219
Finer, Lawrence B., 170
Fingerman, Karen L., 79
Finkelhor, David, 28, 236
Finkelstein, Eric A., 278
Finn, Jeremy D., 254
Fiorina, Carly, 175
Fischman, Josh, 310
Flavin, Jeanne, 130
Fleegler, Eric W., 135
Fleishman, Jeffrey, 177
Fleming John, 99, 312
Florida, Richard, 51
Flower, Shawn M., 130, 137

Niederdeppe, Jeff, 285
Niquette, Mark, 203
Nir, Sarah Maslin, 186
NOAA National Centers, 310
Noeth, Bryan J., 150
Norris, Tina, 191
Nossiter, Adam, 228
Nurmi, Jari-Erik, 70
Nyseth, Hollie, 6

SUBJECT INDEX

A

ability grouping of students, 253–254
abortion, 169–170
    restrictions on, 288–289
absolute poverty, 145
Abu Ghraib prison scandal, 106
acculturation, 184
achieved status, 84–85
achievement, as cultural value, 44
achievement gap, 249–251
    test scores and, 255
activity, as cultural value, 44
activity theory, 242
ACT tests, 256
acute disease, 276
adolescence
    e-cigarette use in, 278
    health status of, 276
    nonmarital childbearing in, 233
    socialization in, 76–77
adulthood, socialization in, 77–78
advergaming, 74–75
advertising, socialization and, 74–75
affirmative action, 257
African Americans
    achievement gaps for, 250
    demographics of, 189–190
    employment rates for, 207–208
    family structure for, 232–235
    health status of, 277
    religion and, 263
    voting rates and, 219–220
age and aging
    conflict theory perspective on,
        242–243
    current issues in, 240–241
    disability and, 274–275
    family structure and, 238–241
    feminist theories concerning, 243
    functionalist perspective on,
        241–242
    health status and, 276
    inequality and, 242–243
    poverty and, 146–147
    religion and, 262–263
    right to die and, 240–241
    sexuality and, 167–168
    sociological perspectives on,
        241–244
    symbolic interactionist perspective
        on, 244
    voting rates and, 219
ageism, 244

agnostics, 261–262
air pollution, 308–310
alcohol abuse, domestic violence and,
    26
alienation, 11
    in bureaucracies, 109
Alzheimer's disease, 276
American Academy of Matrimonial
    Lawyers, 22
American Academy of Pediatrics
    (AAP), 73
American Indian and Alaska Natives
    (AIANs)
    demographics for, 191–192
    genocide of, 183–184
American Medical Association, 285,
    288–289
American Psychiatric Association, 290
Annie Hall (film), 167
anomie, 126–127
antiabortion movement, 169–170
antibiotic resistance, 275
anticipatory socialization, 67
apartheid (South Africa), 183
appearance, impression management
    and, 68–69
applied research, 35
AP tests, 256
Arab Americans, 192–193
arithmetic rate of food supply,
    297–298
Army's Ranger School, 115
arranged marriages, 4, 228
artificial intelligence, 328
ascribed status, 84–85
asexuals, 159
AshleyMmadison.com, 118
Asian Americans
    cohabitation rates of, 230
    demographics on, 190–191
    education levels of, 249–251
    family structure of, 231–235
    religion and, 263
    voting rates for, 219–220
Asians, wealth levels of, 141
assimilation, 184
    language and, 43
atheists, 261–262
attention-deficit-hyperactivity disor-
    der (ADHD), 290–291
authoritarianism
    parenting, 70
    political leadership, 103–104,
        214–215

authoritative parenting, socialization
    and, 70
authority, 20
    types of, 215–216

B

baby boomers, 239–241
back-stage behavior, 90
behavior
    socialization and, 61
    sociological research on, 22
    symbols and, 41
beliefs
    culture and, 40
    religious beliefs, 261–262, 270
    research and, 36–37
bias, social desirability bias, 29
binge drinking, 279–281
biology
    environment and, 63–65
    homosexuality and, 160
biotechnology, 328–329
birth order, social learning theories,
    66
bisexuals, 159
boomerang children/boomerang gen-
    eration, 77–78, 227
bourgeoisie, 10, 155
British Petroleum (BP), 311
brominated vegetable oil (BVO), 327
Brown v. Board of Education, 184
*Brown v. Entertainment Merchants
    Association*, 74
bureaucracies, 108–111
burqa, 41
Burwell v. Hobby Lobby Stores, 288

C

campaign contributions, 217–218
capitalism, 10–11
    characteristics of, 201–202
    crony capitalism, 203
    social inequality and, 128
    social stratification and, 155–156
    welfare capitalism, 202–203
capital punishment, 137
career opportunities
    gender differences in, 252–253
    in sociology, 6–7
casinos, American Indian ownership
    of, 192
caste system, 139–140

causation, 25–26
CBS News, 29
celebrity bogus science, 22
Cesarean sections, prevalence of, 287
chador, 41
charismatic authority, 216
  cults and, 260
cheating, at colleges and universities, 257–258
child abuse and neglect, 236–237
  feminist perspectives on, 243–244
child care, educational institutions and, 248–249
child development
  socialization and, 75–76
  teachers and, 73
child maltreatment, 236–237
children
  global stratification of, 152
  nonmarital childbearing and, 232–233
  in poverty, 146–148
China
  air pollution in, 309
  Elderly Rights Law in, 47
  filial piety in, 49
  mixed economy in, 202–203
  one-child policy in, 296–297
  poverty in, 152
chronic disease, 276
church, defined, 260–261
Church of Jesus Christ of Latter-Day Saints, 228, 268–269
cities, origin and growth of, 300
civil religion, 265
class system. *See also* social class; social stratification
  class structure in America, 142–145, 150–151
  defined, 140
  diversity studies and, 6
  gender and, 175–176
  online interaction and, 96
  social conflict and, 10–11, 15
  stigmatization and, 133–134
climate change, 310
closed-ended questionnaires, 28
closed stratification system, 138–140
coercive organizations, 107–108
coercive power, 215
cohabitation, 230–231
collaborative consumption, 329
collective behavior, 314–322
  types of, 317–318
colleges and universities, current issues for, 256–258
combat, women in, 177
commercialization of sex, 176
common sense, sociology and, 4
communication, in bureaucracies, 109
communism, 202

community outreach programs, crime prevention and, 135
compartmentalization, role conflict and, 87
compromise, role conflict and, 87
compulsory voting, 220
computer-assisted telephone interviewing (CATI), 27–28
computer technology, 328
concentric zone theory, 304
concepts, defined, 22–23
confidentiality, research ethics, 36
conflict
  in families, 235–238
  family and aging and, 242–243
conflict theory, 15
  culture and, 57–58
  deviance and, 128–130
  education and, 249–251
  gender and sexuality and, 174–175
  groups and organizations and, 113–114
  health care and medicine, 285–286
  laws, power and social control, 128–130
  politics and power and, 222
  race and ethnicity and, 194–195
  religion and, 267–268
  social stratification and, 155–156
  urbanization and, 304–305
  work and economy and, 211–212
conformity
  as cultural value, 44
  of groups, 104–106
conglomerates, 203–204
Congress (U.S.), 219–220
conspicuous consumption, 143–144
contact hypothesis, race and ethnicity and, 196–197
contagion theory, 315–316
content analysis, 31–32
continuity theory, 244
control group, 32–33
control theory, 134–135
control variable, 23
convergence theory, 316–317
conversational dominance, 91
corporate crimes, 123–124
corporate welfare, 155
corporations
  defined, 203
  economy and, 203–204
correlation, 25–26
cost-benefit studies, 35
countercultures, 50–51
courtship, 228
creationism, myth of, 264
credential inflation, 251
credentialism, 250–251
crime
  classification of, 121–125

  control of, 134–137
  defined, 119
  functionalist perspectives on, 125–128
  measurement of, 121–122
  offenders and victims, 121–122
  prevention and intervention, 134–135
  victimless crimes, 125
  women offenders, 130–131
  women victims of, 131–132
crime control model, 135–136
CrimeSolutions.gov, 135
criminal deviance, 121
criminal justice system, 134–137
Crisis magazine, 13
critical thinking, 6–7
  research and, 22
crony capitalism, 203
cross-dressers, 162
cross-sectional research, 25
crowds, 321–322
crude birth rate, 294
crude death rate, 294–295
cults, 259–260
cultural capital, 52–54, 140
cultural dynamics, 54
cultural imperialism, 53–54
cultural integration, 54, 247
cultural lag, 56
cultural relativism, 49–50
cultural universals, 48–49
culture
  building blocks of, 40–47
  characteristics of, 39–40
  defined, 38
  education and, 247
  ethnocentrism and, 49–50
  external pressures on, 55–56
  family and, 226–228
  gender and sexuality and, 160–163, 171–173
  high culture, 52–54
  ideal *vs.* real culture, 49
  language and, 42
  laws and, 47
  marriage patterns and, 4
  mass media and, 53
  material and nonmaterial, 40
  multigenerational households and, 240
  norms and, 45–48
  personal space and, 95–96
  pop culture, 52–54
  of rape, 131–132
  rituals and, 48
  sanctions and, 47–48
  sexual orientation and, 160
  socialization and, 61
  sociological perspectives on, 56–59
  symbols and, 40–41

culture (*continued*)
    technology and, 54–56
    touch and, 94
    values and, 44–45
    variations in, 50–52
    violence and, 63
culture boundedness, 51–52
culture of poverty perspective,
    147–148
culture shock, 51–52
cumulative wealth, social stratification
    and, 140–141
cyberbullying, 98
cybercrime, 124
cyberstalking, 98

D

DARE (Drug Abuse Resistance
    Education) program, evaluation
    of, 35
data
    collection methods, 27–29, 34
    field research, 31
    presentation of findings, 27
    secondary analysis of, 33–34
dating
    cohabitation and, 230
    interracial, 197–198
    as marriage market, 228
Davis-Moore thesis, social stratifica-
    tion and, 154
deductive reasoning, 23–24
de facto segregation, 184
deforestation, 310
degradation ceremonies, 80
deindustrialization, 205
de jure segregation, 184
dementia, 276
democracy
    as cultural value, 44
    defined, 213–214
democratic leadership, 104
demographics. *See also* population
    statistics
    aging and, 239–241
    child maltreatment, 236–237
    defined, 292–293
    education and, 248–249
    female-headed households, 234–235
    health status and, 275–276
    intimate partner violence and,
    235–236
    marriage and divorce rates and,
    229–230
    nonmarital childbearing and, 233
    population changes, 292–299
    population composition and struc-
    ture, 296–297
    social mobility and, 149–150
    voter turnout and, 218

demographic transition theory,
    298–299
denominations (religious), 260
dependency theory, inequality and,
    153
dependent variable, 23
depression, 276
developmental differences, nature
    and nurture debate and, 62–63
deviance
    anomie and social strain theory,
    126–128
    characteristics of, 119–121
    classification of, 121–122
    conflict theory and, 128–130
    control of, 134–137
    cultural differences in, 120
    defined, 118–121
    differential association theory, 132
    feminist theories on, 130–132
    functionalist perspectives on,
    125–128
    labeling theory and, 132–133
    laws, power and social control,
    128–130
    medicalization of, 133–134
    social inequality and, 128
    symbolic interactionism and,
    132–134
*Diagnostic and Statistical Manual
    of Mental Disorders (DSM)*,
    133–134, 289–290
dictatorships, 214
differential association theory, 132
diffusion of culture, 54
direct learning, 65–66
direct observational understanding,
    11–12
disability, defined, 274–275
disasters, 320
discouraged workers, 207
discovery, culture and, 54–55
discrimination
    prejudice and, 186–187
    racial and ethnic, 185–186
    segregation and, 184
diversity
    of American Indians, 191–192
    of Asian Americans, 190–191
    majority and minority groups in
    U.S. and, 187–193
    of Middle Eastern Americans,
    192–193
    sociology and, 6
division of labor, 9
Divorce-Online, 22
divorce rates, 229–230
    myths concerning, 22
domestic violence, 172–173, 235–236,
    243–244
    alcohol abuse and, 26

dominant groups, 182–183
do no harm principle, research
    ethics, 36
double segregation, 250
doublespeak, 88
downward mobility, 149
dowry disputes, 172
dramaturgical analysis, 17, 89–90
dropout rate, 256
Dr. Oz Show, 291
drug abuse
    as crime, 125
    prevention and intervention,
    134–135
dyads, 103
dysfunctions, 14
    deviance and, 125–126

E

earmarks, 218
earnings. *See* wages
    education and, 248–249
    in health care occupations, 288
    for teachers, 256
Ebola epidemic, 317
e-cigarettes, 278
economic capital, 140
economic inequality
    conflict and, 15
    race and ethnicity and, 194–195
economy, 200–223
    child maltreatment and, 236–237
    communism and, 202
    corporations and, 203–204
    defined, 200
    education and, 249
    environment and, 310–312
    family and, 225
    global economic systems and,
    201–203
    marriage and divorce rates and,
    229–230
    mixed economies, 202–203
    socialism and, 202
    social mobility and, 149
    sociological explanations of,
    210–213
    wealth distribution and, 141
ecosystems, 306
edge cities, 301
education. *See also* schools and
    schooling
    benefits of, 247–249
    cohabitation rates and, 230–231
    colleges and universities, 256–258
    conflict theory perspectives,
    249–251
    current issues in, 254–258
    female-headed households and,
    234–235

feminist theories (*continued*)
  social stratification and, 156–157
  urbanization, 305
  work and the economy and, 212–213
feminization of poverty, 147–148
fertility, 293–294
  below-placement rates of, 298–299
fetal alcohol disorders, 276
fetal origins theory, 63
fictive kin, 234
field research, 29–31
financial aid, education and access to, 250–251
Fisher-Price company, 37
folkways, 46–47
food, culture and, 48–49
food deserts, 277
foreign-born population, 181
formal deviance, 120
formal organizations, 107–112
  bureaucracies, 108–111
  conflict theory and, 113–114
  feminist theory concerning, 115
  sociological perspectives on, 112–117
  symbolic interactionist perspective on, 115–116
for-profit colleges, 258
fracking, 307–308
freedom, as cultural value, 44
friends, socialization and, 72
front-stage behavior, 90
functionalism, 13–14
  culture and, 56–57
  deviance and, 125–128
  education and, 247–249
  family structure and aging and, 241–242
  gender and sexuality and, 173–174
  health and, 283–285
  politics and power and, 220–222
  race and ethnicity and, 194
  religion and, 265–267
  social groups and organizations and, 112–113
  social stratification and, 153–154
  urbanization and, 304
  work and the economy and, 210–211
fundamentalism, 264
Fundamentalist Church of Jesus Christ of Latter-Day Saints, 228
funerals, cultural differences and, 50

G

gang rape, in India, 172
gay bashing, 163–164
gender
  advertising and, 74–75
  conflict theory concerning, 174–175
  crime and, 130–131
  cultural components of, 160–163, 171–173
  defined, 159
  developmental and health differences, 62–63
  deviance and, 131–1322
  education and, 164–165, 252–253, 306
  family life and, 164, 243–244
  feminist theories and, 15–17, 175–176
  functionalist perspective on, 173–174
  health and, 276, 287–289
  human capital and, 173
  inequality, 171–172
  language and, 42–43
  life expectancy and, 238–239
  money and, 158
  pay gap and, 152, 165–166, 171–172, 209–210, 212–213, 288
  politics and, 166–167, 172, 222–223
  population statistics on, 175–176
  poverty and, 147–148, 156–157
  roles, stereotypes and sexism, 162–163
  sex reassignment and, 63
  sexuality and, 158–177
  social construction of, 176–177
  social mobility and, 150
  sociological perspectives on, 173–177
  stratification, 164–167
  symbolic interactionism and, 176–177
  touching and, 94
  toys and, 72
  urbanization and, 305
  voter turnout and, 218
  in workplace, 165–166
gendercide, 296
gendered institutions, 164
gendered racism, 195–196
gender expression, 162
gender identity, 160
genderqueer, 162
General Electric, 204
generalized other, 68
General Mills, 319–320
General Motors (GM), 124
genetic engineering, 328–329
genetics, race and, 180–181
genocide, 183–184
gentrification, 302
geography
  health and, 277
  social mobility and, 150

geometric rate of population growth, 297–298
gestures, nonverbal communication, 92–93
Ghana, witch camps in, 120
glass ceiling, 115
glass escalator, 115
global economic systems, 201–203
global feminism, 16
global health and illness, 272–274
globalization
  deindustrialization and, 205
  diversity and, 6
  popular culture and, 53–54
global political systems, 213–215
global social stratification, 151–153
global urbanization, 300
global warming, 309–310
goal displacement, 109
Goldman Sachs, 129
gossip, 317–318
government, defined, 213
government policies and programs, social mobility and, 149
government publications, language in, 43
government shutdowns, 104
grade inflation, 257
graduation rates, at colleges and universities, 258
grandparents, socialization and, 71
Great Recession (2007–2008), 141, 150
  education funding and, 250
  racial differences in recovery from, 189
greenhouse effect, 309–310
greenwashers, 311
grief counseling, 6
groups, 100–107
  classification, 100–101
  conflict theory on, 113–114
  conformity in, 104–106
  defined, 100
  dominant group, 182
  dominant-minority relations continuum, 183–184
  ethnic group, 180
  feminist theory concerning, 115
  leadership of, 103–104
  majority and minority groups, 186–193
  racial-ethnic group, 180
  racial group, 180
  size and structure, 103–104
  social networks, 106–107
  sociological perspectives on, 112–117
  symbolic interactionist perspective on, 115–116
groupthink, 105–106
gun homicide, research on, 21

H

habitus, social class and, 151
hacking, 118
Hallmark Corporation, 42
harassment, online, 98
Harlow studies, 62
Harry Potter and the Philosopher's Stone, 149
hate crime, 123
hate groups, 51
Hawthorne effect, 33, 110
health, defined, 272
health care
  bureaucracy, 109
  conflict theory and, 285–286
  feminist theories and, 286–289
  fraud, 124
  functionalist perspective on, 283–285
  gender stratification in jobs relating to, 288
  global disparities in, 273–274
  profit and waste in, 285–286
  technology and, 330
  U.S.-global comparisons, 281–283
health status
  functionalist perspective on, 283–285
  global disparities in, 273–274
  global health and illness, 272–274
  medicalization of health, 289–290
  medicine and, 272–291
  nature and nurture debate and, 62–63
  sexuality and, 167–168
  social class and, 145
  social determinants of, 275–276
  sociological perspectives on, 283–291
  symbolic interactionist perspective on, 289–291
  in United States, 274–281
  urban legends concerning, 318
helicopter parenting, 77
hermaphrodites. *See* intersexuals
heroin use, growth of, 280–281
heterogamy, 226
heterosexism, 163–164
heterosexuals, 159
hidden curriculum, 250–251
high culture, 52–54
hijab/hejab, 41
Hindu, 48
Hispanic Americans
  cohabitation rates of, 230
  growth in U.S. of, 187–188
  voting rates for, 219–220
Hispanic health paradox, 277
HIV infections, 281
Homeboy Industries, 137

homelessness, 148
  field research on, 30
homework, test scores and, 255
homicide, culture and, 63
homophobia, 163–164
homosexuality
  defined, 159
  intolerance of, 163–164
  peer acceptance and, 72
  precancel of, 159
  religious acceptance of, 266–268
  same-sex marriage and, 170–171
  violence against males and, 173
honor killings, 172
hooking up, 168–169
housing, urbanization and, 302–303
human capital, gender role divisions and, 173–174
humanitarianism, as cultural value, 44
hunger
  population and, 298
  poverty and, 152–153
hypotheses, 27
  defined, 22–23

I

ideal culture, 49
ideal types, social groups and, 102
ideology, politics and, 217
illicit drugs, 280
imitation, social learning theory, 65–66
immigration
  functionalist view of, 194
  population change and, 295–296
  race and ethnicity and, 181–182
  upward mobility and, 149
impression management, 68–69, 90
incarceration, crime control and, 136–137
incest taboo, 225
incidence, social epidemiology and, 273
income. *See also* wages
  education and, 248–249, 254
  global inequalities in, 151–153
  health inequality and, 273–274
  by race and ethnic group, 188–192
  schooling and, 250–251
  social stratification and distribution of, 140–141
  voting rates and, 219
independent variable, 23
India, caste system in, 140
indirect lairing, 65–66
individual choice, 3–4
individual discrimination, 185
individualism, as cultural value, 45
inductive reasoning, 23–24

Industrial Revolution, 10
inequality
  aging and, 242–243
  education and, 249–251
  family structure and, 242–243
  gender and sexual inequality, 174–175
  global universality of, 153
  racial-ethnic inequality, 193–198
  social and economic inequality and, 15, 128
infancy, socialization in, 75–76
infant mortality rate, 294–295
  defined, 152
informal deviance, 120
informed consent, research ethics, 36
informed decision making, 6
in-groups, 102–103
inherited wealth, 140
  upper-upper class and, 143–144
innovation
  culture and, 54
  education and, 247
  strain theory and, 127
Institute for Clinical and Economic Review, 285
institutional discrimination, 186
institutionalization
  socialization and, 62
  social movements, 326–327
instrumental tasks, authoritarian leadership and, 104
intellectual property theft, 122
interactionism, 16–17
intergenerational mobility, 148–149
interlocking directorates, 204
internalization, 61
internal migration, 296
international migration, 295–296
International Red Cross and Red Crescent, 41
Internet
  pornography on, 176
  social networks on, 107
Internet Crime Complaint Center, 124
interracial marriage, 197–198, 226
intersexuals, 160
interviews, 28–29
intimate partner violence (IPV), 235–236, 243–244
intragenerational mobility, 148–149
invention, culture and, 54
Investigation Discovery, 128
involuntary resocialization, 80
Iraq War, groupthink dynamic and, 106
iron law of oligarchy, 109
Islamic societies
  symbols in, 41
  women and, 268–269

Middle Eastern Americans, 192–193
migration, population change and, 295–296
millennials, religion and, 262–263
minimum wage, 206–207
minorities
    defined, 183–184
    education inequality and, 249–251
    employment for, 208–210
miscegenation, 197
misconceptions, sociology vs., 4
mixed authority forms, 216
mixed economy, 202–203
mob, 322
modernization theory, 153
    inequality and, 153
monarchies, 214–215
monitoring studies, 35
    technology and, 330
monogamy, 228
monopoly, 201–202
mores, 46–47
mortality, population change and, 294–295
Mosuo ethnic group, 227
motherhood penalty, 212–213
multiculturalism, 51
multigenerational households, 71, 240–241
multinational corporations, 204
multiple-nuclei theory, 304
multiracial diversity, growth of, 197
myths
    about vaccines and autism, 22
    sociological research and, 21–22
    sociology vs., 4

N

nanotechnology, 329
National Alliance to End Homelessness, 148
National Association for the Advancement of Colored People (NAACP), 13
National Crime Victimization Survey, 121–122
National Rifle Association, 21, 324
Native Americans. See American Indians
nature and nurture debate, 62–65
needs assessment studies, 35
negative assimilation, 233
negative sanctions, 47–48
negotiation, role conflict and, 87
neolocal residence pattern, 227
neo-Malthusian theory, 298–299
New American Bible, 43–44
new religious movement (NRM), 259–260
new social movements theory, 324–325
new urban sociology, 304–305

New York Times, 29
noncriminal deviance, 121
nones (religiously unaffiliated persons), 262–263
nonmarital childbearing, 232–233
nonmarital sex, 39–40
nonmaterial culture, 40
nonmaterial facts, 9
nonparticipant observation, 29–30
nonprobability sample, 24–25
nonverbal communication, 91–96
    eye contact, 93
    facial expressions, 93
    personal space and, 94–96
    sexual harassment and, 166
    silence, 91–92
    touch, 93–94
    visual cues, 92–93
normative organizations, 107
norms, culture and, 45–48
norovirus, 275
nouveau riche, 143–144
nuclear family, 226–227
nurse practitioners, 289

O

Obergefell v. Hodges, 170–171
obesity, 278–279
    nature/nurture debate and, 64
objective approach to socioeconomic status, 142
objective data, 4
observation, social learning theory, 65–66
observational studies, 29–30
occupations
    earnings and, 209–210
    gender stratification in, 288
    health and, 277
    prestige of, 141–142
    sex segregation in, 165
Occupy Wall Street (OWS), 11, 326
offshoring, 205–206
old-age dependency ratio, 239–240
oligopoly, 202
online interaction, 96–99
    benefits and costs of, 97–99
    connections and anxieties, 98
    demographics of, 96–97
    harassment and humiliation, 98–99
    social networks and, 107
    worker productivity and, 113
online surveys, 29
open-ended questionnaires, 28
open stratification system, 138–140
Oracle Software, 138
Ordain Women group, 269
organization, social movements and, 325–326
organized crime, 124–125

out-groups, 102–103
overgeneralizations, sociological research and, 21
Oyo Nyimba Kabamba Iguru, 215

P

panic, 318–319
parental leave policies, 243
parenting
    by black males, 234–235
    child maltreatment and, 236–237
    educational outcomes and, 254
    functionalist perspective on, 241–242
    gender stratification and, 164
    helicopter parents, 76–77
    nonmarital childbearing, 232–233
    roles in, 78–79
    socialization and, 70–71
participant observation, 29–30
part-time work, 207
patriarchy
    family system, 227, 243–244
    politics and power and, 222–223
    religion and, 267
    women as victims of, 131–132
patrilocal residence pattern, 227
peer groups, socialization and, 71–62
perception, self-awareness and, 66
performance reviews, 112
performing behavior, 65–66
peripheral theory of urbanization, 304
permissive parenting, socialization and, 70
personal space, nonverbal communication and, 94–96
petit bourgeoisie, 10
Pew Research Center, 220
Pez candy dispensers, 320–321
physical contact, sexual harassment and, 166
physical disability, as master status, 85
physician-assisted suicide, 240
physicians
    as gatekeepers, 284–285
    gender gap in earnings of, 288
Plain Writing Act, 43
play, socialization and, 71–72
pluralism
    in politics, 220–221
    race and ethnicity and, 184
police, crime prevention and, 135
political action committees (PACS), 217–218
political parties, 216–217
politics
    abortion and, 169–170
    defined, 200
    gender gap in, 166–167, 172, 222–223

politics (continued)
  global political systems, 213–215
  power and authority in, 215–216
  sociological perspectives on,
    220–223
  in United States, 216–220
pollution, water supply and, 307–308
polyandry, 228
polygamy, 228
polygyny, 228
popular culture, 52–54
  socialization and, 73–75
population pyramids, 297
population statistics, 292–313. *See
    also* demographics
  aging population, 239–241
  changes in population, 292–299
  composition and structure,
    296–297
  define, 293
  foreign-born population, 181
  gender and, 175–176
  growth in population, 297–298
  multiracial diversity and, 197
  prison population, 136–137
  sampling, 24–25
pornography, 176
positive population checks, 298
positive sanctions, 47–48
postmodern theory, 19
posttraumatic stress disorder, 275
poverty, 145–148
  aging and, 243–244
  causes of, 147–148
  costs of, 146
  education and schooling and, 250
  extreme poverty, 152–153
  female-headed households and,
    234–235
  feminization of, 147–148, 156–157
  social mobility and, 149–150
  structural characteristics of,
    147–148
  in United States, 146–148
poverty line, 145–146
power
  defined, 215
  deviance and, 128–129
  family structure and, 242–243
  social stratification and, 142
  sociological perspectives on,
    220–223
  space and, 95
  in United States, 216–220
power elite perspective, political
    power and, 221–222
practicality, as cultural value, 44
Precious (film), 237
pregnancy
  costs of, 174
  decrease in rates of, 169

health care and, 287–289
  nonmarital childbearing and, 233
prejudice, 185
  discrimination and, 186–187
  religion and, 267
prejudiced discrimination, 186
premarital cohabitation, 231
premarital sex, cultural practices and,
    177
Presbyterian Mission Agency, 269
prescription pain medication,
    280–281
prestige
  social stratification and, 141–142
  for teachers, 256
prevalence, social epidemiology and,
    273
prevention, of crime, 134–135
preventive population checks, 298
primary deviance, 133
primary group, 101
primary labor market, 194–195
priorities, role conflict and, 87
prison population, 136–137
privacy, online interaction and, 99
privatization, water supply, 308
privilege
  education and, 251
  preservation of, 140
probability sample, 24–25
pro-choice movement, 169–170
procreation, 241–242
  family and, 225
profane, defined, 259
progress, as cultural value, 44
proletariat, 10, 155
pro-life movement, 169–170
propaganda, 321
property, crimes of, 122
Protestant ethic, 266
*Protestant Ethic and the Spirit of
    Capitalism* (Weber), 11
public, defined, 320–321
public mourning, 322
public opinion, 321
public policy, sociology and, 6
punishment, crime prevention and,
    135–136

Q

qualitative research, 25
quantitative research, 25
questionnaires, 28

R

race and racism, 178–198
  achievement gap and, 249–250
  child maltreatment and, 236–237
  class structure and, 144–145

conflict theory and, 15, 194–195
  diversity studies and, 6
  dominant and minority groups and,
    182–184
  education and, 249–251
  family structure and, 243–244
  feminist theories and, 195–196
  feminist theory and, 16
  functionalist perspective on, 194
  gender and, 195–196
  health status and, 277
  income levels and, 188–189
  internal migration and, 296
  interracial relationships, 197–198
  language and, 43
  life expectancy and, 238–239
  majority and minority groups and,
    186–193
  nonmarital childbearing and, 233
  police activity and, 135
  poverty and, 147–148
  prison population and, 136–137
  racial-ethnic friction, 184–186
  racism defined, 184–185
  religion and, 263
  social exchange theory and, 90–91
  social mobility and, 150
  social significance of, 180–181
  sociology and, 13, 193–198
  suspicion of research and, 37
  symbolic interactionism and,
    196–197
  voting rates and, 219–220
  wealth distribution and, 141
racial-ethnic groups, 180
  education levels and, 250–253
  friction in, 184–186
racial group, defined, 180
radical feminism, 16
random digit dialing, 27–28
random sample surveys, 27–28
rape
  on campus, 253
  incidence of, 172
  women as victims of, 131–132
rape culture, 131–132
rational-legal authority, 216
reactivity, field research, 30
real culture, 49
reality, social construction of, 88–89
rebellion, strain theory and, 127
recreation, family and, 225
reference groups, 69, 103
refugees, migration of, 295
rehabilitation, crime control and, 137
reinforcement, social learning theory,
    65–66
relational statuses, 84
relationships, online interaction and,
    97–98
relative deprivation theory, 324

status set, 84
STEM fields (science, technology, engineering and math), gender gap in, 165, 252–253
stereotypes
 in films, 237
 gender stereotypes, 162–163
 racial/ethnic stereotypes, 185
stigma
 deviance and, 120
 medicalization of health and, 290–291
strain theory, 126–128, 316
streaming of students, 253–254
street crime, 121–122
stress, job satisfaction and, 207–208
structural factors
 in poverty, 147–148
 social mobility, 149
structural functionalism, 13–14
structural strain theory, 316–317
structured interviews, 28
structures, social systems, 14
student engagement, 254
student loan debt, 258
subculture, 50
subjective approach to socioeconomic status, 142
subjective understanding, 11
subordination, religion and, 267
subsidies, 203
substance abuse, 279–280
suburbanization, 301
success, as cultural value, 44
suffrage movement, 327
sugar consumption, 279
suicide
 childhood trauma and, 64
 Durkheim's research on, 9–10
 gender patterns and, 276
 myths about, 21
 physician-assisted suicide, 240
Suicide (Durkheim), 9
Supreme Court (U.S.), 218
surrogates, grandparents as, 71
surveys, 27–29
 victimization survey, 121–122
sustainable development, 310–311
symbolic interactionism, 16–19
 culture and, 40, 58–59
 deviance and, 132–134
 education and, 253–254
 family structure and aging and, 244
 gender and sexuality and, 176–177
 health and illness and, 289–291
 organizations and groups and, 115–116
 race and ethnicity and, 196–197
 religion and, 269–270
 social interaction and, 88–90
 socialization and, 66–69

social stratification and, 157
 urbanization and, 305–306
 work and economy and, 213
symbols, 18, 40–41
 defined, 40–41
 in religion, 269–270

T

taboos, 47
tattoos, 121, 137
tax credits and deductions, 203
tax policies
 corporate welfare and, 155
 education and, 247–248, 250
teachers
 international comparisons of, 256
 socialization and, 72–73
teaching, educational outcomes and, 255
technology
 bad urbanism and, 305
 benefits and costs, 329–330
 cultural change and, 54–56
 ethics and, 331
 health and medicine, 330
 health care costs and, 282–283
 social change and, 327–328
telephone interviews, 28
test scores
 achievement gaps and, 255
 ACTs, SATs, and AP Courses, 256
 international test scores, 255–256
theory, defined, 7
Thomas Theorem, 89
3-D printing, 329–330
time dimension in research, 25
topic selection in research, 26
Top-Toy Group, 72
total fertility rate (TFR), 294–295
total institutions, 80
totalitarianism, 214
touch, nonverbal communication, 93–94
tracking, of students, 253–254
tradition, 20
traditional authority, 215
traffic congestion, 302
transgender, 160–161
transnational conglomerates, 204
transnational corporations, 204
transsexuals, 161–162
triads, 103
tweeting, 98–99
two-party systems, 217

U

underclass, 144–145
underemployed workers, 207–208

undocumented immigrants, 181–183
unemployment, 5
 discourged workers and, 207
 hypotheses about, 23
Uniform Crime Report (UCR), 121–122
uninvolved parenting, socialization and, 70
United States
 aging in, 238–241
 air pollution in, 309–310
 class structure in, 142–145
 cultural values in, 44–45
 decline in social mobility in, 149
 employment and working conditions in, 207–208
 European American population in, 187
 family structure in, 229–235
 health care coverage in, 281–283
 health status in, 274–281
 immigration in, 181–182
 internal migration in, 296
 Latino population in, 187–188
 LGBT acceptance in, 163–164
 politics and power in, 216–220
 poverty rate in, 146–148
 racial and ethnic classifications in, 181, 186–193
 racial and ethnic diversity in, 179–181
 religion in, 261–265
 test scores in, 255–256
 urbanization in, 300–303
unstructured interviews, 28
upper class
 characteristics of, 143–144
 legacy students from, 251
upper-middle class, 144
upper-upper class, 143–144
urban ecology, 304
urbanization, 299–306
 conflict theory perspective on, 304–305
 consequences of, 302
 feminist perspectives, 305
 functionalist perspective on, 304
 global trends in, 300
 sociological perspectives on, 303–306
 symbolic interactionism and, 305–306
 in United States, 300–303
urban legends, 318
urban sprawl, 302
U.S. Census Bureau, surveys by, 28, 34
U.S. Flag Code, 41
utilitarian organizations, 107

CHAPTER REVIEW 1
Thinking Like a Sociologist

CHAPTER 1 SUMMARY

1-1 What Is Sociology?

Sociology is the systematic study of human behavior in society, behavior that takes place among individuals, small groups (such families), large organizations (such as Apple), and entire societies (such as the United States). Sociology goes well beyond common sense and conventional wisdom because it examines claims and beliefs critically, considers many points of view, and enables us to move beyond established ways of thinking.

1-2 What Is a Sociological Imagination?

The *sociological imagination* emphasizes the connection between personal troubles (biography) and structural (public and historical) issues in understanding the social world. *Microsociology* examines the relationships between individuals, whereas *macrosociology* focuses on large-scale patterns and processes that characterize society as a whole. Macro-level systems shape society, often limiting our personal options on the micro level.

1-3 Why Study Sociology?

Regardless of your major, this course will help you (1) make more informed decisions, (2) understand diversity, (3) shape social policies and practices, (4) think critically, and (5) expand your career opportunities.

1-4 Some Origins of Sociological Theory

Sociologists use *theories* to produce knowledge and to offer solutions to everyday social problems. Some of the most influential theorists have included Auguste Comte, Harriet Martineau, Émile Durkheim, Karl Marx, Max Weber, Jane Addams, and W. E. B. Du Bois. All of these early thinkers shaped contemporary sociological theories.

1-5 Contemporary Sociological Theories

Sociologists typically use more than one theory to explain human behavior. Four major perspectives help us understand society:

• *Functionalism maintains that society is a complex system of interdependent parts that work together to ensure a society's survival.* Critics contend that functionalism is so focused on order and stability that it often ignores social change, social inequality, and social conflict.

• *Conflict theory sees disagreement and the resulting changes in society as natural, inevitable, and even desirable.* Critics argue that conflict theory ignores the importance of harmony and cooperation.

• *Feminist theories*, which build on conflict theory, *emphasize women's social, economic, and political inequality*. Critics claim that these theories are too narrowly focused.

• *Symbolic interaction focuses on individuals' everyday behavior through the communication of knowledge, ideas, beliefs, and attitudes*. Critics maintain that this theoretical perspective overlooks the impact of macro-level factors on our everyday behavior.

Example: Critical Thinking versus Common Sense and Conventional Wisdom

When thinking critically, it's important to differentiate between common sense/conventional wisdom and data-based facts. Here are a few examples:

Myth: Older people make up the largest group of poor people.

Fact: Children younger than 6, not older people, make up the largest group of poor people.

Myth: Divorce rates are higher today than ever before.

Fact: Divorce rates are lower today than they were between 1980 and 1990.

Now, based on the material you read in this chapter, construct your own "myth" and "fact."

KEY TERMS

sociology the systematic study of human behavior in society.

sociological imagination the ability to see the relationship between individual experiences and larger social influences.

microsociology examines the patterns of individuals' social interaction in specific settings.

macrosociology examines the large-scale patterns and processes that characterize society as a whole.

theory a set of statements that explains why a phenomenon occurs.

empirical information that is based on observations, experiments, or other data collection rather than on ideology, religion, intuition, or conventional wisdom.

social facts aspects of social life, external to the individual, that can be measured.

social solidarity social cohesiveness and harmony.

division of labor an interdependence of different tasks and occupations, characteristic of industrialized societies, that produces social unity and facilitates change.

capitalism an economic system in which the ownership of the means of production is in private hands.

alienation the feeling of separation from one's group or society.

value free separating one's personal values, opinions, ideology, and beliefs from scientific research.

functionalism (structural functionalism) maintains that society is a complex system of interdependent parts that work together to ensure a society's survival.

CHAPTER REVIEW 1

dysfunctions social patterns that have a negative impact on a group or society.

manifest functions purposes and activities that are intended and recognized; they're present and clearly evident.

latent functions purposes and activities that are unintended and unrecognized; they're present but not immediately obvious.

conflict theory examines how and why groups disagree, struggle over power, and compete for scarce resources.

feminist theories examine women's social, economic, and political inequality.

symbolic interaction theory (*interactionism*) examines people's everyday behavior through the communication of knowledge, ideas, beliefs, and attitudes.

social interaction a process in which people take each other into account in their own behavior.

TEST YOUR LEARNING

1. _____ looks at the relationship between individual characteristics; _____ examines the relationships between institutional characteristics.
   a. Microsociology; macrosociology
   b. Macrosociology; microsociology
   c. Metasociology; macrosociology
   d. Metasociology; microsociology

2. Which social class, as identified by Karl Marx, includes the ruling elite who own the means of production?
   a. Capitalists
   b. Communists
   c. Power elite
   d. Proletariat

3. James sees Julie and Sam laughing in the hallway and assumes that Julie likes Sam. James is using Weber's
   a. explanatory understanding.
   b. surveillance understanding.
   c. common understanding.
   d. direct observational understanding.

4. Jeremy views society as a system of interrelated parts, whereas Thomas sees society as composed of groups competing for scarce resources. Jeremy would be considered a _____ theorist; Thomas would be seen as a _____ theorist.
   a. symbolic interactionist; functionalist
   b. conflict; functionalist
   c. functionalist; symbolic interactionist
   d. functionalist; conflict

5. Many people buy designer clothes that they can't afford. The clothes are an example of a status symbol that reflects a
   a. latent function.
   b. manifest function.
   c. dysfunction.
   d. social system.

6. *True or False* In the definition of sociology, "systematic" means behavior that is built into the larger social structure of society.

7. *True or False* Émile Durkheim saw sociology as the scientific study of two aspects of society: social statics and social dynamics.

8. *True or False* Jane Addams was an early sociologist who published extensively on topics such as social disorganization, immigration, and urban neighborhoods.

9. *True or False* Much contemporary functionalism grew out of the work of Auguste Comte and Émile Durkheim.

10. *True or False* Feminist theorists see society as cooperative and harmonious.

11. What does C. Wright Mills mean when he says that there's a connection between personal troubles and structural issues? Use an example, but not unemployment, to explain and illustrate the sociological imagination concept.

12. Which theory do you think better explains how the United States operates—functionalist, conflict, feminist, or symbolic interactionist? Why?

1. a 2. a 3. d 4. d 5. a 6. False 7. False 8. False 9. True 10. False

TABLE 1.2   LEADING CONTEMPORARY PERSPECTIVES IN SOCIOLOGY

THEORETICAL PERSPECTIVE	FUNCTIONALISM	CONFLICT	FEMINIST	SYMBOLIC INTERACTION
Level of Analysis	**Macro**	**Macro**	**Macro and Micro**	**Micro**
Key Points	• Society is composed of interrelated, mutually dependent parts. • Structures and functions maintain a society's or group's stability, cohesion, and continuity. • Dysfunctional activities that threaten a society's or group's survival are controlled or eliminated.	• Life is a continuous struggle between the haves and the have-nots. • People compete for limited resources that are controlled by a small number of powerful groups. • Society is based on inequality in terms of ethnicity, race, social class, and gender.	• Women experience widespread inequality in society because, as a group, they have little power. • Gender, ethnicity, race, age, sexual orientation, and social class—rather than a person's intelligence and ability—explain many of our social interactions and lack of access to resources. • Social change is possible only if we change our institutional structures and our day-to-day interactions.	• People act on the basis of the meaning they attribute to others. Meaning grows out of the social interaction that we have with others. • People continuously reinterpret and reevaluate their knowledge and information in their everyday encounters.
Key Questions	• What holds society together? How does it work? • What is the structure of society? • What functions does society perform? • How do structures and functions contribute to social stability?	• How are resources distributed in a society? • Who benefits when resources are limited? Who loses? • How do those in power protect their privileges? • When does conflict lead to social change?	• Do men and women experience social situations in the same way? • How does our everyday behavior reflect our gender, social class, age, race, ethnicity, sexual orientation, and other factors? • How do macro structures (such as the economy and the political system) shape our opportunities? • How can we change current structures through social activism?	• How does social interaction influence our behavior? • How do social interactions change across situations and between people? • Why does our behavior change because of our beliefs, attitudes, values, and roles? • How is "right" and "wrong" behavior defined, interpreted, reinforced, or discouraged?
Example	• A college education increases one's job opportunities and income.	• Most low-income families can't afford to pay for a college education.	• Gender affects decisions about a major and which college to attend.	• College students succeed or fail based on their degree of academic engagement.

CHAPTER 2 SUMMARY

2-1 How Do We Know What We Know?

Much of our knowledge is based on tradition and authority. In contrast, sociologists rely on *research methods* to get information about a particular topic.

2-2 Why Is Sociological Research Important in Our Everyday Lives?

In contrast to tradition and authority, sociological research challenges overgeneralizations, exposes myths, helps explain why people behave as they do, influences social policies, and sharpens critical thinking skills about issues that affect our everyday lives.

2-3 The Scientific Method

The *scientific method* incorporates careful data collection, exact measurement, accurate recording and analysis of findings, thoughtful interpretation of results, and, when appropriate, a generalization of the findings to a larger group. A research question or a *hypothesis* examines the association between an *independent variable* and a *dependent variable*. Sociologists use both *qualitative* and *quantitative* approaches, and are always concerned about the *reliability* and *validity* of their measures.

2-4 The Research Process: The Basics

Research involves choosing a topic, summarizing the pertinent research, formulating a hypothesis or asking a research question, describing the data collection methods, collecting the data, presenting the findings, and analyzing and explaining the results.

2-5 Some Major Data Collection Methods

Five data collection methods are especially useful in sociology (see Table 2.3 on the next page). In designing their studies, sociologists weigh the advantages and limitations of each data collection method. *Evaluation research*, which uses one or more data collection methods, examines whether a social intervention has produced the intended result. Because sociologists don't conduct research in a cultural vacuum, many groups use the findings to change current policies and practices.

Example: The Uneasy Relationship between Research and Practice

Many supporters of the DARE (Drug Abuse Resistance Education) program were unhappy when more than 30 research studies showed that DARE had negligible long-term effects in reducing teen drug use. About 75 percent of U.S. school districts revised the DARE curriculum, but continued to use it because they believed that the program built a positive relationship between police, students, parents, and educators (see text).

2-6 Ethics and Social Research

Sociological research demands a strict code of ethics to avoid mistreating participants. For example, participants must give informed consent and must not be harmed, humiliated, abused, or coerced; researchers must honor their guarantees of privacy, confidentiality, and/or anonymity. Still, sociologists often encounter pressure from policy makers and others to limit their research to non-controversial topics.

KEY TERMS

research methods organized and systematic procedures to gain knowledge about a particular topic.

scientific method a body of objective and systematic techniques for investigating phenomena, acquiring knowledge, and testing hypotheses and theories.

concept an abstract idea, mental image, or general notion that represents some aspect of the world.

variable a characteristic that can change in value or magnitude under different conditions.

independent variable a characteristic that has an effect on the dependent variable.

dependent variable the outcome that may be affected by the independent variable.

control variable a characteristic that is constant and unchanged during the research process.

hypothesis a statement of the expected relationship between two or more variables.

deductive reasoning an inquiry process that begins with a theory, prediction, or general principle that is then tested through data collection.

inductive reasoning an inquiry process that begins with a specific observation, followed by data collection, a conclusion about patterns or regularities, and the formulation of hypotheses that can lead to theory construction.

reliability the consistency with which the same measure produces similar results time after time.

validity the degree to which a measure is accurate and really measures what it claims to measure.

population any well-defined group of people (or things) about whom researchers want to know something.

sample a group of people (or things) that's representative of the population researchers wish to study.

probability sample each person (or thing) has an equal chance of being selected because the selection is random.

nonprobability sample there is little or no attempt to get a representative cross section of the population.

qualitative research examines nonnumerical material and interprets it.

CHAPTER REVIEW 2

quantitative research focuses on a numerical analysis of people's responses or specific characteristics.

causation a relationship in which one variable is the direct consequence of another.

correlation the relationship between two or more variables.

survey a data collection method that includes questionnaires, face-to-face or telephone interviews, or a combination.

questionnaire a series of written questions that ask for information.

interview a researcher directly asks respondents a series of questions.

field research data collected by observing people in their natural surroundings.

content analysis a data collection method that systematically examines some form of communication.

experiment a controlled artificial situation that allows researchers to manipulate variables and measure the effects.

experimental group the participants who are exposed to the independent variable.

control group the participants who aren't exposed to the independent variable.

secondary analysis examination of data that have been collected by someone else.

evaluation research studies whether a social intervention has produced the intended result.

TEST YOUR LEARNING

1. Sociological research is important in our daily lives for a number of reasons. Which of the following is NOT one of those reasons?
   a. It exposes myths.
   b. It challenges the findings of psychology researchers.
   c. It affects social policy.
   d. It sharpens our critical thinking skills.

2. What is a hypothesis?
   a. A theory based on one's opinions.
   b. A measure of how accurate a study is.
   c. A statement of a relationship between two or more variables.
   d. An analysis of multiple studies that researchers use to make generalizations.

3. Alexandra notices that she performs best on tests that are given in the afternoon. She then begins to collect data of her test grades and asks close friends and classmates to do the same. Alexandra is using
   a. intuition.
   b. inferences.
   c. deductive reasoning.
   d. inductive reasoning.

4. Which of the following examines and interprets nonnumerical data?
   a. Qualitative research
   b. Surveys
   c. Secondary analysis
   d. Quantitative research

5. There are three golden rules in sociological research. Which of the following is NOT one of them?
   a. Do no harm.
   b. Be fully transparent with the subject.
   c. Get the subject's informed consent.
   d. Protect the subject's confidentiality.

6. *True or False* Social researchers continuously question the quality of existing studies.

7. *True or False* If researchers use a nonprobability sample, they can generalize the results to a larger population.

8. *True or False* Validity is the consistency with which the same measure produces similar results time after time.

9. *True or False* Observing second graders interact with one another in their classroom is an example of field research.

10. *True or False* Correlation is pretty much the same as causation.

11. State a hypothesis about your GPA, the dependent variable. What three independent variables might you use? Explain why.

12. Suppose you want to examine the effects of student employment on class attendance. Which data collection method(s) would you use? Why?

1. b 2. c 3. d 4. a 5. b 6. True 7. False 8. False 9. True 10. False

TABLE 2.3 SOME DATA COLLECTION METHODS IN SOCIOLOGICAL RESEARCH

Method	Example	Advantages	Disadvantages
Surveys	Sending questionnaires and/or interviewing students on why they succeeded in college or dropped out	Questionnaires are fairly inexpensive and simple to administer; interviews have high response rates; findings are often generalizable	Mailed questionnaires may have low response rates; respondents tend to be self-selected; interviews are usually expensive
Secondary analysis	Using data from the National Center for Education Statistics (or similar organizations) to examine why students drop out of college	Usually accessible, convenient, and inexpensive; often longitudinal and historical	Information may be incomplete; some documents may be inaccessible; some data can't be collected over time
Field research	Observing classroom participation and other activities of first-year college students with high and low grade-point averages (GPAs)	Flexible; offers deeper understanding of social behavior; usually inexpensive	Difficult to quantify and to maintain observer/subject boundaries; the observer may be biased or judgmental; findings are not generalizable
Content analysis	Comparing the transcripts of college graduates and dropouts on variables such as gender, race/ethnicity, and social class	Usually inexpensive; can recode errors easily; unobtrusive; permits comparisons over time	Can be labor intensive; coding is often subjective (and may be distorted); may reflect social class bias
Experiments	Providing tutors to some students with low GPAs to find out if such resources increase college graduation rates	Usually inexpensive; plentiful supply of subjects; can be replicated	Subjects aren't representative of a larger population; the laboratory setting is artificial; findings can't be generalized

CHAPTER 3 SUMMARY

3-1 Culture and Society

Culture is learned, transmitted from one generation to another, adaptive, and always changing. A *society* shares a culture and sees itself as a social unit. People construct a *material culture* (e.g., buildings) and *nonmaterial culture* (e.g., rules for behavior) that influence each other (such as forbidding smoking in public buildings).

3-2 The Building Blocks of Culture

The following are some of the fundamental building blocks of culture:

• *Symbols* take many forms, distinguish one culture from another, can change over time, and can unify or divide a society.

• *Language* makes us human; it can change over time, and affects perceptions about gender, race, social class, and ethnicity.

• *Values* provide general guidelines for behavior; they sometimes conflict, are usually emotion laden, vary across cultures, and change over time.

• *Norms*—whether folkways, mores, or laws—regulate our behavior; they vary across cultures, and range from mild to severe sanctions.

• *Rituals* transmit and reinforce norms that unite people and strengthen their relationships.

Example: Sanctions for Violating the Dead

Sanctions are more severe for violating laws than folkways. Legacy.com, which carries a death notice or obituary for almost all of the roughly 2.4 million Americans who die each year, dedicates at least 30 percent of its budget to weeding out comments (a relatively mild punishment) that "diss the dead" (Urbina, 2006). In contrast, in many states, vandalizing a tombstone is a property crime that can result in a fine of up to $1,000, a jail sentence of up to a year, or both.

3-3 Some Cultural Similarities

Customs and specific behaviors vary across countries, but *cultural universals*, such as food taboos, are common to all societies. *Ideal culture* often differs from *real culture*. *Ethnocentrism* has its benefits but can also lead to conflict and discrimination. *Cultural relativism*, the opposite of ethnocentrism, encourages cross-cultural understanding and respect.

3-4 Some Cultural Variations

Subcultures and *countercultures* account for some of the complexity within a society. Subcultures differ from the larger society in some ways, whereas countercultures oppose or reject some of the dominant culture's basic beliefs, values, and norms. In *multiculturalism*, several cultures coexist without trying to dominate one another. People who encounter an unfamiliar way of life or environment may experience *culture shock* because everyone is culture bound to some degree.

3-5 High Culture and Popular Culture

High culture refers to the cultural expression of a society's highest social classes, whereas *popular culture*—which is widely shared among a population—includes music, social media, sports, hobbies, fashions, the food we eat, the people with whom we spend time, the gossip we share, and the jokes we tell. *Cultural capital* affects social class boundaries, but our everyday life can contain elements of both popular and high culture. Popular culture is typically spread through *mass media*, including television and the Internet, and has enormous power in shaping our attitudes and behavior.

3-6 Cultural Change and Technology

Some societies are relatively stable because of *cultural integration*, but all societies change over time because of diffusion, innovation, invention, discovery, and external pressures. *Cultural lags* can create confusion, ambiguity about what's right and wrong, conflict, and a feeling of helplessness. They also expose contradictory values and behavior.

3-7 Sociological Perspectives on Culture

See Table 3.3 on the next page.

KEY TERMS

culture the learned and shared behaviors, beliefs, attitudes, values, and material objects that characterize a particular group or society.

society a group of people who share a culture and defined territory.

material culture the physical objects that people make, use, and share.

nonmaterial culture the ideas that people create to interpret and understand the world.

symbol anything that stands for something else and has a particular meaning for people who share a culture.

language a system of shared symbols that enables people to communicate with one another.

values the standards by which people define what is good or bad, moral or immoral, proper or improper, desirable or undesirable, beautiful or ugly.

norms specific rules of right and wrong behavior.

folkways norms that involve everyday customs, practices, and interaction.

mores norms that people consider very important because they maintain moral and ethical behavior.

taboos strong prohibitions of any act that's forbidden because it's considered extremely offensive.

laws formally defined norms about what is legal or illegal.

sanctions rewards for good or appropriate behavior and/or penalties for bad or inappropriate behavior.

rituals formal and repeated behaviors that unite people.

cultural universals customs and practices that are common to all societies.

ideal culture the beliefs, values, and norms that people say they hold or follow.

real culture people's actual everyday behavior.

ethnocentrism the belief that one's own culture, society, or group is inherently superior to others.

cultural relativism the belief that no culture is better than another and should be judged by its own standards.

CHAPTER REVIEW 3

subculture a group within society that has distinctive norms, values, beliefs, lifestyle, or language.

counterculture a group within society that openly opposes and/or rejects some of the dominant culture's norms, values, or laws.

multiculturalism (sometimes called *cultural pluralism*) the coexistence of several cultures in the same geographic area, without one culture dominating another.

culture shock confusion, disorientation, or anxiety that accompanies exposure to an unfamiliar way of life.

high culture the cultural expression of a society's highest social classes.

popular culture the beliefs, practices, activities, and products that are widespread among a population.

cultural capital resources such as knowledge, verbal and social skills, education, and other assets that give a group advantages.

mass media forms of communication designed to reach large numbers of people.

cultural imperialism the cultural values and products of one society influence or dominate those of another.

cultural integration the consistency of various aspects of society that promotes order and stability.

cultural lag the gap that occurs when material culture changes faster than nonmaterial culture.

TEST YOUR LEARNING

1. Society and culture are mutually
   a. dependent.
   b. exclusive.
   c. destructive.
   d. diversified.

2. ____ stand for something else and have a particular meaning for people who share a culture.
   a. Mores
   b. Symbols
   c. Values
   d. Norms

3. At a party, George was eating nachos and salsa. He took a bite of his chip and then dipped it back into the salsa bowl. His friend Brett gave him a disgusted look. Which of the following did George break?
   a. Mores
   b. A taboo
   c. A folkway
   d. A law

4. Which of the following is an example of a subculture?
   a. A group of adolescent Goths
   b. Roman Catholics
   c. A 40-and-over women's bridge club
   d. All of the above
   e. None of the above

5. Anaz is an 8-year-old Iranian girl who loves the female fashions she sees in U.S. films and television shows. Lately she's been questioning why her mother wears a burka. Anaz's questioning of her family's traditions could best be explained by
   a. cultural relativism.
   b. multiculturalism.
   c. cultural imperialism.
   d. ethnocentrism.

6. *True or False* Culture is adapting and always changing.

7. *True or False* Language helps communicate ideas, but it's not considered a symbol.

8. *True or False* In most societies, the real culture matches the ideal culture.

9. *True or False* The legal controversy over file sharing and downloading music off the Web is an example of cultural lag.

10. *True or False* Symbolic interaction relies on a macro approach in examining culture.

11. List five examples of material culture and five of nonmaterial culture that affect you. Explain (a) how they intersect and (b) how they shape your attitudes and behavior.

12. Provide an example of each type of norm (folkways, mores, and laws) that you or someone you know has violated. Describe the negative sanctions for each violation.

1. a 2. b 3. c 4. d 5. c 6. True 7. False 8. False 9. True 10. False

TABLE 3.3 SOCIOLOGICAL PERSPECTIVES OF CULTURE

THEORETICAL PERSPECTIVE	FUNCTIONALIST	CONFLICT	FEMINIST	SYMBOLIC INTERACTIONIST
Level of Analysis	**Macro**	**Macro**	**Macro and Micro**	**Micro**
Key Points	• Similar beliefs bind people together and create stability. • Sharing core values unifies a society and promotes cultural solidarity.	• Culture benefits some groups at the expense of others. • As powerful economic monopolies increase worldwide, the rich get richer and the rest of us get poorer.	• Women and men often experience culture differently. • Cultural values and norms can increase inequality because of sex, race/ethnicity, and social class.	• Cultural symbols forge identities (that change over time). • Culture (such as norms and values) helps people merge into a society despite their differences .
Examples	• Speaking the same language (English in the United States) binds people together because they can communicate with one another, express their feelings, and influence one another's attitudes and behaviors.	• Much of the English language reinforces negative images about gender ("slut"), race ("honky"), ethnicity ("jap"), and age ("old geezer") that create inequality and foster ethnocentrism.	• Using male language (e.g., "congressman," "fireman," and "chairman") conveys the idea that men are superior to and dominant over women, even when women have the same jobs.	• People can change the language they create as they interact with others. Many Americans now use "police officer" instead of "policeman," and "single person" instead of "bachelor" or "old maid."

CHAPTER 4 SUMMARY

4-1 Socialization: Its Purpose and Importance

Socialization, a lifelong process, fulfills four key purposes: It establishes our social identity, teaches us roles, controls our behavior (through *internalization*), and transmits culture to the next generation. The research on children who are isolated or institutionalized shows that socialization is critical to our social and physical development.

4-2 Nature and Nurture

Biologists tend to focus on the role of heredity (or genetics) in human development. In contrast, most social scientists, including sociologists, underscore the role of learning, socialization, and culture. This difference of opinion is often called the *nature–nurture debate*. Sociobiologists argue that genetics (nature) can explain much of our behavior, whereas most sociologists maintain that socialization and culture (nurture) shape even biological inputs.

4-3 Sociological Explanations of Socialization

Sociologists have offered many explanations of socialization, but two of the most influential, both at the micro level, have been social learning and symbolic interaction theories (see Table 4.2 on the next page). The theories help us understand social development processes, but also have limitations.

4-4 Primary Socialization Agents

Parents are the first and most important *socialization agents*, but siblings, grandparents, and other family members also play important roles. Other significant socialization agents include play and peer groups, teachers and schools, popular culture, and the media. Electronic media and advertising are especially powerful socialization sources.

Example: Are Parents Raising Self-Centered Children?

Many parents tell their offspring that they're beautiful, can achieve anything they want, and are very intelligent. In fact, only about 5 percent of American kids can be considered "gifted" (have significantly higher than average intellectual or other abilities), even though many are enrolled in gifted classes. Some educators maintain that telling average—or even above average—children that they're superior does them a disservice: It gives them false expectations on how the world will treat them, encourages being self-centered, and increases anger and unhappiness when they don't succeed in college or the workplace (Deveny, 2008).

4-5 Socialization Throughout Life

As we progress through the life course, we learn culturally approved norms, values, and roles. Infants are born with an enormous capacity for learning that parents and other caregivers can enrich and shape. In adolescence, these and other socialization agents teach children how to form relationships, to get along with others, and to develop their social identity through play and peer groups. In adulthood, people must learn new roles that include singlehood, marriage, parenthood, divorce, work, and experiencing the death of a loved one. Socialization continues in later life when many people learn still new roles such as grandparents, retirees, older workers, and being widowed.

4-6 Resocialization

Resocialization can be voluntary or involuntary. Voluntary resocialization includes entering a religious order, joining a religious cult, seeking treatment in a drug abuse rehabilitation facility, or serving in the military. The changes can be short-term and pleasant, but also long, difficult, and intense. Most involuntary resocialization takes place in *total institutions*, where people are isolated from the rest of society, stripped of their former identities, and required to conform to new rules and behavior.

KEY TERMS

socialization the lifelong process through which people learn culture and become functioning members of society.

internalization the process of learning cultural behaviors and expectations so deeply that we accept them without question.

social learning theories maintain that people learn new attitudes, beliefs, and behaviors through social interaction, especially during childhood.

role models people we admire and whose behavior we imitate.

self an awareness of one's social identity.

looking-glass self a self-image based on how we think others see us.

role taking learning to take the perspective of others.

significant other someone whose opinions we value and who influences our thinking, especially about ourselves.

anticipatory socialization learning how to perform a role that a person will occupy.

generalized other the norms, values, and expectations of society as a whole that affect a person's behavior.

impression management the process of providing information and cues to others to present oneself in a favorable light while downplaying or concealing one's less appealing characteristics.

reference groups groups of people who shape an individual's self-image, behavior, values, and attitudes in different contexts.

socialization agents the individuals, groups, or institutions that teach us how to participate effectively in society.

multigenerational households homes in which three or more generations live together.

CHAPTER REVIEW 4

peer group people who are similar in age, social status, and interests.

resocialization the process of unlearning old ways of doing things and adopting new attitudes, values, norms, and behavior.

total institutions settings where people are isolated from the rest of society, stripped of their former identities, and required to conform to new rules and behavior.

TEST YOUR LEARNING

1. Mom taught Emma that girls can do anything that boys can do, including being good in math and science. This is an example of the ___ process.
   a. multicultural
   b. feminist
   c. socialization
   d. hereditary

2. "Human development is fairly fixed." This statement is an example of which side of the nature/nurture debate?
   a. Nature side of the debate
   b. Nurture side of the debate
   c. Neither the nature nor nurture side of the debate
   d. Both the nature and nurture side of the debate

3. According to social learning theory, the greatest impact of social interaction occurs during
   a. childhood.
   b. adolescence.
   c. early adulthood.
   d. middle adulthood.

4. What are the three stages of Mead's role-taking theory?
   a. Pre-work stage, work stage, post-work stage
   b. Anal stage, phallic stage, post-phallic stage
   c. Mirror stage, active stage, passive stage
   d. Preparatory stage, play stage, game stage

5. According to Mead, children learn how to perform roles that they'll occupy in the future. Mead referred to this process as
   a. a developing self.
   b. anticipatory socialization.
   c. role taking.
   d. the generalized other.

6. **True or False** Institutionalization is the process of learning cultural behaviors and expectations so deeply that we accept them without question.

7. **True or False** Charles Horton Cooley proposed that the looking-glass self develops in five phases.

8. **True or False** According to Erving Goffman, social life mirrors a theatre.

9. **True or False** Healthy child development is most likely in authoritative homes.

10. **True or False** Resocialization is the process of presenting ourselves in a favorable light while downplaying or concealing our less appealing characteristics.

11. Think about your early socialization. Were social learning theories, symbolic interaction theories, or both important in your socialization? Explain and illustrate your answer using specific examples.

12. Have you ever engaged in impression management? Describe when, where, how, and why.

1. c 2. a 3. a 4. d 5. b 6. False 7. False 8. True 9. True 10. False

TABLE 4.2	KEY ELEMENTS OF SOCIALIZATION THEORIES
Social Learning Theories	**Symbolic Interaction Theories**
• Social interaction is important in learning appropriate and inappropriate behavior.	• The self emerges through social interaction with significant others.
• Socialization relies on direct and indirect reinforcement.	• Socialization includes role taking and controlling the impression we give to others.
Example: Children learn how to behave when they are scolded or praised for specific behaviors.	*Example:* Children who are praised are more likely to develop a strong self-image than those who are always criticized.

CHAPTER 5 SUMMARY

5-1 Social Structure

Social interaction, central to all social activity, affects people's behavior. Our interaction is part of the *social structure*, which guides our actions and gives us a feeling that life is orderly and predictable. Every society has a social structure that encompasses statuses and roles.

5-2 Status

A *status* is a social position that a person occupies in a society. Every person has many statuses that form her or his *status set*, which include both ascribed and achieved statuses. An *ascribed status* is a social position that a person is born into and can't control, change, or choose (e.g., age, race, and ethnicity). An *achieved status* is a social position that a person attains through personal effort or assumes voluntarily (e.g., college student or wife). A *master status* is usually immediately apparent, makes the biggest impression, affects others' perceptions, and, consequently, often shapes a person's entire life.

Because we hold many statuses, some clash. People experience *status inconsistency* when they occupy social positions that are ranked differently (such as being a low-paid college professor).

5-3 Role

A *role* defines how we're expected to behave in a particular status, but people vary considerably in fulfilling the responsibilities associated with their roles. These differences reflect *role performance*, the actual behavior of a person who occupies a status. A *role set* encompasses different roles attached to a single status (e.g., a parent who's a teacher, volunteer, and PTA member). Playing many roles often leads to *role conflict* because it's difficult to meet the requirements of two or more statuses, and to *role strain*, the stress that arises because of incompatible demands among roles within a single status.

To deal with role conflict and role strain, some people deny that there's a problem. More effective ways to minimize role conflict and role strain include compromising, negotiating, setting priorities, compartmentalizing our roles, refusing to take on more roles, and exiting one or more current roles.

Example: Exiting a Marriage

Divorce is a good example of role exit, but it often involves a long process of five stages that may last several decades (Bohannon 1971):

• The "emotional divorce" begins when one or both partners feel disillusioned or unhappy.

• The "legal divorce" is the formal dissolution of the marriage during which the partner who doesn't want the divorce may try to stall the end of the marriage.

• During the "economic divorce" stage, the partners may argue about who should pay past debts and unforeseen expenses (like moving costs).

• The "coparental divorce" stage involves parents' agreeing on issues such as child support and visitation rights.

• During the "community divorce" stage, partners inform friends, family, and others that they are no longer married.

• Finally, the couple goes through a "psychic divorce" in which the partners separate from each other emotionally. One or both spouses never complete this stage because they can't let go of their pain and anger—even if they remarry.

5-4 Explaining Social Interaction

See Table 5.2 on next page.

KEY TERMS

social interaction the process by which we act toward and react to people around us.

social structure an organized pattern of behavior that governs people's relationships.

status a social position that a person occupies in a society.

status set a collection of social statuses that a person occupies at a given time.

ascribed status a social position that a person is born into.

achieved status a social position that a person attains through personal effort or assumes voluntarily.

master status a status that overrides other statuses and forms an important part of a person's social identity.

status inconsistency the conflict that arises from occupying social positions that are ranked differently.

role the behavior expected of a person who has a particular status.

role performance the actual behavior of a person who occupies a status.

role set array of roles attached to a particular status.

role conflict difficulties in playing two or more contradictory roles.

role strain difficulties due to conflicting demands within the same role.

self-fulfilling prophecy a situation in which if we define something as real and act on it, it can, in fact, become real.

ethnomethodology the study of how people construct and learn to share definitions of reality that make everyday interactions possible.

dramaturgical analysis examines social interaction as if occurring on a stage where people play different roles and act out scenes for the audiences with whom they interact.

social exchange theory proposes that individuals seek through their interactions to maximize their rewards and minimize their costs.

nonverbal communication messages that are sent without using words.

social media websites that enable users to create, share, and/or exchange information and ideas.

social networking site a website that connects people who have similar personal or professional interests.

5-5 Nonverbal Communication

Our *nonverbal communication* includes gestures, facial expressions, eye contact, and silence. Touching and how we use space are also important forms of nonverbal communication because they send powerful messages about our feelings and power. Nonverbal communication, which varies from society to society, can lead to cross-cultural misinterpretation and misunderstanding.

5-6 Online Interaction

Many people interact in *cyberspace*, an online world of computer networks that includes *social media* and *social networking sites*. Internet usage varies by sex, age, race, ethnicity, and social class. Online interaction can be impersonal, socially isolating, and jeopardizes our privacy, but can also save time, foster closer ties among family members and friends, and facilitate working from home.

TEST YOUR LEARNING

1. Jenna is the youngest of three girls in her family. Being the youngest sister is an example of a(n)
   a. achieved status.
   b. ascribed status.
   c. problematic status.
   d. interchangeable status.

2. Jake has a full-time course load and works 30 hours a week. Given the demands of both work and school, Jake is likely to experience
   a. role set.
   b. role strain.
   c. role conflict.
   d. role exchange.

3. According to ____, social interaction is based on trying to maximize one's benefits while minimizing costs.
   a. Ethnomethodology theory
   b. Feminist theory
   c. Conflict theory
   d. Social exchange theory

4. Gestures, eye contact, and silence are all examples of
   a. emotional language.
   b. punishment.
   c. healthy relationship interaction.
   d. nonverbal communication.

5. Which one of the following statements is false?
   a. Almost equal numbers of women and men use the Internet.
   b. The higher a family's income, the greater the likelihood that its members are Internet users.
   c. A major disadvantage of cyberspace communication is that it's more impersonal than face-to-face interaction.
   d. White Americans are the most wired group in the United States.

6. *True or False* In sociology, status signifies prestige.

7. *True or False* A master status is based on one's achieved status.

8. *True or False* Cyberstalking is more common among adolescents than adults.

9. *True or False* If we define something as real and act on it, it can, in fact, become real. This is known as the self-fulfilling prophecy.

10. *True or False* Being an African American college student is an example of having two ascribed statuses.

11. List your social statuses. Which ones are ascribed, and which ones are achieved? Do you have one or more master statuses, and how do they affect you? After describing the difference between role conflict and role strain, choose an example from each and explain how you cope with or have resolved a role conflict or role strain.

12. Describe three benefits and three costs of online interaction that you've encountered. Include any of the topics you read about in the chapter, but illustrate your answer with specific examples.

1. b 2. c 3. d 4. d 5. d 6. False 7. False 8. False 9. True 10. False

TABLE 5.2	SOCIOLOGICAL EXPLANATIONS OF SOCIAL INTERACTION
Perspective	**Key Points**
Symbolic Interactionist	• People create and define their reality through social interaction. • Our definitions of reality, which vary according to context, can lead to self-fulfilling prophecies.
Social Exchange	• Social interaction is based on a balancing of benefits and costs. • Relationships involve trading a variety of resources, such as money, youth, and good looks.
Feminist	• Females and males act similarly in many interactions but may differ in communication styles and speech patterns. • Men are more likely to use speech that's assertive (to achieve dominance and goals), whereas women are more likely to use language that connects with others.

CHAPTER 6 SUMMARY

6-1 Social Groups

A *social group* (such as friends or work groups) gives us a common identity and a sense of belonging. Social groups include *primary groups* (such as family members) that shape our social and moral development and *secondary groups* (such as the students in your sociology class) that pursue a specific goal or activity.

Members of an *in-group* share a sense of identity and "we-ness." In contrast, *out-groups* are viewed and treated negatively because they're seen as having values, beliefs, and other characteristics that differ from those of the in-group.

We also have *reference groups* that influence who we are, what we do, and who we'd like to be in the future. Groups often form a *social network* that may be tightly knit and interact every day or may include large numbers of people whom we don't know personally and with whom we interact only rarely or indirectly.

Example: Secondary Groups Can Replace Primary Groups

In 1864, an alcoholic who had ruined a promising career on Wall Street because of his constant drunkenness cofounded Alcoholics Anonymous (AA), a program that would enable people to stop drinking by undergoing a spiritual awakening, and seeking help from a buddy to stay sober. Initially, AA was a secondary group that encouraged its members to attend regular meetings and discuss their accomplishments in staying sober. Over the years, however, AA became a primary group for many members: It offers support to a relatively small group of people who engage in face-to-face interaction over an extended period, especially when their family and friends have rejected them.

6-2 Formal Organizations

We depend on a variety of *formal organizations* to provide goods and services in a stable and predictable way. Utilitarian, normative, and coercive organizations differ in their general characteristics and membership, but a single formal organization can fall into all three categories.

Bureaucracies (such as your college), are supposed to accomplish goals and tasks in the most efficient and rational way possible, but many experience shortcomings such as weak reward systems, rigid rules, goal displacement, alienation, communication problems, and dehumanization.

6-3 Sociological Perspectives on Social Groups and Organizations

See Table 6.3 on the next page.

6-4 Social Institutions

A *social institution* meets a society's basic survival needs. Social institutions are abstractions, but they have an organized purpose, weave together norms and values, and, consequently, guide behavior. Because social institutions are linked to one another, they can tell us a lot about how a society functions and how we're connected to one another.

KEY TERMS

social group two or more people who share some attribute and interact with one another.

primary group a small group of people who engage in intimate face-to-face interaction over an extended period.

secondary group a large, usually formal, impersonal, and temporary collection of people who pursue a specific goal or activity.

ideal types general traits that describe a social phenomenon rather than every case.

in-group people who share a sense of identity and "we-ness" that typically excludes and devalues outsiders.

out-group people who are viewed and treated negatively because they're seen as having values, beliefs, and other characteristics different from those of an in-group.

reference group people who shape our behavior, values, and attitudes.

dyad a group with two members.

triad a group with three members.

authoritarian leader gives orders, assigns tasks, and makes all major decisions.

democratic leader encourages group discussion and includes everyone in the decision-making process.

laissez-faire leader offers little or no guidance to group members and allows them to make their own decisions.

groupthink a situation in which in-group members make faulty decisions because of group pressures, rather than critically testing, analyzing, and evaluating ideas and evidence.

social network a web of social ties that links individuals or groups to one another.

CHAPTER REVIEW 6

formal organization a complex and structured secondary group designed to achieve specific goals in an efficient manner.

bureaucracy a formal organization designed to accomplish goals and tasks in an efficient and rational way.

goal displacement a preoccupation with rules and regulations rather than achieving the organization's objectives.

alienation a feeling of isolation, meaninglessness, and powerlessness.

iron law of oligarchy the tendency of a bureaucracy to become increasingly dominated by a small group of people.

glass ceiling workplace attitudes or organizational biases that prevent women from advancing to leadership positions.

glass escalator men who enter female-dominated occupations and receive higher wages and faster promotions than women.

social institution an organized and established social system that meets one or more of a society's basic needs.

TEST YOUR LEARNING

1. A(n) ____ is a small group of people who engage in frequent and intimate face-to-face interaction.
   a. Out-group
   b. Reference group
   c. Secondary group
   d. Primary group

2. ____ conducted experiments that used "teachers" who administered electric shocks to "learners."
   a. Solomon Asch
   b. Stanley Milgram
   c. Philip Zimbardo
   d. Irving Janis

3. Max Weber outlined the characteristics of an efficient and productive bureaucracy. Which of the following was NOT one of those characteristics?
   a. High degree of specialization
   b. Explicit rules and regulations
   c. Qualifications-based employment
   d. Decentralized authority

4. Regarding social groups and organizations, symbolic interaction maintains that
   a. some people benefit more than others.
   b. cooperation works.
   c. men benefit more than women.
   d. people define and shape their situations.

5. ____ contend that organizations are based on vast differences in power and control.
   a. Functionalists
   b. Conflict theorists
   c. Symbolic interactionists
   d. Exchange theorists

6. **True or False** A high school football team is a good example of a primary group.

7. **True or False** Groupthink occurs most often when in-group members discuss a diversity of ideas.

8. **True or False** People join normative organizations because of shared interests and to pursue goals that they consider personally worthwhile or rewarding.

9. **True or False** According to the iron law of oligarchy, bureaucracies have a tendency to become increasingly dominated by a small group of people.

10. **True or False** The glass ceiling refers to workplace behavior, not attitudes.

11. Describe yourself using the following concepts: primary group, secondary group, in-group, out-group, reference group, and social network. Illustrate your self-description with specific examples.

12. In the United States, we live in a "McDonaldized" society. What does this concept mean? What are some of the benefits and costs of a McDonaldized society? Illustrate your answer with specific examples from your everyday life.

1. d 2. b 3. d 4. d 5. b 6. False 7. False 8. True 9. True 10. False

TABLE 6.3	VOLUNTEERING IN THE UNITED STATES, 2014
Percentage of adults who volunteer	25
Total number of volunteers age 16 and older	63 million
Median annual hours per volunteer	50
Total dollar value of volunteer time	$173 billion

Sources: Based on Corporation for National & Community Service, 2015, and BLS News Release, 2015.

CHAPTER 7 SUMMARY

7-1 What Is Deviance?

Deviance, the violation of social norms, includes *crime*. Perceptions of deviance and crime vary across and within societies, can change over time. People with authority or power decide what's right or wrong, and a *stigma* accompanies deviant behavior.

Example: Deviance and College Drinking

According to many college presidents, alcohol abuse is the most serious problem on campus. Alcohol abuse leads to alcohol poisoning, blackouts, sexual assault, violent behavior, injuries, and academic problems. Because drinking laws are rarely enforced, some college presidents have proposed that the drinking age be lowered from 21 to 18.

Others argue that doing so would increase traffic fatalities and drinking problems. Young people can get a driver's license at 16 and vote and enlist in the military at 18. Should they be the ones, then, to decide whether drinking laws should be changed?

7-2 Types of Deviance and Crime

Sociologists study *noncriminal deviance*—such as suicide, alcoholism, lying, mental illness, and adult pornography—and *criminal deviance*, behavior that violates laws. Two of the most important sources of U.S. crime statistics are the FBI's *Uniform Crime Report*, and *victimization surveys*.

The media usually focuses on street crimes, but Americans are much more likely to be victimized by *hate crimes*, *white-collar crimes*, *corporate crimes*, *cybercrime*, and *organized crime*. *Victimless crimes* violate laws, but the parties involved don't consider themselves victims.

7-3 to 7-6 Sociological Explanations of Deviance and Crime

See Table 7.2 on the next page.

7-7 Controlling Deviance and Crime

The purpose of *social control* is to eliminate, or at least reduce, deviance and crime. Those who have authority or power administer formal social control. Informal social control is internalized from childhood.

The *criminal justice system* relies on three major approaches in controlling crime: prevention and intervention, punishment, and rehabilitation. A *crime control model* supports a tough approach toward criminals in sentencing, imprisonment, and capital punishment. In contrast, many people believe that *rehabilitation* can change offenders into productive and law-abiding citizens.

KEY TERMS

deviance a violation of social norms.

crime a violation of society's formal laws.

stigma a negative label that devalues a person and changes her or his self-concept and social identity.

victimization survey interviews people about being crime victims.

hate crime a criminal act motivated by hostility toward a particular race, religion, ethnicity, sexual orientation, gender, or disability.

white-collar crime illegal activities committed by high-status people in the course of their occupations.

corporate crimes (also called *organizational crimes*) illegal acts committed by executives to benefit themselves and their companies.

cybercrime (also called *computer crime*) illegal activities that are conducted online.

organized crime activities of individuals and groups that supply illegal goods and services for profit.

victimless crimes acts that violate laws but those involved don't consider themselves victims.

anomie the condition in which people are unsure how to behave because of absent, conflicting, or confusing social norms.

strain theory posits that people may engage in deviant behavior when they experience a conflict between goals and the means available to obtain the goals.

patriarchy a hierarchical system in which cultural, political, and economic structures are controlled by men.

rape culture an environment in which sexual violence is prevalent, pervasive, and perpetuated by the media and popular culture.

CHAPTER REVIEW 7

differential association theory asserts that people learn deviance through interaction, especially with significant others.

labeling theory posits that society's reaction to behavior is a major factor in defining oneself or others as deviant.

primary deviance the initial act of breaking a rule.

secondary deviance rule-breaking behavior that people adopt in response to others' reactions.

medicalization of deviance diagnosing and treating a violation of social norms as a medical disorder.

social control the techniques and strategies that regulate people's behavior in society.

control theory proposes that deviant behavior decreases when people have strong social bonds with others.

criminal justice system government agencies that are charged with enforcing laws, judging offenders, and changing criminal behavior.

crime control model proposes that crime rates increase when offenders don't fear apprehension or punishment.

TEST YOUR LEARNING

1. Which of the following statements is false?
   a. Deviance can be a trait or a behavior.
   b. Informal deviance violates laws.
   c. Deviance is usually accompanied by social stigmas.
   d. Deviance varies across situations.

2. Who of the following is most likely to be the victim of a crime?
   a. A black man
   b. A black woman
   c. A white man
   d. A white woman

3. When 3-year-old Kyle colored the kitchen table blue with his new crayons, his mother frowned and scolded him. Kyle's mother used ____ to control his behavior?
   a. punishment
   b. negative sanctions
   c. behavior modification
   d. positive sanctions

4. Merton proposed four deviant modes by which people adapted to social strain. Which of the following was NOT one of Merton's modes?
   a. Innovation
   b. Retreatism
   c. Ritualism
   d. Recidivism

5. What's a fundamental question that a conflict theorist would ask regarding crime?
   a. "Why do some people commit crimes whereas others don't?"
   b. "Why are some acts defined as criminal whereas others aren't?"
   c. "Why do men commit more violent crimes than women?"
   d. "How does one's social context impact deviant behavior?"

6. *True or False* Female crime rates have decreased.

7. *True or False* Differential association theory claims that people learn deviant behaviors through interaction with others.

8. *True or False* Labeling theory contends that society's reaction to a behavior is a major factor in defining oneself or others as deviant.

9. *True or False* Capital punishment decreases crime.

10. *True or False* The crime control model maintains that rehabilitation is the best way to decrease the frequency of crime.

11. What's the difference, if any, between deviance and crime? Include the positive and negative functions, if any, of deviance and crime in your answer.

12. Which sociological perspective do you think provides the best explanation of deviance and crime? Explain why, illustrating your answer with specific examples.

1. b 2. a 3. b 4. d 5. b 6. False 7. True 8. True 9. False 10. False

TABLE 7.2 SOCIOLOGICAL EXPLANATIONS OF DEVIANCE

Theoretical Perspective	Level of Analysis	Key Points
Functionalist	Macro	• Deviance is both functional and dysfunctional. • Anomie increases the likelihood of deviance. • People are deviant when they experience blocked opportunities to achieve the culturally approved goal of economic success.
Conflict	Macro	• There's a strong association between capitalism, social inequality, power, and deviance. • The most powerful groups define what's deviant. • Laws rarely punish the illegal activities of the powerful.
Feminist	Macro and micro	• There's a large gender gap in deviant behavior. • Women's deviance reflects their general oppression due to social, economic, and political inequality. • Many women are offenders or victims because of patriarchal beliefs and practices and living in a rape culture.
Symbolic Interactionist	Micro	• Deviance is socially constructed. • People learn deviant behavior from significant others such as parents and friends. • If people are labeled or stigmatized as deviant, they're likely to develop negative self-concepts and engage in criminal behavior.

CHAPTER 8 SUMMARY

8-1 Social Stratification Systems and Bases

Social stratification is a society's ranking of people who have different access to valued resources. A *closed stratification system* differs from an *open stratification system* because the latter allows movement from one social class to another. Stratification includes *wealth, prestige,* and *power*. People are more likely to experience status inconsistency if they rank differently on these three dimensions, such as a football player who has great wealth but little power.

8-2 Social Class in America

A good indicator of social class is *socioeconomic status (SES)*, an overall rank based on a person's income, education, and occupation. Using SES and other variables (such as values, power, and *conspicuous consumption*), most sociologists agree that there are at least four social classes in the United States: upper, middle, working, and lower. These groups can be divided further into upper-upper, lower-upper, upper-middle, lower-middle, and the working class. The lower class includes the *working poor* and the *underclass*. A major outcome of social stratification is *life chances*.

Example: Restaurant Menus and Stratification

Two sociologists—in Iowa and Virginia—asked their students in introductory sociology classes to do a content analysis (see Chapter 1) of 10 menus that represented a sampling of restaurants by social class. The students found that the restaurants that catered to upper-class clientele had higher than average entrée prices, described the entrées in foreign languages, used fancy sauces, recommended expensive wines, and had few illustrations. Middle-class menus emphasized "value for the dollar," presented photos of entrées with "bountiful plates overflowing with appetizing food," and popular items such as quesadillas. Menus at lower-class restaurants featured low prices ($3 to $10 entrées), the items were numbered, none of the entrées had "pretentious names," and the typesetting was simple (Wright and Ransom, 2005). In effect, then, even menus denote social class and social status.

8-3 Poverty

Absolute poverty is a more serious social problem than *relative poverty* because millions of Americans live below the *poverty line*. Poverty varies by age, gender, family structure, race, and ethnicity. One explanation for poverty maintains that people are poor because of individual failings; another contends that societal structures create and sustain poverty.

8-4 Social Mobility

Social mobility can be *horizontal, vertical, intragenerational,* or *intergenerational*. There has been a sharp decline in the share of middle-income adults since 1971 because many have slid into a lower class. Structural, demographic, and individual factors affect a person's social mobility. Much social mobility depends on structural, demographic, and family background factors, all of which are interrelated.

8-5 Global Stratification

Global inequality is widespread, and all societies are stratified, but some countries are much wealthier than others, and inequality is increasing across the globe. Historically, currently, and across all nations, women and children experience the greatest poverty. Sociologists use *modernization theory, dependency theory,* and *world-system theory* to explain why inequality is universal.

8-6 Sociological Explanations: Why There Are Haves and Have-Nots

See Table 8.1 on the next page.

KEY TERMS

social stratification a society's ranking of people based on their access to valued resources such as wealth, power, and prestige.

slavery system people own others as property and have almost total control over their lives.

caste system people's positions are ascribed at birth and largely fixed.

class system people's positions are based on both birth and achievement.

social class people who have a similar standing or rank in a society based on wealth, education, power, prestige, and other valued resources.

wealth economic assets that a person or family owns.

income the money a person receives, usually through wages or salaries, but can also include other earnings.

prestige respect, recognition, or regard attached to social positions.

power the ability to influence or control the behavior of others despite opposition.

socioeconomic status (SES) an overall ranking of a person's position in society based on income, education, and occupation.

working poor people who work at least 27 weeks a year but whose wages fall below the official poverty level.

underclass people who are persistently poor and seldom employed, residentially segregated, and relatively isolated from the rest of the population.

life chances the extent to which people have positive experiences and can secure the good things in life because they have economic resources.

absolute poverty not having enough money to afford the basic necessities of life.

relative poverty not having enough money to maintain an average standard of living.

CHAPTER REVIEW 8

poverty line the minimal income level that the federal government considers necessary for basic subsistence (also called the *poverty threshold*).

feminization of poverty the disproportionate number of the poor who are women.

social mobility movement from one social class to another.

intragenerational mobility movement up or down a social class over one's lifetime.

intergenerational mobility movement up or down a social class over two or more generations.

global stratification worldwide inequality patterns that result from differences in wealth, power, and prestige.

infant mortality rate the number of babies under age 1 who die per 1,000 live births in a given year.

Davis–Moore thesis the functionalist view that social stratification benefits a society.

meritocracy a belief that social stratification is based on people's accomplishments.

bourgeoisie those who own and control capital and the means of production.

proletariat workers who sell their labor for wages.

corporate welfare subsidies, tax breaks, and assistance that the government has created for businesses.

TEST YOUR LEARNING

1. Slavery and castes are ____ systems.
   a. capitalist
   b. open stratification
   c. closed stratification
   d. socialist

2. A person's socioeconomic status (SES) is based on
   a. income, education, and occupation.
   b. wealth, prestige, and power.
   c. wealth, prestige, and lifestyle.
   d. education, income, and age.

3. The economic gap between the wealthy and the poor
   a. fluctuates from year to year.
   b. is about the same as during the 1990s.
   c. is decreasing.
   d. is increasing.

4. Which of the following is NOT a major structural factor in social mobility?
   a. Changes in the economy
   b. Immigration patterns
   c. Consumer confidence
   d. Number of available positions in given occupations

5. ____ claim that social stratification ultimately benefits society.
   a. Functionalists
   b. Conflict theorists
   c. Feminist theorists
   d. Symbolic interactionists

6. ***True or False*** One of the criticisms of conflict theory is that it ignores structural factors in explaining stratification.

7. ***True or False*** Conspicuous consumption displays one's social status and enhances one's prestige.

8. ***True or False*** Intergenerational mobility refers to moving up or down the class hierarchy over one's lifetime.

9. ***True or False*** The Davis–Moore thesis is a symbolic interactionist perspective.

10. ***True or False*** According to world-system theory, the global economic system helps richer countries stay rich while poorer countries remain poor.

11. Describe your current social class and explain why you place yourself in this class. Ten years from now, do you expect to experience upward or downward mobility? Explain why.

12. Should public assistance recipients be required to work, regardless of their education level? If no, why not? If yes, should the government subsidize preschool and afterschool care for their children while the parent(s) is/are at work? In your answer, incorporate the sociological perspectives on social stratification.

1. c 2. a 3. d 4. c 5. a 6. False 7. True 8. False 9. False 10. True

TABLE 8.1	SOCIOLOGICAL EXPLANATIONS OF SOCIAL STRATIFICATION	
Perspective	**Level of Analysis**	**Key Points**
Functionalist	Macro	• Fills social positions that are necessary for a society's survival • Motivates people to succeed and ensures that the most qualified people will fill the most important positions
Conflict	Macro	• Encourages workers' exploitation and promotes the interests of the rich and powerful • Ignores a wealth of talent among the poor
Feminist	Macro and micro	• Constructs numerous barriers in patriarchal societies that limit women's achieving wealth, status, and prestige • Requires most women, not men, to juggle domestic and employment responsibilities that impede upward mobility
Symbolic Interactionist	Micro	• Shapes stratification through socialization, everyday interaction, and group membership • Reflects social class identification through symbols, especially products that signify social status

CHAPTER 9 SUMMARY

9-1 Sex, Gender, and Culture

Sex refers to biological characteristics, whereas *gender* refers to learned attitudes and behaviors. *Gender roles* differ depending on whether people perceive themselves as masculine or feminine and because a society expects women and men to think and behave differently. Many Americans still have *gender stereotypes* about how people will look, act, think, and feel based on their sex.

Our sexual identity incorporates a *sexual orientation* that can be *homosexual*, *heterosexual*, *bisexual*, or *asexual*. Many biological theories maintain that sexual orientation has a strong genetic basis, but social constructionists argue that culture, not biology, plays a large role in forming people's *gender identity*.

9-2 Contemporary Gender Stratification

Sexism is widespread because of *gender stratification*, which can lead to inequality in the family (particularly child care and housework), education, the workplace (as in *occupational sex segregation* and a *gender pay gap*), and politics. *Sexual harassment* and workplace bullying are also common in the workplace.

9-3 Sexuality

Sexual attitudes and behavior can vary from situation to situation and change over time, including why we have sex. Contrary to some stereotypes, adolescents aren't sexually promiscuous and older people aren't asexual. All of us have internalized *sexual scripts*, and *sexual double standards* persist.

Example: Gender Roles and Hooking Up—Are Women the Losers?

Hooking up refers to physical encounters, no strings attached,

and can mean anything from kissing and genital fondling to oral sex and sexual intercourse. The prevalence of hooking up has increased only slightly since the late 1980s, but is now more common than dating at many high schools and colleges. Hooking up has its advantages. It's much cheaper than dating, and requires no commitment of time or emotion. In addition, hookups remove the stigma from those who can't get dates but can experience sexual pleasure, and they make people feel sexy and desirable. Hooking up also has disadvantages, especially for women, because they generally get a bad reputation as being "easy" (see Chapter 9). In effect, then, and despite the advantages of hooking up, the sexual double standard persists.

9-4 Some Current Social Issues about Sexuality

Two of the most controversial and politically contested issues continue to be *abortion* and *same-sex marriage*. Almost equal percentages of Americans support or condemn abortion. Those who favor same-sex marriage argue that people should have the same legal rights regardless of sexual orientation; those who oppose same-sex marriage contend that such unions are immoral and contrary to religious beliefs.

9-5 Gender and Sexuality across Cultures

Worldwide, women have fewer rights and opportunities than men. In all countries and regions, the greatest gender gaps are in economic opportunity and participation and political leadership. People in some countries are more accepting of homosexuality than in the past, but *heterosexism* prevails.

9-6 Sociological Explanations of Gender and Sexuality

See Table 9.4 on the next page.

KEY TERMS

sex the biological characteristics with which we are born.

gender learned attitudes and behaviors that characterize women and men.

intersexuals people whose medical classification at birth isn't clearly either male or female.

sexual identity an awareness of ourselves as male or female and how we express our sexual values, attitudes, and feelings.

sexual orientation a preference for sexual partners of the same sex, of the opposite sex, of both sexes, or neither sex.

homosexuals those who are sexually attracted to people of the same sex.

heterosexuals those who are sexually attracted to people of the opposite sex.

bisexuals those who are sexually attracted to both sexes.

asexuals those who lack any interest in or desire for sex.

gender identity a perception of oneself as either masculine or feminine.

transgender an umbrella term for people whose gender identity and behavior differ from the sex to which they were assigned at birth.

gender expression the way a person communicates gender identity to others through behavior, clothing, hairstyles, voice, or body characteristics.

gender roles the characteristics, attitudes, feelings, and behaviors that society expects of females and males.

gender stereotypes expectations about how people will look, act, think, and feel based on their sex.

sexism an attitude or behavior that discriminates against one sex, usually females, based on the assumed superiority of the other sex.

heterosexism belief that heterosexuality is the only legitimate sexual orientation.

homophobia a fear and hatred of lesbians and gays.

gender stratification people's unequal access to wealth, power, status, prestige, and other valued resources because of their sex.

occupational sex segregation (sometimes called *occupational gender segregation*) the process of channeling women and men into different types of jobs.

gender pay gap the overall income difference between women and men in the workplace (also called the *wage gap*, *pay gap*, and *gender wage gap*).

sexual harassment any unwanted sexual advance, request for sexual favors, or other conduct of a sexual nature that makes a person uncomfortable and interferes with her or his work.

sexual script specifies the formal and informal norms for acceptable or unacceptable sexual behavior.

sexual double standard a code that permits greater sexual freedom for men than women.

abortion the expulsion of an embryo or fetus from the uterus.

same-sex marriage (also called *gay marriage*) a legally recognized marriage between two people of the same biological sex and/or gender identity.

TEST YOUR LEARNING

1. ____ refers to learned attitudes and behaviors, whereas ____ refers to biological characteristics with which we are born.
   a. Sex; gender
   b. Gender; sex
   c. Sex; gender roles
   d. Sexual identity; gender stereotypes

2. Generally, the gender pay gap increases as the level of educational attainment
   a. increases.
   b. decreases.
   c. neither of the above because it depends on how hard a person works
   d. neither a nor b because such data are not available

3. Sean, age 16, lies to his friends about the frequency of his sexual activity and new sexual exploits every week. Sean is
   a. being a normal male.
   b. following a sexual script.
   c. falling into a pathologic life course.
   d. challenging gender norms.

4. ____ is the belief that heterosexuality is superior to and more natural than homosexuality or bisexuality.
   a. Heterobiology
   b. Homosexism
   c. Heterosexism
   d. Bisexism

5. ____ posit that gender inequality and sexuality are socially constructed.
   a. Functionalists
   b. Conflict theorists
   c. Feminist theorists
   d. Symbolic interactionists

6. *True or False* Sexism can only be enacted by men against women.

7. *True or False* Sexual harassment is the fastest growing type of employment discrimination.

8. *True or False* U.S. abortion rates have been increasing.

9. *True or False* A majority of Americans support same-sex marriage.

10. *True or False* Conflict theorists believe that capitalism increases gender inequality.

11. How do sex and gender differ? How do they affect each other? "Our sexual behavior is spontaneous." Explain why you agree or disagree with this statement.

12. Describe how gender stratification affects women and men in family life, education, the workplace, and politics. Provide examples to illustrate your answer.

1. b 2. a 3. b 4. c 5. d 6. False 7. True 8. False 9. True 10. True

TABLE 9.4	SOCIOLOGICAL EXPLANATIONS OF GENDER AND SEXUALITY	
Theoretical Perspective	**Level of Analysis**	**Key Points**
Functionalist	Macro	• Gender roles are complementary, equally important for a society's survival, and affect human capital. • Agreed-on sexual norms contribute to a society's order and stability.
Conflict	Macro	• Gender roles give men power to control women's lives. • Most societies regulate women's, but not men's, sexual behavior.
Feminist	Macro and micro	• Women's inequality reflects their historical and current domination by men, especially in the workplace. • Many men use violence—including sexual harassment, rape, and global sex trafficking—to control women's sexuality.
Symbolic Interactionist	Micro	• Gender is a social construction that emerges and is reinforced through everyday interactions. • The social construction of sexuality varies across cultures because of societal norms and values.

CHAPTER 10 SUMMARY

10-1 U.S. Racial and Ethnic Diversity

Perhaps the most multicultural country in the world, the United States includes about 150 distinct ethnic or racial groups among more than 312 million inhabitants. By 2025, only 58 percent of the U.S. population will be white—down from 86 percent in 1950.

10-2 The Social Significance of Race and Ethnicity

Race refers to physical characteristics, whereas an *ethnic group* identifies with a common national origin or cultural heritage. A *racial-ethnic group* has both distinctive physical and cultural characteristics.

10-3 Our Changing Immigration Mosaic

In 1900, almost 85 percent of immigrants came from Europe; now immigrants come primarily from Asia and Latin America. Many Americans are ambivalent about immigrants, especially those who are in the country illegally. Many scholars argue, however, that, in the long run, both legal and undocumented immigrants bring more benefits than costs.

10-4 Dominant and Minority Groups

A *dominant group* has more economic and political power than a *minority*. The latter may be larger in number than a dominant group but is often subject to differential and unequal treatment because of its physical, cultural, or other characteristics. Patterns of dominant-minority group relations include *genocide*, *segregation*, *acculturation*, *assimilation*, and *pluralism*.

10-5 Some Sources of Racial-Ethnic Friction

Racism justifies and preserves the social, economic, and political interests of dominant groups. *Prejudice* is an attitude; *discrimination* is an act that occurs at both the individual and institutional level. All of us can be prejudiced, but minorities are typically targets of *stereotypes* and *ethnocentrism* that often lead to *scapegoating*.

Example: Stuff White People Like

The popular blog *Stuff White People Like* has generated clones (e.g., *Stuff Educated Black People Like* and *Stuff Asian People Like*). Why are these sites so popular? Many fans say that the descriptions are funny because they're true. According to some critics, however, by poking fun at privileged upper-middle-class white people, the sites fuel stereotypes instead of having painfully frank discussions about U.S. race and racism (Sternbergh, 2008). Do you agree or disagree?

10-6 Major Racial and Ethnic Groups in the United States

European Americans, who settled the first colonies, are declining in population, whereas Latinos now comprise 17 percent of the population. Other large racial and ethnic populations are African Americans (13 percent), Asian Americans (6 percent), and American Indians (almost 2 percent). Middle Eastern Americans, who comprise less than 0.05 percent of the population, come from more than 30 countries. All of these groups have experienced prejudice and discrimination, but they have enhanced U.S. society and culture.

10-7 Sociological Explanations of Racial-Ethnic Inequality

See Table 10.4 on the next page.

10-8 Interracial and Interethnic Relationships

About 97 percent of Americans report being only one race, but the numbers of biracial children are increasing because of interracial dating and marriage. The rise of racial-ethnic intermarriage reflects many micro and macro factors such as greater interethnic and interracial contact, changing attitudes, and acculturation.

KEY TERMS

racial group people who share visible physical characteristics that members of a society consider socially important.

ethnic group people who identify with a common national origin or cultural heritage.

racial-ethnic group people who have distinctive physical and cultural characteristics.

dominant group any physically or culturally distinctive group that has the most economic and political power, the greatest privileges, and the highest social status.

minority a group of people who may be treated differently and unequally because of their physical, cultural, or other characteristics.

genocide the systematic effort to kill all members of a particular ethnic, religious, political, racial, or national group.

segregation the physical and social separation of dominant and minority groups.

acculturation the process of adopting the language, values, beliefs, and other characteristics of the host culture.

assimilation conforming to the dominant group's culture, adopting its language and values, and intermarrying with that group.

pluralism minority groups maintain many aspects of their original culture while living peacefully with the host culture.

racism a set of beliefs that one's own racial group is inherently superior to other groups.

prejudice an attitude that prejudges people, usually in a negative way.

stereotype an oversimplified or exaggerated generalization about a group of people.

CHAPTER REVIEW 10

scapegoats individuals or groups whom people blame for their own problems or shortcomings.

discrimination behavior that treats people unequally or unfairly because of their group membership.

individual discrimination harmful action on a one-to-one basis by a dominant group member against someone in a minority group.

institutional discrimination unequal treatment and fewer opportunities because of a society's everyday laws, policies, practices, and customs.

gendered racism the overlapping and cumulative effects of inequality due to racism *and* sexism.

contact hypothesis posits that the more people get to know members of a minority group personally, the less likely they are to be prejudiced against that group.

miscegenation marriage or sexual relations between a man and a woman of different races.

TEST YOUR LEARNING

1. What's the difference between race and ethnicity?
   a. Race is a negative term that refers to difference; ethnicity is the politically correct term that refers to difference.
   b. Race refers to people's physical characteristics; ethnicity refers to people's cultural heritage.
   c. Race refers to people around the world; ethnicity refers only to a group of people within a specific nation.
   d. There is no difference between race and ethnicity; the terms are interchangeable.

2. Today most immigrants to the United States come from which of the following countries?
   a. China and Mexico
   b. Mexico and Canada
   c. England and Canada
   d. Japan and Russia

3. _____ is the physical and social separation of dominant and minority groups.
   a. Racism
   b. Internal colonialism
   c. Segregation
   d. Institutional discrimination

4. Tyrone and Jeanette were up for the same job. Tyrone told his friends that he didn't get the job not because Jeanette was more qualified but because "women are always hired first." This is an example of
   a. scapegoating.
   b. pluralism.
   c. stereotyping.
   d. prejudice.

5. Laws against miscegenation were overturned nationally in
   a. 1807.
   b. 1867.
   c. 1907.
   d. 1967.

6. ***True or False*** On the continuum of dominant and minority group relations, genocide reflects the least tolerance, whereas pluralism illustrates the greatest tolerance.

7. ***True or False*** Ethnocentrism is the belief that one's own racial group is naturally inferior to others.

8. ***True or False*** Asian Americans are the fastest growing U.S. minority group.

9. ***True or False*** American Indians are on the verge of vanishing from the American landscape.

10. ***True or False*** The lower the educational level, the greater the likelihood that a person will have an interracial marriage.

11. What's the difference between individual discrimination and institutional discrimination? Is one more powerful than the other in its effect on minority groups? If so, which one and why? If not, why not?

12. Choose any two minorities discussed in this chapter. Compare them in terms of factors such as population size, diversity, constraints, and economic success.

1. b 2. a 3. c 4. a 5. d 6. True 7. False 8. False 9. False 10. False

Theoretical Perspective	Level of Analysis	Key Points
Functionalist	Macro	Immigration provides needed workers; acculturation and assimilation increase social solidarity; racial-ethnic inequality can be dysfunctional, but benefits dominant groups.
Conflict	Macro	There's ongoing strife between dominant and minority groups; powerful groups maintain their advantages primarily through economic exploitation; race is a more important factor than social class in perpetuating racial-ethnic inequality.
Feminist	Macro and micro	Minority women suffer from the combined effects of racism and sexism; gendered racism occurs within and across racial-ethnic groups.
Symbolic Interactionist	Micro	Because race and ethnicity are socially constructed, social interaction can increase or reduce racial and ethnic hostility; antagonistic attitudes toward minorities, which are learned, can be lessened through cooperative interracial and interethnic contacts.

TABLE 10.3 SOCIOLOGICAL EXPLANATIONS OF RACIAL-ETHNIC INEQUALITY

CHAPTER 11 SUMMARY

11-1 Global Economic Systems

The *economy* determines how a society produces, distributes, and consumes goods and services. *Capitalism* frequently spawns *monopolies* and *oligopolies* that dominate the market and discourage competition. *Socialism* and *communism* promise cooperation and collective ownership of property, but all countries have mixed economies that allow private profits.

11-2 Corporations and the Economy

A *corporation*, usually created for profit, often forms a *conglomerate*. Both corporations and conglomerates are governed by *interlocking directorates* that have become more powerful than ever because of the growth of *transnational corporations* and *transnational conglomerates*.

11-3 Work in U.S. Society Today

Work produces goods or services. Many Americans have been casualties of *deindustrialization* and *globalization*. Others have lost their jobs to *offshoring*. Widespread *downsizing* has resulted in unemployment, *underemployment*, discouraged workers, and low-paid jobs.

11-4 Sociological Explanations of Work and the Economy

See Table 11.3 on the next page.

11-5 Global Political Systems

In a *democracy*, ideally, citizens have a high degree of control over the state. *Totalitarianism* controls people's lives; *authoritarianism* generally permits some degree of individual freedom. A *monarchy* is the oldest type of authoritarian regime.

11-6 Politics, Power, and Authority

Politics includes *power* and *authority*. *Legitimate power* comes from having a role, position, or title that people accept as legal and appropriate; *coercive power* relies on force or threat of force to impose one's will on others. Authority can be based on *tradition, charisma, rational-legal power*, or a combination of these sources.

11-7 Politics and Power in U.S. Society

A *political party* tries to influence and control government. Whereas political parties include diverse individuals, *special-interest groups* are usually made up of people who are similar in social class and political objectives. The most powerful special-interest groups are *political action committees* and *lobbyists*. The voting rate is much higher among some groups than others. Situational and structural factors, such as voter registration, requiring proof of citizenship, and convenience can also encourage or discourage voting.

11-8 Sociological Perspectives on Politics and Power

See Table 11.7 on the next page.

---

interlocking directorate the same people serve on the boards of directors of several companies or corporations.

transnational corporation (also called a *multinational corporation* or an *international corporation*) a large company that's based in one country but operates across international boundaries.

transnational conglomerate (also called a *multinational conglomerate*) a corporation that owns a collection of different companies in various industries in a number of countries.

work a physical or mental activity that produces goods or services.

deindustrialization a process of social and economic change resulting from the reduction of industrial activity, especially manufacturing.

globalization the growth and spread of investment, trade, production, communication, and new technology around the world.

offshoring sending work or jobs to another country to cut a company's costs at home.

discouraged workers people who stop searching for employment because they believe that job hunting is futile.

underemployed people who have part-time jobs but want full-time work or whose jobs are below their experience, skill, and education levels.

motherhood penalty (also called *motherhood wage penalty* or *mommy penalty*) a pay gap between women who are and aren't mothers.

government a formal organization that has the authority to make and enforce laws.

democracy a political system in which, ideally, citizens have control over the state and its actions.

totalitarianism the government controls almost every aspect of people's lives.

authoritarianism the state controls the lives of citizens but permits some degree of individual freedom.

monarchy power is allocated solely on the basis of heredity and passes from generation to generation.

power the ability of a person or group to influence others, even if they resist.

authority the legitimate use of power.

traditional authority power based on customs that justify the ruler's position.

charismatic authority power based on exceptional individual abilities and characteristics that inspire devotion, trust, and obedience.

KEY TERMS

economy determines how a society produces, distributes, and consumes goods and services.

politics individuals and groups acquire and exercise power and authority and make decisions.

capitalism an economic system based on private ownership of property and competition in producing and selling goods and services.

monopoly domination of a particular market or industry by one person or company.

oligopoly domination of a market by a few large producers or suppliers.

socialism an economic system based on the public ownership of the production of goods and services.

communism a political and economic system in which property is communally owned and all people are considered equal.

welfare capitalism (also called *state capitalism*) an economic system that combines private ownership of property, market competition, and a government's regulation of many programs and services.

corporation an organization that has legal rights, privileges, and liabilities apart from those of its members.

conglomerate a corporation that owns a collection of companies in different industries.

CHAPTER REVIEW 11

rational-legal authority power based on the belief that laws and appointed or elected political leaders are legitimate.

political party an organization that tries to influence and control government by recruiting, nominating, and electing its members to public office.

special-interest group (also called an *interest group*) a group of people that seeks or receives benefits or special treatment.

political action committee (PAC) a special-interest group that raises money to elect one or more candidates to public office.

lobbyist someone hired by a special-interest group to influence legislation on the group's behalf.

pluralism a political system in which power is distributed among a variety of competing groups in a society.

power elite a small group of influential people who make the nation's major political decisions.

TEST YOUR LEARNING

1. _____ is a market system dominated by a few large producers or suppliers.
   a. Capitalism
   b. Monopoly
   c. Oligopoly
   d. Socialism

2. Felipe serves on the board of directors of several U.S. companies. He's part of
   a. an interlocking directorate.
   b. a transnational corporation.
   c. a conglomerate.
   d. a monopoly.

3. Which of the following is NOT a major reason for the surge in women's employment since the 1970s?
   a. Men's falling wages
   b. Women's increased education
   c. Rising cost of home ownership
   d. Decrease in sexism

4. According to Max Weber, which one of the following types of authority is most characteristic of the majority of U.S. presidents?
   a. Charismatic
   b. Absolute
   c. Traditional
   d. Rational-legal

5. _____ claim that the U.S. political structure is pluralistic.
   a. Functionalists
   b. Conflict theorists
   c. Feminist theorists
   d. Symbolic interactionists

6. **True or False** Compared with low-wage U.S. jobs, high-wage jobs are relatively safe from offshoring.

7. **True or False** Deindustrialization has spread because of globalization.

8. **True or False** Asian Americans have higher voting rates than Latinos.

9. **True or False** Some of the best-paid lobbyists are retired corporate executives (CEOs).

10. **True or False** In the United States, most Republicans believe that the federal government should provide social programs for its citizens.

11. Have you personally benefited or been harmed by globalization? Be specific in explaining your answer.

12. Suppose someone asked you to explain U.S. political power. Are you more likely to use a functionalist, conflict, or feminist perspective? A combination? Or none? Explain your answer.

1. c 2. a 3. a 4. d 5. a 6. False 7. True 8. False 9. False 10. False

---

TABLE 11.3 SOCIOLOGICAL EXPLANATIONS OF WORK AND THE ECONOMY

Theoretical Perspective	Level of Analysis	Key Points
Functionalist	Macro	Capitalism benefits society; work provides an income, structures people's lives, and gives them a sense of accomplishment.
Conflict	Macro	Capitalism enables the rich to exploit other groups; most jobs are low-paying, monotonous, and alienating; productivity isn't always rewarded.
Feminist	Macro and micro	Gender roles structure women's and men's work experiences differently and inequitably.
Symbolic Interactionist	Micro	How people define and experience work in their everyday lives affects their workplace behavior and relationships with coworkers and employers.

TABLE 11.7 SOCIOLOGICAL EXPLANATIONS OF POLITICAL POWER

	Functionalism: A Pluralist Model	Conflict Theory: A Power Elite Model	Feminist Theories: A Patriarchal Model
Who has political power?	The people	Rich upper-class people—especially those at top levels in business, government, and the military	White men in Western countries; most men in traditional societies
How is power distributed?	Very broadly	Very narrowly	Very narrowly
What is the source of political power?	Citizens' participation	Wealthy people in government, business corporations, the military, and the media	Being white, male, and very rich
Does one group dominate politics?	No	Yes	Yes
Do political leaders represent the average person?	Yes, the leaders speak for a majority of the people.	No, the leaders are most concerned with keeping or increasing their personal wealth and power.	No, the leaders are rarely women who have decision-making power.

CHAPTER 12 SUMMARY

12-1 What Is a Family?

Among other activities, the members of a *family* care for one another and any children. Worldwide, however, families vary in characteristics such as structure (a *nuclear* or an *extended* family), living arrangements (*patrilocal*, *matrilocal*, or *neolocal*), who has authority (*matriarchal*, *patriarchal*, or *egalitarian*), and how many marriage mates a person can have (*monogamy* or *polygamy*).

12-2 How U.S. Families Are Changing

The United States has one of the highest marriage and divorce rates in the world. *Divorce* is easier to obtain than in the past because all states now have *no-fault divorce* laws, but divorce rates vary by race, ethnicity, and social class. The number of single people has risen greatly, primarily because many people are postponing marriage. There has also been a striking increase in *cohabitation* and nonmarital births. Female-headed single-parent families are increasingly common.

12-3 Family Conflict and Violence

We're more likely to experience *intimate partner violence (IPV)* than assault by a stranger. IPV causes are complex and often the product of multiple individual, demographic, and societal factors. *Child maltreatment* is widespread, and parents

are the most common offenders. Similarly, in *elder abuse*, most of the offenders are adult children, spouses, or other family members. Both micro- and macro-level reasons help explain IPV, child maltreatment, and elder abuse and neglect.

12-4 Our Aging Society

How people define "old" varies across societies. There are also significant differences between the *young-old*, the *old-old*, and the *oldest-old* in their health, ability to live independently, and to work. Historically and currently, women live longer than men, but life expectancy varies by race, ethnicity, and social class. Our *old-age dependency ratio* has increased, placing a larger burden on the working-age population. Some important current aging issues include the rise of *multigenerational households*, physician-assisted suicide, and competition for scarce resources.

Example: "He Gets Prettier; I Get Older"

When comparing her own public image with that of her actor husband, the late Paul Newman, actress Joanne Woodward once remarked, "He gets prettier; I get older." Was she right? If aging gracefully is acceptable, why do many companies tout antiaging products? And why are the products targeted primarily at women?

12-5 Sociological Explanations of Family and Aging

See Table 12.3 on the next page.

KEY TERMS

family an intimate group consisting of two or more people who (1) have a committed relationship, (2) care for one another and any children, and (3) share close emotional ties and functions.

incest taboo cultural norms and laws that forbid sexual intercourse between close blood relatives.

marriage a socially approved mating relationship that people expect to be stable and enduring.

endogamy (often used interchangeably with *homogamy*) cultural practice of marrying within one's group.

exogamy (often used interchangeably with *heterogamy*) cultural practice of marrying outside one's group.

nuclear family a family form composed of married parents and their biological or adopted children.

extended family a family form composed of parents, children, and other kin.

patrilocal residence pattern newly married couples live with the husband's family.

matrilocal residence pattern newly married couples live with the wife's family.

neolocal residence pattern a newly married couple sets up its own residence.

boomerang generation young adults who move back into their parents' home or never leave it in the first place.

matriarchal family system the oldest females control cultural, political, and economic resources and, consequently, have power over males.

patriarchal family system the oldest males control cultural, political, and economic resources and, consequently, have power over females.

egalitarian family both partners share power and authority fairly equally.

marriage market prospective spouses compare the assets and liabilities of eligible partners and choose the best available mate.

arranged marriage parents or relatives choose the children's spouses.

monogamy one person is married exclusively to another person.

serial monogamy individuals marry several people, but one at a time.

polygamy a man or woman has two or more spouses.

CHAPTER REVIEW 12

cohabitation two unrelated people aren't married but live together and are in a sexual relationship.

dating cohabitation a couple that spends a great deal of time together decides to move in together.

premarital cohabitation a couple lives together before getting married.

fictive kin nonrelatives who are accepted as part of a family.

intimate partner violence (IPV) abuse that occurs between people in a close relationship.

child maltreatment (also called *child abuse*) a broad range of behaviors that place a child at serious risk or result in serious harm, including physical and sexual abuse, neglect, and emotional mistreatment

elder abuse (sometimes called *elder mistreatment*) any knowing, intentional, or negligent act by a caregiver or other person that causes harm to people age 65 or older.

life expectancy the average expected number of years of life remaining at a given age.

baby boomers people born between 1946 and 1964.

old-age dependency ratio the number of people age 65 and older who aren't in the labor force relative to the number of working-age adults ages 18 to 64.

activity theory proposes that many older people remain engaged in numerous roles and activities, including work.

exchange theory posits that people seek through their social interactions to maximize their rewards and minimize their costs.

ageism discrimination against older people.

continuity theory posits that older adults can substitute satisfying new roles for those they've lost.

TEST YOUR LEARNING

1. This chapter defines a *family* in terms of three criteria. Which of the following is NOT one of the criteria?
   a. Relationships are legally sanctioned by the state
   b. Live together in a committed relationship
   c. Care for one another and any children
   d. Share close emotional ties and bonds

2. The family serves many important functions in a society. Which of the following functions does the family NOT serve?
   a. Sexual regulation
   b. Social placement
   c. Peer approval
   d. Economic security

3. Worldwide, the most common residence pattern is
   a. patriarchal.
   b. matriarchal.
   c. neolocal.
   d. egalitarian.

4. Which of the following is NOT a common factor in intimate partner violence (IPV)?
   a. Income level
   b. Employment status
   c. Drug abuse
   d. Number of children

5. _____ use both a macro and a micro approach in examining families and aging.
   a. Functionalists
   b. Conflict theorists
   c. Feminist theorists
   d. Symbolic interactionists

6. *True or False* Jack marries several people, but one at a time. Jack is a serial monogamist.

7. *True or False* U.S. divorce rates have decreased since 1980.

8. *True or False* College-educated women are less likely to cohabit than women with a high school diploma.

9. *True or False* Most child maltreatment offenders are relatives and unrelated caregivers such as foster parents and boyfriends.

10. *True or False* Baby boomers are the fastest growing segment of the U.S. population.

11. "U.S. families are falling apart." Explain whether you agree or disagree with this statement, using functionalist, conflict, feminist, and symbolic interactionist perspectives.

12. What do you think your life will be like at age 85? How might your options be different because of your sex, social class, race, and ethnicity? Support your answer with specific examples.

1. a 2. c 3. a 4. d 5. c 6. True 7. True 8. False 9. True 10. False

TABLE 12.3	SOCIOLOGICAL PERSPECTIVES ON FAMILIES AND AGING	
Theoretical Perspective	**Level of Analysis**	**Key Points**
Functionalist	Macro	Families are important in maintaining societal stability and meeting family members' needs. Older people who are active and engaged are more satisfied with life.
Conflict	Macro	Families promote social inequality because of social class differences. Many corporations view older workers as disposable.
Feminist	Macro and micro	Families both mirror and perpetuate patriarchy and gender inequality. Women have an unequal burden in caring for children as well as older family members and relatives.
Symbolic Interactionist	Micro	Families construct their everyday lives through interaction and subjective interpretations of family roles. Many older family members adapt to aging and often maintain previous activities.

CHAPTER 13 SUMMARY

13-1 Education and Society

U.S. *education* and *schooling* have changed in four important ways: Universal education has expanded, community colleges have flourished, public higher education has burgeoned, and student diversity has greatly increased.

13-2 Sociological Perspectives on Education

See Table 13.1 on the next page.

13-3 Some Current Issues in U.S. Education

Compared with their counterparts in a number of other countries, many U.S. students are performing poorly in elementary and high schools—especially in mathematics and the sciences. U.S. teachers' salaries are low, many are poorly prepared, out of field, and have less control over curricula than ever before. A growing number of critics contend that standardized tests (such as SATs and ACTs) are little more than gatekeeping tools that exclude lower socioeconomic students from access to higher education, and that the tests scores don't predict college success. Despite grade inflation, high school and college dropout rates are high, and cheating is widespread.

13-4 Religion and Society

Some form of religion exists in all societies and cultures. *Religion* unites believers into a community. Every known society distinguishes between *sacred* and *profane (secular)* activities. Religion, *religiosity*, and spirituality differ; for example, people who describe themselves as religious may not attend services.

Example: Sacred vs. Secular

The corporate mission of Chick-fil-A (a franchise that prepares sandwiches),

as stated on a plaque at company headquarters, is "to glorify God." Chick-fil-A is the only national fast-food chain that closes on Sunday so employees can go to church, and prospective employees are asked about their religious activities. Some franchise operators are delighted with the religious emphasis; others believe that a business should stay out of its workers' personal lives (Schmall, 2007). What do *you* think?

13-5 Religious Organization and Major World Religions

People express their religious beliefs most commonly through organized groups, including *cults* (also called *new religious movements [NRMs]*), *sects*, *denominations*, and *churches*. Some NRMs, which usually are organized around a *charismatic leader*, have been short-lived, whereas others have become established religions with highly organized bureaucracies. Worldwide, the largest religious group is Christians, followed by Muslims, but no religious group comes close to being a global majority. The third largest group is people who don't have a religious affiliation.

13-6 Religion in the United States

Religion in the United States is complex and diverse. About half of U.S. adults have changed their religion since childhood, many opting for no religion at all. Mainline Protestant groups have declined whereas evangelicals have surged. Religious participation varies by gender, age, race, ethnicity, and social class. Some sociologists maintain that *secularization* is increasing rapidly in the United States; others contend that this claim is greatly exaggerated, especially as witnessed by the growth of *fundamentalism* and the prevalence of *civil religion*.

13-7 Sociological Perspectives on Religion

See Table 13.5 on the next page.

KEY TERMS

education a social institution that transmits attitudes, knowledge, beliefs, values, norms, and skills.

schooling formal training and instruction provided in a classroom setting.

achievement gap the difference in academic performance that shows up in grades, standardized test scores, and college completion rates.

hidden curriculum school practices that transmit nonacademic knowledge, values, attitudes, norms, and beliefs.

credentialism an emphasis on certificates or degrees to show that people have certain skills, educational attainment levels, or job qualifications.

tracking (also called *streaming* or *ability grouping*) assigning students to specific educational programs and classes on the basis of test scores, previous grades, or perceived ability.

meritocracy a system that rewards people because of their individual accomplishments.

religion a social institution that involves shared beliefs, values, and practices related to the supernatural.

sacred anything that people see as awe-inspiring, supernatural, holy, and not part of the natural world.

profane the ordinary and everyday elements of life that aren't related to religion.

religiosity the ways people demonstrate their religious beliefs.

cult a religious group that is devoted to beliefs and practices that are outside of those accepted in mainstream society.

new religious movement (NRM) term used instead of *cult* by most sociologists.

CHAPTER REVIEW 13

charismatic leader someone that followers see as having exceptional or superhuman powers and qualities.

sect a religious group that has broken away from an established religion.

denomination a subgroup within a religion that shares its name and traditions and is generally on good terms with the main group.

church a large established religious group that has strong ties to mainstream society.

secularization a process in which religion loses its social and cultural influence.

fundamentalism the belief in the literal meaning of a sacred text.

civil religion (sometimes called *secular religion*) integrating religious beliefs into secular life.

Protestant ethic a belief that hard work, diligence, self-denial, frugality, and economic success will lead to salvation in the afterlife.

false consciousness an acceptance of a system of beliefs that prevents people from protesting oppression.

TEST YOUR LEARNING

1. _____ maintain that the education system creates and perpetuates social inequality.
   a. Functionalists
   b. Conflict theorists
   c. Feminist theorists
   d. Symbolic interactionists

2. An emphasis on certificates or degrees is called
   a. meritocracy.
   b. tracking.
   c. the hidden curriculum.
   d. credentialism.

3. Today, the largest educational achievement gap is between
   a. males and females.
   b. blacks and whites.
   c. social classes.
   d. none of the above.
   e. all of the above.

4. A _____ is a religious group that is devoted to beliefs and practices that are outside of those accepted in mainstream society.
   a. cult
   b. sect
   c. denomination
   d. church

5. According to _____, religion isn't innate but taught.
   a. functionalists
   b. conflict theorists
   c. feminist theorists
   d. symbolic interactionists

6. *True or False* Cultural innovation is one of the latent functions of education.

7. *True or False* U.S. high school dropout rates have been increasing.

8. *True or False* Both religion and religiosity are social institutions.

9. *True or False* The Protestant work ethic is a belief that hard work, diligence, self-denial, frugality, and economic success will lead to salvation in the afterlife.

10. *True or False* One of the criticisms of functionalist theories on religion is that they often ignore the role that religion plays in creating social cohesion and harmony.

11. How does education reinforce existing inequalities in the United States based on gender, social class, race, and ethnicity? Illustrate your answer with specific examples.

12. Explain how religious and secular rituals differ and overlap. Be specific. Also, why do people who say that they have no religion practice civil religion?

1. b 2. d 3. c 4. a 5. d 6. False 7. False 8. False 9. True 10. False

TABLE 13.1 — SOCIOLOGICAL EXPLANATIONS OF EDUCATION

Theoretical Perspective	Level of Analysis	Key Points
Functionalist	Macro	Contributes to society's stability, solidarity, and cohesion and provides opportunities for upward mobility
Conflict	Macro	Reproduces and reinforces inequality and maintains a rigid social class structure
Feminist	Macro and micro	Produces inequality based on gender
Symbolic Interactionist	Micro	Teaches roles and values through everyday face-to-face interaction and behavior

TABLE 13.5 — SOCIOLOGICAL EXPLANATIONS OF RELIGION

Theoretical Perspective	Level of Analysis	Key Points
Functionalist	Macro	Religion benefits society by providing a sense of belonging, identity, meaning, emotional comfort, and social control over deviant behavior.
Conflict	Macro	Religion promotes and legitimates social inequality, condones strife and violence between groups, and justifies oppression of poor people.
Feminist	Macro and micro	Religion subordinates women, excludes them from decision-making positions, and legitimizes patriarchal control of society.
Symbolic Interactionist	Micro	Religion provides meaning and sustenance in everyday life through symbols, rituals, and beliefs, and binds people together in a physical and spiritual community.

CHAPTER 14 SUMMARY

14-1 Global Health and Illness

Health varies among individuals and societies, but all people experience *disease*. In addressing the question of why some people are healthier than others, health and medical practitioners, researchers, and sociologists examine *social epidemiology*, which includes both the incidence and prevalence of illness or health problems within a population during a specific time period. Worldwide, high income populations live longer, have low infant mortality rates, and spend more on health, but experience "diseases of wealth" such as diabetes, heart disease, and various cancers.

14-2 Health and Illness in the United States

Many people live with a *disability*; others die at an earlier age than expected for a variety of macro- and micro-level reasons. The three most important reasons for illness and early death are environmental, demographic, and lifestyle choices.

After age 65, people are more likely to experience *chronic diseases* rather than *acute diseases*. Women tend to live longer than men but have more chronic diseases, such as arthritis, asthma, cancer, and mental illness, and experience higher depression rates than men.

Among U.S. racial-ethnic groups, black babies have the *highest infant mortality rate*, and Latinos are the least likely to die from illicit drugs and prescription drug abuse. Our lifestyle choices also improve or impair health. The top three preventable health hazards, in order of priority, are smoking, obesity, and substance abuse. Sexually transmitted diseases are another important source of preventable health hazards.

14-3 Health Care: United States and Global

Medicine is a vital part of *health care* in diagnosing, treating, and preventing illness, injury, and other health impairments. The United States is one the richest nations in the world, but only the very wealthy don't have to worry about receiving and paying for the best medical care available. About 10 percent of Americans have no health insurance.

Since 2000, employer-based health insurance coverage has deteriorated. Consequently, workers at both large and small firms have been making higher contributions to premiums and paying higher deductibles or copayments. *Medicare* pays many of the medical costs of Americans age 65 and over, regardless of income. *Medicaid*, another government program, provides medical care for those living below the poverty level.

The United States spends more on health care than any other nation in the world, but ranks only 34th worldwide in life expectancy. Compared with other high-income countries, the United States covers a smaller percentage of the total population, and has poorer health outcomes. Unlike people in other high-income countries, Americans have fewer and shorter doctor visits, fewer days of inpatient hospital care, and pay considerably more for prescription drugs and routine medical services (such as doctor's office visits and normal birth deliveries).

In 2010, Congress passed the Patient Protection and Affordable Care Act, also called the Affordable Care Act (ACA) or "Obamacare." The ACA's goal is to give more Americans age 64 and under access to affordable, quality health insurance, and to reduce the growth in U.S. health care spending.

The uninsured include people who live in states where Medicaid doesn't cover the "nearly poor," who are undocumented immigrants, who haven't enrolled in Medicaid, who have incomes above the federal poverty level but can't afford the programs offered by employers or private insurers, and those who forgo coverage by choice.

KEY TERMS

health the state of physical, mental, and social well-being.

social epidemiology examines how societal factors affect the distribution of disease within a population.

health care the prevention, management, and treatment of illness.

disability any physical or mental impairment that limits a person's ability to perform a basic life activity.

chronic diseases long-term or lifelong illnesses that develop gradually or are present from birth.

acute diseases illnesses that strike suddenly and often disappear rapidly but can cause incapacitation and sometimes death.

Alzheimer's disease a progressive, degenerative disorder that attacks the brain and impairs memory, thinking, and behavior.

substance abuse a harmful overindulgence in or dependence on a drug or other chemical.

medicine a social institution that deals with illness, injury, and other health problems.

sick role a social role that excuses people from normal obligations because of illness.

medical-industrial complex a network of business enterprises that influences medicine and health care.

medicalization a process that defines a nonmedical condition or behavior as an illness, disorder, or disease that requires medical treatment.

CHAPTER REVIEW 14

14-4 Sociological Perspectives on Health and Medicine

See Table 14.3 below.

TEST YOUR LEARNING

1. The sick role is most closely associated with which theoretical perspective?
   a. Functionalist
   b. Conflict
   c. Feminist
   d. Symbolic interactionist

2. ____ is the top preventable lifestyle health hazard.
   a. Obesity
   b. Drug abuse
   c. Sexually transmitted diseases
   d. Smoking

3. Some have hailed the ____ as an important milestone in U.S. health care reform.
   a. AMA
   b. DSM
   c. ACA
   d. HMO

4. Which one of the following statements about U.S. health and medicine is *false*?
   a. Illness and death rates vary considerably, especially by social class.
   b. According to functionalists, the medical profession and health care industry maintain the status quo because of the medical-industrial complex.
   c. Women use more health care services than men do.
   d. According to symbolic interactionists, physicians, mental rights advocates, and parents all benefit from medicalization.

5. One of the most common criticisms of ____ theories is that they often ignore the contributions of medical and health care systems.
   a. functionalist
   b. conflict
   c. feminist
   d. symbolic interaction

6. *True or False* Acute diseases increase as people age.

7. *True or False* Chronically ill people are responsible for almost 50 percent of all U.S. health care spending.

8. *True or False* Compared with other groups, Latinos are the most likely to die from illegal drugs and prescription drug abuse.

9. *True or False* All states have obesity rates equal to or greater than 30 percent.

10. *True or False* Medicalization is a network of business enterprises that influences medicine and health care.

11. Is government-paid health care a basic human right? Or should it be limited to taxpayers? Explain your answer. Be as specific as possible, and incorporate the material in this chapter such as health care in other countries and the ACA.

12. Identify any two limitations of functionalist, conflict, feminist, and symbolic interaction theories in explaining health and medicine. (Extra points if you add a third limitation for each theory that's not discussed in this chapter.)

1. a 2. d 3. c 4. b 5. b 6. False 7. True 8. False 9. False 10. False

TABLE 14.3 SOCIOLOGICAL PERSPECTIVES ON HEALTH AND MEDICINE

Perspective	Level of Analysis	Key Points
Functionalism	Macro	• Health and medicine are critical in ensuring a society's survival and are closely linked to other institutions. • Illness is dysfunctional because it prevents people from performing expected roles. • Sick people are expected to seek professional help and get well.
Conflict	Macro	• There are gross inequities in the health care system. • The medical establishment is a powerful social control agent. • A drive for profit ignores people's health needs.
Feminist	Macro and micro	• Women are less likely than men to receive high-quality health care. • Gender stratification in medicine and the health care industry reduces women's earnings. • Men control women's health.
Symbolic Interaction	Micro	• Illness and disease are socially constructed. • Labeling people as ill increases their likelihood of being stigmatized. • Medicalization has increased the power of medical associations, parents, and mental health advocates and the profits of pharmaceutical companies.

CHAPTER 15 SUMMARY

15-1 Population Changes

Demography examines the interplay among *fertility*, *mortality*, and *migration*. The *crude birth rate*, *total fertility rate*, *crude death rate*, and infant mortality rate measure a population's life expectancy and health.

Push and pull factors affect international and internal migration. Demographers also use *sex ratios* and *population pyramids* to understand a population's composition and structure.

Demographers who believe that population growth is a ticking bomb subscribe to *Malthusian theory*, which argues that the world's food supply won't keep up with population growth. *Demographic transition theory*, in contrast, maintains that population growth is kept in check and stabilizes as countries experience greater economic and technological development.

15-2 Urbanization

Globally and in the United States, cities and *urbanization* mushroomed during the twentieth century and are expected to increase. As more people move from rural to urban areas, many of the world's largest cities are becoming *megacities*. In the United States, urban growth has led to suburbanization, *edge cities*, *exurbs*, *gentrification*, and *urban sprawl*.

Sociologists offer several perspectives on how and why cities change, and the consequences (see Table 15.3 on the next page).

15-3 The Environment

Population growth and urbanization are changing the planet's *ecosystem* and, many argue, endangering plants, animals, and humans. Water and air pollution and global warming are good examples of threats to the ecosystem in the United States and globally. Clean water has been depleted for many reasons, including pollution, privatization, waste, and mismanagement.

Four of the most common sources and causes of air pollution are burning fossil fuels, manufacturing plants that spew pollutants into the air, winds that carry contaminants across borders and oceans, and lax governmental policies. Air pollution, which can lead to the *greenhouse effect*, is a major cause of *climate change* and *global warming*.

The rise of environmental problems has sparked discussions about *sustainable development*. Those who are pessimistic about achieving sustainable development show that, worldwide, the United States has one of the worst records on environmental performance, largely because of the close ties between government officials and corporations.

Others are optimistic about achieving sustainable development and point to examples such as decreases in the emission of major air pollutants and some large U.S. corporations' switching to practices that decrease pollution and energy consumption.

KEY TERMS

demography the scientific study of human populations.

population a group of people who share a geographic territory.

fertility the number of babies born during a specified period in a particular society.

crude birth rate (also called the *birth rate*) the number of live births per 1,000 people in a population in a given year.

total fertility rate (TFR) the average number of children born to a woman during her lifetime.

mortality the number of deaths in a population during a specified period.

crude death rate (also called the *death rate*) the number of deaths per 1,000 people in a population in a given year.

migration the movement of people into or out of a specific geographic area.

sex ratio the proportion of men to women in a population.

population pyramid a graphic depiction of a population's age and sex distribution at a given point in time.

Malthusian theory maintains that population is growing faster than the food supply needed to sustain it.

demographic transition theory maintains that population growth is kept in check and stabilizes as countries experience economic and technological development.

zero population growth (ZPG) a stable population level that occurs when each woman has no more than two children.

urbanization people's movement from rural to urban areas.

megacities metropolitan areas with at least 10 million inhabitants.

CHAPTER REVIEW 15

edge cities business centers that are within or close to suburban residential areas.

exurbs areas of new development beyond the suburbs that are more rural but on the fringe of urbanized areas.

metropolitan statistical area (MSA, also called *metro area*) a central city of at least 50,000 people and urban areas linked to it.

urban sprawl the rapid, unplanned, and uncontrolled spread of development into regions adjacent to cities.

gentrification a process in which upper-middle-class and affluent people buy and renovate houses and stores in downtown urban neighborhoods.

urban ecology studies the relationships between people and urban environments.

new urban sociology view that urban changes are largely the result of decisions made by powerful capitalists and high-income groups.

ecosystem a community of living and non-living organisms that share a physical environment.

global warming the increase in the average temperature of earth's atmosphere.

greenhouse effect the heating of earth's atmosphere because of the presence of certain atmospheric gases.

climate change a change in overall temperatures and weather conditions over time.

sustainable development economic activities that don't threaten the environment.

TEST YOUR LEARNING

1. ____ is the study of human populations.
   a. Sociology
   b. Demography
   c. Ecology
   d. Zero population growth (ZPG)

2. Emma is comparing the number of deaths per 1,000 people in Spain and Russia for 2011. She is measuring
   a. fertility.
   b. mortality.
   c. the crude birth rate.
   d. the crude death rate.

3. Which one of the following is NOT one of the phases of demographic transition theory?
   a. Preindustrial society
   b. Early industrial society
   c. Advanced industrial society
   d. Declining industrial society

4. Maleek and Shonda live in a newly developed area that's rural but on the fringe of an urbanized region. They live in
   a. an exurb.
   b. a suburb.
   c. a gentrified area.
   d. an edge city.

5. Why do many feminist scholars maintain that women suffer more problems than men when living in urban areas?
   a. Women don't have the natural aggressive instinct that men have.
   b. Husbands tend to maintain tighter controls over their wives in urban areas.
   c. Urban areas have typically been designed by men and for men.
   d. All of the above explain why women suffer more problems in urban areas.

6. *True or False* Malthusian theory claims that population growth is kept in check and stabilizes as developing countries experience economic and technological development.

7. *True or False* Population pyramids are graphic depictions of a population's makeup in terms of the age and sex of its members at a given point in time.

8. *True or False* Concentric zone theory emphasizes the development of suburbs around a city but away from its center.

9. *True or False* A sex ratio of 115 means that there are 115 women for every 100 men in a population.

10. *True or False* Global warming begins with the greenhouse effect.

11. Does Malthusian theory or demographic transition theory come closer to your perspective on population growth? Why?

12. How do functionalist, conflict, feminist, and symbolic interaction theories differ in their explanations of urbanization? Be specific in illustrating your answer.

1. b 2. d 3. d 4. a 5. c 6. False 7. True 8. False 9. False 10. True

TABLE 15.3	SOCIOLOGICAL EXPLANATIONS OF URBANIZATION	
Perspective	**Level of Analysis**	**Key Points**
Functionalist	Macro	Cities serve many important social and economic functions, but urbanization can also be dysfunctional.
Conflict	Macro	Driven by greed and profit, large corporations, banks, developers, and other capitalist groups shape cities' growth or decline.
Feminist	Macro and micro	Whether they live in cities or suburbs, women generally experience fewer choices and more constraints than men.
Symbolic Interactionist	Micro	City residents differ in their types of interaction, lifestyles, and perceptions of urban life.

Social Change: Collective Behavior, Social Movements, and Technology

CHAPTER 16 SUMMARY

16-1 Collective Behavior

Social scientists offer several explanations of *collective behavior*. Contagion theory proposes that individuals act emotionally and irrationally due to a crowd's almost hypnotic influence. According to convergence theory, crowds consist of like-minded people who deliberately assemble in a place to pursue a common goal. Emergent norm theory emphasizes the importance of social norms in shaping crowd behavior. Structural strain theory proposes that collective behavior occurs only if six conditions are present.

There are many types of collective behavior, some more short-lived or harmful than others. *Rumors, gossip*, and *urban legends* are typically untrue, but many people believe and pass them on for a number of reasons, such as anxiety or to reinforce a community's moral standards. In contrast, *panic* and *mass hysteria* can have dire consequences, including death. *Fashions* and *fads* are harmless because they usually last only a short time and change over time. In contrast, a *disaster* is an unexpected event due to social, technological, or natural causes that result in widespread damage and destruction.

Publics, public opinion, and *propaganda* also affect large numbers of people. *Crowds* vary in their motives, interests, and emotional level. A casual crowd, for example, has little, if any, interaction, the gathering is temporary, and there is little emotion. On the other hand, protest crowds, especially in *mobs* and *riots*, can wreak considerable havoc on property and result in death.

Example: Crowds Can Be Deadly

On Thanksgiving Day, 2008, crowds started gathering at 9:00 p.m. outside the Walmart store in Valley Stream, New York, for a bargain-hunting ritual known as Black Friday, the day after Thanksgiving. By 4:55 a.m., the crowd had grown to more than 2,000 people and could no longer be held back. Suddenly, according to witnesses, the glass doors shattered and "the shrieking mob surged through in a blind rush for holiday bargains." A 34-year-old male temporary worker, who had been hired for the holiday season, was trampled to death, and four other people, including a 28-year-old woman who was eight months pregnant, were treated for injuries (McFadden and Macropoulos, 2008). Review the types of crowds in this chapter. Which type of crowd do you think this Walmart incident most nearly represents?

16-2 Social Movements

Unlike collective behavior, *social movements* are typically organized and have long-lasting effects. Some of the most common social movements are alternative, redemptive, reformative, resistance, and revolutionary (see Table 16.1 on the next page). Sociologists have offered several explanations for the emergence of social movements, including mass society theory, relative deprivation theory, resource mobilization theory, and new social movements theory. Each theory has strengths and weaknesses in explaining social movements.

Social movements generally go through four stages: emergence, organization, institutionalization, and decline. Decline occurs when a social movement is successful and becomes a part of society's fabric; when the members become distracted because the group loses sight of its original goals and/or their enthusiasm wanes; when the membership fragments because the participants disagree about goals, strategies, or tactics; or when a government quashes dissent. Social movements are important because

KEY TERMS

social change transformations of societies and social institutions over time.

collective behavior the spontaneous and unstructured behavior of a large number of people.

rumor unfounded information that people spread quickly.

gossip rumors, often negative, about other people's personal lives.

urban legends (also called *contemporary legends* and *modern legends*) rumors about stories that supposedly happened somewhere.

panic a collective flight, often irrational, from a real or perceived danger.

mass hysteria an intense, fearful, and anxious reaction to a real or imagined threat by large numbers of people.

fashion a popular way of dressing during a particular time or among a particular group of people.

fad a fashion that spreads rapidly and enthusiastically but lasts only a short time.

disaster an unexpected event that causes widespread damage, destruction, distress, and loss.

public a group of people, not necessarily in direct contact with each other, who are interested in a particular issue.

public opinion widespread attitudes on a particular issue.

propaganda spreading information (or misinformation) to influence people's attitudes or behavior.

crowd a temporary gathering of people who share a common interest or participate in a particular event.

mob a highly emotional and disorderly crowd that uses force, the threat of force, or violence against a specific target.

CHAPTER REVIEW 16

riot a violent crowd that directs its hostility at a wide and shifting range of targets.

social movement a large and organized group of people who want to promote or resist a particular social change.

relative deprivation a gap between what people have and what they think they should have compared with others in a society.

technology the application of scientific knowledge for practical purposes.

they can create or resist change at the individual, institutional, and societal levels.

16-3 Technology and Social Change

Technology also generates changes. Some of the most important technological advances have included computer technology, biotechnology, and nanotechnology—all of which have changed our lives. Technology has both benefits and costs, however. For example, the Internet and other forms of telecommunication technology can bring people together, but can also intrude on our privacy. In addition, technological advances raise numerous ethical questions, such as their greater availability to higher-income people.

TEST YOUR LEARNING

1. According to ____, crowds are highly suggestible and out of control.
   a. contagion theory
   b. convergence theory
   c. emergent norm theory
   d. structural strain theory

2. A major difference between panic and mass hysteria is that
   a. panics are typically not as severe as mass hysteria.
   b. panics don't usually last as long as mass hysteria.
   c. panics stem from imagined events, whereas mass hysteria arises from real events.
   d. panics tend to occur less today, whereas mass hysteria occurs more frequently.

3. Cabbage Patch Kids dolls are an example of a
   a. fashion.
   b. fad.
   c. craze.
   d. disaster.

4. ____ maintains that social movements emerge because there's a gap between what people have and what they think they should have compared with other people.
   a. Mass society theory
   b. Relative deprivation theory
   c. Resource mobilization theory
   d. New social movements theory

5. Which of the following is NOT one of four stages of a social movement?
   a. Emergence
   b. Conflict
   c. Institutionalization
   d. Decline

6. *True or False* During the organization stage of social movements, the movement becomes more bureaucratic.

7. *True or False* Gossip and urban legends are types of rumors.

8. *True or False* Mobs typically last longer than riots.

9. *True or False* Environmentalism is an example of the new social movements theory.

10. *True or False* Technological advances are available to most low-income people.

11. What conditions are necessary to produce social movements? What draws people into a social movement? If you're involved in a social movement now, which one and why?

12. What, personally, have been some of the benefits and costs of technology in your life? Illustrate your answer with specific examples.

1. a 2. b 3. b 4. b 5. b 6. False 7. True 8. False 9. True 10. False

TABLE 16.1 FIVE TYPES OF SOCIAL MOVEMENTS

Movement	Goal	Examples
Alternative	Change some people in a specific way	Alcoholics Anonymous, transcendental meditation
Redemptive	Change some people, but completely	Jehovah's Witnesses, born-again Christians
Reformative	Change everyone, but in specific ways	Gay rights advocates, Mothers Against Drunk Driving (MADD)
Resistance	Preserve status quo by blocking or undoing change	Antiabortion groups, white supremacists
Revolutionary	Change everyone completely	Right-wing militia groups, Communism, Islamic State in Iraq and Syria (ISIS)